T0205740

Lecture Notes in Computer Science 13237

Founding Editors

Gerhard Goos
Karlsruhe Institute of Technology, Karlsruhe, Germany

Juris Hartmanis
Cornell University, Ithaca, NY, USA

Editorial Board Members

Elisa Bertino
Purdue University, West Lafayette, IN, USA

Wen Gao
Peking University, Beijing, China

Bernhard Steffen
TU Dortmund University, Dortmund, Germany

Moti Yung
Columbia University, New York, NY, USA

More information about this series at https://link.springer.com/bookseries/558

Seiichi Uchida · Elisa Barney ·
Véronique Eglin (Eds.)

Document Analysis Systems

15th IAPR International Workshop, DAS 2022
La Rochelle, France, May 22–25, 2022
Proceedings

Editors
Seiichi Uchida
Kyushu University
Fukuoka, Japan

Elisa Barney
Boise State University
BOISE, ID, USA

Véronique Eglin
LIRIS UMR CNRS
Villeurbanne, France

ISSN 0302-9743 ISSN 1611-3349 (electronic)
Lecture Notes in Computer Science
ISBN 978-3-031-06554-5 ISBN 978-3-031-06555-2 (eBook)
https://doi.org/10.1007/978-3-031-06555-2

© Springer Nature Switzerland AG 2022, corrected publication 2022
This work is subject to copyright. All rights are reserved by the Publisher, whether the whole or part of the material is concerned, specifically the rights of translation, reprinting, reuse of illustrations, recitation, broadcasting, reproduction on microfilms or in any other physical way, and transmission or information storage and retrieval, electronic adaptation, computer software, or by similar or dissimilar methodology now known or hereafter developed.
The use of general descriptive names, registered names, trademarks, service marks, etc. in this publication does not imply, even in the absence of a specific statement, that such names are exempt from the relevant protective laws and regulations and therefore free for general use.
The publisher, the authors and the editors are safe to assume that the advice and information in this book are believed to be true and accurate at the date of publication. Neither the publisher nor the authors or the editors give a warranty, expressed or implied, with respect to the material contained herein or for any errors or omissions that may have been made. The publisher remains neutral with regard to jurisdictional claims in published maps and institutional affiliations.

This Springer imprint is published by the registered company Springer Nature Switzerland AG
The registered company address is: Gewerbestrasse 11, 6330 Cham, Switzerland

Preface

Welcome to the 15th IAPR International Workshop on Document Analysis Systems (DAS 2022). DAS 2022 was held in La Rochelle, France, during May 22–25, 2022, and brought together many researchers from Europe and abroad.

With the new remote access facilities, the workshop was not confined to a specific location. In a sense, this was truly a worldwide edition of DAS, taking place around the world in a coordinated fashion, employing a schedule we designed to support participation across a wide range of time zones. Of course, this came with some challenges but also with interesting opportunities that caused us to rethink the way of fostering social and scientific interaction in this new medium. It also allowed us to organize an environmentally friendly event, extend the reach of the workshop, and facilitate participation from literally anywhere in the world for those with an interest in our field and an Internet connection. We truly hope we managed to make the most out of a difficult situation.

DAS 2022 continued the long tradition of bringing together researchers, academics, and practitioners in the research field of document analysis systems. In doing so, we built upon the previous workshops held over the years in Kaiserslautern, Germany (1994); Malvern, PA, USA (1996); Nagano, Japan (1998); Rio de Janeiro, Brazil (2000); Princeton, NJ, USA (2002); Florence, Italy (2004); Nelson, New Zealand (2006); Nara, Japan (2008); Boston, MA, USA (2010); Gold Coast, Australia (2012); Tours, France (2014); Santorini, Greece (2016); Wien, Austria (2018); and Wuhan, China (2020).

As with previous editions, DAS 2022 was a rigorously peer-reviewed and 100% participation single-track workshop focusing on issues and approaches in document analysis and recognition. The workshop comprised presentations by invited speakers, oral and poster sessions, and a pre-workshop tutorial, as well as distinctive DAS discussion groups.

This year we received 93 submissions in total, 78 of which were in the regular paper track and 15 in the short paper track. All regular paper submissions underwent a rigorous single-blind review process where the vast majority of papers received three reviews. The reviewers were selected from the 80 members of the Program Committee, judging the originality of work, the relevance to document analysis systems, the quality of the research or analysis, and the overall presentation. Of the 78 regular submissions received, 52 were accepted for presentation at the workshop (67%). Of these, 31 papers were designated for oral presentation (40%) and 21 for poster presentation (27%). All short paper submissions were reviewed by PC chairs. Of the 15 short papers received, all 15 were accepted for poster presentation at the workshop (100%). The accepted regular papers are published in this proceedings volume in the Springer Lecture Notes in Computer Science series. Short papers appear in PDF form on the DAS conference website.

The final program included six oral sessions, two poster sessions, and the discussion group sessions. There were also two awards announced at the conclusion of the workshop: the IAPR Best Student Paper Award and the IAPR Nakano Best Paper Award. We offer our deepest thanks to all who contributed their time and effort to make DAS 2022 a first-rate event for the community.

In addition to the contributed papers, the program also included two invited keynote presentations by distinguished members of the research community: Andreas Dengel from the German Research Center for Artificial Intelligence (DFKI, Germany) and Adam Jatowt from the University of Innsbruck (Austria).

We furthermore would like to express our sincere thanks to the tutorial organizer, Himanshu Sharad Bhatt from American Express AI Labs, for sharing his valuable scientific and technological insights. Special thanks are also due to our sponsors IAPR, the L3i Laboratory, AriadNext, Esker, IMDS, GoodNotes, Yooz, MyScript, ITESOFT, TEKLIA, VIALINK, and the Région Nouvelle Aquitaine and Communauté d'Agglomération de La Rochelle, whose support, especially during challenging times, was integral to the success of DAS 2022.

The workshop program represented the efforts of many people. We want to express our gratitude, especially to the members of the Program Committee for their hard work in reviewing submissions. The publicity chairs, Richard Zanibi (USA) and Joseph Chazalon (France), helped us in many ways, for which we are grateful. We also thank the discussion group chairs, Michael Blumenstein (Australia) and Umapada Pal (India), for organizing the discussion groups and the tutorial chairs, Rafael Dueire Lins (Brazil) and Alicia Fornes (Spain), for organizing the tutorial. A special thank you goes to the publication chair, Cheng-Lin Liu (China), who was responsible for the proceedings at hand. We are also grateful to the local organizing committee who made great efforts in arranging the program, maintaining the web page, and setting up the meeting platform with support for remote attendance. The workshop would not have happened without the great support from the hosting organization, La Rochelle University.

Finally, the workshop would have not been possible without the excellent papers contributed by authors. We thank all the authors for their contributions and their participation in DAS 2022! We hope that this program will further stimulate research and provide practitioners with better techniques, algorithms, and tools. We feel honored and privileged to share the best recent developments in the field of document analysis systems with you in these proceedings.

April 2022

Seiichi Uchida
Elisa Barney Smith
Véronique Eglin

Organization

General Chair

Jean-Marc Ogier	La Rochelle University, France

Conference Chairs

Jean-Christophe Burie	La Rochelle University, France
Mickaël Coustaty	La Rochelle University, France
Antoine Doucet	La Rochelle University, France

Program Committee Chairs

Seiichi Uchida	Kyushu University, Japan
Elisa Barney Smith	Boise State University, USA
Véronique Eglin	INSA Lyon, France

Industrial Chairs

Vincent Poulain d'Andecy	Yooz, France
Robin Mélinand	MyScript, France

Tutorial Chairs

Rafael Dueire Lins	Universidade Federal de Pernambuco, Brazil
Alicia Fornes	Universitat Autònoma de Barcelona, Spain

Discussion Group Chairs

Michael Blumenstein	University of Technology Sydney, Australia
Umapada Pal	Indian Statistical Institute, India

Publication Chair

Cheng-Lin Liu	Institute of Automation, Chinese Academy of Sciences, China

Publicity Chairs

Richard Zanibi	Rochester Institute of Technology, USA
Joseph Chazalon	LRDE, France

Program Committee

Alireza Alaei	Southern Cross University, Australia
Eric Anquetil	INSA Rennes, France
Apostolos Antonacopoulos	University of Salford, UK
Xiang Bai	Huazhong University of Science and Technology, China
Ujjwal Bhattacharya	Indian Statistical Institute, India
Michael Blumenstein	University of Technology Sydney, Australia
Emanuela Boros	La Rochelle University, France
Jorge Calvo-Zaragoza	University of Alicante, Spain
Vincent Christlein	University of Erlangen-Nuremberg, Germany
Christian Clausner	University of Salford, UK
Florence Cloppet	Paris Cité Université, France
Bertrand Coüasnon	INSA Rennes, France
Veronique Eglin	INSA Lyon, France
Maud Ehrmann	EPFL, Switzerland
Miguel Ferrer	Universidad de Las Palmas de Gran Canaria, Spain
Gernot Fink	TU Dortmund University, Germany
Andreas Fischer	University of Fribourg, Switzerland
Volkmar Frinken	University of California, Davis, USA
Basilis Gatos	Institute of Informatics and Telecommunications, Greece
Ahmed Hamdi	La Rochelle University, France
Nicholas Howe	Smith College, USA
Donato Impedovo	UNIBA, Italy
Masakazu Iwamura	Osaka Prefecture University, Japan
Brian Kenji Iwana	Kyushu University, Japan
Adam Jatowt	University of Innsbruck, Austria
C. V. Jawahar	IIIT, Hyderabad, India
Lianwen Jin	South China University of Technology, China
Dimosthenis Karatzas	Universitat Autonoma de Barcelona, Spain
Ergina Kavallieratou	University of the Aegean, Greece
Florian Kleber	TU Wien, Austria
Camille Kurtz	Université de Paris, France
Bart Lamiroy	Université de Reims Champagne-Ardenne, France
Aurélie Lemaitre	Université de Rennes, France

Laurence Likforman-Sulem	Institut Polytechnique de Paris, France
Cheng-Lin Liu	Institute of Automation, Chinese Academy of Sciences, China
Marcus Liwicki	Luleå University of Technology, Sweden
Josep Llados	Universitat Autònoma de Barcelona, Spain
Daniel Lopresti	Lehigh University, USA
Georgios Louloudis	Institute of Informatics and Telecommunications, Greece
Andreas Maier	University of Erlangen-Nuremberg, Germany
Muhammad Imran Malik	DFKI, Germany
Angelo Marcelli	Universita' degli Studi di Salerno, Italy
Volker Märgner	Technische Universität Braunschweig, Germany
Simone Marinai	University of Florence, Italy
Harold Mouchère	Université de Nantes, France
Hung Tuan Nguyen	Tokyo University of Agriculture and Technology, Japan
Journet Nicholas	Université de Bordeaux, France
Wataru Ohyama	Saitama Institute of Technology, Japan
Shinichiro Omachi	Tohoku University, Japan
Shivakumara Palaiahnakote	University of Malaya, Malaysia
Thierry Paquet	Université de Rouen, France
Liangrui Peng	Tsinghua University, China
Giuseppe Pirlo	Bari University, Italy
Ashok C. Popat	Google, USA
Ioannis Pratikakis	Democritus University of Thrace, Greece
Irina Rabaev	Shamoon College of Engineering, Beer Sheva, Israel
Jean-Yves Ramel	Université de Tours, France
Oriol Ramos-Terrades	Universitat Autònoma de Barcelona, Spain
Frédéric Rayar	Université de Tours, France
Kaspar Riesen	University of Applied Sciences and Arts Northwestern Switzerland, Switzerland
Verónica Romero	Universitat de València, Spain
Marçal Rusiñol	AllRead Machine Learning Technologies, Spain
Robert Sablatnig	TU Wien, Austria
Joan Andreu Sanchez	Universitat Politècnica de València, Spain
Marc-Peter Schambach	Siemens Logistics GmbH, Germany
Mathias Seuret	University of Fribourg, Switzerland
Faisal Shafait	National University of Sciences and Technology, Pakistan
Nicolas Sidere	University of La Rochelle, France
Nikolaos Stamatopoulos	Institute of Informatics and Telecommunications, Greece

Jun Sun	Fujitsu R&D Center Co., Ltd., Japan
Salvatore Tabbone	Université de Lorraine, France
Kengo Terasawa	Future University Hakodate, Japan
Iuliia Tkachenko	Université Lyon 2, France
Ernest Valveny	Universitat Autònoma de Barcelona, Spain
Ekta Vats	Uppsala University, Sweden
Christian Viard-Gaudin	Université de Nantes, France
Mauricio Villegas	Omni:us, Germany
Nicole Vincent	Université de Paris, France
Berrin Yanikoglu	Sabanci University, Turkey
Anna Zhu	Wuhan University of Technology, China

Local Organizing Committee

Beatriz Martínez Tornés	La Rochelle University, France
Carlos Gonzalez	La Rochelle University, France
Damien Mondou	La Rochelle University, France
Dominique Limousin	La Rochelle University, France
Emanuela Boros	La Rochelle University, France
Guillaume Bernard	La Rochelle University, France
Ibrahim Souleiman Mahamoud	La Rochelle University, France
Kais Rouis	La Rochelle University, France
Lady Viviana Beltran Beltran	La Rochelle University, France
Latifa Bouchekif	La Rochelle University, France
Marina Dehez–Clementi	La Rochelle University, France
Mélanie Malinaud	La Rochelle University, France
Muhammad Muzzamil Luqman	La Rochelle University, France
Musab Al-Ghadi	La Rochelle University, France
Nathalie Renaudin-Blanchard	La Rochelle University, France
Nicholas Journet	University of Bordeaux, France
Nicolas Sidère	La Rochelle University, France
Souhail Bakkali	La Rochelle University, France
Théo Taburet	La Rochelle University, France
Zuheng Ming	La Rochelle University, France

Sponsors

IAPR, L3i Laboratory, La Rochelle Université, AriadNext, Esker, GoodNotes, IMDS, MyScript, Yooz, ITESOFT, TEKLIA, VIALINK, La Région Nouvelle Aquitaine, Communauté d'Agglomération de La Rochelle, La Ville de La Rochelle and CIRCULARSEAS.

Academic Sponsors

Platinum Sponsors

Gold Sponsors

Silver Sponsors

With the Support of

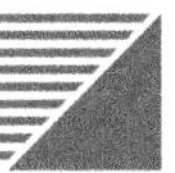

Sponsors

Contents

Historical Document Analysis + CSAWA

Handwriting Text Recognition

Applications in Handwriting

Open-Source Software and Benchmarking

Poster Session 1

Poster Session 2

Document Analysis Systems and Applications

Font Shape-to-Impression Translation

Masaya Ueda[1(✉)], Akisato Kimura[2], and Seiichi Uchida[1]

[1] Kyushu University, Fukuoka, Japan
masaya.ueda@human.ait.kyushu-u.ac.jp
[2] NTT Communication Science Laboratories, NTT Corporation, Chiyoda City, Japan

Abstract. Different fonts have different impressions, such as elegant, scary, and cool. This paper tackles part-based shape-impression analysis based on the Transformer architecture, which is able to handle the correlation among local parts by its self-attention mechanism. This ability will reveal how combinations of local parts realize a specific impression of a font. The versatility of Transformer allows us to realize two very different approaches for the analysis, i.e., multi-label classification and translation. A quantitative evaluation shows that our Transformer-based approaches estimate the font impressions from a set of local parts more accurately than other approaches. A qualitative evaluation then indicates the important local parts for a specific impression.

Keywords: Font shape · Impression analysis · Translator

1 Introduction

Different fonts, or typefaces, have different impressions. Each font just has a specific shape style and this style is converted to some special impression in our mind. At MyFonts.com, each font is tagged with several impression words. We can see those words as a result of a shape-to-impression translation by humans.

It is still not clear why and how different font shape styles give different impressions. As will be reviewed in Sect. 2, many subjective experiments have proved that font shapes surely affect their impression, readability, and legibility. However, most of those experiments just confirm that a specific font (such as Helvetica) has a specific impression (such as reliable). In other words, they do not reveal more detailed relationships or general trends between font shapes and impressions.

To understand the shape-impression relationship, we need more objective experiments using a large-scale dataset. Fortunately, Chen et al. [2] have published a shape-impression dataset by using the content of MyFonts.com. This is a very large dataset containing 18,815 fonts and 1,824 vocabularies of *impression tags* attached to each font. Figure 1 shows several examples of fonts and their impression tags. Some tags directly express font style types such as *Sans-Serif* and *Script*, more shape-related properties such as *Bold* and *Oblique* and more abstract impressions such as *Elegant* and *Scary*.

© Springer Nature Switzerland AG 2022
S. Uchida et al. (Eds.): DAS 2022, LNCS 13237, pp. 3–17, 2022.
https://doi.org/10.1007/978-3-031-06555-2_1

Bodoni

HERONS

decorative, sensitive, legible, feminine, magazine, business-text, neutral, serif, elegant, narrow

Myriad

HERONS

heavy, information, legible, apple, transport, sans-serif, traffic, humanist

Roadblock

HERONS

angle, bold, headline, oblique, retro, italic, display, sans-serif

Elegeion-script

HERONS

elegant, swash, script

Fig. 1. Fonts and their impressions (from MyFonts dataset [2]). `Bodoni`, `Myriad`, etc. are font names.

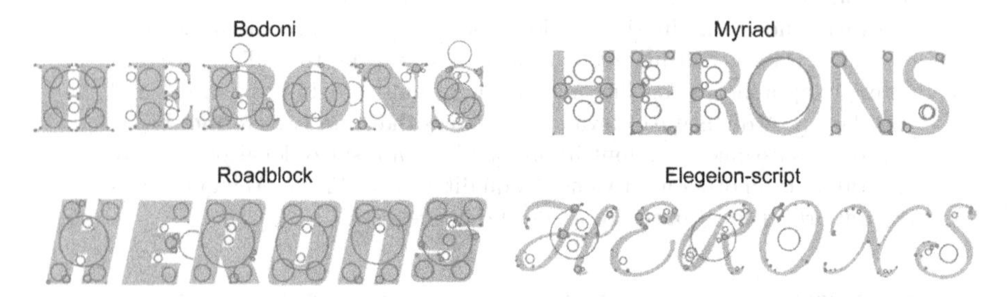

Fig. 2. SIFT keypoints. The center of each circle represents the location of a keypoint and the radius represents the scale at the keypoint. Roughly speaking, the SIFT descriptor at each keypoint represents the local shape around the circle.

If we can reveal the relationships between font shapes and those impression tags, it will be very meaningful for not only several practical applications but also more fundamental research. Example practical applications include (1) font selection or recommendation systems that provide a suitable font according to typographer's ambiguous requests by impression. (e.g., [2,3]) and (2) font generation systems that can accept the impression words as the constraints on the generation (e.g., [13]). From the viewpoint of more fundamental research, the shape-impression relationship is still unrevealed and will give important evidence to understand the human perception mechanism.

To understand those relationships between font shapes and impressions, one of the promising approaches is *part-based approach*, where a font image is decomposed into a set of local parts and then the individual parts and their combinations are correlated with impressions. This is because the shape of 'A' is comprised of two shape factors—the letter shape (so-called 'A'-ness [9]) and the font style. The part-based approach can discard the letter shape, while retaining various impression clues from local shapes, such as serif, curvature, corner shape, stroke width, etc.

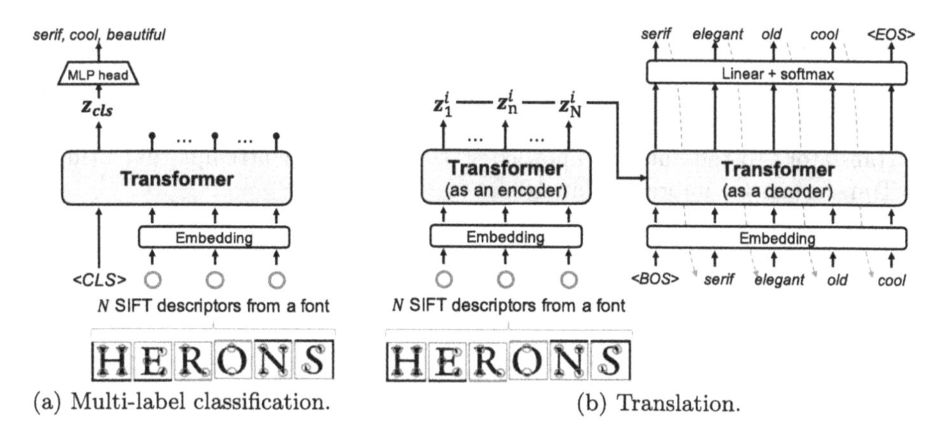

(a) Multi-label classification. (b) Translation.

Fig. 3. Two approaches of part-based shape-impression relationship analysis with Transformers.

This paper proposes a novel method for part-based shape-impression analysis that fully utilizes Transformer [20], which is a recent but already well-known deep neural network architecture. We first train Transformer to output impressions for a given set of local shapes. Then, we analyze the trained Transformer in various ways to understand the important local shapes for a specific impression. The advantages of Transformer for our analysis are threefold.

1. Transformer is a versatile model and offers us two different approaches. As shown in Fig. 3, the classification approach (a) accepts N local descriptors by SIFT [12] as its input elements and outputs the probability of each of K impression classes. In the translation approach (b), one Transformer as an encoder accepts N SIFT descriptors as its input and then encodes them into latent vectors (called "key" and "value"), and then the latent vectors are fed to another Transformer as a decoder that outputs a set of impression words like a translation result.
2. Transformer can accept a variable number of input elements. Since the number of local shapes from a single font image is not constant, this property is suitable for our task.
3. The most important advantage is its *self-attention* mechanism. Self-attention determines a weight for every input element by considering other input elements. Therefore, if we input local shapes to Transformer as multiple input elements, their correlation is internally calculated and used for the task. For example in the classification approach, the correlation among the local shapes that are important for a correct impression class will be boosted through the self-attention mechanism.

We also introduce explainable-AI (XAI) techniques [18,22] for a deep understanding of the importance of local shapes on a specific impression. In the experiment, we reveal the important parts for a specific impression by several different techniques.

Our contributions are summarized as follows:

- We propose two methods for part-based font-impression analysis using Transformer. They are a multi-label impression classifier and a shape-to-impression translator. To the authors' knowledge, this is the first attempt that utilizes Transformer for impression analysis.
- Using a large font-impression dataset, we experimentally prove that our Transformer-based methods enable us to realize more reliable and flexible analysis than the previous approaches.
- Our proposed method reveals local parts that well describe a specific impression with the help of XAI.

2 Related Work

2.1 Subjective Impression Analysis of Fonts

The analysis of the relationships between font shapes and impressions has a long history from the 1920's [5,16]. Shaikh and Chaparro [17] defined the impression of 40 fonts by collecting answers from 379 people. In 2014, O'Donovan et al. [14] have published their dataset with 200 fonts with impression tags. In this pioneering work on data-driven impression analysis, impression tags were only 37 vocabularies. More recently, Chen et al. [2] published a far larger font-impression dataset by using `MyFonts.com` and then proposed a font retrieval system using impressions as a query. As will be detailed in Sect. 3, about 20,000 fonts are annotated with about 2,000 impression tags in their dataset. Although it contains noisy annotations, its rich variation is useful for analyzing the relationship between font shape and impression. For example, Matsuda et al. [13] utilize it for a font generation with specific impressions.

2.2 Objective Impression Analysis of Fonts

To the best of our knowledge, only a single attempt [19] has been made so far for shape-impression analysis by using a large dataset and an objective methodology. Although it also takes a part-based approach like ours, it has still a large room for improvement. One of the most significant issues is that all local shape descriptors extracted from a letter image are treated *totally independently* in the framework of DeepSets [21]. More specifically, DeepSets converts individual local shape descriptors to discriminative feature vectors independently, and then just adds those vectors into a single vector. Compared to [19], ours utilizes Transformer which can deal with the combinatorial relationships among the local descriptors by its self-attention mechanism, and therefore, as shown by the later experimental results, realizes a more accurate analysis.

2.3 Transformer

Transformer [20] is a multi-input and multi-output network and was originally developed for various tasks in natural language processing (NLP). It can be used as an encoder and also a decoder in some NLP tasks [15], such as language translation. It is also used as a classifier in some NLP tasks [1,6], such as sentence sentiment classification. Transformer has also been applied to image classification tasks [10]. Vision Transformer (ViT) [7] is the most well-known application of Transformer to image classification. In ViT, an input image is first divided into small patches and then fed to Transformer. In this research, we use ViT as one of the comparative methods. Transformer is also applied to image captioning [4]. Although Transformer has been employed in a vast number of applications from its development, its usefulness in impression analysis tasks has not been proved so far.

3 Dataset and Local Descriptor

As the font-impression dataset, we used the MyFonts dataset provided by Chen et al. [2], which contains 18,815 fonts. As shown in Fig. 1, multiple impression tags are attached to each font by font experts and non-experts (including the customers of `MyFonts.com`). The vocabulary of tags is 1,824. By following [19], we remove minor tags with less than 100 occurrences. Consequently, we use $M = 18,579$ fonts and $K = 483$ impression tags. In the following experiments, we randomly split the dataset into three font-disjoint subsets for train, validation, and test, with the ratio of 0.8, 0.1, and 0.1.

As noted in Sect. 1 and shown in Fig. 1, some impression tags express a typical font style type, such as *Sans-Serif* and *Script*, or a more shape-related property, such as *Bold* and *Oblique*, or a more abstract impression, such as *Elegant* and *Scary*. There is no clear taxonomy among them, and we will simply refer to them as impression tags in this paper.

In this paper, we apply SIFT [12] for extracting local shape descriptors from six letter images of 'H', 'E', 'R', 'O', 'N', and 'S', as shown in Fig. 2. These letters are often employed in typographic work or font design because they contain almost all elements of English alphabets, such as horizontal strokes, vertical strokes, diagonal lines ('N' and 'R'), intersections ('H' and 'R'), curves ('R' and 'S'), corners ('E' and 'R'), and a circle ('O'). Using those six letters (instead of all the alphabets) can limit the number of SIFT vectors and improve the efficiency of training and testing Transformers.

As local shape descriptors, several successors of SIFT have been proposed so far. For example, SURF was proposed as an efficient approximation of SIFT. BRIEF and ORB are also faster versions of SIFT and they provide binary features (for Hamming distance-based fast matching); namely, they do not represent the local shapes in a direct way. We, therefore, decided to use SIFT mainly, following [19].

4 Shape-Impression Relationship Analysis by Multi-label Classification Approach

4.1 Transformer as a Multi-label Classifier

For our purpose of the part-based shape-impression relation analysis, we use Transformer as a multi-label impression classifier. If Transformer is trained to estimate the impression tags for a set of local shape descriptors extracted from a font image, the trained Transformer should know some shape-impression relationships internally. We, therefore, can understand the local shapes that are important for a specific impression by visualizing and analyzing the trained Transformer. Moreover, this analysis result will reflect the combinatorial relationship among the local shapes because of the self-attention mechanism in Transformer.

Figure 3 (a) shows the multi-label impression classification by Transformer. Each of N SIFT descriptors extracted from six letter images (as noted in Sect. 3) is first embedded into another feature space by a single fully-connected (FC) layer and then the resulting N feature vectors and a dummy vector, called class-token $\langle \texttt{CLS} \rangle$, are fed into Transformer as $N+1$ inputs. Note that the number N is variable with font types. After going through a transformer layer (which is comprised of self-attention and FC) L times, we have the Transformer output corresponding to $\langle \texttt{CLS} \rangle$. This output is finally fed to a multi-layer perceptron (MLP)-head with a sigmoid function to have the probability of each of K impressions.

The above model is similar to ViT [7] but very different at three points. First, ours input local descriptors (which are extracted irregularly like Fig. 2), instead of regular square patches. Second, ours does not employ the position encoding of the input elements. This is because the locations of the SIFT keypoints where local descriptors are extracted will incur the letter shape (i.e., 'A'-ness [9]) and therefore disturb our part-based impression analysis. Without the position encoding, N SIFT descriptors are treated as a set with N elements; consequently, we do not need to pay attention to the order of the N features when we input them to Transformer. The third and a rather minor difference is that ours have a variable number N of inputs, whereas ViT always accepts the same number (i.e., the number of patches).

4.2 Implementation Details

The multi-label classification model in Fig. 3 (a) is comprised of five transformer layers internally (i.e., $L = 5$). The self-attention module is organized as so-called multi-head attention with five heads. Each 128-dimensional descriptor input is embedded into 128-dimensional space by an FC layer. The maximum number of the input descriptors is 300. If more than 300 keypoints are detected from the six letters ("HERONS"), 300 descriptors are randomly selected. Otherwise, the zero-padding is applied to have 300 inputs in total. MLP head accepts the output of the Transformer and outputs a K-dimensional class likelihood vector via an FC layer. Binary Cross Entropy (BCE) is used as the multi-label classification loss. The Adam optimizer [11] is used with a learning rate of 0.001.

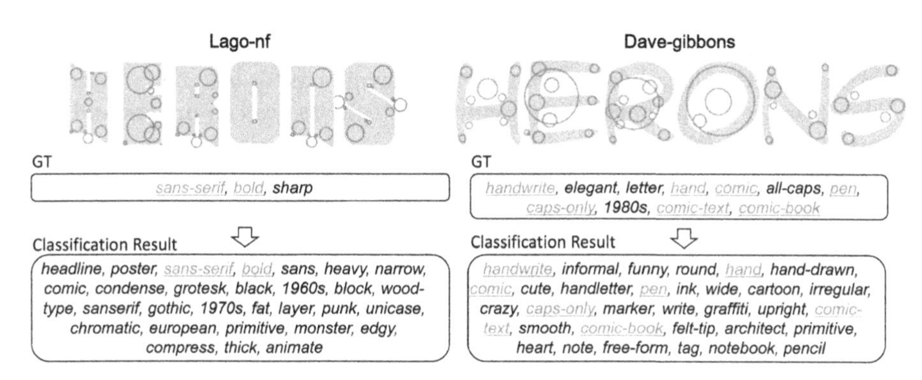

Fig. 4. Example results of multi-label classification. Two fonts are from the test set.

4.3 Classification Examples

Figure 4 shows two test examples of the multi-label classification. Many impressions in the ground-truth are correctly found in the result. Both classification results include significantly more impressions than the ground-truth. A closer look at those excessively detected impressions will find that they are often very similar to one of the labeled impressions. For example, *cartoon* and *pencil* for `Dave-gibbons` are similar to the labeled impressions *comic* and *pen*, respectively. The impression labels of the MyFonts dataset are attached by unspecified people and there is no guarantee that all possible impressions among K vocabularies are fully attached and organized. This situation is the so-called *missing labels* condition. The impression *cartoon* and *pencil* might be missing labels; they could be included in the ground-truth, but none attached them.

4.4 Shape-Impression Relation Analysis by Group-Based Occlusion Sensitivity

To understand the importance of local shapes on a specific impression, we analyze the trained Transformer by using a classical but reliable explainable-AI (XAI) technique, called *occlusion sensitivity*. Occlusion sensitivity was first introduced to an image recognition task in [22]. Its idea is to measure the change of the class likelihood (BCE for our case) when an input part is removed, i.e., occluded. If the removal of an input part drastically decreases the likelihood of a certain class, we consider that the part is very important for the class. In [22], a gray image patch is superimposed on the input image to remove a part.

In our case, the naive application of occlusion sensitivity is to remove a local descriptor from the N inputs and observe the change of the class likelihood. However, this naive application is not meaningful; this is because removing a descriptor from the N inputs often does not affect the likelihood, because very similar descriptors exist in the remaining $N-1$. In other words, if there are similar descriptors, effective occlusion is no longer possible by this one-by-one removal.

We, therefore, group the N descriptors (i.e., local shapes) into Q clusters and remove all the descriptors belonging to a certain cluster $q \in [1, Q]$, from the N descriptors. By this *group-based occlusion sensitivity*, we can understand the importance of a particular type of local shape for a specific impression. For each of K impressions and each of Q local shape types, we measure this sensitivity and then have a Q-dimensional sensitivity vector for each impression. For clustering, k-means is performed on all SIFT descriptors extracted from all the training font images. The Q centroids are treated like "visual words" and SIFT descriptors extracted from a certain font image are quantized into Q types.

5 Shape-Impression Relationship Analysis by Translation Approach

5.1 Transformer as a Shape-to-Impression Translator

Inspired by the fact that cascading two Transformers realizes a language translator [20], we now consider another shape-impression relation analysis approach with Transformers. Figure 3 (b) shows the overall structure of the shape-to-impression translator, which roughly follows the structure of [20]. Two Transformers work as an encoder and a decoder, respectively. In our task, we expect that the decoder outputs the correct impressions by feeding a set of N SIFT descriptors to the encoder.

The encoder is the same as the multi-label classifier of Fig. 3 (a), except it has N outputs and no class-token input. More precisely, after embedding N SIFT descriptors into a certain feature space by an FC layer, the resulting N vectors are fed to the encoder Transformer. Then, the encoder outputs N corresponding latent vectors. Note that the n-th latent vector conveys the information from not only the n-th SIFT descriptor but also the remaining $N - 1$ SIFT descriptors, because of the self-attention mechanism in the encoder.

By utilizing the N latent vectors, the decoder Transformer outputs impression words one by one. More precisely, the special token ⟨BOS⟩ (beginning-of-sentence) is first input to the decoder and then the decoder outputs the first $(K + 3)$-dimensional impression likelihood vector. The extra three dimensions correspond to three special tokens, ⟨BOS⟩, ⟨EOS⟩ (end-of-sentence), and ⟨PAD⟩[1] (padding), respectively. The impression with the maximum likelihood in the K vector elements is determined as the first impression. Second, the first impression word is fed to the decoder as the second input, and then the second impression is output. By repeating this process until ⟨EOS⟩ is output, we have a sequence of impressions.

[1] ⟨PAD⟩ token is used when we train the decoder. ⟨PAD⟩ tokens are added to the end of the ground-truth (i.e., the sequence of the labeled impressions) multiple times until the length of the ground-truth reaches the maximum output length.

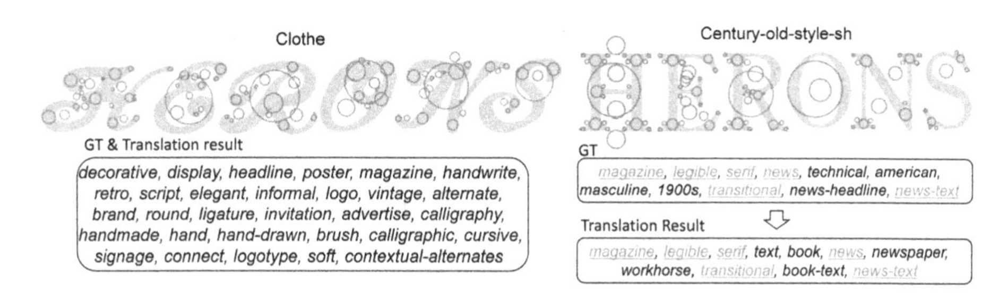

Fig. 5. Example results of shape-to-impression translation. Two fonts are from the test set.

5.2 Implementation Details

For training the decoder of the shape-to-impression translator, we need to specify the unique order of impressions as its ideal output, although they have no pre-defined order. We, therefore, rank the K impressions according to their frequency; the most popular impression is the top and the least popular impression is the bottom. Consequently, the decoder tends to outputs impressions in the descending order of their popularity.

Both Transformers in Fig. 3 (b) are organized by five transformation layers ($L = 5$). The self-attention module is organized by single-head attention. The embedding process for the N input descriptors, the optimizer, and the learning rate are the same as for the multi-label classifier in the previous section. Only for the input of the decoder, the sinusoidal position encoding [20] is used. We employ the beam search during translation. Cross-Entropy is used as the loss function.

5.3 Translation Examples

Figure 5 shows two test examples of the shape-to-impression translator. Surprisingly, for the test font `Clothe`, the translator could output all 30 labeled impressions perfectly. For `Century-old-sty`, the translator cannot give the perfect result. As noted in Sect. 4.3, the ground-truth of the MyFonts dataset often has missing labels and therefore some impressions similar to a labeled impression are often detected excessively.

5.4 Shape-Impression Relation Analysis Using Integrated Gradients

To evaluate the shape-impression relationship with the translation approach, we use a gradient-based XAI method, called *Integrated Gradients* (IG) [18], which has a high versatility for explaining arbitrary networks and has thus been recently utilized in various applications [4,8]. Although our translation approach employs a complex encoder-decoder framework, IG is still applicable to it for understanding its input (local shape) and output (impression) relationship.

More precisely, in our case, IG is used to evaluate the decoder output gradients against one of N local shapes. The key idea of IG is to evaluate multiple gradients that are measured by gradually changing the input vectors from a zero vector (called "baseline" in [18]) to the original SIFT descriptors. The multiple gradients are then aggregated (i.e., integrated) into the integrated gradient, i.e., IG. If an input local shape has a large IG to a specific impression, the input is important for the impression.

6 Experimental Results

6.1 Quantitative Evaluation of the Trained Transformer

Although our main purpose is to analyze the relationship between local shapes and impressions through the trained Transformers, we first need to confirm that those Transformers show reasonable performance in the multi-label classification task and the translation task. In other words, if the trained Transformers' performance is poor on these tasks, the relationship learned in the Transformers is not reliable. We, therefore, first conducted a quantitative evaluation of the trained Transformers.

Evaluation Metrics

F1@100, F1@200, and F1@all are evaluation metrics for multi-label classification tasks. They are simply the average of the F1 scores of the most frequent 100, 200, and all $K(=483)$ impressions, respectively. For the classification approach of Sect. 4, the multi-label classification is first made by applying the threshold 0.5 to the class likelihood values of K impressions. If the likelihood of the k-th impression class is larger than 0.5, we determine that the set of N local shapes show the impression k. Then, the F1 score is calculated for K individual impressions and then finally averaged. For the translation approach of Sect. 5, we calculate the K F1 scores by comparing the output impression sequence (i.e., the translation result) to the correct set of impressions for each font.

mean average precision (mAP) is also an evaluation metric for multi-label classification tasks. It is the average of the average precisions of all K impressions. The average precision for the k-th impression is calculated by using the list of all M fonts ranked in the descending order of the likelihood of the k-th impression. More precisely, it is calculated as $(\sum_{m=1}^{M_k} m/r_m)/M_k$, where M_k is the number of fonts labeled with k, $r_m \in [1, M]$ is the rank of the font with the mth largest likelihood among M_k. Since the translator approach does not provide the impression likelihood, we cannot calculate its mAP.

Table 1. Quantitative evaluation result.

Inputs	Multi-label classification	Translator	ViT [7]	DeepSets [19]
	SIFT	SIFT	Patch	SIFT
F1@100↑	**0.301**	0.264	0.264	0.279
F1@200↑	**0.221**	0.186	0.185	0.194
F1@all↑	**0.145**	0.117	0.109	0.110
mAP ↑	**0.135**	N/A	0.115	0.115

Results and Comparisons. In Table 1, the impression estimation accuracies of the proposed two approaches are compared with the performance by two existing methods, ViT [7] and DeepSets [19]. This table clearly shows that our classification approaches with local shape inputs outperform the existing methods. Comparison between our classification approach and ViT indicates the SIFT descriptors[2] are more suitable than the regular patches, for capturing the important local shapes. More importantly, the fact that our classification approach outperforms DeepSets (with the SIFT descriptors) suggests that the correlation among local shapes by the self-attention in Transformer is important to estimate the font impressions. (As noted in Sect. 2.2, DeepSets treats local shapes totally independently.) The translator shows a similar performance with the existing methods.

Figure 6 shows the F1 scores of three methods (DeepSets-based method [19], the multi-label classifier, and the shape-to-impression translator) for 30 impressions by parallel coordinate plots. The 30 impressions are selected from font-style type impressions (such as *sans-serif*), shape-related property impressions (such as *round*), and abstract impressions (such as *elegant*). Figure 6 also shows examples of font images.

Although it is difficult to find strong trends from Fig. 6, several weak trends can be found as follows. (1) The multi-label classifier shows higher F1 than (or equal F1 to) the DeepSets-based method for most impressions. (2) The multi-label classifier shows significantly higher F1 for the impression whose discriminative shape needs to be described by a combination of local parts. Multiple narrow gaps (with abrupt stoke ends) of *stencil*, square-shaped serifs of *slab-serif*, and densely-distributed (sparsely-distributed) local parts of *narrow* (*wide*) are examples. This might be because of the positive effect of self-attention in Transformer. (3) Frequent impressions tend to get higher F1 values and those values are rather stable among the three methods.

Although it is also difficult to find general trends in the relationship between the multi-label classifier and the translator, the latter often outperforms the

[2] We have tried the SURF descriptors instead of the SIFT descriptors to show the justification to select SIFT as local shape descriptors. We found no significant differences between them. More precisely, the multi-label classifier using SURF achieved about 0.16-point higher mAP and 0.05-point lower F1@all than SIFT.

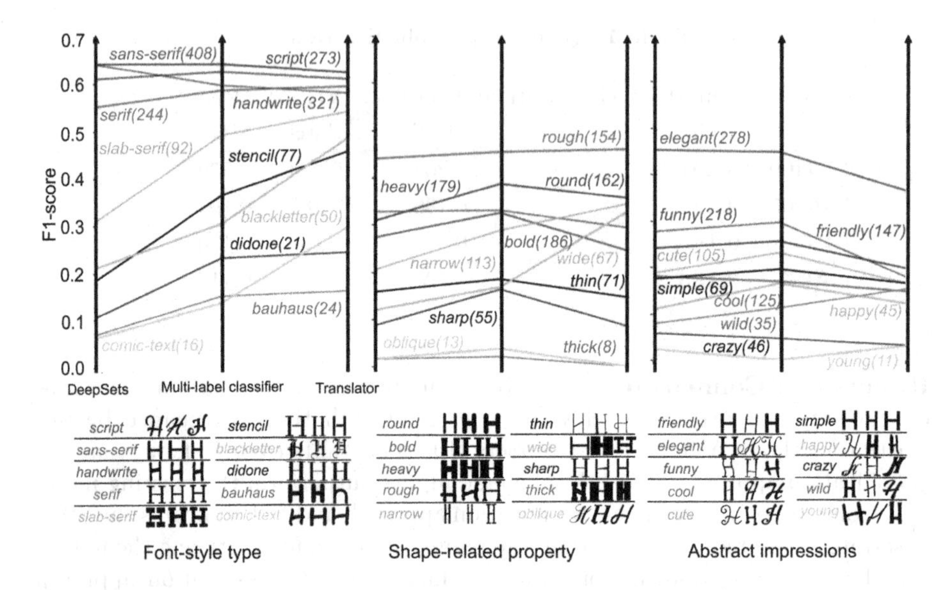

Fig. 6. F1 scores of three methods for 30 impressions are represented by parallel coordinate plots. The number of the test fonts with the impression is parenthesized. Three font image examples for each impression are also shown.

former for minor but strong impressions, such as *blackletter* and *comic-text*. This will be because the fonts with such strong impressions tend to have a stable impression set. In fact, the stability is beneficial for the translator. Since the translator recursively outputs impressions in order of popularity, major but unexpected impressions will interfere with the output of subsequent minor impressions. Therefore, if a font has a stable impression set, the translator can output minor but specific impressions, such as *blackletter*.

6.2 Analysis Results of the Shape-Impression Relationship

Analysis with the Multi-label Classification Approach. Figure 7 shows the local shapes that are important for specific impressions, such as *round* and *elegant*, by the multi-label classifier with the group-based occlusion sensitivity. The bar charts illustrate the sensitivity (i.e., the importance) of $Q = 64$ local shape types for each of the four impressions. Since these bar charts show the *difference* from the median sensitivity, they have a minus element at the qth bin when the qth local shape type is less frequent than usual in the impression. Large positive peaks in the bar chart indicate the very important types. For example, for *round*, the 10th and 11th shape types are the most important. On the font images (from the test set), the local parts belonging to the important types are superimposed as circles, whose colors correspond to the types.

The following provides a brief interpretation of individual results. *Round* has its peaks at $q = 10, 11$ that correspond to round corners or round spaces.

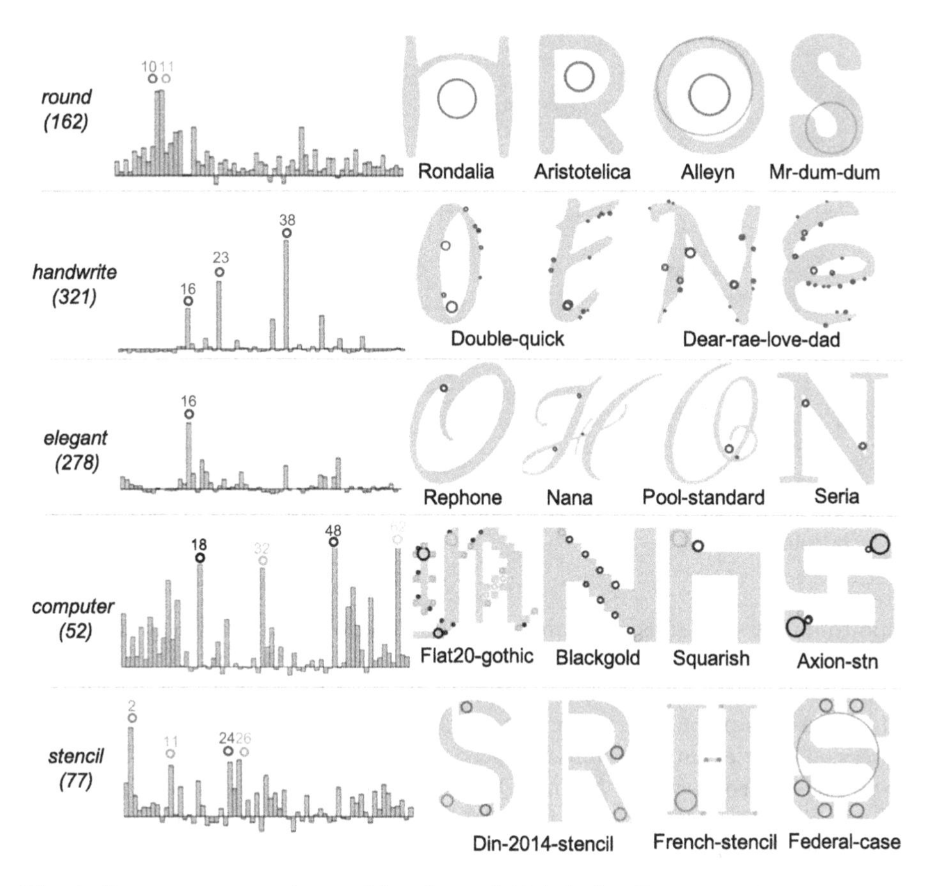

Fig. 7. Important parts detected by the multi-label classification approach, for four impressions. The font names is shown below its image. The parenthesized number is the number of fonts with the impression.

Handwrite has a peak at $q = 16$ that corresponds to sharp and asymmetric (i.e., organic) curves and two peaks at $q = 23, 38$ that correspond to rough shapes mimicking brush stroke. *Elegant* also has a peak at $q = 16$. However, it does not have peaks at $q = 23, 38$; this indicates that *Elegant* is organized by some organic and non-rough (i.e., smooth) curved strokes. *Computer* shows several peaks; among them the local shape types of $q = 32, 48$ are often found together. This indicates that self-attention could successfully enhance the co-occurrence of those types for *Computer*. *Stencil* has several peaks at $q = 2, 24, 26$ that correspond to the abrupt end of a constant-width stroke. Since *Stencil* often contains the abrupt-end shapes (for mimicking the actual stencil letters), excessive existence of those local shapes is important for its impression.

Analysis with the Translator. Figure 8 shows important local parts given by IG on two translation results. Compared to the multi-label classifier, the

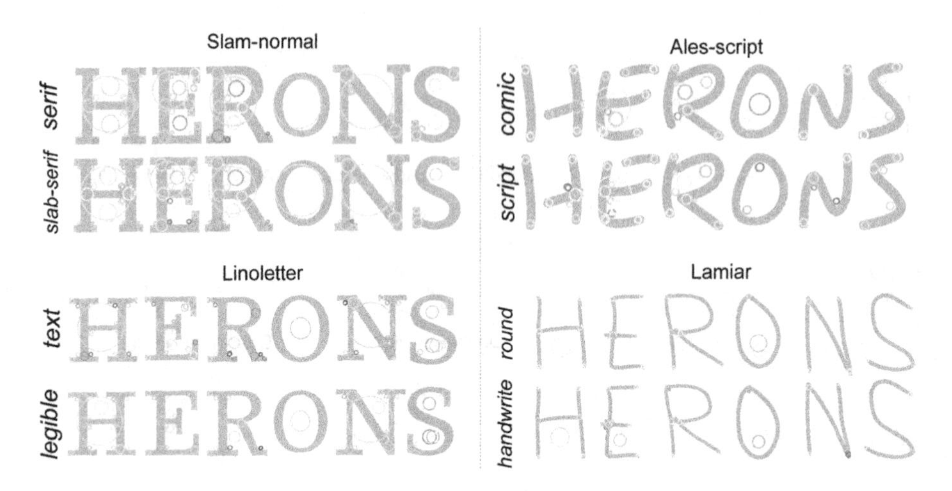

Fig. 8. Important local parts by the translator with IG. More important parts are darker in color.

shape-to-impression translator can explain its important parts without any quantization and grouping by using IG. The darker circle has more IG, i.e., more important. In `Slam-normal`, the serif parts of 'R' contributes to *serif* and *slab-serif*. In `Ales-script`, the important local parts are different according to the impressions. For example, *comic-text* needs a large space of 'O' and 'R,' whereas *script* needs sharp curves of 'N' and 'O.' These results show that the shape which seems to be unique for a specific impression is important to translate into them. In `Linoletter`, many of serif parts contribute to *text*, and the round space of 'S' contributes to *legible*. In `Lamiar`, the enclosed area of 'O' is important for *round* and the sharp stroke ends of 'E', 'R', and 'O' are important for *handwrite*. These results coincide with the results by the multi-label classifier shown in Fig. 7.

7 Conclusion and Future Work

In this paper, part-based font-impression analysis is performed using Transformer. The versatility of Transformer offers us to realize two analysis methods: a multi-label impression classifier and a shape-to-impression translator. Using a large font-impression dataset, we experimentally prove that the multi-label classifier could achieve better impression estimation performance; this means it can learn the trends between local shapes and impressions more accurately. We also revealed important local parts for specific impressions by using the trained Transformer with explainable-AI techniques.

Future work will focus on practical applications of the proposed methods to font selection or recommendation systems, font generation systems, and so on. Our analysis results on shape-impression relationships will be validated by collaborations with experts of cognitive psychology.

References

1. Brown, T., et al.: Language models are few-shot learners. In: NeurIPS (2020)
2. Chen, T., Wang, Z., Xu, N., Jin, H., Luo, J.: Large-scale tag-based font retrieval with generative feature learning. In: ICCV (2019)
3. Choi, S., Aizawa, K., Sebe, N.: FontMatcher: font image paring for harmonious digital graphic design. In: ACM IUI (2018)
4. Cornia, M., Stefanini, M., Baraldi, L., Cucchiara, R.: Meshed-memory transformer for image captioning. In: CVPR (2020)
5. Davis, R.C., Smith, H.J.: Determinants of feeling tone in type faces. J. Appl. Psychol. **17**(6), 742–764 (1933)
6. Devlin, J., Chang, M.W., Lee, K., Toutanova, K.: BERT: pre-training of deep bidirectional transformers for language understanding. arXiv preprint arXiv:1810.04805 (2018)
7. Dosovitskiy, A., et al.: An image is worth 16×16 words: transformers for image recognition at scale. In: ICLR (2020)
8. He, S., Tu, Z., Wang, X., Wang, L., Lyu, M., Shi, S.: Towards understanding neural machine translation with word importance. In: EMNLP-IJCNLP (2019)
9. Hofstadter, D.R.: Metamagical Themas: Questing for the Essence of Mind and Pattern. Basic Books, Inc. (1985)
10. Khan, S., Naseer, M., Hayat, M., Zamir, S.W., Khan, F.S., Shah, M.: Transformers in vision: a survey. arXiv preprint arXiv:2101.01169 (2021)
11. Kingma, D.P., Ba, J.: Adam: a method for stochastic optimization. In: ICLR (2015)
12. Lowe, D.G.: Distinctive image features from scale-invariant keypoints. Int. J. Comput. Vis. **60**(2), 91–110 (2004)
13. Matsuda, S., Kimura, A., Uchida, S.: Impressions2Font: generating fonts by specifying impressions. In: Lladós, J., Lopresti, D., Uchida, S. (eds.) ICDAR 2021. LNCS, vol. 12823, pp. 739–754. Springer, Cham (2021). https://doi.org/10.1007/978-3-030-86334-0_48
14. O'Donovan, P., Lībeks, J., Agarwala, A., Hertzmann, A.: Exploratory font selection using crowdsourced attributes. ACM TOG **33**(4), 92 (2014)
15. Otter, D.W., Medina, J.R., Kalita, J.K.: A survey of the usages of deep learning for natural language processing. IEEE Trans. Neural Netw. Learn. Syst. **32**(2), 604–624 (2021)
16. Poffenberger, A.T., Franken, R.: A study of the appropriateness of type faces. J. Appl. Psychol. **7**(4), 312–329 (1923)
17. Shaikh, D., Chaparro, B.: Perception of fonts: perceived personality traits and appropriate uses. In: Digital Fonts and Reading, chap. 13. World Scientific (2016)
18. Sundararajan, M., Taly, A., Yan, Q.: Axiomatic attribution for deep networks. In: ICML (2017)
19. Ueda, M., Kimura, A., Uchida, S.: Which parts determine the impression of the font? In: Lladós, J., Lopresti, D., Uchida, S. (eds.) ICDAR 2021. LNCS, vol. 12823, pp. 723–738. Springer, Cham (2021). https://doi.org/10.1007/978-3-030-86334-0_47
20. Vaswani, A., et al.: Attention is all you need. In: NeurIPS (2017)
21. Zaheer, M., Kottur, S., Ravanbhakhsh, S., Póczos, B., Salakhutdinov, R., Smola, A.J.: Deep sets. In: NeurIPS (2017)
22. Zeiler, M.D., Fergus, R.: Visualizing and understanding convolutional networks. In: Fleet, D., Pajdla, T., Schiele, B., Tuytelaars, T. (eds.) ECCV 2014. LNCS, vol. 8689, pp. 818–833. Springer, Cham (2014). https://doi.org/10.1007/978-3-319-10590-1_53

TrueType Transformer: Character and Font Style Recognition in Outline Format

Yusuke Nagata(✉), Jinki Otao, Daichi Haraguchi, and Seiichi Uchida

Kyushu University, Fukuoka, Japan
yusuke.nagata@human.ait.kyushu-u.ac.jp

Abstract. We propose TrueType Transformer (T^3), which can perform character and font style recognition in an outline format. The outline format, such as TrueType, represents each character as a sequence of control points of stroke contours and is frequently used in born-digital documents. T^3 is organized by a deep neural network, so-called Transformer. Transformer is originally proposed for sequential data, such as text, and therefore appropriate for handling the outline data. In other words, T^3 directly accepts the outline data without converting it into a bitmap image. Consequently, T^3 realizes a resolution-independent classification. Moreover, since the locations of the control points represent the fine and local structures of the font style, T^3 is suitable for font style classification, where such structures are very important. In this paper, we experimentally show the applicability of T^3 in character and font style recognition tasks, while observing how the individual control points contribute to classification results.

Keywords: Outline format · Font style recognition · Transformer

1 Introduction

In a born-digital document, text data is embedded into the document as various vector formats, including the *outline format*, instead of the bitmap image format and the ASCII format (i.e., character codes plus font names). Figure 1 illustrates the outline format. Each character is represented as single or multiple contours, and each contour is represented by a sequence of *control points*. Even for the same letter 'A,' its contour shape is largely variable, depending on its typeface (i.e., font style). In other words, the outline format conveys the font style information by itself. Much graphic design software (such as Adobe Illustrator) have a function of converting a character image in a specific font into the outline format.

The outline format is often used by graphic design experts because of the following two reasons.

© Springer Nature Switzerland AG 2022
S. Uchida et al. (Eds.): DAS 2022, LNCS 13237, pp. 18–32, 2022.
https://doi.org/10.1007/978-3-031-06555-2_2

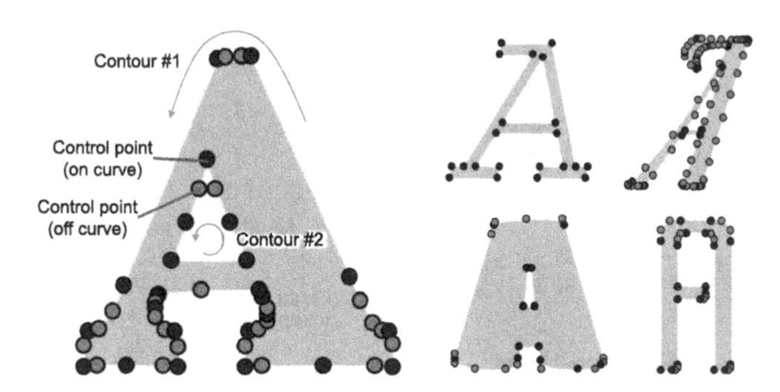

Fig. 1. An outline format. Each stroke contour is represented as a sequence of *control points*. A control point is *on curve* or *off curve*. In the former case, the stroke outline passes through the control point, whereas the control point controls the contour curvature in the latter case. Roughly speaking, TrueType follows this format.

Reproducibility: Imagine that an expert designs a web advertisement with texts in a special font X and sends it to another expert who does not have the font X in her/his design environment. If the texts are not in the outline format (i.e., the texts are represented as their character codes plus font names), a different font will be used for showing the texts as the substitution of X. In contrast, if they are in the outline format, exactly the same text appearance is reproduced.

Flexibility: Experts often modify font shapes for a specific purpose. Especially, the contour shapes of characters in logotypes and posters are often slightly modified from their original font shape in order to enhance their uniqueness and conspicuousness. Kerning (i.e., space between the adjacent characters) is also tuned by ignoring the default kerning rule of the font. The outline format has the flexibility to represent texts with the unique shapes and kernings.

In this paper, we attempt to recognize characters in their outline format by a Transformer-based neural network, called TrueType Transformer (T^3), *without* converting them into a bitmap image format. Figure 2 shows the overview of T^3. In fact, there are two practical merits to recognizing characters directly in the outline format.

- First, we do not need to be careful of the image resolution. If it is too low, the recognition performance becomes poor; if it is too high, we need to waste unnecessary computation costs. Recognizing characters in the outline format is free from this unnecessary hyperparameter, i.e., the resolution.
- The second and more important merit is that the outline format will carry very detailed information about font styles. Font styles are always designed very carefully by font designers, and they pay great attention about the curvatures of contours and fine structures, such as serifs. In the image format, those design elements are often not emphasized. In the outline format, the location and the numbers of control points show the design elements explicitly. Consequently, to understand the font styles, the outline format will be more suitable.

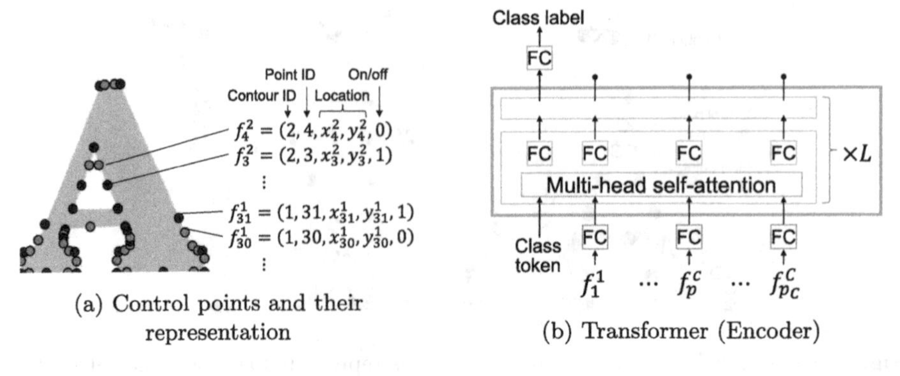

(a) Control points and their representation

(b) Transformer (Encoder)

Fig. 2. The overview of the proposed TrueType Transformer (T^3). (Color figure online)

Despite these merits, there are two issues in handling the outline format for recognition tasks. First, characters in the outline format have a different number of control points, and thus their input dimension is also variable. Second, the control points should not be treated as just a point set. For example, suppose we have a character consisting of two contours (such as 'A' on the left side of Fig. 1). By cascading the two sequences obtained from both contours, a single control point sequence can be obtained. In this case, in order to understand, for example, the stroke thickness, it is not enough to use the relationship between adjacent points on the sequence; instead, we need to know the positional relationship between points that are spatially close but distant in the sequence. Therefore, the mutual relationships between all the control points are often important to understand the shape in the outline format.

Transformer used in T^3 is very suitable for dealing with the above issues on the recognition tasks with the outline format. For the first issue, Transformer has sufficient flexibility in the input length varieties; in fact, Transformer has been originally developed for natural language texts, which have no fixed length. For the second issue, the self-attention mechanism inside Transformer can learn the importance of the individual relationships among all inputs, i.e., control points.

The purpose of this paper is twofold. The first purpose is to prove that characters are recognizable in the outline format. As noted above, outline-based recognition has the merits for dealing with born-digital documents. To the best knowledge of the authors, this is the first attempt at outline-based character recognition. For this purpose, we consider two different tasks; character class recognition and font style recognition. The former is the orthodox character recognition task and the latter is to classify character data into one of font style classes, such as *Serif* and *Display*. The latter task needs to be more careful of local shapes and thus is more suitable for understanding the merit of the outline-based recognition.

The second purpose is to analyze the importance of individual control points for the recognition tasks. This is important for a scientific question to understand discriminative parts for a specific character class or font style class. The analysis is formed by visualizing the self-attention weight, which directly shows the importance of the control points. The method called attention roll-out [1] is employed for the visualization. We will observe how the important control points are different between the character recognition task and the font style recognition task.

We summarize our main contributions as follows.

- To the authors' best knowledge, this is the first attempt to tackle character and font style recognition tasks in an outline format. Recognizing in the outline format is practically useful for dealing with born-digital documents.
- To conduct the above recognition tasks, we propose TrueType Transformer (T^3) that accepts a set of control points of the character outline as its input.
- T^3 showed higher accuracy than two image-based models (ResNet and ViT [7]) in the font style recognition task, which needs to be more careful of fine outline shapes.
- We analyze the learned attention weight to understand the various trends of important control points for a specific character class or font style class. The analysis results, for example, show that important control points are very different between the recognition tasks and also among the four style classes.

2 Related Work

2.1 Transformer

Transformer [20] began to be used and has updated the state-of-the-art records [6,12] in the field of natural language processing (NLP). The transformer initially consists of an encoder and decoder and can perform a language translation. One of the main characteristics of the transformer is that it is possible to process sequential data, i.e., textual data. This characteristic is compatible with our outline data because the outline data is regarded as sequential data as we described in the above Sect. 1.

In recent years, the transformer also has been used in the field of computer vision and image classification [10]. Vision Transformer (ViT) [7] is one of the most famous applications of Transformer for image classification. Unlike transformer for NLP, Transformer for image classification often accept the constant number of input (because the input image size is fixed), and only a transformer encoder is used.

The visualization of attention in Transformer is useful to understand not only the internal behavior of Transformer but also the important elements among all the inputs. Abnar *et al.* [1] proposed attention rollout to visualize attention weight by integrating the features extracted by each transformer layer. Chefer *et al.* [4] proposed the visualization method of attention weight based on a specific formulation while maintaining the total relevancy in each layer.

In this paper, we propose T^3 that utilize the characteristics of Transformer to recognize character class and font style from the outline. In the experiments, we visualize the attention weights by using the above rollout for understanding important control points (i.e., control points that have a larger attention weight) in character and font style recognition tasks.

2.2 Font Analysis by Using Vector Format

To the authors' best knowledge, there is no attempt at character recognition in the outline format. In fact, for many years, the mainstream of font analysis has used bitmap image format (or raster format). However, since digital font data is provided in the TrueType format or OpenType format, it is quite natural to deal with them in the vector format.

Nowadays, we can find several font-related attempts by using other vector formats, such as scalable vector graphics (SVG). Especially for font generation and conversion tasks, focus on the advantage of vector format because it can be free from resolution or scale. Cambell et al. [2] have done pioneering work for outline-based font generation, where a fixed-dimensional font contour model is represented as a low-dimensional manifold. More recently, Lopes et al. [13] proposed a VAE-based model to extract the font style feature from images and SVG decoder for converting the feature into the SVG format. Carlier et al. proposed DeepSVG [3], which is a font generation model that outputs font images in the SVG format. Im2Vec [14] by Reddy et al. generate a vector path from an image input. DeepVecFont [22] by Wang et al. can address font generation in both image and sequential formats by an encoder-decoder framework.

Unlike the above font generation methods, we mainly focus on recognition tasks by using an outline format with an arbitrary number of control points. For this purpose, we utilize Transformer, which characteristics perfectly fit our problem, as noted in Sect. 1. Moreover, by using the learned attention in Transformer, we analyze the important control points for character class recognition and font style recognition.

3 TrueType Transformer (T^3)

3.1 Representation of Outline

As shown in Fig. 2 (a), each control point is represented as a five-dimensional vector f_p^c, where $c \in \{1, \ldots, C\}$ is the contour ID and C is the number of contours in a character data. The point ID $p \in \{1, \ldots, P^c\}$ shows the order in the c-th contour[1]. Consequently, the character data is represented by a set of $N = \sum_c P^c$ vectors and fed to Transformer in the order of $\mathbf{f} = f_1^1, \ldots, f_{P^1}^1, f_1^2, \ldots, f_p^c, \ldots, f_{P^C}^C$.

The five elements of f_p^c are the contour ID c, the font ID p, the location (x_p^c, y_p^c), and the on/off flag $\in \{0, 1\}$. If the on/off flag is 1, it is an "*on-curve*" point; otherwise, an "*off-curve*." A contour is drawn to pass on-curve points. An off-curve point controls the curvature of the contour. As shown in Fig. 1, a character contour sometimes does not have any off-curve point; in this case, the contour is drawn only with straight line segments.

[1] In the outline format, the start point and the endpoint of each contour are located at exactly the same position, that is, $f_1^c = f_{P^C}^c$.

3.2 Transformer Model for T^3

Figure 2 (b) shows the proposed TrueType Transformer (T^3), which performs a recognition task by using an outline data $\mathbf{f}$. T^3 is based on a Transformer encoder model, like BERT [6]. A sequence $\mathbf{f}$ of N control points is first fed to Transformer via a single $5 \times D$ fully-connected (FC) layer (depicted as a green box in (b)). To the resulting N D-dim vectors, another D-dim dummy vector, called *Class token*, is added. Class token is generated as a random and constant vector.

The succeeding process with Transformer is outlined as follows. The $(N+1)$ D-dim input vectors are fed to the multi-head self-attention module, which calculates the correlation of each pair of $(N+1)$ inputs by M different self-attention modules and then outputs $(N+1)$ D-dim vectors by using the correlations as weights. Consequently, $(N+1)$ input vectors are "transformed" to other $(N+1)$ output vectors by the multi-head self-attention module and an FC layer, while keeping their dimensionality D. This transformation operation is repeated L times in Transformer, as shown in Fig. 2(b). The final classification result is given by feeding the leftmost output vector (corresponding to Class token) to a single $D \times K$ FC layer, where K is the number of classes. This FC layer is the so-called MLP-head and implemented with the soft-max operation.

As noted above, the structure of T^3 follows BERT [6], but has an important difference in the *position encoding* procedure [8]. In most Transformer applications, including BERT and ViT, position encoding is mandatory and affects the performance drastically [21]. T^3, however, does not employ position encoding. This is because the input vector f_p^c already contains position information as contour ID c and point ID p. These IDs specify the order of N inputs and therefore it is possible to omit any extra position encoding.

4 Experiment

4.1 Dataset

Google Fonts[2] were used in our experiments. One reason for using them is that they are annotated with five font style categories (*Serif*, *Sans-Serif*, *Display*, *Handwriting*, and *Monospace*). Another reason is that Google Fonts are one of the most popular font sets for font analysis research [15,17,18]. From Google Fonts, we selected the Latin fonts that were used in STEFANN [15]. We then omitted *Monospace* fonts because of two reasons. First, they contain only 75 fonts. Second, they are mainly characterized by not their style but their kerning rule (i.e., space between adjacent letters); therefore, some *Monospace* fonts have a style of another class, such as *Sans-Serif*. Consequently, we used 456 *Serif* fonts, 859 *Sans-Serif*, 327 *Display*, and 128 *Handwriting*, respectively. Those fonts are split into three font-disjoint subsets; that is, the training set with 1,218 fonts, the validation set with 135, and the test set with 471.

[2] https://github.com/google/fonts.

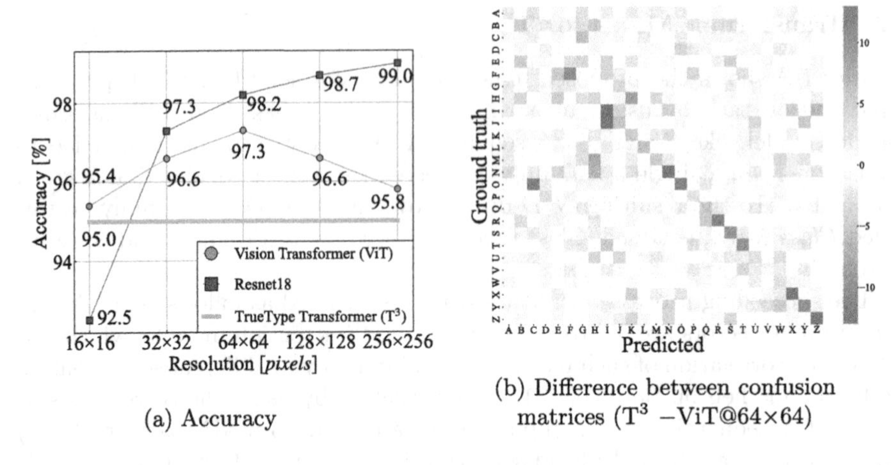

(a) Accuracy

(b) Difference between confusion matrices (T^3 −ViT@64×64)

Fig. 3. Character recognition result. In (b), the darker green non-diagonal components and the darker red diagonal components mean that T^3 have less misrecognition. (Color figure online)

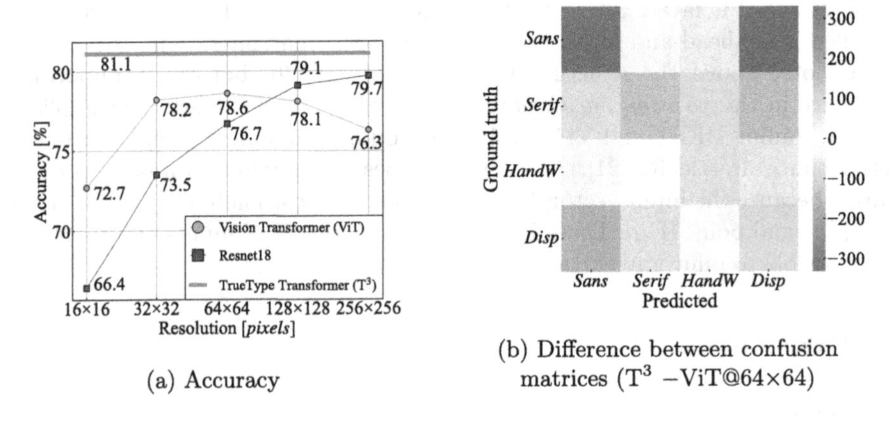

(a) Accuracy

(b) Difference between confusion matrices (T^3 −ViT@64×64)

Fig. 4. Font style recognition result.

4.2 Implementation Details

We set the hyperparameters of Transformer as follows. The internal dimensionality of each feature vector is set at $D = 100$. The number of multiple-heads in the self-attention module is $M = 5$. The number of the transformation layers was $L = 6$. As the last step of the multi-head self-attention in Transformer, each set of M D-dimensional vectors (corresponding to each input) are converted into a single D-dimensional vector by a single FC layer. Consequently, the multi-head attention module accepts $(N + 1)$ D-dimensional vectors and outputs $(N + 1)$ D-dimensional vectors. The FC layer in the yellow box in Fig. 2 (b) converts D-dimensional vectors into 1024 and then $D = 100$. During training, Adam [11] is used as the optimizer with a learning rate of 0.0001 to minimize the cross-entropy loss.

As comparative methods, we used two image-based recognition methods; one is ResNet18 [9], and the other is ViT [7]. From the outline format, we can generate bitmap images in arbitrary resolutions. We, therefore, examine different resolutions (16×16, 32×32, 64×64, 128×128, and 256×256) in the later experiments. In ViT, a whole image is divided into 16 square patches; therefore, the patch size depends on the resolution. For example, for 64×64 images, each patch is 16×16. Different from T^3, ViT employs position encoding proposed in [7]. We did not use the pretrained ViT because it was pretrained with natural images from ImageNet-21k and JFT-300M instead of binary (i.e., black and white) images.

4.3 Quantitative Comparison Between Outline-Based and Image-Based Recognition Methods

Accuracy in the Character Recognition Task. Figure 3 (a) shows the character recognition accuracy by T^3, ResNet18, and ViT, at different image resolutions. Since T^3 is free from image representation, its accuracy is shown as a flat horizontal line in each graph. T^3 achieved 95.0%; although it is not higher than the image-based methods (except for ResNet@16×16), it is also true that the accuracy by T^3 is not far inferior to the image-based methods. This result proves that characters can be directly recognized in their outline format. Moreover, T^3 does not need to be careful of the extra hyper-parameter, i.e., image resolution, whereas the image-based methods need[3].

For a more detailed comparison, Fig. 3 (b) shows the difference of confusion matrices between T^3 and ViT@64×64. This result indicates that both methods have their own advantages. For example, T^3 has fewer misrecognition of 'E'↔'F' and 'I'↔'J.' This observation suggests that T^3 can capture fine differences between visually-similar classes. (This point is confirmed in the experimental results of font style recognition, which requires to discriminate visually-similar but slightly-different samples.) On the other hand, T^3 has more misrecognition of 'H'→'K' and 'O'→'C.' Two letters 'H' and 'K' are very different in their appearance, but often not in their control points – just removing several control points from 'H,' it becomes similar to 'K.' Consequently, outline and image bitmap show their own advantages, and thus their complementary usage will further improve the accuracy.

[3] The accuracy of ResNet monotonically increases with the resolution; this is a strong advantage of ResNet. However, at the same time, the higher resolution induces more input elements. A character data in the 256×256 image format has 65,536 elements. In contrast, the outline format has only $\bar{p} \times 5$ elements, where $\bar{p}$ is the average number of the control points and about 45 in our dataset. If we use the median, $\bar{p} \sim 30$. Therefore, T^3 needs only 0.2% elements of the 256×256 image format. Even compared with the 16×16 image, T^3 needs about 60%.

Accuracy in Font Style Recognition Task. The advantage of T^3 is more obvious in the font style recognition task, as shown in Fig. 4 (a). In this result, T^3 outperforms the image-based methods regardless of the image resolution; even with the 256×256 resolution, image-based methods could not outperform T^3. This is because the outline format can represent fine structures of font shapes. In fact, the outline format can explicitly represent the sharpness of stroke corners, tiny serif structures, fluctuations of curvatures, etc.

Figure 4 (b) shows the difference of confusion matrices between T^3 and ViT@64×64. This result shows that the misrecognition of *Sans-Serif*→*Display* (the darkest green element) is drastically decreased by T^3. In fact, there are many ambiguous fonts between *Sans-Serif* and *Display*, and thus the style classification based on their appearance is difficult. However, in the outline format, it becomes easier to classify them. This is because, as we will see later, fonts in *Display* often have many control points to realize their unique shape. Therefore, by utilizing the number of the control points for the final classification inside Transformer, T^3 can distinguish *Display* from *Sans-Serif* and vice versa.

4.4 Qualitative Comparison Between Outline-Based and Image-Based Recognition Methods

Samples Correctly Recognized by T^3. To understand the characteristics of T^3 via its successful results, we randomly selected 100 samples that were correctly recognized by T^3 and misrecognized by ResNet and/or ViT, and show them in Fig. 5. In (a), we can observe a clear trend that T^3 could recognize the character class of very thin font images. This is reasonable because very thin strokes might be overlooked by the image-based methods, whereas they have no large difference from thicker strokes for our outline-based method. In other words, in the outline format, almost the same number of control points will be used regardless of the stroke thickness. This property realizes the robustness of T^3 to thin strokes.

Compared to Fig. 5 (a), the samples in (b) did not show any clear trend. However, for example, the shapes of those *Sans-Serif* fonts have very sharp corners and, therefore, T^3 can give the correct recognition results for them if T^3 has the ability to emphasize the sharp corner. In Sect. 4.5, we will conduct a more detailed analysis of the style recognition results by using the learned attention. We then can confirm that T^3 has the ability to emphasize the unique characteristics of each font style class.

The Number of Control Points in *Display*. For *Display* style, having a large number of control points is an important clue for the correct recognition. Each character sample of *Display* has 166 control points on average when its style is correctly recognized as *Display*; however, misrecognized samples only have 47 control points on average. Fonts of *Display* often have a complicated outline or a rough (i.e., non-smooth) outline and thus need to use many control points. Therefore, it is reasonable that T^3 internally uses the number of points for discriminating *Display* from the others.

(a) Character recognition task.

(b) Font style recognition task.

Fig. 5. Samples correctly recognized by T^3. In (b), samples are divided into four blocks according to their style class; from top left to bottom right, *Sans-Serif*, *Serif*, *Handwriting*, and *Display*.

4.5 Analysis of Learned Attention

Visualization of Learned Attention. In this section, we visualize the importance of the individual control points for the classification result, for understanding (1) how T^3 makes its classification and (2) where are the most important parts for each class. For the visualization, we use learned attention. Specifically, we employ *attention rollout* [1] to visualize attention weight, which estimates the contribution of each input to the final classification result by using LRP (layer-wise relevance propagation)-like calculation. Attention rollout does not visualize the learned self-attentions (i.e., correlation among N inputs) in each of L transformation layers, but visualizes how each of N inputs affects the final output by going through all the internal layers.

It should be emphasized that the point-wise visualization is an important advantage of the outline format over the image format. If we visualize important parts for image data by using XAI techniques, such as Grad-CAM, the result is somewhat blurry and fuzzy. Other XAI techniques, such as LRP, can show important pixels, but the effect of spatial smoothing in convolutional neural networks (CNN) is inevitable for them, and thus their result often becomes less reliable. T^3 deals with a limited number (about 30) of control points, and thus the explanation of their importance becomes more reliable and sharper.

Important Points for Character Recognition. As shown in the middle column in Fig. 6, the control points around the lower part in characters are often contributed more than the upper part to character recognition regardless of font styles. A classic typography book [16] claimed that the legibility of characters is lost drastically by hiding their lower part. Our result coincides with this claim. For 'I,' 'T,' and 'Y,' the control points in the upper part are also important; however, they are still a minority compared to the lower parts.

Fig. 6. Visualized attention to important control points. For each style class, the left column shows character images and their all control points. The middle and right columns plot the control points with the six largest attention in character and font style recognition, respectively. Since the start point and the endpoint of each contour are located at exactly the same position in the outline format, sometimes only five points appear. Blue and red points are on-curve and off-curve, respectively, and their darkness is relative to their importance, i.e., the attention weight. In each style class, 26 fonts are randomly selected and then used to print 26 characters 'A'–'Z'. (Color figure online)

Important Points for Font Style Recognition. The rightmost column of each style in Fig. 6 shows that the location of important control points is very different among the four styles.

- Important control points for *Sans-Serif* show several particular trends. First, the important points are often located at the ends of straight lines. This trend reflects that the fonts of *Sans-Serif* often contain sharp corners formed by two straight line segments. In other words, curves (specified by off-curve points) are not important to characterize *Sans-Serif*. Second, the control points that represent a constant stroke width become important. For example, 'E,' 'F,' 'T,' and 'Z' in (a) have four important points for their top horizontal stroke; they can represent two parallel line segments, which suggest a constant stroke width. The points of 'Q' also represent that the stroke width is constant at both left and right sides. The four points at the top of 'H' indicate the constant stroke width in its own way.
- For *Serif*, the control points with larger attention locate around serifs. More interestingly, these points are not scattered but gather each other. This will be because a single point is not enough to recognize the shape of a serif. Since Transformer can deal with the correlation among the inputs, this collaborative attention among the neighboring control points was formed[4].
- The important control points for *Handwriting* have a trend similar to *Serif*; they gather around the endpoint (or the corner) of strokes. This may be because to capture the smooth and round stroke endpoint shapes that are unique to *Handwriting*.
- *Display* shows various cases (which are also observed in the other styles) because the fonts of *Display* have less common trends among themselves. Sometimes a special serif shape (such as 'K' in (d)) is a clue and sometimes a rough curve (such as 'A' and 'R' in (d)). As noted before, having a large number of control points is also an important clue to characterize complex font shapes of *Display*, besides the location of important points.

Comparison Between Two Recognition Tasks. It is natural to suppose that the important points will be different in the character recognition and style recognition tasks. In the character recognition task, several control points that contribute to its legibility will have larger attention. In contrast, in the style recognition task, several control points that form decorative parts (such as serif) will have larger attention. As expected, a comparison between the middle and rightmost columns in Fig. 6 reveals that the important points are often totally different in both tasks. For example, in the samples 'A' of *Handwriting* and *Display* their important points locate totally upside down.

[4] When applying an XAI technique to CNN results, the importance of the neighboring pixels are often similar because of the smoothing effect of spatial convolution. In contrast, Transformer does not have any function of spatial convolution, and the relationship among all N inputs is treated equally regardless of their locations. Consequently, this gathering phenomenon in Fig. 6 proves that Transformer learns that those control points have a strong relationship to form a style.

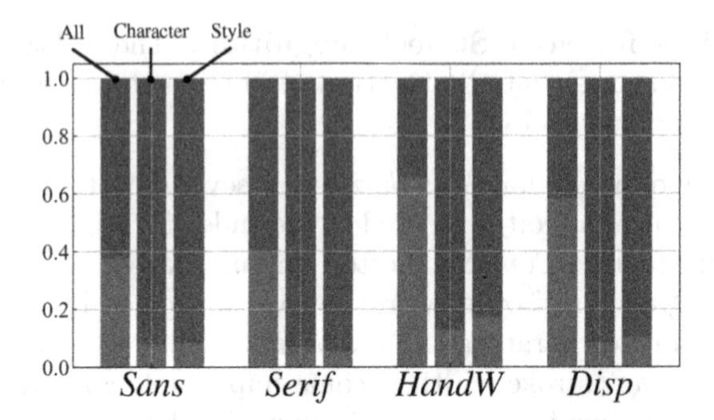

Fig. 7. The ratio between on-curve points (blue) and off-curve points (red) for each style class. "All," "Character," and "Style" bars show the ratios among all control points, top six control points for character recognition, and them for font style recognition, respectively. (Color figure online)

Importance Difference Between On-curve and Off-curve Points. A careful observation of Fig. 6 reveals that on-curve (blue) points are treated more importantly than off-curve (red) points. Except for a small number of samples, on-curve points dominate the important points with the six largest attentions in most samples. This is not because on-curve points are much more than off-curve points—in fact, the numbers of on-curve and off-curve points do not differ significantly. Figure 7 shows the ratio between on-curve and off-curve points for each style. Specifically, the leftmost bar for each style shows the ratio, and the ratio of off-curve points is at least about 40% (at *Sans-Serif* and *Serif*). For *Handwriting* and*Display*, its ratio is more than 50%.

Figure 7 also shows the ratio of on-curve and off-curve among the most important six points on both recognition tasks. As observed above, the ratio of off-curve points decreases to about 10%. This fact indicates that off-curve points are not very important for recognition tasks. For the character recognition task, this fact is convincing because the rough character shape described only by the on-curve points will be enough to recognize the character class. For the font style recognition task, however, this fact is somewhat unexpected because a fine curvature seems to be important as the clue of the style. Although the off-curve points might become more important for finer style classification tasks (such as font impression classification [5,19]), the fine curvature was not important, at least, for the current four style classes.

5 Conclusion

This paper reports a trial of character and font style recognition in an outline format. The outline format has different information from the bitmap image format, even when they seem almost identical in their appearance as images. This difference realizes several advantages of the outline format over the image format. For example, as we proved experimentally, the outline format is suitable for font style recognition since it can explicitly represent the fine structure of font shapes. Moreover, in the character recognition task, where the image format also has a high recognition accuracy, a comparable accuracy can be obtained with the outline format.

For the outline-based recognition, we developed TrueType Transformer (T^3), which is a Transformer-based classification model. Different from an image-based classification model, such as ResNet, it can directly accept a set of an arbitrary number of control points that define the outline, i.e., the font shape. Moreover, by visualizing the attention weight learned in T^3, we can understand which control points are important for determining the recognition results. Various analyses have been made and their results indicate the strategy of T^3 in recognizing different font styles using control points.

Future work will focus on the following attempts. First, we apply T^3 to a recognition task of font impressions (such as *elegant* and *funny*) to see its performance on this more delicate recognition task. In addition, we will observe the importance of off-curve points, which control subtle curvatures of the outline. Second, we extend T^3 to convert the outline data into its deformed version. In other words, we can utilize Transformer as a shape transformer. More precisely, we formulate this shape transformation task in a Transformer-based encoder-decoder system (just like a language translator [20]). In these future attempts, we will be able to utilize the quantitative and qualitative analysis results given in this paper.

References

1. Abnar, S., Zuidema, W.: Quantifying attention flow in transformers. In: Proceedings of the 58th Annual Meeting of the Association for Computational Linguistics (ACL) (2020)
2. Campbell, N.D.F., Kautz, J.: Learning a manifold of fonts. ACM Trans. Graph. (ToG) **33**, 1–11 (2014)
3. Carlier, A., Danelljan, M., Alahi, A., Timofte, R.: DeepSVG: a hierarchical generative network for vector graphics animation. In: Advances in Neural Information Processing Systems (NeurIPS) (2020)
4. Chefer, H., Gur, S., Wolf, L.: Transformer interpretability beyond attention visualization. In: Proceedings of the IEEE/CVF Conference on Computer Vision and Pattern Recognition (CVPR) (2021)
5. Chen, T., Wang, Z., Xu, N., Jin, H., Luo, J.: Large-scale tag-based font retrieval with generative feature learning. In: Proceedings of the IEEE/CVF International Conference on Computer Vision (ICCV) (2019)
6. Devlin, J., Chang, M.W., Lee, K., Toutanova, K.: BERT: pre-training of deep bidirectional transformers for language understanding. arXiv preprint arXiv:1810.04805 (2018)
7. Dosovitskiy, A., et al.: An image is worth 16 × 16 words: transformers for image recognition at scale. In: Proceedings of the 8th International Conference on Learning Representations (ICLR) (2020)
8. Dufte, P., Schmitt, M., Schütze, H.: Position information in transformers: an overview. arXiv preprint arXiv:2102.11090 (2021)

9. He, K., Zhang, X., Ren, S., Sun, J.: Deep residual learning for image recognition. In: Proceedings of the IEEE Conference on Computer Vision and Pattern Recognition (CVPR) (2016)
10. Khan, S., Naseer, M., Hayat, M., Zamir, S.W., Khan, F.S., Shah, M.: Transformers in vision: a survey. arXiv preprint arXiv:2101.01169 (2021)
11. Kingma, D.P., Ba, J.: Adam: a method for stochastic optimization. In: Proceedings of the 3rd International Conference on Learning Representations (ICLR) (2015)
12. Liu, Y., et al.: Roberta: a robustly optimized BERT pretraining approach. arXiv preprint arXiv:1907.11692 (2019)
13. Lopes, R.G., Ha, D., Eck, D., Shlens, J.: A learned representation for scalable vector graphics. In: Proceedings of the IEEE/CVF International Conference on Computer Vision (ICCV) (2019)
14. Reddy, P., Gharbi, M., Lukac, M., Mitra, N.J.: Im2Vec: synthesizing vector graphics without vector supervision. In: Proceedings of the IEEE/CVF Conference on Computer Vision and Pattern Recognition (CVPR) (2021)
15. Roy, P., Bhattacharya, S., Ghosh, S., Pal, U.: STEFANN: scene text editor using font adaptive neural network. In: Proceedings of the IEEE/CVF Conference on Computer Vision and Pattern Recognition (CVPR) (2020)
16. Spencer, H.: The Visible Word: Problems of Legibility. Visual Communication Books (1969)
17. Srivatsan, N., Barron, J., Klein, D., Berg-Kirkpatrick, T.: A deep factorization of style and structure in fonts. In: Proceedings of the 2019 Conference on Empirical Methods in Natural Language Processing and the 9th International Joint Conference on Natural Language Processing (EMNLP-IJCNLP) (2019)
18. Srivatsan, N., Wu, S., Barron, J., Berg-Kirkpatrick, T.: Scalable font reconstruction with dual latent manifolds. In: Proceedings of the 2021 Conference on Empirical Methods in Natural Language Processing (EMNLP) (2021)
19. Ueda, M., Kimura, A., Uchida, S.: Which parts determine the impression of the font? In: Lladós, J., Lopresti, D., Uchida, S. (eds.) ICDAR 2021. LNCS, vol. 12823, pp. 723–738. Springer, Cham (2021). https://doi.org/10.1007/978-3-030-86334-0_47
20. Vaswani, A., et al.: Attention is all you need. In: Advances in Neural Information Processing Systems (NeurIPS) (2017)
21. Wang, B., et al.: On position embeddings in BERT. In: Proceedings of the 9th International Conference on Learning Representations (ICLR) (2021)
22. Wang, Y., Lian, Z.: DeepVecFont: synthesizing high-quality vector fonts via dual-modality learning. ACM Trans. Graph. (TOG) **40**, 1–15 (2021)

Unified Line and Paragraph Detection by Graph Convolutional Networks

Shuang Liu[1(✉)], Renshen Wang[2], Michalis Raptis[2], and Yasuhisa Fujii[2]

[1] University of California, San Diego, USA
s3liu@eng.ucsd.edu
[2] Google Research, Mountain View, USA
{rewang,mraptis,yasuhisaf}@google.com

Abstract. We formulate the task of detecting lines and paragraphs in a document into a unified two-level clustering problem. Given a set of *text detection boxes* that roughly correspond to words, a text line is a cluster of boxes and a paragraph is a cluster of lines. These clusters form a two-level tree that represents a major part of the layout of a document. We use a graph convolutional network to predict the relations between text detection boxes and then build both levels of clusters from these predictions. Experimentally, we demonstrate that the unified approach can be highly efficient while still achieving state-of-the-art quality for detecting paragraphs in public benchmarks and real-world images.

Keywords: Text detection · Document layout · Graph convolutional network

1 Introduction

Document layout extraction is a critical component in document analysis systems. It includes pre-OCR layout analysis [1] for finding text lines and post-OCR layout analysis [2,3] for finding paragraphs or higher level entities. These tasks involve different levels of document entities – line-level and paragraph-level – both are important for OCR and its downstream applications, and both have been extensively studied. Yet, to the best of authors' knowledge, there has been no study that tries to find text lines and paragraphs at the same time.

Graph convolutional networks (GCNs) are becoming a prominent type of neural networks due to their capability of handling non-Euclidean data [4]. They naturally fit many problems in OCR and document analysis, and have been applied to help form lines [5–7], paragraphs [3] or other types of document entities [8]. Besides the quality gain from these GCN models, another benefit from these approaches is that we can potentially combine all the machine learning tasks and build a single, unified, multi-task GCN model.

In this paper, we propose to apply a multi-task GCN model in OCR text detection to find both text lines and paragraphs in the document image.

Work done during the first author's internship at Google Research.

© Springer Nature Switzerland AG 2022
S. Uchida et al. (Eds.): DAS 2022, LNCS 13237, pp. 33–47, 2022.
https://doi.org/10.1007/978-3-031-06555-2_3

Compared to separated models dedicated to specific tasks, this unified approach has some potential advantages:

- System performance. Running a single step of model inference is usually faster than running multiple steps.
- Quality. A multi-stage system often suffers from cascading errors, where an error produced by a stage will cause ill-formed input for the next stage. Pruning stages can reduce the chance for this type of errors.
- Maintainability. The overall system complexity can be reduced, and it is easier to retrain and update a unified model than multiple models.

There are also potential drawbacks on this approach. For example, some post-OCR steps [3] need to be moved to pre-OCR, which reduces the available input signals to the model.

The rest of this paper is organized as follows. Section 2 reviews the related work. Section 3 presents our proposed method, where the problem formulation and the modeling solution are discussed in details. Section 4 discusses potential drawbacks and limitations of this approach. Section 5 presents experimental results. Finally, Sect. 6 concludes the paper with suggestions for future work.

2 Related Work

2.1 Text Line Detection

The first step of OCR is text detection, which is to find the text words or lines in the input image. For documents with dense text, lines are usually chosen over words for more reliable performance [9]. Due to the high aspect ratios of text lines, an effective way to find them is to first detect small character-level boxes, and then cluster them into lines using algorithms like Text Flow [10].

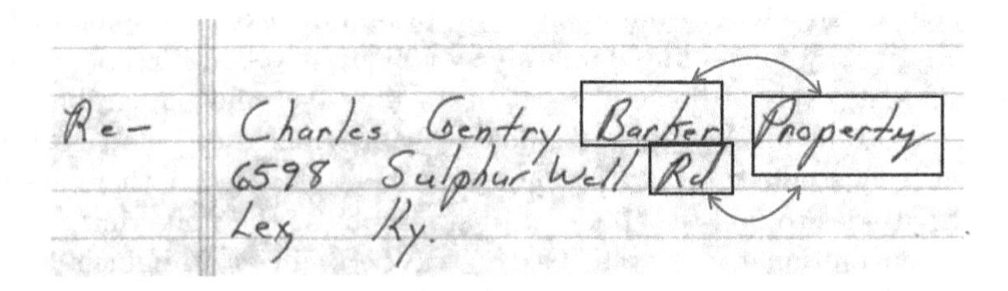

Fig. 1. A difficult handwriting example for Text Flow clustering—the upper positive edge has a larger distance between the pair of words than that of the lower negative edge, which may cause the algorithm to choose the wrong edge.

While Text Flow is effective for most printed documents, handwritten text can be challenging for the network flow model. As shown in Fig. 1, handwritten words can be sparser and more scattered than printed ones, and may cause difficulty for Text Flow, as well as for connected component based post-processing like [11]. A sophisticated graph clustering algorithm based on superpixels is proposed in [12] for better performance with historical/handwritten documents, yet with potential limitations on line curvilinearity and word spacings.

We think graph neural network based approaches like [5–7] are in a better position to solve the handwritten text detection problem, or more difficult layout problems in general.

2.2 Paragraph Detection

PubLayNet [13] provides a large data set containing paragraphs as well as baseline object detection models trained to detect these paragraphs. While the image based detection models can work well on this data set, it is shown in [3] that they are less adaptable to real-world applications with possible rotations and perspective distortions, and require more training data and computing resources. For example, the detection models in [13] are around 1 GB in size, while the GCN models based on OCR outputs in [3] are under 200 KB each.

So once again, graph neural network based approaches have significant advantages in the layout problem. Note that in [3], two GCN models take OCR bounding boxes as input to perform operations on lines. If these lines are formed by another GCN model, there will be totally three GCN models in the system, which leaves opportunities for unification and optimization.

3 Proposed Method

In this paper, we propose a unified graph convolutional network model to form both text lines and paragraphs from word-level detection results.

3.1 Pure Bounding Box Input

Our GCN model takes only geometric features from bounding boxes like [3], rather than using RRoI (rotated region of interest) features from input images [7]. Not only does it greatly reduce the model size and latency, but it also makes the model generalize better across domains—as verified by the synthetic-to-real-world test in [3].

Not coincidentally, a street view classification model proposed in [14]—which can be viewed as another type of layout analysis—is discovered to perform better with pure bounding boxes.

We use word-level boxes that are obtained by grouping character-level boxes. Since the difficulty of Text Flow shown in Fig. 1 mainly resides in the connections between words, character-to-word grouping is easier and can be performed with a geometric heuristic algorithm at a high accuracy.

3.2 Problem Statement

The input of our algorithm will be a set of n rotated rectangular boxes as well as m undirected edges between them.

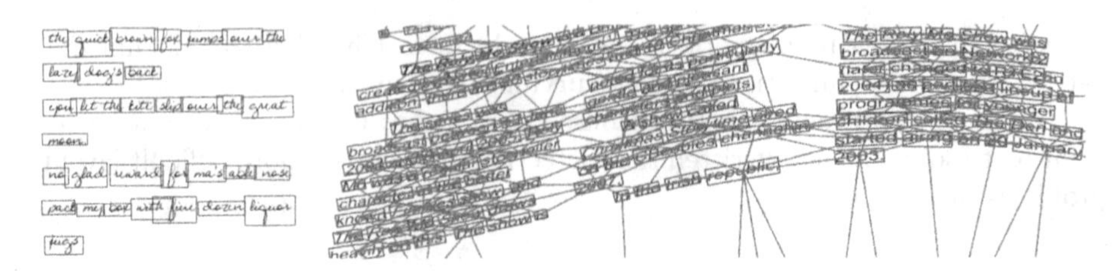

Fig. 2. Left: GCN boxes (blue) which do not necessarily correspond to words or characters. Right: GCN edges (purple) over GCN boxes (blue). (Color figure online)

- Each box is represented by five real numbers $(x_i, y_i, w_i, h_i, a_i)$ where (x_i, y_i) is the coordinate of its upper left corner, w_i is its width, h_i is its height, and a_i is its angle. We will call these *GCN boxes*, since they will be used as nodes by a GCN. These boxes correspond to word-level text regions in a document; an example is shown on the left of Fig. 2—note that each GCN box does not necessarily correspond to a word or a character in the usual sense. While our approach is largely agnostic to the specification of the boxes, the approximate word level boxes discussed in Sect. 3.1 are preferred for their good accuracy and efficiency.
- The edges connect pairs of boxes that are close to each other. We will call these edges *GCN edges* since they will be used as edges by a GCN. An example is shown on the right of Fig. 2. We describe how the edges are constructed in Sect. 3.4.

The output of our algorithm should be a two-level clustering of the input: ideally, the first level clusters GCN boxes into lines, and the second level clusters lines into paragraphs.

3.3 Main Challenge

For efficiency considerations, the GCN edges should be constructed in a way such that its size grows linearly with the number of GCN boxes—we will call such a GCN a linear-sized GCN. We discuss a way of constructing a linear-sized graph for a GCN in Sect. 3.4. However, this requirement introduces a problem: as we will explain shortly, paragraph is a global property, while a linear-sized GCN makes only local predictions.

Consider a naive attempt that is to simply predict for each GCN edge, whether the two GCN boxes connected by the edge are from the same paragraph. Figure 3 shows that this would not work as we had expected—Humans typically infer paragraph structures by tracing all the way back to the beginning of each line, no matter how long the line is; on the other hand, each edge in a linear-sized GCN can only receive information from a neighbourhood of the edge.

We overcome this challenge by making the GCN to predict only local relations and later using these local relations to perform global inference. See Sect. 3.5 for a detailed discussion.

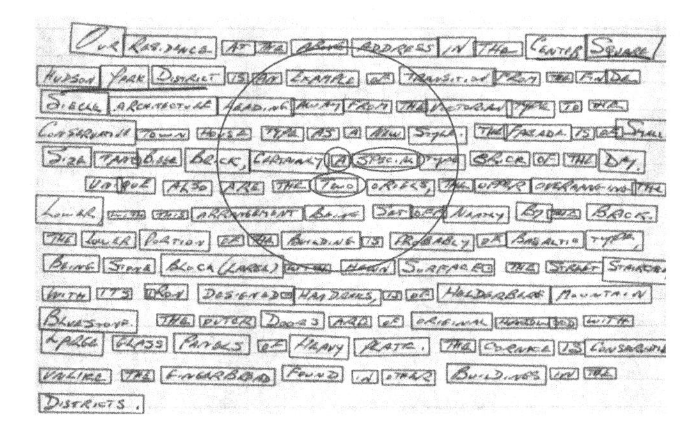

Fig. 3. Consider the three words in orange circles. To predict any property that relates to the box that contains "A", a linear-sized GCN can only use information in its neighborhood. For example, a GCN by itself, using information only in the red circle, can determine whether "A" is in the same line as "SPECIAL", but cannot determine whether "A" is in the same paragraph as "Two".

3.4 β-skeleton Graph with 2-Hop Connections

The β-skeleton graph used in [3] is a good candidate as an efficient graph for layout tasks. However, in some cases with a combination of low line spacing and high word spacing, the graph constructed on word boxes cannot guarantee connectivity within all lines. As shown on the left of Fig. 4, the middle line has an extra large space between the two words, and the β-skeleton edge in this space is a vertical cross-line connection rather than a horizontal edge we need for line-level clustering.

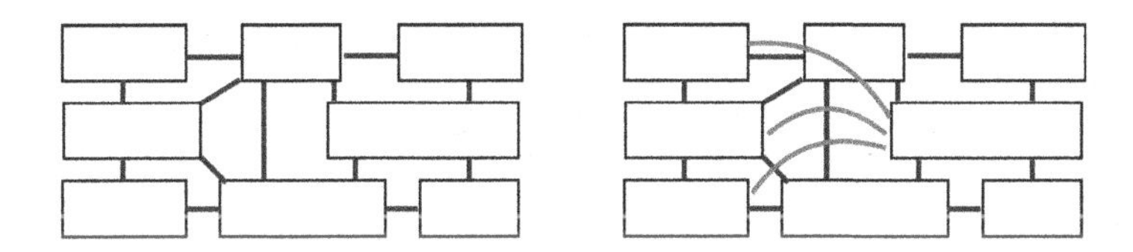

Fig. 4. Left: β-skeleton graph on a set of word boxes [3]. Right: Added 2-hop edges so that neighboring words within the same line are connected.

The solution we use is to add 2-hop connections to the graph. For each node in the β-skeleton graph, we check each of its neighbor's neighbor and add a hop edge if the two boxes fall in certain distance and angle constraints. This provides the necessary connectivity while still maintaining reasonable sparsity (shown in Fig. 4 on the right). Also see the graph on the right of Fig. 2 for an example from synthetic documents.

3.5 GCN Predictions

Let b_1 and b_2 be two GCN boxes that are connected by a GCN edge. We use a GCN to make the following binary predictions.

- **box_left**(b_1, b_2): 1 if b_1 and b_2 are in the same line and b_1 is adjacent to b_2 to the left, 0 otherwise (Fig. 5).

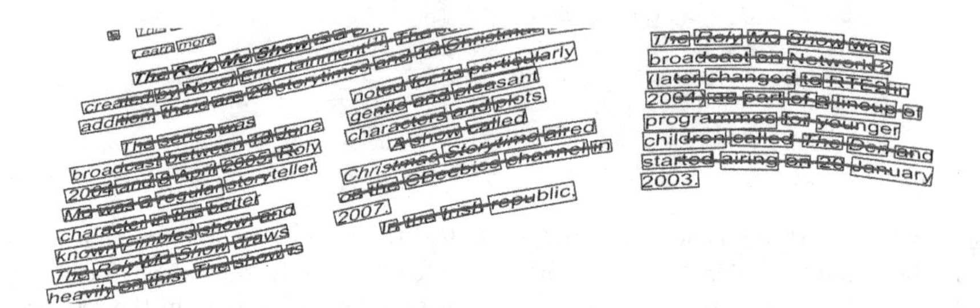

Fig. 5. GCN boxes and **box_left** labels. If there is a yellow edge connecting two GCN boxes b_1 on the left and b_2 on the right, it means **box_left**$(b_1, b_2) = 1$. (Color figure online)

- **box_above**(b_1, b_2): (only makes valid prediction when b_1 is the first box in a line) 1 if b_1 and b_2 are in the same paragraph and b_1 is in a line that is directly above the line b_2 is in (Fig. 6).

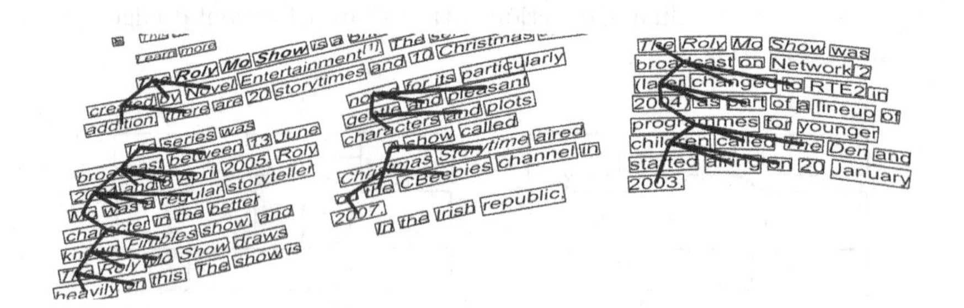

Fig. 6. GCN boxes and **box_above** labels. If there is a green edge connecting two GCN boxes b_1 on the top and b_2 on the bottom, it means **box_above**$(b_1, b_2) = 1$. (Color figure online)

- **box_below**(b_1, b_2): (only makes valid prediction when b_1 is the first box in a line) 1 if b_1 and b_2 are in the same paragraph and b_1 is in a line that is directly below the line b_2 is in (Fig. 7).

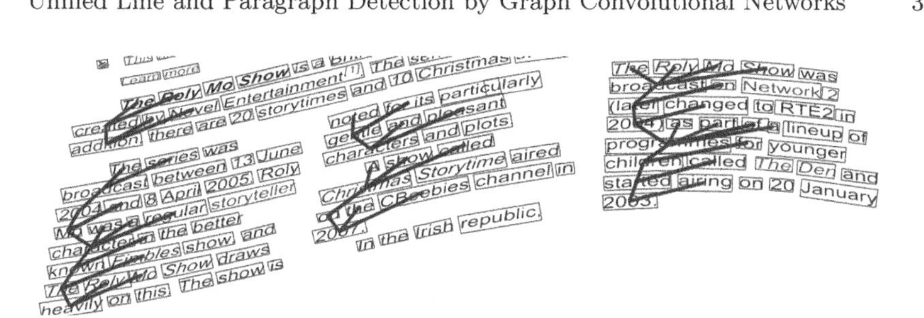

Fig. 7. GCN boxes and **box_below** labels. If there is a blue edge connecting two GCN boxes b_1 on the bottom and b_2 on the top, it means **box_below**$(b_1, b_2) = 1$. (Color figure online)

A curious reader may wonder whether it is necessary to have both **box_above** and **box_below**. When GCN edges has size linear in the size of GCN boxes, we cannot guarantee that the first GCN boxes in two adjacent lines are always connected by a GCN edge, especially when there is a large indentation. By utilizing *both* **box_above** and **box_below**, we can be assured that the first GCN box in each line is connected to *some* GCN box in any adjacent line. Figure 8 gives an example where **box_below** misses such a connection but **box_above** does not; Fig. 9 gives an example where **box_above** misses such a connection but **box_below** does not.

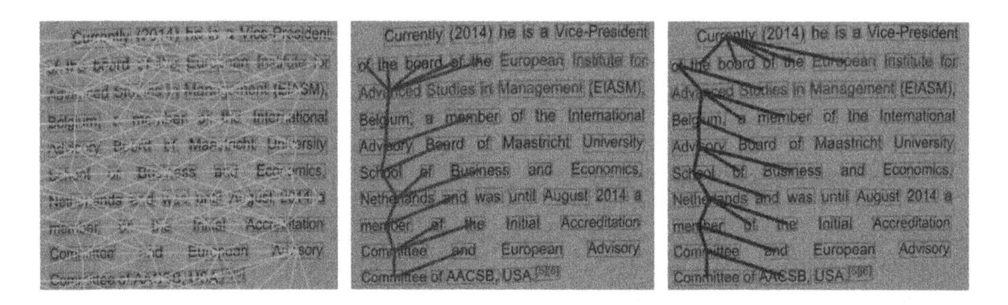

Fig. 8. The left image shows the graph edges. **box_below** could miss the connection between the first two lines (middle image), but this could be complemented by **box_above** (right image).

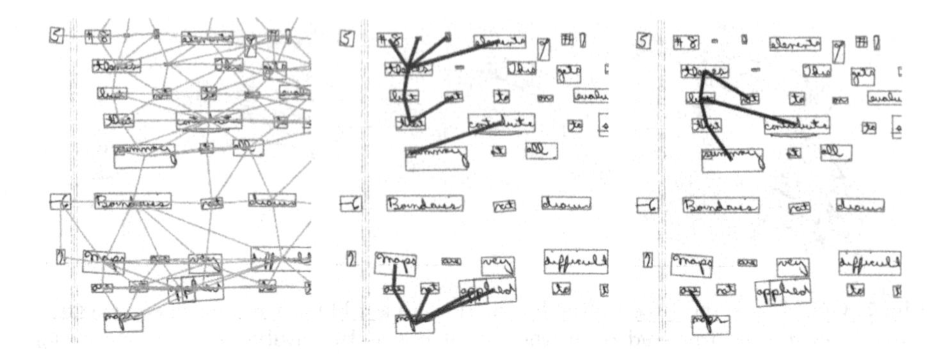

Fig. 9. The left image shows the graph edges. **box_above** could miss the connection between the first two lines of paragraphs (right image), but this could be complemented by **box_below** (middle image).

3.6 Forming Lines

Initially all GCN boxes belong to a separate line. Iteratively, if **box_left** $(b1, b2) = 1$, then we merge b_1 and b_2 into the same line. A GCN box b is identified as the first box of a line if there is no GCN box b' such that **box_left**$(b', b) = 1$.

3.7 Forming Paragraphs

Initially all lines belong to a separate paragraph. Iteratively, if **box_above**$(b1, b2) = 1$ or **box_below**$(b1, b2) = 1$, then we merge b_1's line and b_2's line into the same paragraph.

3.8 Overall System Pipeline

The end-to-end pipeline of our OCR system with the proposed unified GCN model is shown in Fig. 10. It is similar to the pipeline in [15], with extra paragraph outputs directly produced in the text detector, so the layout post-processing can be greatly simplified, resulting in better system efficiency.

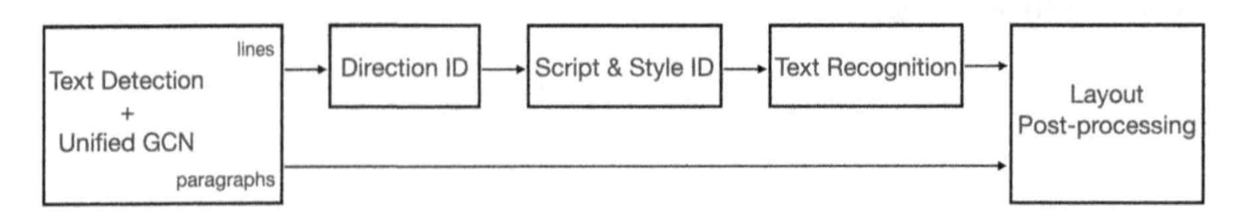

Fig. 10. The overall OCR pipeline with the unified GCN model producing both lines and paragraphs at text detection.

4 Limitations

4.1 Single-Line Paragraphs

In rare circumstances, there are many consecutive equal-lengthed paragraphs and a linear-sized GCN would fail to identify the signal of a paragraph start from local information (Fig. 11). This is rather a limitation of linear-sized GCNs and is beyond the scope of the current project.

This is a contrived example. In this example, all paragraphs are
single-line paragraphs, except this one.
This is the second paragraph of this contrived example.
This is the third paragraph of this contrived example.
This is the fourth paragraph of this contrived example.
This is the fifth paragraph of this contrived example.
This is the sixth paragraph of this contrived example.
This is the seventh paragraph of this contrived example.
This is the eighth paragraph of this contrived example.
This is the ninth paragraph of this contrived example.
This is the last paragraph of this contrived example.

Fig. 11. In rare cases, the limited perceptive field of a linear-sized GCN (red) could make it unable to determine whether a GCN box (blue) starts a paragraph. (Color figure online)

4.2 Document Rotations

Since our approach is purely geometric, we focus on document images that are rotated no more than 45°. GCN models are shown in [3] to be much more robust than object detection models for rotated/distorted inputs.

However, when the rotation angle gets close to 90° or even 180°, the document orientation becomes uncertain, which will affect the **box_above** and **box_below** predictions for paragraph-level clustering at the left side of text regions.

As shown in Fig. 12, if a document image is rotated by 90°, it may be impossible, even for a human, to recover the original orientation correctly using only geometrical information (bounding boxes). In a typical OCR engine, text directions can be obtained from a "direction ID" model [15] which runs on detected text lines (see Fig. 10). But here we don't have text lines yet, and will run into a "chicken-and-egg" dependency cycle to rely on line based text direction models. So if directions are needed, we may add a direction prediction to each character-level detection box, and take majority votes to decide the document orientation.

5 Experiments

We experiment with the proposed GCN model on both the open PubLayNet dataset [13] and our own annotated paragraph set from real-world images. These

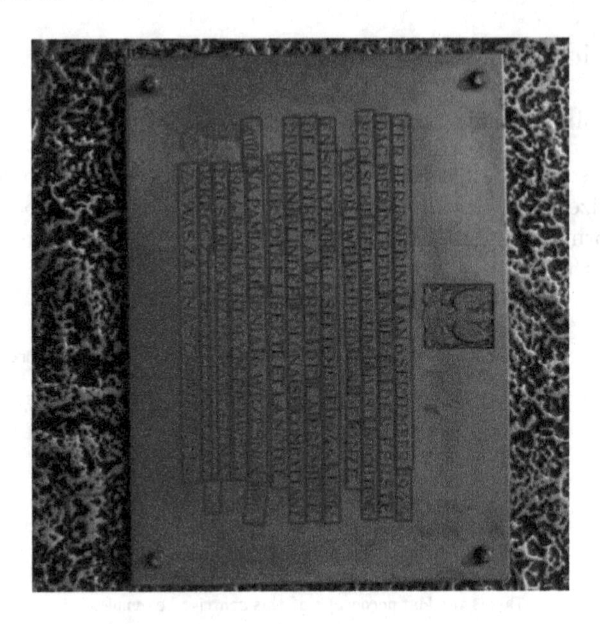

Fig. 12. Image example rotated by 90°, where it is hard to tell whether the texts are rotated clockwise or counter-clockwise purely from geometric information.

datasets can evaluate the clustering quality on output paragraphs, and on output lines as well since any mistake in line-level clustering will affect the parent paragraph.

The model input is from the OCR text detector behind the Google Cloud Vision API of DOCUMENT_TEXT_DETECTION[1] version 2021. Following the input layer are 5 steps of "message passing" in [16] as the backbone and 3 binary classification heads for the outputs described in Sect. 3.5 – similar to the "line clustering" model in [3] but with 3 independent node-to-edge prediction heads.

In each message passing step, the node-to-edge message kernel has a hidden layer of size 64 and output layer of size 32, the edge-to-node aggregation is by average pooling, the node-to-next-step and the final prediction heads have the same (64, 32) sizes as the node-to-edge message kernel. The resulting GCN model is under 200 KB in parameter size, which is negligible compared to a typical OCR engine or an image based layout model.

5.1 PubLayNet Results

We first train the unified GCN model using labels inferred from the PubLayNet training set, and evaluate the paragraph results on the PubLayNet validation set against the $F1_{var}$ metric introduced in [3]. The result comparison with a few other methods are shown in Table 1.

---

[1] https://cloud.google.com/vision/docs/fulltext-annotations.

Table 1. Paragraph $F1_{var}$ score comparisons.

Model	$F1_{var}$
Tesseract [1]	0.707
Faster-RCNN-Q [3]	0.945
OCR + Heuristic [3]	0.364
OCR + 2-step GCNs [3]	0.959
Unified GCN	0.912

Compared to the results of 2-step GCNs, the only obvious loss pattern is caused by a situation discussed in Sect. 4.1, which appears frequently in the title sections of this dataset, as shown in the top-right of Fig. 13. The paragraphs here are not indicated by indentation or vertical spacing, but by changes in font size and style which are not easily captured by early stage text detection. The 2-step approach can do better in this case because the line clustering model has more useful input features such as the total length of each line.

Fig. 13. Evaluation examples from PubLayNet [13]. Paragraph-clustering results are mostly perfect except for list items (not annotated as paragraphs in ground truth) and the title section on top-right (not separating the title and author names due to lack of indentations).

5.2 Real-World Evaluation Results

For evaluation on real-world images, we train the unified GCN model with both the augmented web synthetic data from [3] and human annotated data. The metric is F1@IoU0.5 since there is no ground truth of paragraph line count to support the variable IoU threshold. Results are in Table 2.

Table 2. Paragraph F1-scores tested on the real-world test set.

Model	Training data	F1@IoU0.5
OCR + Heuristic	–	0.602
Faster-RCNN-Q [3]	Annotated data (pre-trained on PubLayNet)	0.607
OCR + 2-step GCNs [3]	Augmented web synthetic + Annotated	0.671
Unified GCN	Augmented web synthetic + Annotated	0.659

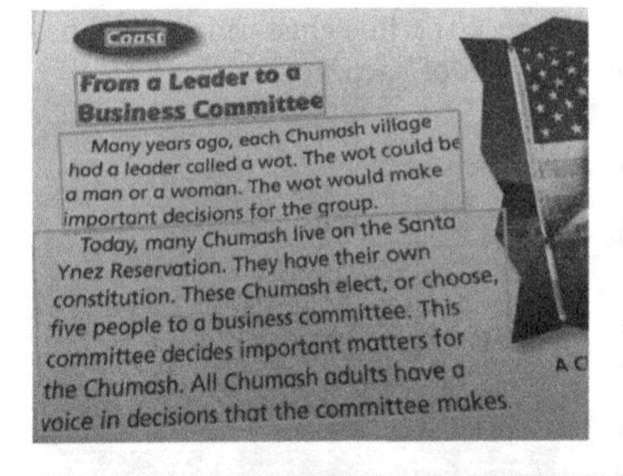

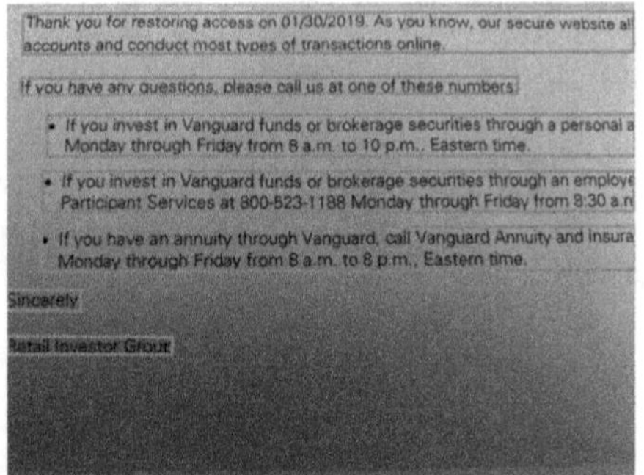

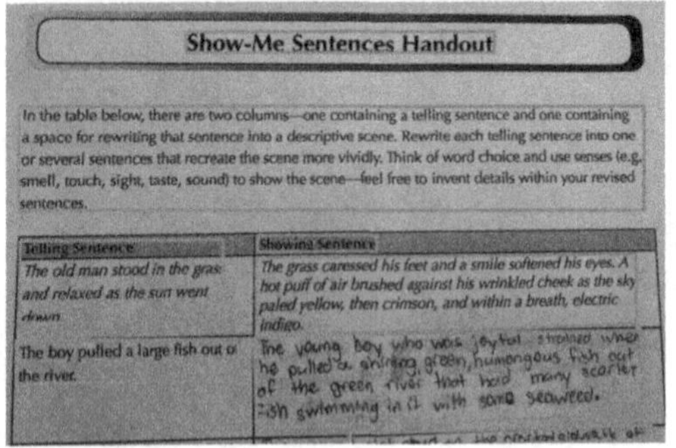

Fig. 14. Success examples among real-world images.

Figure 14 and 15 show success and failure examples from this set. The loss pattern from Sect. 4.1 is still here in the left image of Fig. 15. But such cases are quite rare in the real world as told by the smaller score difference than in Table 1. The overall quality of the unified GCN output is very close to that from the 2-step approach.

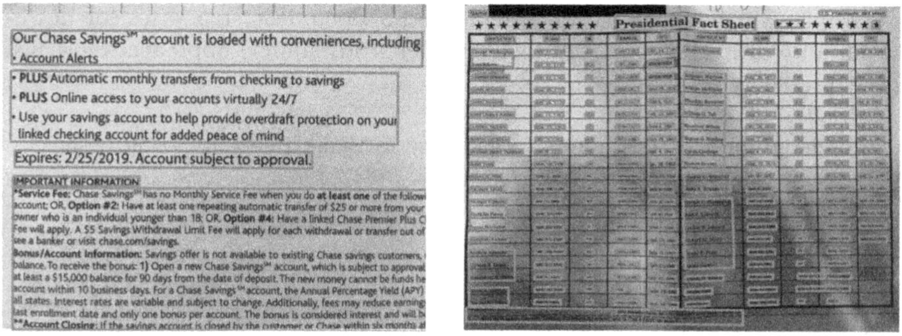

Fig. 15. Failure examples among real-world images. Left: paragraphs without indentations or extra vertical spacing may not be detected. Right: tables will need extra handling beyond the scope of this work.

6 Conclusions and Future Work

We demonstrate that two OCR related layout tasks can be performed by a single unified graph convolutional network model. Compared to a multi-step pipeline, this unified model is more efficient and can produce paragraphs at virtually the same quality in real world images.

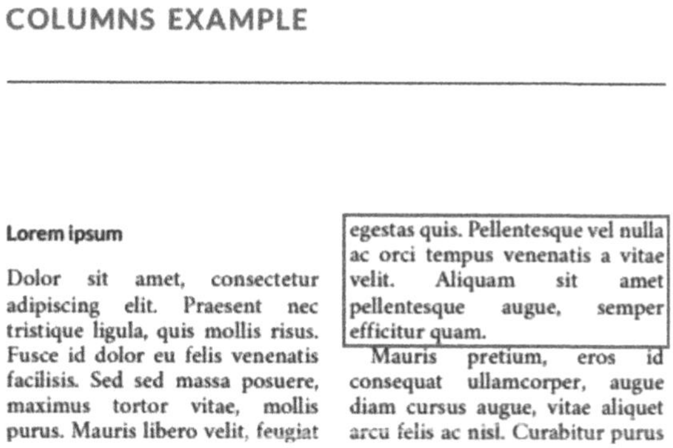

Fig. 16. The region contained in the red rectangle is physically a paragraph, but needs to be combined into the one in the left text column to form a complete semantic paragraph. (Color figure online)

We believe the number of clustering levels is not limited to two, since document layout tasks are extremely diverse in nature. Paragraphs can further be clustered into text columns or sections, which may belong to even higher level blocks. Figure 16 shows a physical paragraph, or part of a semantic paragraph which spans across multiple text columns. Our current approach stops at physical paragraphs. But if we can obtain text columns as layout entities, semantic paragraphs will be available to provide better document structures for downstream applications. Future work will further study unified approaches for such layout problems.

Acknowledgements. The authors would like to thank Reeve Ingle and Ashok C. Popat for their helpful reviews and feedback.

References

1. Smith, R.: An overview of the Tesseract OCR engine. In: Proceedings of the 9th International Conference on Document Analysis and Recognition, pp. 629–633 (2007)
2. Yang, X., Yumer, E., Asente, P., Kraley, M., Kifer, D., Lee Giles, C.: Learning to extract semantic structure from documents using multimodal fully convolutional neural network. In: Proceedings of the IEEE/CVF Conference on Computer Vision and Pattern Recognition, pp. 5315–5324 (2017)
3. Wang, R., Fujii, Y., Popat, A.C.: Post-OCR paragraph recognition by graph convolutional networks. In: Proceedings of the IEEE/CVF Winter Conference on Applications of Computer Vision, pp. 493–502 (2022)
4. Wu, Z., Pan, S., Chen, F., Long, G., Zhang, C., Yu, P.S.: A comprehensive survey on graph neural networks. IEEE Trans. Neural Netw. Learn. Syst. **32**, 4–24 (2020)
5. Zhang, S.-X., Zhu, X., Yang, C., Wang, H., Yin, X.-C.: Adaptive boundary proposal network for arbitrary shape text detection. In: Proceedings of the IEEE/CVF International Conference on Computer Vision, pp. 1305–1314 (2021)
6. Ma, C., Sun, L., Zhong, Z., Huo, Q.: ReLaText: exploiting visual relationships for arbitrary-shaped scene text detection with graph convolutional networks. Pattern Recogn. **111** (2021)
7. Zhang, S.-X., et al.: Deep relational reasoning graph network for arbitrary shape text detection. In: Proceedings of the IEEE/CVF Conference on Computer Vision and Pattern Recognition, pp. 9699–9708 (2020)
8. Lee, C.-Y., et al.: ROPE: reading order equivariant positional encoding for graph-based document information extraction. In: Proceedings of the 59th Annual Meeting of the Association for Computational Linguistics and the 11th International Joint Conference on Natural Language Processing (ACL-IJCNLP), pp. 314–321 (2021)
9. Diaz, D., Qin, S., Ingle, R., Fujii, Y., Bissacco, A.: Rethinking text line recognition models. arxiv, abs/2104.07787 (2021)
10. Tian, S., Pan, Y., Huang, C., Lu, S., Yu, K., Tan, C.L.: Text flow: a unified text detection system in natural scene images. In: Proceedings of the IEEE/CVF International Conference on Computer Vision, pp. 4651–4659 (2015)
11. Lladós, J., Lopresti, D., Uchida, S.: Page layout analysis system for unconstrained historic documents. In: Proceedings of the International Conference on Document Analysis and Recognition, pp. 492–506 (2021)
12. Grüning, T., Leifert, G., Strauß, T., Michael, J., Labahn, R.: A two-stage method for text line detection in historical documents. Int. J. Document Anal. Recogn. (IJDAR) **22**, 285–302 (2019)
13. Zhong, X., Tang, J., Jimeno-Yepes, A.: PubLayNet: largest dataset ever for document layout analysis. In: 2019 International Conference on Document Analysis and Recognition, ICDAR 2019, Sydney, Australia, 20–25 September 2019, pp. 1015–1022. IEEE (2019)

14. Zhao, K., Liu, Y., Hao, S., Lu, S., Liu, H., Zhou, L.: Bounding boxes are all we need: street view image classification via context encoding of detected buildings. IEEE Trans. Geosci. Remote Sens. **60**, 1–17 (2022)
15. Reeve Ingle, R., Fujii, Y., Deselaers, T., Baccash, J., Popat, A.C.: A scalable handwritten text recognition system. In: Proceedings of the International Conference on Document Analysis and Recognition, pp. 17–24 (2019)
16. Gilmer, J., Schoenholz, S.S., Riley, P.F., Vinyals, O., Dahl, G.E.: Neural message passing for quantum chemistry. In: Proceedings of the 34th International Conference on Machine Learning, ICML 2017, vol. 70, pp. 1263–1272. JMLR.org (2017)
17. Monnier, T., Aubry, M.: docExtractor: an off-the-shelf historical document element extraction. In: Proceedings of the International Conference on Frontiers in Handwriting Recognition, pp. 91–96 (2020)

The Winner Takes It All: Choosing the "best" Binarization Algorithm for Photographed Documents

Rafael Dueire Lins[1,2,3(✉)], Rodrigo Barros Bernardino[1,3], Ricardo Barboza[3], and Raimundo Oliveira[3]

[1] Universidade Federal de Pernambuco, Recife, PE 50730-120, Brazil
{rdl,rbb4}@cin.ufpe.br
[2] Universidade Federal Rural de Pernambuco, Recife, PE 52171-900, Brazil
[3] Universidade do Estado do Amazonas, Manaus, AM, Brazil
rsbarboza@uea.edu.br

Abstract. The recent Time-Quality Binarization Competitions have shown that no single binarization algorithm is good for all kinds of document images and that the time elapsed in binarization varies widely between algorithms and also depends on the document features. On the other hand, document applications for portable devices have space and processing limitations that allow to implement only the "best" algorithm. This paper presents the methodology and assesses the time-quality performance of 61 binarization algorithms to choose the most time-quality efficient one, under two criteria.

Keywords: Smartphone applets · Document binarization · DIB-dataset · Photographed documents · Binarization competitions

1 Introduction

Today, half of the population of the world has a smartphone with a built-in digital camera, according with the June 2021 report from the consulting firm Strategy Analytics[1]. Such devices are incredibly versatile and even low cost ones have good quality cameras that allow digitizing document images, that are widely used in a large number of everyday situations that in a recent past photocopying was used.

Binarization, or *thresholding*, is the name given to the conversion process of a color image into its black-and-white (or monochromatic) version. Binary images make most documents more readable and save toner for printing. save storage space [47], communication bandwidth. Binarization also works as a file compression strategy, as the size of binary images is often orders of magnitudes smaller than the original gray or color images. Thresholding is a key preprocessing step for document OCR, classification and indexing. The recent Time-Quality

[1] https://www.strategyanalytics.com/.

© Springer Nature Switzerland AG 2022
S. Uchida et al. (Eds.): DAS 2022, LNCS 13237, pp. 48–64, 2022.
https://doi.org/10.1007/978-3-031-06555-2_4

Binarization Competitions [25,27,30,31] have shown that no single binarization algorithm is good for all kinds of document images and that the time elapsed in binarization varies widely between algorithms and also depends on the document features.

Portable devices are limited in space and users are eager for outputs. Thus, being able to pinpoint which algorithm would fast provide a good quality binary image, capable of being embedded in applications in a certain smartphone model is a valuable information. **This paper assesses 61 binarization algorithms to choose the one that presents the best time-quality trade-off to be implemented in embedded applications in smartphones**. The universe of the tested algorithms is formed by "classical" and recently published binarization algorithms: Akbari_1 [1], Akbari_2 [1], Akbari_3 [1], Bataineh [3], Bernsen [5], Bradley [6], Calvo-Z [7], CLD [41], CNW [40], dSLR [46], DeepOtsu (SL) [12], DiegoPavan (DP) [50], DilatedUNet [27], DocDLink [57], Doc-UNet (WX) [30], ElisaTV [2], Ergina_G [49], Ergina_L [18], Gattal [9], Gosh [4], Howe [13], Huang [14], HuangBCD (AH_1) [28], HuangUnet (AH_2) [28], iNICK (KS_1) [42], Intermodes [38], ISauvola [11], IsoData [53], Jia-Shi [15], Johannsen [16], KSW [17], Li-Tam [22], Lu-Su [32], Mean [10], Mello-Lins [34], Michalak [27], Michalak21_1 (MO_1) [28], Michalak21_2 (MO_3) [28], Michalak21_3 (MO_3) [28], Min-Error [20], Moments [52], Niblack [36], Nick [19], Otsu [37], Percentile [8], Pun [39], RenyEntropy [43], Sauvola [44], Shanbhag [45], Singh [48], Su-Lu [51], Triangle [56], Vahid (RNB) [28], WAN [35], Wolf [54], Wu-Lu [33], Yen-CC [55], YinYang [27], YinYang21 (JB), [27], Yuleny [30].

The test set used for such an assessment is part of the IAPR DIB dataset (https://dib.cin.ufpe.br) including "real-world" offset, laser, and deskjet printed text documents. Such documents were photographed at two different places, with four different models of smartphones widely used today, with their in-built strobe flash *on* and *off*. The methodology presented here may be used to find the most suitable algorithm for other devices, or the same smartphone models under different setups.

2 Quality-Time Evaluation Methods

Two quality measures were used to evaluate the performance of the 61 binarization algorithms assessed here. The first one, made use of Google Vision to perform Optical Character Recognition (OCR) on the documents and applies the Levenshtein distance ($[L_{dist}]$) to the correct number of characters in the document transcription ($\#\ char$). The error rate is calculated as:

$$([L_{dist}] = (\#char - L_{dist})/\#char.) \quad (1)$$

The second quality measure, P_{err}, compares the proportion between the black-to-white pixels in the scanned and photographed binary documents [24]. One expects that although the photographed and scanned documents have different resolutions, the number of black pixels in a photographed and its scanned version of a document will be balanced, somehow. Thus,

$$([P_{err}] = 100 * \frac{abs(PW_{GT} - PW_{bin})}{PW_{GT}}) \tag{2}$$

where PW_{GT} and PW_{bin} are the proportion of white pixels in the ground-truth and the binarized image, respectively, and $abs()$ obtains the absolute value of the difference. The quality evaluation was done in the context of each measure separately. They were ranked using the mean value for the whole dataset.

The processing time evaluation provides the order of magnitude of the time elapsed for binarizing the whole datasets. The training-times for the AI-based algorithms were not computed. The processing device was CPU: Intel(R) Core(TM) i7-10750H CPU @ 2.60 GHz, with 32 GB RAM and a GPU GeForce GTX 1650 4 GB The algorithms were implemented using two operating systems and different programming languages, for specific hardware platforms such as GPUs:

- **Windows 10 (version 1909), Matlab:** Akbari_1, Akbari_2, Akbari_3, CLD, CNW, ElisaTV, Ergina-Global, Ergina-Local, Gattal, Ghosh, Howe, iNICK, Jia-Shi, Lu-Su, Michalak, MO_1, MO_2, MO_3
- **Linux Pop!_OS 20.10:**
  - **C++ (GCC 10.3):** Bataineh, Bernsen, ISauvola, Niblack, Nick, Otsu, Sauvola, Singh, Su-Lu, WAN, Wolf
  - **Python 2.7:** SL
  - **Java 14:** YinYang, JB, Bradley, daSilva-Lins-Rocha, Huang, Intermodes, IsoData, Johannsen-Bille, Kapur-SW, Li-Tam, Mean, Mello-Lins, Min-Error, Minimum, Moments, Percentile, Pun, RenyEntropy, Shanbhag, Triangle, Wu-Lu, Yen
  - **Python 3.6:** AH_1, AH_2, Calvo-Z, DP, DilatedUNet, DocDLink, WX, RNB, Yuleny

The algorithms were executed on different operating systems (OS), but on the same hardware. For those that could be executed on both OS types, the processing times for each OS was measured and no significant difference was noticed. This behaviour was also observed and reported in [27]. The mean processing time was used in the analysis. The primary purpose is to provide the order of magnitude time of the processing time elapsed. The **SL** algorithm (DeepOtsu) would take weeks to process the images using a CPU; therefore, a NVIDIA Tesla K80 has been used to accelerate the processing. However, an approximation of the CPU processing time is used as reference in order to compare with the other algorithms, each of which was processed using a CPU on the specific platform.

3 Test Set

Document images acquired using mobile phones are harder to binarize if compared with the use of scanners. The distance between the document and the capturing device and the illumination may vary significantly. Other external light

body can see at once that 3 straight lines, taken at random, divide the plane into 7 parts (look at the only finite part, the triangle included by the 3 lines). Scarcely anybody is able to see, even straining his attention to the utmost, that 5 planes, taken at random, divide space into 26 parts. Yet it can be rigidly proved that the right number is actually 26, and the proof is not even long or difficult.

Carrying out our plan, we check each step. Checking our step, we may rely on intuitive insight or on formal rules. Sometimes the intuition is ahead, sometimes the formal reasoning. It is an interesting and useful exercise to do it both ways. *Can you see clearly that the step is correct?* Yes, I can see it clearly and distinctly. Intuition is ahead; but could formal reasoning overtake it? *Can you also* PROVE *that it is correct?*

Trying to prove formally what is seen intuitively and to see intuitively what is proved formally is an invigorating mental exercise. Unfortunately, in the classroom there is not always enough time for it. The example, discussed in sections 12 and 14, is typical in this respect.

Condition is a principal part of a "problem to find." See PROBLEMS TO FIND, PROBLEMS TO PROVE, 3. See also TERMS, NEW AND OLD, 2.

A condition is called *redundant* if it contains superfluous parts. It is called *contradictory* if its parts are mutually opposed and inconsistent so that there is no object satisfying the condition.

Thus, if a condition is expressed by more linear equations than there are unknowns, it is either redundant or contradictory; if the condition is expressed by fewer equations than there are unknowns, it is insufficient to determine the unknowns; if the condition is expressed by just as many equations as there are unknowns it is

Fig. 1. Samples of the images clustered by device (Motorola G9 Plus, iPhone SE 2, Samsung A10S, Samsung S20) and set-up of the strobe flash (top-line "off", bottom-line "on").

sources and the activation or not of the strobe flash may interfere in the quality of the obtained image. The kind of document images used here are representative of the kind of images that ordinary people take photos of and correspond to the kinds of documents people often used to take photocopies a few years ago. Typically, such documents are text ones with a plain background, printed in either plain white printer or recycled paper. The test set used here, samples of which are presented in Fig. 1, is formed by nine documents offset printed book pages, and deskjet and laser printed documents. Very seldom, people take a photo of a historic documents. If that is the case, such a document image tends not to be binarized in the camera itself as historic documents tend to have a darker background, some show back-to-front interference [29] and physical noises [23]. Such images are part of The IAPR DIB-dataset, which encompasses nine documents obtained from four different models of portable cell-phones, widely used today. Besides the device model, the documents in this set were clustered according to having the in-built strobe-flash set as "*on*" or "*off*".

4 Results

Four models of smartphones of different manufacturers were used is this study. The choice of the devices was made to cover mid-price range models of different manufacturers in such a way to be representative of the smartphones used by the majority of the population. The technical specifications of their front cameras are presented in Table 1.

Table 1. Summary of specifications of the front camera of the devices studied

	Moto G9	iPhone SE2	Galaxy S20	Galaxy A10S
Megapixels	48	12	64	13
Flash	Dual LED	Quad-LED	Dual LED	Dual LED
Aperture	f/1.8	f/1.8	f/2.0	f1.8
Sensor size	1/2 in	–	1/1.72 in	–
Pixel size	0.8 μm	–	0.8 μm	–

In this paper, for each device tested, three algorithms will be selected:

- (i) the best binarization algorithm for printing, screen reading, or less storage space or claims for less communication bandwidth for transmission.
- (ii) the best for OCR processing binarization algorithm.
- (iii) the overall "winner" - the algorithm that provides the best quality-time trade-off for any sort of binarization application.

The developers of applications for other device models should use the methodology presented here to make a criterious choice of which algorithm(s) to use.

4.1 Motorola Moto G9

The Moto G9 device used in this assessment is an Android 10 smartphone developed by Motorola Mobility[2], a subsidiary of Lenovo. It was first released in August 2020. The analysis of the data presented on Table 2 for the document images in the testset acquired with this smartphone, shows that several algorithms perform very well in terms of the quality of the generated monochromatic image. Two of them compete for the podium in analysing the general quality time trade-off: Michalack [27] and MO_1 [25], both developed by Hubert Michalak and Krzysztof Okarma at the West Pomeranian University of Technology, Poland.

- (i) best for printing: $\mathbf{MO_1}$: the difference of the P_{err} with flash on and off makes it slightly better than Michalak, as both are top fast among the top quality algorithms.
- (ii) best for OCR: **Michalak**
- (iii) global "winner": **Michalak**: it is just as fast as MO_1, but has better quality measures than MO_1.

[2] https://www.motorola.com/we/compare-smartphones.

Table 2. Best binarization algorithms using Motorola G9 Plus

Motolora G9 Plus						
	OFF			ON		
#	**Alg.**	P_{err}	**Time (s)**	**Alg.**	P_{err}	**Time (s)**
1	**Michalak**	0.92	0.06	**KS$_1$**	0.55	3.42
2	**MO$_3$**	0.94	1.41	**MO$_1$**	0.59	0.05
3.00	**Bradley**	0.95	0.41	**Gosh**	0.70	145.16
4	**MO$_1$**	0.97	0.06	**Yasin**	0.74	1.75
5	**ElisaTV**	1.06	11.59	**ElisaTV**	0.83	11.2
6	**Yasin**	1.14	2.03	**MO$_3$**	0.86	1.34
7	**DilatedUNet**	1.17	188.27	**Bradley**	0.91	0.40
8	**MO$_2$**	1.19	3.09	**Michalak**	0.97	0.05
9	**Gosh**	1.24	143.09	**Singh**	1.00	0.44
10	**WX**	1.25	281.66	**Nick**	1.12	0.21
11	**KS$_2$**	1.42	3.80	**Su-Lu**	1.22	2.17
12	**DocDLink**	1.43	300.18	**DilatedUNet**	1.24	187.73
13	**KS$_1$**	1.68	3.72	**Wolf**	1.32	0.29
14	**ISauvola**	1.72	0.53	**WX**	1.64	281.16
15	**Su-Lu**	1.74	2.19	**MO$_2$**	1.65	3.00
#	**Alg.**	$[L_{dist}]$	**Time (s)**	**Alg.**	$[L_{dist}]$	**Time (s)**
1	**KS$_2$**	0.98	3.80	**AH$_1$**	0.98	398.98
2	**MO$_3$**	0.98	1.41	**AH$_2$**	0.98	91.2
3	**Bradley**	0.98	0.41	**KS$_2$**	0.98	3.69
4	**Michalak**	0.98	0.06	**MO$_3$**	0.98	1.34
5	**RNB**	0.98	46.17	**SL**	0.98	13666.25
6	**WAN**	0.98	1.36	**Michalak**	0.98	0.05
7	**ISauvola**	0.97	0.53	**Bradley**	0.98	0.40
8	**MO$_2$**	0.97	3.09	**RNB**	0.98	45.58
9	**MO$_1$**	0.97	0.06	**WAN**	0.97	1.35
10	**ElisaTV**	0.97	11.59	**MO$_2$**	0.97	3.00
11	**JB**	0.97	1.79	**JB**	0.97	1.73
12	**KS$_1$**	0.97	3.72	**KS$_1$**	0.97	3.42
13	**Gosh**	0.97	143.09	**MO$_1$**	0.97	0.05
14	**YinYang**	0.97	2.08	**ISauvola**	0.97	0.52
15	**Bataineh**	0.97	0.16	**ElisaTV**	0.97	11.2

Figure 2 presents the results of the binarization produced by the top two algorithms for two of the document images produced by the Moto G9 smartphone.

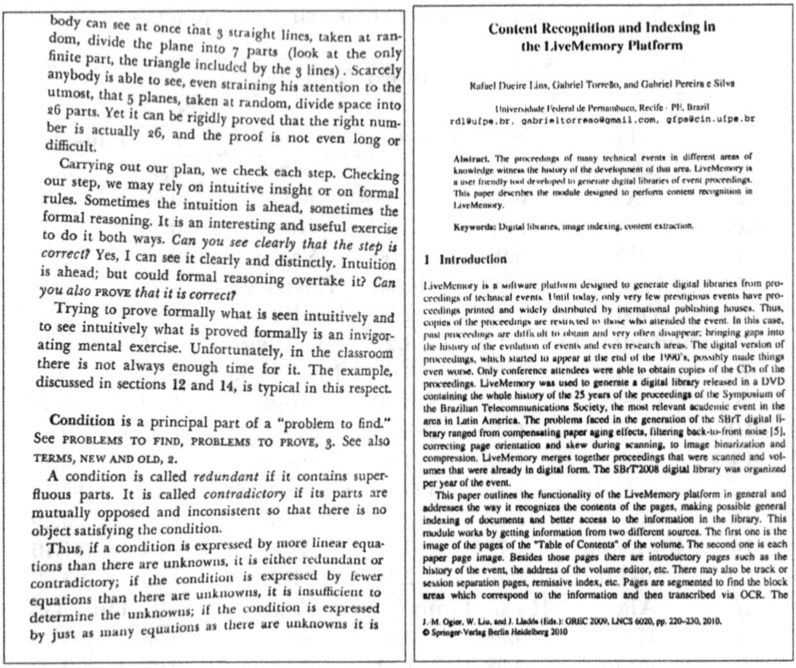

body can see at once that 3 straight lines, taken at random, divide the plane into 7 parts (look at the only finite part, the triangle included by the 3 lines). Scarcely anybody is able to see, even straining his attention to the utmost, that 5 planes, taken at random, divide space into 26 parts. Yet it can be rigidly proved that the right number is actually 26, and the proof is not even long or difficult.

Carrying out our plan, we check each step. Checking our step, we may rely on intuitive insight or on formal rules. Sometimes the intuition is ahead, sometimes the formal reasoning. It is an interesting and useful exercise to do it both ways. *Can you see clearly that the step is correct?* Yes, I can see it clearly and distinctly. Intuition is ahead; but could formal reasoning overtake it? *Can you also* PROVE *that it is correct?*

Trying to prove formally what is seen intuitively and to see intuitively what is proved formally is an invigorating mental exercise. Unfortunately, in the classroom there is not always enough time for it. The example, discussed in sections **12** and **14**, is typical in this respect.

Condition is a principal part of a "problem to find." See PROBLEMS TO FIND, PROBLEMS TO PROVE, 3. See also TERMS, NEW AND OLD, 2.

A condition is called *redundant* if it contains superfluous parts. It is called *contradictory* if its parts are mutually opposed and inconsistent so that there is no object satisfying the condition.

Thus, if a condition is expressed by more linear equations than there are unknowns, it is either redundant or contradictory; if the condition is expressed by fewer equations than there are unknowns, it is insufficient to determine the unknowns; if the condition is expressed by just as many equations as there are unknowns it is

Content Recognition and Indexing in the LiveMemory Platform

Rafael Dueire Lins, Gabriel Torreão, and Gabriel Pereira e Silva

Universidade Federal de Pernambuco, Recife - PE, Brazil
rdl@ufpe.br, gabrieltorreao@gmail.com, gfps@cin.ufpe.br

Abstract. The proceedings of many technical events in different areas of knowledge witness the history of the development of that area. LiveMemory is a user friendly tool developed to generate digital libraries of event proceedings. This paper describes the module designed to perform content recognition in LiveMemory.

Keywords: Digital libraries, image indexing, content extraction.

1 Introduction

LiveMemory is a software platform designed to generate digital libraries from proceedings of technical events. Until today, only very few prestigious events have proceedings printed and widely distributed by international publishing houses. Thus, copies of the proceedings are restricted to those who attended the event. In this case, past proceedings are difficult to obtain and very often disappear; bringing gaps into the history of the evolution of events and even research areas. The digital version of proceedings, which started to appear at the end of the 1990's, possibly made things even worse. Only conference attendees were able to obtain copies of the CDs of the proceedings. LiveMemory was used to generate a digital library released in a DVD containing the whole history of the 25 years of the proceedings of the Symposium of the Brazilian Telecommunications Society, the most relevant academic event in the area in Latin America. The problems faced in the generation of the SBrT digital library ranged from compensating paper aging effects, filtering back-to-front noise [5], correcting page orientation and skew during scanning, to image binarization and compression. LiveMemory merges together proceedings that were scanned and volumes that were already in digital form. The SBrT'2008 digital library was organized per year of the event.

This paper outlines the functionality of the LiveMemory platform in general and addresses the way it recognizes the contents of the pages, making possible general indexing of documents and better access to the information in the library. This module works by getting information from two different sources. The first one is the image of the pages of the "Table of Contents" of the volume. The second one is each paper page image. Besides those pages there are introductory pages such as the history of the event, the address of the volume editor, etc. There may also be track or session separation pages, remissive index, etc. Pages are segmented to find the block areas which correspond to the information and then transcribed via OCR. The

J.-M. Ogier, W. Liu, and J. Lladós (Eds.): GREC 2009, LNCS 6020, pp. 220–230, 2010.
© Springer-Verlag Berlin Heidelberg 2010

Fig. 2. Result of the binarization with $\mathbf{MO}_1$ algorithm of an offset printed **book** page with the strobeflash **on** (left) and, using **Michalak**, a **deskjet** printed document with the stobeflash **off** (right), both acquired using the **Motorola Moto G9**

4.2 Samsung A10S

The smartphone Samsung Galaxy A10S[3] was released around August 2019 and became the second top selling device worldwide in December 2019 and it is still on sale today [4] It originally runs an Android 9.0 (Pie), upgradable to Android 11, One UI 3.1. The two assessments made here with the 61 binarization algorithms yielded the data shown on Table 3 for the top 15 algorithms and allow to point as global results:

[3] https://www.gsmarena.com/samsung_galaxy_a10s-9793.php.
[4] https://www.91mobiles.com/hub/best-selling-phone-q3-2019-iphone-xr-11-samsung-galaxy-a10-a50/?pid=33347.

Table 3. Best binarization algorithms using Samsung A10S

Samsung A10S						
	OFF			ON		
#	**Alg.**	P_{err}	**Time (s)**	**Alg.**	P_{err}	**Time (s)**
1	**Michalak**	0.76	0.05	**Michalak**	0.76	0.03
2	**MO_2**	0.91	1.95	**MO_2**	0.91	1.86
3	**MO_1**	0.92	0.04	**MO_1**	0.92	0.03
4	**MO_3**	0.92	0.87	**MO_3**	0.92	0.8
5	**Bradley**	0.94	0.24	**Bradley**	0.94	0.24
6	**Bernsen**	1.06	1.98	**Bernsen**	1.06	1.96
7	**ElisaTV**	1.16	6.13	**ElisaTV**	1.16	6.09
8	**DocDLink**	1.24	173.78	**Yasin**	1.24	1.29
9	**Yasin**	1.24	1.46	**DocDLink**	1.24	173.34
10	**ISauvola**	1.25	0.31	**ISauvola**	1.25	0.31
11	**Gosh**	1.27	80.84	**Gosh**	1.27	80.66
12	**Howe**	1.32	37.38	**Howe**	1.32	37.27
13	**WX**	1.35	174.81	**WX**	1.35	174.31
14	**Wolf**	1.38	0.18	**Wolf**	1.38	0.18
15	**KS_2**	1.4	3.26	**KS_2**	1.4	3.31
#	**Alg.**	$[L_{dist}]$	**Time (s)**	**Alg.**	$[L_{dist}]$	**Time (s)**
1	**RNB**	0.98	27.77	**RNB**	0.98	27.86
2	**KS_2**	0.98	3.26	**AH_2**	0.98	56.78
3	**ElisaTV**	0.98	6.13	**KS_2**	0.98	3.31
4	**JB**	0.98	1.24	**ElisaTV**	0.98	6.09
5	**ISauvola**	0.98	0.31	**JB**	0.98	1.23
6	**Bradley**	0.98	0.24	**ISauvola**	0.98	0.31
7	**AH_2**	0.98	59.22	**AH_1**	0.98	257.38
8	**$Akbari_1$**	0.98	15.27	**Bradley**	0.98	0.24
9	**Jia-Shi**	0.98	15.19	**$Akbari_1$**	0.98	15.18
10	**MO_3**	0.98	0.87	**Jia-Shi**	0.98	15.22
11	**Michalak**	0.98	0.05	**MO_3**	0.98	0.8
12	**WAN**	0.98	0.82	**Michalak**	0.98	0.03
13	**KS_1**	0.97	3.49	**WAN**	0.98	0.83
14	**YinYang**	0.97	1.41	**KS_1**	0.97	3.38
15	**Gosh**	0.97	80.84	**SL**	0.97	11627.4

- (i) best for printing and transmitting: **Michalak** – it has the best P_{err} either with flash on or off
- (ii) best for OCR: **Michalak** – it is the fastest among the smallest $[L_{dist}]$, with value 0.98
- (iii) overall winner: **Michalak** – it is the best either for OCR or printing and transmitting applications

The result of the binarization of two of the test images in the dataset used here processed by Michalak algorithm may be seen in Fig. 3.

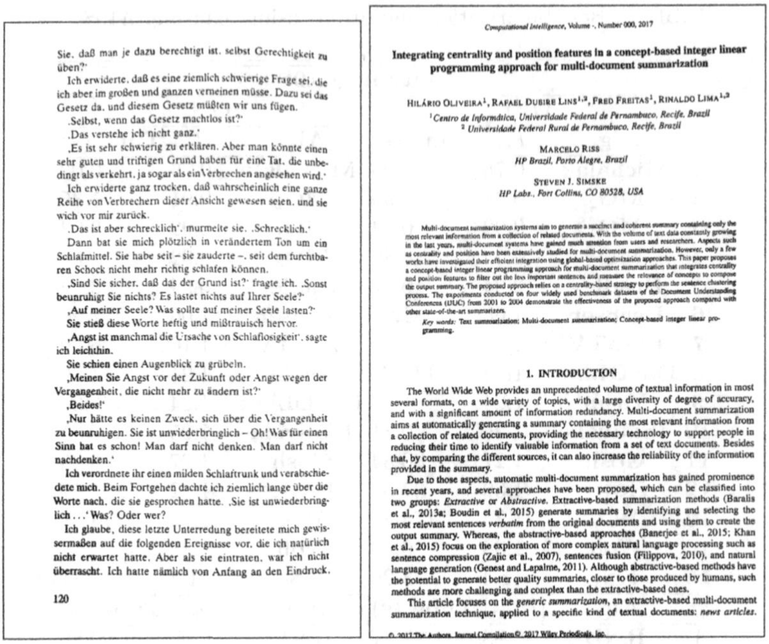

Sie, daß man je dazu berechtigt ist, selbst Gerechtigkeit zu üben?'

Ich erwiderte, daß es eine ziemlich schwierige Frage sei, die ich aber im großen und ganzen verneinen müsse. Dazu sei das Gesetz da, und diesem Gesetz müßten wir uns fügen.

,Selbst, wenn das Gesetz machtlos ist?'

,Das verstehe ich nicht ganz.'

,Es ist sehr schwierig zu erklären. Aber man könnte einen sehr guten und triftigen Grund haben für eine Tat, die unbedingt als verkehrt, ja sogar als ein Verbrechen angesehen wird.'

Ich erwiderte ganz trocken, daß wahrscheinlich eine ganze Reihe von Verbrechern dieser Ansicht gewesen seien, und sie wich vor mir zurück.

,Das ist aber schrecklich', murmelte sie. ,Schrecklich.'

Dann bat sie mich plötzlich in verändertem Ton um ein Schlafmittel. Sie habe seit – sie zauderte –, seit dem furchtbaren Schock nicht mehr richtig schlafen können.

,Sind Sie sicher, daß das der Grund ist?' fragte ich. ,Sonst beunruhigt Sie nichts? Es lastet nichts auf Ihrer Seele?'

,Auf meiner Seele? Was sollte auf meiner Seele lasten?'

Sie stieß diese Worte heftig und mißtrauisch hervor.

,Angst ist manchmal die Ursache von Schlaflosigkeit', sagte ich leichthin.

Sie schien einen Augenblick zu grübeln.

,Meinen Sie Angst vor der Zukunft oder Angst wegen der Vergangenheit, die nicht mehr zu ändern ist?'

,Beides!'

,Nur hätte es keinen Zweck, sich über die Vergangenheit zu beunruhigen. Sie ist unwiederbringlich – Oh! Was für einen Sinn hat es schon! Man darf nicht denken. Man darf nicht nachdenken.'

Ich verordnete ihr einen milden Schlaftrunk und verabschiedete mich. Beim Fortgehen dachte ich ziemlich lange über die Worte nach, die sie gesprochen hatte. ,Sie ist unwiederbringlich ...' Was? Oder wer?

Ich glaube, diese letzte Unterredung bereitete mich gewissermaßen auf die folgenden Ereignisse vor, die ich natürlich nicht erwartet hatte. Aber als sie eintraten, war ich nicht überrascht. Ich hatte nämlich von Anfang an den Eindruck,

120

Computational Intelligence, Volume -, Number 000, 2017

Integrating centrality and position features in a concept-based integer linear programming approach for multi-document summarization

HILÁRIO OLIVEIRA[1], RAFAEL DUEIRE LINS[1,2], FRED FREITAS[1], RINALDO LIMA[1,2]
[1] *Centro de Informática, Universidade Federal de Pernambuco, Recife, Brazil*
[2] *Universidade Federal Rural de Pernambuco, Recife, Brazil*

MARCELO RISS
HP Brazil, Porto Alegre, Brazil

STEVEN J. SIMSKE
HP Labs., Fort Collins, CO 80528, USA

Multi-document summarization systems aim to generate a succinct and coherent summary containing only the most relevant information from a collection of related documents. With the volume of text data constantly growing in the last years, multi-document systems have gained much attention from users and researchers. Aspects such as centrality and position have been extensively studied for multi-document summarization. However, only a few works have investigated their efficient integration using global-based optimization approaches. This paper proposes a concept-based integer linear programming approach for multi-document summarization that integrates centrality and position features to filter out the less important sentences and measure the relevance of concepts to compose the output summary. The proposed approach relies on a centrality-based strategy to perform the sentence clustering process. The experiments conducted on four widely used benchmark datasets of the Document Understanding Conferences (DUC) from 2001 to 2004 demonstrate the effectiveness of the proposed approach compared with other state-of-the-art summarizers.

Key words: Text summarization; Multi-document summarization; Concept-based integer linear programming.

1. INTRODUCTION

The World Wide Web provides an unprecedented volume of textual information in most several formats, on a wide variety of topics, with a large diversity of degree of accuracy, and with a significant amount of information redundancy. Multi-document summarization aims at automatically generating a summary containing the most relevant information from a collection of related documents, providing the necessary technology to support people in reducing their time to identify valuable information from a set of text documents. Besides that, by comparing the different sources, it can also increase the reliability of the information provided in the summary.

Due to those aspects, automatic multi-document summarization has gained prominence in recent years, and several approaches have been proposed, which can be classified into two groups: *Extractive* or *Abstractive*. Extractive-based summarization methods (Baralis et al., 2013a; Boudin et al., 2015) generate summaries by identifying and selecting the most relevant sentences *verbatim* from the original documents and using them to create the output summary. Whereas, the abstractive-based approaches (Banerjee et al., 2015; Khan et al., 2015) focus on the exploration of more complex natural language processing such as sentence compression (Zajic et al., 2007), sentences fusion (Filippova, 2010), and natural language generation (Genest and Lapalme, 2011). Although abstractive-based methods have the potential to generate better quality summaries, closer to those produced by humans, such methods are more challenging and complex than the extractive-based ones.

This article focuses on the *generic summarization*, an extractive-based multi-document summarization technique, applied to a specific kind of textual documents: *news articles*.

© 2017 The Authors. Journal Compilation © 2017 Wiley Periodicals, Inc.

Fig. 3. Result of the binarization with **Michalak** algorithm of an offset printed **book** page with the strobeflash **on** (left) and a **deskjet** printed document with the strobeflash **off** (right), adcquired using the **Samsung A10S**

4.3 Samsung S20

The Samsung Galaxy S20 is another Android-based smartphone designed and manufactured by Samsung. It is the successor model to the successful Galaxy S10 and it was released on 11 February 2020 [21]. The analysis of the data presented on Table 4 allows one to pinpoint the "best" algorithms in the in terms of image quality-time and OCR-performance and time, and the overall "winner" as:

Table 4. Best binarization algorithms using Samsung S20

Samsung S20						
	OFF			ON		
#	**Alg.**	P_{err}	**Time (s)**	**Alg.**	P_{err}	**Time (s)**
1	**MO_1**	0.91	0.05	**Gattal**	0.66	55.68
2	**MO_3**	0.92	1.09	**IsoData**	0.72	0.13
3	**Bradley**	0.96	0.31	**Otsu**	0.74	0.02
4	**Michalak**	0.99	0.05	**MO_1**	0.79	0.04
5	**DilatedUNet**	1.06	151.65	**Li-Tam**	0.84	0.13
6	**WX**	1.13	279.6	**Yasin**	0.92	1.47
7	**Howe**	1.26	49.79	**Gosh**	0.95	102.95
8	**DocDLink**	1.27	228.22	**MO_3**	0.96	0.98
9	**Gosh**	1.28	120.9	**ElisaTV**	0.97	7.46
10	**KS_1**	1.28	3.79	**Wolf**	1.02	0.22
11	**Wolf**	1.28	0.23	**KS_1**	1.05	3.39
12	**Yasin**	1.28	1.75	**Michalak**	1.05	0.04
13	**Singh**	1.29	0.34	**Bradley**	1.05	0.29
14	**MO_2**	1.33	2.49	**Singh**	1.06	0.32
15	**Nick**	1.37	0.16	**$Ergina_L$**	1.06	0.62
#	**Alg.**	$[L_{dist}]$	**Time (s)**	**Alg.**	$[L_{dist}]$	**Time (s)**
1	**MO_3**	0.98	1.09	**$Ergina_G$**	0.98	0.44
2	**RNB**	0.98	36.34	**KSW**	0.98	0.13
3	**KS_2**	0.98	3.47	**Yen-CC**	0.98	0.13
4	**Michalak**	0.98	0.05	**Bradley**	0.98	0.29
5	**ISauvola**	0.98	0.41	**MO_3**	0.98	0.98
6	**JB**	0.98	1.43	**SL**	0.98	10319.87
7	**Bradley**	0.98	0.31	**ElisaTV**	0.98	7.46
8	**WAN**	0.98	1.07	**IsoData**	0.98	0.13
9	**ElisaTV**	0.98	7.68	**Wolf**	0.98	0.22
10	**Bataineh**	0.98	0.12	**Su-Lu**	0.98	1.62
11	**YinYang**	0.98	1.64	**AH_2**	0.98	72.09
12	**DocDLink**	0.97	228.22	**RNB**	0.98	34.71
13	**MO_1**	0.97	0.05	**AH_1**	0.98	319.31
14	**MO_2**	0.97	2.49	**RenyEntropy**	0.98	0.13
15	**AH_2**	0.97	75.01	**MO_1/ Michalak**	0.98	0.04

- (i) best for printing: **MO_1**.
- (ii) best for OCR: **Michalak**
- (iii) the overall "winner": **MO_1**, P_{err} and the L_{dist} are reasonably small.

Figure 4 shows the monochromatic version of two of the images in this dataset.

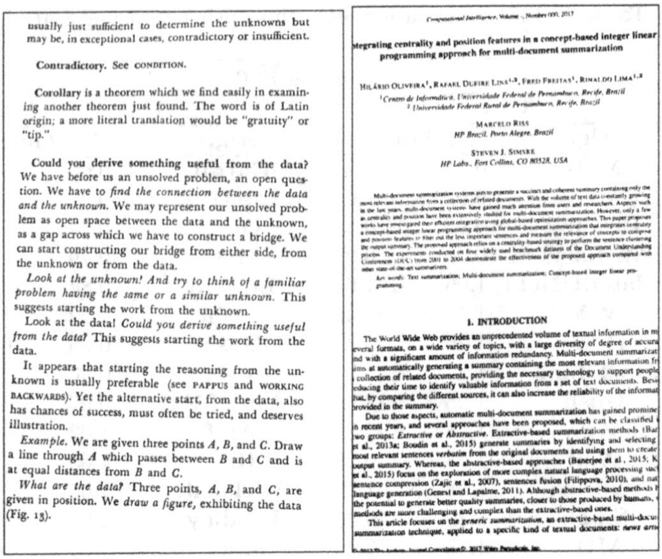

usually just sufficient to determine the unknowns but may be, in exceptional cases, contradictory or insufficient.

Contradictory. See CONDITION.

Corollary is a theorem which we find easily in examining another theorem just found. The word is of Latin origin; a more literal translation would be "gratuity" or "tip."

Could you derive something useful from the data? We have before us an unsolved problem, an open question. We have to *find the connection between the data and the unknown.* We may represent our unsolved problem as open space between the data and the unknown, as a gap across which we have to construct a bridge. We can start constructing our bridge from either side, from the unknown or from the data.

Look at the unknown! And try to think of a familiar problem having the same or a similar unknown. This suggests starting the work from the unknown.

Look at the data! *Could you derive something useful from the data?* This suggests starting the work from the data.

It appears that starting the reasoning from the unknown is usually preferable (see PAPPUS and WORKING BACKWARDS). Yet the alternative start, from the data, also has chances of success, must often be tried, and deserves illustration.

Example. We are given three points *A*, *B*, and *C*. Draw a line through *A* which passes between *B* and *C* and is at equal distances from *B* and *C*.

What are the data? Three points, *A*, *B*, and *C*, are given in position. We *draw a figure*, exhibiting the data (Fig. 13).

Computational Intelligence, Volume -, Number 000, 2017

tegrating centrality and position features in a concept-based integer linear programming approach for multi-document summarization

HILÁRIO OLIVEIRA[1], RAFAEL DUEIRE LINS[1,2], FRED FREITAS[1], RINALDO LIMA[1,2]
[1]*Centro de Informática, Universidade Federal de Pernambuco, Recife, Brazil*
[2]*Universidade Federal Rural de Pernambuco, Recife, Brazil*

MARCELO RISS
HP Brazil, Porto Alegre, Brazil

STEVEN J. SIMSKE
HP Labs., Fort Collins, CO 80528, USA

1. INTRODUCTION

Fig. 4. Result of the binarization with $\mathbf{MO}_1$ algorithm of an offset printed **book** page with the strobe flash **on** (left) and a **deskjet** printed document with the stobeflash **off** (right), acquired using the **Samsung S20**

4.4 Apple iPhone SE

The second-generation iPhone SE (also known as the iPhone SE 2 or the iPhone SE 2020) is a smartphone designed and developed by Apple Inc. It was released on April, 2020 and became one of the top selling smartphone models in 2020 (24.2 million devices sold).[5] It continues today as one of the top sold mid-price devices. Table 5 presents the results for the assessment of this dataset. As one can see, several "classical" binarization algorithms appear high-up in the quality P_{err} rank, with very efficient time figures. Taking the formula $Perr_{off}\ x\ T_{off} + Perr_{on}\ x\ T_{on}$ as the way to decide the best algorithm for printing, MO_1 appears top with 0.0972, closely followed by Michalak (0.1194) and Otsu (0,1316). Thus, in this category the winner is MO_1. The global results are:

- (i) best for printing and transmitting: $\mathbf{MO}_1$
- (ii) the best for OCR: $\mathbf{MO}_1$
- (iii) the overall winner: $\mathbf{MO}_1$

[5] https://www.gizmochina.com/2021/02/25/most-shipped-smartphones-2020-omdia/.

Table 5. Best binarization algorithms using Apple iPhone SE

Apple iPhone SE 2						
	OFF			ON		
#	Alg.	P_{err}	Time (s)	Alg.	P_{err}	Time (s)
1	**Yasin**	0.72	1.96	**IsoData**	0.60	0.12
2	**Nick**	0.79	0.17	**Otsu**	0.60	0.02
3	**Sauvola**	0.79	0.17	**Sauvola**	0.73	0.18
4	**Singh**	0.79	0.30	**Gattal**	0.74	54.59
5	**Gosh**	0.79	88.74	**Gosh**	0.77	85.64
6	**JB**	0.88	1.27	**Yasin**	0.81	1.55
7	**YinYang**	0.94	1.70	**MO_1**	0.81	0.04
8	**Wolf**	0.95	0.23	**Singh**	0.81	0.29
9	**KS_1**	0.96	4.23	**Wolf**	0.84	0.24
10	**ElisaTV**	1.04	5.00	**Nick**	0.84	0.17
11	**Su-Lu**	1.04	1.77	**JB**	0.85	1.27
12	**MO_1**	1.08	0.06	**ElisaTV**	0.90	3.44
13	**KS_3**	1.21	4.70	**YinYang**	0.94	1.78
14	**Michalak**	1.31	0.06	**Michalak**	1.02	0.04
15	**Bradley**	1.36	0.34	**KS_1**	1.03	3.30
#	Alg.	$[L_{dist}]$	Time (s)	Alg.	$[L_{dist}]$	Time (s)
1	**KS_1**	0.98	4.23	**YinYang**	0.98	1.78
2	**$Akbari_1$**	0.98	21.76	**SL**	0.98	10,310.89
3	**Jia-Shi**	0.98	20.74	**Yasin**	0.97	1.55
4	**Singh**	0.98	0.30	**KS_2**	0.97	3.39
5	**Wolf**	0.98	0.23	Singh	0.97	0.29
6	**Wu-Lu**	0.98	0.13	**Nick**	0.97	0.17
7	**Bataineh**	0.98	0.13	**KS_3**	0.97	4.65
8	**AH_1**	0.98	277.31	**Bataineh**	0.97	0.13
9	**ElisaTV**	0.98	5.00	**RNB**	0.97	33.9
10	**Calvo-Z**	0.98	9.83	**$Ergina_G$**	0.97	0.43
11	**MO_2**	0.98	2.56	**Howe**	0.97	55.39
12	**RNB**	0.98	33.45	**Li-Tam**	0.97	0.13
13	**Nick**	0.98	0.17	**MO_2**	0.97	2.28
14	**MO_1**	0.98	0.06	**$Ergina_L$**	0.97	0.59
15	**Bradley**	0.98	0.34	**DocDLink**	0.97	191.72
37	**Yen-CC**	0.97	0.13	MO_1	0.97	0.04

Figure 5 presents two of the imagesin this dataset binarized with the overall "winner".

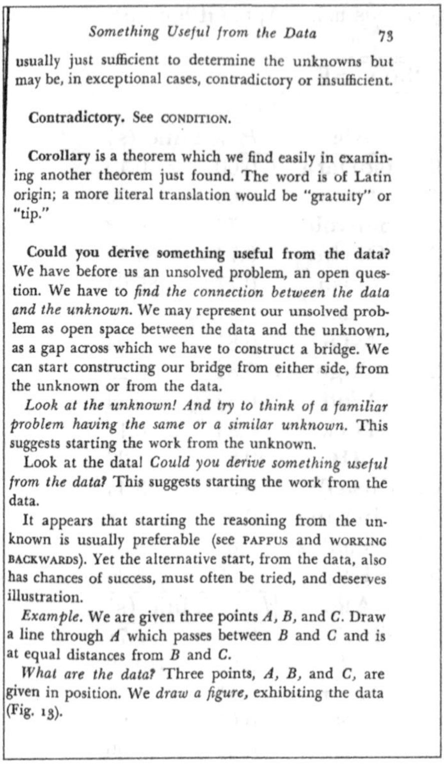

Something Useful from the Data 73

usually just sufficient to determine the unknowns but may be, in exceptional cases, contradictory or insufficient.

Contradictory. See CONDITION.

Corollary is a theorem which we find easily in examining another theorem just found. The word is of Latin origin; a more literal translation would be "gratuity" or "tip."

Could you derive something useful from the data? We have before us an unsolved problem, an open question. We have to *find the connection between the data and the unknown.* We may represent our unsolved problem as open space between the data and the unknown, as a gap across which we have to construct a bridge. We can start constructing our bridge from either side, from the unknown or from the data.

Look at the unknown! And try to think of a familiar problem having the same or a similar unknown. This suggests starting the work from the unknown.

Look at the data! *Could you derive something useful from the data?* This suggests starting the work from the data.

It appears that starting the reasoning from the unknown is usually preferable (see PAPPUS and WORKING BACKWARDS). Yet the alternative start, from the data, also has chances of success, must often be tried, and deserves illustration.

Example. We are given three points *A, B,* and *C.* Draw a line through *A* which passes between *B* and *C* and is at equal distances from *B* and *C.*

What are the data? Three points, *A, B,* and *C,* are given in position. We *draw a figure,* exhibiting the data (Fig. 13).

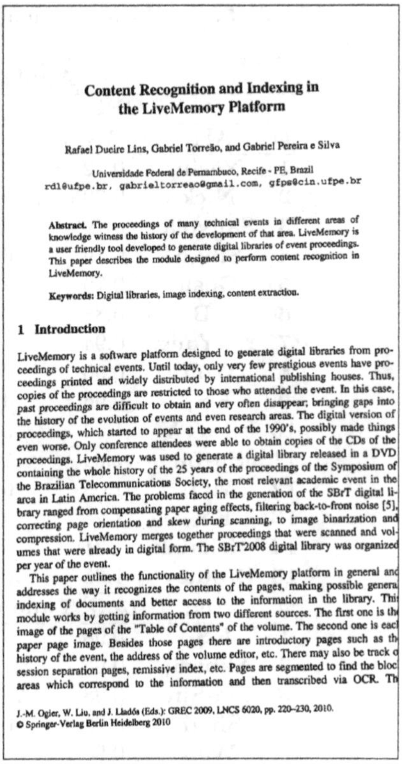

Content Recognition and Indexing in the LiveMemory Platform

Rafael Dueire Lins, Gabriel Torreão, and Gabriel Pereira e Silva

Universidade Federal de Pernambuco, Recife - PE, Brazil
rdl@ufpe.br, gabrieltorreao@gmail.com, gfps@cin.ufpe.br

Abstract. The proceedings of many technical events in different areas of knowledge witness the history of the development of that area. LiveMemory is a user friendly tool developed to generate digital libraries of event proceedings. This paper describes the module designed to perform content recognition in LiveMemory.

Keywords: Digital libraries, image indexing, content extraction.

1 Introduction

LiveMemory is a software platform designed to generate digital libraries from proceedings of technical events. Until today, only very few prestigious events have proceedings printed and widely distributed by international publishing houses. Thus, copies of the proceedings are restricted to those who attended the event. In this case, past proceedings are difficult to obtain and very often disappear; bringing gaps into the history of the evolution of events and even research areas. The digital version of proceedings, which started to appear at the end of the 1990's, possibly made things even worse. Only conference attendees were able to obtain copies of the CDs of the proceedings. LiveMemory was used to generate a digital library released in a DVD containing the whole history of the 25 years of the proceedings of the Symposium of the Brazilian Telecommunications Society, the most relevant academic event in the area in Latin America. The problems faced in the generation of the SBrT digital library ranged from compensating paper aging effects, filtering back-to-front noise [5], correcting page orientation and skew during scanning, to image binarization and compression. LiveMemory merges together proceedings that were scanned and volumes that were already in digital form. The SBrT'2008 digital library was organized per year of the event.

This paper outlines the functionality of the LiveMemory platform in general and addresses the way it recognizes the contents of the pages, making possible genera indexing of documents and better access to the information in the library. Thi module works by getting information from two different sources. The first one is th image of the pages of the "Table of Contents" of the volume. The second one is eac paper page image. Besides those pages there are introductory pages such as th history of the event, the address of the volume editor, etc. There may also be track o session separation pages, remissive index, etc. Pages are segmented to find the bloc areas which correspond to the information and then transcribed via OCR. Th

J.-M. Ogier, W. Liu, and J. Lladós (Eds.): GREC 2009, LNCS 6020, pp. 220–230, 2010.
© Springer-Verlag Berlin Heidelberg 2010

Fig. 5. Result of the binarization with $\mathbf{MO}_1$ algorithm of an offset printed **book** page with the strobeflash **on** (left) and, a **deskjet** printed document with the stobeflash **off** (right), acquired using the **iPhone SE 2**

5 Conclusions

Smartphones have drastically changed the way of life of people worldwide with their omnipresence, growing computational power and high-quality embedded cameras. Photographing documents is now a simple way of digitizing everyday documents and book pages for later referencing and even meeting legal requirements in many countries. Document binarization plays a key role in many document processing pipelines, besides yielding smaller documents for storing and sending via networks better readable and more economic to print. Recent document binarization competitions [27,30,31] show that no single algorithm is the best for all kinds of documents. Each smartphone model has a camera with different features making the binarization of photographed document images a challenging task.

Applications that run on smartphones need to be light due to the hardware limitations of a device that needs to execute several processes simultaneously. Thus a binarization algorithm to be used in an embedded smartphone application must have an excellent quality-time balance. This paper presents a methodology to choose such an algorithm. Four popular smartphone models of three different manufacturers were quality-time assessed using 61 of the possibly best binarization algorithms of today, pointing out the "best" algorithm for printing, the "best" algorithm for OCR applications, and the global "winner" for each of those devices.

The recent paper [26] shows that feeding binarization algorithms with the image, their RGB-components or the grayscale converted image yield to differences in their quality-time performance. That analysis would multiply the number of the assessed algorithms by five, thus is left as one of the lines for further work.

Acknowledgements. The authors grateful to all those who made their binarization code available.

The research reported in this paper was mainly sponsored by the RD&I project between the Universidade do Estado do Amazonas and Transire Eletrônicos and Tec Toy S.A. through the Lei de Informática/SUFRAMA. Rafael Dueire Lins was also partly sponsored by CNPq – Brazil.

References

1. Akbari, Y., Britto, A.S., Jr., Al-Maadeed, S., Oliveira, L.S.: Binarization of degraded document images using convolutional neural networks based on predicted two-channel images. In: ICDAR (2019)
2. Barney Smith, E.H., Likforman-Sulem, L., Darbon, J.: Effect of pre-processing on binarization. In: Document Recognition and Retrieval XVII, p. 75340H (2010)
3. Bataineh, B., Abdullah, S.N.H.S., Omar, K.: An adaptive local binarization method for document images based on a novel thresholding method and dynamic windows. Pattern Recogn. Lett. **32**(14), 1805–1813 (2011)
4. Bera, S.K., Ghosh, S., Bhowmik, S., Sarkar, R., Nasipuri, M.: A non-parametric binarization method based on ensemble of clustering algorithms. Multimed. Tools Appl. **80**(5), 7653–7673 (2020). https://doi.org/10.1007/s11042-020-09836-z
5. Bernsen, J.: Dynamic thresholding of gray-level images. In: International Conference on Pattern Recognition, pp. 1251–1255 (1986)
6. Bradley, D., Roth, G.: Adaptive thresholding using the integral image. J. Graph. Tools **12**(2), 13–21 (2007)
7. Calvo-Zaragoza, J., Gallego, A.J.: A selectional auto-encoder approach for document image binarization. Pattern Recogn. **86**, 37–47 (2019)
8. Doyle, W.: Operations useful for similarity-invariant pattern recognition. J. ACM **9**(2), 259–267 (1962)
9. Gattal, A., Abbas, F., Laouar, M.R.: Automatic parameter tuning of k-means algorithm for document binarization. In: 7th ICSENT, pp. 1–4. ACM Press (2018)
10. Glasbey, C.: An analysis of histogram-based thresholding algorithms. Graph. Models Image Process. **55**(6), 532–537 (1993)
11. Hadjadj, Z., Meziane, A., Cherfa, Y., Cheriet, M., Setitra, I.: ISauvola: improved Sauvola's algorithm for document image binarization. In: Campilho, A., Karray, F. (eds.) ICIAR 2016. LNCS, vol. 9730, pp. 737–745. Springer, Cham (2016). https://doi.org/10.1007/978-3-319-41501-7_82

12. He, S., Schomaker, L.: DeepOtsu: document enhancement and binarization using iterative deep learning. Pattern Recogn. **91**, 379–390 (2019)
13. Howe, N.R.: Document binarization with automatic parameter tuning. IJDAR **16**(3), 247–258 (2013)
14. Huang, L.K., Wang, M.J.J.: Image thresholding by minimizing the measures of fuzziness. Pattern Recogn. **28**(1), 41–51 (1995)
15. Jia, F., Shi, C., He, K., Wang, C., Xiao, B.: Degraded document image binarization using structural symmetry of strokes. Pattern Recogn. **74**, 225–240 (2018)
16. Johannsen, G., Bille, J.: A threshold selection method using information measures. In: International Conference on Pattern Recognition, pp. 140–143 (1982)
17. Kapur, J., Sahoo, P., Wong, A.: A new method for gray-level picture thresholding using the entropy of the histogram. Comput. Vis. Graph. Image Process. **29**(1), 273–285 (1985)
18. Kavallieratou, E., Stathis, S.: Adaptive binarization of historical document images. In: Proceedings - International Conference on Pattern Recognition, vol. 3, pp. 742–745 (2006)
19. Khurshid, K., Siddiqi, I., Faure, C., Vincent, N.: Comparison of Niblack inspired binarization methods for ancient documents. In: SPIE, p. 72470U (2009)
20. Kittler, J., Illingworth, J.: Minimum error thresholding. Pattern Recogn. **19**(1), 41–47 (1986)
21. Knapp, M.: Samsung Galaxy S20 price and deals: here's where you can get it in the US. In: TechRadar (2020). Accessed 12 Feb 2020
22. Li, C., Tam, P.: An iterative algorithm for minimum cross entropy thresholding. Pattern Recogn. Lett. **19**(8), 771–776 (1998)
23. Lins, R.D.: A taxonomy for noise in images of paper documents - the physical noises. In: Kamel, M., Campilho, A. (eds.) ICIAR 2009. LNCS, vol. 5627, pp. 844–854. Springer, Heidelberg (2009). https://doi.org/10.1007/978-3-642-02611-9_83
24. Lins, R.D., Almeida, M.M.D., Bernardino, R.B., Jesus, D., Oliveira, J.M.: Assessing binarization techniques for document images. In: DocEng 2017, pp. 183–192 (2017)
25. Lins, R.D., Bernardino, R.B., Barney Smith, E., Kavallieratou, E.: ICDAR 2019 time-quality binarization competition. In: ICDAR, pp. 1539–1546 (2019)
26. Lins, R.D., Bernardino, R.B., da Silva Barboza, R., Lins, Z.D.: Direct binarization a quality-and-time efficient binarization strategy. In: Proceedings of the 21st ACM Symposium on Document Engineering, DocEng 2021. ACM (2021)
27. Lins, R.D., Bernardino, R.B., Simske, S.J.: DocEng 2020 time-quality competition on binarizing photographed documents. In: DocEng 2020, pp. 1–4. ACM (2020)
28. Lins, R.D., Bernardino, R.B., Smith, E.B., Kavallieratou, E.: ICDAR 2021 competition on time-quality document image binarization. In: Lladós, J., Lopresti, D., Uchida, S. (eds.) ICDAR 2021. LNCS, vol. 12824, pp. 708–722. Springer, Cham (2021). https://doi.org/10.1007/978-3-030-86337-1_47
29. Lins, R.D., Guimarães Neto, M., França Neto, L., Galdino Rosa, L.: An environment for processing images of historical documents. Microprocess. Microprogram. **40**(10–12), 939–942 (1994)
30. Lins, R.D., Kavallieratou, E., Barney Smith, E., Bernardino, R.B., de Jesus, D.M.: ICDAR 2019 time-quality binarization competition. In: ICDAR, pp. 1539–1546 (2019)
31. Lins, R.D., Simske, S.J., Bernardino, R.B.: DocEng 2021 time-quality competition on binarizing photographed documents. In: DocEng 2021, pp. 1–4. ACM (2020)
32. Lu, S., Su, B., Tan, C.L.: Document image binarization using background estimation and stroke edges. IJDAR **13**(4), 303–314 (2010)

33. Lu, W., Songde, M., Lu, H.: An effective entropic thresholding for ultrasonic images. In: 14th ICPR, vol. 2, pp. 1552–1554 (1998)
34. Mello, C.A.B., Lins, R.D.: Image segmentation of historical documents. In: Visual 2000 (2000)
35. Mustafa, W.A., Abdul Kader, M.M.M.: Binarization of document image using optimum threshold modification. J. Phys.: C. Ser. **1019**(1), 012022 (2018)
36. Niblack, W.: An Introduction to Digital Image Processing. Strandberg Publishing Company (1985)
37. Otsu, N.: A threshold selection method from gray-level histograms. IEEE Trans. Syst. Man Cybern. **9**(1), 62–66 (1979)
38. Prewitt, J.M.S., Mendelsohn, M.L.: The analysis of cell images. Ann. N. Y. Acad. Sci. **128**(3), 1035–1053 (2006)
39. Pun, T.: Entropic thresholding, a new approach. Comput. Graph. Image Process. **16**(3), 210–239 (1981)
40. Saddami, K., Afrah, P., Mutiawani, V., Arnia, F.: A new adaptive thresholding technique for binarizing ancient document. In: INAPR, pp. 57–61. IEEE (2018)
41. Saddami, K., Munadi, K., Away, Y., Arnia, F.: Effective and fast binarization method for combined degradation on ancient documents. Heliyon (2019)
42. Saddami, K., Munadi, K., Muchallil, S., Arnia, F.: Improved thresholding method for enhancing Jawi binarization performance. In: ICDAR, vol. 1, pp. 1108–1113. IEEE (2017)
43. Sahoo, P., Wilkins, C., Yeager, J.: Threshold selection using Renyi's entropy. Pattern Recogn. **30**(1), 71–84 (1997)
44. Sauvola, J., Pietikäinen, M., Pietikainem, M.: Adaptive document image binarization. Pattern Recogn. **33**(2), 225–236 (2000)
45. Shanbhag, A.G.: Utilization of information measure as a means of image thresholding. CVGIP: Graph. Models Image Process. **56**(5), 414–419 (1994)
46. Silva, J.M.M., Lins, R.D., Rocha, V.C.: Binarizing and filtering historical documents with back-to-front interference. In: ACM SAC 2006, pp. 853–858 (2006)
47. da Silva, J.M.M., Lins, R.D.: Color document synthesis as a compression strategy. In: ICDAR (ICDAR), pp. 466–470 (2007)
48. Singh, T.R., Roy, S., Singh, O.I., Sinam, T., Singh, K.M.: A new local adaptive thresholding technique in binarization. IJCSI Int. J. Comput. Sci. Issues **08**(6), 271–277 (2011)
49. Sokratis, V., Kavallieratou, E., Paredes, R., Sotiropoulos, K.: A hybrid binarization technique for document images. In: Biba, M., Xhafa, F. (eds.) Studies in Computational Intelligence, pp. 165–179. Springer, Cham (2011). https://doi.org/10.1007/978-3-642-22913-8_8
50. Souibgui, M.A., Kessentini, Y.: DE-GAN: a conditional generative adversarial network for document enhancement. IEEE Trans. Pattern Anal. Mach. Intell. 1 (2021)
51. Su, B., Lu, S., Tan, C.L.: Binarization of historical document images using the local maximum and minimum. In: 8th IAPR DAS, pp. 159–166. ACM Press (2010)
52. Tsai, W.H.: Moment-preserving thresolding: a new approach. Comput. Vis. Graph. Image Process. **29**(3), 377–393 (1985)
53. Velasco, F.R.: Thresholding using the isodata clustering algorithm. Technical report, OSD or Non-Service DoD Agency (1979)
54. Wolf, C., Jolion, J.M., Chassaing, F.: Text localization, enhancement and binarization in multimedia documents. In: Object Recognition Supported by User Interaction for Service Robots, vol. 2, pp. 1037–1040. IEEE Computer Society (2003)
55. Yen, J.C., Chang, F.J.C.S., Yen, J.C., Chang, F.J., Chang, S.: A new criterion for automatic multilevel thresholding. Trans. Image Process. **4**(3), 370–378 (1995)

56. Zack, G.W., Rogers, W.E., Latt, S.A.: Automatic measurement of sister chromatid exchange frequency. J. Histochem. Cytochemi. **25**(7), 741–753 (1977)
57. Zhou, L., Zhang, C., Wu, M.: D-LinkNet: LinkNet with pretrained encoder and dilated convolution for high resolution satellite imagery road extraction. In: Computer Vision and Pattern Recognition (2018)

A Multilingual Approach to Scene Text Visual Question Answering

Josep Brugués i Pujolràs, Lluís Gómez i Bigordà(✉),
and Dimosthenis Karatzas

Computer Vision Center, Universitat Autònoma de Barcelona, Barcelona, Spain
lgomez@cvc.uab.cat

Abstract. Scene Text Visual Question Answering (ST-VQA) has recently emerged as a hot research topic in Computer Vision. Current ST-VQA models have a big potential for many types of applications but lack the ability to perform well on more than one language at a time due to the lack of multilingual data, as well as the use of monolingual word embeddings for training. In this work, we explore the possibility to obtain bilingual and multilingual VQA models. In that regard, we use an already established VQA model that uses monolingual word embeddings as part of its pipeline and substitute them by FastText and BPEmb multilingual word embeddings that have been aligned to English. Our experiments demonstrate that it is possible to obtain bilingual and multilingual VQA models with a minimal loss in performance in languages not used during training, as well as a multilingual model trained in multiple languages that match the performance of the respective monolingual baselines.

Keywords: Scene text · Visual question answering · Multilingual word embeddings · Vision and language · Deep learning

1 Introduction

Visual Question Answering (VQA) is a semantic task where given a question about an image, an algorithm that combines Natural Language Processing (NLP) and Computer Vision (CV) must infer the answer. When the answer can be found by reading the text present in the image, the task is called Scene Text VQA (ST-VQA). To respond the question the algorithm can use any information found in the image, such as objects, colours, or text. The query can be arbitrary and can encompass many sub-problems in Computer Vision, such as:

- Object recognition: "What is in the image?"
- Object detection: "Are there any persons in the image?"
- Attribute classification: "What colour is the woman's shirt?"
- Scene classification: "Is it raining?"
- Counting: "How many persons are in the image?"
- Activity recognition: "Is the dog running?"
- Spatial recognition: "Are the man and the woman holding hands?"

© Springer Nature Switzerland AG 2022
S. Uchida et al. (Eds.): DAS 2022, LNCS 13237, pp. 65–79, 2022.
https://doi.org/10.1007/978-3-031-06555-2_5

ST-VQA models can be used in very powerful applications, such as for blind people assistance, where a visually impaired person asks a question out loud, and the VQA algorithm uses the transcription of the question and the picture from the user's camera to answer the question. Another interesting application is for travellers that do not speak the language of the country they are visiting and ask questions about things they do not understand. These sorts of models have a lot of advantages, but some limitations too.

Firstly, a huge database is needed to train a VQA model. The database needs to contain images, whose collection can be automated with tools such as web crawlers or by taking images from other datasets. The more problematic part is the data annotation, i.e. to write one or more question-answer pairs for each image, which takes a lot of time and effort to do and to revise.

Secondly, the majority of datasets available for the VQA task are monolingual, i.e. they only contain question and answer pairs in one language, typically English. This is a huge problem in some cases, such as on the application for travellers mentioned above, where the scene text information is in one language, i.e. the language of the place where they are, and the question is given on another language, i.e. the language that the traveller speaks. Some of the possible solutions to get the upper hand on this problem are either to train a monolingual model for each language needed, which needs data annotated in another language and increases the resources needed or to use automatic translation tools to translate the question to English and return back the translated answer.

In this paper, we explore the possibilities of using a third solution to these problems. In that regard, our main focus is to explore the feasibility of using bilingual or multilingual word embeddings instead of the currently used monolingual embeddings in order to obtain VQA models that perform well in more than one language. Word embeddings are learnt word representations in a certain vector space, such that words with similar meaning have a similar representation. By aligning the space of one or more source languages into the space of a target language, we obtain a bilingual or multilingual word embedding. That means that the same word in the source language(s) and the target language should be assigned the same or a close representation.

To perform our experiments, we use an existing method for the ST-VQA task [8] that currently uses monolingual word embeddings in its implementation, and replace them with FastText [3] and BPEmb [12] multilingual embeddings that have been aligned to the English word embedding. We train our models with data from the English-only ST-VQA dataset, and translate the questions into three languages (Catalan, Spanish, and Chinese) using a machine translation model to simulate a multilingual scenario. We have also collected a small custom test set with questions and answers annotated in Catalan and Spanish languages in order to compare the performance between genuine text and machine-translated text. Each model is respectively trained with a different combination of languages, and the performance is analysed in several target languages separately. The process is repeated with several types of configurations to analyse the performance of different embedding models and alignment methods.

2 Related Work

2.1 Word Embeddings

A word embedding is a learned representation for text, where each individual word is represented as a real-valued vector in a predefined vector space: each word is mapped to one vector. The main benefit of dense representations is their generalisation power [7], as they capture and exploit the semantic similarities in the language, which has proved to be useful in many NLP tasks.

The *word2vec* family of architectures was introduced by Mikolov et al. [17,19]. Originally, they proposed two different architectures to obtain a *word2vec* model: the Continuous Bag of Words (CBOW) [19] and the Skip-gram [17]. Both use context to learn the representations: the embeddings are learned by looking at nearby words in a given corpus so that words that always share the same context will all end up with similar representations, e.g. home appliances (like *"refrigerator"*, *"fridge"*, *"oven"*, etc.) will have similar representations.

Since its inception, different modifications of the original *word2vec* architectures, as well as new ones, have arisen with the aim to improve the word representations. Some of them add the possibility to obtain out-of-vocabulary (OOV) representations, i.e. to obtain the representations of words that do not appear on the training vocabulary. Joulin et al. [16] introduced FastText, a method similar to the CBOW that was on par with state-of-the-art, deep-learning models at the time. The advantage of FastText is that it was several orders of magnitude faster to train than its deep learning counterparts. FastText is built on top of linear models with a rank constraint and a fast loss approximation. The model starts off with word representations that are averaged into a text representation, i.e. a hidden layer that is shared among features and classes, which is fed to a classifier. To improve the speed of the model, two solutions were proposed. On the one hand, a hierarchical softmax based on the Huffman coding tree is used. On the other hand, to avoid the computational cost of taking into account the word order, they use bags of n-grams[1] to maintain efficiency without losing accuracy. A feature hashing [24] is used to have an efficient mapping of the n-grams. Later, the same authors introduced modifications to the FastText architecture to exploit sub-word information [3]. The main idea is to learn representations for character n-grams instead of word n-grams and to represent words as the sum of the n-gram vectors. This modification allows FastText to obtain representations for words that did not appear in the training data.

Bilingual and Multilingual Embeddings. Once monolingual word embeddings have been learned with any of the methods presented above, a source language embedding can be aligned to a target language embedding, so that they lay in a common vector space. Such alignment can be performed either in a supervised way, i.e. with a bilingual dictionary of word translation pairs, or in an unsupervised way.

[1] An n-gram is a contiguous sequence of n items from a given sample of text or speech.

Initially, Mikolov et al. [18] proposed a method to automate the process of generating and extending dictionaries and phrase tables, which can be used to translate single words between languages. By using monolingual embeddings learned by CBOW and Skip-gram models and a small bilingual dictionary, they propose to learn a transformation matrix **W** such that **W** approximates each word x_i of the source language to its correspondence z_i of the target language. The transformation matrix is learned by the following optimization problem:

$$\min_{\mathbf{W}} = \sum_{i=1}^{n} ||\mathbf{W}\mathbf{x}_i - \mathbf{z}_i||^2 \tag{1}$$

Similar to Mikolov et al. [18], Faruqui et al. [6], Gouws et al. [10], Smith et al. [21] and Jawanpuria et al. [14] resort to learning a transformation matrix **W** by means of an optimisation process. Faruqui et al. [6] employ Canonical Correlation Analysis (CCA) to maximise the correlation ρ between the projected vectors from the two languages. Gouws et al. [10] introduce BilBOWA (Bilingual Bag-of-Words without alignments), an efficient model for learning bilingual representations of words that does not require word-aligned parallel training data, i.e. the bilingual representation is learned without the need of having learnt monolingual embeddings first. Smith et al. [21] introduce the idea that the transformation between languages should be orthogonal, as such transformations are more robust to noise. In order to mitigate the effect of the hubness problem[2], they propose to use the inverted softmax instead of the softmax when estimating the confidence in the prediction. Jawanpuria et al. [14] propose a novel geometric approach to learn bilingual mappings given monolingual embeddings and a bilingual dictionary. Their model can be easily generalised to more than two languages to obtain multilingual word embeddings.

Heinzerling et al. [12] present a set of monolingual and multilingual word embeddings based on Byte Pair Encoding (BPE), a variable-length encoding that views texts as a sequence of symbols and iteratively merges the most frequent symbol pair into a new symbol. By using this approach, such embeddings can be used to obtain representations for out-of-vocabulary words, similar to FastText. To learn multilingual embeddings, they concatenate the data from the Wikipedia corpus of each language and perform the training process with all the data at the same time.

Conneau et al. [5] and Chen et al. [4] propose an adversarial training [9] approach to learn the mapping between 2 languages without the need of supervision. In the work of Conneau et al. [5], for every input sample, the discriminator and the matrix **W** are trained successively with stochastic gradient updates. The transformation **W** is orthogonal following the approach of Smith et al. [21], and they optimise the Cross-domain Similarity Local Scaling (CSLS) loss. Chen et al. [4], to learn a multilingual embedding without any type of supervision, employ a series of language discriminators, one per language, which they are in turn a binary classifier trained to identify how likely a given vector is from a language.

[2] The hubness problem is caused by words that are the closer word of too many words.

Table 1. Summary of the most important scene text VQA datasets.

VizWiz [11]	ST-VQA [2]	Text VQA [20]	EST-VQA [23]
2018	2019	2019	2020
English	English	English	Chinese, English
31,000 img	22,020 img	28,408 img	25,239 img.
70,000 ques	30,471 ques	45,336 ques	28,158 ques.
58,789 ans	30,471 ans	453,360 ans	28,158 ans

Joulin et al. [15] utilise a relaxation of the orthogonality constraint from [21] with the CSLS loss from [5], which lead to a convex formulation that can be solved using manifold optimisation tools. Their approach needs to have a bilingual dictionary.

2.2 Scene Text Visual Question Answering

The Scene Text VQA (ST-VQA) task has recently emerged as a natural evolution of the generic VQA task [1,11]. In ST-VQA, the answer to the question is found in the image in the form of text. Because of that, VQA models started incorporating an Optical Character Recognition (OCR) module, and some new datasets emerged [2,20,23]. Gomez et al. [8] introduced a model for the ST-VQA task that incorporates an attention mechanism that attends to multi-modal features conditioned to the question. Such a mechanism is based on pointer networks [22], and is trained to "point" to the region of the image that contains the text that answers the question. Singh et al. [20] presented LoRRa, another attention-based model that is able to either form the answer from a predefined vocabulary or by pointing to a text token in the image. Hu et al. [13] presented the Multimodal Multi-Copy Mesh (M4C) model, a Transformer-based multi-modal fusion mechanism coupled with a dynamic pointer network that is able to answer from text found on the image or from a predefined vocabulary. The current state-of-the-art models [25] for scene text VQA are mostly based on the M4C architecture.

Table 1 sums up the most important datasets for scene text VQA and presents the most important characteristics of each one of them. In VizWiz [11], the images were taken by blind people, questions are spoken and sometimes not answerable. ST-VQA [2] and EST-VQA [23] are datasets where the answer to the question appears always in the image as text, i.e. they are datasets specifically for the ST-VQA task. For our experiments we utilise the VQA architecture presented by Gomez et al. [8] and the ST-VQA [2] and EST-VQA [23] datasets. Notice that although EST-VQA is a multilingual dataset the only available baselines use monolingual word embeddings.

3 Methodology

In this section, our multilingual Visual Question Answering model is presented. First, we briefly justify the choice of the three multilingual word embeddings that will be used.

3.1 Word Embeddings

Multilingual FastText embeddings [3] are used because of having both a state-of-the-art performance and an easy-to-use python API. The API has full integration with the pre-trained models released by its authors and contains functionality to align a source language embedding to a target language embedding using the supervised method of Joulin et al. [15], which enables us to perform some of the bilingual experiments carried on this paper. In the alignment process, each dictionary is formed by the 6500 most common words in English and their translation. Another important fact to take into account is that FastText is able to obtain vector representations for out-of-vocabulary (OOV) words thanks to its sub-word approach. This is useful because not all words that appear in the dataset appear also in the training corpora. Without this capability, a huge amount of words would end up having the same representation. The provided FastText embeddings are either trained on Wikipedia (Wiki) or Common Crawl (CC) corpora and are available in 157 languages[3]. We have used the monolingual embeddings trained in both Wiki and CC, and we have performed the alignments ourselves.

BPEmb [12] has been used for several reasons. Firstly, to test a different approach to obtain the embeddings to compare with FastText. Secondly, BPEmb does not provide bilingual alignments, but a multilingual embedding obtained with the alignment of 274 languages to English. This has enabled us to test the difference in performance between bilingual and multilingual alignments. Thirdly, similarly to FastText it also allows the obtention of vector representations for OOV words. It provides an API to load the models and interact with them. The authors provide 275 monolingual embeddings and a multilingual embedding[4], all of them trained with data from Wikipedia.

Smith et al. [21] is also used in this work to obtain bilingual alignments. As it has been seen in Sect. 2, the method uses an orthogonal transformation matrix W that maps the source language to the target language. The authors provide transformations that align 78 languages to English by using the pre-trained monolingual FastText embeddings, as well as the code[5] to align more languages. We have chosen to use this approach as we will be able to compare directly two different techniques to align the same embeddings.

In all three cases, the languages that have been used in our experiments are English, Catalan, Spanish and Chinese. When aligning FastText embeddings, Catalan, Spanish and Chinese have been aligned to English.

---

[3] https://fasttext.cc/.
[4] https://bpemb.h-its.org/.
[5] https://github.com/babylonhealth/fastText_multilingual.

3.2 Visual Question Answering Architecture

We use the architecture presented by Gomez et al. [8]. As Fig. 1 shows, the model is composed of four different modules: a scene text feature extractor, a visual feature extractor, a question encoder, and an answer predictor. The model constructs a grid of multi-modal features by concatenating the extracted features from the first two modules. Each section in the grid contains textual and visual features at each spatial location. The latter stage attends to this multi-modal grid conditioned to the encoded question on the third module. Its output is the probability of each cell in the grid to contain the answer to the question. We refer the reader to the original paper for more details.

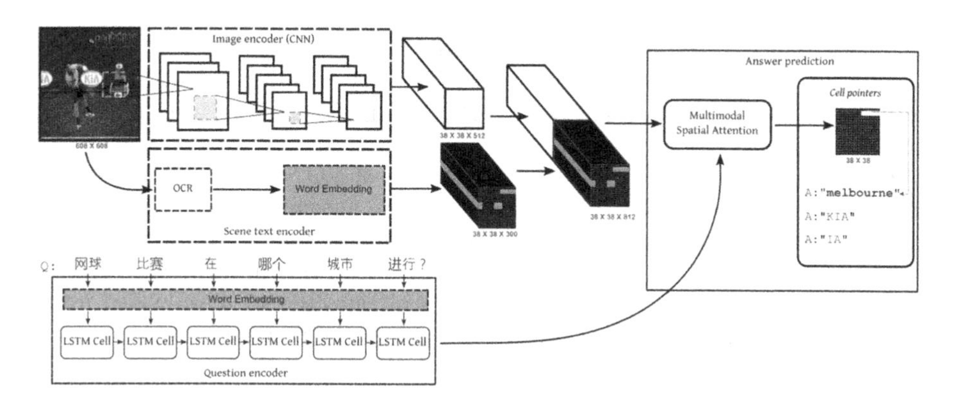

Fig. 1. Scene text VQA model [8], composed with the combination of a visual feature extractor, a scene text feature extractor, a question encoder, and an attention-based answer predictor.

The trainable modules of this architecture, i.e. the question encoder and the attention-based answer predictor, have been re-implemented in Keras/TensorFlow 2 in order to bring the codebase into a more modern and easy-to-use framework. After that, using our implementation of the model we replaced the monolingual FastText word embeddings in the scene text feature extractor and the question encoder (light blue boxes in Fig. 1) by our multilingual embeddings introduced in Sect. 3.1. Notice that the proposed approach is model agnostic and we can expect similar performance comparisons with any existing ST-VQA model.

4 Experiments

In this section, the different experiments that have been performed during the development of our work are explained. First we give some details about the experimental set-up: the used datasets, evaluation metrics, and implementation details.

4.1 Datasets

The **ST-VQA dataset** is formed by 22,020 images, from which 30,471 question/answer pairs are annotated: 26,308 for training and the rest for evaluation. The answer to any of the questions on the dataset is always a set of text tokens found on the image. It is important to note that alongside the images, questions and answers, the authors provide the OCR tokens for each of the images which were obtained using the Google OCR commercial API. To simulate a multilingual scenario we translated the English questions of this dataset into three languages (Catalan, Spanish, and Chinese) using the Google Translate API.

The **EST-VQA** dataset was released with the focus on being the first dataset that tackles the monolingual problem [23], as it contains question and answer pairs in Chinese and English. Apart from that, alongside the images, it provides the location of the answer inside the image. Different to ST-VQA, it does not provide pre-computed OCR tokens. The collection contains a total of 25,239 images and 28,158 question-answers pairs, of which 15,152 are in English and 13,006 in Chinese.

Custom datasets: To test the performance of VQA models with Catalan and Spanish data that has not been obtained from automatic translated data, 2 small datasets have been created. In that regard, the Catalan dataset contains 66 images and 66 questions, while the Spanish dataset contains 60 images and 63 questions. The images have been either taken by ourselves or downloaded from the Web. Figure 2 shows four samples from each of the used datasets.

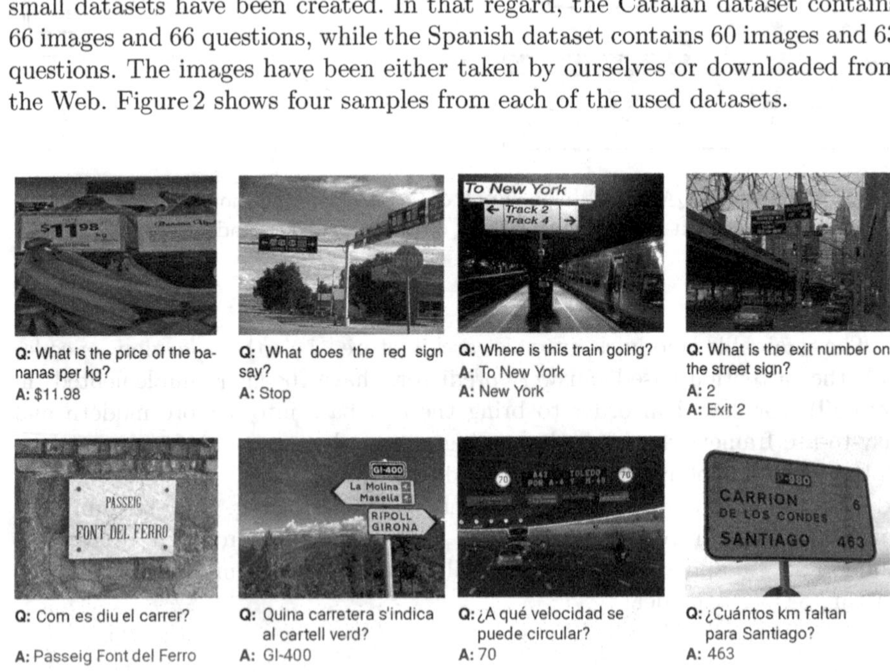

Fig. 2. Sample of data of ST-VQA (top) and custom (bottom) datasets.

4.2 Evaluation Metrics

The Average Normalised Levenshtein Similarity, ANLS, is the official metric in the ST-VQA dataset challenge [2]. The main motivation behind it is to soft-penalise possible OCR errors, so that a 0 score is not given to answers where the OCR token has a small error, e.g. "hella" instead of "hello". The Levenshtein score for a given answer is a number between 0 and 1 and indicates how similar it is to a groud-truth annotation based on their edit distance.

4.3 Implementation Details

Table 2 summarize the training parameters used in all our experiments for the VQA models. In total, 35 VQA models were trained in the ST-VQA dataset. Each model took around 36 h to train on a 12 GB Titan X Pascal GPU. Note that no data augmentation on the images was used, as the original authors of the VQA model found that it did not improve the results for them [8].

Table 2. Summary of the training parameters for the VQA models.

Model	Batch Size	Learning Rate	Decay Rate	Optimiser	Epochs
VQA	32	0.0003	0.99	Adam	100

4.4 VQA Experiments

Our firsts experiments consisted in training our implementation of the VQA model using monolingual word embeddings on the ST-VQA dataset. For that, we employed the original English annotations and their machine translation to Catalan, Spanish and Chinese. The goal behind these experiments was to ensure that our implementation was correct and to obtain the baseline results for each combination of language and embedding.

Table 3 shows the ANLS of the models trained with FastText and BPEmb monolingual embeddings on the ST-VQA dataset. As expected the maximum performance is obtained when the model is tested with the same language as it was trained. This is due to the embeddings not being aligned so that the same words in different languages produce different vector representations. It can also be seen that using machine translation tools does not affect the performance of the non-English languages. We can also see that in general our implementation outperforms the results of the original model. This can be attributed to the use of more modern, higher quality FastText and BPEmb embeddings.

We appreciate in Table 3 that when testing the models in different languages than the one used in training, the performance drops but the models are still producing decent results. This can be explained by the fact that in the majority of images in the ST-VQA dataset the number of OCR tokens is small (between 1 and 5), thus sometimes the model can point to the correct answer just by chance.

Table 3. Monolingual performance (ANLS) on the ST-VQA dataset on English (en), Catalan (ca), Spanish (es) and Chinese (zh), using FastText and BPEmb word embeddings.

		FastText				BPEmb			
		en	ca	es	zh	en	ca	es	zh
Tested on	en	**0.34**	0.16	0.15	0.16	**0.33**	0.22	0.21	0.21
	ca	0.17	**0.32**	0.18	0.16	0.19	**0.33**	0.19	0.19
	es	0.18	0.18	**0.33**	0.17	0.20	0.22	**0.34**	0.19
	zh	0.18	0.20	0.17	**0.33**	0.19	0.18	0.19	**0.31**

Once the monolingual baselines were obtained, the second goal was to test if an embedding from one language that was aligned to English was performing similarly to their respective non-aligned monolingual embedding or, to the contrary, performance was degraded. Table 4 presents the results from models trained with Catalan, Spanish and Chinese FastText embeddings aligned to English by using Smith et al. [21] and Joulin et al. [15] methods. In the case of English, we make the evaluations on the monolingual model, as the transformation to apply to obtain the aligned representation is the identity, and there is no need to train the model.

Table 4. Bilingual performance (ANLS) on the ST-VQA dataset in English (en), Catalan (ca), Spanish (es) and Chinese (zh), using FastText and aligned with Joulin et al. or Smith et al. Green, red, and blue results indicate an improvement with respect to the monolingual models (details in the main text).

		FastText							
		Aligned with Joulin et al.				Aligned with Smith et al.			
		en	ca	es	zh	en	ca	es	zh
Tested on	en	**0.34**	0.27	0.28	0.25	**0.34**	0.29	0.30	0.22
	ca	**0.27**	**0.32**	0.15	0.12	**0.26**	**0.33**	0.29	0.19
	es	**0.27**	0.14	**0.33**	0.12	**0.28**	0.28	**0.34**	0.20
	zh	**0.23**	0.14	0.15	**0.33**	0.16	0.16	0.17	**0.32**

We appreciate in Table 4 that the aligned models perform similarly to their respective monolingual counterparts on the language that has been used for training, but they consistently improve on English (marked in green in the table), between **35%** and **75%** depending on the case. They still do not match the results of the English monolingual model but this is expected since the translation between languages using these embeddings is not perfect. Secondly, we can observe (in blue in the table) that in the case of testing the models trained with English data, the performance increases for all languages. The big difference between these two groups of models is that the embeddings aligned with Smith et al. and trained on non-English languages produce performance gains in all languages (marked in red) except for Chinese, while in the embeddings aligned with Joulin et al. such behaviour is not happening. Notice that Spanish and Catalan are two languages with small lexical distance and the aligned embeddings should be able to exploit that. The smaller or null gain in performance in some Chinese tests can be explained due to the different alphabet and sentence structure, which in turn make it difficult to obtain good alignments.

Table 5 shows the ANLS obtained by training the models with the multilingual BPEmb embedding aligned with Heinzerling et al. We appreciate that they clearly outperform the models obtained by aligning the FastText embeddings, and with the additional advantage of only needing one embedding for all languages. We appreciate that the models' performance is the same as the monolingual counterparts when trained and tested in the same language, while the score on the other languages improves at a better rate than before (marked in green). In the cases of the model trained in Chinese, it only generalises well on English, and not Catalan and Spanish (marked in red).

Table 5. Multilingual performance (ANLS) on the ST-VQA dataset on English (en), Catalan (ca), Spanish (es) and Chinese (zh), using BPEmb embeddings aligned with Heinzerling et al.

		BPEmb			
		Aligned on Heinzerling et al.			
		en	ca	es	zh
Tested on	en	**0.35**	0.30	0.28	0.24
	ca	0.30	**0.35**	0.30	0.15
	es	0.28	0.32	**0.34**	0.14
	zh	0.24	0.23	0.22	**0.32**

As the final experiment on the ST-VQA dataset, Table 6 shows the results of training with data from English together with data from other languages.

As it can be seen in the results, the performance of Catalan, Spanish and Chinese is maintained in their respective monolingual models (marked in bold black) while the English performance increases, matching the score of the English monolingual models (marked in green), and the rest of languages maintain the same performance as the models trained in one language (marked in orange). The same behaviour is observed when training with all four languages (marked in blue), where all of them are able to match their baseline score. This is an important result as it demonstrates that using multilingual embeddings can potentially allow us to perform multilingual VQA with a single model. By using aligned embeddings, languages that share the same alphabet perform decently, especially with the embeddings aligned with the methods of Smith et al. and

Table 6. Bilingual and multilingual performance (ANLS) on the ST-VQA dataset in English (en) + Catalan (ca), English (en) + Spanish (es) and English (en) + Chinese (zh), using FastText embeddings aligned with Smith et al., and BPEmb embeddings aligned with Heinzerling et al. Blue results show that results on the model trained in 4 languages match the baseline performances.

		FastText			BPEmb			
		Aligned with Smith et al.			Aligned with Heinzerling et al.			
		en + ca	en + es	en + zh	en + ca	en + es	en + zh	en + ca + es + zh
Tested on	en	0.34	0.34	0.34	0.34	0.34	0.34	**0.34**
	ca	**0.33**	0.29	0.26	**0.33**	0.31	0.27	**0.34**
	es	0.32	**0.33**	0.29	0.30	**0.34**	0.28	**0.34**
	zh	0.21	0.21	**0.32**	0.15	0.14	**0.31**	**0.31**

Heinzerling et al. Finally, we also trained the model with all the languages at a time with the multilingual BPEmb embedding. As it can be seen, all languages match their baseline score (marked in blue), thus accomplishing the goal to obtain a model that achieves good performance in all languages and, in this case, by using only one single word embedding.

Table 7 summarises the best scoring models from all the previous experiments and their average ANLS across all languages. It can be seen that by using a multilingual embedding and training the model in 4 languages, all of them match their monolingual performance, with the advantage of: (1) having only one model that performs well on all languages, instead of four models; and (2) the model needs only one embedding instead of four, reducing the resources needed. This might be of great importance in low resource scenarios such as mobile applications, where storage and computational power is limited. Figure 3 shows qualitative results on the ST-VQA dataset with models trained with multilingual BPEmb embeddings.

Table 7. Summary of the best ANLS scores per language and type of embedding used in training. In the case of bilingual and multilingual embeddings, we show the ones trained in 1 language (where we exclude the English model) and the ones trained in 2 and 4 languages.

	(en)	(ca)	(es)	(zh)	Average	# models
Monolingual embeddings	0.34	0.33	0.34	0.33	0.335	4
Bilingual embeddings (excluding English models)	0.30	0.33	0.34	0.33	0.325	4
Bilingual embeddings (trained in 2 languages)	0.34	0.33	0.33	0.32	0.33	3
Multilingual embeddings (trained in 1 language)	0.30	0.35	0.34	0.32	0.3275	4
Multilingual embeddings (trained in 2 languages)	0.34	0.33	0.34	0.31	0.33	3
Multilingual embeddings (trained in 4 languages)	0.34	0.34	0.34	0.31	0.3325	1

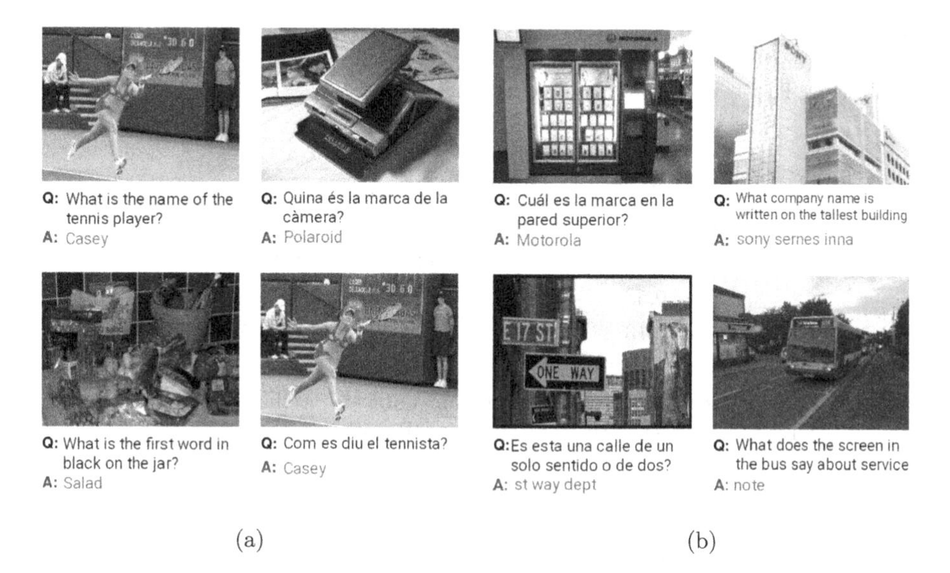

Fig. 3. Qualitative results on the ST-VQA dataset. Models were trained with multilingual BPEmb embeddings. (a) shows English and Catalan results and (b) Spanish and Chinese results. For easier reading, Chinese questions have been written in English.

In Table 8 we provide results on the EST-VQA dataset using FastText and BPEmb embeddings aligned with Heinzerling et al. and Smith et al., respectively. The Chinese and English translated data has been obtained by translating the original dataset using Google Translate. The overall better performance in this experiment is because we assume a perfect OCR output (i.e. these are Upper-Bound results). Since in this dataset the OCR performance is extremely low, we have used the ground-truth annotations as OCR outputs. It can be seen that the best performing models are with non-translated data (marked in green). It is also observed that training in English translated from Chinese or vice-versa does not affect the performance when evaluated with non-translated data, but in the reverse case, the performance drops (marked in red). We also appreciate that the models trained with BPEmb multilingual embeddings trained with both languages (blue) match their BPEmb monolingual performance on the model trained.

Finally, Table 9 shows the results on the Catalan and Spanish custom datasets with the monolingual models trained on the ST-VQA dataset. The difference in performance with respect to the models evaluated with the ST-VQA dataset may be due to differences in the types of questions between the two datasets. It is clear that using original data in one language still performs well when tested in a model trained with machine-translated data. Finally, it can be observed that the performance between different types of embeddings is big compared to the models evaluated in the ST-VQA datasets, where all embeddings were performing similarly. This indicates that some embedding alignments are more affected by the gap between automatically translated data and genuine data.

Table 8. Monolingual and multilingual performance (ANLS) on the EST-VQA dataset on English (en-or), Chinese (zh-or), English translated (en-tr) and Chinese translated (zh-tr) using FastText and BPEmb embeddings.

		FastText				BPEmb		
		zh-or	en-or	zh-tr	en-tr	zh-or	en-or	en-or + zh-or
Tested on	zh-or	0.80	0.50	0.65	0.12	**0.71**	0.31	**0.68**
	en-or	0.30	0.71	0.36	0.52	0.56	**0.62**	**0.62**
	zh-tr	0.53	0.56	0.70	0.27	0.65	0.49	**0.65**
	en-tr	0.13	0.43	0.18	0.54	0.52	0.43	**0.56**

Table 9. Performance (ANLS) on the custom datasets of the monolingual models trained on ST-VQA dataset with FastText and BPEmb embeddings.

		FastText		BPEmb	
		ca	es	ca	es
Tested on	ca	0.41	–	0.30	–
	es	–	0.43	–	0.23

5 Conclusions

In this work, we have adapted an existing VQA monolingual model [8] to a newer multilingual framework by substituting its monolingual word embeddings with aligned word embeddings. Based on the results obtained we can draw the following conclusions: (1) It is possible to train multilingual Visual Question Answering models by only using data from one language and aligned word embeddings, although the performance on the languages not used during training is lower than their respective monolingual baseline. (2) By using the BPEmb multilingual embedding and training the model with several languages at once, a single multilingual VQA model matches the monolingual baselines.

Acknowledgment. This work has been supported by: Grant PDC2021-121512-I00 funded by MCIN /AEI/10.13039/501100011033 and the European Union NextGenerationEU/PRTR;
Project PID2020-116298GB-I00 funded by MCIN/ AEI /10.13039/501100011033; Grant PLEC2021-007850 funded by MCIN/AEI/10.13039/501100011033 and the European Union NextGenerationEU/PRTR.

References

1. Antol, S., et al.: VQA: visual question answering. In: ICCV (2015)
2. Biten, A.F., et al.: Scene text visual question answering. In: ICCV (2019)
3. Bojanowski, P., Grave, E., Joulin, A., Mikolov, T.: Enriching word vectors with subword information (2017)

4. Chen, X., Cardie, C.: Unsupervised multilingual word embeddings. arXiv preprint arXiv:1808.08933 (2018)
5. Conneau, A., Lample, G., Ranzato, M., Denoyer, L., Jégou, H.: Word translation without parallel data. arXiv preprint arXiv:1710.04087 (2017)
6. Faruqui, M., Dyer, C.: Improving vector space word representations using multilingual correlation. In: Proceedings of the EC-ACL (2014)
7. Goldberg, Y., Hirst, G.: Neural Network Methods in Natural Language Processing. Morgan & Claypool Publishers (2017). 9781627052986 (zitiert auf Seite 69) (2017)
8. Gómez, L., et al.: Multimodal grid features and cell pointers for scene text visual question answering. CoRR abs/2006.00923 (2020)
9. Goodfellow, I., et al.: Generative adversarial nets. In: NIPS (2014)
10. Gouws, S., Bengio, Y., Corrado, G.: BilBOWA: fast bilingual distributed representations without word alignments. In: ICML (2015)
11. Gurari, D., et al.: VizWiz grand challenge: answering visual questions from blind people. In: CVPR (2018)
12. Heinzerling, B., Strube, M.: BPEmb: tokenization-free pre-trained subword embeddings in 275 languages. In: LREC (2018)
13. Hu, R., Singh, A., Darrell, T., Rohrbach, M.: Iterative answer prediction with pointer-augmented multimodal transformers for textVQA. In: CVPR (2020)
14. Jawanpuria, P., Balgovind, A., Kunchukuttan, A., Mishra, B.: Learning multilingual word embeddings in latent metric space: a geometric approach. Trans. ACL (2019)
15. Joulin, A., Bojanowski, P., Mikolov, T., Jégou, H., Grave, E.: Loss in translation: learning bilingual word mapping with a retrieval criterion. In: EMNLP (2018)
16. Joulin, A., Grave, E., Bojanowski, P., Mikolov, T.: Bag of tricks for efficient text classification (2016)
17. Mikolov, T., Chen, K., Corrado, G., Dean, J.: Efficient estimation of word representations in vector space. arXiv preprint arXiv:1301.3781 (2013)
18. Mikolov, T., Le, Q.V., Sutskever, I.: Exploiting similarities among languages for machine translation (2013)
19. Mikolov, T., Sutskever, I., Chen, K., Corrado, G.S., Dean, J.: Distributed representations of words and phrases and their compositionality. In: NIPS (2013)
20. Singh, A., et al.: Towards VQA models that can read. In: CVPR (2019)
21. Smith, S.L., Turban, D.H.P., Hamblin, S., Hammerla, N.Y.: Offline bilingual word vectors, orthogonal transformations and the inverted softmax. CoRR abs/1702.03859 (2017)
22. Vinyals, O., Fortunato, M., Jaitly, N.: Pointer networks (2017)
23. Wang, X., et al.: On the general value of evidence, and bilingual scene-text visual question answering. In: CVPR (2020)
24. Weinberger, K., Dasgupta, A., Langford, J., Smola, A., Attenberg, J.: Feature hashing for large scale multitask learning. In: ICML (2009)
25. Yang, Z., et al.: Tap: text-aware pre-training for text-VQA and text-caption. In: CVPR (2021)

[illegible]

Information Extraction and Applications

Sequence-to-Sequence Models for Extracting Information from Registration and Legal Documents

Ramon Pires[1,2(✉)], Fábio C. de Souza[1,3], Guilherme Rosa[1,3], Roberto A. Lotufo[1,3], and Rodrigo Nogueira[1,3]

[1] NeuralMind Inteligência Artificial, São Paulo, Brazil
{ramon.pires,fabiosouza,guilherme.rosa,roberto, rodrigo.nogueira}@neuralmind.ai
[2] Institute of Computing, University of Campinas, Campinas, Brazil
[3] School of Electrical and Computer Engineering, University of Campinas, Campinas, Brazil

Abstract. A typical information extraction pipeline consists of token- or span-level classification models coupled with a series of pre- and post-processing scripts. In a production pipeline, requirements often change, with classes being added and removed, which leads to nontrivial modifications to the source code and the possible introduction of bugs. In this work, we evaluate sequence-to-sequence models as an alternative to token-level classification methods for information extraction of legal and registration documents. We finetune models that jointly extract the information and generate the output already in a structured format. Post-processing steps are learned during training, thus eliminating the need for rule-based methods and simplifying the pipeline. Furthermore, we propose a novel method to align the output with the input text, thus facilitating system inspection and auditing. Our experiments on four real-world datasets show that the proposed method is an alternative to classical pipelines. The source code is available at https://github.com/neuralmind-ai/information-extraction-t5.

Keywords: Information extraction · Sequence-to-sequence · Legal texts

1 Introduction

Current commercial information extraction (IE) systems consist of individual modules organized in a pipeline with multiple branches and merges controlled by manually defined rules. Most of such modules are responsible for extracting or converting a single piece of information from an input document [6,12,24]. In production pipelines, the requirements and specifications often change, with new types of documents being added and more information being extracted. This

© Springer Nature Switzerland AG 2022
S. Uchida et al. (Eds.): DAS 2022, LNCS 13237, pp. 83–95, 2022.
https://doi.org/10.1007/978-3-031-06555-2_6

leads to higher maintenance costs due to an ever larger number of individual components.

Although recent pretrained language models significantly improved the effectiveness of IE systems, modern pipelines are made of various of such models, each responsible for extracting or processing a small portion of the document. Due to the large computational demands of these models, it can be challenging to deploy a low-latency low-cost IE pipeline.

In this work, we study the viability of a framework for information extraction based on a single sequence-to-sequence (seq2seq) model finetuned on a question answering (QA) task, for extracting and processing information from legal and registration documents. The main advantage of this framework is that a single model needs to be trained and maintained. Thus, it can be shared by multiple projects with different requirements and types of documents. Figure 1 depicts an illustration comparing a classical IE pipeline with the proposed one.

We validate this single-model framework on four datasets that represent demands from real customers. We summarize our contributions as follows:

- We show that a single seq2seq model is a competitive alternative to a classical pipeline made of multiple extraction and normalization modules (see Sect. 5.3 for an ablation study). Our model is trained end-to-end to generate the structured output in a single decoding pass.
- We evaluate different pretraining and intermediate finetuning strategies and show that target language tokenization and pretraining are by far the most important contributors to the final performance. Further pretraining on the legal domain and finetuning on a similar task have minor contributions.
- One of the limitations of generative models is that the location of extracted tokens in the input text can be ambiguous. We propose a novel method to address this problem (Sect. 3.3).

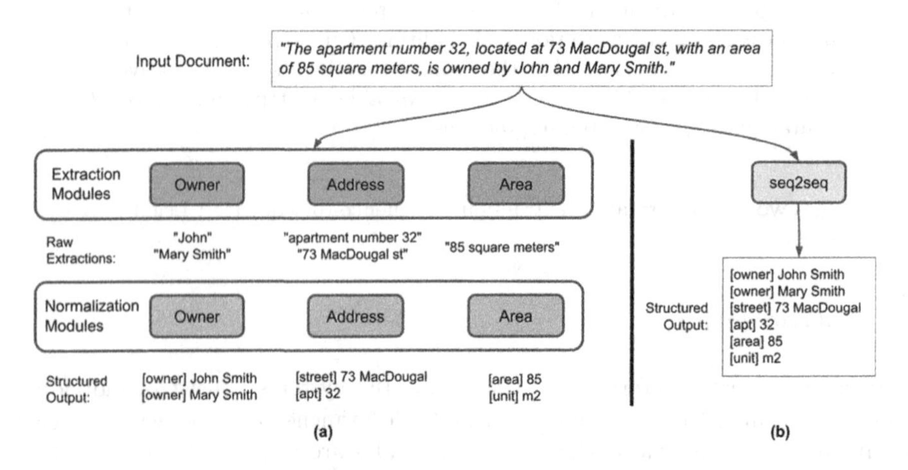

Fig. 1. a) A classical IE pipeline consisting of multiple extraction and normalization modules. **b)** Our proposed IE framework that converts the input document into structured data using a single seq2seq model.

2 Related Work

A go-to choice of methods for information extraction are token-level classification models, which predict a label for each element in the input sequence [5,8,10,11, 13,16,18,21]. Seq2seq models are typically used to solve complex language tasks like machine translation, text summarization, and question answering [14,20, 25,30]. This special class of language models has shown excellent performance on tasks that are commonly tackled by token-level or span-level classification methods. For instance, T5 [22] is a seq2seq model that achieves the state of the art on various extractive question-answering and natural language inference tasks. Others have confirmed that this single model can replace multiple task-specific NLP models [15].

We are not the first to apply seq2seq models to IE tasks. Chen et al. [9] proposed a seq2seq model that can recognize previously unseen named entities. Athiwaratkun et al. [4] finetuned a T5 model to jointly learn multiple tasks such as named entity recognition (NER), slot labeling and intent classification. Hybrids of token-level classification and seq2seq models were also used to extract information from legal documents [1]. Others use a seq2seq model for nested and discontinuous NER tasks [27,29]. Most of the datasets used in these works contain shorter input texts than the ones used here.

Although IE systems have a long history in the scientific literature, there are few studies that analyze their use in commercial NLP pipelines. There are legal NER datasets similar to the ones used in this work [2,3,17,19], but they rarely reflect the complexities of production pipelines, such as processing scanned documents with OCR errors and extracting nested and discontiguous entities.

3 Methodology

This section first describes input and output formats used by our models. Then, we present compound QAs, and introduce sentence IDs and answers in canonical (normalized) format. Figure 2 depicts one example of a Registry of Property, along with some human-annotated questions and answers.

3.1 Questions and Answers

Our method uses questions coupled with contexts as input and answers as output. In this section, we describe how we prepare QAs and contexts for extracting information from fields of different legal and registration documents.

We seek to format the answers by adding clues for each category of information. Clues are indispensable for designing compound answers (see Sect. 3.2). To maintain consistency and standardization, those clues are reused for related fields, regardless of the document type. We format the answers by preceding the core answer with the clue in square brackets followed by a colon.

Due to the quadratic cost with respect to the sequence length of transformer models, we have to divide long documents into windows (e.g., of 10 sentences)

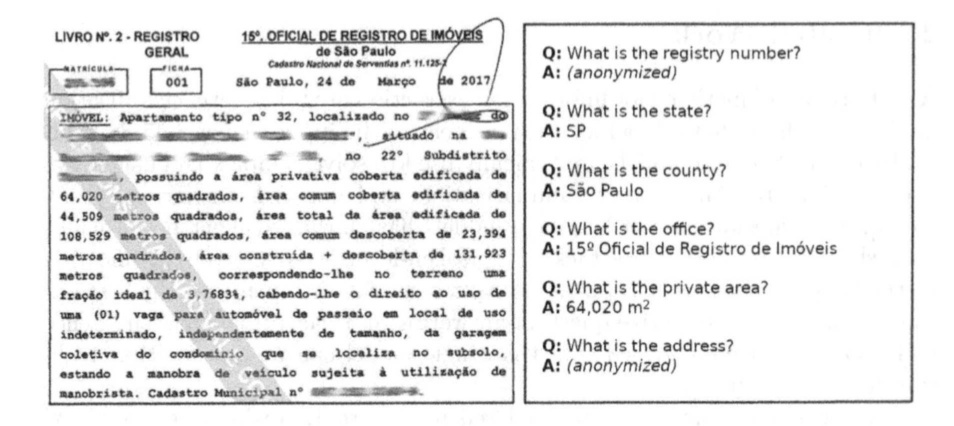

Fig. 2. Example of Registry of Property (left) along with its questions and answers annotated by a human (right). Some answers are omitted to preserve anonymity.

with a certain overlap and repeat the questions in each window. In some cases, the model should not provide an answer because the necessary information is not present in the current window. Thus, we finetune it to output "N/A" (an abbreviation for not available) for those question and context pairs. At inference time, the extracted information is the one that achieves the highest probability among all windows, except for N/A responses. The extracted information is empty only if the model has generated "N/A" in all windows.

3.2 Compound QAs

Some subsets of fields are often closely related or even appear connected. The classical pipeline is cumbersome as it requires the model to analyze each document oftentimes for extracting information of the same scope. An example is the *address* field, whose extraction requires recovering, in general, seven individual subfields: street, number, complement, district, city, state, and zip code.

We propose a novel alternative for extracting all that information at once by using compound QAs. We represent compound answers by preceding each field response with the respective field indicator (clue) in square brackets, and concatenating them all in a fixed, pre-set order (e.g., value and unit, in sequence). We also replace the set of questions by a single, more generic question. Table 1 presents examples of compound QAs (column *Comp* marked).

Compound QAs can be valuable to extract discontinuous information (like nested NER) [27,29]. The user may request, for example, the complete list of heirs along with their respective registration numbers and the fraction. The use of compound QAs would be beneficial as it requires a single shot IE for each heir.

Table 1. Examples of different formats of contexts, questions and answers. When the column *Sent* is marked (✓), question and answer pairs use the "sent IDs" context. When not marked, the "original" context is used.

	Contexts			
Original	Apartment type nº 32, located on the 10th floor of the Central Building, situated at 1208 Santos Dumont St., having a private covered built area of 64,020 square meters, a common covered built area of 44,509 square meters...			
Sent IDs	[SENT1] Apartment type nº 32, [SENT2] located on the 10th floor of the Central Building, [SENT3] situated at 1208 Santos Dumont St., [SENT4] having a private covered built area of 64,020 square meters, [SENT5] a common covered built area of 44,509 square meters...			
	Questions and Answers	**Comp**	**Sent**	**Raw**
(1)	(Q) What is the value of the private area? (A) [value]: 64.02 (Q) What is the unit of the private area? (A) [unit]: m^2	–	–	–
(2)	(Q) What is the private area? (A) [value]: 64.02 [unit]: m^2	✓	–	–
(3)	(Q) What is the value of the private area? (A) [SENT4] [value]: 64.02 (Q) What is the unit of the private area? (A) [SENT4] [unit]: m^2	–	✓	–
(4)	(Q) What is the private area? (A) [SENT4] [value]: 64.02 [SENT4] [unit]: m^2	✓	✓	–
(5)	(Q) What is the value of the private area and how does it appear in the text? (A) [value]: 64.02 [text] 64,020 (Q) What is the unit of the private area and how does it appear in the text? (A) [unit]: m^2 [text] square meters	–	–	✓
(6)	(Q) What is the private area and how does it appear in text? (A) [value]: 64.02 [text] 64,020 [unit]: m^2 [text] square meters	✓	–	✓
(7)	(Q) What is the value of the private area and how does it appear in the text? (A) [SENT4] [value]: 64.02 [text] 64,020 (Q) What is the unit of the private area and how does it appear in the text? (A) [SENT4] [unit]: m^2 [text] square meters	–	✓	✓
(8)	(Q) What is the private area and how does it appear in text? (A) [SENT4] [value]: 64.02 [text] 64,020 [SENT4] [unit]: m^2 [text] square meters	✓	✓	✓

3.3 Sentence IDs and Canonical Format

Commercial IE systems are part of important decision making pipelines, and as such, they need to be constantly monitored and audited. One way to monitor the quality of predictions is to know the location in the input text from which the information was extracted. However, the location cannot be trivially inferred from the output of seq2seq models if the information appears more than once in the text. To address this limitation, we propose the use of *sentinel tokens* that represent sentence IDs and allow the generative model to reveal the location of its prediction in the original sequence. Table 1 shows examples of QAs with sentence IDs (column *Sent* marked).

The questions are not changed for requesting sentence IDs. The document text, however, must be split into sentences following some criterion (e.g., paragraph, form field, number of tokens, lines). Then, each sentence is preceded by its own sentinel token using the pattern [SENTx], in which x represents the sentence ID, starting from 1 and increasing for each new sentence. The answers are preceded by the sentinel token of the sentence it came from. If the answer is compound, each component has its sentence ID preceding the clue and response.

Often, certain types of information appear in a document in a variety of formats. For example, a date may be expressed in a day-month-year format, such as "20 November 2021", or abbreviated with some sort of separator like "20/11/2021" or "20-11-2021". For a commercial system, it is required to extract those types of information in a standardized, canonical format to harmonize varied formats and ensure accuracy in all situations. Our IE system is able to directly extract those particular fields in the canonical format.

However, the canonical format might not match the span in the original text, and thus the use of sentence IDs may not be enough for locating the extracted information in the original document. To maintain this functionality, we complement the use of sentence IDs by extracting the raw format that the information appears in the document. For those varied-format fields, we include a complement to the question— *"and how does it appear in the text?"*—and change the answers to encompass the canonical-format response (with its respective clue) together with a reserved token *"[text]"* and the raw format of the response as a suffix. Table 1 depicts examples of QAs that use the raw format along with the canonical format (column *Raw* marked).

4 Experimental Setup

We evaluated the effectiveness of the models on four different datasets in Portuguese. Each dataset represents a specific type of document. We finetune each model (Sect. 4.1) on the four datasets (Sect. 4.2), instead of training individually one model for each type of document. We measure the effectiveness using exact matching (EM), which corresponds to the accuracy; and token-based F1.

4.1 Models

In this section we describe the pretrained and finetuned models used in this work. Some of those models were taken from the shelf, while for others we adapted to our particular task and domain.

T5: Encoder-decoder model that is a Transformer [28] with a combination of relevant insights from a systematic study [22]. T5 achieved state-of-the-art results in many NLP benchmarks. We work with texts in Portuguese, so this English model acts as a weak baseline.

PTT5: It consists of a T5 model, without changes in the architecture, pretrained on a large Brazilian Portuguese corpus [7]. We employ PTT5 Base which showed to be not only more efficient but also more effective than the PTT5 Large [7].

PTT5-Legal: To adapt PTT5 to in-domain text, we further pretrain it on a large corpus that comprises two out of the four document types we explore in this work: registries of property and legal publications. This corpus contains approximately 50 million tokens, which is a fraction of 2.7 billion tokens used to pretrain PTT5.

PTT5-QA: We also intend to improve the ability of the model to extract information from diverse documents by question-answering, regardless of the domain of the data. For that, we finetune PTT5 on the SQuAD v1.1 dataset [23] before finetuning it on our datasets.

PTT5-Legal-QA: Finally, we investigate if the unsupervised in-domain pretraining followed by supervised task-aware finetuning provides effectiveness gains on the task of extracting relevant information from documents.

4.2 Datasets

We finetune and evaluate the models on four datasets that contain legal documents in Portuguese. Table 2 presents statistics of the datasets.[1]

NM-Property: is composed by the opening section of property documents that discriminate a property or land, containing essential information for legal identification, such as location or qualification of the owners (if an individual or legal entity). The documents are scanned and susceptible to OCR errors due to poor image quality. The dataset has 3191 documents for training, 799 for validation and 242 for testing.

NM-Certificates: represents a set of documents issued by the competent agency of the tax office whose function is to prove that a person, organization or property has or does not have debts, that is, that there are no collection of actions in relation to this natural or legal person, nor in relation to real estate. The dataset has 760 documents for training, 191 for validation and 311 for testing.

NM-Publications: is comprised of documents issued by the Brazilian government gazette (*Diario Oficial*, similar to U.S. Federal Register). The documents contain legal notices that have been authorised or that are required by law to be published. The dataset has 1600 documents for training, 401 for validation and 500 for testing.

NM-Forms: consists of forms for opening a bank account filled with personal information. In general, most of the fields are filled. The scanned documents present common OCR errors, such as confusing vertical lines in text boxes as characters. The dataset has 240 documents for training, 60 for validation and 282 for testing.

4.3 Training and Inference

For input preprocessing, we prepare sliding windows with an overlap of 50%. Thus, each window—that can be interspersed by sentence ID tokens or not—concatenated with the question occupies a maximum of 512 tokens. Regarding

[1] As the legal and registration documents are of private nature, the datasets cannot be made available.

Table 2. Statistics of the datasets.

Dataset	Chars/doc	Fields	Train	Valid	Test
NM-Property	3011.53	17	3191	799	242
NM-Certificates	4914.39	10	760	191	311
NM-Publications	1895.76	3	1600	401	500
NM-Forms	1917.14	25	240	60	282
Total	–	55	5791	1451	1334

the use of sentence IDs, we use line breaks as a criterion to divide the document text into sentences. We finetune the models using AdamW optimizer with a constant learning rate of 1e−4 and weight decay of 5e−5. Training takes approximately 22,500 optimization steps with batches of 64 samples. The model we select for testing is the one with the highest EM over the validation set. At test time, the information is extracted using beam search with 5 beams. The maximum output length is set to 256 tokens, but none of our test examples reached this limit.

5 Results

The first set of experiments involves finetuning the pretrained models presented in Sect. 4.1 for information extraction using individual questions and answers as well as the original document text as context. Rows 1 to 6 in Table 3 present the results.

Table 3. Main results. "Pre-ft" means finetuning on the English SQuAD 1.1 dataset before finetuning on the target datasets.

	Pre-train	Pre-ft	Comp	Sent	Raw	Dataset					
							Prop	Cert	Publ	Form	Avg
(1)	Eng	–	–	–	–	EM	57.5	39.4	64.3	69.1	57.6
						F1	79.0	73.5	86.7	89.4	82.2
(2)	Port (Eng vocab)	–	–	–	–	EM	68.2	40.2	64.4	70.9	60.9
						F1	84.3	74.1	86.6	90.3	83.8
(3)	Port	–	–	–	–	EM	84.1	79.9	95.9	93.8	88.4
						F1	92.2	91.4	98.0	97.2	94.7
(4)	Port	✓	–	–	–	EM	83.6	79.2	95.9	94.2	88.2
						F1	91.8	91.3	98.0	97.4	94.6
(5)	Port+Legal	–	–	–	–	EM	84.5	80.5	96.1	94.1	88.8
						F1	92.3	91.8	98.1	97.3	94.9
(6)	Port+Legal	✓	–	–	–	EM	84.1	82.1	96.8	94.4	89.4
						F1	92.2	92.9	98.4	97.5	95.3
(7)	Port+Legal	✓	✓	–	–	EM	81.7	84.4	97.2	94.8	89.5
						F1	90.8	93.4	98.6	97.6	95.1
(8)	Port+Legal	✓	✓	✓	✓	EM	81.4	85.3	96.9	94.1	89.4
						F1	90.4	93.2	98.4	97.2	94.8

We experimented with distinct T5 Base models: the original T5 Base with English tokenizer (row 1), with further pretraining on Portuguese corpus [7] but using the English tokenizer (row 2), and PTT5 Base (row 3). We notice that the adoption of a Portuguese tokenizer by PTT5 provided an error reduction in EM of 70.3% over the previous experiment that used the same pretraining dataset (rows 2 vs 3).

Afterward, we used the PTT5 model finetuned on the question-answering task, and further finetuned it on the four datasets (PTT5-QA) (row 4). We observe that adapting the model for QA on the SQuAD dataset did not provide improvements over the large-scale pretraining.

We then applied the PTT5 model pretrained over a corpus of legal documents (PTT5-Legal), and finetuned it on the four datasets. The unsupervised pretraining brought the best result over the NM-Properties dataset, and a minor improvement on the average of the four datasets (row 5). Finally, we explored the PTT5 model that incorporated the finetuning on question-answering after in-domain pretraining (PTT5-Legal-QA). This model achieved the best average EM and F1 (row 6).

5.1 Experiments for Compound QAs

In this section, we investigate whether the use of compound QAs, described in Sect. 3.2, brings benefits in effectiveness or at least reaches comparable performance to the use of individual QAs. We applied compound QAs for address, land area, private area and built area from NM-Property; and for address from NM-Forms. To compare results, we post-process the compound answers by splitting them into individual sub-responses, each one including the clue in square brackets and the extracted information.

As shown in row 7 of Table 3, compound QAs yields better results than using individual QAs, reaching superior performance on three datasets. We emphasize, however, that the average results are comparable. This confirms that the use of compound QAs brings efficiency and does not affect the model effectiveness.

However, the EM results for NM-Property decreased from 84.1% to 81.7%. Most errors occur for the address field, in scenarios in which the response is N/A (often not annotated), but the model outputs an answer. One of the reasons is that the NM-Property concentrates a large number of registries in rural or allotted areas, in which some information is missing. We noticed that the four most negatively impacted fields are from NM-Property's address: complement, street, district and number. On the other hand, five out of seven address subfields of NM-Form had improvements.

5.2 Experiments for Sentence IDs and Canonical Format

Herein, we investigate whether exploring sentence IDs and using the raw format along with the canonical format text described in Sect. 3.3, provides comparable results to the previous set of experiments.

We employed canonical format extraction for areas, dates, ID numbers, states and certificate results. Specifically, we use it for 8 NM-Property fields, 4 NM-Certificate fields and 7 NM-Forms fields. To compare results, we post-process the compound answers by splitting them into individual sub-responses. Each sub-response involves the clue in square brackets and the information, preceded by sentence ID in brackets, and succeeded by the text in raw format. Results in row 8 of Table 3 show that this method does not outperform the one without sentence IDs and raw-text extractions. Nonetheless, the model robustness for extracting information on varied legal and registration documents was not affected.

5.3 Comparison with BERT on a NER Task

In this experiment, we compare the extraction capabilities of our proposed framework to a named entity recognition (NER) system, a common part of classical IE pipelines. We use the NER implementation of Souza et al. [26], which uses BERT and casts NER as a token-level classification task. Since NER models cannot predict nested or overlapping entities and requires all outputs to be tied to input tokens, we filter out all classification fields and commonly overlapping field classes from our dataset. We also remove entire documents if they contain overlapping entities of the remaining fields. The reduced dataset retains 38% of the documents and 42 of the 55 fields. We then finetune both the NER and the PTT5-Legal-QA models on the reduced training set and evaluate on the reduced test set.

Following CoNLL03 evaluation procedures, the evaluation is performed at entity-level using exact match. For a given entity, the prediction is marked as correct only if the system predicts the correct entity label and location (start and end offsets) in the input text. Thus, this experiment validates exclusively the performance of our output alignment method detailed in Sect. 5.2.

Table 4. NER ablation results.

Model	Params	Precision	Recall	F1-score (micro)
BERT-Large	330M	90.2	92.9	91.5
T5-base (ours)	220M	91.6	89.6	90.6

Table 4 shows the results. Our seq2seq framework shows a slightly lower but competitive performance on the task, even though this ablation setup gives an edge to the NER system by removing impossible cases for NER, in which it would be penalized with false negatives. This task also does not evaluate answer post-processing, which is one of seq2seq's strengths and would have to be done by an extra component.

6 Conclusion

We validated the use of a single seq2seq model for extracting information from four different types of legal and registration documents in Portuguese. The model is trained end-to-end to output structured text, thus replacing parts of rule-based normalization and post-processing steps of a classical pipeline. The compound QAs also replaces a relationship extraction step that is required when fields are extracted independently in classical pipelines, such as by a named entity recognition model. We found that target language (Portuguese) pretraining and tokenization are the most important adaptations to increase effectiveness, while pretraining on in-domain (legal) texts and finetuning on a large question-answering dataset marginally improve results. Finally, we propose a method to align answers with the input text, thus allowing seq2seq models to be more easily monitored and audited in IE pipelines.

References

1. de Almeida, M., Samarawickrama, C., de Silva, N., Ratnayaka, G., Perera, A.S.: Legal party extraction from legal opinion text with sequence to sequence learning. In: 2020 20th International Conference on Advances in ICT for Emerging Regions (ICTer), pp. 143–148. IEEE (2020)
2. Angelidis, I., Chalkidis, I., Koubarakis, M.: Named entity recognition, linking and generation for Greek legislation. Front. Artif. Intell. Appl. 1 (2018)
3. Luz de Araujo, P.H., de Campos, T.E., de Oliveira, R.R.R., Stauffer, M., Couto, S., Bermejo, P.: LeNER-Br: a dataset for named entity recognition in Brazilian legal text. In: Villavicencio, A., et al. (eds.) PROPOR 2018. LNCS (LNAI), vol. 11122, pp. 313–323. Springer, Cham (2018). https://doi.org/10.1007/978-3-319-99722-3_32
4. Athiwaratkun, B., dos Santos, C., Krone, J., Xiang, B.: Augmented natural language for generative sequence labeling. In: Proceedings of the 2020 Conference on Empirical Methods in Natural Language Processing (EMNLP), pp. 375–385 (2020)
5. Baevski, A., Edunov, S., Liu, Y., Zettlemoyer, L., Auli, M.: Cloze-driven pretraining of self-attention networks. In: Proceedings of the 2019 Conference on Empirical Methods in Natural Language Processing and the 9th International Joint Conference on Natural Language Processing (EMNLP-IJCNLP), pp. 5363–5372 (2019)
6. Bommarito, M.J., II., Katz, D.M., Detterman, E.M.: LexNLP: natural language processing and information extraction for legal and regulatory texts. In: Research Handbook on Big Data Law. Edward Elgar Publishing (2021)
7. Carmo, D., Piau, M., Campiotti, I., Nogueira, R., Lotufo, R.: PTT5: pretraining and validating the T5 model on Brazilian Portuguese data. arXiv preprint arXiv:2008.09144 (2020)
8. Carreras, X., Marquez, L., Padro, L.: Learning a perceptron-based named entity chunker via online recognition feedback. In: Proceedings of the Seventh Conference on Natural Language Learning at HLT-NAACL 2003, pp. 156–159 (2003)
9. Chen, L., Moschitti, A.: Learning to progressively recognize new named entities with sequence to sequence models. In: Proceedings of the 27th International Conference on Computational Linguistics, pp. 2181–2191 (2018)

10. Chen, Q., Zhuo, Z., Wang, W.: BERT for joint intent classification and slot filling. arXiv preprint arXiv:1902.10909 (2019)
11. Chieu, H.L., Ng, H.T.: Named entity recognition with a maximum entropy approach. In: Proceedings of the Seventh Conference on Natural Language Learning at HLT-NAACL 2003, pp. 160–163 (2003)
12. Chieze, E., Farzindar, A., Lapalme, G.: An automatic system for summarization and information extraction of legal information. In: Francesconi, E., Montemagni, S., Peters, W., Tiscornia, D. (eds.) Semantic Processing of Legal Texts. LNCS (LNAI), vol. 6036, pp. 216–234. Springer, Heidelberg (2010). https://doi.org/10.1007/978-3-642-12837-0_12
13. Chiu, J.P., Nichols, E.: Named entity recognition with bidirectional LSTM-CNNs. Trans. Assoc. Computat. Linguist. **4**, 357–370 (2016)
14. Dong, L., et al.: Unified language model pre-training for natural language understanding and generation. In: Advances in Neural Information Processing Systems, vol. 32, pp. 13063–13075 (2019)
15. Du, Z., et al.: All NLP tasks are generation tasks: a general pretraining framework. CoRR abs/2103.10360 (2021)
16. Florian, R., Ittycheriah, A., Jing, H., Zhang, T.: Named entity recognition through classifier combination. In: Proceedings of the Seventh Conference on Natural Language Learning at HLT-NAACL 2003, pp. 168–171 (2003)
17. Huang, W., Hu, D., Deng, Z., Nie, J.: Named entity recognition for Chinese judgment documents based on BiLSTM and CRF. EURASIP J. Image Video Process. **2020**(1), 1–14 (2020)
18. Lample, G., Ballesteros, M., Subramanian, S., Kawakami, K., Dyer, C.: Neural architectures for named entity recognition. In: Proceedings of the 2016 Conference of the North American Chapter of the Association for Computational Linguistics: Human Language Technologies, pp. 260–270 (2016)
19. Leitner, E., Rehm, G., Schneider, J.M.: A dataset of German legal documents for named entity recognition. In: Proceedings of the 12th Language Resources and Evaluation Conference, pp. 4478–4485 (2020)
20. Lewis, M., et al.: BART: denoising sequence-to-sequence pre-training for natural language generation, translation, and comprehension. In: Proceedings of the 58th Annual Meeting of the Association for Computational Linguistics, pp. 7871–7880 (2020)
21. Li, X., Sun, X., Meng, Y., Liang, J., Wu, F., Li, J.: Dice loss for data-imbalanced NLP tasks. In: Proceedings of the 58th Annual Meeting of the Association for Computational Linguistics, pp. 465–476 (2020)
22. Raffel, C., et al.: Exploring the limits of transfer learning with a unified text-to-text transformer. J. Mach. Learn. Res. **21**(140), 1–67 (2020). http://jmlr.org/papers/v21/20-074.html
23. Rajpurkar, P., Zhang, J., Lopyrev, K., Liang, P.: SQuAD: 100,000+ questions for machine comprehension of text. In: Proceedings of the 2016 Conference on Empirical Methods in Natural Language Processing, Austin, Texas, pp. 2383–2392. Association for Computational Linguistics, November 2016. https://doi.org/10.18653/v1/D16-1264. https://aclanthology.org/D16-1264
24. Sarawagi, S.: Information Extraction. Now Publishers Inc. (2008)
25. Song, K., Tan, X., Qin, T., Lu, J., Liu, T.Y.: Mass: masked sequence to sequence pre-training for language generation. In: International Conference on Machine Learning, pp. 5926–5936. PMLR (2019)

26. Souza, F., Nogueira, R., Lotufo, R.: BERTimbau: pretrained BERT models for Brazilian Portuguese. In: Cerri, R., Prati, R.C. (eds.) BRACIS 2020. LNCS (LNAI), vol. 12319, pp. 403–417. Springer, Cham (2020). https://doi.org/10.1007/978-3-030-61377-8_28
27. Straková, J., Straka, M., Hajic, J.: Neural architectures for nested NER through linearization. In: Proceedings of the 57th Annual Meeting of the Association for Computational Linguistics, Florence, Italy, pp. 5326–5331. Association for Computational Linguistics, July 2019. https://doi.org/10.18653/v1/P19-1527. https://www.aclweb.org/anthology/P19-1527
28. Vaswani, A., et al.: Attention is all you need. In: Advances in Neural Information Processing Systems, pp. 5998–6008 (2017)
29. Yan, H., Gui, T., Dai, J., Guo, Q., Zhang, Z., Qiu, X.: A unified generative framework for various NER subtasks. In: Proceedings of the 59th Annual Meeting of the Association for Computational Linguistics and the 11th International Joint Conference on Natural Language Processing (Volume 1: Long Papers), pp. 5808–5822. Association for Computational Linguistics, August 2021. https://doi.org/10.18653/v1/2021.acl-long.451. https://aclanthology.org/2021.acl-long.451
30. Zhang, J., Zhao, Y., Saleh, M., Liu, P.: Pegasus: pre-training with extracted gap-sentences for abstractive summarization. In: International Conference on Machine Learning, pp. 11328–11339. PMLR (2020)

Contrastive Graph Learning with Graph Convolutional Networks

G. Nagendar(✉) and Ramachandrula Sitaram

247.ai, Bangalore, India
{Gattigorla.Nagendar,NagaVenkata.R}@247.ai

Abstract. We introduce a new approach for graph representation learning, which will improve the performance of graph-based methods for the problem of key information extraction (KIE) from document images. The existing methods either use a fixed graph representation or learn a graph representation for the given problem. However, the methods which learn graph representation do not consider the nodes label information. It may result in sub-optimal graph representation and also can lead to slow convergence. In this paper, we propose a novel contrastive learning framework for learning the graph representation by leveraging the information of node labels. We present a contrastive graph learning convolutional network (CGLCN), where the contrastive graph learning framework is used along with the graph convolutional network (GCN) in a unified network architecture. In addition to this, we also create a labeled data set of receipt images (Receipt dataset), where we do the annotations at the word level rather than at the sentence/group of words level. The Receipt dataset is well suited for evaluating the KIE models. The results on this dataset show the superiority of the proposed contrastive graph learning framework over other baseline methods.

Keywords: Contrastive learning · Graph neural networks (GNN) · Key information extraction · Graph convolutional networks (GCN)

1 Introduction

Optical character recognition (OCR) is the process of identifying text characters from scanned documents, which includes text detection and recognition. The scanned documents can be historical documents, receipts, bills, and invoices. Unlike only identifying the text characters from the document images, the key information extraction (KIE) from document images recently gained a lot of attention [1,2]. It involves associating a label to each of the recognized word. Key information (entities) extraction from document images can play an important role in many applications like converting document information into a structured format, efficient archiving, fast indexing, document analytics, and so on. For example, in invoices, information like invoice number, total amount, and date gives richer and meaningful information about the invoice. Recently, there are

© Springer Nature Switzerland AG 2022
S. Uchida et al. (Eds.): DAS 2022, LNCS 13237, pp. 96–110, 2022.
https://doi.org/10.1007/978-3-031-06555-2_7

many deep learning-based KIE approaches [8,9,11,12,40] that have emerged and outperformed traditional rule-based [3] and template-based [4,5] methods. These traditional methods [3–5] are not robust against real-world settings, like the document images that are captured using mobile devices.

The deep learning-based KIE methods can be roughly categorized into three types: sequence-based [8,9], graph-based [11,12], and grid-based [10]. Among these methods, the graph-based methods, in particular, graph convolutional networks (GCN) [6,7] are popularly used for the KIE problem. These methods [11,12] model each document image as a graph, where the text segments (words or group of words) are represented as nodes. GCNs capture the structural information within the node's neighborhood by following a neighborhood aggregation scheme. However, the performance of graph-based methods [11,12] majorly rely on the graph structure representation of given data. The node's neighborhood aggregation is computed using the graph structure. Recently, few methods have been proposed to learn a graph structure for the given data [13–15]. Henaff *et al.* [13] proposed a graph learning framework using a fully connected network. In [14], Li *et al.* proposed another framework for graph learning using a distance metric. Jiang *et al.* [15] proposed a graph learning convolutional network (GLCN) framework, which integrates both graph learning and graph convolution in a unified network architecture. Along with graph convolution, it also learns graph structure. Among these methods, GLCN [15] significantly outperforms other graph learning frameworks [13,14].

The current graph representation learning methods [13–15] use the distance between the node features for checking whether two nodes are nearer or farther in the graph learning. These methods do not use the label information of the nodes in learning the graph representation. However, since the node features obtained in the initial steps are not specific to the given problem, these methods [13–15] result in sub-optimal graph representation and also lead to slow convergence. This is mainly because of the fact that, in the initial steps, the nodes that belong to different entities/classes may be nearer in the initial feature space. It can be avoided by leveraging the label information of the nodes.

In this paper, we propose a novel contrastive learning [18–20] framework for learning the graph representation, which assist graph-based methods [11,12] to improve performance. Compared to other graph learning techniques [13–15], where the label information is not used in graph construction, the proposed contrastive learning framework allows to leverage this label information effectively. The proposed graph learning framework improves the performance of GCN based key information extraction methods [11,12,15]. We use contrastive loss [18,19] to learn a class discriminative node representations by projecting them into an embedding space. Since we are considering the label information, the proposed framework results in an embedding space where elements of the same class are more closely aligned compared to other graph learning methods [13–15]. The graph representation is learned in this embedding space. The main idea for the proposed contrastive learning framework is to learn a projection network by pulling the set of nodes belonging to the same class (entity) together in an embedding space while simultaneously pushing apart clusters of nodes from different classes by leveraging the label information. In the embedding space, the samples in the same class will get a higher edge weight, while the samples belonging to different classes will get a smaller edge weight. In this way, an optimal graph representation is learned using contrastive loss in the embedding space. We present both supervised and semi-supervised contrastive loss functions for learning the graph representation. In addition to this, we also present a contrastive graph learning convolutional network (CGLCN), where both contrastive graph learning and graph convolution are integrated into a unified network architecture. It can be trained in a single optimization manner. Also, different from other KIE models [11,12], we use multi-modal feature fusion using block super diagonal tensor decomposition [31] for combining the textual, visual, and positional features. We also created a Receipt dataset of grocery bills, with annotations at the word level. The images are captured in different view points and contain some amount of perspective distortion.

The main contributions of this paper can be summarized as follows

- We propose a novel graph representation learning framework using contrastive learning. Both graph learning and graph convolution are integrated into a unified network architecture.
- We also present a supervised and semi-supervised contrastive loss functions for the graph learning.
- The proposed CGLCN is evaluated on a real-world dataset, which shows the superiority of our model over other baseline models.

2 Related Work

In recent years, with the advantage of deep learning methods, key information extraction (KIE) from document images have gained encouraging improvement. Before the deep learning methods, the early works mainly used rule-based [3] or template matching methods [4,5]. In general, these methods tend to fail on unseen templates and might lead to poor performance on real application scenarios. In real applications, the captured document images contain non-frontal views and some amount of perspective distortion. In KIE, significant progress has been made with the development of deep learning techniques. Most of these approaches formulate it as a token classification problem. The sequence-based approaches serialize the document into a 1D text sequence, then use sequence tagging methods [30,42]. The sequence-based methods like LayoutLM [8] and LAMBERT [9] model the layout structure of documents based on the pre-training process of a BERT-like model. These methods pre-train on a large-scale dataset, making them less sensitive to the serialization step. The graph-based methods [11,12,41], represent the document image as a graph with the text segments as nodes in the graph. Several representation techniques [13–15] are used for the graph representation. The KIE problem is characterised as a node classification problem [11,12]. In SPADE [41], a dependency graph is constructed for the given document. In VRD [11] and PICK [12], graph convolutional networks (GCN) [6]

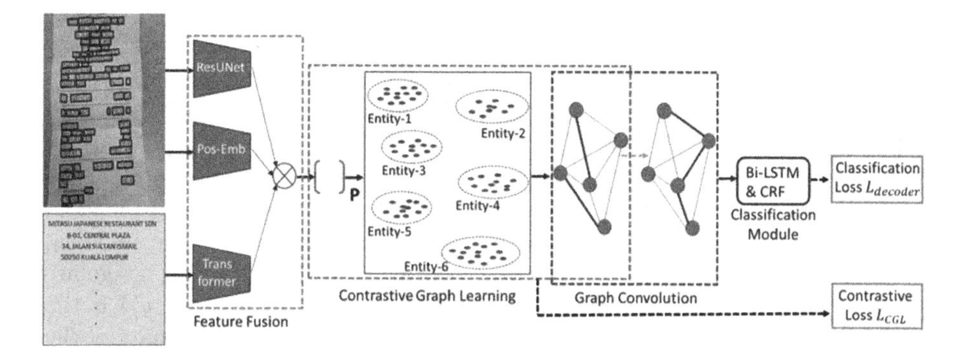

Fig. 1. The overall architecture of the proposed CGLCN model. The input for our model is an image containing text bounding boxes and their corresponding transcript. It contains graph learning, graph convolution, and classification modules. The features are combined using feature fusion techniques.

are used to get a richer representation for the nodes using message passing techniques. Our proposed contrastive graph learning framework can be used in the graph-based KIE methods [11,12] for the improved graph representation.

Contrastive Learning. Contrastive learning enables learning representations by contrasting positive against negative pairs [17]. It allows to learn class discriminative features. Contrastive learning is popularly used in self-supervised learning [23] and unsupervised visual representation learning [24,25]. Recently, it has been successfully used for the self-supervised learning on graph structure data [21,22]. Contrastive learning techniques are not previously explored for graph representation learning. In this work, we use contrastive learning techniques [18–20] for learning the graph representation, where contrastive loss is used to train the discriminative node embeddings.

3 Methodology

We present a contrastive graph learning framework for the graph-based key information extraction (KIE) methods [11,12]. The overall architecture of the proposed contrastive graph learning convolutional network (CGLCN) is shown in Fig. 1. In the first module, we extract the textual, visual, and positional features from the text bounding boxes. These features correspond to the node attributes in the graph representation. The extracted features are combined using the feature fusion techniques [31]. The node attributes are used for learning the graph representation using contrastive learning. We then use GCN for computing the node embeddings. The GCN contains one input layer, multiple hidden (convolutional) layers and the node embeddings are obtained from the final convolutional layer. Finally, in the classification module, we use Bi-LSTM [27] and CRF [28] for assigning the entity labels to the node attributes.

3.1 Graph Representation

We model the problem of key information extraction from document images as a graph node classification task. Here, we represent a document image as a undirected dynamic graph $G = \{V, E\}$, where $V = \{v_1, v_2, \ldots, v_n\}$ is the set of nodes and $E = [e_{ij}]$ is the edge (adjacent) matrix, which contains the edge weights between the nodes. $X = [x_1, x_2, \ldots, x_n] \in \mathbb{R}^{n \times m}$ is the feature matrix, where x_i is the m dimensional feature vector corresponding to the node v_i. The detected text regions (text bounding boxes) tr_i in the document image are represented as nodes in the graph, where, $tr_i = (x_1^i, y_1^i, x_2^i, y_2^i)$, (x_1^i, y_1^i) is the top left most coordinate and (x_2^i, y_2^i) is the bottom right most coordinate of the bounding box. We extract the following visual, textual, and positional features from the text bounding boxes, which are represented as the node attributes,

- *Visual Features*: The CNN features obtained from the cropped text bounding box region are used as the corresponding visual features of the node. We use ResUnet [33,34] for extracting the visual features.
- *Textual Features*: The text features are extracted from the recognized text obtained from the cropped text bounding box. We use transformers [26] for extracting the text features.
- *positional Features*: For positional features, we use top left most coordinate of the text bounding box and its height. We use positional embedding of the bounding box location.

Now, each node in the graph is represented as $v_i = (vf_i, tf_i, pf_i)$, where vf_i, tf_i, and pf_i represents the visual, textual, and positional features respectively.

Multimodal Feature Fusion. The node attributes play an important role in learning the graph representation. As discussed in the previous section, we are using textual, visual, and positional features for representing the nodes in a graph. Each of these features carries a different type of information. The standard way of feature fusion techniques like concatenation and weighted sum may not yield the best results. The fusion techniques should leverage the interactions between these feature representations. In this paper, we use the feature fusion technique proposed in BLOCK [31], which is a multimodal fusion framework based on the block-superdiagonal tensor decomposition [32]. It is primarily used for the fusion of textual and visual features for Visual Question Answering (VQA) and Visual Relationship Detection (VRD).

3.2 Contrastive Graph Learning

In this section, we present a framework for learning the graph representation using contrastive learning. The objective is to learn a projection network using contrastive loss [18–20] such that the cluster of nodes belonging to the same class (entity) will be pulled together in the projection space, while the cluster of nodes from different classes (entities) will be simultaneously pushed apart.

This is done by effectively leveraging label information. The contrastive loss is applied to the outputs of the projection network. We normalize the outputs of the projection network to lie on a unit hypersphere. This enables to use the inner product as the similarity measure between the nodes in the projection space and we use it as the corresponding edge weight in the edge matrix E. The edge matrix gives an optimal graph representation for the given nodes. Since the nodes which belong to the same entity are closer to each other in the projection space, their corresponding edge weight (inner product) will be higher compared to the nodes which belong to different entities (clusters).

Projection Network P. We use a projection network $P : \mathbb{R}^m \rightarrow \mathbb{R}^l$, where m is the dimension of the node attributes, for projecting the node attributes into a projection space. l is the dimension of projection space and $l < m$. We use a multi-layer perceptron [16] with a single hidden layer as the projection network, and GeLU [35] is used as the activation function. The projection network is trained using a contrastive loss function.

The main advantage of the proposed approach is that we can use our framework for learning graph representation under supervised and semi-supervised settings. We now present the supervised and semi-supervised contrastive loss functions for training the projection network. The semi-supervised contrastive loss function enables to use both labeled and unlabeled nodes for training the projection network.

Supervised Contrastive Loss Function. Let $V_P = \{P(x_1), P(x_2), \ldots, P(x_n)\}$ be the projection of node attributes in V for the projection network P. For supervised contrastive learning [18], the contrastive loss function is given as,

$$L_{sup} = \sum_{x \in V_P} \frac{-1}{|\pi(x)|} \sum_{u \in \pi(x)} log \frac{exp(\langle x, u \rangle / \tau)}{\sum_{v \in V_P \setminus \{x\}} exp(\langle x, v \rangle / \tau)} \tag{1}$$

where, $\pi(x) = \{z \in V; y_z = y_x\}$, y_x is the class label for the node x, $|\pi(x)|$ is the cardinality of $\pi(x)$ and $\tau \in \mathbb{R}^+$ is the temperature parameter. $\langle .,. \rangle$ denotes the inner product between node attributes to measure their similarity in the projection space.

Semi-supervised Contrastive Loss Function. In semi-supervised contrastive learning [19,20], in addition to labeled document images, we also use unlabeled document images for learning the graph representation. Here, we use unlabeled data only for learning the graph representation. Using semi-supervised contrastive learning, we are able to achieve similar performance as supervised contrastive learning (Table 5) with a fewer number of labeled documents. In the first step, we cluster the nodes (text bounding boxes) from the unlabeled document images using a clustering algorithm. Since the densities of actual clusters vary a lot, the clusters obtained from this step may not be reliable clusters.

There is a possibility of outliers in these clusters, which means the class labels for these nodes are different from the cluster representative label. These outliers are treated as unclustered nodes, and we remove these outlier nodes from their cluster by leveraging their context in the graph. We introduce a reliability criterion for recognizing the outliers in the clusters. In a cluster, a node is considered as an outlier if none of its neighbors in the graph are present in that cluster. We use the graph representation obtained from the previous iteration for computing the neighbors of a node.

For semi-supervised contrastive learning [19,20], the contrastive loss function is given as,

$$L_{semi-sup} = - \sum_{x \in V_P^l \cup V_P^{ul} \setminus V_P^{ol}} log \frac{exp(\langle x, c \rangle / \tau)}{\sum_{i=1}^{n_l} exp(\langle x, m_i \rangle / \tau) + \sum_{i=1}^{n_{ul}} exp(\langle x, m_i^{ul} \rangle / \tau)} \tag{2}$$

where, V_P^l and V_P^{ul} are the projection of node attributes for the projection network P from the labeled and unlabeled documents, respectively. V_P^{ol} is the projection of outlier node attributes, n_l is the number of classes (entities) in the labeled data and n_{ul} is the number of clusters obtained from the clustering algorithm over the unlabeled data. $\tau \in \mathbb{R}^+$ is the temperature parameter. m_i and m_i^{ul} are the ith class/cluster mean of labeled and unlabeled nodes respectively. If $x \in V_P^l$ belongs to k^{th} class, then $c = m_i$ and if $x \in V_P^{ul}$ belongs to k^{th} cluster, then $c = m_i^{ul}$. The major difference between the supervised (Eq. 1) and semi-supervised (Eq. 2) contrastive loss functions is the addition of a new term in the denominator ($\sum_{i=1}^{n_{ul}} exp(\langle x, m_i^{ul} \rangle / \tau)$) for the semi-supervised formulation, which corresponds to the unlabeled data. This term acts as a regularizer in the semi-supervised formulation.

4 Graph Convolution

The graph representation learned using contrastive learning (Sect. 3.2) is used along with the graph convolutional network (GCN) [6] for computing the node embeddings. The node embeddings obtained from the GCN are the problem specific node attributes. These node attributes are fed into the classification (decoder) module for classifying the nodes into one of the information categories. We use Bi-LSTM [27] and CRF [28] as the decoder and, negative log-likelihood as the loss function in the decoder module.

4.1 Loss

The overall loss in CGLCN constitutes the loss from the graph learning module (Eq. 1 or Eq. 2) and the loss from the classification (decoder) module. We take the weighted combination of these 2 losses as the final loss, which is given as,

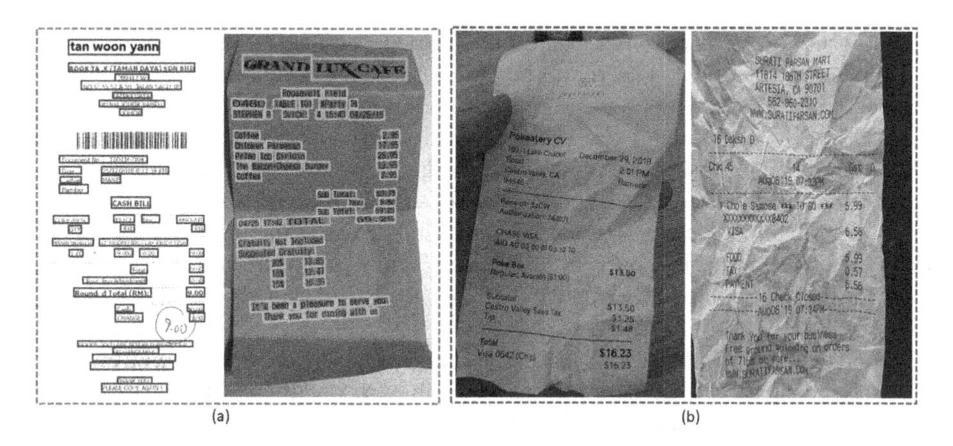

Fig. 2. (a) Annotations from SROIE (left) and Receipt datasets (right). (b) Few sample images from the Receipt dataset.

$$L_{CGLCN} = L_{CGL} + \lambda L_{decoder} \quad (3)$$

where, L_{CGL} is the loss from the contrastive graph learning module, which can either supervised or semi-supervised contrastive loss function. $L_{decoder}$ is the loss from the final classification module. λ is the trade off parameter.

5 Experiments

In this section, we validate the performance of proposed contrastive graph representation learning framework for the key information extraction task.

5.1 Datasets

To evaluate various components of the proposed contrastive learning framework, we test it on the following datasets.

SROIE Dataset [1]**.** This dataset is part of the 2019 ICDAR-SROIE [1] competition. It contains the scanned receipt images, which do not have any perspective distortions. The training set contains 626 images, and the test set contains 347 images. All the receipts are labeled with 4 entities (categories), company, address, date, and total. It mainly contains English characters and digits. This dataset provides the text bounding boxes and their corresponding transcript. Some of the text bounding boxes are at a word level, and some of them contain multiple words. An example receipt image and the overlaid text bounding boxes are shown in Fig. 2(a) (left).

Receipt Dataset. The receipt dataset is our in-house dataset, which contains 1612 images. These images are downloaded from the internet using different keywords. These are captured from different views and contain some amount of perspective distortion. We randomly selected 1300 images for training and 312 images for testing. The templates for test images differ from the training images. For the SROIE dataset [1], the text bounding boxes are not at the word level (Fig. 2(a) (left)), and the key information extraction from those types of text bounding boxes is easier compared to the word level text bounding boxes. Also, producing those types of bounding boxes in real world setting is not quite easy. The document images captured in real application setting contains different types of layouts, perspective distortions, and capture from different views. This limits the applicability of SROIE dataset for evaluating the key information extraction methods.

For the Receipt dataset, we do the annotation at the word level. A word level bounding box is drawn for all the text in each document image and annotated their corresponding text. We then label each bounding box to one of 12 key information categories, including company name, address, date, item name, item price, total, subtotal, quantity, phone number, time, tax and individual item price. The annotated text bounding boxes for the Receipt dataset are given in Fig. 2(a) (right). Few sample images containing non-frontal views and folds are shown in Fig. 2(b). Since our dataset has word level annotations and is captured from different views containing perspective distortions, it is well suited for evaluating the performance of key information extraction methods. We use the same evaluation settings described in [1].

CORD Dataset [39]. It includes images of Indonesian receipts and consists of 30 fields under 4 categories. We use the train/dev/test split of 800/100/100 respectively as proposed in [39].

5.2 Implementation Details

The proposed framework is implemented in PyTorch and trained on a NVIDIA GPU with 12 GB memory. Our model is trained from scratch using Adam [38] optimizer. We use a batch size of 4 during training. The learning rate is set to 10^{-4}. The encoder module of transformers [26] is used for text embeddings. The dimensionality for visual, textual and positional features are set to 256. We set the number of convolutional layers in CGLCN to 2 and train CGLCN for a maximum of 60 epochs. The averaged F1 score is used as the evaluative metric.

5.3 Baseline Methods

To verify the performance of the proposed contrastive graph learning framework for the key information extraction (KIE), we compare CGLCN with the graph-based methods GLCN [15], VRD [11] and PICK [12]. All these graph-based methods use graph convolutional networks [6] for computing the node embeddings. In these methods, the text bounding boxes in the document image are modeled as nodes in a graph. In both VRD [11] and PICK [12], nodes are represented using textual and visual features. These methods are summarized as follows,

- VRD [11]. It uses a fully connected graph for the graph representation. The node embeddings obtained from the GCN are fed into a standard BILSTM as the decoder for information extraction.
- GLCN [15]. Graph representation is learnt from the given data. We use textual, visual, and positional features as node attributes. It use MLP as the decoder.
- PICK [12]. The graph representation used in this work is inspired by the GLCN [15]. The node embeddings obtained from the GCN are fed into the Bi-LSTM and CRF layer for information extraction.

5.4 Comparison with Baseline Methods

We first compare our proposed CGLCN with the baseline methods (Sect. 5.3) on SROIE dataset in Table 1. To evaluate the superiority of the proposed contrastive learning for graph representation, we replace the graph representation module in GLCN [15], VRD [11] and PICK [12] with the proposed contrastive graph learning module and denote it as cont-GLCN, cont-VRD and cont-PICK respectively. The results are given in Table 2. We report the accuracy in terms of F1 score. As we can see, the contrastive counterpart of baseline methods perform reasonably well compared to their actual method. Also, we report an improvement of 1.8%, 2.4% and 1.6% in terms of F1 score over the baseline methods GLCN, VRD and PICK respectively. We also report the accuracy for our CGLCN model, where we use textual, visual, and positional features as compared to other baseline models. Also, different from these models, we use multimodal feature fusion using block super diagonal tensor decomposition [31] for combining the textual, visual, and positional features. The CGLCN model outperforms all other baseline models.

Table 1. Comparison with the baseline methods on SROIE dataset in terms of F1 score.

Method	F1 score
GLCN [15]	90.8
cont-GLCN	92.4
VRD [11]	85.6
cont-VRD	87.8
PICK [12]	93.7
cont-PICK	95.1
CGLCN (proposed)	**96.5**

To further evaluate the proposed CGLCN in the presence of perspective distortions and viewpoint changes, we compare with the baseline methods on the Receipt dataset in Table 2. For all the methods, we also report the category wise accuracy and it is given in terms of the F1 score. Similar to the SROIE dataset, we compare the baseline methods with their corresponding contrastive version (cont-GLCN, cont-VRD, cont-PICK). From the results, we can observe that the proposed contrastive graph learning framework significantly improves the performance of all the baseline methods. We report the maximum improvement of 2.5% in the F1 score for the VRD [11], where a fully connected graph is used for graph representation. We report an improvement of 1.8% and 1.7% over the GLCN and PICK models. As we notice from Table 2, CGLCN outperforms all other baseline methods. We also observe the improvement in the performance over all the 12 information categories. Note that the images in the Receipt dataset are not scanned images, and these are captured from different viewpoints and

Table 2. Comparison with the baseline methods on Receipt dataset in terms of F1 score. Here, we compare the accuracy over all the 12 information categories (entities), which includes, company name (Comp Nam), address (Addr), date, item name, item price (Item Pr), total, sub total, quantity (Quant), phone number (Ph-Num), time, tax and individual item price (Ind Itm-Pr).

Method	Comp Nam	Addr	Date	Item Nam	Item Pr	Total	SubTotal	Quant	Ph-Num	Time	Tax	Ind Itm-Pr	Avg
GLCN [15]	76.2	82.0	85.1	87.4	84.3	72.6	69.5	86.4	82.4	86.3	68.6	71.5	79.3
cont-GLCN	78.2	83.5	86.7	88.6	85.7	74.6	73.1	88.2	84.3	88.2	69.8	73.0	81.1
VRD [11]	71.2	75.1	83.6	80.9	85.2	75.2	75.7	84.7	82.7	87.5	66.1	65.3	77.6
cont-VRD	73.6	77.5	85.4	83.9	87.1	77.8	77.3	87.4	84.2	90.2	68.8	69.1	80.1
PICK [12]	75.6	82.7	85.7	84.9	86.1	77.8	76.4	85.9	83.1	91.3	69.4	70.4	80.8
cont-PICK	77.9	85.1	87.5	87.2	88.3	79.5	78.0	87.3	83.7	92.8	71.0	72.4	82.5
CGLCN (proposed)	**79.4**	**85.9**	**88.1**	**90.6**	**88.6**	**81.4**	**79.4**	**90.7**	**84.8**	**95.6**	**71.6**	**73.8**	**84.1**

contain some amount of perspective distortion. This shows the robustness of our model, and is well suited for real-world applications.

We compare CGLCN with the state-of-the-art methods on CORD dataset in Table 3. The proposed CGLCN is comparable with LayoutLMv2 [8], TILT [40], DocFormer [43] and performing well compared to LAMBERT [9]. Note that, the CGLCN is taking fewer number of parameters compared to LayoutLMv2, TILT and DocFormer for achieving comparable performance. Also, CGLCN do not involve any pre-training.

Table 3. Comparison with the state-of-the-art methods on CORD dataset.

Method	#param(M)	F1 score
LAMBERT [9]	125	94.41
LayoutLMv2 [8]	426	96.01
TILT [40]	780	96.33
DocFormer [43]	536	**96.99**
CGLCN	308	95.83

Table 4. Comparison with the baseline methods, where text bounding boxes and the corresponding text is obtained from the text detection and recognition models.

Method	GT	T-Det	T-Recg	End-to-End
GLCN [15]	79.3	78.4	74.8	73.2
VRD [11]	77.6	76.9	73.0	71.7
PICK [12]	80.8	80.0	76.5	75.1
CGLCN	**84.1**	**83.2**	**81.0**	**79.7**

In real applications, the text bounding boxes and their corresponding text is obtained from the text detection and recognition models. In general, these models may induce some detection and recognition errors. To further evaluate the CGLCN in the presence of these errors, for the test images, we take the text bounding boxes and their corresponding text obtained from the detection [36] and recognition [37] models. The results are given in Table 4. In the Table, the results under GT are obtained using ground truth text bounding boxes and their corresponding text. In T-Det, only the text bounding boxes are obtained from the detection model [36]. In T-Recg, the text in the GT bounding boxes are obtained from the recognition model [37]. In End-to-End setting, both the text bounding boxes and the corresponding text is taken from the text detection and recognition model. The proposed CGLCN outperforms all other baseline methods even if the text bounding boxes and the corresponding text is obtained from the detection and recognition models. For the T-Det, we observe a minimal drop in the performance for all the methods. This is mainly due to the word level annotations for the Receipt dataset, which suggests that it is well suited for evaluating the KIE models. For the T-Recg, we observe 3.1% performance drop in terms of the F1 score. This is mostly due to the recognition error in the numerical entities like date, item price, tax, subtotal, and total, where some digits are misrecognized as characters. Under End-to-End setting, we notice a smaller drop in the performance for CGLCN compared to all the baseline methods. It is more effective in handling text spotting errors than other baseline methods.

5.5 Supervised vs. Semi-supervised Contrastive Graph Learning

This section evaluates the performance of proposed semi-supervised and supervised contrastive graph learning techniques. The results are given in Table 5. We show the performance of both semi-supervised and supervised models over a varying number of training images. For the semi-supervised model, the images that are not part of the training set are used as the unlabeled images. We do not use the node (text bounding boxes) labels in those images for learning the graph representation. As we can see, the semi-supervised model performs better compared to supervised model. As we increase the number of training images, the semi-supervised model outperform the supervised model. For the semi-supervised model, using 1300 samples, it is able to achieve the performance of the supervised model on 1500 training samples. The semi-supervised framework is well suited for the problems where we have unlimited unlabeled data and limited labeled data.

Table 5. Performance of semi-supervised and supervised contrastive graph learning over varying number of training images in terms of F1 score.

# Training images	700	1000	1300	1500
Semi-Supervised	79.9	82.3	84.0	-
Supervised	79.8	81.6	83.0	84.1

5.6 Ablation Studies

To evaluate the various components of proposed CGLCN, we conduct a few ablation studies on the Receipt dataset.

The effect of textual, visual and positional features on CGLCN over Receipt dataset are demonstrated in Table 6. The results show that the textual features play an important role compared to visual and positional features. We notice a 6.7% drop in the F1 score without textual features.

Table 6. Effect of textual, visual, positional and graph learning modules of CGLCN on Receipt dataset.

Method	F1 score
Without textual features	77.4
Without visual features	82.0
Without positional features	83.3
Without graph learning (GL)	74.3
Without adding GL features	83.2
Without feature fusion	82.3
CGLCN	**84.1**

The textual features capture the main context of the node compared to visual and positional features. The positional features are contributing minimally to the CGLCN compared to textual and visual features. The positional features assist the graph learning module in learning the graph representation. We also show the effect of the graph learning module in Table 6 under without graph learning (GL). As we can see, the graph learning module has an essential role in our proposed CGLCN. Compared to all other modules, it results in a significant drop in the F1 score (9.8%). The graph learning module helps significantly in capturing the structural information within nodes neighborhoods. In addition to this, we also show the importance of the features obtained from the graph learning (GL) module for the GCN in Table 6, under 'without adding GL features'. Here, we only use the features obtained from the feature fusion for computing the final node embeddings in GCN and we notice a 0.9% drop in the F1 score. In another ablation study, we examine the significance of feature fusion for combining textual, visual, and positional features. The results are shown under 'without feature fusion' in Table 6. Here, instead of the feature fusion discussed in Sect. 3.1, we simply concatenate the textual, visual, and positional features. We notice a drop in the F1 score from 84.1% to 82.3% without feature fusion. This suggests the importance of feature fusion techniques for key information extraction.

We also analyze the impact of the number of layers in GCN. We obtain the F1 score of 82.4%, 84.1%, and 83.6% for the number of layers 1, 2, and 3, respectively. We achieve the best performance for 2 layers. It tends to overfit if we set the number of layers to more than 2. In GCN, the probability of overfitting will tend to increase with the increase in the number of layers. The number of layers in GCN is task-specific.

6 Conclusion and Future Work

In this paper, we proposed a graph representation learning framework using contrastive learning. We also present contrastive graph learning convolutional network (CGLCN), where the contrastive graph learning framework and graph convolutional network (GCN) are integrated into an unified network architecture. The proposed contrastive graph learning framework can be used in any graph-based learning problem. In this paper, we validated for the problem of key information extraction. The experimental results on our Receipt dataset demonstrate the effectiveness of contrastive graph learning over other graph representation learning frameworks. In the future, we explore the contrastive graph learning framework for other tasks, such as graph clustering, pose estimation, image classification and image cosegmentation.

References

1. Huang, Z., et al.: ICDAR2019 competition on scanned receipt OCR and information extraction. In: ICDAR (2019)

2. Lample, G., Ballesteros, M., Subramanian, S., Kawakami, K., Dyer, C.: Neural architectures for named entity recognition. In: NAACL (2016)
3. Esser, D., Schuster, D., Muthmann, K., Berger, M., Schill, A.: Automatic indexing of scanned documents: a layout-based approach. In: DRR (2012)
4. Cesarini, F., Francesconi, E., Gori, M., Soda, G.: Analysis and understanding of multi-class invoices. DAS **6**, 102–114 (2003)
5. Simon, A., Pret, J.-C., Johnson, A.P.: A fast algorithm for bottom-up document layout analysis. In: PAMI (1997)
6. Kipf, T.N., Welling, M.: Semi-supervised classification with graph convolutional networks. In: ICLR (2017)
7. Veličković, P., Cucurull, G., Casanova, A., Romero, A., Liò, P., Bengio, Y.: Graph attention networks. In: ICLR (2017)
8. Xu, Y., et al.: LayoutLMv2: multi-modal pre-training for visually-rich document understanding. arXiv (2020)
9. Garncarek, Ł, et al.: LAMBERT: layout-aware language modeling for information extraction. In: Lladós, J., Lopresti, D., Uchida, S. (eds.) ICDAR 2021. LNCS, vol. 12821, pp. 532–547. Springer, Cham (2021). https://doi.org/10.1007/978-3-030-86549-8_34
10. Lin, W., et al.: ViBERTgrid: a jointly trained multi-modal 2D document representation for key information extraction from documents. In: Lladós, J., Lopresti, D., Uchida, S. (eds.) ICDAR 2021. LNCS, vol. 12821, pp. 548–563. Springer, Cham (2021). https://doi.org/10.1007/978-3-030-86549-8_35
11. Liu, X., Gao, F., Zhang, Q., Zhao, H.: Graph convolution for multimodal information extraction from visually rich documents. In: NAACL (2019)
12. Yu, W., Lu, N., Qi, X., Gong, P., Xiao, R.: PICK: processing key information extraction from documents using improved graph learning-convolutional networks. In: ICPR (2020)
13. Henaff, M., Bruna, J., LeCun, Y.: Deep convolutional networks on graph-structured data. arXiv (2015)
14. Zhu, F., Huang, J., Li, R., Wang, S.: Adaptive graph convolutional neural networks. In: AAAI (2018)
15. Jiang, B., Zhang, Z., Lin, D., Tang, J., Luo, B.: Semi-supervised learning with graph learning-convolutional networks. In: CVPR (2019)
16. Hastie, T., Tibshirani, R., Friedman, J.: The Elements of Statistical Learning. Springer Series in Statistics, Springer, New York (2001)
17. Hadsell, R., Chopra, S., LeCun, Y.: Dimensionality reduction by learning an invariant mapping. In: CVPR (2006)
18. Khosla, P., et al.: Supervised contrastive learning. In: NIPS (2020)
19. Ge, Y., Zhu, F., Chen, D., Zhao, R., Li, H.: Self-paced contrastive learning with hybrid memory for domain adaptive object Re-ID. In: NIPS (2020)
20. Zhang, Y., Zhang, X., Qiu, R.C., Li, J., Xu, H., Tian, Q.: Semi-supervised contrastive learning with similarity co-calibration. CoRR abs/2105.07387 (2021)
21. You, Y., Chen, T., Sui, Y., Chen, T., Wang, Z., Shen, Y.: Graph contrastive learning with augmentations. In: NIPS (2020)
22. You, Y., Chen, T., Shen, Y., Wang, Z.: Graph contrastive learning automated. In: ICML (2021)
23. Chen, T., Kornblith, S., Norouzi, M., Hinton, G.: A simple framework for contrastive learning of visual representations. arXiv (2020)
24. He, K., Fan, H., Wu, Y., Xie, S., Girshick, R.: Momentum contrast for unsupervised visual representation learning. In: CVPR (2020)

25. Chen, T., Kornblith, S., Norouzi, M., Hinton, G.: A simple framework for contrastive learning of visual representations. In: ICML (2020)
26. Vaswani, A., et al.: Attention is all you need. In: NIPS (2017)
27. Graves, A., Schmidhuber, J.: Framewise phoneme classification with bidirectional LSTM networks. In: IJCNN (2005)
28. Lafferty, J., McCallum, A., Pereira, F.C.: Conditional random fields: probabilistic models for segmenting and labeling sequence data. In: ICML (2001)
29. Xu, Y., Li, M., Cui, L., Huang, S., Wei, F., Zhou, M.: LayoutLM: pre-training of text and layout for document image understanding. In: SIGKDD (2020)
30. Devlin, J., Chang, M.-W., Lee, K., Toutanova, K.: BERT: pre-training of deep bidirectional transformers for language understanding. In: NAACL (2019)
31. Ben-younes, H., Cadene, R., Thome, N., Cord, M.: BLOCK: bilinear superdiagonal fusion for visual question answering and visual relationship detection. In: AAAI (2019)
32. De Lathauwer, L.: Decompositions of a higher-order tensor in block terms part II: definitions and uniqueness. In: SIMAX (2008)
33. Zhang, Z., Liu, Q.: Road extraction by deep residual U-Net. In: GRSL (2017)
34. Diakogiannis, F.I., Waldner, F., Caccetta, P., Wu, C.: ResUNet-a: a deep learning framework for semantic segmentation of remotely sensed data. In: ISPRS (2020)
35. Hendrycks, D., Gimpel, K.: Gaussian error linear units (GELUs). arXiv (2016)
36. Liao, M., Wan, Z., Yao, C., Chen, K., Bai, X.: Real-time scene text detection with differentiable binarization. In: AAAI (2020)
37. Shi, B., Bai, X., Yao, C.: An end-to-end trainable neural network for image-based sequence recognition and its application to scene text recognition. In: PAMI (2017)
38. Kingma, D.P., Ba, J.: Adam: a method for stochastic optimization. arXiv (2014)
39. Park, S., et al.: CORD: a consolidated receipt dataset for post-OCR parsing. In: Document Intelligence Workshop at NeurIPS (2019)
40. Powalski, R., Borchmann, Ł., Jurkiewicz, D., Dwojak, T., Pietruszka, M., Pałka, G.: Going full-tilt boogie on document understanding with textimage-layout transformer. arXiv (2021)
41. Hwang, W., Yim, J., Park, S., Yang, S., Seo, M.: Spatial dependency parsing for semi-structured document information extraction. In: ACL-IJCNLP (2021)
42. Ma, X., Hovy, E.: End-to-end sequence labeling via bi-directional LSTM-CNNsCRF. In: ACL (2016)
43. Appalaraju, S., Jasani, B., Kota, B.U., Xie, Y., Manmatha, R.: DocFormer: end-to-end transformer for document understanding. In: ICCV (2021)

Improving Information Extraction on Business Documents with Specific Pre-training Tasks

Thibault Douzon[1,2](✉), Stefan Duffner[1], Christophe Garcia[1], and Jérémy Espinas[2]

[1] INSA Lyon, CNRS, LIRIS, Lyon, France
{thibault.douzon,stefan.duffner,christophe.garcia}@insa-lyon.fr
[2] Esker, Lyon, France
jeremy.espinas@esker.com

Abstract. Transformer-based Language Models are widely used in Natural Language Processing related tasks. Thanks to their pre-training, they have been successfully adapted to Information Extraction in business documents. However, most pre-training tasks proposed in the literature for business documents are too generic and not sufficient to learn more complex structures. In this paper, we use LayoutLM, a language model pre-trained on a collection of business documents, and introduce two new pre-training tasks that further improve its capacity to extract relevant information. The first is aimed at better understanding the complex layout of documents, and the second focuses on numeric values and their order of magnitude. These tasks force the model to learn better-contextualized representations of the scanned documents. We further introduce a new post-processing algorithm to decode BIESO tags in Information Extraction that performs better with complex entities. Our method significantly improves extraction performance on both public (from 93.88 to 95.50 F1 score) and private (from 84.35 to 84.84 F1 score) datasets composed of expense receipts, invoices, and purchase orders.

Keywords: Business documents · Document understanding · Information extraction · Pre-training · BIESO Decoding · Transformer

1 Introduction

Business documents are paper-sized files containing useful information about interactions between companies. They may take the form of invoices, purchase orders, various reports, and agreements. The exact layout of a document depends on the issuer, but the contained information is conventionally structured. For example invoices and purchase orders share the same header, table, footer structure that almost all issuers have adopted. Because such documents trace every transaction made by companies, they are the key to business process automation. With the emergence of modern resource planning systems, accurate Information Extraction (IE) has become one of the core problems of Document Intelligence.

© Springer Nature Switzerland AG 2022
S. Uchida et al. (Eds.): DAS 2022, LNCS 13237, pp. 111–125, 2022.
https://doi.org/10.1007/978-3-031-06555-2_8

Initially, information extraction was done by human operators, but software solutions have been developed since the early days of document analysis to tackle the problem. Their intent was to ease the work of human operators with hard-coded extraction rules. Unfortunately, these rules needed to be adapted for each and every layout of documents. This limitation has led to the rise of Machine Learning (ML) models for automatic document IE.

First ML approaches relied on Optical Character Recognition (OCR) systems to provide the textual content of the document. This transition from image to text allowed for standard Natural Language Processing (NLP) methods to be applied by adopting a Sequence Labeling problem. The amount of labeled data necessary to train accurate NLP models has always been a problem. Business documents are inherently private which strongly limits the quantity of publicly available data. Thus, only companies selling business process automation software are able to collect larger amounts of such data. Moreover, they often rely on their customer to implicitly label the documents.

Most recent proposals often include a pre-training step, where the model is trained on a pretext task. Those pretext tasks are self-supervised problems that teach the model many useful "skills" for manipulating the data. Usually, these tasks are as broad as possible, teaching the model common sense about language, grammar, and global structure.

In this work, we focus on LayoutLM [31], a pre-trained Transformer that is specialized in business documents. As shown in Fig. 1, it reuses the same Trans-

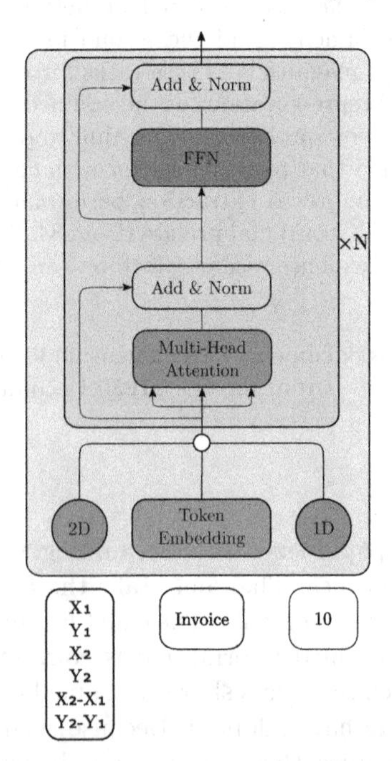

Fig. 1. LayoutLM architecture. Token embeddings are enriched with 1D positional encoding and 2D spatial encoding specific to this architecture. The number of blocks N varies from 12, for the base model, to 24, for the large one.

former layer with multi-head attention with the addition of a 2D positional encoding. Its larger version achieved state-of-the-art performance in both document classification and information extraction. However, the required hardware to train it can be repelling. In this paper, we propose new pre-training tasks specific to business documents that will provide additional skills to the model. We also propose a new decoding post-processing algorithm that prevents many errors made by the model due to ambiguities. Combined, our contributions[1] allow for the base LayoutLM model to perform on par with the large version.

2 Related Work

2.1 Information Extraction

Rule-based approaches [10] have been supplanted by Deep Learning models in the last decade. Document IE first capitalized on the state of the art in Named Entity Recognition for NLP [8]. Recurrent Neural Networks with Long-Short Term Memories were first used to encode documents at a word level [16,23], allowing a simple classifier to predict each word's associated label. Instead of a softmax and cross-entropy loss, a Conditional Random Field [25] model has been used in addition to `BIESO` tags. Other architectures have also been proposed to better adapt to the specificity of the document. For example, graphs [3,12,13,33] and convolutions over a grid [1,7,11] constrained the model based on the words' positional information. Because most architectures relied on textual representations, they benefited from pre-trained word embeddings like Word2Vec [14] or GloVe [17].

With the emergence of Transformers [26] and text encoders like BERT [2], attention-based document analysis models [5,30,31] evolved quickly, which resulted in a large improvement of state-of-the-art performance. In line with [7] which included both textual and visual representations, multi-modal Transformers [11,30] superseded conventional textual models.

In parallel to the rise of Transformers, end-to-end IE models tried to reduce the labeling cost. First using RNNs with attention layers [15,22], then shifting to Transformers [19]. Adopting at the same time the Question Answering [4] (QA) format, instead of the usual Sequence Labeling, provided more flexibility on the predicted labels.

2.2 Pre-training

Semi-supervised training and pre-trained models were popularised in NLP with enriched word embeddings [14,17,18]. With the emergence of Transformers, large pre-trained models have been proposed [29]. Thanks to their pre-training, they can efficiently adapt to various tasks [27,28] and data types. In general, these models are pre-trained on large unlabeled datasets in a self-supervised manner. This self-supervision removes parts of the burden of data labeling [24] and leverages the huge quantities of available data.

[1] Code available here: https://github.com/thibaultdouzon/business-document-pre-training.

A wide variety of pre-training tasks have been proposed. General-purpose tasks aiming at learning the language and grammar were used first. Auto-regressive tasks [20] and Masked Language Modeling [2] are still frequently used in new pre-trained models as they have proven to be effective in most situations. In addition to incremental improvements [21,32], some new pre-training tasks were designed to align representations of multi-modal inputs [19,30].

3 Models

3.1 Architecture

We used the well-established LayoutLM architecture [31] which itself is based on BERT Transformer [2]. More specifically, we chose the base model[2] with 12 layers and 512 dimensions for token embeddings. This model is computationally much more efficient compared to the larger version while still giving very good performance.

Transformer models work on tokens that are in between characters and words. LayoutLM and BERT both use the WordPiece algorithm. We use the same tokenizer as LayoutLM in order to compare our performance with the base LayoutLM model. It uses a vocabulary size of 30000, and we limit the sequence length to 512 tokens, including the special tokens `[CLS]` and `[SEP]`. This limitation due to GPU memory consumption of self-attention operations often forces us to cut documents in multiple pieces of 512 tokens and process them separately.

Contrary to RNNs, all positions in the sequence are equivalent in a Transformer model. To provide information about position inside the sequence, a linear positional encoding [26] is added for each token. Then LayoutLM adapted this positional encoding to a 2D version that can represent the positions of words on a page.

For both pre-training tasks and fine-tuning, we use a simple dense layer to map each token's final internal representation to the dimension of the prediction space. A softmax layer is applied to produce the final model confidence scores. For training, the cross-entropy loss is used on the model confidence scores.

3.2 ConfOpt Post-processing

We model the Information Extraction task as sequence tagging on tokens. Predictions are done at the token level and then aggregated by a lightweight post-processing step to give the model's final prediction. In all experiments, we use `BIESO` tagging. That is, each field to extract is composed of a sequence of target tags of the following types: `B` for the beginning of the entity, `I` for inside, `E` for its end, or otherwise `S` for a single token entity. `O` is used for any token that is outside any target label. `BIESO` is widely used in IE as it provides structure to the target sequence that helps the model.

Instead of the trivial post-processing which consists of simply following the maximum confidence of the model, we decided to decode a model's prediction by solving a basic optimization problem. We will refer to this method as ConfOpt in the

[2] Pre-trained weights available here: https://huggingface.co/microsoft/layoutlm-base-uncased.

remaining of the paper. The predicted sequence for a target label is the sequence that maximizes model confidence over the whole input sequence. There is a constraint to decode a prediction: it must match the following regular pattern: `(BI*E)` | `S` where * denotes zero or many occurrences and | denotes an alternative.

This optimisation problem can be solved with a dynamic programming approach. The model's predictions for one target label can represented as a $4 \times N$ dimensional matrix where N is the sequence length and 4 comes from the 4 tags `B,I,E,S`. By noting $C_{T,0}$ the model's confidence in `T` tag at position 0 and $P_{T,i}$ the best prediction confidence ending at token i with tag `T`, the objective is to determine $S = \max_{\substack{0 \le i < N \\ T \in \{E,S\}}} P_{T,i}$ where

$$P_{B,i} = C_{B,i} \ ; \ P_{I,i} = C_{I,i} + \max \begin{cases} P_{B,i-1} \\ P_{I,i-1} \end{cases}$$

$$P_{S,i} = C_{S,i} \ ; \ P_{E,i} = C_{E,i} + \max \begin{cases} P_{B,i-1} \\ P_{I,i-1} \end{cases}$$

One drawback of this post-processing is dealing with no prediction and non-unique predictions. It can be solved with an empirically determined threshold below which no predictions are made. Though in this paper this is not further studied because fields are mandatory in a document and always unique.

4 Pre-training

Transformer models provide great performance when first pre-trained on pretext tasks on very large unlabelled datasets. This pre-training is most of the time done in a self-supervised manner in order to avoid the labeling cost. LayoutLM uses Masked Visual-Language Modeling [31] which is adapted from BERT's Masked Language Modeling [2]. It teaches the model how text and documents are formed at a token level. In practice, at each training step, 15% of the tokens are randomly chosen and replaced by either a `[MASK]` token, a random token, or not replaced at all. The model tries to guess which token is the most probable right replacement at those positions.

For all pre-training tasks when a document is too long to be processed at once, we randomly select a continuous span of words of maximum size and provide it to the model instead. We expect the model to learn useful features on various parts of documents thanks to the long training. For very short documents, the input is padded to the maximum size.

We introduce two new specific pre-training tasks in addition to Masked Visual-Language Modeling (MVLM). The first one, Numeric Ordering teaches the model how to compare and order numbers. The second one, Layout Inclusion focuses on words in the 2D plane and their relative positioning. We chose to avoid regression tasks, even though their implementation would have been simpler. For example, simply removing the 2D positioning of some tokens, and

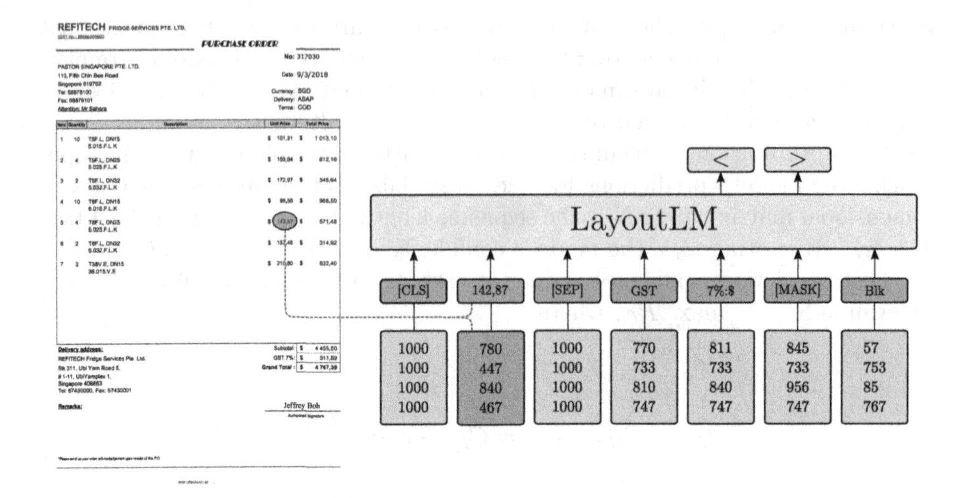

Fig. 2. A pre-training example with Numeric Ordering task. A random token containing a number is selected, then the target is to predict whether other numbers are smaller or bigger. Some random noise can be added by masking tokens' textual or spatial representations. Only a small part of the document's input is represented in this illustration.

asking the model to predict tokens' position is an alternative to what we propose. But this does not behave well for a token that could appear either at the top or the bottom of the document: the model would learn its mean position – the middle – where the token would never appear. In the following, we will describe the two pre-training tasks in detail.

4.1 Numeric Ordering Task

Numeric Ordering (NO) focuses on numeric figures in the document and their relative values. Contrary to MLM which only relies on self-supervised data, NO relies on a handcrafted number parser to find and parse all numbers that appear in a document. Because business documents are mostly made of decimal numbers written with digits, we ignore those written out in words. The numeric value of each token is determined by parsing beforehand each word in the document, looking for numbers and ignoring irrelevant characters.

As shown in Fig. 2, the model must predict for every numeric figure in the document if its parsed value is smaller, equal or greater than a randomly selected number among the document. The loss is only computed on tokens starting a new word, but tokens continuing a word are important to determine the value represented by a word.

We want the model not only to reason on the textual features, but also on the spatial context surrounding each figure in the document. Therefore, we randomly mask the textual representations of 15% of the numbers in the document and replace them with the `[MASK]` token as shown in Fig. 2. For the same reason, we

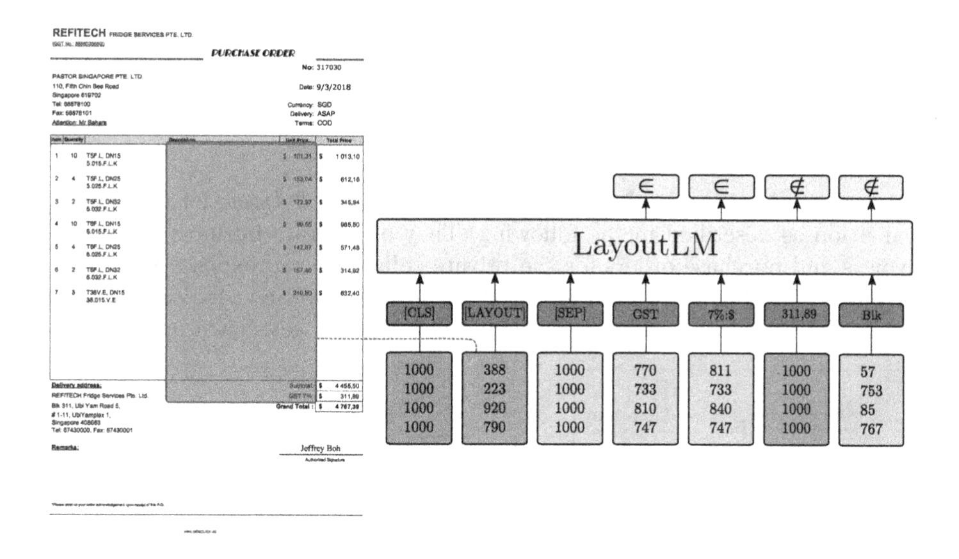

Fig. 3. A pre-training example with Layout Ordering task. Coordinates of the purple rectangle are drawn uniformly. Random noise is added by masking the 2D position of some tokens. Only a small part of the document is represented.

also mask the spatial encoding of 15% of the numbers and make sure both text and position are not masked at the same time. All masked positions are replaced with (1000, 1000, 1000, 1000).

4.2 Layout Inclusion Task

We introduce another pre-training task focusing on the 2D positional encoding, which we called Layout Inclusion (LI). Its purpose is to provide a better understanding of document layouts and complex structures. In fact, most business documents, including invoices, purchase orders, and expense receipts, contain tables where the meaning of tokens is mostly driven by their position relative to headers.

As shown in Fig. 3, Layout Inclusion is formatted like a question answering prompt: a question followed by the content of the document. The question is simply a special token `[LAYOUT]` positioned at random coordinates (x_1, y_1, x_2, y_2). The model must then classify every token in the document into 2 groups: either `inside` or `outside` of the question token. More precisely, the target answer is whether the middle point of a document token is inside or outside the rectangle described by the coordinates of the question.

Again, the objective is for the model to not only reason on the 2D positions of tokens but also use their textual embedding. In order to force the model to use both representations, we randomly replace 15% of documents token positions with (1000, 1000, 1000, 1000). In case of a random position replacement, the target value is still computed based on the real position of the token, and the model must make its prediction based on the token's text and the neighboring tokens using the classical 1D positional encoding.

5 Datasets

We used 2 different collections of documents to build 3 datasets for training and evaluation as described in the following. They all contain business documents: invoices and purchase orders for the private collection and expense receipts for the public one. The largest dataset used for pre-training isn't labeled, document samples with their target fields for the others datasets are shown in Fig. 4.

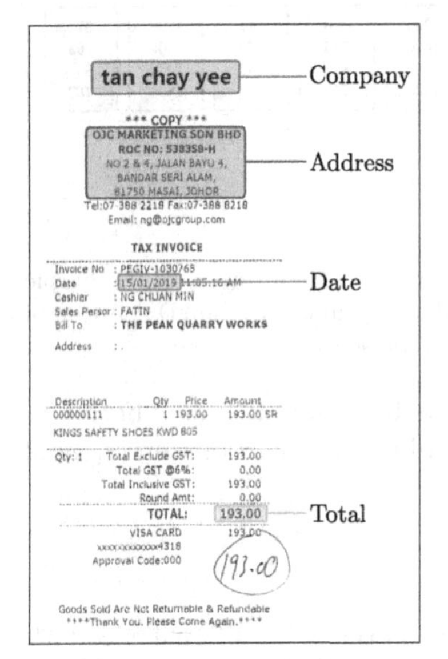

(a) Receipt from SROIE.

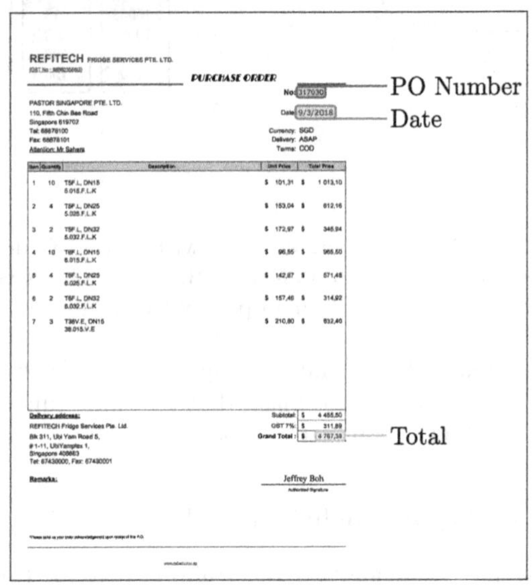

(b) Purchase Order from BDC-PO.

Fig. 4. A document sample for each training dataset annotated with the expected predictions. For BDC-PO, we replaced the document with a fictive one due to privacy reasons.

5.1 Business Documents Collection

The Business Documents Collection (BDC) is a large private dataset composed of 100k invoices and 300k purchase orders. Those real documents were submitted and processed on a commercial document automation solution in the last 3 years. It contains English-only documents divided into 70000 different issuers. All documents sharing the same issuer usually use the same information template. Therefore, we limited the maximum number of documents of the same issuer to 50. It is important to keep the number of similar layouts in the collection low and the variety of examples high. We used this collection for pre-training language models on business documents that are closer to our final objective than RVL-CDIP [9].

Textual and positional information have been extracted using a commercial OCR system. It achieves excellent accuracy on properly scanned documents and provides accurate word positions. We also use the provided read order to determine the order of tokens when feeding the network. This order determines the 1D positional encoding given to each token that complements the 2D positional encoding.

Because we only used this collection for pre-training models on self-supervised tasks, most documents do not have extraction labeling. Only a subset composed of purchase orders is labeled for the IE task.

5.2 Business Documents Collection – Purchase Orders

We selected a subset of the Business Documents Collection to build a labeled dataset of English purchase orders called BDC-PO. It contains almost 9000 different issuers split into training, validation, and test set. In order to not introduce bias for models pre-trained on the BDC, we removed from BDC all documents emitted by a supplier contained in the test set. This means that document layouts contained in the test set have never been seen before by the model at pre-training or training time.

Long purchase orders are rare but can sometimes be longer than 20 pages. If we wanted to train models and make predictions on such documents, we would have to evaluate the model on dozens of inputs for one document. Instead, we chose to limit documents to one page and crop the remaining. It only concerns roughly 25% of the dataset and sometimes impacts the prediction because labels are missing from the input.

The extraction task consists of 3 fields: document number, delivery date, and total amount. Those fields were chosen because they are mandatory for most customers and thus are well labeled at the word level by the end-user. We controlled the labeling quality at the issuer level and rejected from the dataset some issuers with undesirable labeling practices.

5.3 ICDAR 2019 – Scanned Receipts

We also trained and evaluated our model on the public Scanned Receipts OCR and Information Extraction [6] (SROIE) dataset that was published for ICDAR 2019. We focus on the third task which consists in extracting information from the documents. SROIE contains Malaysian receipts split into 626 train and 347 test documents. Unfortunately, we do not have control over the composition of the test set, and most of the test layouts also occur in the training set.

We used the OCR text provided with the dataset instead of using our own OCR system. As others have pointed out [31], it contains numerous little errors that negatively affect the final performance. For a fair comparison with the leaderboard, we manually fixed them such that the expected string appears in the input, at least. These fixes mostly concern addresses and company names. It almost exclusively involves fixing errors related to white-spaces and commas.

6 Experiments

All experiments were performed on a single machine equipped with two Nvidia RTX A6000 with 48Go of video memory each. This allowed us to boost the batch size up to 32 per device on a base transformer model. To further increase the batch size, we also aggregated 6 batches together before propagating the gradient for a total batch size of 192. We used the Adam optimizer with a learning rate of $1e{-}5$ and 200 linear warm-up steps as it improved our model's convergence. We used 1500 training steps for SROIE and 3000 steps for BDC-PO. Finally, we ran each fine-tuning 10 times in each setup to get a precise idea of the performance of the models and the variability of the results. For the different pre-training scenarios, we performed only two runs and the best model was kept.

6.1 Post-processing

This first set of experiments aims at comparing the post-processing used to decode the sequence produced by the model. We want to determine whether our proposed ConfOpt algorithm is competitive with other decoding methods. We decided to use the LayoutLM base model and compare the proposed ConfOpt against two other decoding algorithms as shown in Table 1.

We named Ad-Hoc the basic decoding using the label with maximal confidence for each token. When decoding with this method, a `B` tag starts a new entity, a `I` tag continues the previous entity, a `E` closes the previous entity, and a `S` tag produces a new entity and closes it right away. Ad-Hoc and ConfOpt use the same model weights in this experiment as they do not introduce any trainable parameters.

The second decoding algorithm uses a Conditional Random Field (CRF) [8, 25] that processes LayoutLM's predictions. In this particular case, we did not use the classical cross-entropy loss but the score provided by the CRF layer. Because the CRF required specific training and did not optimize the same loss, its weights are different from the two other post-processing methods.

Table 1. Performance comparison on SROIE and BDC-PO between multiple post-processing algorithm. Score is computed on the exact match between the prediction and the target string.

Post processing	Fine tuning (F1 score)	
	SROIE	BDC-PO
Ad-Hoc	93.88 ± 0.59	84.35 ± 0.12
CRF	94.01 ± 0.55	84.40 ± 0.16
ConfOpt	$\mathbf{94.94 \pm 0.38}$	$\mathbf{84.57 \pm 0.10}$

We evaluated these algorithms on both SROIE and BDC-PO. The results in Table 1 show a tiny improvement using a CRF instead of the Ad-Hoc post-processing (0.13 and 0.05 F1 points) but those differences are always within one standard deviation range. We would need more evidence to conclude on the effect of adding a CRF layer for the post-processing.

On both datasets, using ConfOpt significantly increases performance (1.06 and 0.22 F1 points) compared to the Ad-Hoc post-processing, even though the model is strictly identical. In light of these results, we decided to use the ConfOpt for the next experiment.

6.2 Business Document-Specific Pre-training

We conducted another set of experiments in order to study the effects of the new business data-specific pre-training tasks on the model performance. At the same time, we controlled the performance gap obtained by pre-training with the basic MVLM task on the same new dataset. Both comparisons are insightful to decide whether it is useful to pre-train on clients' data and/or with data-specific pre-training tasks.

For the pre-training part, we always initialize the model's weights with the base version [31]. We pre-train models for 20 epochs on 80% of BDC. When using multiple pre-training tasks at the same time, we chose to provide batches of single tasks to the model. Gradient aggregation over multiple batches helps smoothing the update between different tasks. We pre-trained 2 models on the BDC, one with MVLM only and another with MVLM+NO+LI.

Table 2. Model performance when fine-tuning on BDC-PO

Pre training		F1 score	Accuracy per field		
Task(s)	Dataset		PO number	Total	Date
MVLM	RVL-CDIP	84.57 ± 0.10	89.98	89.10	93.59
MVLM	BDC	84.77 ± 0.12	90.61	89.33	93.59
MVLM+NO+LI	BDC	$\mathbf{84.84 \pm 0.08}$	**90.71**	**89.36**	**93.83**

Table 3. Model performance when fine-tuning on SROIE. Models name ending with a * are our contribution. The second part contains published scores of the original LayoutLM and LayoutLMv2 as a comparison.

Architecture	Pre training		F1 score	Accuracy per field			
	Task(s)	Dataset		Company	Address	Total	Date
LayoutLM base *	MVLM	RVL-CDIP	94.94 ± 0.38	92.91	90.81	89.25	99.48
LayoutLM base *	MVLM	BDC	95.18 ± 0.23	**93.72**	91.00	89.48	**99.68**
LayoutLM base *	MVLM+NO+LI	BDC	$\mathbf{95.50 \pm 0.22}$	93.60	**91.41**	**90.89**	99.57
LayoutLM base [31]	MVLM	RVL-CDIP	94.38				
LayoutLM large [31]	MVLM	RVL-CDIP	95.24				
LayoutLMv2 large [30]	MVLM+TIA+TIM	RVL-CDIP	**97.81**				

We evaluated each pre-trained model on both datasets, the results are available in Table 2 for BDC-PO and Table 3 for SROIE. Each cell contains the means of 10 runs with different seeds and the standard deviation is provided for the F1 score. There are a few interesting things to notice.

The first important remark is the importance of the pre-training dataset. Pre-training on BDC significantly improves performance on both SROIE and BDC-PO, even though the pretext training task is the same as what was used for LayoutLM. BDC is more homogeneous and focuses on invoices and purchase orders. Contrary to our expectations, we observe a greater improvement on SROIE than on BDC-PO (0.24 vs 0.2 F1 points). But the overall improvement by using BDC can be explained because RVL-CDIP contains a broader panel of document types and is not specialized like BDC. Even though BDC does not contain expense receipts, its global structure is similar to invoices.

Next, we can compare the pre-training tasks. Introducing Numeric Ordering (NO) and Layout Inclusion (LI) tasks also improves the performance over the previously pre-trained model. We observe a 0.32 F1 point improvement on SROIE but only 0.07 on BDC-PO. We suspect the small improvement introduced by the new tasks can be explained because most useful skills to process purchase orders were learned by pre-training on such documents. The new pre-training tasks help more for generalizing on new types of documents.

We also can look at the results on a field per-field basis. We observe that using BDC over RVL-CDIP improved the recognition of all fields except for the dates in BDC-PO. If introducing new training tasks did not improve all fields, we notice that some fields were greatly enhanced like the total amount in SROIE (1.41 F1 points difference). We expected to observe a greater improvement in the total field with the new pre-training tasks. But it does not seem to improve performance much on BDC-PO's total.

Finally it is interesting to compare on Table 3 our results with the published scores of LayoutLM and LayoutLMv2. Our pre-trained model with NO and LI tasks performs better than LayoutLM large which contains 3 times more parameters. However, LayoutLMv2 – which uses both textual and visual information – performance level is still unreachable for a textual-only model.

7 Conclusion

In this work, we showed significant improvements are accessible without introducing more trainable parameters and computational complexity. Only using the base transformer architecture, we achieved a performance that is comparable to the large version which contains 3 times more parameters. Pre-trained models can be further specialized through in-domain datasets and specific pretext training tasks. We demonstrated that by introducing a new collection of business documents and training tasks focusing on documents' layout and number understanding. We showed that performance improvements can be imputed to both pre-training tasks (Numeric Ordering and Layout Inclusion) and new pre-training dataset.

In the future, we will investigate on IE as a Question Answering problem. It has already been proposed in the past [4] as an alternative to Sequence Labeling when fine-tuning models. It should improve the model's generalization capabilities and enable few-shot learning. But nowadays all models are pre-trained, and we would like to study the impact on generalization of a QA-only pre-training.

References

1. Denk, T.I., Reisswig, C.: BERTgrid: contextualized embedding for 2D document representation and understanding. arXiv:1909.04948 [cs], September 2019
2. Devlin, J., Chang, M.W., Lee, K., Toutanova, K.: BERT: pre-training of deep bidirectional transformers for language understanding. arXiv:1810.04805 [cs], May 2019
3. Gal, R., Ardazi, S., Shilkrot, R.: Cardinal graph convolution framework for document information extraction. In: Proceedings of the ACM Symposium on Document Engineering 2020, pp. 1–11. ACM, Virtual Event, CA, USA, September 2020. https://doi.org/10.1145/3395027.3419584. https://dl.acm.org/doi/10.1145/3395027.3419584
4. Gardner, M., Berant, J., Hajishirzi, H., Talmor, A., Min, S.: Question answering is a format; when is it useful? arXiv:1909.11291 [cs], September 2019
5. Garncarek, L., Powalski, R., Stanisławek, T., Topolski, B., Halama, P., Graliński, F.: LAMBERT: layout-aware language modeling using BERT for information extraction. arXiv:2002.08087 [cs], March 2020
6. Huang, Z., et al.: ICDAR2019 competition on scanned receipt OCR and information extraction. In: 2019 International Conference on Document Analysis and Recognition (ICDAR), pp. 1516–1520, September 2019. https://doi.org/10.1109/ICDAR.2019.00244. ISSN 2379-2140
7. Katti, A.R., et al.: Chargrid: towards understanding 2D documents. arXiv:1809.08799 [cs], September 2018
8. Lample, G., Ballesteros, M., Subramanian, S., Kawakami, K., Dyer, C.: Neural architectures for named entity recognition. arXiv:1603.01360 [cs], April 2016
9. Lewis, D., Agam, G., Argamon, S., Frieder, O., Grossman, D., Heard, J.: Building a test collection for complex document information processing. In: Proceedings of the 29th Annual International ACM SIGIR Conference on Research and Development in Information Retrieval, pp. 665–666 (2006)
10. Li, Y., Krishnamurthy, R., Raghavan, S., Vaithyanathan, S., Jagadish, H.V.: Regular expression learning for information extraction. In: Proceedings of the 2008 Conference on Empirical Methods in Natural Language Processing, Honolulu, Hawaii, pp. 21–30. Association for Computational Linguistics, October 2008. https://www.aclweb.org/anthology/D08-1003
11. Lin, W., et al.: ViBERTgrid: a jointly trained multi-modal 2D document representation for key information extraction from documents. In: Lladós, J., Lopresti, D., Uchida, S. (eds.) ICDAR 2021. LNCS, vol. 12821, pp. 548–563. Springer, Cham (2021). https://doi.org/10.1007/978-3-030-86549-8_35
12. Liu, X., Gao, F., Zhang, Q., Zhao, H.: Graph convolution for multimodal information extraction from visually rich documents. arXiv:1903.11279 [cs], March 2019

13. Lohani, D., Belaïd, A., Belaïd, Y.: An invoice reading system using a graph convolutional network. In: Carneiro, G., You, S. (eds.) ACCV 2018. LNCS, vol. 11367, pp. 144–158. Springer, Cham (2019). https://doi.org/10.1007/978-3-030-21074-8_12
14. Mikolov, T., Chen, K., Corrado, G., Dean, J.: Efficient estimation of word representations in vector space. arXiv:1301.3781 [cs], September 2013
15. Palm, R.B.: End-to-end information extraction from business documents, p. 99 (2019)
16. Palm, R.B., Winther, O., Laws, F.: CloudScan - a configuration-free invoice analysis system using recurrent neural networks. arXiv:1708.07403 [cs], August 2017
17. Pennington, J., Socher, R., Manning, C.: Glove: global vectors for word representation. In: Proceedings of the 2014 Conference on Empirical Methods in Natural Language Processing (EMNLP), Doha, Qatar, pp. 1532–1543. Association for Computational Linguistics (2014). https://doi.org/10.3115/v1/D14-1162. http://aclweb.org/anthology/D14-1162
18. Peters, M.E., et al.: Deep contextualized word representations. arXiv:1802.05365 [cs], March 2018
19. Powalski, R., Borchmann, Ł, Jurkiewicz, D., Dwojak, T., Pietruszka, M., Pałka, G.: Going full-TILT boogie on document understanding with text-image-layout transformer. In: Lladós, J., Lopresti, D., Uchida, S. (eds.) ICDAR 2021. LNCS, vol. 12822, pp. 732–747. Springer, Cham (2021). https://doi.org/10.1007/978-3-030-86331-9_47
20. Radford, A., Narasimhan, K., Salimans, T., Sutskever, I.: Improving language understanding by generative pre-training, p. 12 (2018)
21. Raffel, C., et al.: Exploring the limits of transfer learning with a unified text-to-text transformer. arXiv:1910.10683 [cs, stat], July 2020
22. Sage, C., Aussem, A., Eglin, V., Elghazel, H., Espinas, J.: End-to-end extraction of structured information from business documents with pointer-generator networks. In: EMNLP 2020 Workshop on Structured Prediction for NLP, Punta Cana (online), Dominican Republic, November 2020. https://hal.archives-ouvertes.fr/hal-02958913
23. Sage, C., Aussem, A., Elghazel, H., Eglin, V., Espinas, J.: Recurrent neural network approach for table field extraction in business documents. In: 2019 International Conference on Document Analysis and Recognition (ICDAR), pp. 1308–1313. IEEE (2019)
24. Sage, C., et al.: Data-efficient information extraction from documents with pre-trained language models. In: Barney Smith, E.H., Pal, U. (eds.) ICDAR 2021. LNCS, vol. 12917, pp. 455–469. Springer, Cham (2021). https://doi.org/10.1007/978-3-030-86159-9_33
25. Sutton, C., McCallum, A.: An introduction to conditional random fields. arXiv:1011.4088 [stat], November 2010
26. Vaswani, A., et al.: Attention is all you need. arXiv:1706.03762 [cs], June 2017
27. Wang, A., et al.: SuperGLUE: a stickier benchmark for general-purpose language understanding systems. arXiv:1905.00537 [cs], February 2020
28. Wang, A., Singh, A., Michael, J., Hill, F., Levy, O., Bowman, S.R.: GLUE: a multi-task benchmark and analysis platform for natural language understanding. arXiv:1804.07461 [cs], February 2019
29. Wolf, T., et al.: HuggingFace's transformers: state-of-the-art natural language processing. arXiv:1910.03771 [cs], July 2020
30. Xu, Y., et al.: LayoutLMv2: multi-modal pre-training for visually-rich document understanding. arXiv:2012.14740 [cs], May 2021

31. Xu, Y., Li, M., Cui, L., Huang, S., Wei, F., Zhou, M.: LayoutLM: pre-training of text and layout for document image understanding. In: Proceedings of the 26th ACM SIGKDD International Conference on Knowledge Discovery & Data Mining, pp. 1192–1200, August 2020. https://doi.org/10.1145/3394486.3403172. http://arxiv.org/abs/1912.13318
32. Yang, Z., Dai, Z., Yang, Y., Carbonell, J., Salakhutdinov, R., Le, Q.V.: XLNet: generalized autoregressive pretraining for language understanding. arXiv:1906.08237 [cs], January 2020
33. Yu, W., Lu, N., Qi, X., Gong, P., Xiao, R.: PICK: processing key information extraction from documents using improved graph learning-convolutional networks. arXiv:2004.07464 [cs], April 2020

How Confident Was Your Reviewer? Estimating Reviewer Confidence from Peer Review Texts

Prabhat Kumar Bharti[1(✉)], Tirthankar Ghosal[2], Mayank Agrawal[1], and Asif Ekbal[1]

[1] Department of Computer Science and Engineering, Indian Institute of Technology Patna, Daulatpur, India
{prabhat_1921cs32,mayank265,asif}@iitp.ac.in

[2] Institute of Formal and Applied Linguistics, Faculty of Mathematics and Physics, Charles University, Prague, Czech Republic
ghosal@ufal.mff.cuni.cz

Abstract. The scholarly peer-reviewing system is the primary means to ensure the quality of scientific publications. An area or program chair relies on the reviewer's confidence score to address conflicting reviews and borderline cases. Usually, reviewers themselves disclose how confident they are in reviewing a certain paper. However, there could be inconsistencies in what reviewers self-annotate themselves versus how the preview text appears to the readers. This is the job of the area or program chair to consider such inconsistencies and make a reasonable judgment. Peer review texts could be a valuable source of Natural Language Processing (NLP) studies, and the community is uniquely poised to investigate some inconsistencies in the paper vetting system. Here in this work, we attempt to automatically estimate how confident was the reviewer directly from the review text. We experiment with five data-driven methods: Linear Regression, Decision Tree, Support Vector Regression, Bidirectional Encoder Representations from Transformers (BERT), and a hybrid of Bidirectional Long-Short Term Memory (BiLSTM) and Convolutional Neural Networks (CNN) on Bidirectional Encoder Representations from Transformers (BERT), to predict the confidence score of the reviewer. Our experiments show that the deep neural model grounded on BERT representations generates encouraging performance.

Keywords: Peer reviews · Confidence prediction · Deep neural network

1 Introduction

The scholarly peer review system is deemed as the foremost process for research validation [18]. However, the peer review process has long been criticized for being

---

The original version of this chapter was revised: one crucial acknowledgement was missing. This has now been corrected. The correction to this chapter is available at https://doi.org/10.1007/978-3-031-06555-2_53

© Springer Nature Switzerland AG 2022, corrected publication 2022
S. Uchida et al. (Eds.): DAS 2022, LNCS 13237, pp. 126–139, 2022.
https://doi.org/10.1007/978-3-031-06555-2_9

arbitrary [2] biased [5,7,23] etc. In case of large conferences or popular venues, submissions are increasing at an exponential rate [3,8,22]. It is difficult to find expert reviewers to ensure a proper evaluation of the manuscript. Editors/chairs[1] had to resort to invite novice reviewers [13] to counter the reviewing workload for this voluntary, yet critical job [12]. Quite expectantly the quality of the reviews goes down, authors remain dissatisfied after receiving a low-quality, arbitrary review. Usually the editors rely on the recommendation given by the reviewers in form of some scores or categories like "weak reject" or "strong accept" to decide the fate of the paper. However, there might be cases where reviewers of the same paper can have conflicting opinions or the recommendation scores can be miscalibrated as well [4]. In such borderline or conflicting cases, editors usually look up to another artifact: *the background of the reviewer or the confidence score of the reviewer*. Sometimes, the editors are aware of the background of the reviewer and they appropriately measure quality in their reviews (*decide which reviewer's opinion has higher significance in comparison to the other*).

In conferencing systems, there is usually an option to self-annotate how confident was the reviewer in their reviews, either on a scale or on some categorical values. The editors use that score to mitigate conflicting or borderline decisions. The reviewer's confidence comes as an useful indicator to gauge the reviewer's knowledge and strength of the review. The reviewer's confidence score reinforces their conviction on their judgement and serves as an useful decision artifact. In our research, we are intrigued by the question: *Can we automatically predict the reviewer's conviction on their judgement directly from the review text using NLP/ML techniques?*

With this objective, we would like to predict the reviewer's confidence score for a given peer review text, which may be used as a signal to measure the review's quality. Therefore, we propose hybrid bidirectional LSTM and CNN architecture based on Bidirectional Encoder Representations from Transformers (BERT) to anticipate the reviewer's confidence score. To the best of our knowledge, this is the very first time where the peer review text has been used for the specified task. Additionally, extensive experiments have demonstrated that the proposed model outperforms several competing baselines. However, the automatic estimation of peer review quality is not straightforward. Our ultimate goal is to augment human judgment about the quality of a review.

Our key contributions in this paper are:

1. This paper proposed a novel hybrid bidirectional LSTM and CNN architecture grounded on BERT that leverages peer review text as input. We compare our studies with Linear Regression (LR), Decision Tree (DT), Support Vector Regression (SVR) and Bidirectional Encoder Representations from Transformers (BERT).
2. We predict the confidence score of the reviewer based on review text so that the area chair can automatically recognize the review quality. In addition, we use a semantic-level representation of the peer review text to improve our prediction performance.

[1] In this manuscript, editors/chairs are used interchangeably.

3. The proposed model will assist the area or program chair to create an automatic judgment of review quality.
4. Additionally, we evaluate our proposed model using data from International Conference on Learning Representations (ICLR) 2018, 2019 and 2021 in a cross-year fashion to verify its efficacy.
5. At the end, we establish statistical-driven baselines with an experimental dataset to evaluate mean absolute error (MAE), root mean square error (RMSE) and (R^2) coefficient of determination.

2 Related Work

Peer Review: Peer review is probably the most widely-accepted method to accumulate scientific knowledge [15] in the current era. Peer review research faces a significant obstacle due to the lack of confidentiality and proprietary peer review texts and metadata. In response, some scholarly venues e.g., International Conference on Learning Representations (ICLR) have begun publicly making peer review data available to encourage transparency and restore trust in the peer-review process.

In AI applications, predicting the reviewer's confidence is a practical and important task that can enhance efficiency in paper review. Furthermore, comparing reviewer recommended scores with machine recommended scores also aims to enhance consistency in the assessment procedures and outcomes.

Confidence Score Prediction: Most existing research portrays review rating prediction as a multi-class classification/regression task [16]. They use supervised machine learning models with review texts to predict rating. Furthermore, because features are so important, most studies extract useful features such as context-level features [20] and user features [10] to improve the prediction performance. Feature engineering, on the other hand, is time-consuming and labour-intensive. Various deep learning-based models are being proposed to automatically determine text features [1] based on the development of neural networks and their wide application. However, the existing deep learning models learn continuous representations (like words, phrases, sentences, and documents)[6,11,17,19,24,25]. Although deep learning models can learn extensive feature representations automatically, they cannot efficiently capture the hierarchical relationship inherent to the review text.

Unlike other reviews (e.g., product reviews, movie reviews, restaurant reviews, etc.), peer reviews are hierarchical and usually longer; most of these models do not leverage up-to-date representation techniques such as BERT. The BERT [9] language representation is built based on the cutting-edge neural technology. NLP tools provide an extensive range of language features that can significantly improve the performance of NLP tasks. It is possible to perform downstream tasks directly with the models that return context-aware embeddings.

Moreover, the embedded algorithms can be tailored relatively inexpensively for any given task by adding small extensions to the core BERT architecture, allowing more significant gains in task performance, however, at the expense of feature interpretability. Finally, we provide BERT with hybrid BiLSTM and CNN architecture for the confidence score prediction task.

3 Methodology

Here, we describe the problem statement for reviewer confidence score prediction in the context of the proposed model, as well as the details of the proposed model.

3.1 Problem Statement

Quantitative analysis of peer reviews as an NLP/IR resource can help understand peer review texts and provide editors with valuable insights about review quality. If the reviewer has less confidence about her review, the area chair would want to pay less attention to such reviews while making the final decision. Based on the answers to the above question, we hypothesize that the quality of reviews can be improved by clearly bringing out the reviewers' understanding of the work, which help editors determine the reviewers, most familiar with the work.

The following task is defined in this context:

Suppose there are reviewer set U for paper P and the confidence score matrix $C \in C^{|U| \times |P|}$, where the entry C_{uP} indicates the confidence score of reviewer $u \in U$ towards paper P. For a reviewer u, the review written by u can be represented as $R = \{\mathrm{S}_1, \cdots, \mathrm{S}_\mathrm{n}\}$, for paper P. Where $\mathrm{S}_1, \cdots, \mathrm{S}_\mathrm{n} \rightarrow$ are the sentences in the review texts.

Based on the preceding considerations. We want to learn a predictive function that takes the review text as input and predicts the reviewer's confidence score based on the input.

$$f\left(\{R\}_{n=1}^{n}\right) \rightarrow C_{uP}$$

It can be used as a signal to measure the quality of the review and help program chairs to automatically estimate the review quality based on the reviewer's confidence score.

3.2 Framework

The architecture for the proposed confidence score prediction is shown in Fig. 1, we first divide the review into sentences and then pass it through BERT [9]. The BERT produces sentence embedding of 768 dimensions. Those 768 dimensions are then passed through the BiLSTM. After this, we stack all the vectors to apply CNN. Then the output of the CNN after max-pooling operation passed through a dense layer of 256 units, which is ultimately passed through a dropout layer to predict the reviewer's confidence score. We train the model using mean squared error Loss function.

1. **Review Encoding:** Review encoding was performed using BERT [9]. Sentences of the review are input, and a sentence embedding with dimensions $d = 768$ is the output. The final review representation is given by $R = S_1 \oplus S_2 \oplus S_3 \oplus \ldots \oplus S_n, R \in {}^{n \times d}$, where $\oplus$ is the concatenation operator and n is the maximum number of sentences.

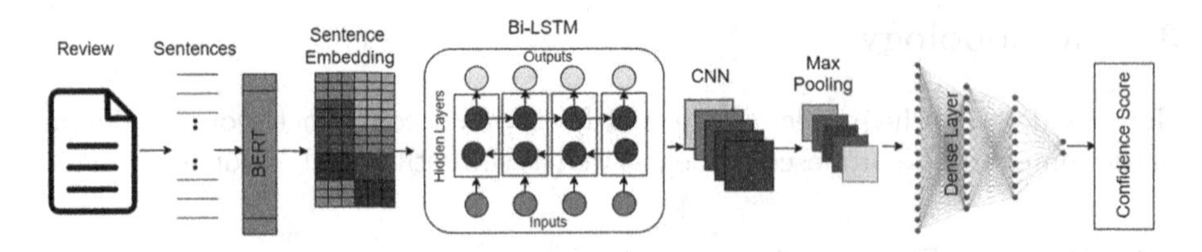

Fig. 1. Bidirectional Encoder Representations from Transformers (BERT) using hybrid bidirectional LSTM and CNN architecture for prediction of confidence score.

2. **Deep Learning-Based Review-Feature Extraction:** We use the hybrid of BiLSTM and CNN to extract features from the sentence embeddings. As shown in Fig. 1, the sentence embedding generated by BERT passed through the BiLSTM. Then, we stack all the vectors and apply convolution operation with K different sets of CNN filters $[f_1, f_2, \ldots, f_k]$ on $W_{f_k^t} \in R^{t \times d}$, where t is the sliding window length, The convolutional outputs of filter f_k^t over window h^{th} is given as.
$$y_h^{f_k^t} = g\left(W_{f_k^t} \odot R_{h:h+t-1} + b_{f_k^t}\right)$$
where $\odot$ is inner product, $R_{h:h+t-1}$ means h^{th} window of $(R_{t \times emb})$, where R_{emb} represents embedding dimension of review sentences generated by BERT. g() is the non-linear function, $b_{f_k^t}$ is the bias for the filter f_k^t. Afterwards, we do a max-pooling operation to obtain the most significant features. $\hat{f_k^t} = \max\left(y_1^{f_k^t}, y_2^{f_k^t}, y_3^{f_k^t}, \ldots\right)$ for a review R, And then, we finally obtain the review cnn representation i.e. $R_{cnn} = (\hat{f_k^t}, 1 \leq t \leq Z, and\ 1 \leq f \leq K)$. To some extent, it can capture the semantic features of the review text.
3. **Prediction Layer:** The obtained feature vectors from review is ultimately passed through a dense layer with 256 units. The final representation is:
$$X_{R_{\text{final}}} = f_{\text{MLP}}\left(\Theta_{\text{predict }[R_{cnn}]}\right)$$
Where $\Theta_{predict[R_{cnn}]}$ the parameters of the a multilayer perceptron (MLP). The following equation defines the prediction layer as:
$$\mathrm{C_{uP}} = \mathrm{ReLU}\left(w_d \cdot [X_{R_{\text{final}}}] + b_d\right)$$
4. **Training:** We define confidence score prediction error over training samples using the mean square error (MSE) loss function:
$$L(\theta) = \frac{1}{|\mathcal{D}|} \sum_{R \in D} \left(C_{uP} - \hat{C}_{uP}\right)^2$$
Where D denotes the set of instances for training, and $\hat{C}_{uP}$ is the normalized real confidence score assigned by the reviewer u to the paper P. Our model parameters are learned by optimizing the $L(\theta)$ loss. To optimize our model, we use the Adam optimizer [14]. Detailed information on the implementation is available in Sect. 4.1.

3.3 Baseline Models

We used three well-known machine learning and one state-of-the-art BERT baseline to compare the performance of our proposed model (Fig. 2).

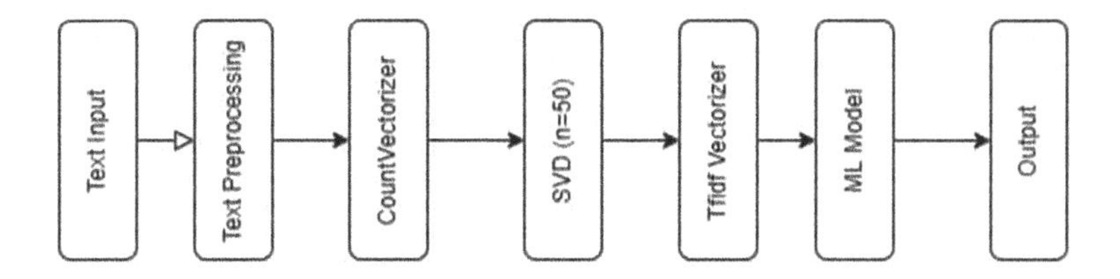

Fig. 2. Flowchart of machine learning models to predict the confidence score.

- **Traditional machine learning models.** First, basic regex is used to clean the input text. Next, stemming is performed using PorterStemmer. After that, the text corpus is fed to the CountVectorizer for feature extraction. The CountVectorizer produces a sparse matrix of the text corpus. Here, we have chosen the n_gram range to be 1,1, which means only unigrams are allowed. Next, the sparse matrix produced by the CountVectorizer is passed to the term frequency-inverse document frequency (TF-IDF) transformer, which is mainly used to get the TF-IDF vectors for information retrieval. Finally, a single value decomposition algorithm has been used to reduce the dimensionality of the vectors. This scales down the vectors to 50. Then the output from the SVD is passed through LR, DT and SVR. We provide the hyperparameter details of the machine learning models in our git repository.
- **Bidirectional Encoder Representations from Transformers (BERT)** We compared our proposed model with a transformer variant of BERT [9] for the contextual embedding of review sentences. We initially divided the review into sentences in our confidence score prediction task and passed them into the pre-trained BERT model. After fine-tuning the pre-trained BERT model, we added a fully connected layer with ReLU activation and a dropout rate of 0.3. Finally, the model was trained using mean squared error loss. And we keep the training testing ratio identical across the task for a fair comparison.

4 Data Description and Experimental Setup

A dataset of 11.2k reviews submitted to the ICLR The conference and their scores for 2018, 2019 and 2021 were curated using OpenReview Platform. The dataset includes confidence scores on a 5-point scale. With a value of 5, this review indicates that the reviewer understands the paper and reads it carefully. Value 1 indicates that the reviewer does not understand the paper because her expertise does not match the paper's topic. The distributions of scores of the datasets are shown in Fig. 3. The detailed year-wise statistics are given in Table 1. ICLR reviews are publicly available through the official OpenReview API[2].

Table 1. Analysis of data and statistics

Conference edition	Reviews	Avg length of reviews (in terms of words)	Avg length of reviews (in terms of sentences)	Testing sample taken (20%)
ICLR-2018	2967	365.12	22.62	593
ICLR-2019	4764	394.83	24.49	953
ICLR -2021	3290	436.49	26.26	658

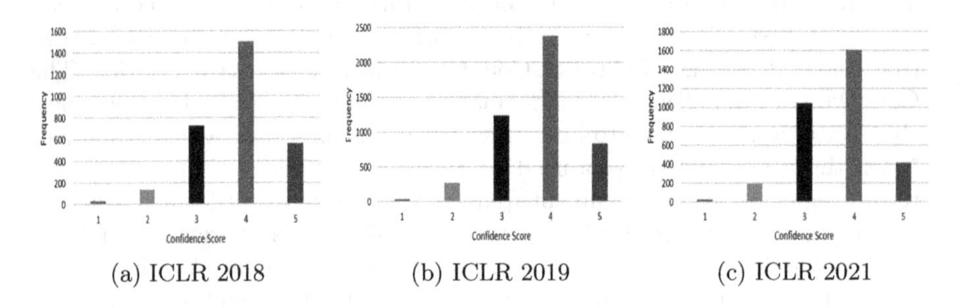

(a) ICLR 2018 (b) ICLR 2019 (c) ICLR 2021

Fig. 3. Score distributions of ICLR-2018, 2019 and 2021 peer reviews

In terms of our experimental setup, we utilize three popular criteria to evaluate the proposed model and compare it with the baselines. We use the following metrics to determine the efficacy of the models: root mean square error (RMSE), mean absolute error (MAE), and the coefficient of determination (R^2). MAE and RMSE measure the variation between the predicted and actual values. Therefore, lower values of MAE and RMSE are desirable, and worth of (R^2) is used to assess how well a regression model matches the observed data. The following equations illustrate RMSE, MAE, and (R^2).

$$RMSE = \sqrt{\frac{1}{n}\sum_{i=1}^{n}(y_i - x_i)^2}$$

$$MAE = \frac{1}{n}\sum_{i=1}^{n}|y_i - x_i|$$

[2] https://openreview.net/.

$$R^2 = 1 - \frac{\sum_{i=1}^{n} (y_i - x_i)^2}{\sum_{i=1}^{n} (y_i - \bar{x})^2}$$

Where: $y_i \Rightarrow$ predicted output, $x_i \Rightarrow$ expected output, n $\Rightarrow$ number of samples.

Table 2. The hyper-parameters details of proposed model.

	Parameter name	Value
Proposed model	Embedding dimension	768
	CNN (Kernel size, Number of filters)	8, 4, 2, 100
	LSTM (OutputUnit)	64
	Dense (Hidden units)	256
	Activation function	ReLU
	Dropout rate	0.3
	Output Layer (Type Hidden Units)	1
	Epochs	20
	Batch size	32
	Optimizer	Adam
	Loss Function	Mean Square Error
	Learning-Rate	$5e^{-5}$

4.1 Implementation Details

We use 80% of the data for training and 20% for evaluation in all experiments. While training, we oversampled to mitigate the imbalanced data distribution problem. Our model is built utilizing Keras on top of TensorFlow-2.4.1. We also use the Adam optimizer [14] to train our model, with batch size 32 and dropout set to 0.3 [21]. The hyper-parameters details for the proposed model are presented in Table 2. We make our codes available[3] to enable the readers to replicate our findings.

5 Experimental Results and Discussion

As shown in Fig. 3, data distribution is highly imbalanced; 50% of data in all the years have a confidence score of 4. Due to the imbalance data distribution, the machine learning models are highly biased on the training set, therefore, yields non-promising results for all the testing parameters. Moreover, the machine learning models do not consider the sense of the words, which is important in finding out the confidence score of the reviewer. The BERT model returns contextualized embeddings for sentences directly used for the mentioned task. As shown in the Table 4, the proposed model outperforms all the other comparing baselines.

[3] https://github.com/PrabhatkrBharti/Rev.Conf.git.

Table 3. Performance comparisons of linear regression, decision tree, SVR, BERT, and proposed model using different statistical criteria during the training stage.

	Model types	ICLR 2018			ICLR 2019			ICLR 2021		
		RMSE	MAE	R^2	RMSE	MAE	R^2	RMSE	MAE	R^2
Baselines	Linear Regression	0.059	0.028	0.961	0.039	0.002	0.997	4.073	3.041	0.989
	Decision Tree	0.001	0.043	0.019	0.912	0.209	0.83	0.012	0.009	0.998
	SVR	0.351	0.206	0.822	0.313	0.994	0.34	0.342	0.212	0.818
	BERT	0.581	0.439	0.392	0.607	0.467	0.407	0.585	0.427	0.373
Proposed model		0.397	0.318	0.751	0.369	0.284	0.798	0.359	0.286	0.784

5.1 Evaluation Criteria

The training and test performance results are presented in Table 3 and Table 4, respectively. It is visible from Table 4 that the proposed model have much less RMSE and MAE than the other machine learning models for all the years. As shown in Table 4, SVR still performs better than the other two machine learning models, but a gap remains between SVR and proposed model with a quiet margin 35%, 25%, 57% (ICLR 2018), 38%, 30%, 56% (ICLR 2019) and 34%, 39%, 55% (ICLR 2021) in term of RMSE, MAE, R^2, respectively.

Contrary to our proposed model, the BERT baseline does not perform well, as shown in Table 4. Moreover, the vectorization of the inputs using BERT yields almost similar results during testing and training. In addition, the hybrid bidirectional LSTM and CNN architecture demonstrates that it can guide the proposed model to make a good prediction for the mentioned task. As a final result, our proposed model outperforms all baseline models substantially for the ICLR datasets (2018, 2019, and 2021).

The proposed model has an R^2 score greater than 0.60 for all the years, whereas the maximum R^2 score for any machine learning model is 0.08. This suggests that the proposed model is well-suited with the observed confidence scores in order to predict its values.

Table 4. Performance comparisons of linear regression, decision tree, SVR, BERT, and proposed model using different statistical criteria during the testing stage.

	Model types	ICLR 2018			ICLR 2019			ICLR 2021		
		RMSE	MAE	R^2	RMSE	MAE	R^2	RMSE	MAE	R^2
Baselines	Linear Regression	0.944	0.739	−0.358	1.099	0.859	−0.708	1.034	0.828	−0.696
	Decision Tree	1.055	0.767	−0.696	1.116	0.812	−0.759	1.061	0.778	−0.786
	SVR	0.762	0.577	0.115	0.804	0.625	0.085	0.766	0.628	0.045
	BERT	0.591	0.451	0.362	0.617	0.497	0.374	0.605	0.437	0.369
Proposed model		**0.406**	**0.324**	**0.689**	**0.418**	**0.323**	**0.654**	**0.423**	**0.334**	**0.600**

Contrary to this, the R^2 values of the decision tree and linear regression are negative, indicating that they do not follow the trend. It is visible in the error values presented in Table 4. Furthermore, the SVR fails to deliver a high R^2 score indicating that it fails to adhere to the data trend.

Table 5. Results for cross-year experiments # ICLR means the proposed model is trained on ICLR dataset.

Proposed model	# ICLR 2018		# ICLR 2019		# ICLR 2021	
	ICLR 2019	ICLR 2021	ICLR 2018	ICLR 2021	ICLR 2018	ICLR 2019
RMSE	0.423	0.443	0.418	0.408	0.403	0.381
MAE	0.334	0.339	0.321	0.309	0.316	0.295
R^2	0.617	0.575	0.626	0.626	0.648	0.638

5.2 Cross-Year Experiments

To test the robustness of our proposed model. We perform cross-year experiments and evaluate the RMSE, MAE, and R^2 scores. The proposed model was trained on the ICLR 2018 dataset and tested on the ICLR 2019 and 2021 datasets, vice-versa. Table 5 shows that the proposed model performed equally well for cross-year experiments.

5.3 Ablation Study

To evaluate the effectiveness of our proposed framework. As shown in Table 6. We designed two variants, w/o BiLSTM (proposed model without BiLSTM) and w/o CNN (proposed model without CNN), to perform an ablation study and analyze the results of our framework using RMSE, MAE, R^2. If either BiLSTM or CNN is removed, we observe the drop in performance across all datasets. These results indicate that both BiLSTM and CNN-based approaches would efficiently guide the framework to make good predictions. In addition, combining these hybrid approaches (BiLSTM and CNN) further enhances performance, which shows that the two are complementary.

Table 6. Effects of the proposed model's internal structure.

Model types	ICLR 2018			ICLR 2019			ICLR 2021		
	RMSE	MAE	R^2	RMSE	MAE	R^2	RMSE	MAE	R^2
Proposed model	0.406	0.324	0.689	0.418	0.323	0.654	0.423	0.334	0.6
Proposed model w/o LSTM	0.536	0.385	0.451	0.515	0.414	0.418	0.495	0.367	0.416
Proposed model w/o CNN	0.576	0.439	0.395	0.565	0.457	0.382	0.562	0.41	0.386

5.4 Non Parametric (Levene's) Test

We also carried out Levene's test to assess the homogeneity of variance between the predicted output and required output. Since the dataset is heavily tailed towards confidence score 4 (see Fig. 3), we use the adjusted-mean (trimmed mean) version to calculate the p-value for the test. If the p-value between the two distributions is less than 0.05, the distributions are non-homogeneous. In Table 7, the output produced by the proposed model is homogeneous with the required output. The linear regression model and the SVR model have very low p-value (practically zero), but the decision tree and BERT have better values than these two models. This score is calculated using the scipy library[4]. It also proves the efficacy of our proposed model.

Table 7. Non Parametric (Levene's test) for linear regression, decision tree, SVR, BERT and proposed model

Model types	P-value for Levene's test for Year wise		
	ICLR 2018	ICLR 2019	ICLR 2021
Linear Regression	1.19 × 10–66	1.92 × 10–125	1.55 × 10–141
SVR	1.42 × 10–66	8.432 × 10–108	1.06 × 10–110
Decision Tree	0.0084	0.017	0.044
BERT	0.0612	0.180	0.163
Proposed model	0.0945	0.264	0.278

5.5 Observations

It is also observed in Table 3 and Table 4 that all the machine learning models are highly biased on the training set, therefore, yields non-promising results while testing. The vectorization of the inputs using BERT has almost similar testing and training results. Table 3 shows that the linear regression and decision tree outperforms the other two models by a margin. The skewness of the data also contributes to this problem since most of the time, the models predict the value 4 and thus get better performance for the training set. On the other hand, due to its vectorization and gradual learning, the proposed model got a better score than the rest and performed better in the testing data, resulting from optimal training. Table 4 clearly shows that the machine learning models perform poorly on the testing set. But the SVR model performs better than the other two machine learning models because it considers extreme cases, reducing error risk, while the linear regression and decision tree work on clear boundary algorithms.

[4] https://docs.scipy.org/doc/scipy/reference/generated/scipy.stats.levene.html.

5.6 Error Analysis

As a result of analyzing the error cases, we could determine very clearly where our model fails.

Imbalance Data Distribution: The proposed model has performed well for all the test sets. But it was seen that the proposed model could not predict a perfect confidence score for the reviews with a confidence score of 1. The possible reason could be the very low availability of reviews with a confidence score of 1 (see Fig. 3).

Table 8. Misclassified actual labels

Confidence Score	ICLR 2018			ICLR 2019			ICLR 2021		
	No. of wrong predictions	Total predictions	% of wrong predictions	No. of wrong predictions	Total predictions	% of wrong predictions	No. of wrong predictions	Total predictions	% of wrong predictions
1	34	34	100%	40	40	100%	26	26	100%
2	43	137	31%	118	275	42%	60	199	30%
3	153	728	21%	383	1239	31%	338	1042	33%
4	224	1507	15%	352	2378	13%	303	1610	19%
5	187	561	23%	227	832	27%	75	413	18%

Misclassified Actual Labels: We train the proposed model using the standard training validation approach. Using the proposed model, we predict the output for all the data points in the dataset. We remove all the reviews whose confidence scores are in the range of 0.5 of the expected output. Using these data points, we analyze the error. On analyzing the data, we found that the proposed model cannot correctly predict the confidence score for reviews with confidence score 1. The reason is the unavailability of data, and the reviews are very short. The reviews with a confidence score of 1 only contribute to less than 1% of the entire dataset (Table 8).

6 Conclusion and Future Work

Peer review quality evaluation is a critical problem for scientific health, adding influence to the gatekeeper of scientific knowledge. Many studies and experiments have been conducted to improve the peer review system and enhance the quality of peer reviews, focusing on improving the quality of peer reviews. However, none of the previous research focused on the automatic evaluation of peer review quality towards gauging reviewer knowledge from peer-review texts. In this research, we aim to explore this paradigm. And, we proposed a hybrid BiLSTM and CNN architecture grounded on BERT baseline that leverages review texts to predict the reviewer's confidence score. Statistical testing of the proposed model has consistently shown that it outperforms the baselines by a wide margin. Thus, an experimental dataset has the potential to pose as a benchmark experimental testbed to leverage current NLP techniques to address peer-review quality, leading to more transparency and trust in scientific research validation. In the future, we intend to explore how we can broaden the scope of our work by modeling the linguistic properties of the review content as they frame uncertainty and conviction.

Acknowledgments. The first author, Prabhat Kumar Bharti, acknowledges Quality Improvement Programme, an initiative of All India Council for Technical Education (AICTE), Government of India, for fellowship support. Tirthankar Ghosal is funded by Cactus Communications, India (Award # CAC-2021-01) to carry out this research. The fourth author Asif Ekbal receives the Visvesvaraya Young Faculty Award. Thanks to the Digital India Corporation, Ministry of Electronics and Information Technology, Government of India for funding this research.

Ethics Compliance. In no way does this work attempt to replace human reviewers; instead, a good use case of this research is to focus and enhance the quality of the review process is to leverage Human AI collaboration to predict the confidence scores of the reviewers. It will eventually help the editors with review quality.

References

1. Bengio, Y., Courville, A., Vincent, P.: Representation learning: a review and new perspectives. IEEE Trans. Pattern Anal. Mach. Intell. **35**(8), 1798–1828 (2013)
2. Bornmann, L., Daniel, H.D.: Reliability of reviewers' ratings when using public peer review: a case study. Learned Publishing **23**(2), 124–131 (2010)
3. Bornmann, L., Mutz, R.: Growth rates of modern science: a bibliometric analysis based on the number of publications and cited references. J. Am. Soc. Inf. Sci. **66**(11), 2215–2222 (2015)
4. Brezis, E.S., Birukou, A.: Arbitrariness in the peer review process. Scientometrics **123**(1), 393–411 (2020). https://doi.org/10.1007/s11192-020-03348-1
5. Chavalarias, D., Ioannidis, J.P.: Science mapping analysis characterizes 235 biases in biomedical research. J. Clin. Epidemiol. **63**(11), 1205–1215 (2010)
6. Conneau, A., Schwenk, H., Barrault, L., Lecun, Y.: Very deep convolutional networks for text classification. arXiv preprint arXiv:1606.01781 (2016)
7. Cortes, C., Lawrence, N.D.: Inconsistency in conference peer review: revisiting the 2014 neurips experiment. arXiv preprint arXiv:2109.09774 (2021)
8. De Bellis, N.: Bibliometrics and Citation Analysis: From the Science Citation Index to Cybermetrics. Scarecrow Press (2009)
9. Devlin, J., Chang, M.W., Lee, K., Toutanova, K.: BERT: pre-training of deep bidirectional transformers for language understanding. arXiv preprint arXiv:1810.04805 (2018)
10. Gao, W., Yoshinaga, N., Kaji, N., Kitsuregawa, M.: Modeling user leniency and product popularity for sentiment classification. In: Proceedings of the Sixth International Joint Conference on Natural Language Processing, pp. 1107–1111 (2013)
11. Huang, B., Carley, K.M.: Parameterized convolutional neural networks for aspect level sentiment classification. arXiv preprint arXiv:1909.06276 (2019)
12. Huisman, J., Smits, J.: Duration and quality of the peer review process: the author's perspective. Scientometrics **113**(1), 633–650 (2017)

13. Kelly, J., Sadeghieh, T., Adeli, K.: Peer review in scientific publications: benefits, critiques, & a survival guide. Ejifcc **25**(3), 227 (2014)
14. Kingma, D.P., Ba, J.: Adam: a method for stochastic optimization. arXiv preprint arXiv:1412.6980 (2014)
15. Kronick, D.A.: Peer review in 18th-century scientific journalism. JAMA **263**(10), 1321–1322 (1990)
16. Pang, B., Lee, L.: Seeing stars: exploiting class relationships for sentiment categorization with respect to rating scales. arXiv preprint cs/0506075 (2005)
17. Pennington, J., Socher, R., Manning, C.D.: Glove: global vectors for word representation. In: Proceedings of the 2014 Conference on Empirical Methods in Natural Language Processing (EMNLP), pp. 1532–1543 (2014)
18. Price, S., Flach, P.A.: Computational support for academic peer review: a perspective from artificial intelligence. Commun. ACM **60**(3), 70–79 (2017)
19. Qiao, C., et al.: A new method of region embedding for text classification. In: ICLR (Poster) (2018)
20. Qu, L., Ifrim, G., Weikum, G.: The bag-of-opinions method for review rating prediction from sparse text patterns. In: Proceedings of the 23rd International Conference on Computational Linguistics (Coling 2010), pp. 913–921 (2010)
21. Srivastava, N., Hinton, G., Krizhevsky, A., Sutskever, I., Salakhutdinov, R.: Dropout: a simple way to prevent neural networks from overfitting. J. Mach. Learn. Res. **15**(1), 1929–1958 (2014)
22. Tabah, A.N.: Literature dynamics: studies on growth, diffusion, and epidemics. Ann. Rev. Inf. Sci. Technol. (ARIST) **34**, 249–86 (1999)
23. Tomkins, A., Zhang, M., Heavlin, W.D.: Reviewer bias in single-versus double-blind peer review. Proc. Natl. Acad. Sci. **114**(48), 12708–12713 (2017)
24. Wang, B.: Disconnected recurrent neural networks for text categorization. In: Proceedings of the 56th Annual Meeting of the Association for Computational Linguistics (Volume 1: Long Papers), pp. 2311–2320 (2018)
25. Zhang, Y., Wallace, B.: A sensitivity analysis of (and practitioners' guide to) convolutional neural networks for sentence classification. arXiv preprint arXiv:1510.03820 (2015)

Historical Document Analysis + CSAWA

Recognition and Information Extraction in Historical Handwritten Tables: Toward Understanding Early 20th Century Paris Census

Thomas Constum[1](✉), Nicolas Kempf[1], Thierry Paquet[1], Pierrick Tranouez[1], Clément Chatelain[2], Sandra Brée[3], and François Merveille[4]

[1] LITIS EA4108, University of Rouen Normandy, Rouen, France
{thomas.constum1,nicolas.kempf,thierry.paquet, pierrick.tranouez}@univ-rouen.fr

[2] LITIS EA4108, INSA Rouen Normandy, Saint-Étienne-du-Rouvray, France
clement.chatelain@insa-rouen.fr

[3] LARHRA, UMR 5190, CNRS, Lyon, France
sandra.bree@msh-lse.fr

[4] Campus Condorcet - The Grand Équipement Documentaire, Aubervilliers, France
francois.merveille@campus-condorcet.fr

Abstract. We aim to build a vast database (up to 9 million individuals) from the handwritten tabular nominal census of Paris of 1926, 1931 and 1936, each composed of about 100,000 handwritten simple pages in a tabular format. We created a complete pipeline that goes from the scan of double pages to text prediction while minimizing the need for segmentation labels. We describe how weighted finite state transducers, writer specialization and self-training further improved our results. We also introduce through this communication two annotated datasets for handwriting recognition that are now publicly available, and an open-source toolkit to apply WFST on CTC lattices.

Keywords: Handwriting recognition · Document layout analysis · Self-training · Table analysis · WFST · Semi-supervised learning

1 Introduction

In the digital age, many handwritten corpora containing very interesting information for historians still remain unexploited because it would be too costly and time-consuming to analyze them by hand entirely. The POPP project (Project for the OCRing of the Paris Population census) aims to build a vast database (12 million individuals) from the handwritten tabular nominal census of Paris of the years 1926, 1931, 1936 and 1946, each composed of about 100,000 handwritten

Project supported by CollEx-Persée (AAP19_20), with the financial collaboration of the TGIR Progedo and the Grand Équipement Documentaire Campus Condorcet.

© Springer Nature Switzerland AG 2022
S. Uchida et al. (Eds.): DAS 2022, LNCS 13237, pp. 143–157, 2022.
https://doi.org/10.1007/978-3-031-06555-2_10

simple pages in a tabular format (Fig. 3a shows a double page). In this paper, the first three census are considered since they have the same table structure.

A similar table structure was used for the population census at the national level in France at the same epoch. This material is a primary source of information for historian demographers, urban historians and more generally researchers in the humanities and social sciences. This fixed structure that spans over the entire corpus facilitates the recognition process of handwritten information. Indeed, every cell contains an expected specific type of information (a name, a date, an address, etc.) that can be modeled thanks to a dictionary or a regular expression that is used to drive the recognition process based on beam search Viterbi alignment [5,14,16].

The core of the processing chain is the handwriting recognition module. Even if significant progress has been made these last years thanks to the use of deep neural network architectures, performance can degrade quickly on heterogeneous data, or heterogeneous writing styles, showing generalization difficulties of such architectures. The size and thoroughness of the training dataset have proved to be a major factor in getting high performance recognition systems [10], thus the interest to generate additional synthetic or augmented data for training the network [22].

Less explored in the handwriting recognition literature, semi-supervised learning offers an interesting framework to exploit large amounts of un-annotated datasets during training. Considering the large amount of data available to us with the three handwritten Paris census, this is a path we decided to explore, more precisely self-training. Evaluation results show a significant performance increase with this framework, reaching nearly mono-writer performance. We analyze the performance in relation to the size of the network architecture.

In this communication, we present the main components of the whole processing chain of the handwritten tabular nominal census and the recognition performance obtained on a subset of the corpus that was manually annotated for training as well as evaluation purposes. The main components are based on deep neural networks for table detection, table line detection and line recognition, whereas syntax-driven handwritten field recognition uses Weighted Finite State Transducers (WFST). Finally, a set of coherency rules that are derived from the fixed structure of the census tables is exploited to fill in missing fields as much as possible. In the end, the extraction results exported through CSV files are made available to the research community of historian demographers. Some components of the processing chain are made freely available to the research community, notably the Kaldi-based WFST decoder[1] that has been re-factorized for easy integration in a Python programming framework. We also provide free access to the POPP annotated handwritten corpus to the Document Image Analysis research community.[2] This paper is organized as follows.

Section 2 is devoted to the presentation of the corpus. Section 3 provides an overview of the processing pipeline. Section 4 is devoted to the pre-processing

[1] https://gitlab.com/projet-popp/sigra/.
[2] https://github.com/Shulk97/POPP-datasets/.

stage, including table detection, page classification, table dewarping and table line detection. Section 5 is devoted to the handwriting recognition module presentation, including the deep NN optical model, self-training and the presentation of the recognition results. Section 6 is devoted to domain knowledge leveraging and content verification, including syntactical and lexicon-based language models, decoding, normalization of the recognized fields and logical deduction rules. Section 7 provides some information regarding processing time all along the pipeline.

2 Corpus and Ground-Truthed Datasets

2.1 Presentation of the Census

Three census of the population of Paris are considered in this paper: 1926, 1931 and 1936. Each census is divided into 20 boroughs *(arrondissements* in French), themselves divided into 4 districts each (80 districts in total). Each census consists of approximately 100,000 pages and 3 million lines, and we estimate the number of writers for one census to be between 80 and 500. The raw dataset stored by the Archives of Paris consists in a set of double pages (see Fig. 3a) made of every pages of the census booklets, including the front and back covers. The pages that are interesting for us are the pages containing tables filled with information that describes each individual. As depicted on Fig. 3b, these tables contain around 30 rows and each row describes one individual. The tables contain 15 columns. The first 5 columns are actually filled with only 3 types of information: street names, street numbers and household numbers. The 10 other columns contain information about each individual, such as name, place of birth and occupation. In this communication, we focus our attention on these 10 columns.

The challenging aspect of this dataset lies in the handwritten nature of the table information. They have been written by the multiple enumerators that were involved in gathering the information when visiting each home. Because of the handwritten aspect of the data, the table layout is often not respected. Sometimes, words are written across two columns, and in some cases, some words are written between rows. Figure 1 shows an overview of a page and the challenges that can arise with this dataset. Moreover, since the tables were filled by many writers, some columns were not filled out in the same way. For example, the column *place of birth* is sometimes filled out with the city and the *département* of birth (French administrative division), with the *département* only, or with the country of birth only.

2.2 Annotation of Two Datasets

In order to tune the text recognition model, a first training dataset was necessary. Although the table structure is extremely stable for every page of the census, there is a large variability in the writing styles, background color, and ink type.

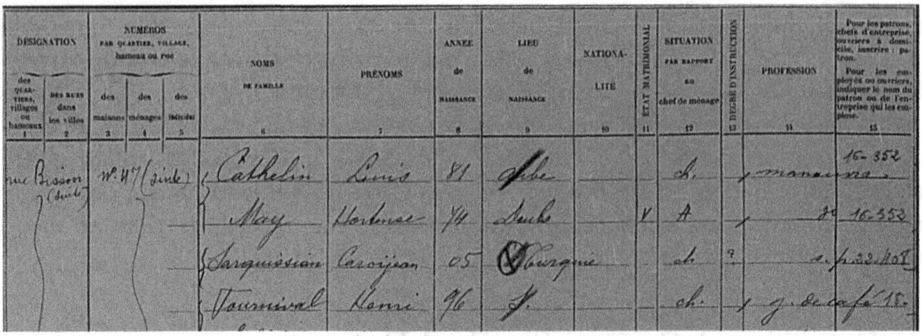

Fig. 1. Headers and first rows of a table from 1926. The corresponding annotation of the first row is "Cathelain/Louis/81/Aube/¤/¤/ch./¤/manoeuvre/?16-352".

Therefore, we annotated one double page for each of the 80 districts of the census of 1926, so as to create a generic dataset as representative as possible. Since this dataset already contains a significant diversity of writing styles, there is no need to create year-specific annotated datasets. Indeed, these three census have been scanned under the same conditions, the printed table template is the same and there is no year-specific degradation. This first dataset thus contains 160 pages made of 4800 handwritten lines, split into training (80%), validation (10%) and test (10%) sets. The split has been conducted at the double page level so that lines from a given district are located in only one subset. This dataset was manually transcribed line by line using the specific character "/" as a *logical* column separator and the specific character "¤" to indicate empty columns. To describe words written between lines, we defined the symbol "!" and "?" indicating a word respectively written below and above the regular line. These out of line words are not used yet, but are annotated for future use. Figure 1 presents an example of annotation.

Moreover, to conduct some mono-writer experiments, we annotated another dataset consisting of 49 pages (1470 lines) from a single district named *Belleville* located in the 20th *arrondissement* and written by a single writer[3]. Among these 49 pages, 39 pages were used for training, 5 pages were used for validation and the last 5 pages were used for testing.

We make available to the research community the images and labels of these two datasets[4].

3 Processing Pipeline

The processing pipeline is depicted on Fig. 2, below. It is composed of four main stages: pre-processing, handwriting recognition, domain knowledge integration

[3] The complete Belleville census was written by three writers but their writing style are very similar and can be therefore considered as one unique writing style.

[4] https://github.com/Shulk97/POPP-datasets/.

and content verification. The pre-processing stage is devoted to table detection, image dewarping and table row detection. The second stage is devoted to handwriting recognition applied on the detected rows. Language models are introduced during the third stage to constrain the handwriting recognition stage to allow content extraction and normalization using Weighted Finite State Transducers (WFST). We also apply a rejection rule based on grammars defined for each column. Finally, the stage of content verification introduces a set of coherency rules that are derived from the fixed structure of the census tables to fill in some missing fields as much as possible. In the end, the extraction results exported through CSV files are made available to the research community of historian demographers.

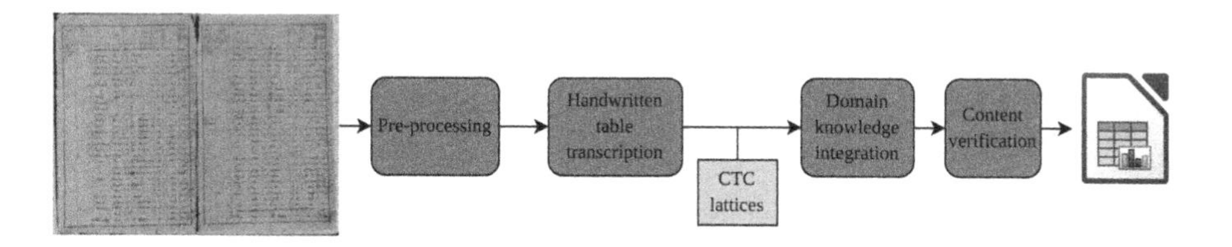

Fig. 2. Organization of the pipeline.

4 Layout Analysis and Information Extraction

4.1 Segmentation and Dewarping of Tables

The first step in the processing chain is to detect the two tables in the double page scans. Deep Convolutional Neural Networks (CNN) have proved to be the best pixel-wise predictors for this task [15, 19]. The *dhSegment* approach [15] was chosen since it obtains competitive results on Pagenet [25] and its source code is open source[5]. First, we created a dataset made of 80 double pages annotated for page segmentation. The network was trained using the same parameters as in [15] with 90% of the dataset and obtained a mean intersection over union (mIoU) of 0.987 on the remaining 10% of the dataset. We then used the newly trained model to generate new labels and we also annotated a few extra images manually that were found difficult for the model. This new labelled dataset is composed of 223 double pages for training and 40 pages for testing. We finally trained the model on this new dataset and obtained a mIoU of 0.9924 on the test set. Figure 3b shows the segmentation result obtained from the image in Fig. 3a.

After segmenting the tables, the last step is to use the corners of the tables to dewarp them by applying an affine transformation. The final result of this step is a table image with a null inclination angle. Figure 3c shows the two dewarped tables obtained from the segmentation result of Fig. 3b. Once dewarped, we can

[5] https://github.com/dhlab-epfl/dhSegment.

easily crop the irrelevant parts of the image, namely the headers and the 5 first columns. The structure of these columns is indeed not respected as can be seen in Fig. 1 and they contain address information only.

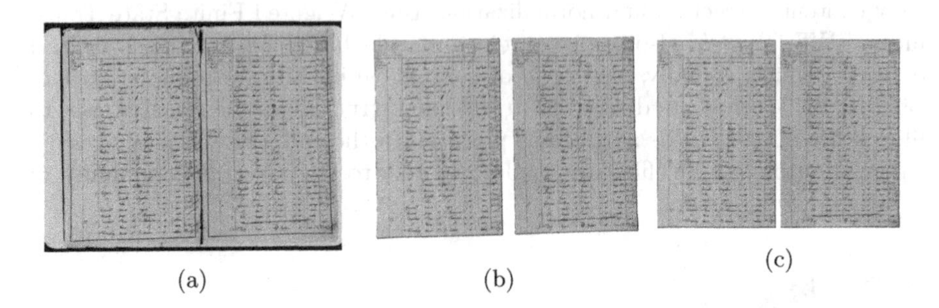

Fig. 3. (a) Example of a double page from the 1926 census. (b) Segmentation results. (c) Dewarped tables.

4.2 Page Classification

As mentioned in Subsect. 2.1, we need to detect the pages with tables and discard the other types of pages. Given the amount of data, we trained a classifier to automatically categorize pages as relevant, irrelevant and badly segmented. For this simple classification task we used the pre-trained MobileNetV2 architecture [20] with a width multiplier of 1. We bootstrapped training the system with 200 pages which were annotated by hand. Then, we generated a larger training dataset by generating predictions on unlabeled pages that were then verified manually to obtain a dataset of 1000 annotated pages with 80% of data for training, 10% for validation and 10% for testing. The network was then trained again and we obtained an accuracy of 97.8% on the validation set and 98.5% on the test set. In order to evaluate the performance of the segmentation model, we segmented the 100000 pages of the 1926 census and then classified each one with the classifier. By counting the number of pages classified as badly segmented by the classifier, we obtained an evaluated ratio of 98% of correctly segmented pages.

4.3 Segmentation of Tables into Rows

In order to segment our tables into rows, we first used baseline detection using the ARU-Net [12] pre-trained on cBAD. We chose ARU-Net because it has competitive results on cBAD and has the advantage of having the source code and pre-trained weights publicly available. Thanks to the detected baselines, we were then able to reconstruct the table rows by regrouping baselines with similar vertical location. Then, we used this first method to quickly annotate data for segmentation. This time, we trained a dhSegment architecture to segment the baselines of the rows by using 260 annotated pages for training and 32 pages for validation and obtained a mIoU of 0.80 on validation. Though this result could be improved, it is enough to localize the rows by taking the median vertical position of each detected baseline. Finally, to obtain the segmented lines, we simply have to deduce the bounding box from the baseline, knowing that the height of the lines is constant.

Most of the time, the logical separation in columns follows the separation materialized by the printed lines of the table template. However it is not the case when words are overlapping columns. That is why we introduced in the ground truth the column separator '/' (Subsect. 2.2) so that the model learns to predict the logical separation into columns beside predicting the handwritten text. Thanks to this column separator symbol, we avoid the need to segment the rows of the table into columns.

5 Handwriting Recognition

5.1 Architecture of the Optical Model

State of the art neural network architectures for handwritten text recognition are typically made of convolutional layers followed by recurrent layers, either MDLSTM [27] or BLSTM [13,17]. However, some recent studies showed very competitive results using fully convolutional networks [7–9,30,31]. This kind of architecture has the advantage of being much faster to train than recurrent architectures because it can take full advantage of the parallelization capabilities offered by GPUs. Moreover, several recent publications report on segmentation free approaches [3,4,6,7,21,30] which do not introduce any explicit line segmentation stage. These network architectures have the ability to be trained at paragraph level, and learn to detect and recognize text lines at the same time, thus avoiding the need for any segmentation ground truth.

In this case, line segmentation is not a difficult step because once the table is dewarped, the lines of text are straight, without slope, parallel and equidistant. Thus, we use in this work a line recognition model. Indeed, paragraph text recognition models can have difficulties to converge especially when we try to modify the width or the depth of the model as we did in Subsect. 5.3. For our experiments we chose to use the line text architecture described in [7] because of its recurrent-free aspect and also because the source code and training weights on reference datasets are publicly available on Github[6]. This architecture is modular since the encoder of the recognition part can be directly connected to an attention module to perform handwriting recognition on paragraphs.

5.2 Results of the Optical Model

We first trained our model on the generic dataset and obtained a CER of 7.08% and a WER of 19.05% on the test set. This first result is correct, although the

[6] https://github.com/FactoDeepLearning/VerticalAttentionOCR.

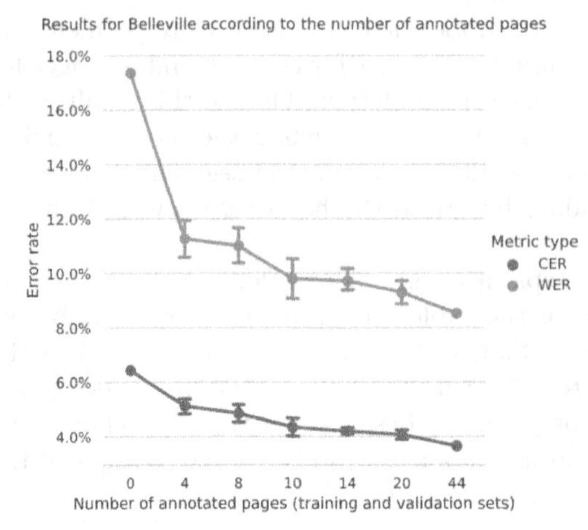

# pages (lines) training and validation	Number of trainings performed
0 (0)	no fine-tuning
4 (120)	5
8 (240)	5
10 (300)	4
14 (420)	3
20 (600)	2
44 (1320)	1

Fig. 4. Left: Recognition results on the *Belleville* dataset for different numbers of annotated data. Right: Number of trainings performed for each experiment.

error rate is higher than the state of the art results on datasets like IAM or RIMES with respectively 2.3% [17] and 4.87% [13] of CER on the test set. This indicates that the difficulty represented by this dataset constitutes an interesting challenge for the Document Image Analysis research community.

Regarding mono-writer recognition, we first evaluated the model trained with the generic dataset on the *Belleville* dataset and obtained 6.42% of CER. Then, by specializing the model on the whole *Belleville* dataset, we improve the CER to 3.65%, i.e. 43% of relative improvement, which shows the great improvements that can be expected from writer specialization. Then, to evaluate the gain with respect to the number of annotated pages, we also tuned the system using only a fraction of the training and validation sets while testing on the complete test dataset. In order to have more reliable results, we performed cross-validation. For example, to evaluate the performance obtained using 13 pages for training and 1 page for validation, we performed 3 trainings using each time randomly chosen pages that had not been used before. The results are summarized in Fig. 4. With only 4 annotated pages (3 for training and 1 for validation), we reach an average CER of 5.11%. Then, the more we annotate data of a target dataset, the less improvement we get by new annotated lines which confirms the results obtained in [2]. The next challenge to address will be to minimize the number of new lines needed to benefit from writer specialization.

5.3 Self-training

Self-training is a technique related to semi-supervised learning [26] that consists in using a model trained on labelled data, called a teacher, to generate pseudo-labels that are then used for training another model, called a student. This kind of technique is especially useful when a very large amount of unlabelled data is available, which is the case in our project. Self-training has been well explored in the computer vision field [28,29,32], but few publications have investigated self-training for handwriting recognition [23].

In this study, we followed the Noisy Student Training scheme described in [28] which injects noise during the student training. Noise injection makes the student model better than the teacher because its task is harder: it has to reproduce the prediction of the teacher model while having noise applied on the input (using data augmentation) and on the model (using dropout). The student can then become a teacher for another iteration of the process. Student networks can have exactly the same architecture as their teacher, or they can have a wider and deeper architecture, possibly with some recurrent layers or other components. This training process can be repeated until there is no further improvement.

The unlabelled dataset consists in this case of 2.4 million line images selected randomly from the 1926 census. First we trained the initial model (model 0) in a supervised mode on the generic dataset, then we entered the self-training mode for 4 iterations (more iterations did not provide any improvement). Student 1 and 2 have a similar architecture as model 0, while architecture B of student 3 was 1.5 deeper and 1.25 wider, following the scaling method described in [24]. Indeed, a student 2*bis* with the initial architecture trained on the predictions of student 2 showed nearly no improvements on the validation set and regressed on the test set. Thus, this model was not used for the next iterations and we used student 3 instead. This new architecture allowed us to further improve the results. This result is particularly interesting because such an architecture is not able to converge using supervised learning on the generic dataset. Finally, we experimented the influence that a BLSTM layer could have on the results. Indeed, this type of layer is most efficient when a large volume of training data is available. Moreover, it is shown in the literature that LSTM layers are very much suited to model contextual information when on top of convolutional layers. Thus, we trained a student 4 using the predictions of student 3. This final model has an architecture C made of the encoder of architecture B and a BLSTM-based decoder. We illustrate this architecture in Fig. 5. This final architecture brought an improvement of almost 1% of CER on test set compared to student 3, which demonstrates that a BLSTM-based decoder can still be useful compared to a fully convolutional decoder when a large amount of data is available. By using self-training, we have thus improved the CER from 7.08% to 4.52%, which

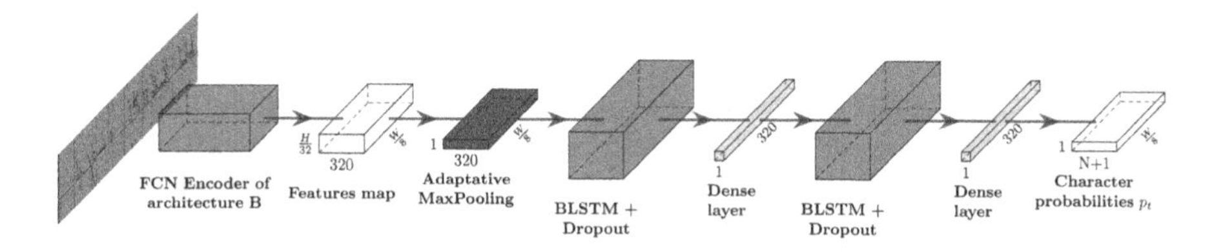

Fig. 5. Schema of architecture C.

represents 36% of relative improvement compared to supervised learning. The Table 1 summarizes the results obtained at each experiment.

We also evaluated this self-training process on a single writer dataset. We used the weights of the best model described above to initialize training on the Belleville dataset and then performed inference on the complete unlabelled Belleville district data to generate pseudo-labels. Then we re-trained the model on these pseudo-labels and obtained a CER of 2.66% and a WER of 6.37% on the test set. Compared to the results on the Belleville dataset using supervised learning (CER of 3.65% and a WER of 8.53%), this represents a relative improvement of 30% on the CER. This result indicates that self-learning can improve recognition performance even when the results are already good. Moreover, it shows that noisy student training can also be beneficial when combined with writer specialization.

Table 1. Recognition results of the self-training experiments

Model	Dataset	CER (%) validation	WER (%) validation	CER (%) test	WER (%) test
Initial (Model 0)	Generic	6.86	18.66	7.08	19.05
Student 1	Generic	6.07	17.12	6.12	17.12
Student 2	Generic	5.94	16.80	5.97	16.83
Student 2*bis*	Generic	5.89	16.68	6.02	16.89
Student 3	Generic	5.64	15.98	5.43	15.50
Student 4	Generic	**5.01**	**14.53**	**4.52**	**13.57**
Student 4 specialized	Mono-writer	**2.14**	**5.24**	**2.66**	**6.37**

6 Leveraging Domain Knowledge

6.1 Language Models

Each column, once located at some beginning and ending positions in the lattice thanks to the recognition of the column separator symbol ('/') is thereafter decoded using the beam search Viterbi algorithm [5,14] under the constraints of the expected content in this column. As the census tables share the same structure over years, it is possible to model the expected content of each column either with lists of words (in the case of family names, surnames, marital status, position in a household, etc.) or with regular expressions (in the case of date of birth, place of birth, occupation, etc.). We created the dictionaries and formal grammars, using different resources. For example, a list of names and surnames was built using the French National Statistics Institute (INSEE) deceased database of people in France since 1970[7], historical researches [11] and official documents[8]

[7] Deceased people database since 1970 (INSEE, in French): https://www.insee.fr/fr/information/4190491.

[8] http://www.toponymiefrancophone.org/divfranco/Bougainville/Liste_generale.aspx?nom=liste_pays.

to generate French town names and administrative *département* names. Some country names and foreign towns names were also included as much as possible. Moreover, when necessary, we have created lists of abbreviations used by census agents to compose location names, or occupations. For example, the names of *départements* such as *"Hautes-Pyrénées"* or *"Hautes-Charentes"* can be abbreviated as *"Hte Pyrénées"* or *"H. Charentes"*. Using Finite State Transducers (FST), we combined lists of words and abbreviations in regular expressions to detect every combinations of these elements, and constrain the recognition process of each individual column.

Weighted Finite State Transducers have been proposed for speech recognition [14,16], and are similarly useful for HTR/OCR. However, the freely available libraries that implement WFST and their integration into a recognition engine are not very well documented while they require many steps. Thus, we developed a framework called SIGRA (SImple ctc GRAmmar toolkit) that regroups two modules:

- "py-ctc-wfst-composer": a WFST generator for decoding CTC probability lattices
- "py-ctc-wfst-decoder": a module that allows to use a WFST to decode CTC lattices

Based on three existing open source components, Thrax[18], OpenFST[1] and Kaldi[16], this framework is user-friendly, with a Python API that implements the required pipeline in RAM instead of writing multiple temporary files on disk, which is time-consuming. This framework is open source and available on Gitlab[9]. Each grammar is thus defined and compiled into a Weighted Finite State Transducer using Thrax and exported in the OpenFST format using the "py-ctc-wfst-composer" module. Then, the obtained WFSTs are decoded by the "py-ctc-wfst-decoder" module that uses Kaldi to explore the optical lattice, and find the best recognition path over the lattice.

Finally, by introducing the grammar decoding stage, multiple positive aspects are expected: 1- should the optical model misrecognize a character, lattice decoding may recover the correct character 2- simple but effective *rejection rules* can be implemented when too much discrepancy is detected between the optical best path and the grammar best path. The rejection rules are based on a threshold on the Levenshtein distance between the output string of the grammar and the output string of the optical model. Each threshold has been defined individually for each column. 3- abbreviations can be detected and normalized afterwards to a unique form.

Family names are not effectively rejected by the rejection rule because the list of surnames is huge and a family name with a misrecognized character can often be another valid family name. However, this is not a serious issue because it does not prevent to find individuals with close surnames and thus to remove the ambiguity on the individuals on the basis of other information such as year of birth, place of birth, occupation, marital status etc. Also notice that the

[9] https://gitlab.com/projet-popp/sigra/.

recognition hypothesis of the rejected fields are often close to the results even if rejected. This is why they are still stored in the final CSV exported files but indicated as rejected by using the symbol $ as an opening and closing tag, so that they can be used later on by historians, even if the result was not validated by the system. To illustrate the benefit of using grammars, we have introduced four metrics as follows:

- *Field Rejection Rate (FRR)*: ratio of rejected fields over the total number of fields
- *Accepted Field Error Rate (AFER)*: ratio of incorrectly recognized fields among the accepted fields over the total number of accepted fields.
- *OCR-Field Error Rate (OCR FER)*: percentage of erroneous fields at the output of the OCR (no rejection applied).
- *Grammar Field Error Rate (Grammar FER)*: percentage of erroneous fields provided by the grammar (no rejection applied).

Figure 6 shows the performance on test set of the generic dataset regarding these metrics. The rejection strategy is overall effective because the AFER is lower than the OCR-FER and the Grammar-FER. However, the results are very different depending on the column. The lower error rates are obtained on columns with restricted lexicons such as “marital status”, “birth year” or “nationality” whereas higher error rates are obtained on columns with complex or infinite lexicons such as “names” or “occupation”.

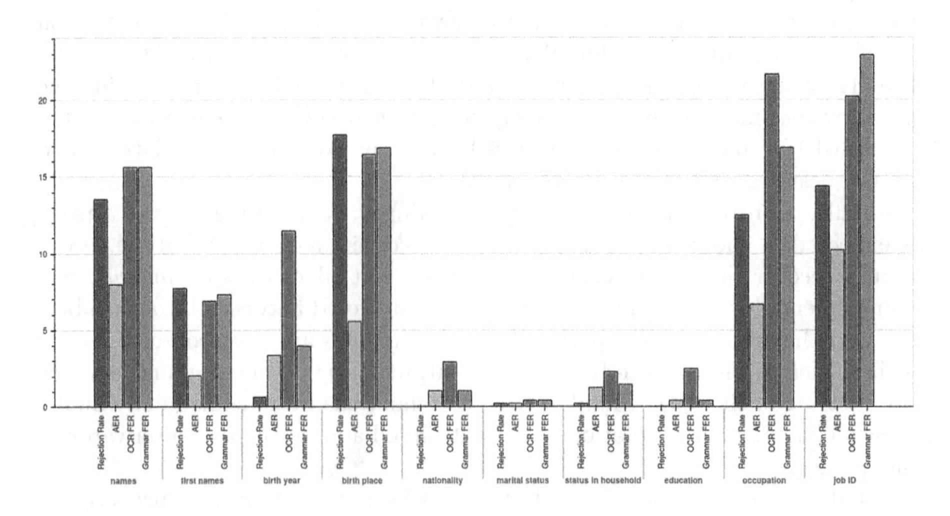

Fig. 6. Field performance metrics on the generic test dataset for each column category

6.2 Normalization and Logical Deductions

In this project, we wanted not only to obtain the textual content in the census but also to make this content usable by historians. That is why after the text recognition, we apply a step of normalization that standardizes the columns combined with logical deductions. This step is necessary because as explained in Subsect. 2.1, the columns were not filled in the same way depending on the census agents. This normalization is made possible by the grammars that detect all existing forms for a given term. Beside normalizing the content of the cells, we also applied logical deductions to enrich the database and facilitate the work of the historians. For example, some writers use the Latin formula *ditto* in a cell to indicate that its content is similar to that of the cell above. That is why, we also perform tabular operations such as deducing the corresponding value of each *ditto*.

7 Processing Time

In this part, we detail the time of each processing step to process the whole 1926 census, namely 50000 double pages, 100000 simple pages and 3 million lines. Processing times were measured on a machine with an Intel Core I7 CPU and an Nvidia Tesla V100 GPU with 16 Gb of memory. The total processing time of the whole pipeline amounts to 67 h approximately. However, since each double page is processed independently, the processing chain may easily be parallelized among several machines. A total of 8.68 h were spent on page segmentation, 0.52 h on page classification, 24.16 h on line segmentation, 22 h on model predictions, and 12 h on decoding, rejection step and post-processing. Among the different steps of the processing pipeline, the line segmentation step is the most time consuming. Adapting our architecture to perform paragraph-based recognition could thus save a considerable amount of time by eliminating the line segmentation step.

8 Conclusion

We have successfully extracted handwritten alphanumeric information from three census of the city of Paris representing a total of 300,000 pages and 9 million individuals. The processing chain relies on state-of-the-art components based on deep neural networks. The contribution of self-training allowed us to benefit from the unannotated corpus to improve by 2.5% CER (40% of relative improvement) the performance of the optical model. Improvements are expected on this point by exploring alternative strategies to the self-training that has been studied here.

Thanks to weighted state automata, it was possible to efficiently constrain and control the recognition process and to reject wrong recognition hypothesis. The knowledge modeling step allowed the integration of external knowledge in the form of demographic and historical databases in order to model the expected content using regular expressions and dictionaries.

The processing chain could easily process the census carried out everywhere in France between the end of the 19th century and the beginning of the 20th century because the same census procedures, the same information, and the same tabular registers were used at that time.

We make available to the community a new labeled manuscript corpus of 6720 lines in total. We also release a flexible library in Python allowing to easily decode recognition lattices under lexical and grammatical constraints by integrating known open source components that have remained difficult to integrate together until now for handwriting recognition.

References

1. Allauzen, C., Riley, M., Schalkwyk, J., Skut, W., Mohri, M.: OpenFst: a general and efficient weighted finite-state transducer library. In: Holub, J., Žd'árek, J. (eds.) CIAA 2007. LNCS, vol. 4783, pp. 11–23. Springer, Heidelberg (2007). https://doi.org/10.1007/978-3-540-76336-9_3
2. Aradillas, J.C., Murillo-Fuentes, J.J., Olmos, P.M.: Boosting offline handwritten text recognition in historical documents with few labeled lines. arXiv:2012.02544 (2020)
3. Bluche, T.: Joint line segmentation and transcription for end-to-end handwritten paragraph recognition. arXiv:1604.08352 (2016)
4. Bluche, T., Louradour, J., Messina, R.: Scan, attend and read: end-to-end handwritten paragraph recognition with MDLSTM attention. arXiv:1604.03286 (2016)
5. Chowdhury, S., Garain, U., Chattopadhyay, T.: A weighted finite-state transducer (WFST)-based language model for online Indic script handwriting recognition. In: International Conference on Document Analysis and Recognition, pp. 599–602 (2011)
6. Coquenet, D., Chatelain, C., Paquet, T.: Handwritten text recognition: from isolated text lines to whole documents. In: ORASIS 2021 (2021)
7. Coquenet, D., Chatelain, C., Paquet, T.: End-to-end handwritten paragraph text recognition using a vertical attention network. arXiv:2012.03868 (2020)
8. Coquenet, D., Chatelain, C., Paquet, T.: Recurrence-free unconstrained handwritten text recognition using gated fully convolutional network. In: International Conference on Frontiers in Handwriting Recognition, pp. 19–24 (2020)
9. Coquenet, D., Soullard, Y., Chatelain, C., Paquet, T.: Have convolutions already made recurrence obsolete for unconstrained handwritten text recognition? In: ICDAR Machine Learning Workshop, Sydney, Australia, pp. 65–70. IEEE (2019)
10. Deng, J., Dong, W., Socher, R., Li, L.J., Li, K., Fei-Fei, L.: ImageNet: a large-scale hierarchical image database. In: 2009 IEEE Conference on Computer Vision and Pattern Recognition, pp. 248–255 (2009)
11. Gay, V.: TRF-GIS Communes (1870–1940) Type: dataset
12. Grüning, T., Leifert, G., Strauß, T., Michael, J., Labahn, R.: A two-stage method for text line detection in historical documents. Int. J. Doc. Anal. Recogn. **22**(3), 285–302 (2019)
13. Michael, J., Labahn, R., Grüning, T., Zöllner, J.: Evaluating sequence-to-sequence models for handwritten text recognition. arXiv:1903.07377 (2019)
14. Mohri, M., Pereira, F., Riley, M.: Weighted finite-state transducers in speech recognition. Comput. Speech Lang. **16**(1), 69–88 (2002)

15. Oliveira, S.A., Seguin, B., Kaplan, F.: dhSegment: a generic deep-learning approach for document segmentation. In: International Conference on Frontiers in Handwriting Recognition, pp. 7–12 (2018)
16. Povey, D., et al.: The Kaldi speech recognition toolkit. In: IEEE 2011 Workshop on Automatic Speech Recognition and Understanding. IEEE Signal Processing Society (2011)
17. Puigcerver, J.: Are multidimensional recurrent layers really necessary for handwritten text recognition? In: International Conference on Document Analysis and Recognition (ICDAR), vol. 01, pp. 67–72 (2017)
18. Roark, B., Sproat, R., Allauzen, C., Riley, M., Sorensen, J., Tai, T.: The OpenGrm open-source finite-state grammar software libraries. In: Proceedings of the ACL 2012 System Demonstrations, Jeju Island, Korea, pp. 61–66. Association for Computational Linguistics, July 2012
19. Ronneberger, O., Fischer, P., Brox, T.: U-Net: convolutional networks for biomedical image segmentation. arXiv:1505.04597 (2015)
20. Sandler, M., Howard, A., Zhu, M., Zhmoginov, A., Chen, L.C.: MobileNetV2: inverted residuals and linear bottlenecks. arXiv:1801.04381 (2019)
21. Schall, M., Schambach, M.P., Franz, M.O.: Multi-dimensional connectionist classification: reading text in one step. In: International Workshop on Document Analysis Systems (DAS), pp. 405–410 (2018)
22. Shorten, C., Khoshgoftaar, T.M.: A survey on image data augmentation for deep learning. J. Big Data **6**(1), 1–48 (2019)
23. Stuner, B., Chatelain, C., Paquet, T.: Self-training of BLSTM with lexicon verification for handwriting recognition. In: 2017 14th IAPR International Conference on Document Analysis and Recognition (ICDAR), vol. 01, pp. 633–638 (2017)
24. Tan, M., Le, Q.V.: EfficientNet: rethinking model scaling for convolutional neural networks. arXiv:1905.11946 (2020)
25. Tensmeyer, C., Davis, B., Wigington, C., Lee, I., Barrett, B.: PageNet: page boundary extraction in historical handwritten documents. arXiv:1709.01618 (2017)
26. Thomas, P.: Semi-supervised learning by Olivier Chapelle, Bernhard Schölkopf, and Alexander Zien (review). IEEE Trans. Neural Netw. **20**, 542 (2009)
27. Voigtlaender, P., Doetsch, P., Ney, H.: Handwriting recognition with large multidimensional long short-term memory recurrent neural networks. In: International Conference on Frontiers in Handwriting Recognition, pp. 228–233 (2016)
28. Xie, Q., Luong, M.T., Hovy, E., Le, Q.V.: Self-training with noisy student improves imagenet classification. arXiv:1911.04252 (2020)
29. Yalniz, I.Z., Jégou, H., Chen, K., Paluri, M., Mahajan, D.: Billion-scale semi-supervised learning for image classification. arXiv:1905.00546 (2019)
30. Yousef, M., Bishop, T.E.: OrigamiNet: weakly-supervised, segmentation-free, one-step, full page text recognition by learning to unfold. In: IEEE/CVF Conference on Computer Vision and Pattern Recognition, pp. 14698–14707. IEEE (2020)
31. Yousef, M., Hussain, K.F., Mohammed, U.S.: Accurate, data-efficient, unconstrained text recognition with convolutional neural networks. Pattern Recogn. **108**, 107482 (2020)
32. Zou, Y., Yu, Z., Liu, X., Kumar, B.V.K.V., Wang, J.: Confidence regularized self-training. arXiv:1908.09822 (2020)

Importance of Textlines in Historical Document Classification

Martin Kišš(✉), Jan Kohút, Karel Beneš, and Michal Hradiš

Brno University of Technology, Brno, Czech Republic
{ikiss,ikohut,ibenes,hradis}@fit.vutbr.cz

Abstract. This paper describes a system prepared at Brno University of Technology for ICDAR 2021 Competition on Historical Document Classification, experiments leading to its design, and the main findings. The solved tasks include script and font classification, document origin localization, and dating. We combined patch-level and line-level approaches, where the line-level system utilizes an existing, publicly available page layout analysis engine. In both systems, neural networks provide local predictions which are combined into page-level decisions, and the results of both systems are fused using linear or log-linear interpolation. We propose loss functions suitable for weakly supervised classification problem where multiple possible labels are provided, and we propose loss functions suitable for interval regression in the dating task. The line-level system significantly improves results in script and font classification and in the dating task. The full system achieved 98.48%, 88.84%, and 79.69% accuracy in the font, script, and location classification tasks respectively. In the dating task, our system achieved a mean absolute error of 21.91 years. Our system achieved the best results in all tasks and became the overall winner of the competition.

Keywords: Historical document classification · Script and font classification · Document origin localization · Document dating

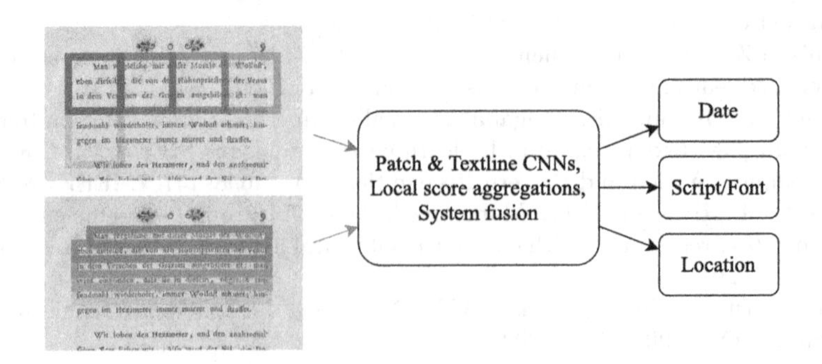

Fig. 1. Proposed historical document classification system.

© Springer Nature Switzerland AG 2022
S. Uchida et al. (Eds.): DAS 2022, LNCS 13237, pp. 158–170, 2022.
https://doi.org/10.1007/978-3-031-06555-2_11

1 Introduction

Visual document classification is a fundamental task when working with historical documents. If one wants to study documents originating from a given location, a given time period or written in a particular font or script, it is first necessary to identify such documents.

This paper describes our approach to the ICDAR 2021 Competition on Historical Document Classification [6] (HDC). The challenge targets three different tasks: (1) Document localization is a simple classification task, where a document needs to be associated with one of the thirteen possible locations of origin. (2) The goal of font and script classification is to identify typefaces present on a document. The training data does not contain any information about location of individual typefaces and, correspondingly, the output is evaluated only at the page level. (3) In the dating task, the goal is to estimate the year of production of a given document page. The ground truth annotations are in the form of intervals (not-sooner-than, not-later-than).

Our approach combines two neural networks, where one operates on square patches of the original documents and the other processes cropped and height-normalized textlines. The local scores are aggregated into page-level output by simple ad-hoc rules. For a given task, the page-level outputs of both networks are combined into a final prediction by a trained fusion (Fig. 1).

While the models for the localization task are trained using the standard cross entropy, we propose two loss functions to deal with the weakly supervised script and font classification. As the labels for the dating task are intervals, we train the models using our proposed interval Huber loss function.

In this paper we show that:

- the proposed loss functions are effective for weakly supervised classification and interval regression,
- the textline processing significantly outperforms patch level processing in the dating and the script and font classification tasks,
- the fusion of both models increases the accuracy of the overall system confirming their complementarity.

2 Related Work

Most recent approaches to visual classification of historical documents are based on the use of convolutional neural networks [1–4,8]. Specifically, these networks are mostly based on the architecture of the ResNet or VGG network [1,2,8] and their task is to classify individual patches or textlines obtained from the original document. The individual local outputs often need to be aggregated to obtain a single output for a given page, which is usually done by averaging [3,4,8]. During training of the network, the input images are often augmented using affine transformations and/or by adding noise.

Christlein et al. [2] proposed a Deep Generalized Max Pooling layer to replace the Global Average Pooling layer at the end of the ResNet architecture, which resulted in accuracy improvement on the CLaMM'16 [4] and CLaMM'17 [3] competition datasets.

Cheikhrouhou et al. [1] used multi-task learning to train a textline model for keyword spotting and script identification. In this approach, the model is a sequence of convolutional, recurrent, and fully connected layers. The output of the last fully connected layer in the main branch is processed by a keyword spotting (KWS) decoder. Local features, obtained from the convolutional part of the network, and global features, obtained from the recurrent part of the network, are combined in the auxiliary branch using a Compact Bilinear Pooling (CBP) and further processed by fully connected layers to classify the script of the input image. The identified script is also used as a secondary input to the KWS decoder to further improve its accuracy. In contrast to this approach, we primarily focus on font and script classification, we compare the textline approach with the patch approach, and we train and evaluate the systems on historical documents.

The most successful approaches in previous historical document classification competitions [3,4] use the ResNet architecture operating on patches, using data augmentations comprising of scaling and adding noise. Apart from CNN-based approaches, some successful systems follow more traditional approach by extracting SIFT descriptors and aggregating them into i-vectors [4] or GMM-supervectors [3].

3 Datasets

For training, validation and testing, we used a dataset published within the ICDAR 2021 Competition on Historical Document Classification [6]. The competition consists of three page-level tasks: 1) font group and script type classification, 2) dating, 3) and place of origin estimation. Overall, the dataset consists of about 70k unique pages. For most of the pages, only one of font, location, script, or date is annotated; for pages from CLaMM'17 [3] both script and date labels are provided. Sizes of datasets for the individual tasks are summarized in Table 1.

Table 1. Number of pages with specific task annotations in the training dataset from ICDAR 2021 Competition on Historical Document Classification.

Task	Training	Validation	Testing
Font	35 382	239	5 506
Script	7 594	419	1 256
Location	5 397	65	325
Date	10 294	1 000	2 516

As the script, font and dating datasets are not partitioned into training–validation subsets, we created our own splits[1]. For script and font datasets,

[1] The splits are publicly available at https://pero.fit.vutbr.cz/hdc_dataset.

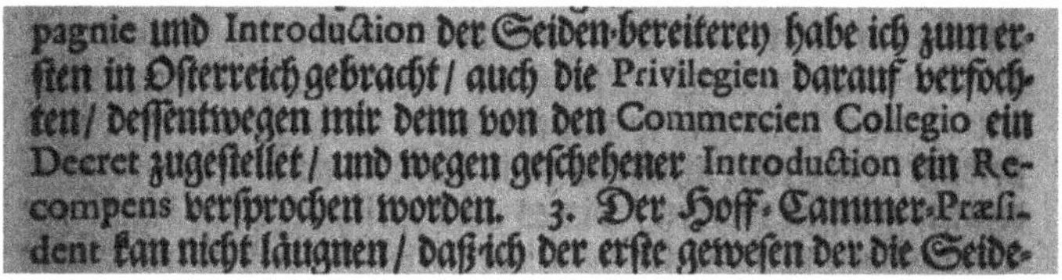

pagnie und Introduction der Seiden-bereiterey habe ich zum er-
sten in Österreich gebracht/ auch die Privilegien darauf verfoch-
ten/ dessentwegen mir denn von den Commercien Collegio ein
Decret zugestellet/ und wegen geschehener Introduction ein Re-
compens versprochen worden. 3. Der Hoff-Cammer-Præsi-
dent kan nicht läugnen/ daß ich der erste gewesen der die Seide-

Fig. 2. An example of a page with mixed fonts.

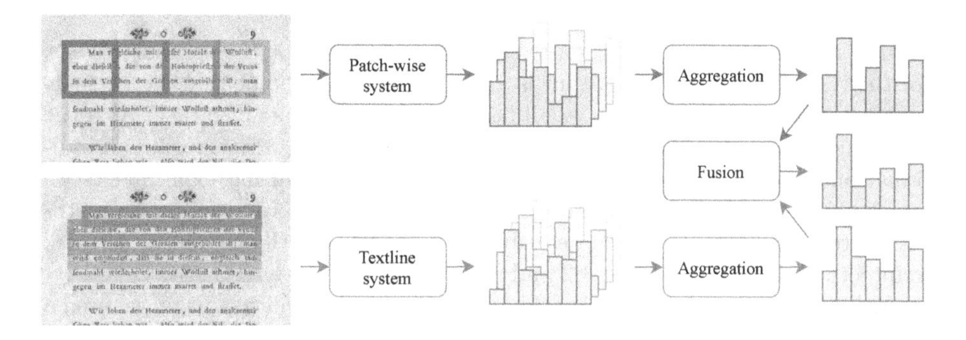

Fig. 3. Schema of the document classification system. An input page is processed by both the patch system (red) and the textline system (blue). The patch and textline outputs are aggregated into two respective page-level results which are finally fused into a single decision (violet). (Color figure online)

we created a validation set with uniform class distribution. This decision was motivated by our expectation that the test set may contain significantly different class distribution (however, this is not the case). For the dating dataset, we randomly selected 1000 pages as the validation set.

While there is only one correct label for each page in the location classification, some pages from the script and font classification tasks contain multiple scripts/fonts and are thus assigned multiple labels in the dataset. This occurs for example in texts written using Fraktur, where some words are written using Antiqua as depicted in Fig. 2. Pages in the dating task are annotated with intervals $\langle a; b \rangle$ representing the estimated range of years when the document was created.

4 Document Classification Systems

Our document classification systems are based on convolutional neural networks processing *cutouts* of the original page images. One system operates on fixed size rectangular image patches while the second system operates on height-normalized crops of automatically detected textlines. The type of the systems' output depends on the target task. For the dating task, the output is a single

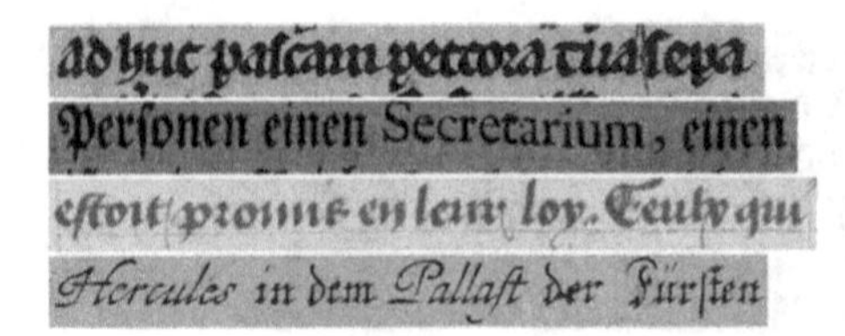

Fig. 4. Examples of detected and cropped textlines from the dataset.

floating point value which is then transformed to represent the estimated year. For the rest of the tasks, the output is a vector of floating point values representing a categorical distribution over all possible classes of the given task. As a the neural networks process each image cutout independently, we aggregate all individual outputs to compute the final page-level output. During experiments we have tested several patch system approaches and local score aggregation methods for textline system and we describe them together with the neural network architectures in detail in Sect. 4.2 and Sect. 4.3, respectively (Fig. 3).

As both systems utilize information about text placement in a page, we used a detector based on ParseNet architecture [5] to detect text regions and individual textlines. The detector was trained on various documents mainly based on the PERO layout dataset [5] and it detected more than 3M textlines in the provided datasets. Examples of the detected textlines are depicted in Fig. 4.

4.1 Loss Functions

Except for the localization task, where standard cross entropy loss function is used, the other tasks presented in the challenge don't provide a clear objective to optimize during training. The difficulty posed in the script and font classification tasks arises from the fact that a single page can be labeled with multiple target classes, e.g. because there are multiple scripts present on it. We propose two loss functions which effectively allow the network to choose one of the page-level labels as relevant for a given cutout.

Both weakly supervised loss functions are based on cross entropy with respect to each of the page-level labels. Loss L_{hard} effectively selects the most probable page-level label (based on network output) and ignores the other labels. Loss L_{soft} considers all page-level labels weighted by the network output probability.

If we denote the network output for a single image cutout x as $f(x)$, the predicted probability of class i as $f(x)_i$, and the set of relevant page-level classes as $\mathcal{T}$, the two loss functions are defined as:

$$L_{hard} = \min_{i \in \mathcal{T}} \left[- \log(f(x)_i)\right] \tag{1}$$

$$L_{soft} = \sum_{i \in \mathcal{T}} - \log(f(x)_i) \cdot f(x)_i. \tag{2}$$

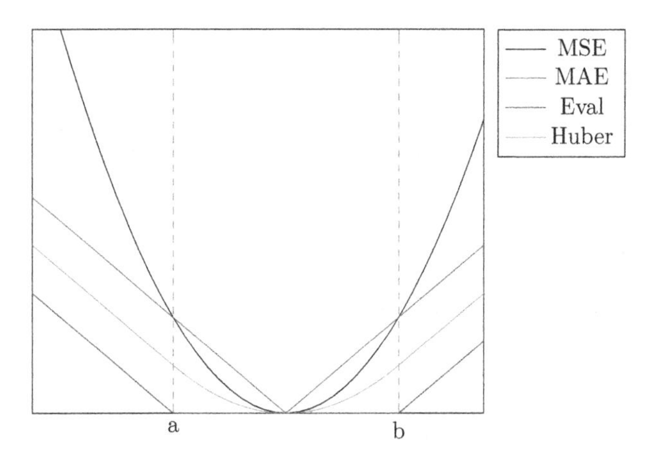

Fig. 5. Different losses used for date regression. *MSE* and *MAE* are the corresponding norm of the distance from the midpoint of the target interval $\langle a; b\rangle$. *Eval* is the evaluation metric, i.e. absolute error from the target interval. Finally, *Huber* is the proposed interval Huber loss, which pulls the output of the system towards the midpoint, but the learning signal gradually weakens inside the interval.

In the dating task, the difficulty arises from the target not being a single point in time (year), but rather an interval $\langle a; b\rangle$. The evaluation metric defined in HDC competition—mean absolute distance from the interval—is differentiable and could directly be used as a loss function. However, it remains open if there really should not be any learning signal provided within the target interval and how do objectives based directly on the interval compare with the more common regression objective functions that drive the output of a system to a single point. Therefore, we compare the evaluation metric with mean absolute and mean squared error from the midpoint of the target interval that force the output of the network deep inside the label interval. Additionally, we introduce a interval Huber loss:

$$L_{Huber}(y, a, b) = \begin{cases} a - y + \frac{r}{2}, & \text{if } y \leq a \\ y - b + \frac{r}{2}, & \text{if } y \geq b \\ \frac{(y - m)^2}{2r}, & \text{otherwise} \end{cases}, \tag{3}$$

where y is the output of the system, $m = (a + b)/2$ is the midpoint of the target interval, and $r = m - a$ is its radius. As shown in Fig. 5, L_{Huber} is linear outside of the ground truth interval as is the evaluation metric, yet it provides a weak learning signal inside the target interval.

4.2 Patch System

The patch-level system relies on a ResNeXt-50 neural network [9] operating on 224×224 patches. The classification of a page is done by aggregating results of four different image scales, while each scale is divided into a grid of non-overlapping patches.

Patches of each scale are aggregated separately. We consider either all patches in a whole page approach (P), or only text region patches in a text region approach (R). Only the ten most confident patches are averaged to give the result for an image scale. We select the most confident scale as the final page output.

We also combined the whole page approach and the text regions approach, whole page + text regions approach (P+R). In this case, the final page aggregation is the average of all eight scale aggregations. The motivation is that the information inside the text regions should be more valuable for the font and the script tasks but still there can be some useful information outside of these text regions.

4.3 Textline System

The textline system operates on arbitrarily long textline images with normalized height of 40 pixels. The neural network of the textline system uses the convolutional part of VGG16 [7] pretrained on ImageNet. The covolutional features are aggregated using global average pooling, and processed by three linear layers interleaved with dropout layers and LeakyReLU activation functions. As each page contains a variable number of textlines, we propose several methods to aggregate these into a single page-level output.

Classification Tasks Aggregations. We propose three methods to obtain aggregated result for the classification tasks. The *mean* method calculates a mean vector from the outputs obtained on individual textlines and the final label is identified as the class with the highest probability. Rest of the methods represent each textline i by its most probable predicted class c_i and its probability p_{i_c}. The *count* method selects most frequently predicted textline class. The *probs* method selects the class with highest cumulative probability by accumulating the predicted probabilities of the winning textline classes.

Dating Task Aggregations. We propose two methods to obtain page-level predictions for the dating task. The first method computes the *mean* of the results obtained on individual textlines and the second selects the *median* value.

4.4 System Fusion

To benefit from the different nature of the two systems, we combine them in two fusions. In both cases, the fusion operates on the page level, i.e. it is agnostic to the local score aggregation method.

The first one is a simple linear fusion of the outputs:

$$\mathbf{y}_{\text{lin}} = \alpha\mathbf{y}_1 + (1 - \alpha)\mathbf{y}_2, \tag{4}$$

where $\mathbf{y}_1$ and $\mathbf{y}_2$ is the full output the patch and textline systems, i.e. two vectors of probabilities of all classes relevant for the task. This approach naturally extends to the date regression, where $\mathbf{y}_{1,2}$ are scalar date estimates. For each of the tasks, we tune the interpolation coefficient α independently to on the relevant validation set.

For the classification tasks, we trained a multiclass logistic regression on the outputs of the individual systems:

$$\mathbf{y}_{\text{log}} = \text{softmax}(\mathbf{W}[\mathbf{y}_1; \mathbf{y}_2] + \mathbf{b}) \tag{5}$$

Here, $[\mathbf{y}_1; \mathbf{y}_2]$ is a concatentation of the class probabilities coming from the individual systems and $\mathbf{W}$ and $\mathbf{b}$ are trained parameters of the fusion. In order to avoid overfitting, we apply a L2 regularization towards aggressive averaging, i.e. towards $\mathbf{W} = \mathbf{1}$ and $\mathbf{b} = \mathbf{0}$. We determine the strength of this regularization by a 10-fold cross-validation on all the available data. Finally, we train the fusion on all the data.

5 Experiments

We conducted experiments with the patch and the textline system on the dataset provided for the HDC competition. Specifically, we trained both systems using both proposed weakly supervised loss functions L_{hard} and L_{soft} for the script and the font tasks. For the dating task, we also trained both systems with all four regression loss functions mentioned in Sect. 4.1. For all tasks, we evaluated the proposed patch system approaches and proposed local score aggregation methods for textline system.

5.1 Experimental Setup

Patch System Training. The patches were randomly cropped out of the pages and augmented with a small random linear transformation, color jitter and gaussian blur. For each task, a system had been trained until it converged, approximately 100k iterations. We used Adam as the optimization method, the learning rate was progressively enlarged for the first 5k iterations until it reached 3×10^{-4}, the dropout was set to 25%. Every 20k iterations, the learning rate and dropout were reduced until they hit values of 3.5×10^{-5} and 0%, respectively.

Textline-Level System Training. Before the training, we removed textlines shorter than 128 pixels to avoid training on false-positive textline detections. During the training, we used data augmentation comprising of defocusing, adding gaussian noise, and color shifting. To mitigate the imbalance of the provided

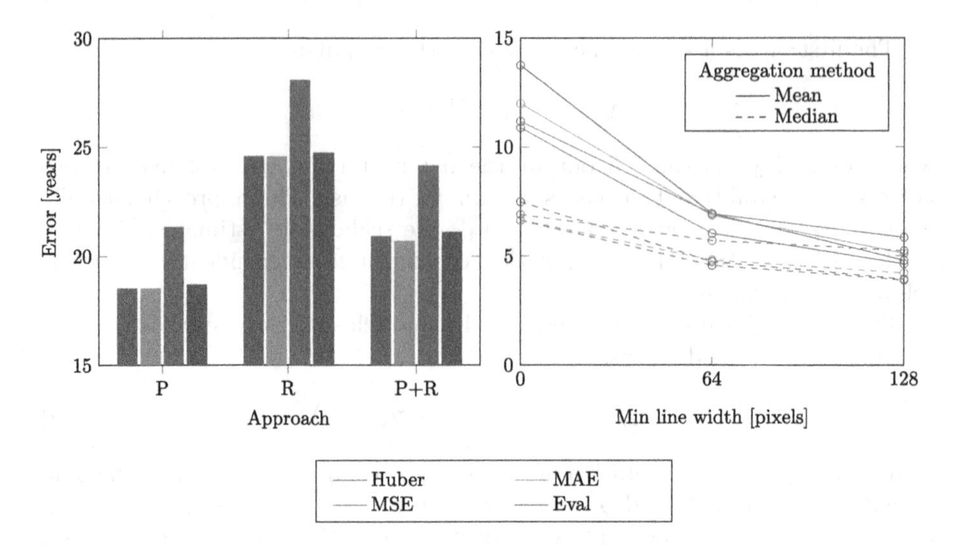

Fig. 6. Comparison of different date regression loss functions, approaches for patch system (left), and aggregation methods for textline system (right). The P, R, P+R stand for whole page approach, text region approach, and whole page + text regions approach, respectively. We report the mean absolute error [years] on the validation data. Note the different scales on the vertical axes.

datasets we weighted contribution of training samples of each class in our loss functions by inverse frequency of samples of the class. Optimization of the model was performed using Adam optimizer with learning rate of 1×10^{-4} for 100k iterations with batch size of 32 and dropout rate 10%.

5.2 Results

Dating Task. For the dating task, we trained models with each proposed date regression loss function and we evaluated them with all the proposed approaches and aggregation methods. Results are depicted in Fig. 6. From these results we decided to use the models trained with the interval Huber loss function. For the patch system we decided to use the whole pages approach (P) and for the textline system we decided to use the median aggregation method for textlines longer than 128 pixels.

Classification Tasks. For the script and font tasks, we trained both systems using both proposed weakly supervised loss functions. Results of these systems are summarized in Table 2. The evaluation was performed on validation sets using the P+R approach for the patch system and mean aggregation method of textlines longer than 64 pixels for the textline system. We used this configurations based on results of preliminary experiments.

The results show that for the patch system the L_{soft} loss function outperforms the L_{hard} loss function by nearly 50% for the font task, while the difference is negligible for the script task. For the textline system the differences for both tasks are even more negligible. From these results we decided to use the patch system trained using the L_{soft} loss function and textline system trained using L_{hard} loss function.

For all classification tasks, we also evaluated all approaches described in Sect. 4.2 and all aggregation methods described in Sect. 4.3 on the validation set. Results of the patch system and textline system are shown in Table 3 and Fig. 7, respectively. Based on the results for the patch system, we decided to use the P+R approach for the font and the script classification tasks and the P approach for the localization task. For the textline system, we selected the mean aggregation method of textlines longer than 64 pixels.

Table 2. Comparison of classification loss functions for training models on dataset with multiple annotations for a single training example.

Loss function	Patch system		Textline system	
	Font	Script	Font	Script
L_{hard}	6.22	19.09	2.93	10.98
L_{soft}	3.32	20.29	2.93	10.74

Table 3. Comparison of different patch system approaches on classification tasks. The P, R, P+R stand for whole page approach, text region approach, and whole page + text regions approach, respectively. We report the error rate on the validation data.

Task	Approach		
	P	R	P+R
Font	5.02	7.11	**2.93**
Script	17.90	25.06	**16.47**
Location	20.00	20.00	**18.46**

System Fusion. We optimized and evaluated both types of system fusion using a 10-fold cross-validation on our validation sets. The results presented in Table 4 are the averages obtained over those 10 splits.

Final Submission. Finally, our systems were evaluated on the test data by the organizers of the challenge, the results are shown in Table 5. Overall, we see that the line system performs much better in classification of script and font type and in dating. This makes sense as we expect that all the necessary information is within the text content, so allowing the model to work with pre-segmented

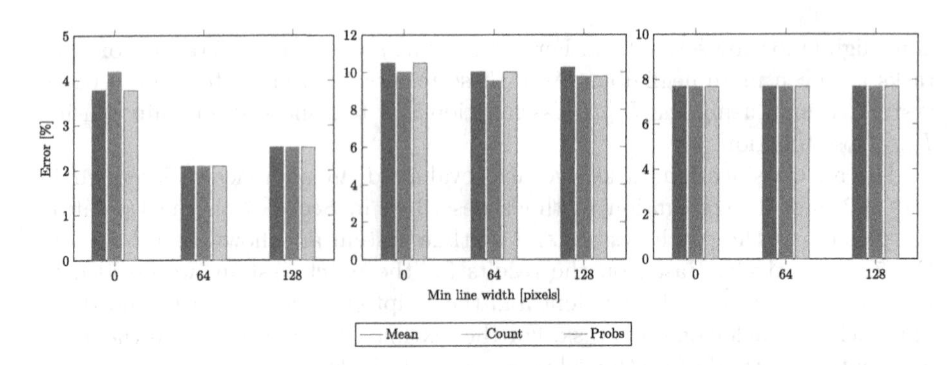

Fig. 7. Comparison of different aggregation methods for textline system on font (left), script (middle), and localization (right) task. We report the error rate on the validation data. Note the different scales on the vertical axes.

Table 4. Average system fusion results. The values are the average of the results obtained from all possible combinations of validation set splits in the cross-validation.

	Font	Script	Location	Date
Linear fusion	1.67	9.55	7.69	4.29
Log-linear fusion	2.09	9.55	7.69	–

textlines allows to focus on extracting discriminative features. On the other hand, the patch system did a better job on localization of documents. We hypothesise that there are some important visual cues outside of the text in this case, such as ornaments, illustrations, or visual separators. For the textline system, documents containing little or no text (e.g. title pages) can also be problematic.

When compared to the other systems participating in the ICDAR 2021 Competition on Historical Document Classification, our system achieved the best results in each task and became the overall winner of the competition.

Checkpoints Averaging. In an attempt to mitigate the effect of spurious changes in validation error coming from random steps of gradient descent, we introduce checkpoint averaging for our models used for the final submission. For the patch system, we took the best performing checkpoint by validation performance and averaged it with checkpoints $\pm 1, 2, 3$ thousand updates apart, for a total of 7 model averaged. For textline system, we averaged the last N checkpoints from the training. Here, N was tuned to optimize validation performance in range $1 < N < 10$.

Table 5. Final results on the ICDAR 2021 Competition on Historical Document Classification test data. Font, script and location are reported as accuracy [%]; for date, mean absolute error is given [years]. All of the presented results have been measured by the challenge organizers. The first four lines represent our submissions. The results of the other teams that participated in the competition [6] are shown below the line. Since each team could have participated with multiple systems, we present here only the best result for each task for these teams.

	Font	Script	Location	Date
Textline system	98.42	88.54	69.85	**21.91**
Patch system	95.68	80.26	75.08	32.45
Linear fusion	98.27	**88.84**	70.77	21.99
Log-linear fusion	**98.48**	88.60	**79.69**	–
Baseline	–	55.22	62.46	–
The North LTU	82.80	74.12	43.69	79.43
CLUZH	95.66	35.25	–	–
NAVER Papago	97.17	–	–	–

6 Conclusion

We proposed a system for visual document classification fusing patch level and textline level approaches. The experiments on the datasets from ICDAR 2021 Competition on Historical Document Classification show that line level processing significantly outperforms patch processing in classification of font and script types and in dating. On the other hand, the patch level approach yields better localization results.

Our proposed weakly supervised classification loss functions proved effective in utilizing multiple page-level classification labels. Similarly for the dating problem, where the labels are given as intervals, specific interval loss functions brought improvement over baseline regression approaches; namely the proposed interval Huber loss yields best results. The fusion provided mediocre improvements. It improved results significantly only in the localization task, the other tasks were dominated by the textline system. The small improvement due to fusion indicates that the patch and line level systems have similar strengths and weaknesses.

We believe the proposed approach could be further improved by combining the task-specific networks into a single multi-task network. Further, text surely contains important cues for dating and localization and we intend to add a text-level system working with automatic text transcriptions, and possibly utilizing exiting historical text corpora.

Acknowledgement. This work has been supported by the Ministry of Culture Czech Republic in NAKI II project PERO (DG18P02OVV055) and by Czech National Science Foundation (GACR) project "NEUREM3" No. 19-26934X.

References

1. Cheikhrouhou, A., Kessentini, Y., Kanoun, S.: Multi-task learning for simultaneous script identification and keyword spotting in document images. Pattern Recogn. **113**, 107832 (2021). https://doi.org/10.1016/j.patcog.2021.107832, https://www.sciencedirect.com/science/article/pii/S0031320321000194
2. Christlein, V., Spranger, L., Seuret, M., Nicolaou, A., Král, P., Maier, A.: Deep generalized max pooling. In: 2019 International Conference on Document Analysis and Recognition (ICDAR), pp. 1090–1096, September 2019. https://doi.org/10.1109/ICDAR.2019.00177, iSSN 2379-2140
3. Cloppet, F., Eglin, V., Helias-Baron, M., Kieu, C., Vincent, N., Stutzmann, D.: ICDAR2017 competition on the classification of medieval handwritings in latin script. In: 2017 14th IAPR International Conference on Document Analysis and Recognition (ICDAR), vol. 01, pp. 1371–1376, November 2017. https://doi.org/10.1109/ICDAR.2017.224, iSSN 2379-2140
4. Cloppet, F., Églin, V., Kieu, V.C., Stutzmann, D., Vincent, N.: ICFHR2016 competition on the classification of medieval handwritings in latin script. In: 2016 15th International Conference on Frontiers in Handwriting Recognition (ICFHR), pp. 590–595, October 2016. https://doi.org/10.1109/ICFHR.2016.0113, iSSN 2167-6445
5. Kodym, O., Hradiš, M.: page layout analysis system for unconstrained historic documents. In: Lladós, J., Lopresti, D., Uchida, S. (eds.) ICDAR 2021. LNCS, vol. 12822, pp. 492–506. Springer, Cham (2021). https://doi.org/10.1007/978-3-030-86331-9_32
6. Seuret, M., et al.: ICDAR 2021 competition on historical document classification. In: Lladós, J., Lopresti, D., Uchida, S. (eds.) ICDAR 2021. LNCS, vol. 12824, pp. 618–634. Springer, Cham (2021). https://doi.org/10.1007/978-3-030-86337-1_41
7. Simonyan, K., Zisserman, A.: Very deep convolutional networks for large-scale image recognition. In: Bengio, Y., LeCun, Y. (eds.) 3rd International Conference on Learning Representations, ICLR 2015, San Diego, CA, USA, 7–9 May 2015, Conference Track Proceedings (2015). http://arxiv.org/abs/1409.1556
8. Tensmeyer, C., Saunders, D., Martinez, T.: Convolutional neural networks for font classification. In: 2017 14th IAPR International Conference on Document Analysis and Recognition (ICDAR), vol. 01, pp. 985–990, November 2017. https://doi.org/10.1109/ICDAR.2017.164, iSSN: 2379-2140
9. Xie, S., Girshick, R., Dollar, P., Tu, Z., He, K.: Aggregated residual transformations for deep neural networks. In: Proceedings of the IEEE Conference on Computer Vision and Pattern Recognition (CVPR), pp. 1492–1500 (2017). https://openaccess.thecvf.com/content_cvpr_2017/html/Xie_Aggregated_Residual_Transformations_CVPR_2017_paper.html

Historical Map Toponym Extraction for Efficient Information Retrieval

Ladislav Lenc[1,2](✉), Jiří Martínek[1,2], Josef Baloun[1,2], Martin Prantl[1,2], and Pavel Král[1,2]

[1] Department of Computer Science and Engineering, Faculty of Applied Sciences, University of West Bohemia, Plzeň, Czech Republic
{llenc,jimar,balounj,perry,pkral}@kiv.zcu.cz
[2] NTIS - New Technologies for the Information Society, Faculty of Applied Sciences, University of West Bohemia, Plzeň, Czech Republic

Abstract. The paper deals with detection, classification and recognition of toponyms in hand-drawn historical cadastral maps. Toponyms are local names of towns, villages and landscape features such as rivers, forests etc. The detected and recognized toponyms are utilized as keywords in an information retrieval system that allows intelligent and efficient searching in historical map collections. We create a novel annotated dataset that is freely available for research and educational purposes. Then, we propose a novel approach for toponym classification based on KAZE descriptor. Next we compare and evaluate several state-of-the-art methods for text and object detection on our toponym detection task. We further show the results of toponym text recognition using popular Tesseract engine.

Keywords: Historical maps · Toponyms · Text detection · OCR · IR · FCN

1 Introduction

Information retrieval (IR) in historical documents is a challenging task. Documents occur mostly as digital images with no text layer that makes an efficient information retrieval very difficult and cumbersome. A special group of such documents are historical maps. The maps contain several interesting pictographic elements: boundary stones and milestones, scale, and other symbols determined by the map legend. In addition, they also include so-called toponyms, i.e. local names of towns, municipalities, villages, but also forests, hills, or paths. Such toponyms are crucial for IR and make up the main keywords that are mostly found in user queries in the IR systems.

We focus on toponym detection, classification and recognition in order to annotate each map sheet with a set of keywords to facilitate efficient searching and IR. The task is solved within an ongoing project that concentrates on seamless connection of cadastral maps and efficient searching in them [13]. We focus on scanned historical cadastral Austro-Hungarian maps from the 19th century provided by Czech Office for Surveying, Mapping and Cadastre (CUZK)[1].

[1] https://www.cuzk.cz/.

© Springer Nature Switzerland AG 2022
S. Uchida et al. (Eds.): DAS 2022, LNCS 13237, pp. 171–183, 2022.
https://doi.org/10.1007/978-3-031-06555-2_12

We create a novel annotated dataset which represents the first main contribution of this work. The dataset contains 800 annotated map sheets and it is freely available for research and educational purposes. The second important contribution consists in proposition of a novel method for map toponym classification based on KAZE descriptor. We also compare and evaluate several state-of-the-art methods for text and object detection on our toponym detection task using the newly created dataset. Finally, we show the results of optical character recognition (OCR) using Tesseract system [22].

The rest of the paper is organized as follows. The next section covers the related work. Our dataset is described in Sect. 3. We present the approaches for text detection, toponym classification and OCR in Sect. 4. Experiments are described in Sect. 5 and Sect. 6 concludes the paper.

2 Related Work

The detection of regions of interest (RoI) is one of essential tasks in the computer vision field. There are many challenges, e.g.: noise, image distortion, video frames. The quality of the detection results is often crucial since it is usually followed by further steps such as OCR or classification and can negatively influence the results of the successive steps.

Many scene text detection approaches are evaluated on well-known benchmark datasets (some of them are Street View Text (SVT) [23], IIIT5k [17], ICDAR2003 [16], ICDAR2013 [12], ICDAR2015 [11]).

Traditional approaches are based on manual feature engineering such as: Stroke Width Transform [5], Maximally Stable Extremal Regions [18,19] or others [10,14]. Individual character recognition approaches are also employed [4,28].

Currently, the mainstream methods in this topic are deep neural network models like Region-based Convolutional Neural Networks (R-CNNs). These networks typically use pre-trained ResNet 50 or VGG 16 backbones to extract features and region proposals. A significant approach for efficient object detection using deep convolutional networks has been proposed in [7]. R-CNN system consists of region proposal, feature extraction and classification modules. Proposed regions are separately processed and classified for presence of the desired objects. The system achieved promising results but the drawbacks are multi-stage training and the computational inefficiency due to the number of regions and thus classifications per single image.

An approach based on Spatial Pyramid Pooling has been proposed by He et al. [9]. The main strength of this approach is the capability to use a specific pooling operation that generates a fixed-length representation regardless of image size or scale.

In 2015, Ross Girshick went further in his research proposing Fast R-CNN [6] with several innovations to deal with the drawbacks and speed-up. Compared to R-CNN, the image features are calculated only once. Based on the regions, the region features are pooled into a fixed size feature space using RoI pooling (which can be understood as 1-level Spatial Pyramid Pooling).

The selective search algorithm for region proposals was still considered a drawback resulting in Faster R-CNN model [21]. Instead of selective search algorithm, the model is extended by Region Proposal Network which is faster.

Additionally, the task of instance segmentation can be solved by mask R-CNN [8]. It adopts the Faster R-CNN scenario and outputs also segmentation mask for each RoI.

You Only Look Once (YOLO) [20] is even faster than Faster R-CNN and allows real-time object detection. Briefly, as says the network name, you only look once at the image to predict the bounding boxes and the class probabilities for these boxes. It uses the grid over the image and predicts the boxes for each cell of the grid. The trade-off is the limited amount of predicted boxes per cell in the grid.

EAST: An Efficient and Accurate Scene Text Detector has been proposed by Zhou et al. [29] in 2017. In 2019, Baek et al. proposed Character Region Awareness for Text Detection (CRAFT) [2] which is able to detect characters from the scene without necessity of annotating such characters and glyphs individually.

Some approaches for text detection and recognition are focused on historical maps (e.g. Weinman [24,25]). This task is very challenging since the maps (especially historical) are very different and heterogeneous. Moreover, the texts are often warped and rotated. Chiang and Knoblock [3] present a general semi-automatic approach that should require minimal user effort for the text recognition.

3 Dataset

The data are annotated for all three sub-tasks that we have to handle: 1) toponym detection, 2) toponym classification and 3) text recognition. We thus need to locate and extract text regions, determine the toponym category and finally recognize the text from it. There are two groups of toponyms:

1. Municipal toponyms (cities, villages, significant buildings);
2. General toponyms (forests, roads, hills, swamps, rivers and other objects).

These two groups can be separated from each other since the municipal toponyms are written with a printed-like font (capital letters) while the general toponyms are handwritten. For simplicity, in the rest of the article, we will refer toponyms as either *printed* (P) or *handwritten* (H). An example of a map sheet with detected and labeled toponyms can be seen in Fig. 1.

There is also one important text element in the map sheet, so-called nomenclature. See the top right corner of Fig. 1 and more detailed in Fig. 2. It determines the location of the map sheet in the coordinate system. The nomenclature is also labeled in our dataset, however, we do not use it in this work.

The created dataset consists of 800 manually annotated map sheets which cover various areas across the Czech Republic[2]. In total, we marked 3700 regions indicating toponyms or nomenclatures (see Table 1). The vast majority of toponyms are thus handwritten since there are much more general objects throughout the maps. The dataset is freely available for research and educational purposes[3].

[2] The Czech Republic was a part of the Austria-Hungary Empire until 1918.

[3] https://corpora.kiv.zcu.cz/nomenclature/.

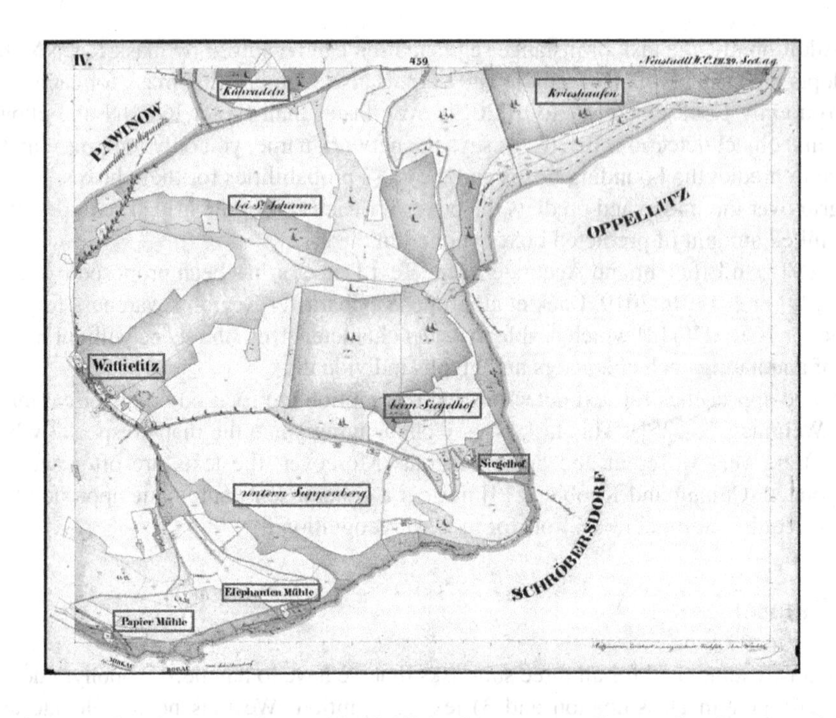

Fig. 1. Example of a map sheet with highlighted toponyms. Green-colored are municipal toponyms with printed-like font and purple-colored are general toponyms with handwritten font. Not labeled curved text specifies information about neighboring map sheet and it is not used in our case. (Color figure online)

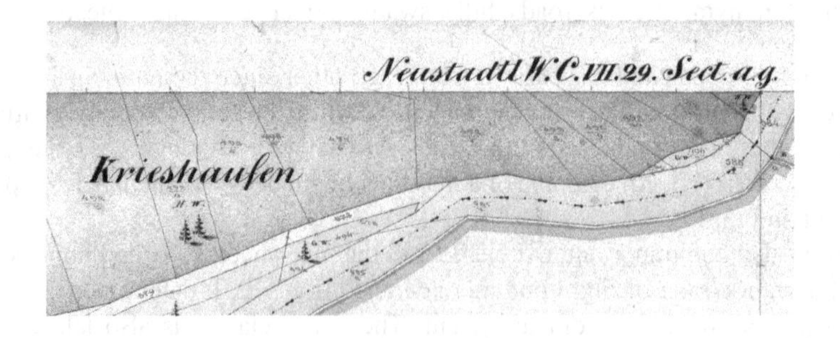

Fig. 2. Example of a nomenclature (right top corner)

Table 1. Numbers of nomenclatures, handwritten and printed toponyms within our dataset.

Dataset	Map sheets	Nomenclatures	Handwritten toponyms	Printed toponyms
Train	650	650	2050	335
Test	100	100	305	41
Dev	50	50	141	28

4 Toponym Processing Approach

Within this section, we describe the approach we used in all our tasks. The overall pipeline is depicted in Fig. 3. Three blue boxes represent models for particular subtasks: text detection, toponym classification and OCR. Note that the text detection model includes toponym classifier in the case of Faster R-CNN and YOLOv5 methods.

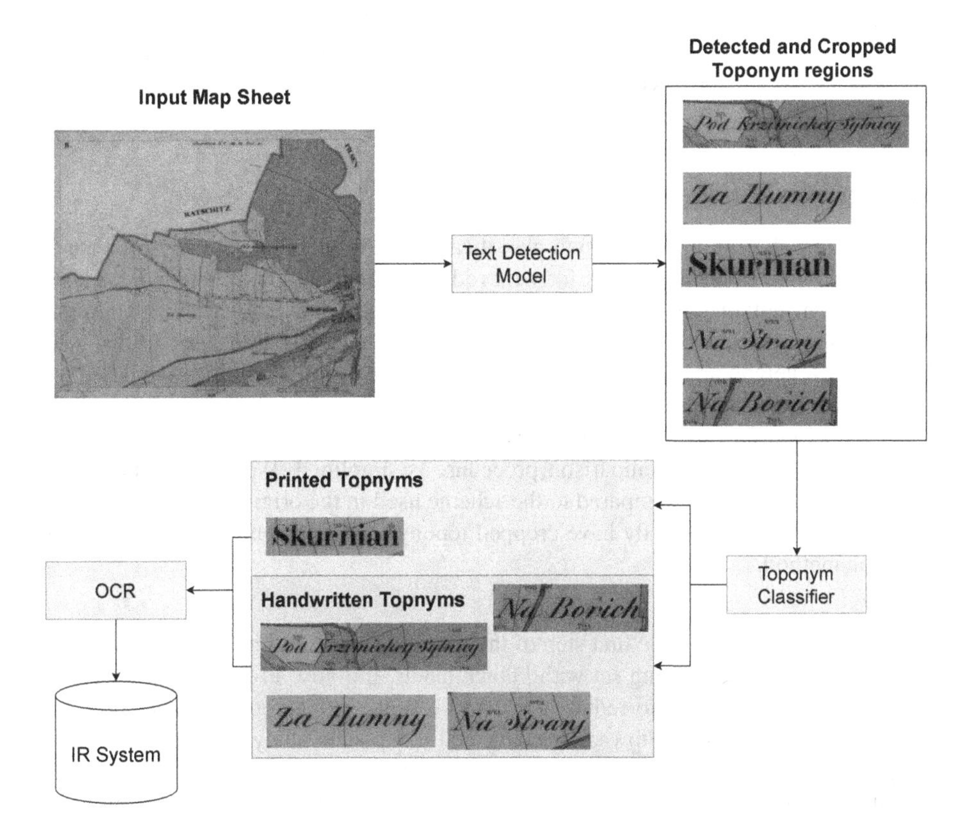

Fig. 3. Overall processing pipeline

4.1 Toponym Detection Methods

As a baseline we use a simple algorithm based on the connected components analysis (CCA). We first apply a median filter, binarize the image and use morphological dilation to connect single letters into one larger component. Then we use the CCA to identify large enough components that should represent the toponyms.

The more sophisticated approaches are based on neural networks and therefore, they must be trained. We use the training part of the created dataset for this task.

The following neural detectors presented in details in Related Work section are evaluated and compared:

- Faster R-CNN model with ResNet-50-FPN backbone;
- EAST: an efficient and accurate scene text detector;
- HP-FCN: High Performance Fully Convolutional Network [26];
- YOLO: You Only Look Once (we use YOLOv5 for our experiments).

YOLOv5 and Faster R-CNN are able to realize the detection and classification together. However, the other models predict only a bounding box (region of interest). In this case, we use for classification a novel algorithm based on KAZE key-point detector described below.

4.2 Toponym Classification Method

The proposed algorithm for toponym classification into printed and handwritten categories is inspired by the algorithm for writer identification based on image descriptors from [27]. The authors utilized SIFT [15] image descriptor in combination with clustering for image representation. The representations were then compared using χ^2 distance and k-nearest neighbours algorithm. They further utilized a finer identification based on contour-directional features to get the final categories.

In the proposed method, we have used faster and smaller KAZE [1] descriptor instead of the SIFT. Our comparison procedure is simplified. We have applied only a single-step comparison compared to the scheme used in the original paper. The algorithm assumes that we already have cropped toponym images obtained by a toponym detection method.

Codebook Generation. The first step of the proposed algorithm is a codebook generation. It is based on a training set with known labels. Initially, all input image regions are pre-processed in order to reduce noise and to remove some unnecessary objects. We perform binarization followed by connected component analysis to preserve only components that are large enough to be part of the text content (glyphs). The minimal size of a component is based on the input resolution and degree of noise (in our case components with area smaller than 100 px are excluded).

The KAZE detector is applied on all pre-processed regions. We thus obtain a set of key-points and corresponding descriptors for each region. All descriptors are put together and the resulting set of descriptor vectors is clustered with K-means algorithm. We work with normalized descriptors (sum up to 1.0). The number of clusters K is a key parameter of this approach. Based on the recommendation from [27] we have tested several values for K and finally set $K = 100$. This process is described in Fig. 4.

Image Representation. Image representation is calculated based on the codebook. For each descriptor vector, we find the closest cluster using Euclidean distance. Even though the descriptor vectors have a relatively high dimension (64 or 128), this simple metric has sufficient results based on our preliminary experiments. We have also tested Cosine

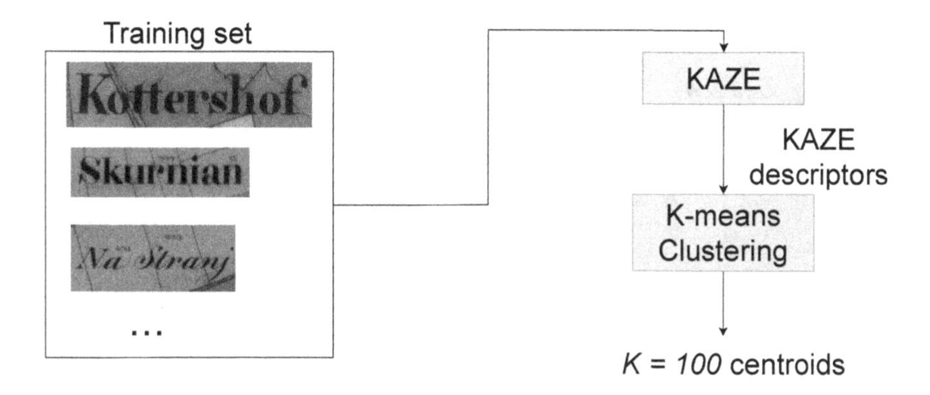

Fig. 4. Codebook generation process

distance, χ^2 and Bhattacharyya distances, but the results were similar or even worse. The representation is then a histogram of size K where each bin represents how many times the given centroid was the closest to a descriptor vector. Finally, the histograms are size-normalized. This process is depicted in Fig. 5.

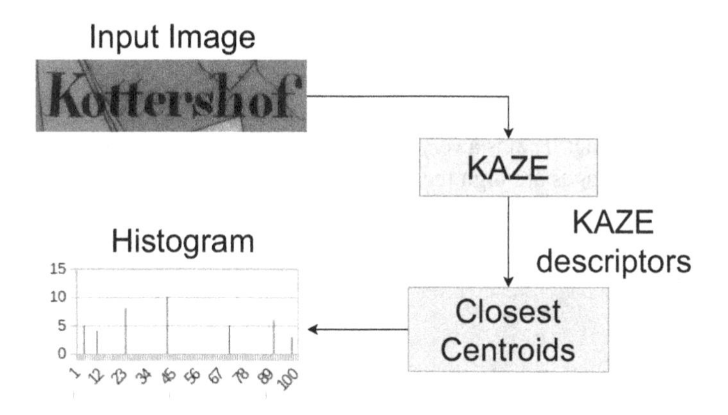

Fig. 5. Image representation creation

Prediction. The prediction of an unknown text region is based on comparison of its representation with representations of known samples (training set). We first pre-process the image and obtain its histogram representation. Then we find N most similar histograms from the training set using Bhattacharyya distance. Based on our experiments, this distance is more suitable for the comparison of histograms than other traditional distance measures. The predicted class is determined as the majority class occurring in the N most similar histograms. Figure 6 shows the prediction phase of our approach.

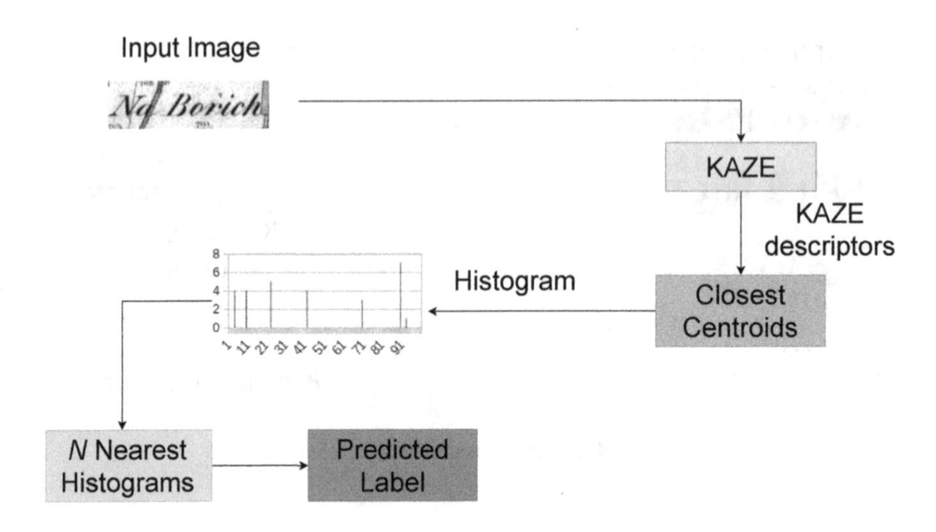

Fig. 6. Test image prediction process

The biggest advantage of this algorithm is the fact that only a small amount of training examples is sufficient for reasonable results (comparing to the neural network models).

4.3 OCR Models

Tesseract 4.0. Tesseract[4] [22] is a very popular and powerful open-source OCR engine. The important advantages are high recognition score and availability of many different language supports. It employs a powerful LSTM based OCR engine and integrates models for 116 additional languages. It is also possible to carry out training of the Tesseract engine and create a custom *tessdata* file which is fine-tuned on particular font and data.

5 Experiments

The experiments are divided into three sections. The first part deals with the toponym detection sub-task. The second experiment measures the success rate of toponym classification. The last set of experiments shows the results of OCR using Tesseract engine.

5.1 Toponym Detection

We evaluate the four different neural network models and compare it with the baseline algorithm based on connected components analysis, see Sect. 4.1. We use the evaluation protocol for the COCO dataset. We calculate precision, recall and F1 score for two

[4] https://github.com/tesseract-ocr/.

Table 2. Evaluation of the toponym detection algorithms on 0.5 and 0.75 IoU levels

Model	IoU@50				IoU@75				Avg AP
	Prec.	Rec.	F1	AP	Prec.	Rec.	F1	AP	
CCA (baseline)	19.5	60.4	29.5	11.3	10.7	33.1	16.2	0.27	2.78
EAST detector	84.5	**89.9**	**87.1**	**77.8**	77.5	**82.4**	**79.9**	**51.3**	**46.7**
HP-FCN	65.4	75.4	70.1	44.4	53.9	62.2	57.8	17.1	20.6
YOLOv5	84.6	79.2	81.8	67.1	76.4	71.7	73.9	39.7	37.1
Faster R-CNN	**87.2**	80.9	83.9	71.2	**80.6**	75.0	77.7	45.4	41.8

intersection over union (IoU) levels (IoU above 0.5 and 0.75). Moreover, we calculate also average precision (AP, area under precision/recall curve). Finally, we provide the average AP at IoU interval 0.5–0.95 with 0.05 step. The overall text detection results are summarized in Table 2.

As we can see, the best results have been obtained by EAST Detector. The Faster R-CNN obtained also very good results. Its advantage compared to the EAST detector is the ability to predict also the label together with the bounding box.

Figure 7 extends the results reported in Table 2 and shows the precision of the utilized models based on the different IoU levels. This figure shows that the best precision is obtained by Faster R-CNN at all IoU levels. Moreover, EAST detector is only slightly worse.

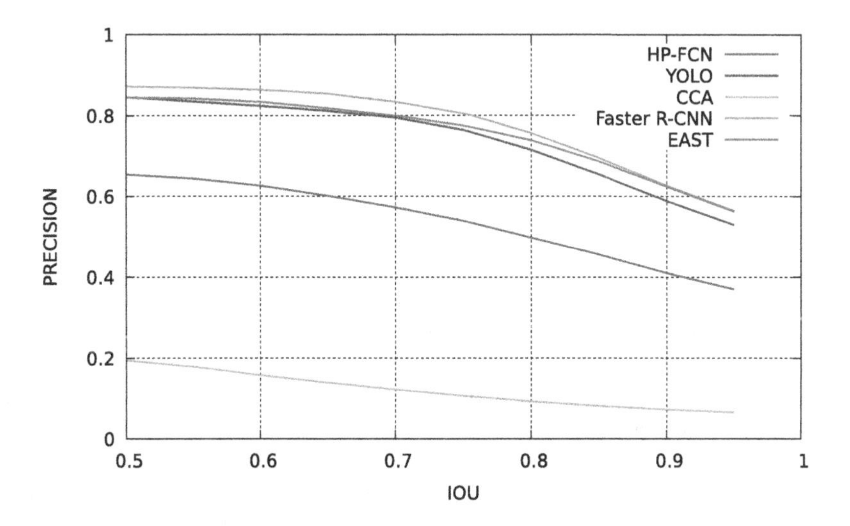

Fig. 7. Precision results based on IoU level

5.2 Toponym Classification

This experiment compares two approaches for toponym classification. The first one is the proposed toponym classification algorithm already described in Sect. 4.2. The classification is done for all toponym detection models. We use only correctly detected toponym regions for evaluation (IoU between predicted and ground truth regions at least 0.5). In such a case, we ran our algorithm and predicted a toponym label (P or H).

The second part of this experiment measures the quality of YOLOv5 and Faster-RCNN models and their built-in classification heads.

We show classification accuracies in Table 3. This table shows that using the proposed classification approach brings comparable results for all tested detection methods which indicates that it is robust against different sizes of the detected regions. It is also superior to the toponym classification accuracy of the YOLOv5 and Faster-RCNN models. The best results are obtained using HP-FCN together with our proposed classification algorithm.

Table 3. Toponym classification results; accuracy (ACC) in %

Detection approach	Classification approach	ACC
CCA (baseline)	Proposed	98.7%
EAST	Proposed	99.1%
HP-FCN	Proposed	**99.2**%
YOLOv5	Proposed	98.8%
Faster R-CNN	Proposed	98.8%
YOLOv5	YOLOv5	97.6 %
Faster R-CNN	Faster R-CNN	98.2 %

5.3 OCR Results

The last experiment measures the quality of the OCR using popular Tesseract engine with three different configurations. As a baseline OCR, we used Tesseract with standard ENG trained data that is available as a default configuration.

Since we have two types of toponyms that are visually different, we trained a separate Tesseract model for each of them. As a result, we created **Tesseract**$_P$ and **Tesseract**$_H$ for printed and hand-written toponyms respectively. During the evaluation step (on 346 bounding boxes in the test set), we measured the character error rate (CER). First, we split all test toponyms (bounding boxes) into printed and handwritten, and we evaluated each Tesseract configuration. Table 4 shows the results.

Naturally, **Tesseract**$_P$ obtained excellent results for printed toponyms and **Tesseract**$_H$ was very successful in handwritten toponyms and vice versa. The best results on the full test set (combined printed and handwritten toponyms) were obtained by **Tesseract**$_H$. Basically, the reason is that the vast majority of toponyms are handwritten. We

Table 4. OCR results with tesseract

	Printed toponyms	Handwritten toponyms	All toponyms
Number of toponyms	41	305	346
Tesseract ENG (baseline)	0.153	0.477	0.437
Tesseract_P (trained)	**0.061**	0.512	0.459
Tesseract_H (trained)	0.076	**0.185**	0.185
Combined tesseract	–	–	**0.171**

also employed a combined Tesseract approach where we pick a Tesseract configuration according to the predicted class (e.g. if a bounding box of a toponym has been previously marked as P, $\textbf{Tesseract}_P$ is used for OCR and similarly for the label H). With such a configuration we obtained the best results (see the last row in Table 4) reaching CER value of 0.171. This result is sufficient for the target application.

6 Conclusions and Future Work

The main goal of this work was to detect, classify and recognize toponyms occurring in historical maps to enable an efficient information retrieval.

We have evaluated several text detection neural network models. We have conducted detection experiments on a newly published dataset and presented the results. In terms of the average precision, the best text detection model is EAST. Its drawback, though, is that it is not able to predict a toponym label per se.

We have also implemented a novel toponym classification algorithm that can be used for regions detected by arbitrary text detector. It is based on KAZE descriptor and the idea behind it comes from writer identification approaches. We have shown that this approach reaches excellent classification results. The drawback of training YOLOv5 and/or Faster R-CNN is the necessity of having a relatively high number of training samples. Our method does not require such a large number of samples and works very well even if only less than 200 toponym images (detected text-regions) are used for training.

Although Faster R-CNN and YOLOv5 obtained very good detection and classification results, the best strategy seems to be in separated training: 1) focus on the best possible text detection and 2) use a separate classification algorithm that can be easily adapted for classification into more classes. Finally, we have conducted the OCR experiments which are crucial for the success rate of the following IR system.

In the future work, we would like to concentrate on the further improvement of the processing pipeline, especially on the OCR part which is very important. We would also like to try a finer categorization of the toponyms (distinguish between rivers and hills etc.) We further plan to utilize our methods on map sheets from a different era and different geographical area.

Acknowledgements. This work has been partly supported from ERDF "Research and Development of Intelligent Components of Advanced Technologies for the Pilsen Metropolitan Area (InteCom)" (no.: CZ.02.1.01/0.0/0.0/17_048/0007267) and by Grant No. SGS-2022-016 "Advanced methods of data processing and analysis".

References

1. Alcantarilla, P.F., Bartoli, A., Davison, A.J.: KAZE features. In: Fitzgibbon, A., Lazebnik, S., Perona, P., Sato, Y., Schmid, C. (eds.) ECCV 2012. LNCS, vol. 7577, pp. 214–227. Springer, Heidelberg (2012). https://doi.org/10.1007/978-3-642-33783-3_16
2. Baek, Y., Lee, B., Han, D., Yun, S., Lee, H.: Character region awareness for text detection. In: Proceedings of the IEEE/CVF Conference on Computer Vision and Pattern Recognition, pp. 9365–9374 (2019)
3. Chiang, Y.-Y., Knoblock, C.A.: Recognizing text in raster maps. GeoInformatica **19**(1), 1–27 (2014). https://doi.org/10.1007/s10707-014-0203-9
4. De Campos, T.E., Babu, B.R., Varma, M., et al.: Character recognition in natural images. VISAPP (2) **7**, 2 (2009)
5. Epshtein, B., Ofek, E., Wexler, Y.: Detecting text in natural scenes with stroke width transform. In: 2010 IEEE Computer Society Conference on Computer Vision and Pattern Recognition, pp. 2963–2970. IEEE (2010)
6. Girshick, R.: Fast R-CNN. In: Proceedings of the IEEE International Conference on Computer Vision (ICCV), December 2015
7. Girshick, R., Donahue, J., Darrell, T., Malik, J.: Rich feature hierarchies for accurate object detection and semantic segmentation. In: Proceedings of the IEEE Conference on Computer Vision and Pattern Recognition, pp. 580–587 (2014)
8. He, K., Gkioxari, G., Dollár, P., Girshick, R.: Mask R-CNN. In: Proceedings of the IEEE International Conference on Computer Vision, pp. 2961–2969 (2017)
9. He, K., Zhang, X., Ren, S., Sun, J.: Spatial pyramid pooling in deep convolutional networks for visual recognition. IEEE Trans. Pattern Anal. Mach. Intell. **37**(9), 1904–1916 (2015)
10. Jung, K., Kim, K.I., Jain, A.K.: Text information extraction in images and video: a survey. Pattern Recogn. **37**(5), 977–997 (2004)
11. Karatzas, D., et al.: ICDAR 2015 competition on robust reading. In: 2015 13th International Conference on Document Analysis and Recognition (ICDAR), pp. 1156–1160. IEEE (2015)
12. Karatzas, D., et al.: ICDAR 2013 robust reading competition. In: 2013 12th International Conference on Document Analysis and Recognition, pp. 1484–1493. IEEE (2013)
13. Lenc, L., Prantl, M., Martínek, J., Král, P.: Border detection for seamless connection of historical cadastral maps. In: Barney Smith, E.H., Pal, U. (eds.) ICDAR 2021. LNCS, vol. 12916, pp. 43–58. Springer, Cham (2021). https://doi.org/10.1007/978-3-030-86198-8_4
14. Liang, J., Doermann, D., Li, H.: Camera-based analysis of text and documents: a survey. Int. J. Doc. Anal. Recognit. (IJDAR) **7**(2), 84–104 (2005)
15. Lowe, D.G.: Distinctive image features from scale-invariant keypoints. Int. J. Comput. Vis. **60**(2), 91–110 (2004). https://doi.org/10.1023/B:VISI.0000029664.99615.94, https://dx.doi.org/10.1023/B:VISI.0000029664.99615.94
16. Lucas, S.M., et al.: ICDAR 2003 robust reading competitions: entries, results, and future directions. Int. J. Doc. Anal. Recognit. (IJDAR) **7**(2–3), 105–122 (2005)
17. Mishra, A., Alahari, K., Jawahar, C.: Scene text recognition using higher order language priors. In: BMVC-British Machine Vision Conference. BMVA (2012)
18. Neumann, L., Matas, J.: A method for text localization and recognition in real-world images. In: Kimmel, R., Klette, R., Sugimoto, A. (eds.) ACCV 2010. LNCS, vol. 6494, pp. 770–783. Springer, Heidelberg (2011). https://doi.org/10.1007/978-3-642-19318-7_60

19. Neumann, L., Matas, J.: Real-time scene text localization and recognition. In: 2012 IEEE Conference on Computer Vision and Pattern Recognition, pp. 3538–3545. IEEE (2012)
20. Redmon, J., Divvala, S., Girshick, R., Farhadi, A.: You only look once: unified, real-time object detection. In: Proceedings of the IEEE Conference on Computer Vision and Pattern Recognition, pp. 779–788 (2016)
21. Ren, S., He, K., Girshick, R., Sun, J.: Faster R-CNN: towards real-time object detection with region proposal networks. Adv. Neural. Inf. Process. Syst. **28**, 91–99 (2015)
22. Smith, R.: An overview of the tesseract OCR engine. In: Ninth International Conference on Document Analysis and Recognition (ICDAR 2007), vol. 2, pp. 629–633. IEEE (2007)
23. Wang, K., Babenko, B., Belongie, S.: End-to-end scene text recognition. In: 2011 International Conference on Computer Vision, pp. 1457–1464. IEEE (2011)
24. Weinman, J.: Geographic and style models for historical map alignment and toponym recognition. In: 2017 14th IAPR International Conference on Document Analysis and Recognition (ICDAR), vol. 1, pp. 957–964. IEEE (2017)
25. Weinman, J., Chen, Z., Gafford, B., Gifford, N., Lamsal, A., Niehus-Staab, L.: Deep neural networks for text detection and recognition in historical maps. In: 2019 International Conference on Document Analysis and Recognition (ICDAR), pp. 902–909. IEEE (2019)
26. Wick, C., Puppe, F.: Fully convolutional neural networks for page segmentation of historical document images. In: 2018 13th IAPR International Workshop on Document Analysis Systems (DAS), pp. 287–292. IEEE (2018)
27. Xiong, Y.J., Wen, Y., Wang, P.S.P., Lu, Y.: Text-independent writer identification using sift descriptor and contour-directional feature. In: 2015 13th International Conference on Document Analysis and Recognition (ICDAR), pp. 91–95 (2015). https://doi.org/10.1109/ICDAR.2015.7333732
28. Yokobayashi, M., Wakahara, T.: Segmentation and recognition of characters in scene images using selective binarization in color space and gat correlation. In: Eighth International Conference on Document Analysis and Recognition (ICDAR 2005), pp. 167–171. IEEE (2005)
29. Zhou, X., et al.: EAST: an efficient and accurate scene text detector. In: Proceedings of the IEEE Conference on Computer Vision and Pattern Recognition, pp. 5551–5560 (2017)

Information Extraction from Handwritten Tables in Historical Documents

José Andrés[1(✉)], Jose Ramón Prieto[1], Emilio Granell[1], Verónica Romero[2], Joan Andreu Sánchez[1], and Enrique Vidal[1]

[1] PRHLT Research Center, Universitat Politècnica de València, Valencia, Spain
{joanmo2,joprfon,emgraro,evidal}@prhlt.upv.es
[2] Departament d'Informàtica, Universitat de València, Valencia, Spain
veronica.romero@uv.es

Abstract. Recently, significant advances have been made in Document Understanding in structured historical documents. However, not much research has been done in information extraction from handwritten structured historical documents. In this paper, we compare two Machine Learning approaches and another approach that is based on heuristic rules to extract information in historical pre-printed forms with handwritten information. We analyze how each approach performs at each step of the extraction process. The proposed approaches improve the heuristic-rule baseline by up to 0.14 F-measure points throughout the information extraction pipeline.

Keywords: Structured handwritten documents · Information extraction · Neural networks

1 Introduction

In the last years, important advances have been achieved in Handwritten Text Recognition (HTR) of historical documents. The current prevalent techniques based on Deep and Recurrent Neural Networks (DRNN) [5] have largely contributed to these achievements. In this way, new problems related to HTR in large collections can now be explored. This paper researches the problem of information extraction from handwritten tables in historical documents.

Information Extraction from tables of historical documents is a very difficult problem given the laxity which the layout was usually considered in the past [8]. This problem is also present even for more recent printed documents for which some interesting solutions have been explored recently [10]. The interest in Information Extraction from historical handwritten tables is enormous since there exist large amounts of huge collections in which the information was registered in tabular format: border records, military registers, hospital

J. Andrés and J.R. Prieto have equally contributed to this paper.

© Springer Nature Switzerland AG 2022
S. Uchida et al. (Eds.): DAS 2022, LNCS 13237, pp. 184–198, 2022.
https://doi.org/10.1007/978-3-031-06555-2_13

records, records related to industrial processes, financial, population, forest, travel records, etc. Extracting the relevant information from these collections would allow researchers to study in more depth the past. Currently adopted solutions for extracting reliable information are mainly based on crowd-sourcing approaches. In these approaches, the crowd does not receive any automatic assistance.[1,2]

Two current limitations to get useful results in these collections are, on the one hand, the complex tabular layout that they have, and on the other hand, the current HTR results, which are not error-free.

The tabular layout can be very free in some situations with lines of running text from left side to right side of the image [19]. In this case, the line extraction process can be very good, and consequently, the HTR results and the extraction of relevant information can be good since they can rely on useful context [20]. The tabular layout can also be composed of pre-printed sheets [21]. Extracting row lines from side to side of the page image has been proven to be not very useful because the linguistic context along the columns in the same row may be not useful [21] and it seems better to extract lines at cell level. In such case, the line detection can be difficult since sometimes only quotes are written to indicate that the value of the preceding row is reproduced. In addition, the HTR results and the extraction of relevant information results cannot be very good since they cannot rely on the HTR context [21].

This paper researches on how to perform information extraction over a set of textlines given its geometric location. Three approaches have been considered for this task. The main idea in the two first ones is to use just the geometrical location to perform information extraction, while the last approach can also extract features from neighboring textlines [15].

The techniques that are researched in this paper are applied to the HisClima dataset [21] which is composed of logbooks of printed forms, and the tables are filled in with handwritten text. Although these forms have a fixed layout, the handwritten text does not strictly respect the cells. In addition, the tables contain difficult vocabulary related with winds, temperatures, atmospheric pressure, etc., which make the recognition process really difficult.

2 Related Work

In recent years, great advances have been made in Document Layout Analysis. Many of these advances have been possible thanks to neural networks, which have taken a leading role. Using a pre-trained encoder, [1] trained a generic deep learning architecture for various layout analysis tasks, such as text line detection and page extraction. In [4] they used Mask Convolutional and Recurrent Network (CRNN) detecting tables as an object detection problem; however, this work only focuses on detecting the table, not its structure. To deal with structure

[1] https://www.zooniverse.org/projects/krwood/old-weather-ww2.
[2] https://fromthepage.com/.

detection, Fully-Convolutional Neural Networks have been used to analyze the tables structures [22].

Graph Neural Networks (GNN) have also started to take on an important role in finding relationships in documents. [18] used a CNN to extract visual features, detecting rows, columns, and cells. Then, they classified the relationships among the different detected objects using GNNs.

However, none of these methods have been applied to handwritten images but assume printed images with high regularity and straightness.

Regardless, progress has been made in the field of Table Understanding *historical handwritten*. A competition was developed for detecting the structure of handwritten tables, [3] track B. One of the teams used a fully CNN and then constructed an adjacency matrix from the detected objects. [13] created a graph from the rows and classified each edge to find rows and columns, doing a subsequent connected component analysis on the initial pruned graph. Edges were classified as nodes in a conjugate graph, created as a pre-process of the initial graph of textlines. However, although the method is powerful, it has certain weaknesses as it depends on the initial graph. In [15], the initial graph was improved, making the method more robust. In this paper, GNN-based neural networks modify the latter research by making it less costly and complex, eliminating conjugate graphs for edge classification.

In the field of Information Extraction (IE) for handwritten historical tables, interesting researches have been published so far. Thus, [8,21] use geometry to select the rows and columns of already known headers. However, it only presents results on ground truth lines. This paper tests methods with automatically detected lines and compares them with those based on neural networks.

3 HisClima Dataset

The HisClima database is a freely available handwritten text database [14] compiled from logbooks of ships that sailed the Arctic ocean from July of 1880 until February of 1881. In this paper we only used one of the logbooks belonging to the Jeannette ship.

These logbooks documents are composed by two different kind of pages, some pages containing tables and other ones containing descriptive text. There is preprinted text in both type of pages. The Jeannette logbook follows this structure and it is composed of 419 pages, 208 correspond to table pages and the other 211 correspond to descriptive text.

Given that the most relevant information for researchers of this kind of documents is included in the tables, that contain a lot of numerical data, we focus only on table pages. These pages are divided into two parts: the upper part is for registering the information in the AM period of each day and the bottom part registers the information referred to the PM period of each day (see Fig. 1). The registered information is related to weather navigation conditions and is plenty of the particular jargon related to this problem: wind type, atmospheric pressure, etc.

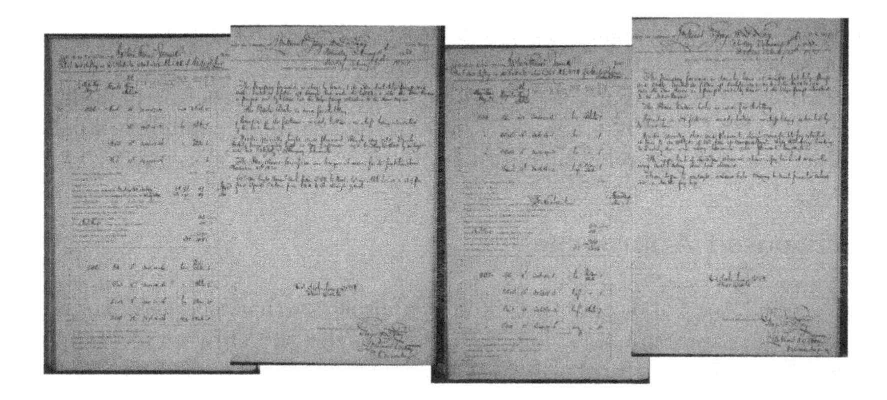

Fig. 1. Example of ship logs with annotation about the weather conditions.

These documents entail some challenges related with different areas such as layout analysis, handwritten recognition and information extraction. One of the main difficulties for automatic layout analysis comes from the fact that the information included in a cell sometimes is replaced with quotation marks (") when the data is the same as the data in the same column in the previous row (or some row above). These quotation marks are very short and sometimes they are difficult to automatically detect. Other layout difficulties that can be found are: data related with a cell that is really written in the upper and lower cells, crossed out column names, words written between cells, and different number of rows completed in every table.

The HisClima database has been endowed with two different types of annotations. First, the layout of each page was annotated to indicate blocks, columns, rows, and baselines. Second, the text was completely transcribed by an expert paleographer, including relevant semantic information. The lines in the table regions were marked at cell level, resulting in a total of 3 525 lines. In addition, the printed and the handwritten text have been labeled with different tags to differentiate between both types of text in the recognition process. Note that the

Table 1. Basic statistics of the HisClima database and average values for the three partitions.

Number of	Train	Validation	Test	Total
Pages	143	15	50	208
Lines	23 617	2 284	7 838	33 739
Running words	46 599	4 604	15 611	66 814
Lexicon	1 287	491	924	1 483
Character set size	76	76	76	76
Rel. information	10 917	1 021	3 533	15 471

printed text is more regular and easy to learn and therefore, it is important to provide separated results for both types of text.

In this paper, the partitions defined in [21] have been used. The main figures are shown Table 1. The last row, *Rel. Information*, shows the number of cells in the tables that contain relevant information.

4 Proposed Approaches

In this section, we will discuss the different approaches that have been employed to face the information extraction task. We consider the problem of extracting textual information in hybrid printed/handwritten tables. We wish to be agnostic with respect to possibly predefined table layouts, but we assume tables to be organized into orthogonal rows and columns. Each column typically has at least one column header h_c, and each row has a row header h_r. The textual content of interest v is contained in cells which are the intersection of columns and rows.

Moreover, to perform information extraction over a tabular image x, a set of possible column headers (one per column) and row headers (one per row) are (semi-automatically) extracted from the ground-truth transcripts of the training table images. In addition, we assume that the regions of x matching h_c, h_r and v are small bounding boxes (BBs), b_c, b_r and b_v respectively, each tightly containing the corresponding textual contents.

The following approaches take as input the extracted textlines and their 1-best transcripts. As output, they return the textual contents of interest v associated to each tuple (h_c, h_r). A diagram showing the necessary steps to go from images to a database filled with the extracted information can be seen in Fig. 2.

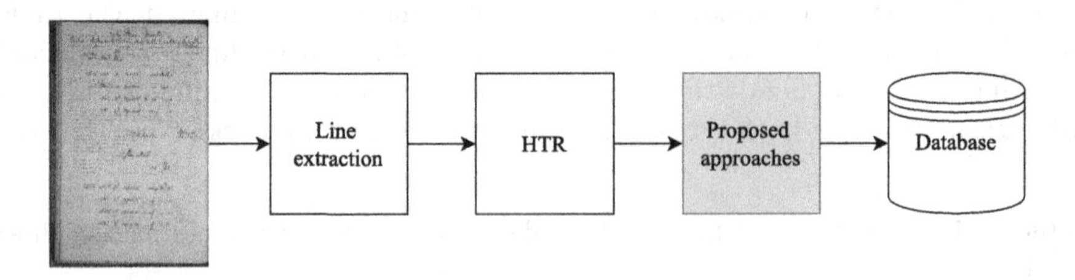

Fig. 2. Overall diagram of the methodological pipeline. It is highlighted in red the phase which accounts for the information extraction techniques described in this section. (Color figure online)

4.1 Heuristic Geometric Information

This approach follows the information retrieval method presented in [21]. This method is based on the geometrical information of a structured multi-word query and the 1-best transcript of the detected lines in a table page.

The retrieval process is carried out in three steps for each table page. First, every column header (h_c) word is retrieved. Second, the row header (h_r) words are retrieved by looking for all the lines that appear below the first column region, delimited by the horizontal span of the column-heading word line. Then, every cell-content word is searched by the combination of the corresponding column and row regions. Finally, for those cells whose content is a double quotation marks, it is replaced by the content of the first previous cell different to the double quotation marks.

4.2 Log-Linear Model

As an alternative technique to solve this problem and building on the foundations of the previous approach, we propose making use of the geometric information employing a combination of robust machine learning models.

In this method, we assume that the structure of a table is given by the positions of its column headers, row headers and cells of interest. Nevertheless, two problems should be addressed before starting to perform information extraction: the multi-line cells and the quotation marks. As discussed before in Sect. 3, there are cells whose content is spread across multiple lines. This is a problem since, when we perform information extraction, we are interested in extracting the whole cell. Similarly, we aim at extracting the textual content that a quotation marks symbol represents, not the quotations themselves. Therefore, this approach entails three consecutive steps: grouping the textlines that belong to the same semantic cell, substituting the quotation marks for the content they represent, and finally, performing information extraction over the resulting groups of textlines. This pipeline can be seen in Fig. 3.

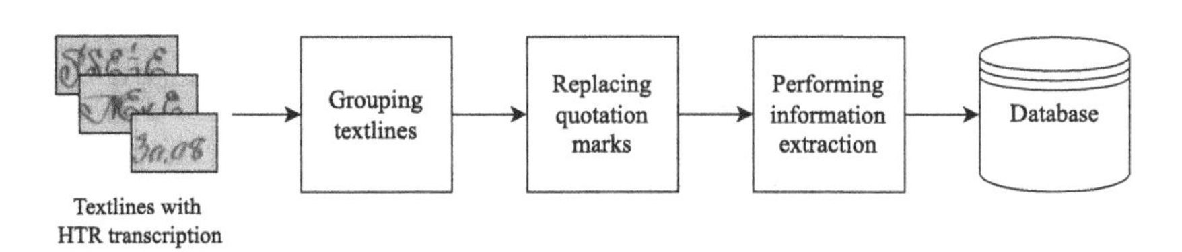

Fig. 3. Pipeline of the log-linear model approach. In the first phase, the textlines are grouped into semantic cells. Then, the quotation marks are replaced by the textual content they represent. Finally, information extraction is performed over the resulting set of semantic cells.

Grouping Textlines. This phase can be divided in two steps. In the first step, we group two textlines b_v and $b_{v'}$ when their probability of being in the same cell is greater than a given threshold t_1. To express this formally, we introduce the binary random variable C, which denotes if b_v and $b_{v'}$ belong to the same semantic cell. Therefore, we will group b_v and $b_{v'}$ when $P(C = 1 \mid b_v, b_{v'}) \geq t_1$.

To learn this probability, an MLP has been employed. As input features we have considered the x coordinate of b_v, the absolute distance between b_v and

$b_{v'}$ in the y-axis and in the x-axis. With the use of the two first attributes, the system learns which is the expected distance in the y-axis between two textlines that belong to the same cell while taking into account their position in the x-axis. Moreover, with the last considered attribute, the system learns which deviation can be expected in the x-axis. Note that the multi-line cells appear only in a concrete image region in this collection. Therefore, the allowed distance in the y-axis in that image region will be larger than in other regions.

Once this first step is finished, we will have the textlines grouped in semantic cells. However, this procedure could leave us with textlines appearing in multiple cells simultaneously, as the network might have trouble grouping the most distant textlines. To overcome this problem, the second step of this phase consists in merging two groups of textlines if they share at least a textline. An example is shown in Fig. 4. The output of this second step is the final set of semantic cells.

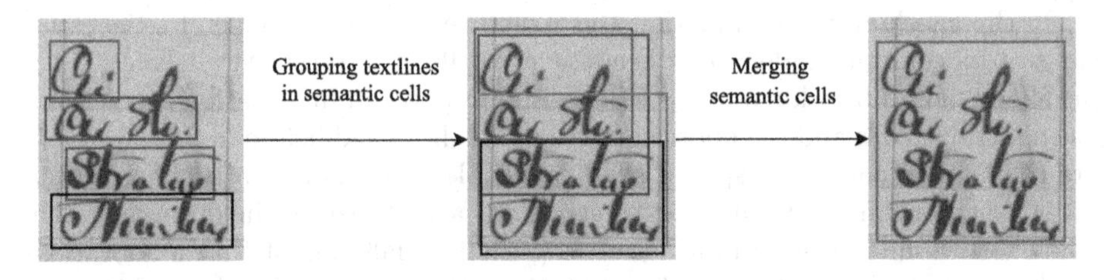

Fig. 4. Example showing the two steps that conform the grouping textlines procedure. At the beginning, we have four different textlines. Then, we apply the first step and group the textlines into semantic cells. As can be seen in the middle figure, the first step might fail to group the most distant textlines. However, when we perform the second step, the semantic cells that share at least one textline are merged, leading us to the semantic cell shown in the last figure.

Replacing Quotation Marks. In this phase, our objective is to substitute the quotation marks for the textual contents they represent. Note that every quotation mark has a unique textual content it refers to. Therefore, we replace the textual content of each quotation marks BB, b_q, with the textual content of the semantic cell, b_v, with largest probability of being the precedent cell of b_q (and consequently, the cell b_q is referring to). This probability is denoted as $P(S = 1 \mid b_v, b_q)$, where S is a binary random variable that denotes if b_v is the semantic cell to which b_q is referencing.

This probability has been approximated by means of an MLP, where the distance in the y-axis and the absolute distance in the x-axis between b_v and b_q have been adopted as input features. By doing this, the model learns at which distance in both axes the referenced cell b_v is expected to be found. It is worth mentioning that the textual contents of b_v might also be another quotation marks. In that case, we would repeat the replacing quotation marks process but employing the previous b_v as current b_q, until achieving a b_v whose textual contents are not a quotation marks. The final output of this phase is the set of semantic cells without quotations.

Performing Information Extraction. Finally, after grouping the textlines and replacing the quotation marks by the textual content they represent, we can perform information extraction over the resulting semantic cells.

In this task, we extract the textual content of interest v when its probability of being relevant to a header tuple (h_c, h_r) is larger than a given threshold t_2. This is formally expressed as extracting v when $P(\mathcal{R} = 1 \mid h_c, h_r, v) \geq t_2$, where $\mathcal{R}$ is a binary random variable which indicates if v is relevant to (h_c, h_r). Note that, following the ideas presented in [24], $P(\mathcal{R} = 1 \mid h_c, h_r, v)$ can be approximated as:

$$P(\mathcal{R} = 1 \mid h_c, h_r, v) \approx \max P(\mathcal{R} = 1 \mid h_c, h_r, v, b_c, b_r, b_v) \tag{1}$$

To model this distribution we have utilized a combination of probabilities, where each of them accounts for a different structural component of a table. They are the following:

The first probability is $P(H_c = 1 \mid b_c)$, where H_c is a binary random variable that denotes if a semantic cell b_c is classified as a column header. The second considered probability is $P(H_r = 1 \mid b_r)$, where H_r is a binary random variable that denotes if b_r is classified as a row header. The third probability that has been used is $P(A_c = 1 \mid b_c, b_v)$, where A_c is a binary random variable that denotes if cell b_v is aligned with a given column header b_c. Finally, the last probability distribution that has been considered is $P(A_r = 1 \mid b_r, b_v)$, where A_r is a binary random variable that denotes if a semantic cell b_v is aligned with respect to a given row header b_r.

With the purpose of modeling these probabilities, different attributes have been considered. For $P(H_c \mid b_c)$, the x-coordinate of b_c has been employed. Analogously, the y-coordinate of b_r has been considered for $P(H_r \mid b_r)$. This choice of attributes is due to the fact that column headers are expected to appear at the top of the page, while row headers are expected to appear at the left side of the table. Moreover, the absolute distance in the x-axis between b_c and b_v has been chosen as attribute for $P(A_c \mid b_c, b_v)$, while the homologous distance in the y-axis has been utilized for $P(A_r \mid b_r, b_v)$. When employing these attributes, $P(A_c \mid b_c, b_v)$ learns which is the expected deviation in the x-axis between two BBs which are considered to be in the same column, while $P(A_r \mid b_r, b_v)$ learns the expected deviation in the y-axis when two BBs are considered to be in the same row. To learn each of these probabilities, a Perceptron has been used.

With the purpose of combining these probabilities, the log-linear framework has been followed. This framework has been used with great success in diverse fields such as statistical machine translation [11] and automatic speech recognition [6] among others. When using it in this context, we find a set of 4 feature functions, $g_1, ..., g_4$, which we want to combine taking into account their corresponding weight, $\lambda_1, ..., \lambda_4$. Note that as feature functions, we have utilized the four previously described probabilities. It is worth mentioning that, in this task we are interested in the textual content of interest v when $P(\mathcal{R} \mid h_c, h_r, v) \geq t_2$, and not necessarily in its actual probability. Therefore, we could get rid of the normalization term of a log-linear model, leading us to:

$$f(\mathcal{R}, h_c, h_r, v, b_c, b_r, b_v) = \exp(\sum_{m=1}^{4} \lambda_m g_m(\mathcal{R}, h_c, h_r, v, b_c, b_r, b_v))$$

Finally, the whole information extraction process described in this approach could be summarized as follows: Firstly, we group the textlines into semantic cells. Secondly, we substitute the quotation marks with the content they represent. Thirdly, employing the list of predefined column headers and row headers described in Sect. 4, we calculate the score of each textual content of interest v for each possible tuple (h_c, h_r). When the achieved score of v for a tuple is greater than a given threshold t_2, we extract v.

4.3 Graph Neural Network

Techniques based on neural network graphs (GNN) have been used to extract sub-structures of each table (rows and columns) from previously detected textlines. The first step is to create a graph by connecting the textlines by their line of sight (LoS) and, in addition, using the improvements on the graph of [15]. These techniques result in a graph depending on the sub-structure to be detected. Then, a score in $[0, 1]$ is obtained for each edge, which can be interpreted as an estimate of the probability that the corresponding edge joins two elements of the same sub-structure.

In this work, the GNN-based techniques of [15] have been slightly modified. Instead of using the conjugate graph, the raw graph is used directly to classify edges using Eq. 2. This result in a more memory-efficient method, with no pre-processing (conjugation) and faster to train since conjugation treats each edge as a node, and this makes it much more expensive when the number of edges grows. In this work, the number of nodes is kept the same, growing the number of edges but avoiding conjugation.

In order to obtain the classification of the edges between the origin (s) and destination (d), Eq. 2 is used, being $\boldsymbol{x}'_s$ and $\boldsymbol{x}'_d$ the embeddings associated to the nodes calculated by the GNN, from origin and destination respectively, and $\boldsymbol{e}_{s,d}$ edge properties given these node embeddings. From $\boldsymbol{x}'_s$ and $\boldsymbol{x}'_d$ another vector is obtained from the result of the absolute value of the subtraction of each component, expressed as $|\boldsymbol{x}'_s - \boldsymbol{x}'_d|$ in the equation. $\boldsymbol{e}_{s,d}$ is the value calculated at the time of creating the graph and remains unchanged. The vector function $Q : \mathbb{R}^{\phi} \times \mathbb{R}^{\gamma} \rightarrow \mathbb{R}$ is implemented as an MLP with a sigmoid activation function, therefore the results can be interpreted as a probability of the edge between s and d nodes, and trained together with the GNN as a whole network.

$$\hat{y}_{s,d} = Q(|\boldsymbol{x}'_s - \boldsymbol{x}'_d|, \boldsymbol{e}_{s,d}) \tag{2}$$

At the same time, with another GNN the textlines headers are detected. In this case, each textline, which is equivalent to a node in the original network, is classified as a binary header detection problem.

With all the information from the three trained GNNs (row, column and header detection), a process is performed to obtain, for each existing cell, the row and the column to which it belongs, as well as the header of that column.

Note that with these GNN-based methods, multi-line cells are naturally detected by intersecting rows and columns, resulting in a cell with all the lines in it.

5 Evaluation Criteria and Metrics

The transcripts quality obtained by the HTR system is assessed using the Levenshtein edit distance [9] with respect to the reference text, at both the word and the character levels. The Character Error Rate (CER) is the Levenshtein edit distance at character level and it can be defined as the minimum number of substitutions, deletions and insertions needed to transform the transcription into the reference text, divided by the number of characters in the reference text. Similarly, Word Error Rate (WER) is this edit distance calculated at word level.

The performance of the different information extraction approaches has been assessed employing the precision (P), recall (R) and its harmonic mean (F_1). In this context, the precision denotes the percentage of the retrieved cells that are correct according to the GT, while the recall denotes the proportion of relevant cells that has been retrieved from all the relevant cells. Finally, the F_1 accounts for the balance between both metrics.

It is worth noting that this task has been evaluated at cell level. Therefore, a match is considered a *true positive* (TP) when the content retrieved by the system is exactly the same as the content found in the GT for that cell. Otherwise, a match is considered a *false positive* (FP). Moreover, when the contents of a cell in the GT are not retrieved, this is considered as a *false negative* (FN).

Finally, we would like to remark that 95% confidence intervals ($\alpha = 0.025$) have been calculated using the bootstrap method with 10 000 repetitions [2].

6 Experimental Framework and Results

Different experiments have been performed to assess the text recognition and the proposed information extraction approaches. In the following sections, we will describe the experimental framework that has been employed and the results that have been achieved.

6.1 Experimental Settings

Handwriting Text Recognition. The text recognition system used in this work is based on the technology described in [16] and is implemented by using the PyLaia [17] toolkit. The input images were pre-processed to correct the slope and inclination.

Optical models are based on Convolutional and Recurrent Neural Networks (CRNN), consisting of a convolutional block and a recurrent block. This model consists of three convolutional layers with filters composed of different maps of characteristics (16, 32 and 48) with kernel sizes of 3×3 pixels and horizontal and vertical dilations of 1 pixel. *LeakyReLU* is used as activation function, and the output of the convolutional layers is fed to a layer of *max pooling* with non-overlapping of 1×1 pixels, only at the output of the first two layers. After that, the recurrent block is made up of three recurrent layers composed of 256 bidirectional long-short term memory units. Finally, a fully connected linear layer is used after the recurrent block. All hyper-parameters, such as the number of convolutional and recurring layers were configured in the validation set. It is important to remark that the optical models for the printed characters and for the handwritten characters were the same.

A language model was estimated as 3-gram of characters directly from the transcripts of the included lines of text on the training and validation partitions using the SRILM [23] toolkit. Just a language model was used, both for the printed part and for the handwritten part.

As input for the three different approaches that are described below, the 1-best transcript obtained using the previously described HTR system has been employed.

Heuristic Geometric Information. In the case of the heuristic geometric information method, the geometric position of every detected line in the page was used.

Log-Linear Model. MLPs and Perceptrons have been employed to tackle this problem. Concretely, the employed MLPs are formed by one hidden layer of 128 units plus the final layer for binary classification.

To optimize every network, we have employed Adam [7] solver with a learning rate of 0.01. To handle class imbalance, a version of weighted cross-entropy has been used as a loss function. In this version, weights per class were computed as $w_k = \frac{s_m}{s_k}$, where s_k denotes the number of samples of the class k and s_m denotes the number of samples of the majority class. As batch size, we have employed 256. As metric for choosing the best model of each network during training, we have employed the F_1 score obtained over the validation dataset.

Finally, the weights of the log-linear model and the employed threshold in the grouping textlines phase (t_1) have been estimated jointly employing the Powell Search algorithm [12], while the threshold used to associate semantic information to a textual content of interest (t_2) has been estimated as the optimal threshold obtained when calculating the F_1 score over the validation dataset.

Graph Neural Network. We have used the same configuration for the three cases (rows, columns and headers), with four layers of 64 filters each in the first steps of the GNN and, finally, the MLP composed of four layers of 64 neurons each, plus the final layer of binary classification.

To optimize the GNNs we used minibatch SGD and Adam solver [7] with a learning rate of 0.01. To reduce false positives, which have a very detrimental effect, a version of weighted cross-entropy has been used as a loss function. In this version, cross-entropy values of the negative class are multiplied by a weight $w_0 = \frac{\alpha}{log\epsilon + p_0}$, $\epsilon \geq 0$, where p_0 is the prior-probability of the negative class, after some tests, the hyper-parameter α has been set to 5 in all experiments. The networks were trained for 6000 epochs. To consider an edge as positive, a threshold of 0.95 has been used on $\hat{y}_{s,d}$, minimizing false positives.

In order to improve the created graph we used, in the case of columns σ_1 were set to 1 and σ_2 to 10 and the other way round in the case of columns. To detect headers we used the original graph.

Line Detection. Line detection was carried out with MaskRCNN implemented with Detectron2 tool [25].

6.2 Text Recognition Results

Table 2 presents the obtained text recognition results for the test partition. As it can be observed, the error in the overall results is quite low (CER = 2.0% and WER = 4.4%). But as noted in previous sections, in this type of experiments is quite relevant to distinguish between the errors in the printed part and in the handwritten part. The GT was annotated with this information. We can observe that the recognition of the handwritten text (CER = 5.7% and WER = 10.4%) represents a greater challenge than the printed text (CER = 1.5% and WER = 1.8%). These results are reported with the GT, which means that all annotated lines were used to compute these values.

Table 2. Results of text recognition. 95% confidence intervals are never larger than 1.1% for manuscript text, 0.4% for printed text and 0.5 % overall.

Text type	Manuscript	Printed	Overall
CER	5.7%	1.5%	2.0%
WER	10.4%	1.8%	4.4%

6.3 Information Extraction Results

Table 3 reports the results obtained when extracting information from the GT lines and from the automatically extracted lines, using the HTR system previously described in both cases. Note that the former experiment simulates the scenario in which a perfect line detector is used. Therefore it can be considered optimistic. The latter experiment simulates a real scenario.

In order to qualify the results obtained by our approaches, we are interested in knowing which is the maximum performance that could be achieved if our systems made no errors obtaining the structure of columns, rows and headers. These results are denoted as oracle in Table 3. On the one hand, when employing GT lines, the only errors present in the oracle are the ones corresponding to HTR. On the other hand, another source of errors is added when employing automatic lines, which is a possible bad line detection. Oracle/Automatic cells are indirectly measuring the error detection line.

Table 3. Information extraction results. 95% confidence intervals are never larger than 0.01 when using GT lines and 0.02 when employing automatic lines.

Lines	Ground thruth			Automatic		
Metric	P	R	F_1	P	R	F_1
Heuristic geometric information	0.79	0.78	0.78	0.64	0.55	0.59
Log-linear model	0.87	0.79	0.83	0.77	**0.69**	**0.73**
Graph neural network	**0.88**	**0.83**	**0.85**	**0.78**	0.67	0.72
Oracle	0.89	0.88	0.89	0.79	0.72	0.76

Results in Table 3 show that the best performance using GT lines has been achieved by the GNN, reaching 0.85 out of the 0.89 possible F_1 points, while the best results with automatic lines have been achieved by the log-linear model, which has obtained 0.73 out of the 0.76 possible F_1 points. Nevertheless, it is worth noting that the difference in F_1 between both approaches is very narrow and therefore, a similar performance in practice is expected.

Moreover, it can be seen that the worst performance has been achieved by the heuristic geometric information approach, achieving 0.78 F_1 points and 0.59 F_1 points when employing GT and automatic lines respectively.

Finally, we would like to remark that if we compare the oracle results with GT and automatic lines, the automatic extraction leads us to a loss of 0.13 F_1 points.

7 Discussion

First of all, it is clear that there is a significant difference in performance between the heuristic approach and the other two employed techniques. This difference is due to the fact that this method is not able to group textlines, and therefore, it is not able of retrieving correctly any multi-line cell. The neural networks methods are able of joining textlines when they belong to the same cell, and therefore have obtained significantly better results. As mentioned before, a similar performance has been achieved by these methods.

The log-linear model approach is much less computationally expensive than the GNN method, as it is formed by simple neural networks, such as MLPs and Perceptrons. However, it presents a major drawback: it needs to choose the appropriate attributes to model the different distributions that have been employed, and therefore, a study should be carried out to choose the appropriate attributes for each new dataset.

On the other hand, the GNN method is much more computationally expensive than the other approaches. Nevertheless, it presents a much wider generalization capability, allowing this approach to be employed in other layout understanding tasks. Moreover, this generalization ability lets the GNN to automatically learn the appropriate attributes without having to manually select them.

8 Reproducibility

We hope this work will help to boost research on how to perform information extraction over handwritten historical tables. For this reason, the source code used in the experiments presented in this paper is freely available on https://github.com/PRHLT/table-ie-das2022, along with the subsets used as training, validation, test[3].

9 Conclusions

This paper has shown a research on information extraction on handwritten historical tabular documents. Two of the employed approaches are based on Machine Learning techniques while the other is based on heuristic rules. Empirical results show that the best results are obtained by the Machine Learning approaches, achieving results close to the maximum reachable considering the HTR and line detection errors.

Taking this information into account, future efforts should be focused on improving the automatic line extraction and the HTR transcriptions. A possible solution to overcome HTR errors is to use *probabilistic indices* [24] instead of the 1-best transcripts given that, where the HTR fails, the *probabilistic indices* might have other accurate hypotheses. As another possible solution to face this issue, an end-to-end system which performs the line detection, HTR transcription and information extraction could be developed.

Acknowledgement. Work partially supported by the Universitat Politècnica de València under grant FPI-I/SP20190010 (Spain), by Generalitat Valenciana under project DeepPattern (PROMETEO/2019/121), by grant PID2020-116813RB-I00a funded by MCIN/AEI/10.13039/501100011033, by the Generalitat Valenciana under the project GV/2021/072, by grant RTI2018-095645-B-C22 funded by MCIN/AEI/10.13039/501100011033 and by "ERDF A way of making Europe".

References

1. Ares Oliveira, S., Seguin, B., Kaplan, F.: DhSegment: a generic deep-learning approach for document segmentation. In: Proceedings of International Conference on Frontiers in Handwriting Recognition, ICFHR, pp. 7–12 (2018)
2. Bisani, M., Ney, H.: Bootstrap estimates for confidence intervals in ASR performance evaluation. In: ICASSP, vol. 1, pp. 409–412 (2004)
3. Gao, L., et al.: ICDAR 2019 competition on table detection and recognition (cTDaR). In: Proceedings of the International Conference on Document Analysis and Recognition, ICDAR, pp. 1510–1515 (2019)
4. Gilani, A., Qasim, S.R., Malik, I., Shafait, F.: Table detection using deep learning. Proceedings of the International Conference on Document Analysis and Recognition, ICDAR, vol. 1, pp. 771–776 (2017)
5. Graves, A., Liwicki, M., Fernández, S., Bertolami, R., Bunke, H., Schmidhuber, J.: A novel connectionist system for unconstrained handwriting recognition. IEEE TPAMI **31**(5), 855–868 (2009)

[3] https://doi.org/10.5281/zenodo.4106887.

6. Heigold, G.: A log-linear discriminative modeling framework for speech recognition (2010)
7. Kingma, D.P., Ba, J.L.: Adam: a method for stochastic optimization. In: 3rd International Conference on Learning Representations, ICLR 2015 - Conference Track Proceedings (2015)
8. Lang, E., Puigcerver, J., Toselli, A.H., Vidal, E.: Probabilistic indexing and search for information extraction on handwritten German parish records. In: 2018 16th International Conference on Frontiers in Handwriting Recognition (ICFHR), pp. 44–49 (2018). https://doi.org/10.1109/ICFHR-2018.2018.00017
9. Levenshtein, V.I.: Binary codes capable of correcting deletions, insertions, and reversals. Sov. Phys. Dokl. **10**(8), 707–710 (1966)
10. Lin, W., et al.: ViBERTgrid: a jointly trained multi-modal 2D document representation for key information extraction from documents. In: Lladós, J., Lopresti, D., Uchida, S. (eds.) ICDAR 2021. LNCS, vol. 12821, pp. 548–563. Springer, Cham (2021). https://doi.org/10.1007/978-3-030-86549-8_35
11. Och, F.J., Ney, H.: Discriminative training and maximum entropy models for statistical machine translation. In: Proceedings of the 40th Annual meeting of the Association for Computational Linguistics, pp. 295–302 (2002)
12. Powell, M.J.: An efficient method for finding the minimum of a function of several variables without calculating derivatives. Comput. J. **7**(2), 155–162 (1964)
13. Prasad, A., Dejean, H., Meunier, J.L.: Versatile layout understanding via conjugate graph. In: Proceedings of the International Conference on Document Analysis and Recognition, ICDAR, pp. 287–294 (2019)
14. PRHLT: HisClima dataset, October 2020. https://doi.org/10.5281/zenodo.4106887
15. Prieto, J.R., Vidal, E.: Improved graph methods for table layout understanding. In: Lladós, J., Lopresti, D., Uchida, S. (eds.) ICDAR 2021. LNCS, vol. 12822, pp. 507–522. Springer, Cham (2021). https://doi.org/10.1007/978-3-030-86331-9_33
16. Puigcerver, J.: Are multidimensional recurrent layers really necessary for handwritten text recognition? In: ICDAR, vol. 01, pp. 67–72, November 2017
17. Puigcerver, J., Mocholí, C.: Pylaia (2018). https://github.com/jpuigcerver/PyLaia
18. Qasim, S.R., Mahmood, H., Shafait, F.: Rethinking table recognition using graph neural networks. In: Proceedings of the International Conference on Document Analysis and Recognition, ICDAR, pp. 142–147 (2019)
19. Romero, V., et al.: The ESPOSALLES database: an ancient marriage license corpus for off-line handwriting recognition. Pattern Recogn. **46**(6), 1658–1669 (2013)
20. Romero, V., Fornés, A., Granell, E., Vidal, E., Sánchez, J.A.: Information extraction in handwritten marriage licenses books. In: Proceedings of the 5th International Workshop on Historical Document Imaging and Processing, HIP 2019, pp. 66–71 (2019)
21. Romero, V., Sánchez, J.A.: The HisClima database: historical weather logs for automatic transcription and information extraction. In: ICPR (2020)
22. Siddiqui, S.A., Fateh, I.A., Rizvi, S.T.R., Dengel, A., Ahmed, S.: DeepTabStR: deep learning based table structure recognition. In: Proceedings of the International Conference on Document Analysis and Recognition, ICDAR, pp. 1403–1409 (2019)
23. Stolcke, A.: SRILM-an extensible language modeling toolkit. In: Proceedings of the 3rd Annual Conference of the International Speech Communication Association (Interspeech), pp. 901–904 (2002)
24. Vidal, E., Toselli, A.H., Puigcerver, J.: A probabilistic framework for lexicon-based keyword spotting in handwritten text images. arXiv preprint arXiv:2104.04556 (2021)
25. Wu, Y., Kirillov, A., Massa, F., Lo, W.Y., Girshick, R.: Detectron2 (2019). https://github.com/facebookresearch/detectron2

Named Entity Linking on Handwritten Document Images

Oliver Tüselmann(✉) and Gernot A. Fink

Department of Computer Science, TU Dortmund University, 44227 Dortmund, Germany
{oliver.tueselmann,gernot.fink}@cs.tu-dortmund.de

Abstract. Named Entity Linking (NEL) is an information extraction task that semantically enriches documents by recognizing mentions of entities in a text and matching them against an entry in a Knowledge Base (KB). This semantic information is fundamentally important for realizing a semantic search. Furthermore, it serves as a feature for subsequent tasks (i.e. Question Answering) as well as for improving the user experience. Current NEL approaches and datasets from the Document Image Analysis community are mainly focusing on machine-printed documents and do not consider handwriting. This is mainly due to the lack of annotated NEL handwriting datasets. To fill this gap, we manually annotated the well known IAM and George Washington datasets with NEL labels and created a synthetic handwritten version of the AIDA-CoNLL dataset. Furthermore, we present an evaluation protocol as well as a baseline approach.

Keywords: Named entity linking · Named entity disambiguation · Document image analysis · Information retrieval · Handwritten documents

1 Introduction

Over the last few years, we observe an increasing interest of the Document Image Analysis (DA) community towards the semantic analysis of documents. Thereby, the understanding of a document is becoming increasingly important. Several tasks from the field of Natural Language Processing (NLP), such as Named Entity Recognition (NER) [3,5,7,39] and Question Answering (QA) [22,23], have already been investigated in the context of document images. Beside these two, Named Entity Linking (NEL) constitutes another task of high interest in the DA community. NEL recognizes entities in unstructured texts and links them to their corresponding entries in a Knowledge Base (KB) (see Fig. 1). KBs (e.g. Wikidata [40], DBpedia [20]) contain information about entities and usually their relations. Entries in a KB are often so-called Named Entities (NEs). These are objects from the real world, such as persons, places, organizations or events. NEL is particularly useful for indexing document images and thus realizing a semantic search. Moreover, the extracted information can contribute to subsequent tasks and can improve the user experience [35].

© Springer Nature Switzerland AG 2022
S. Uchida et al. (Eds.): DAS 2022, LNCS 13237, pp. 199–213, 2022.
https://doi.org/10.1007/978-3-031-06555-2_14

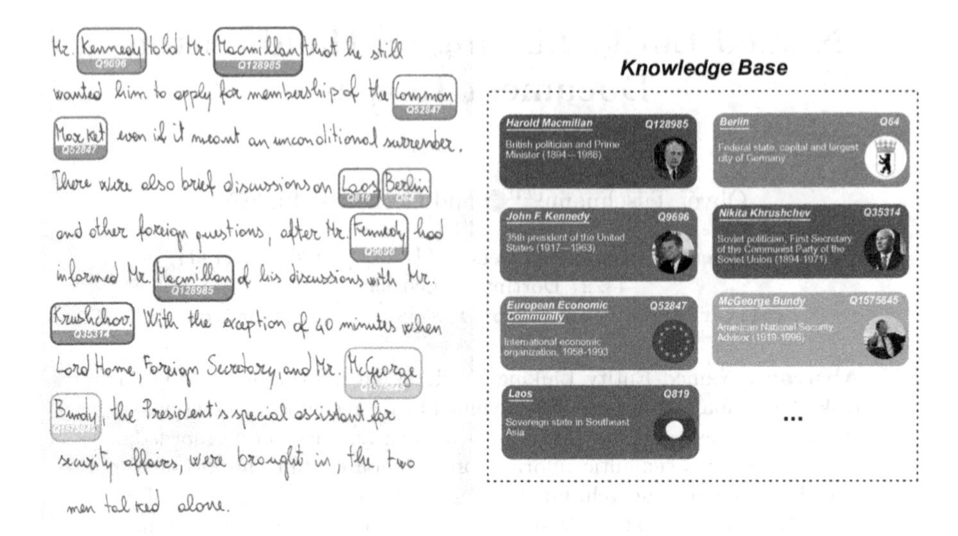

Fig. 1. An example for linking Named Entities against Wikidata on a handwritten document image from the IAM database.

The information extraction task of NEL is an ongoing research topic in the NLP community. Multiple approaches and datasets already exist for this task [35]. In DA, there is also an active community working on the recognition and mapping of entities in historical newspapers [7]. Approaches and datasets mainly focus on machine-printed documents and do not consider handwriting separately. In [22], Mathew et al. recently showed that Question Answering approaches trained on machine-printed data perform rather limited on handwritten document images and they suggested that specialized models are needed. This is mainly due to the fact that handwriting generally has more variability compared to machine-printed text, leading to more errors during recognition. Furthermore, there is a difference between machine and handwritten recognition errors, which would make a simple adaptation less than optimal [22].

To the best of our knowledge, there is only one work dealing with the task of NEL on handwritten document images [13]. This work focuses on a historical, non-public dataset. The lack of publicly available datasets hinders the investigation, development and comparison of NEL approaches on handwritten documents. To tackle this problem, we present three new datasets for this task. The first dataset is synthetically generated based on AIDA-CoNLL [14], the standard benchmark in the textual domain. In contrast, the other two datasets feature real handwriting, that we manually labeled with NEL annotations. We make the datasets publicly available for the community and present an evaluation protocol as well as an initial baseline approach[1].

[1] https://patrec.cs.tu-dortmund.de/cms/en/home/Resources/index.html.

Many places and monuments have been named in honour of Washington PERSON ,
most notably the capital of the United States LOCATION , Washington, D.C. LOCATION
In addition, Washington LOCATION is the only US LOCATION state named after a president.

Fig. 2. A visualization of the mention detection step. The task does not include the prediction of Named Entity tags.

2 Named Entity Linking

The general goal of NEL is to recognize entities in an unstructured text and assign them to a unique identity from a given KB. To achieve a mapping between mentions in a text and entities in a KB, state-of-the-art approaches usually divide the task into a mention detection and an entity disambiguation phase [15]. In the following, we describe the two main steps of NEL approaches in more detail.

In the literature the term Named Entity Linking is often used as a synonym for Named Entity Disambiguation (NED). It is important to distinguish both tasks, as NED skips the mention detection step. Therefore, only gold standard mentions are considered and have to be disambiguated.

Mention Detection. The first stage of a NEL system detects mentions of entities from a given KB in an unstructured text (see Fig. 2). For this task Named Entity Recognition (NER) models are often used [35]. These models do not only recognize NEs but also predict if an identified NE is, e.g., a person, an organization or an event. Even if some NEL systems use these class predictions for their further steps, only the identification of mentions is important in this phase. Traditional NER methods are mainly implemented based on handcrafted rules, orthographic features or dictionaries [42]. Statistical-based methods such as Hidden Markov Models and Conditional Random Fields (CRFs) achieved further progress in this field [41]. In recent years, deep neural network approaches greatly improved the recognition accuracies. Especially combinations of recurrent neural networks, CRFs and pre-trained word embeddings have been successful [18]. For a detailed overview of NER, see [42].

Entity Disambiguation. In most approaches, the disambiguation process is further separated into the candidate generation and entity ranking stages [10,15,19]. The candidate generation step creates a list of possible entities that are associated to an identified mention from the previous step (see Fig. 3). An intuitive realization is to find entities in the KB which textually match the mention [2,35,36]. To counteract the variability of mentions, heuristics like the Levenshtein distance or normalization are used [35]. Since entities are often associated with different names, state-of-the-art methods usually rely on dictionaries with additional aliases for entities [35]. Such dictionaries are often generated from the disambiguation and redirect pages of Wikipedia. There are several specialized resources available online containing aliases and synonyms of entities

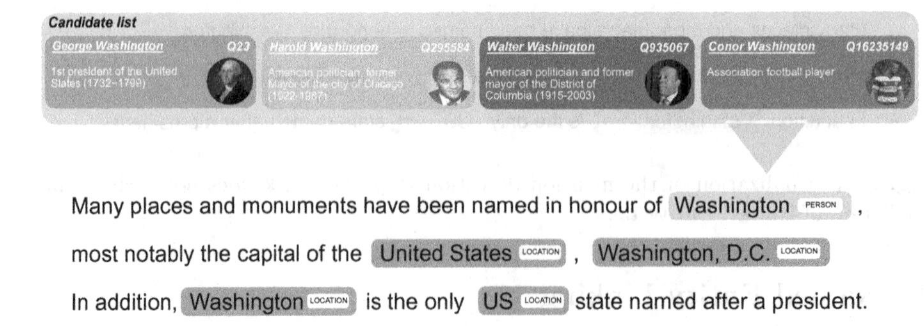

Fig. 3. An example showing the generation of candidates for the first mention of *Washington* in the sample.

[14]. Most systems rely on precalculated prior probabilities of correspondence between mention and entities $p(e|m)$ [10,15,17,29,44]. This probability is commonly computed based on Wikipedia hyperlinks, where the URL provides the entity e and the corresponding text of the link provides the mention m [38].

Entity Ranking is the final step of the traditional NEL pipeline and is usually interpreted as a retrieval problem. The system assigns a score to each entity from the candidate generation step, indicating how well the entity matches the given mention (see Fig. 4). For the disambiguation process it is crucial to capture the semantic information from the context in which the mention appears [35]. Therefore, state-of-the-art approaches usually produce a vector representation for the given mention in its context and also for each candidate entity [2,35]. Finally, the similarity between the mention and entity representation is computed. Traditional NEL approaches typically use handcrafted features to calculate similarities between a mention and its candidate entities [14,31,36]. In recent years, approaches based on neural networks outperformed traditional ones [35]. The general idea is to represent the mention, context and entities as vector representations and to use a neural architecture to compute similarity scores between the given entities and the mention. There are different strategies in the literature to encode entities. A widely used strategy is to map entities into a continuous vector space, such that entity representations are embedded into the same semantic space as words [19,44]. Another strategy uses relations between entities in a KB and graph embedding methods [34]. Lately, neural encoders are used to convert textual descriptions of entities into embeddings and tackle the task with a self-attention model based on a pre-trained BERT model [43].

While it is common to separate the entity recognition and disambiguation step, a few systems provide a joint model [4,17]. It is possible that some mentions do not have their corresponding entity in the given KB. Therefore, NEL approaches should also be able to recognize if a mention does not exist in the KB. This prediction could be interpreted and realized as a classification problem with rejection [35]. For a more detailed overview of textual NEL, see [2,35,36].

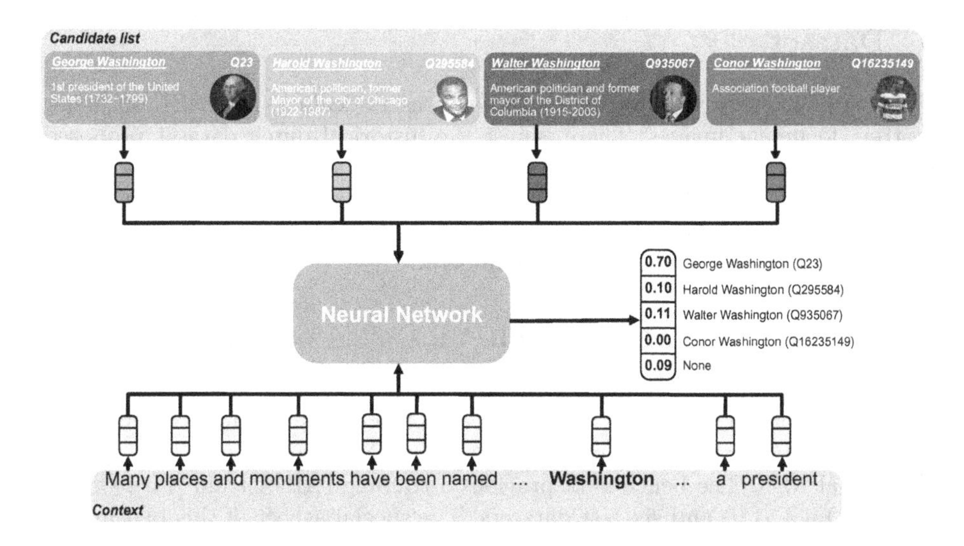

Fig. 4. An overview of the entity ranking step for the first mention of *Washington* given the candidate list generated in Fig. 3.

Document Image Domain. In recent years, several works on the recognition and mapping of entities in historical document images were proposed [8,27]. Traditional approaches use rule based systems [12] or just concentrate on specific types of entities [37]. There are also multiple works that use available NEL approaches from the literature to analyze historical data. For example, Ruiz et al. [32] apply NEL as a sub task for generating an overview of the manually transcribed Bentham dataset. Hereby, approaches working on historic data often have to deal with the problem that most publicly available KBs do not contain most of the entities mentioned in the historic texts [25].

A fundamental work of Pontes et al. [28] evaluates the effects of synthetically generated Optical Character Recognition (OCR) errors in machine-printed documents on the NED process. They show that the performance of NED systems can decrease by 20% when about 15% of the words are not transcribed properly. Lately, Ehrmann et al. provided a large annotated NER and Linking dataset on historical newspapers [7] and organized the CLEF 2020 HIPE Shared Task [8]. Furthermore, Pontes et al. [27] proposed recently a multilingual end-to-end NEL approach based on the model of Kolitsas et al. [17].

In the handwriting domain, Hendriks et al. do a case study in [13] on a handwritten, historic, dutch document corpus. They use a fuzzy matching strategy to detect mentions. Afterwards, they select candidates by using fuzzy string matching and additional context information, e.g., dates. Finally, an active learning approach is used for linking entities to entries in their KB. It is important to note that this work just concentrates on person names and uses a small KB.

3 Dataset

There are currently no publicly available datasets dealing with NEL on handwritten document images. There is just the historical dutch dataset proposed by [13], which is currently not publicly available and is highly specialized for the specific use-case. To evaluate and develop NEL approaches on handwritten data, we provide three new annotated datasets for this task. In the following, we describe each of the datasets in more detail.

3.1 Synthetic HW-AIDA-CoNLL

To train and evaluate NEL models on handwritten document images, we created a synthetic dataset called HW-AIDA-CoNLL. It consists of synthetically generated handwritten pages using the text of AIDA-CoNLL, the standard benchmark for textual NEL. The benchmark provides a training (AIDA-train [14]), validation (AIDA-A [14]) and six test datasets. The special aspect of this benchmark is that there is not only a so-called in-domain test dataset (AIDA-B [14]), which is evaluated on the same type of data as used during training and validation. But there are five out-of-domain datasets (MSNBC [6], CWEB [9], ACE2004 [31], AQUAINT [24] and WIKI [31]), on which the system is not fine-tuned and thus provides an indication of its adaptivity on other domains. Even though the five out-of-domain datasets are very useful for evaluating NED approaches, several NEL publications consider only a subset of them [4,15,17]. We follow this trend and restrict our synthetically generated out-of-domain datasets to MSNBC, CWEB and WIKI.

The AIDA-CoNLL dataset [14] was manually annotated by Hoffart et al. and is based on the CoNLL 2003 dataset [33]. The data is divided into AIDA-train for training, AIDA-A for validation, and AIDA-B for testing. It is one of the biggest manually annotated NEL datasets available, containing 1393 news articles and 27817 linkable mentions. The MSNBC dataset [6] contains 20 news articles from 10 different topics and provides 656 linkable mentions. The datasets was cleaned and updated by Guo et al. [11]. CWEB [9] contains 320 documents and 11154 linkable mentions. Therefore, web pages from the ClueWeb Corpora in English language were annotated. Finally, WIKI [31] is composed of 320 documents and 6821 linkable mentions that correspond to existing hyperlinks in Wikipedia articles. The NEL annotations of CWEB and WIKI are generated automatically, while the others were checked or created manually and are therefore more reliable.

For generating the synthetic handwritten document images of these datasets, we used nearly the same approach as proposed in [22]. We synthetically generated each text document as an image using a handwritten font. The font is randomly sampled from over 300 publicly available True Type fonts that resemble handwriting. Each word of the document is rendered onto a transparent background. The font size is set randomly from a range of 28–52 pts and the intensity of the text stroke is varied between 0 and 50. In the next step, all word images are pasted onto a background image in the same order as in the original passage

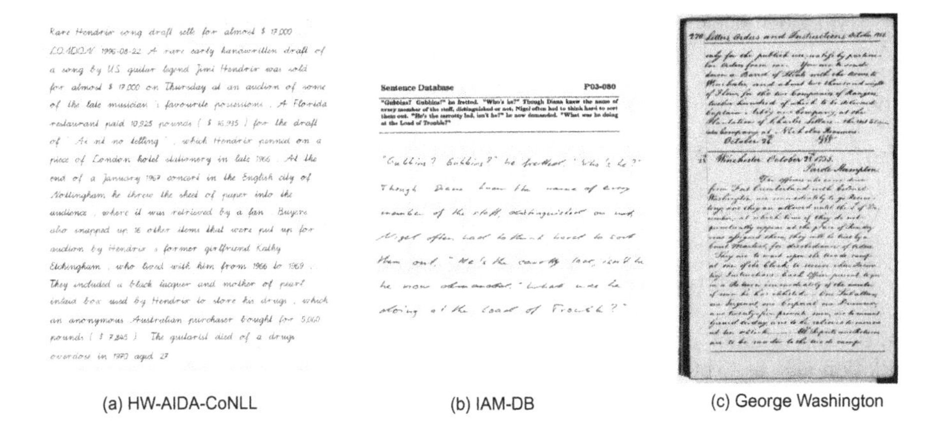

Fig. 5. Examples of document images for the synthetically generated HW-AIDA-CoNLL as well as the IAM-DB and George Washington datasets.

using alpha blending. We randomly sample the background image from a small set of manuscripts like textures. While pasting the words onto the background image, we break the lines whenever a word does not fit on the actual line. To provide variability, we randomly set the page width to a minimum of 800 and a maximum of 1600 pixel. Furthermore, we randomly padded the borders on all four sides of the document. To provide usability for approaches that work on full pages, we additionally divided long documents into multiple pages. Sentence segmentation is given for the documents and there is no overlap between the fonts in the training and test set (Fig. 5).

3.2 IAM-DB

The IAM Database [21] is a major benchmark for several DA tasks. The documents contain modern English sentences and were written by a total of 657 different people. The pages contain text from the genres listed in Fig. 6. The wide range of genres makes this an excellent choice of dataset for NEL. The database consists of 1539 scanned text pages containing a total of 13353 text lines and 115320 words.

Since NEL is a semantic task, we use an optimized partitioning of the IAM-DB data into training, validation and testing as proposed by Tüselmann et al. [39]. For the manual NEL annotations of the IAM-DB, we use Wikidata [40] as our KB. It is one of the largest public databases with over 95 million entries and it is widely used in the literature. The advantage of Wikidata is that its IDs do not change over time as Wikipedia page numbers or titles sometimes do. Given a Wikidata ID, it is possible to extract the page number or title of their corresponding Wikipedia page if needed for an approach.

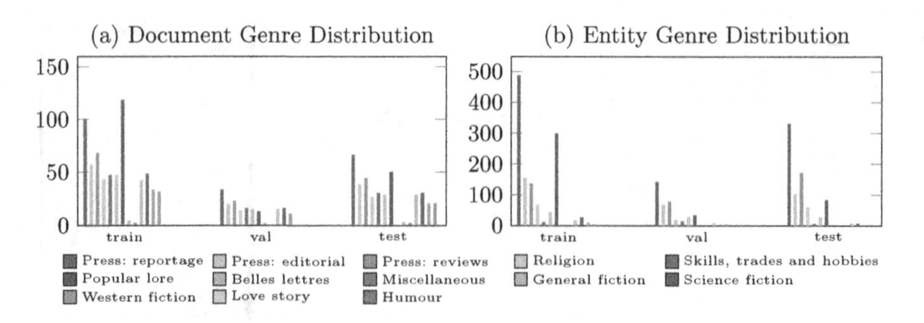

Fig. 6. The amount of document pages (a) and entities (b) per genre in training, validation and test set for the IAM database [39].

Our NEL dataset contains 3650 links between word images and Wikidata IDs. There are a total of 2405 entity-mention pairs in the dataset. These are divided into 1238 pairs in the training set, 785 pairs in the test set and 382 pairs in the validation set. Figure 6 shows that most linkable entities are located in the news articles and only a few in books and novels. This poses a major challenge to the NEL model, as fictitious entities should not be mapped to mismatched entities from the KB.

3.3 George Washington

The George Washington (GW) dataset [30] is a widely used benchmark in DA. It includes 20 pages of correspondences between George Washington and his associates produced in 1755. The documents were written by a single person in historical English. The manual NEL annotations were created in the same way as for the IAM database. The NEL dataset contains 218 links between word images and Wikidata IDs. There are a total of 145 entity-mention pairs in the dataset. These are divided into 105 pairs in the training set, 31 pairs in the test set and 9 pairs in the validation set.

4 Baseline Approach

In this section, we present our baseline approach for NED and Linking. We propose a two-stage system that works on segmented word images. The model transforms a document image into a textual representation and determines the linking of NEs using a textual NEL model (see Fig. 7). The approach is based on a state-of-the-art Handwriting Text Recognizer (HTR) [16] and NEL model [15]. The first step of our approach is to feed all word images into the recognizer in their order of occurrence on the document pages. The recognized text is then processed sequentially by the NEL model. Thereby, it extracts relevant mentions from the text by using a state-of-the-art NER model [1]. Given the relevant mentions, candidates are generated and finally disambiguated. In the following, we will provide further information about each component of this approach.

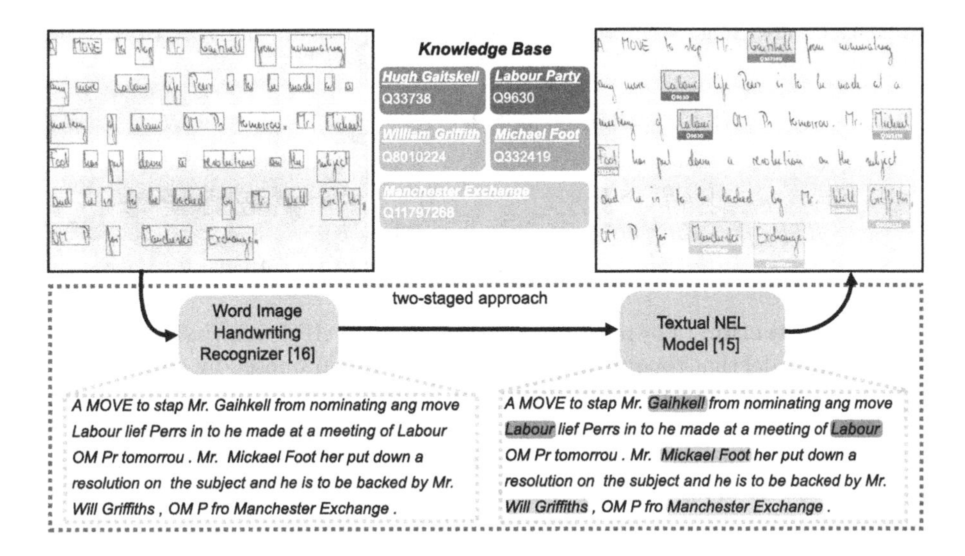

Fig. 7. An overview of our two-staged baseline approach for linking entities on handwritten document images against Wikidata.

4.1 Handwriting Text Recognizer

We use the attention-based sequence-to-sequence HTR model proposed by Kang et al. [16] for transcribing word images. It works on character-level and does not require any information about the language, except for an alphabet. The approach has the advantage that it does not require any dataset-specific pre-processing steps and provides satisfying results for the majority of datasets. Apart from adjusting the size of the input images, the maximum word length and the alphabet, we use the parameters proposed in [16]. As the approach does not use any linguistic resources during recognition, it has the advantage of not penalizing out-of-vocabulary words. However, the use of a lexicon often helps with minor recognition errors. Therefore, we additionally apply a fixed vocabulary consisting of its training, validation and test words on the outputs of the HTR model. In the following, *HTR+D* denotes a model using a dictionary. Table 1 shows the Character Error Rate (CER) and Word Error Rate (WER) for the recognition models on the datasets introduced in Sect. 3. We report similar error rates for the IAM and GW datasets as published in the literature [16]. Improvements are possible with further optimizations and dataset specific adaptations.

4.2 Named Entity Linking

For NEL, we use the Radboud Entity Linker (REL), a state-of-the-art NEL approach proposed by van Hulst et al. [15]. Their model follows the standard entity linking architecture, consisting of mention detection, candidate selection and entity disambiguation.

Table 1. Handwriting recognition (HTR) rates measured in Character Error Rate (CER) and Word Error Rate (WER) for the used datasets.

		AIDA-B	MSNBC	WIKI	CWEB	IAM	GW
HTR	*CER*	6.9	5.7	7.4	8.7	7.1	5.2
	WER	18.3	16.9	18.2	18.7	20.4	14.5
HTR+D	*CER*	5.7	4.4	6.3	8.0	6.4	4.1
	WER	9.2	7.6	9.8	12.7	10.8	6.1

Mention Detection. For mention detection, the FLAIR framework [1] is used. It is a NER model that roughly follows the architecture proposed by Lample et al. [18]. The approach transforms the input words into a vector representation using a pre-trained word embedding model (e.g. BERT [26]). These representations are encoded using a Bidirectional Long Short-Term Memory (BLSTM). Finally, a CRF uses the encoding from the BLSTM and predicts NE tags for each input word.

Candidate Selection. The selection of candidates follows the idea proposed by Le and Titov [19], where for each mention up to seven candidate entities are selected. Four candidates are selected based on mention-entity prior statistics, collected from Wikipedia, CrossWikis [38] and the YAGO dictionary [14]. Furthermore, three candidates are chosen based on their similarity to the 50 word context (c) surrounding the mention. REL uses word and entity embeddings provided by Yamada et al. [44] to estimate the similarity between an entity and the context of a mention. The similarity is obtained for 30 entities with the highest mention-entity prior and is calculated for each of these entities e by $\mathbf{e}^T \sum_{w \in c} \mathbf{w}$, whereby $\mathbf{w}$ and $\mathbf{e}$ are word and entity embeddings.

Entity Disambiguation. REL follows the Ment-norm approach proposed by Le and Titov [19] for mapping mentions to their entries in a given KB. The linking is done by a local score, composed of mention-entity prior and context similarity, as well as the coherence with other linking decisions in the document. The coherence between all entity decisions is done by assuming K latent relations between mentions. To optimize the local and global conditions, max-product loopy belief propagation is used [10]. The final score for an entity of a mention is obtained by a two-layered neural network that combines mention-entity prior information and the max-marginal probability for the entity and its document. Finally, the posterior probabilities of the linked entities are computed by applying a logistic function over the final scores of the neural network.

5 Experiments

We evaluate NED and NEL on the handwritten datasets introduced in Sect. 3 using our two-stage baseline approach. Before we present and discuss the evaluation results, we describe the evaluation protocol in Sec. 5.1.

Table 2. Named Entity Disambiguation (NED) and Linking (NEL) performances measured in InKB micro F1 score based on strong annotation matching. Training was done on AIDA-train.

		AIDA-B	MSNBC	WIKI	CWEB	IAM	GW
NED	*Annotation*	87.9	91.9	75.7	76.7	69.4	22.6
	HTR+D	74.8	80.1	63.1	60.1	54.9	19.4
	HTR	52.0	59.0	47.9	45.8	31.4	12.9
NEL	*Annotation*	79.1	75.4	40.1	47.8	49.8	15.2
	HTR+D	68.0	64.2	33.2	34.9	38.1	12.5
	HTR	56.7	55.7	30.6	33.8	27.5	7.1

5.1 Evaluation Protocol

For NEL, the InKB micro F1 score based on strong annotation matching (see Eq. 3) is often used in the literature to compare approaches. InKB only focuses on those cases in which either the mention has a valid entity in the KB or the NEL approach predicts an entity of the KB. Strong annotation matching implies that a prediction is only correct if it recognizes the mention and links it to the proper entity. Therefore, it is not enough to recognize a part of the mention and match it correctly. For computing the F1 score, it is important to distinguish between micro and macro averaging. The macro F1 computes the F1 score independently for each class and then takes the average, whereas the micro F1 aggregates the contributions of all classes. The evaluation of NED provides a special case as the mentions are provided as input and do not have to be detected. Therefore, the number of predicted mentions is equal to the number of mentions in gold standard, which leads to F1 being equal to precision, recall and accuracy [36].

$$Precision = \frac{\#\text{ of correctly detected and disambiguated mentions}}{\#\text{ of predicted mentions by model}} \tag{1}$$

$$Recall = \frac{\#\text{ of correctly detected and disambiguated mentions}}{\#\text{ of mentions in gold standard}} \tag{2}$$

$$F1 = \frac{2 * Precision * Recall}{Precision + Recall} \tag{3}$$

5.2 Results

In this section, we present and discuss the NED and NEL performances on our created datasets using our baseline approach. To evaluate the impact of the HTR errors, we need the scores of the NED and NEL models on perfect recognition results. In the following, we denote approaches that work on the textual annotations of the word images instead of the HTR results as *Annotation*.

The results of our baseline approach show that NEL is a considerably harder task compared to NED. This is shown in Table 2, where the F1 scores for NEL are significantly lower than for NED on all datasets, even without using handwriting recognition. The NED and NEL models perform well on all textual datasets and the scores are similar to the published results obtained in [15]. The scores show that the IAM dataset is comparatively challenging for both tasks. This is probably due to the small context and the large variety of topics [35].

The recognition errors have a strong impact on the performance of the models. In comparison to handwriting recognition only, the scores improved significantly by adding a dictionary. This is mainly due to the lower WER on the datasets. However, even with small recognition error rates as given for the MSNBC dataset, a strong decrease of the F1 score occurs. Interestingly, REL uses the same model and approach for NED as described in [28]. However, we cannot obtain similar results as observed in their OCR evaluation, even though similar error rates were considered there.

For several reasons, NED and NEL perform particularly poorly on the GW dataset. It is historical data in which only 38% of people and place mentions are indexed in Wikidata. Furthermore, we work on textual annotations that are all in lowercase characters, which makes mention detection and linking considerably harder. For the detection we already used a NER model from FLAIR, which was trained on lowercase letters. The dataset also contains difficult conditions for linking, since there are several abbreviations like *gw* for *George Washington* and often only the last name and military rank of the person is given. In addition, a lot of contextual knowledge is needed as the data commonly involves people and places in the area of Virginia between 1700 and 1800. Finally, the training dataset is very small, which makes fine-tuning difficult.

6 Conclusion

In this work, we introduce the task of Named Entity Linking on handwritten document images. We also propose and publish the first Entity Linking datasets for this task (HW-AIDA-CoNLL, IAM and George Washington) along with a suitable evaluation protocol. The datasets offer two tasks, Named Entity Disambiguation and Named Entity Linking. We investigate a two-stage approach for both tasks on word-segmented handwritten document images in this work. Even though our experiments show that the two tasks are particularly difficult for handwritten document images, we are already achieving promising results with our baseline approach.

References

1. Akbik, A., Bergmann, T., Blythe, D., Rasul, K., Schweter, S., Vollgraf, R.: FLAIR: an easy-to-use framework for state-of-the-art NLP. In: Annual Conference of the North American Chapter of the Association for Computational Linguistics, Minneapolis, MN, USA, pp. 54–59 (2019)
2. Al-Moslmi, T., Ocaña, M.G., Opdahl, A.L., Veres, C.: Named entity extraction for knowledge graphs: a literature overview. IEEE Access **8**, 32862–32881 (2020)

3. Boroş, E., et al.: Robust named entity recognition and linking on historical multilingual documents. In: Working Notes of Conference and Labs of the Evaluation Forum, Thessaloniki, Greece (2020)
4. Cao, N.D., Izacard, G., Riedel, S., Petroni, F.: Autoregressive entity retrieval. In: International Conference on Learning Representations, Vienna, Austria (2021)
5. Carbonell, M., Fornés, A., Villegas, M., Lladós, J.: A neural model for text localization, transcription and named entity recognition in full pages. Pattern Recogn. Lett. **136**, 219–227 (2020)
6. Cucerzan, S.: Large-scale named entity disambiguation based on Wikipedia data. In: Joint Conference on Empirical Methods in Natural Language Processing and Computational Natural Language Learning, Prague, Czech Republic, pp. 708–716 (2007)
7. Ehrmann, M., Romanello, M., Bircher, S., Clematide, S.: Introducing the CLEF 2020 HIPE shared task: named entity recognition and linking on historical newspapers. In: Jose, J.M., et al. (eds.) ECIR 2020. LNCS, vol. 12036, pp. 524–532. Springer, Cham (2020). https://doi.org/10.1007/978-3-030-45442-5_68
8. Ehrmann, M., Romanello, M., Flückiger, A., Clematide, S.: Overview of CLEF HIPE 2020: named entity recognition and linking on historical newspapers. In: Arampatzis, A., et al. (eds.) CLEF 2020. LNCS, vol. 12260, pp. 288–310. Springer, Cham (2020). https://doi.org/10.1007/978-3-030-58219-7_21
9. Gabrilovich, E., Ringgaard, M., Subramanya, A.: FACC1: freebase annotation of ClueWeb corpora, Version 1 (Release date 2013–06-26, Format version 1, Correction level 0) (2013)
10. Ganea, O., Hofmann, T.: Deep joint entity disambiguation with local neural attention. In: Proceedings of Conference on Empirical Methods in Natural Language Processing, Copenhagen, Denmark, pp. 2619–2629 (2017)
11. Guo, Z., Barbosa, D.: Robust entity linking via random walks. In: Proceedings of ACM International Conference on Information and Knowledge Management, Shanghai, China, pp. 499–508 (2014)
12. Heino, E., et al.: Named entity linking in a complex domain: case second world war history. In: Gracia, J., Bond, F., McCrae, J.P., Buitelaar, P., Chiarcos, C., Hellmann, S. (eds.) LDK 2017. LNCS (LNAI), vol. 10318, pp. 120–133. Springer, Cham (2017). https://doi.org/10.1007/978-3-319-59888-8_10
13. Hendriks, B., Groth, P., van Erp, M.: Recognizing and linking entities in old dutch text: a case study on VOC notary records. In: Proceedings of International Conference on Collect and Connect: Archives and Collections in a Digital Age, Leiden, Netherlands, pp. 25–36 (2020)
14. Hoffart, J., et al.: Robust disambiguation of named entities in text. In: Proceedings of Conference on Empirical Methods in Natural Language Processing, Edinburgh, UK, pp. 782–792 (2011)
15. van Hulst, J.M., Hasibi, F., Dercksen, K., Balog, K., de Vries, A.P.: REL: an entity linker standing on the shoulders of giants. In: Proceedings of International ACM SIGIR Conference on Research and Development in Information Retrieval, Xi'an, China, pp. 2197–2200 (2020)
16. Kang, L., Toledo, J.I., Riba, P., Villegas, M., Fornés, A., Rusiñol, M.: Convolve, attend and spell: an attention-based sequence-to-sequence model for handwritten word recognition. In: Brox, T., Bruhn, A., Fritz, M. (eds.) GCPR 2018. LNCS, vol. 11269, pp. 459–472. Springer, Cham (2019). https://doi.org/10.1007/978-3-030-12939-2_32

17. Kolitsas, N., Ganea, O., Hofmann, T.: End-to-end neural entity linking. In: Proceedings of Conference on Computational Natural Language Learning, Brussels, Belgium, pp. 519–529 (2018)
18. Lample, G., Ballesteros, M., Subramanian, S., Kawakami, K., Dyer, C.: Neural architectures for named entity recognition. In: Annual Conference of the North American Chapter of the Association for Computational Linguistics, San Diego, CA, USA, pp. 260–270 (2016)
19. Le, P., Titov, I.: Improving entity linking by modeling latent relations between mentions. In: Annual Meeting of the Association for Computational Linguistics, Melbourne, Australia, pp. 1595–1604 (2018)
20. Lehmann, J., et al.: DBpedia - a large-scale, multilingual knowledge base extracted from Wikipedia. Semant. Web **6**(2), 167–195 (2015)
21. Marti, U., Bunke, H.: The IAM-database: an English sentence database for offline handwriting recognition. Int. J. Doc. Anal. Recognit. **5**(1), 39–46 (2002)
22. Mathew, M., Gomez, L., Karatzas, D., Jawahar, C.V.: Asking questions on handwritten document collections. Int. J. Doc. Anal. Recognit. (IJDAR) **24**(3), 235–249 (2021). https://doi.org/10.1007/s10032-021-00383-3
23. Mathew, M., Karatzas, D., Jawahar, C.V.: DocVQA: a dataset for VQA on document images. In: IEEE Workshop on Applications of Computer Vision, Waikoloa, HI, USA, pp. 2199–2208 (2021)
24. Milne, D., Witten, I.H.: Learning to link with Wikipedia. In: Proceedings of ACM International Conference on Information and Knowledge Management, Napa Valley, CA, USA, pp. 509–518 (2008)
25. Munnelly, G., Pandit, H.J., Lawless, S.: Exploring linked data for the automatic enrichment of historical archives. In: Gangemi, A., et al. (eds.) ESWC 2018. LNCS, vol. 11155, pp. 423–433. Springer, Cham (2018). https://doi.org/10.1007/978-3-319-98192-5_57
26. Peters, M.E., Neumann, M., Iyyer, M., Gardner, M., Clark, C., Lee, K., Zettlemoyer, L.: Deep contextualized word representations. In: Annual Conference of the North American Chapter of the Association for Computational Linguistics, New Orleans, LA, USA, pp. 2227–2237 (2018)
27. Linhares Pontes, E., et al.: Entity linking for historical documents: challenges and solutions. In: Ishita, E., Pang, N.L.S., Zhou, L. (eds.) ICADL 2020. LNCS, vol. 12504, pp. 215–231. Springer, Cham (2020). https://doi.org/10.1007/978-3-030-64452-9_19
28. Linhares Pontes, E., Hamdi, A., Sidere, N., Doucet, A.: Impact of OCR quality on named entity linking. In: Jatowt, A., Maeda, A., Syn, S.Y. (eds.) ICADL 2019. LNCS, vol. 11853, pp. 102–115. Springer, Cham (2019). https://doi.org/10.1007/978-3-030-34058-2_11
29. Raiman, J., Raiman, O.: DeepType: Multilingual entity linking by neural type system evolution. In: Proceedings of AAAI Conference on Artificial Intelligence, New Orleans, LA, USA, pp. 5406–5413 (2018)
30. Rath, T.M., Manmatha, R.: Word spotting for historical documents. Int. J. Doc. Anal. Recognit. **9**(2–4), 139–152 (2007)
31. Ratinov, L., Roth, D., Downey, D., Anderson, M.: Local and global algorithms for disambiguation to Wikipedia. In: Annual Meeting of the Association for Computational Linguistics, Portland, Oregon, pp. 1375–1384 (2011)
32. Ruiz, P., Poibeau, T.: Mapping the Bentham corpus: concept-based navigation. J. Data Min. Digit. Humanit. (2019)

33. Sang, E.F.T.K., Meulder, F.D.: Introduction to the CoNLL-2003 shared task: language-independent named entity recognition. In: Proceedings of Conference on Computational Natural Language Learning, Edmonton, Canada, pp. 142–147 (2003)
34. Sevgili, Ö., Panchenko, A., Biemann, C.: Improving neural entity disambiguation with graph embeddings. In: Annual Meeting of the Association for Computational Linguistics, Florence, Italy, pp. 315–322 (2019)
35. Sevgili, Ö., Shelmanov, A., Arkhipov, M.Y., Panchenko, A., Biemann, C.: Neural entity linking: a survey of models based on deep learning. CoRR abs/2006.00575 (2020)
36. Shen, W., Wang, J., Han, J.: Entity linking with a knowledge base: issues, techniques, and solutions. IEEE Trans. Knowl. Data Eng. **27**(2), 443–460 (2015)
37. Smith, D.A., Crane, G.: Disambiguating geographic names in a historical digital library. In: Constantopoulos, P., Sølvberg, I.T. (eds.) ECDL 2001. LNCS, vol. 2163, pp. 127–136. Springer, Heidelberg (2001). https://doi.org/10.1007/3-540-44796-2_12
38. Spitkovsky, V.I., Chang, A.X.: A cross-lingual dictionary for English Wikipedia concepts. In: Proceedings of International Conference on Language Resources and Evaluation, Istanbul, Turkey, pp. 3168–3175 (2012)
39. Tüselmann, O., Wolf, F., Fink, G.A.: Are end-to-end systems really necessary for NER on handwritten document images? In: Lladós, J., Lopresti, D., Uchida, S. (eds.) ICDAR 2021. LNCS, vol. 12822, pp. 808–822. Springer, Cham (2021). https://doi.org/10.1007/978-3-030-86331-9_52
40. Vrandečić, D., Krötzsch, M.: Wikidata: a free collaborative knowledgebase. Commun. ACM **57**(10), 78–85 (2014)
41. Wen, Y., Fan, C., Chen, G., Chen, X., Chen, M.: A survey on named entity recognition. In: Liang, Q., Wang, W., Liu, X., Na, Z., Jia, M., Zhang, B. (eds.) CSPS 2019. LNEE, vol. 571, pp. 1803–1810. Springer, Singapore (2020). https://doi.org/10.1007/978-981-13-9409-6_218
42. Yadav, V., Bethard, S.: A survey on recent advances in named entity recognition from deep learning models. In: Proceedings of International Conference on Computational Linguistics, Santa Fe, NM, USA, pp. 2145–2158 (2018)
43. Yamada, I., Shindo, H.: Pre-training of deep contextualized embeddings of words and entities for named entity disambiguation. CoRR abs/1909.00426 (2019)
44. Yamada, I., Shindo, H., Takeda, H., Takefuji, Y.: Joint learning of the embedding of words and entities for named entity disambiguation. In: Proceedings of Conference on Computational Natural Language Learning, Berlin, Germany, pp. 250–259 (2016)

Pattern Analysis Software Tools (PAST) for Written Artefacts

Hussein Mohammed(✉), Agnieszka Helman-Wazny(✉), Claudia Colini(✉), Wiebke Beyer, and Sebastian Bosch(✉)

Cluster of Excellence: Understanding Written Artefacts, Universität Hamburg, Hamburg, Germany
{hussein.adnan.mohammed,agnieszka.helman-wazny,claudia.colini, wiebke.beyer,sebastian.bosch}@uni-hamburg.de
https://www.csmc.uni-hamburg.de/written-artefacts.html

Abstract. The research of ancient written artefacts results in an ever-increasing amount of digital data in different forms, ranging from raw images of artefacts, to automatically generated data from advanced acquisition techniques. The manual analysis of this data is typically time consuming, and can be subject to human error and bias. Therefore, we present in this work a set of Pattern Analysis Software Tools (PAST), which are dedicated to the automatic analysis of visual and tabular patterns in the research data from the study of ancient written artefacts. These software tools have been developed to facilitate a more efficient study of written artefacts and to help the scholars benefit from the rapid advancements in the fields of pattern analysis and artificial intelligence. Furthermore, these tools can provide new insights which can only be emerged from the statistical analysis of research data. Each tool in PAST is developed and tested in close collaboration with experts from relevant fields of research, and presented here with actual use cases in order to demonstrate its usability and applicability to real research questions.

Keywords: Pattern analysis · Software development · Manuscript research

1 Introduction

Increasing amount of research data is generated by scholars in the Humanities every day. This data can be raw images of digitised artefacts, automatically generated measurements from material-analysis equipment, or even manually-generated data which reflects observations or calculations of certain features in ancient artefacts. The manual analysis of this data typically requires a lot of time and effort; furthermore, it can be subject to human error and bias. Therefore, automating the process of analysing, classifying, and detecting patterns in research data is becoming an essential need to sustain the productivity of scholars and benefit from the technological advancements in the field of pattern recognition. Furthermore, this automation can provide new insights in some cases for researchers through the statistical analysis of data from different artefacts.

Having proper and reliable interpretation of the generated results from automated pattern recognition methods often require experts from both computer science and the

© Springer Nature Switzerland AG 2022
S. Uchida et al. (Eds.): DAS 2022, LNCS 13237, pp. 214–229, 2022.
https://doi.org/10.1007/978-3-031-06555-2_15

Humanities. In many cases, such close collaboration is not only desired, but rather necessary to have a fruitful outcome. Nevertheless, frequent exchange of data, results, and opinions can be restricted due to geographical limitations, time constraints, or copyright issues. Therefore, having an easy-to-use off-line implementation of pattern recognition methods can help overcoming these difficulties and offer a suitable platform for quick tests to be carried out by experts from different fields independently.

Several computational systems have been published to analyse different aspects of digitised manuscripts [3, 11, 12]. Most of these systems provide valuable contributions to the research community in general. Nevertheless, certain attributes made most of them hard to use and benefit from by scholars in the Humanities. The majority of these systems are on-line applications, which require the scholars to upload their research data. This could be problematic with regards to copyright issues in some cases. Furthermore, some of these systems are designed for several tasks at once, this can make them difficult to use and confusing for the scholars. In addition, many of these systems are designed, developed, and tested only by computer scientists, which can create a gap between what scholars expect and what the system generate as final results.

In this work, we present a set of Pattern Analysis Software Tools (PAST) which are dedicated to the analysis of both visual and tabular patterns in the research data generated from the study of ancient written artefacts. All of these tools are developed as off-line applications which does not require installation or any additional dependencies. Furthermore, these tools have been designed, developed, and tested in close collaboration with experts from relevant fields of research in the Humanities.

The PAST was developed as off-line Razor Pages web applications using the open-source .NET Core framework from Microsoft (https://github.com/dotnet/core). All of these software tools have been made freely available and published under the Creative Commons Attribution-NonCommercial 4.0 International Public License.

The Graphical User Interface (GUI) of the PAST software tools is designed to be easy-to-use and allows the user to perform the analysis in steps. The instructions for each step can be found at the bottom of the corresponding page. Furthermore, a general guideline is provided in a "How To" Section of each software tool.

This paper is organised as follows: Every software tool will be described in a dedicated section, which includes the general description, the basic functionality, and a use case from the research of ancient written artefacts. The final Section contains our conclusions.

2 The Handwriting Analysis Tool (HAT)

The HAT [16] is a software tool based on the method in [22], and can be used to analyse handwriting images. Multiple and different handwriting styles can be analysed concurrently and sorted according to their similarity to a questioned or unknown style. A similarity score can be calculated for each predefined style to create a relative comparison between them with respect to an unknown style.

A state-of-the-art training-free classifier [22] has been used in this tool in order to offer the possibility to analyse handwriting styles without the need for training data, which can not be provided in many cases. This classifier has been tested against several degradation types in [15], and it calculates the distances between detected local features in handwriting images as follows:

$$Dist_N(d,c) = \frac{Dist(d,c)}{K_c}, \tag{1}$$

$Dist_N(d,c)$ is the normalised distance between the detected features d in the test image and class c using the distance calculation presented in [22]. K_c is the number of features from the labelled samples in class c, and $Dist(d,c)$ is the Local Naïve Bayes Nearest-Neighbour (NBNN) [14], which has been reformulated in [22] as follows:

$$Dist(d,c) = \sum_{i=1}^{n} \left[\left(\| d_i - \phi(\mathrm{NN}_c(d_i)) \|^2 - \| d_i - \mathrm{N}_{k+1}(d_i) \|^2 \right) \right], \tag{2}$$

where

$$\phi(\mathrm{NN}_c(d_i)) = \begin{cases} \mathrm{NN}_c(d_i) & \text{if } \mathrm{NN}_c(d_i) \leq \mathrm{N}_{k+1}(d_i) \\ \mathrm{N}_{k+1}(d_i) & \text{if } \mathrm{NN}_c(d_i) > \mathrm{N}_{k+1}(d_i), \end{cases}$$

and $\mathrm{N}_{k+1}(d_i)$ is the neighbour $(k+1)$ of d_i. In a similar way to the work in [14], we used the distance to the $k+1$ nearest neighbours ($k = 10$) as a "background distance" to estimate the distances of classes which were not found in the k nearest neighbours.

2.1 Basic Functionality

The Graphical User Interface (GUI) of the HAT allows the user to perform the analysis in the following steps: selecting the unlabelled handwriting styles to be analysed, the labelled handwriting styles to be considered, and finally the method's parameters; see Fig. 1.

The calculated scores are relative similarity measures. The HAT calculates how similar an unknown style is to a given known style relative to the other known styles as follows:

Let D_s be the absolute value of the distance to the handwriting style s, and $Sum_D = \sum_{s=1}^{n} D_s$, where n is the number of known styles. The Score S for a given style s is $S_s = \frac{D_s}{Sum_D} \times 100$.

The results are displayed automatically in separated tables, where all the styles are ranked according to their relative similarity to the unknown handwriting; see Fig. 2. They can also be stored as spreadsheet files.

2.2 Use Case

In recent years scholars have made some attempts at analysing handwriting styles in Tibetan manuscripts [5,25]. From a range of auxiliary tools available they have turned to palaeography in order to define styles of Tibetan writing [23]. Such attempts, together with analyses of other physical features, can help to link documents in meaningful ways or determine the age of the manuscripts or documents that are currently undated [8,9].

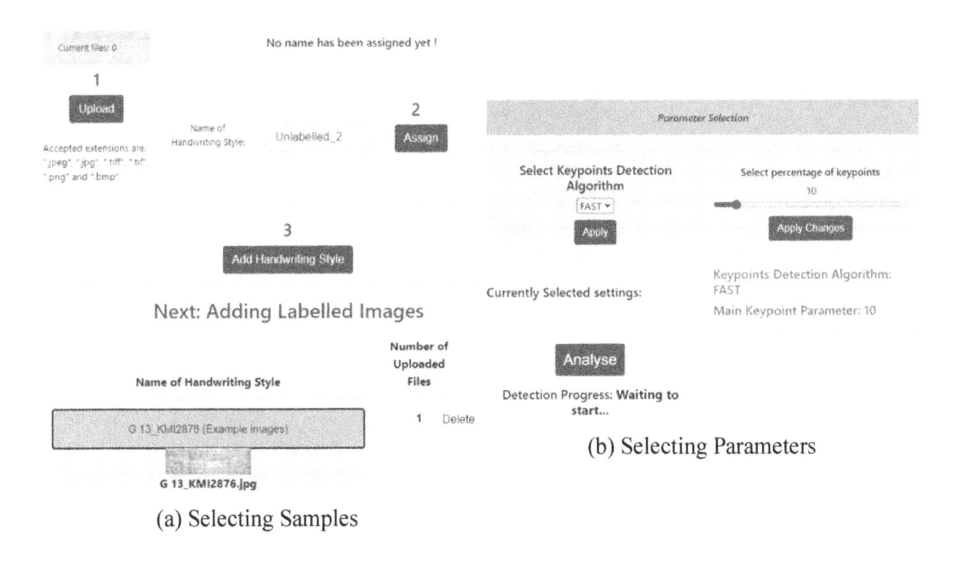

(a) Selecting Samples

(b) Selecting Parameters

Fig. 1. Two steps from the HAT are shown here. Part (a) shows the sample selection step, while part (b) shows the parameter selection step.

The Mustang archives afford a rare opportunity to test the extent to which we can or cannot identify the same scribal hands. Most of these documents were produced by a small community of local scribes using a very narrow selection of materials and tools and unrestricted by dogmatic script standards, as would have been the case with official documents produced in Central Tibet. Five documents have been selected from these archives, which are signed with the same name (*Gyaltsen*), suggesting that they might be written by the same scribal hand. Three of these documents (G13, 24, and 50) are signed by Gyaltsen, while the other two (G102 and 150) are signed by "the man (or the son) Gyaltsen" (bu rgyal mtshan). See Fig. 2.

Although the signatory have the same name, the handwriting did not seem identical in these documents. Therefore, we used HAT in order to analyse the different handwriting styles. The documents G102 and 150 had the highest similarity values, suggesting that they are written by the same person. G102 is an agreement between a married couple, while G150 is about two brothers who are sharing a wife (polyandrous marriage being a common arrangement in Mustang). We might well conclude that the two documents were written by the same person, especially since the scribe in both cases identifies himself as "the man Gyaltsen" (bu rgyal mtshan).

3 The Visual-Pattern Detector (VPD)

The VPD [19] is an efficient and easy-to-use software tool for pattern detection, which is based on the proposed method in [21]. This tool can be used to recognise and allocate visual patterns (such as words, drawings and seals) automatically in digitised manuscripts. The recall-precision balance of detected patterns can be controlled

Results for G 150:

Style Name	Score (Relative Similarity)
G 102	41.5 %
G 13	27.1 %
G 24	19.4 %
G 50	12 %

(a) Calculated Score by HAT.

(b) Manuscript G150 from the Mustang archives.

Fig. 2. Example of the results produced by HAT. The used samples are cropped images of handwriting without any preprocessing.

visually, and the detected patterns can be saved as annotations on the original images or as cropped images depending on the needs of users.

The general approach of the method used in this tool is based on the voting of every detected local feature for a proposed centre of a pattern hypothesis. A detection matrix $M^d(L_{i,c})$ per class is created for the image, where the vote of each feature in the matrix is calculated from the distance to features of the corresponding class using the Normalised Local NBNN distance calculation presented in Eq. 1 as follows:

$$M^d(L_{i,c}) = M^d(L_{i,c}) + Dist_N(d_i, c), \tag{3}$$

where $M^d(L_{i,c})$ is the detection matrix of class c.

Each detection matrix is convolved with a kernel in order to produce the final detections. The detection kernel can be described as follows:

$$K_c^{d_i}(x, y) = \begin{cases} 1 \text{ if } Offset_x^2 + Offset_y^2 < R_c \\ 0 \; otherwise, \end{cases} \tag{4}$$

where $K_c^{d_i}(x, y)$ is the detection kernel of class c for the detected feature d_i centred at location (x, y). $Offset_x$ and $Offset_y$ are the differences in the x- and y-axis between the kernel centre and the current location (x, y) respectively.

3.1 Basic Functionality

The GUI of the VPD allows the user to perform the detection process in steps: selecting patterns to be detected, the images to be searched, and finally the detection parameters; see Fig. 3.

The detection results can be generated in a wide range of formats so that different requirements that scholars may have can be met. The detection threshold can be controlled intuitively by visually inspecting the three best and worst detection results from the set of considered detections; see Fig. 4.

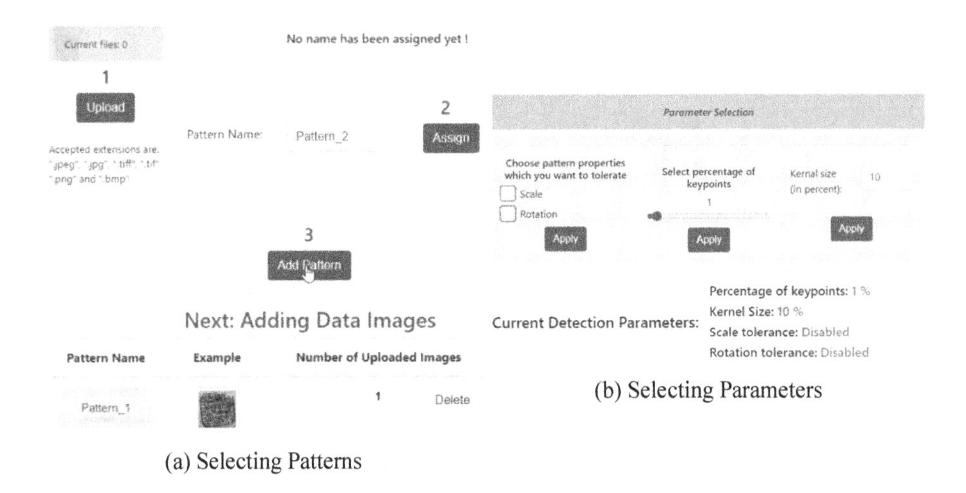

(a) Selecting Patterns

(b) Selecting Parameters

Fig. 3. Two steps from the VPD are shown here. Part (a) shows the patterns selection step, while part (b) shows the parameter selection step.

3.2 Use Case

While various approaches can be used for the authentication of documents from the Mustang archives, the identification and analysis of seals can be among the most effective ones. Although there is a growing body of published research on the subject of Tibetan seals [1,26,29], their use is on borderline topics such as the general Mustang merits.

The seals were intended to certify ownership or validate the document. Analysing them can help us shed a light on features such as the chosen shape of seals, the choice of language (Tibetan, Nepali) and script (which includes examples of Arabic and Roman), the existence and relevance of certain motifs, and the extent to which the content of the seal was implicated in the identity of the owner [9].

A total of 330 manuscripts have been digitised in order to be analysed. However, it is a challenging and time consuming task to manually search for similar seals in all of them. Therefore, we used the VPD in order to automate this process and speed it up. An example of this automated search is the different versions of the Geling community seal seen in Fig. 4, which were detected by VPD in the documents MA/Geling/Tib/2, MA/Geling/Tib/24, and MA/Geling/Tib/43 from the Mustang Archives. This detected pattern is a square seal printed in black ink and outlined by the same size frame, with 3 double columns or rows of indistinct lettering, possibly pseudo-Zhangzhung script.

4 The Line Detection Tool (LDT)

The main goal of the LDT [17] is to analyse images of writing supports in order to detect lines (such as sieve imprints or papyri fibres) and estimate their density. These detected lines form a pattern, which can be used as a distinctive feature of the writing support.

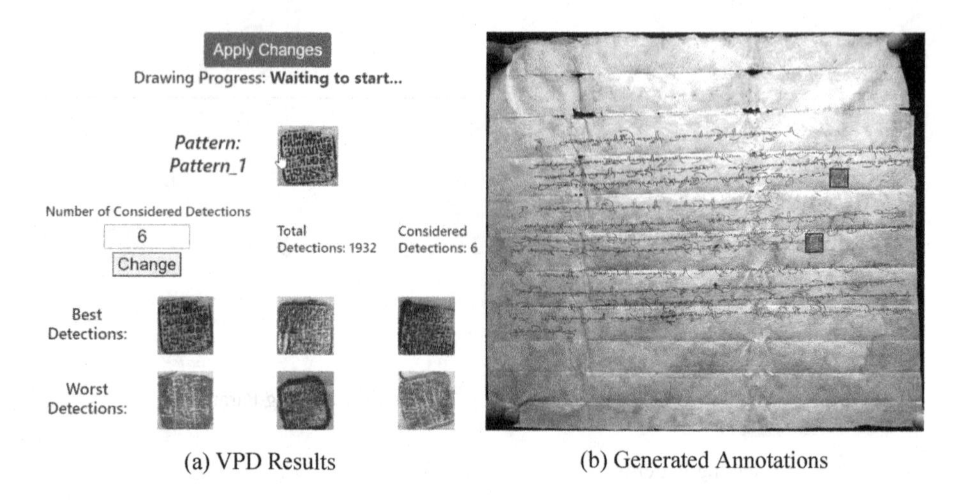

(a) VPD Results (b) Generated Annotations

Fig. 4. Example of the results that can be generated by the VPD. The images are reproduced from the Mustang collection.

The contrast of the selected images is first enhanced using the Contrast Limited Adaptive Histogram Equalisation (CLAHE) [24], then a vertical projection is calculated. These projections are smoothed using a Gaussian filter in order to construct a histogram like the one in Fig. 5, part (a). Lines are detected from the resulting histogram as follows:

$$L_{col} = \begin{cases} 1 & \text{if } H_{min} < (H_{col} \times T_{min}) \text{ and } H_{col} > (H_{max} \times T_{max}) \\ 0 & otherwise, \end{cases} \quad (5)$$

where

$$H_{col} = \sum_{i=1}^{n} I(col, i). \quad (6)$$

L_{col} is the line to be detected at the column *col* in the image. H_{max} and H_{min} are the maximum and minimum values of the histogram correspondingly. T_{max} and T_{min} are threshold values which can be changed by the user. H_{col} is the histogram value at the column *col*, and $I(col, i)$ is the pixel value of image I at the (col, i) position.

This tool can calculate the detection accuracy if ground-truth information is provided, which includes the number of lines and the length of digitised region as a part of the file name. The accuracy is calculated as follows:

$$Accuracy = \left[\frac{GTN - |GTN - CN|}{GTN} \right] \times 100 \quad (7)$$

where GTN is the ground-truth number of lines, and CN is the calculated number of lines.

4.1 Basic Functionality

The GUI of LDT allows the user to perform line detection on multiple images concurrently. The results are presented in a table, which can be saved as a spreadsheet file. The detections can be visualised and saved as images; see Fig. 5. Furthermore, the main parameters of the algorithm can be modified by the user.

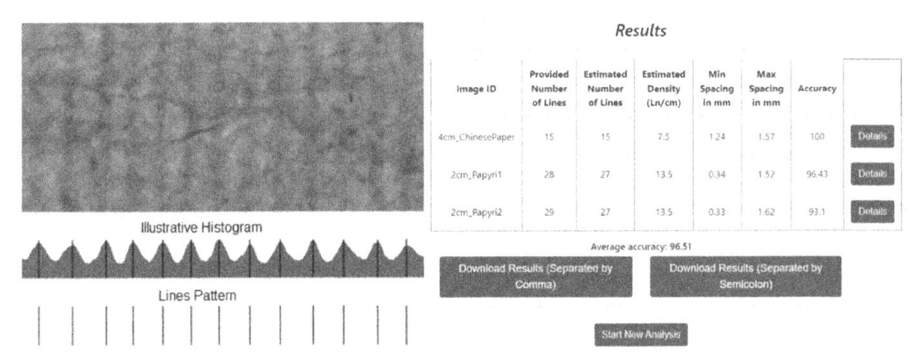

(a) A histogram is generated from the smoothed projections.

(b) The generated results along with performance evaluation.

Fig. 5. Example of the results produced by LDT.

4.2 Use Case

Depending on what type of material was attached to the frame in the paper-making process, we can observe a slight difference in texture and imprint of the sieve on the paper [6, 10]. The imprint of a textile sieve clearly differs from that made of bamboo (laid down in a regular pattern), reed, or other grasses (laid in an irregular pattern). This pattern, sealed in the structure of the paper, allows us to distinguish between handmade woven paper and laid paper, which can be characterized by the density of laid lines. Eventually, using laid-lines pattern allows to group papers which may potentially be made using the same moulds.

The process of making moulds and sieves by hand made them as unique as the papers themselves. Thus, papers made by different moulds will not have identical patterns of laid and chain lines. On the other hand, papers shaped by the same mould should have a nearly identical laid and chain lines imprint. In order to categorize various Asian papers, however, sufficient amount of data needs be collected and analysed as a reference material. Therefore, LDT has been used to mark, measure, and compare the idiosyncratic pattern of laid lines in a collection of Central Asian papers as seen in Fig. 5 in a hunt for mould-mates.

5 The XRF-Data Analysis Tool (XRF-DAT)

The main goal of the XRF-DAT [18] is to analyse tabular data generated by X-ray Fluorescence (XRF) spectroscopy in order to ease and speed up the processing and evaluation of data obtained when analysing written artefacts; in particular their inks, pigments and writing supports. XRF spectroscopy allows the elemental composition of materials to be determined, since each element re-emits photons at characteristic energies after being irradiated with X-rays. It is important to mention that X-rays penetrate the written artefacts, thus when analysing inks and pigments, the detector also collects data from the writing support.

The measurement outcome is a spectrum expressing the energy of photons on the x-axis (in keV) and the intensity of the detection signal on the y-axis (in pulses or counts per second). In the native XRF-spectrometers software (we use the ARTAX spectra by Bruker), an operator manually identifies the elements present in the spectrum as signals, runs the internal deconvolution of these signals and exports the output as *.csv* or *.xls* files. The deconvoluted net intensity for each element is the most important value, which is then used to determine the relative amount of elements in the analysed material.

From these files, an operator would need to perform a long series of time-consuming operations that vary depending on the written artefact, the identified elements and the way the measurements were acquired. The most common operations are: sorting the detected elements by their atomic number, calculating the average of selected spectra, subtracting the writing support from the ink or pigment measurements and normalising by key elements.

The operator can use XRF-DAT to automate these processes in a modular and transparent way, by selecting which data to use (with the limitation that they have to be in *.csv* format) and which operation to perform. In addition, different measurements can be compared using the distance measure provided by XRF-DAT, which is calculated as follows:

$$Dist(M_a, M_b) = \sqrt{\sum_{e=1}^{n}(Net_e^N(M_a) - Net_e^N(M_b))^2}, \tag{8}$$

where

$$Net_e^N(M) = \frac{Net_e}{\sum_{i=1}^{n} Net_i(M)}. \tag{9}$$

M_a and M_b are the measurements a and b correspondingly. Net_e is the net spectra of element e, and $Net_e^N(M)$ is Net_e normalised by the summation of all net values in measurement M. n is the number of elements in measurement M.

5.1 Basic Functionality

The GUI of XRF-DAT allows the user to select a set of measurement tables in order to be analysed. This tool offers three possible functions to choose from:

- Applying the operations mentioned in Sect. 5 automatically.
- Generating scatter plot tables such as the one in Fig. 8.
- Calculating the distance between different measurements. See Fig. 6, part b.

The settings for each of these functionalities can be changed by the user, and the results can be saved as a *.csv* file.

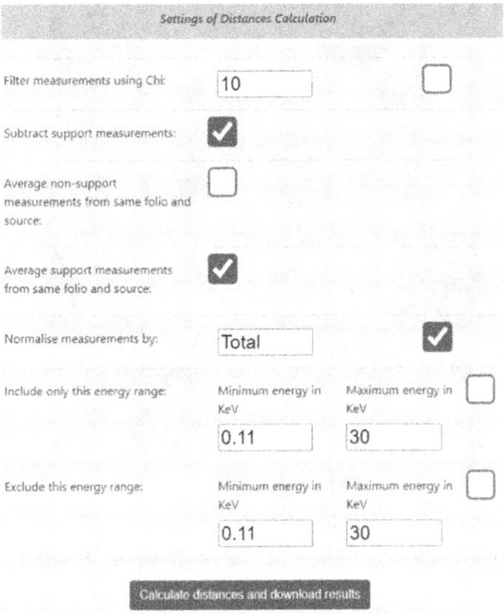

Settings of Distances Calculation

Filter measurements using Chi: 10

Subtract support measurements:

Average non-support measurements from same folio and source:

Average support measurements from same folio and source:

Normalise measurements by: Total

Include only this energy range: Minimum energy in KeV 0.11 Maximum energy in KeV 30

Exclude this energy range: Minimum energy in KeV 0.11 Maximum energy in KeV 30

Calculate distances and download results

(a) The settings of distance calculation.

Results for P6_ink_8	address verso	
FileName		L2_Distances
P6_ink_2	letter recto top	0.04
P6_ink_10	address verso	0.11
P6_ink_7	address verso	0.11
P6_ink_1	letter recto top	0.17
P6_ink_3	letter recto top	0.19
P6_ink_4	letter recto bottom	0.25
P6_ink_6	letter recto bottom	0.27
P6_ink_5	letter recto bottom	0.31
P6_ink_13	account verso	0.86
P6_ink_12	account verso	0.97
P6_ink_11	account verso	1.01
P6_ink_14	account verso	1.05

(b) Distance calculated using XRF-DAT with respect to the measurement P6˙ink˙8. The blocks of the two different inks are separated by a horizontal bold line.

Fig. 6. One of the three functionalities provided by XRF-DAT.

5.2 Use Case

The fragment *P.Hamb.Arab.6* is an incomplete letter on papyrus dating 315 h./927-8 CE, on whose verso an account was later recorded (Fig. 7). The written artefact was analysed by Olivier Bonnerot, Claudia Colini, Kyle Ann Huskin and Ivan Shevchuk following the protocol for ink analysis in use at the CSMC [4]. The inks used in these two occurrences are different, as the letter is written with a mixed carbon-iron-gall ink, while the account is penned with a black one, of the carbon-based type. The XRF analysis was used to assess the elemental composition of inks and papyrus and to verify if the ink of the letter coincide with the one of the address, written on the upper margin of the verso.

We analysed 14 spots for the inks: their location is stated in Fig. 6 part b, second column. The writing support was also analysed on 14 spots, selected close to the ink spots, as papyrus, and archaeological written artefacts in general, show an extremely heterogeneous elemental composition, influencing the measurements of inks. After the identification of the elements in the XRF spectra and their deconvolution with ARTAX software, the 28 .csv files with net intensity values were analysed by XRF-DAT.

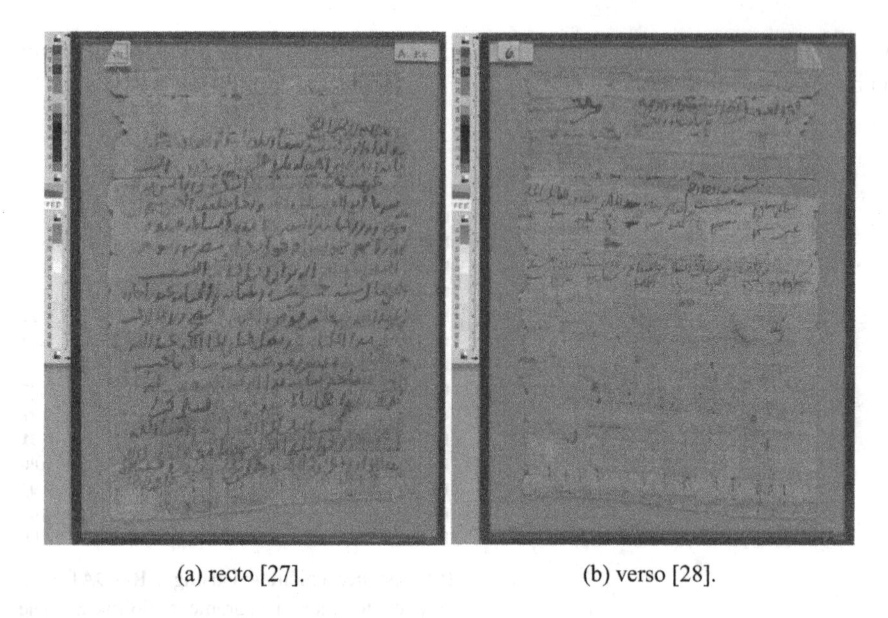

(a) recto [27]. (b) verso [28].

Fig. 7. The fragment *P.Hamb.Arab.6*. Copyright of Staats- und Universitätsbibliothek Hamburg Carl von Ossietzky.

The software was set to give a table (Fig. 8) with final intensity values for each elements of the measured inks, which result from subtracting the net intensity values of the papyrus from the corresponding ink spots. Having the results in one table is beneficial for a quick comparison of the inks and, in fact, the odd values of ink spot 9 (Fig. 8, marked yellow) could be quickly explained by looking at the writing support measurements, in the second half of Fig. 8, as a contamination affecting the papyrus spot 9 (Fig. 8, marked green). Therefore spot 9 was excluded from the rest of the evaluation.

File_Name		P	S	Cl	K	Ca	Ti	Mn	Fe	Ni	Cu	Zn	Sr
P6_ink_3	letter recto top	201	576	0	4613	4040	245	42	14741	42	0	0	36
P6_ink_6	letter recto bottom	234	642	0	642	10743	1713	182	27920	0	0	55	131
P6_ink_9	account verso	1	0	0	0	5608	0	0	0	0	0	0	0
P6_ink_10	address verso	361	378	0	4199	1045	73	166	9845	66	0	0	108
P6_ink_11	account verso	0	59	0	328	1692	0	0	1382	0	0	0	504
P6_p_3		9	468	6101	8957	21355	798	895	4710	115	205	406	857
P6_p_6		61	403	6962	8769	18374	755	915	4464	259	226	374	919
P6_p_9	contamination	52	1079	6145	13022	19102	1218	1216	35752	280	218	582	1031
P6_p_10		31	408	6371	9356	20996	864	838	7001	198	237	490	859
P6_p_11		23	468	5280	8109	25594	773	926	4645	364	362	579	803

Fig. 8. Part of the scattered plot table generated by XRF-DAT after subtraction, with comments. The ink spot 9 is marked in yellow, while the contaminated papyrus spot 9 is marked in green. (Color figure online)

Calculating the fingerprint of the ink [7] will be inconsequential due to the values shown in Fig. 8. Therefore, we used the L2 distance described in Eq. 8 using the total net as a normalisation factor as described in Eq. 9. The values in Fig. 6 part b, clearly shows strong differences between the spots measured on the account and those of the letter and address, but also particular similarities between the inks of the address and the letter, especially the upper part. It is very likely, therefore, that the ink used to write the address is the same one used for the letter, but the analysis also shows that, despite the subtraction of the support, the location of the spots on the papyrus still plays a role.

6 The Artefact-Feature Analysis Tool (AFAT)

The main goal of AFAT [20] is to calculate statistical information from manually generated tabular data which consists of distinctive features of artefacts. The value of each feature in these tables is represented by a positive integer, which describes the particular variant of this feature in an artefact. Furthermore, each variant can have any number of variations, represented by an alphabetical letter. The combination of variant and variation can be used to describe the observed version of a given feature in an artefact. The distance between feature-artefact values is calculated as follows:

$$D_A(a,b) = \frac{\sum_{i=1}^{n} D_i^F(a,b)}{n}, \tag{10}$$

where

$$D_i^F(a,b) = \begin{cases} 1 & \text{if } vt_i^a \neq vt_i^b \\ \dfrac{1}{vn_i} & \text{if } vt_i^a = vt_i^b \text{ and } vn_i^a \neq vn_i^b, \\ 0 & \text{otherwise,} \end{cases} \tag{11}$$

and $D_i^F(a,b)$ is the distance between feature i in artefacts a and b. vt_i^a and vt_i^b are the variants of feature i in artefacts a and b correspondingly. vn_i^a and vn_i^b are the variations of feature i in artefacts a and b correspondingly. Finally, vn_i is the total number of variants in feature i.

Once the distances are calculated between all artefacts in a table, statistical information are produced by AFAT, such as the mean, standard deviation, and detected outliers. Furthermore, the distance between tables of artefact-features is calculated by averaging the total distance between the artefacts in different tables.

6.1 Basic Functionality

The GUI of AFAT allows the user to select a set of artefact-features tables in order to be analysed. Statistical information is automatically calculated and displayed for each table. Furthermore, the distance is measured between all tables. See Fig. 9. All results can be saved as *.csv* files.

6.2 Use Case

One difficulty in the identification of cuneiform hands is that this ancient script consists of three dimensional signs, pressed in clay tablets. A two dimensional image of such tablets can be useful for a manual analysis, but the uneven shadows cast by the signs,

Results for Style4					
Artefact	Mean Distance	Standard Deviation	Average Mean	Average SD	Outliers
LA236	0.16	0.1	0.24	0.09	LA245
LA244	0.16	0.1			LA533
LA235	0.22	0.09			
LA245	0.34	0.08			
LA237	0.25	0.12			
LA247	0.21	0.08			
LA533	0.34	0.1			

(a) Statistical information for an artefact-features table.

Results for Style4	
Artefact-Features Table	Distance
Style1	0.53
Style2	0.57
Style6	0.7
Style5	0.73
Style3	0.79

(b) The distances from one table to all other tables.

Fig. 9. Example of the results provided by AFAT.

and the textured surface of such tablets make it futile to apply computational methods such as OCR and Handwriting Text Recognition (HTR). In addition, annotated training samples are extremely limited and learned features can not be generalised in many cases. For instance, the signs of the Old Assyrian period have not yet been standardised. Thus different constructions leads to having different variants and variations.

Wiebke Beyer from the CSMC developed an approach based on sign variants and variations in order to distinguish between different handwriting styles in cuneiform tablets. A selection of diagnostic signs from the individual manuscripts is examined manually and the variant or variation used is recorded in a table (for more details see [2]). This approach is adopted in the AFAT as described in Sect. 6. See Fig. 10.

Once these tables are composed, variants and variations of different tablets need to be compared. Manual comparison is prone to errors and very hard to scale for a large number of tablets. Therefore, AFAT has been used to automate the comparison and calculate statistical information as described in Sect. 6. Furthermore, it opened new possibilities by measuring the distance between different sets of tablets (tables) as well.

Initial tests have shown that AFAT can identify related tables and also outliers in a meaningful way. This helps to quickly and efficiently group many tables written by the same hand or to verify manually performed analyses. Furthermore, the comparison of several groups of handwritings is a very helpful feature, as any similarities and differences between the handwritings of individuals can be substantiated with numbers. Thus, the tool is not only used to check and sort a large number of tablets, but also to determine relative connections between their features.

(a) An image of a cuneiform tablet, reproduced from [13].

	A	B	C	D	E	F
1		Feature1	Feature2	Feature3	Feature4	Feature5
2	Artefact1	2b	1		2a	1
3	Artefact2	1a	2c	4a		1
4	Artefact3	2b	2a	1b	2a	2b
5	Artefact4	2a	1	4e	2a	1
6	Artefact5	3		4a	2a	2a

(b) An artificial example of a constructed table to be used in AFAT.

Fig. 10. Example of manually-constructed table from observations.

7 Conclusion

We presented a set of five Pattern Analysis Software Tools (PAST), which are dedicated to the automatic analysis of visual and tabular patterns in the research data from the study of ancient written artefacts. These tools are developed as off-line applications, which does not require installation or any additional dependencies. Furthermore, the GUI is easy to use and annotated with sufficient user instructions. PAST tools are developed in close collaboration with scholars of relevant fields from manuscript research. In addition, use cases are presented for each of the software tools in order to demonstrate its usability and applicability to actual research questions.

Several optimisations are planned for each of the software tools as future works. These optimisations involve general enhancements for all tools such as error reporting and format diversifying of input/output files, and tool specific enhancements such as integrating an image cropping mechanism for HAT, VPD and LDT, adding a clustering feature for XRF-DAT and AFAT, and calculating statistical information from measurements for XRF-DAT.

We think that the close collaboration between computer scientists and experts from manuscript research is not only desired, but rather necessary to have useful systems and proper interpretation of the results. Therefore, we made all PAST tools freely available for non-commercial purposes, and we hope that this interdisciplinary collaboration will help reinforcing the synergy across different fields of research in order to expand our knowledge about the past.

Acknowledgements. The research for this work was funded by the Deutsche Forschungsgemeinschaft (DFG, German Research Foundation) under Germany's Excellence Strategy - EXC 2176 'Understanding Written Artefacts: Material, Interaction and Transmission in Manuscript

Cultures', project no. 390893796. The research was conducted within the scope of the Centre for the Study of Manuscript Cultures (CSMC) at Universität Hamburg.

In addition, we thank Katrin Janz-Wenig and the staff of the SUB, Hamburg, for facilitating our investigations on P.Hamb.Arab.6 and Olivier Bonnerot, Kyle Ann Huskin and Ivan Shevchuk, who participated in the analysis.

References

1. Bertsch, W.: A Tibetan official seal of Pho lha nas. Tibet J. **29**(1), 3–8 (2004)
2. Beyer, W.: How did they learn writing? A palaeographic case study. In: Kulakoğlu, F., Kryszat, G., Michel, C. (eds.) Cultural Exchange and Current Research in Kültepe and its Surroundings, 1–4 August 2019, Subartu, vol. 46. Brepols, Kültepe (2021)
3. Caceres, A., Weber, A., Schomaker, L.: Monk in practice: indexing heterogeneous handwritten collections (2020). https://doi.org/10.5281/zenodo.3860811, http://2020.dhbenelux.org/. 7th Digital Humatities Benelux 2020, DH Benelux 2020; Conference date: 03-06-2020 Through 05-06-2020
4. Colini, C., Shevchuk, I., Huskin, K.A., Rabin, I., Hahn, O.: A new standard protocol for identification of writing media, pp. 161–182. De Gruyter (2021). https://doi.org/10.1515/9783110753301-009
5. Davis, T.: Beyond anonymity: palaeographic analyses of the Dunhuang manuscripts. J. Int. Assoc. Tibetan Stud. **3**(1) (2007)
6. Durkin-Meisterernst, D., Friedrich, M., Hahn, O., Helman-Ważny, A., Nöller, R., Raschmann, S.C.: Scientific methods for philological scholarship: pigment and paper analyses in the field of manuscriptology. J. Cult. Herit. **17**, 7–13 (2016). https://doi.org/10.1016/j.culher.2015.06.004
7. Hahn, O.: Analyses of iron gall and carbon inks by means of x-ray fluorescence analysis: A non-destructive approach in the field of archaeometry and conservation science. **31**(1), 41–64 (2010). https://doi.org/10.1515/rest.2010.003
8. Helman-Ważny, A., Ramble, C.: Tibetan documents in the archives of the tantric lamas of Tshognam in Mustang, Nepal: an interdisciplinary case study. Rev. d'Etudes Tibétaines **39**, 266–341 (2017)
9. Helman-Wazny, A., Ramble, C.: The mustang archives: analysis of handwritten documents via the study of papermaking traditions in Nepal (2021)
10. Helman-Ważny, A.: More than meets the eye: fibre and paper analysis of the Chinese manuscripts from the silk roads. STAR: Sci. Technol. Archaeol. Res. **2**(2), 127–140 (2016). https://doi.org/10.1080/20548923.2016.1207971
11. Kahle, P., Colutto, S., Hackl, G., Mühlberger, G.: Transkribus - a service platform for transcription, recognition and retrieval of historical documents. In: 2017 14th IAPR International Conference on Document Analysis and Recognition (ICDAR), vol. 04, pp. 19–24 (2017). https://doi.org/10.1109/ICDAR.2017.307
12. Kiessling, B., Tissot, R., Stokes, P., Stökl Ben Ezra, D.: eScriptorium: an open source platform for historical document analysis. In: 2019 International Conference on Document Analysis and Recognition Workshops (ICDARW), vol. 2, p. 19 (2019). https://doi.org/10.1109/ICDARW.2019.10032
13. Larsen, M.T.: The archive of the šalim-aššur family, the first two generations. Volume 1: the archive of the Šalim-Aššur family (2010). Ankara: Türk Tarih Kurumu, p. 507: tablet no 176 (kt 94/k 1461)
14. McCann, S., Lowe, D.G.: Local naive bayes nearest neighbor for image classification. In: 2012 IEEE Conference on Computer Vision and Pattern Recognition, pp. 3650–3656, June 2012. https://doi.org/10.1109/CVPR.2012.6248111

15. Mohammed, H., Märgner, V., Stiehl, H.S.: Writer identification for historical manuscripts: analysis and optimisation of a classifier as an easy-to-use tool for scholars from the humanities. In: 2018 16th International Conference on Frontiers in Handwriting Recognition (ICFHR), pp. 534–539, August 2018. https://doi.org/10.1109/ICFHR-2018.2018.00099
16. Mohammed, H.: Handwriting Analysis Tool (HAT), February 2020. https://doi.org/10.25592/uhhfdm.900
17. Mohammed, H.: Line Detection Tool (LDT), June 2020. https://doi.org/10.25592/uhhfdm.1042
18. Mohammed, H.: X-Ray Fluorescence Data Analysis Tool (XRF-DAT), June 2020. https://doi.org/10.25592/uhhfdm.969
19. Mohammed, H.: Visual-Pattern Detector (VPD), February 2021. https://doi.org/10.25592/uhhfdm.8832
20. Mohammed, H.: Artefact-Features Analysis Tool (AFAT), January 2022. https://doi.org/10.25592/uhhfdm.9778
21. Mohammed, H., Märgner, V., Ciotti, G.: Learning-free pattern detection for manuscript research. Int. J. Doc. Anal. Recogn. (IJDAR) **24**, 1–13 (2021)
22. Mohammed, H., Märgner, V., Konidaris, T., Stiehl, H.S.: Normalised local naïve bayes nearest-neighbour classifier for offline writer identification. In: 2017 14th IAPR International Conference on Document Analysis and Recognition (ICDAR), pp. 1013–1018. IEEE (2017)
23. Morris, R.N.: Forensic Handwriting Identification: Fundamental Concepts and Principles. Academic press (2020)
24. Pizer, S.M., et al.: Adaptive histogram equalization and its variations. Comput. Vis. Graph. Image Process. **39**(3), 355–368 (1987)
25. van Schaik, S.: Towards a Tibetan palaeography: Developing a typology of writing styles in early Tibet. In: Manuscript Cultures: Mapping the Field, pp. 299–338. De Gruyter (2014)
26. Schuh, D.: Grundlagen tibetisch siegelkunde. Monument Tibeticum Guide **5**(3) (1981)
27. Staats- und Universitätsbibliothek Hamburg Carl von Ossietzky: P.Hamb.Arab. 6, https://resolver.sub.uni-hamburg.de/kitodo/HANSh954
28. Staats- und Universitätsbibliothek Hamburg Carl von Ossietzky: P.Hamb.Arab. 6, https://resolver.sub.uni-hamburg.de/kitodo/HANSh871
29. Walsh, E.: XXI. Examples of Tibetan Seals: Supplementary Note. J. Roy. Asiatic Soc. **47**(3), 465–470 (1915)

TEI-Based Interactive Critical Editions

Simon Schiff[1](✉), Sylvia Melzer[2], Eva Wilden[2], and Ralf Möller[1]

[1] Institute of Information Systems, University of Lübeck, Ratzeburger Allee 160, 23562 Lübeck, Germany
{schiff,moeller}@ifis.uni-luebeck.de
[2] Centre for the Study of Manuscript Cultures, Universität Hamburg, Warburgstraße 26, 20354 Hamburg, Germany
{sylvia.melzer,eva.wilden}@uni-hamburg.de
http://www.ifis.uni-luebeck.de,
https://www.written-artefacts.uni-hamburg.de

Abstract. A critical edition is the reconstitution of a text based on a survey of the available witnesses (manuscripts and quotations), resulting in a text and all its attested variants. It is usually created with high effort by scholars in the humanities, possibly separated by chronological or geographical boundaries, over several years. During the editing process, scholars in the humanities prefer to work with any tools and documents in any format they are familiar with. Independently of any boundary or tool in use, one primary goal usually is to produce a traditional print edition. However, working towards such a print edition is very time-consuming, and we argue that there is potential for reducing the time required without the need to forgo preferred tools and methods. Our contribution consists of providing different interactive views of the data as an additional offer to perform analyses. We enable scholars in the humanities to easily transform documents into the well-known TEI format and to store them in a database to create services on demand. The database management system Heurist, a tool for the humanities, allows for searching, automatic linking parts of the documents, interaction with the documents, and exporting the documents into a printable version of the edition. Our approach is validated by direct collaboration with scholars who are currently working on a critical edition. Our solution allows these scholars to efficiently work on a critical edition, independent of chronological or geographical borders, while they still use their preferred tools and document formats.

Keywords: TEI · Critical edition · Databasing on demand · Annotation · Cross-disciplinary collaboration

The research for the extended abstract was funded by the Deutsche Forschungsgemeinschaft (DFG, German Research Foundation) under Germany's Excellence Strategy - EXC 2176 'Understanding Written Artefacts: Material, Interaction and Transmission in Manuscript Cultures', project no. 390893796.

© Springer Nature Switzerland AG 2022
S. Uchida et al. (Eds.): DAS 2022, LNCS 13237, pp. 230–244, 2022.
https://doi.org/10.1007/978-3-031-06555-2_16

1 Introduction

A critical edition is the reconstruction of a text including all witnesses. It is completely as possible and usually created by humanities scholars over several years. They work with their preferred tools and document formats, possibly separated by chronological and geographical borders. Usually, the goal is to have at the end an edition in a printable format, which is preferred by most of the readers. Due to the different tools, formats and borders, some tasks involved in the creation of the critical edition are time-consuming and cost-intensive. For instance, tasks such as transcribing texts from pictures of manuscripts are very time-consuming if humanities scholars have to merge their transcribed texts of any format, produced by different tools, into one that could be printed. However, there is a growing interest in having access to digital editions, in contrast to a printed edition. By digital editions, we do not refer to, for instance, scans of printed editions. Instead, we refer to digital editions that follow some principles, such as the FAIR (**F**indable, **A**ccessible, **I**nteroperable, **R**euse) principles. Creating a digital output in addition to the print format output is additional costly work that must be done either manually by humanities researchers or by an IT expert who processes the documents semi-automatically.

Our objective is to enable humanities scholars to work with their preferred tools and document formats across chronological and geographical borders and yet to achieve their desired results. We aim to achieve that by providing a web application as an add-on in addition to the tools that are already in use. The web application should only be an add-on, as we aim to support humanities scholars without trying to change their preferred tools and document formats in use. The web application allows humanities scholars to transform documents, in any format, into the well-known text encoding initiative (TEI) format. Documents transformed into the TEI format are machine-readable and thus are combined optionally and then exported into any desired format to be printed or published online. In addition, a database is created on demand, that allows for simple search queries as well as classification of an object by assignment of several independent terms (faceted searches) to analyze data. A database would not only represent the data for one text, but would allow the data to be analyzed as a whole in large quantities in less time.

The database management tool Heurist [9] offers a management for access rights, an environment for sharing and collaborating online with other users, and modification of the database scheme without programming.

We give in Sect. 2 a detailed overview of critical editions, including their creation, use and present why it is challenging but useful to publish them online, following the FAIR principles. The related work in Sect. 3 gives an insight into current research and tools. In Sect. 4, we present the work by Eva Wilden who is a leading expert in classical Tamil literature and poetics, and currently working on the critical edition of a Tamil poetic anthology on the basis of 18 manuscripts. The preferred tool for producing a critical edition in a printable format is Microsoft Word. We worked together across disciplines for testing our developed web application using Microsoft Word DOCX documents as input. We present our web application step by step in the sections following.

First, in Sect. 5, we present how to enable humanities researchers to transform documents from their preferred format into TEI without the help of an IT expert. Documents encoded in TEI are designed to be machine-readable and not human-readable, therefore we present in Sect. 6 a viewer, part of our web application, for documents that stem from our transformation of documents into TEI. Depending on the context, one possibly needs a database, which we create on demand from the TEI documents, for a more extensive data analysis. The databasing on demand process is described in Sect. 7. TEI documents visualized by our web application are combined with linked sources such as images, audio, or other websites. As a use case, we present in Sect. 8 our annotation system, which allows humanities scholars to annotate images that are linked and visualized by our viewer. In addition, the annotation system provides a login for access control and a chat below each annotation for remarks. Finally, we present the use of our web application in Sect. 9, briefly summarize our paper in Sect. 10, and give an outlook for future work. Among other things, we aim to extend our web application such that it could be adapted by humanities scholars, without the help of an IT expert.

2 Preliminaries

A critical edition is a scholarly attempt at reconstructing the transmission history of a text by following up the various textual witnesses over the centuries, and thus reproducing the trajectory the text made through various manuscript and print versions into to the modern days, as depicted in Fig. 1.

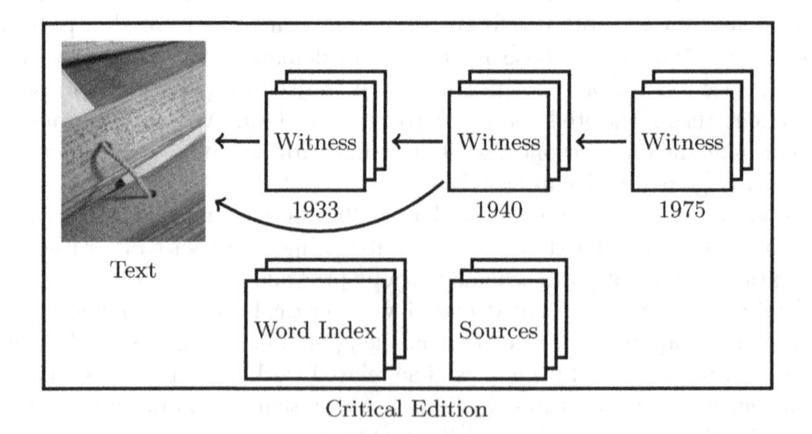

Fig. 1. Contents of a critical edition

The task comprises the collection and decipherment of a set of sources in manuscript and print as completely as possible, a comparison of the differences between these sources, and their documentation. This is achieved by the constitution of a core text and a critical apparatus that records the variants and distinguishes what is meaningful from what is lost, damaged or simply erroneous. Additionally, critical editions contain word indexes with morphological and semantic analysis of the words and their occurrences in the texts. The advantage of this type of edition is that its author presents a hypothesis about the constitution of the text while at the same time giving to his or her potential readers all the elements necessary to revise or even reject it and form conclusions of their own.

Readers often prefer critical editions in printed format. However, there is a growing interest of critical editions in a digital format [1]. Having a critical edition in digital format alone is not sufficient and one should follow specific principles such as the FAIR principles.

Editions are easier to find by using web search engines if they are digitally available online. An author of an edition that is easy to find has a higher chance of being cited than others. Depending on the format of the edition and the viewer that visualizes the contents of the edition the implementation of a faceted search engine helps to satisfy the information needs of various readers. Following links by clicking on them to find linked elements is not as time-consuming as going through a book by hand. One might need to pay a lot to access a complete edition in printed form and only some parts, such as pictures of manuscripts, are the reason for high prices. If an edition is digitized, one has a more fine-grained control as to which parts can be accessed by whom. In particular, specific parts can be made publicly available. Results are combined with other results automatically on demand, if the edition is in a machine-processable format (cf. Sect. 7). In addition, editions in such a format are reused for various kind of research purposes, such as in the field of artificial intelligence (AI) [2].

Following the FAIR principles during the creation of a critical edition is a challenging task, as even the creation of a critical edition in any format is a cost-intensive process lasting several years. Additionally, many humanities scholars who work on a critical edition are possibly separated by chronological and geographical boundaries and with preferences for different tools and document formats. We argue that the solution is not to intervene in the processes, tools, and document formats humanities scholars prefer to work with. Instead, we provide a web application that allows humanities scholars to upload their documents of any format and transform them to TEI, which is a common and standardized format in the digital humanities. As depicted in Fig. 2, we create Antlr4 [12] grammars for the generation of parsers for parsing different document formats, such as Microsoft Word DOCX, Markdown, or TXT, where parts of the documents are written in a specific syntax we know beforehand. Antlr4 is a tool that generates the source code for a parser from a grammar. The source code provides an application programming interface (API) to access parts of the texts, depending on the grammar. Parsed documents are transformed into TEI and

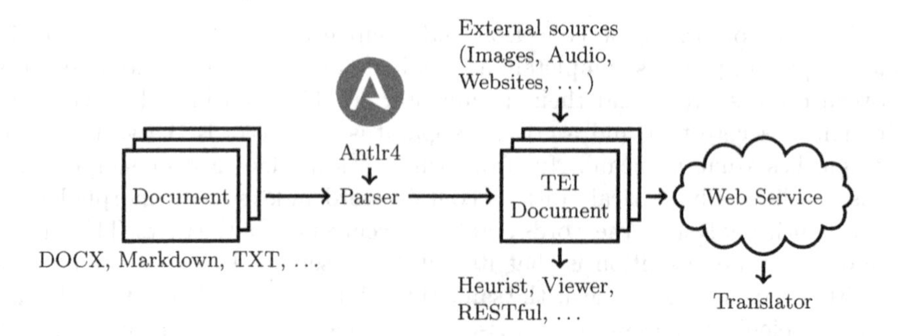

Fig. 2. Processing of word documents

stored in a repository. The documents can be linked with external sources such as images, audio, or websites and exported into any format, visualized with a viewer, accessed using for instance RESTful or a webservice, and exported into a Heurist database instance. RESTful provides an API for tools to access data over the web.

Heurist [9] is an open-source web-based database management system for the humanities and allows researchers without prior IT knowledge to develop data models, store and search their data, and publish it on an automatically-generated website. A further highlight of Heurist is the possibility of displaying the data according to the WYSIWYG (What You See Is What You Get) principle which plays an important role in the humanities. The combination of, on the one hand, storing the data and, on the other hand, representing it in a humanities-friendly way with a powerful search function for extensive data analysis with large amounts of data argues for the use of a database. If the data is already available, it can be manually inserted into a database. However, with an increasing amount of data, it is advisable to carry out this time-consuming work with the help of a computer. Heurist has a CSV import function that could be used to transform data into the CSV format. If the data is in TEI, the EpiDoc extensible stylesheet language transformations (XSLT) stylesheets [4] can be used to convert the TEI files to CSV. When using these stylesheets, the text in the CSV file is stored in HTML encoded form. These stylesheets also can be easily customized. Importing such a CSV file displays the text as in HTML encoded on the web page generated with Heurist. The transformation process is either manual or automatic. The automatic process is called databasing on demand.

3 Related Work

In the field of manuscript research, the amount of data has grown significantly, however at present it cannot be taken into account by search engines in part because the data is available only in printed form. The evaluation is therefore mostly done manually or with a high resource effort by implementing customized algorithms. However, existing algorithms and tools can easily be made to use more powerful, customized applications that can be derived from existing approaches through the systematic harmonization of processes and tools.

For proteomics researchers, the creation of a custom database was made as easy as possible by developing a so-called Database on Demand (DoD) [13]. The idea of DoD is to generate a custom database according to a set of user-selectable criteria using commonly used database templates as source. The database PRIDE [14] is one DoD to provide users to build a database in five steps. However, the DoD in this case only works for data for one - the proteomics - area. In the field of e.g. the Humanities, this concept of a DoD could also be useful. With DOD's approach, it is then possible not only to simply implement an information system such as presented in [10], but also to link these into a network in such a way that even simple, federated, or more complex searches are possible [11]. When presenting the data from the humanities scholars, it is still important to note that text formatting is also a constraint and very important for the humanities scholars. Text formatting is explicitly supported with the TEI format. Since TEI is a machine format that has yet to be converted into a human-readable format. Efforts have been made to implement TEI viewers such as the Edition Visualization Technology (ETV)[1], but these are only usable for very specific TEI schemes. For example, an viewer was developed for epigraphers, which requires a slimmed-down TEI variant, the EpiDoc [5] scheme, as input. While the texts are correctly displayed according to appropriate conditions, e.g. the Leiden conventions [6], the tool called EpiDoc Front-End Services (EFES) [8] must be installed locally before a data set can be viewed. With EFES the texts are displayed, but only one text at a time. An evaluation of all texts or other data is not possible. Therefore, an approach that is based on the combination of a database and the corresponding formatting of the texts is highly recommended when it comes to efficiently carrying out an analysis with the existing data. In addition, the storage of data also has the advantage here of being FAIR (**F**indable, **A**ccessible, **I**nteroperable, **R**eusable) from an ethical point of view [7] and that of good research practice [3].

4 Critical Edition - Critical Texts of Cankam Literature

The point of departure for the current experiment is a traditional critical edition of average complexity, from one of the smaller disciplines within the humanities, namely Indology, or, to be more precise, Tamilistics. Tamil is the language of classical literature in South India; the text selected is an anthology of love poetry that dates back roughly to the beginning of the Christian era. The anthology consists of 400 poems between 13 and 33 lines of length, accompanied by miniature commentaries for each piece and a more detailed old anonymous commentary for the first 90 poems. The part already completed (and available in print) comprises book 1, that is, the first 120 poems plus the old commentary [15]. The critical edition is based on eighteen manuscripts (palm-leaf and paper), three

[1] http://evt.labcd.unipi.it/.

editions from the early and middle 20th century and numerous quotations from the commentaries of an extensive tradition of poetics. The edition consists of a metrical Tamil text, a word-split transcription, a critical apparatus, an interlinear English version and an English translation, with heavy annotation for every layer. On the basis of the text and the variants reported in the critical apparatus a word index cum concordance for every word and all its derivations is produced which also contains an analysis and a translation as well as references to the existing lexical works. At the basis of such a work is a complex web of cross-references on various levels. The Tamil text itself needs to be seen in an intertextual context of parallels within the corpus of literary texts; many of its elements reoccur in the commentary. The critical apparatus relates directly to the manuscript and print sources that contain the corresponding variants. The word-split transcript separates metrical items into lexical items, the interlinear version adds their morphological and semantic analysis. The translation adds an interpretation of the whole both in terms of syntax and in terms of a piece of poetry that follows a set of conventions, has parallels and has been a long-term subject of scholarly discourse. The annotation adds direct references to parallels and ongoing scholarly discussions. The word-index contains a detailed morphological and semantic analysis of each item in relation to various external tools such as dictionaries, grammars, commentaries and scholarly articles. In book format this work consists of three volumes that, including the bibliography, make up a total of about 1100 pages. The tool that was used to produce this edition, in DOCX format, was simply Microsoft Word.

5 Transforming Documents into TEI

Editions contain natural language text paragraphs and sections written in a specific syntax. The latter are, for example, translations of Tamil poems with footnotes or word-indexes with associated information including occurrences of the words in the poems and external sources such as dictionaries. A word-index entry of the word *acaitta* is depicted in Fig. 4 on the left hand side. The associated information of the word acaitta is separated by a tab, including the information *pey. p.a.* and the occurrences *155.14* and *301.22* meaning that the word acaitta occurs at poem 155 in line 14 and poem 301 in line 22. A reader, who has access to a printed edition or digital edition violating the FAIR principles, has to search for the poems 155 and 301 manually, if he or she wants to gather further information about the context, in which the word acaitta is being used. A website could help a reader to reduce the time required to find specific information, by providing links for each occurrence of the words in the indexes, such that the reader can click on them to be redirected to the corresponding poems. As depicted in Fig. 3, the word acai in the word-index on the right hand side of the picture appears in line 5 of poem 40 on the left hand side. In our work, we focus on processing Microsoft Word DOCX documents, as we collaboratively work together with Eva Wilden, who is using primarily Microsoft Word. Direct accessing translations or word-indexes in the editions written in DOCX format, for implementing a website is

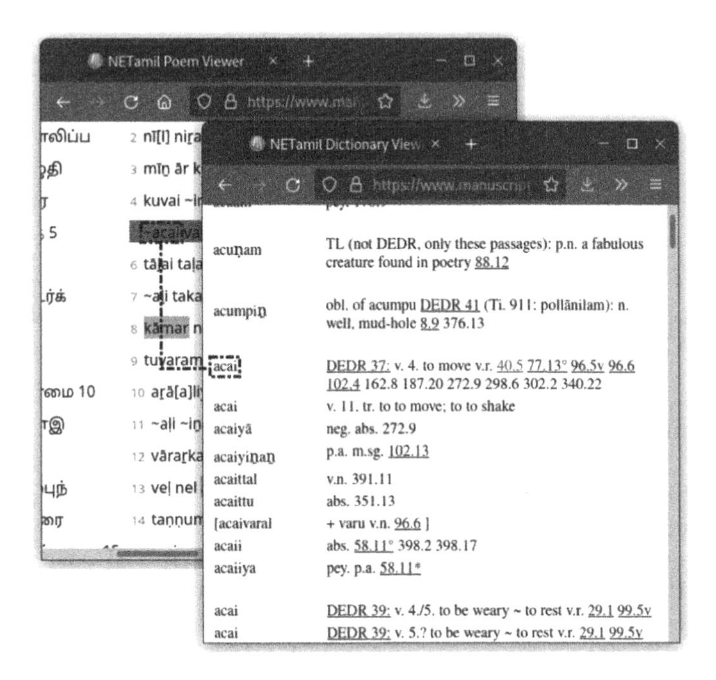

Fig. 3. Poem linked with word-index

impossible. Therefore, we provide a web application, implemented in Java, for uploading DOCX documents by humanities scholars as a service, as depicted in Fig. 2. The application analyzes the uploaded documents and searches for parts that are written in a specific syntax, such as word indexes. If a word index is found, it is processed and a web service is created, access is via the internet to implement a website.

More specifically, a DOCX document is a zip archive standardized as Office Open XML (OOXML), containing among other files a document.xml file. We created a parser, using Antlr4 [12], for parsing and processing the contents of the document.xml file. The result is a document in TXT format containing the contents of the DOCX document without any XML tags, with annotated parts such as sections or tables. The TXT document is again parsed with a parser, we have generated, using Antlr4. A parser is generated, as a Java code snipped, by Antlr4 from a grammar, as depicted in Fig. 4 on the right hand side. The grammar consists of four rules and in this case, the rule *entry* matches with the vocabulary index on the left hand side that is a sequence of tokens acaitta (text), a tab (tab), pey. p.a. (text), 155.14 (occurrence), and 301.22 (occurrence). Each token matches with the corresponding rules text, tab, and occurrence. If the syntax of the documents changes, the grammar can be easily adapted, without the necessity to change anything else part of our web application. We instantiate Java objects from the parsed TXT file and deserialize them as TEI documents.

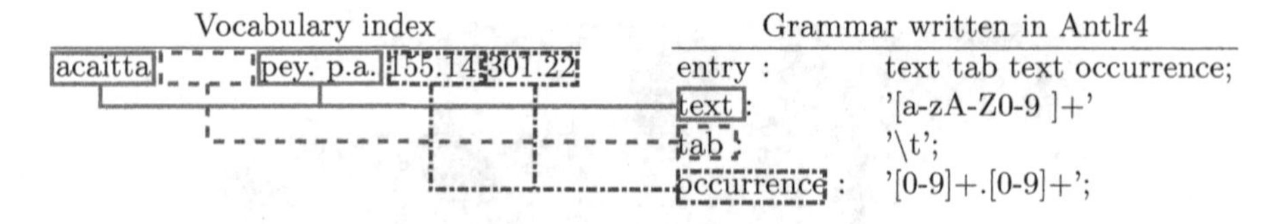

Fig. 4. Parsing a text using Antlr4

A webservice could be created on demand, using the web services description language (WSDL), that accesses the created TEI documents.

6 Interactive Critical Editions

As depicted in Fig. 2 and described in Sect. 5, our web application supports the work of a humanities scholar by converting his Word documents into TEI when parts of the document are written in a certain syntax. A TEI document, that stems from the transformation, is not very pleasant to read for humans, but in a well known scheme readable by our web applications or in general by a machine. Therefore, we create viewer for all kind of TEI documents, our web application creates, using HTML5 together with JavaScript and CSS. The viewer has several advantages over a document written in Word or a printed book. For instance, the viewer allows for (i) a search engine, (ii) linking of various elements such as words in word indexes with their poems at where they appear, (iii) restricting access to only authorized humans, (iv) export the visualized TEI document into various formats, and (v) merging of TEI documents with other TEI documents or other resources such as images mentioned in the documents. The latter can save humanities scholars weeks or even month of work. For instance, as depicted in Fig. 5, two vocabulary indexes need to be merged. Vocabulary index I and II are tiny excerpts of two very large word indexes, containing thousands of words that need to be merged. Doing that by hand would take several weeks of work and is a very laborious and cost-intensive task. Instead, our web application allows for merging the indexes automatically on demand. The merged vocabulary index is again a TEI document, that could be exported into various formats or visualized by our viewer. A humanities researcher does not need to search for the same word in two different documents anymore.

7 Databasing on Demand

The advent of more flexible and precise tools in recent years has enabled the way to evaluate critical editions as well as new methods for manuscript evaluation, thus expanding the field of the humanities. In addition, a large amount of data accumulates over time, which can only be evaluated by machines in a short time and are available to be included in the analysis processes.

Vocabulary index I		Vocabulary index II		I & II merged	
acai	DEDR 37: v. 4. to move - v.r. 40.5 77.13° 96.5v	acai	DEDR 37: v. 4. to move - v.r. 77.13° 187.20	acai	DEDR 37: v. 4. to move - v.r. 40.5 77.13° 96.5v 187.20
acaiyā	neg. abs. 272.9	acai	v. 11. tr. to move; to shake	acaiyā	neg. abs. 272.9
				acai	v. 11. tr. to move; to shake

Vocabulary index I		Vocabulary index II		I & II merged	
acai	DEDR 43: v. 11. to tie – v.r. 54.7	acai	DEDR 43: v. 11. to tie	acai	DEDR 43: v. 11. to tie – v.r. 54.7
acaii	abs. 188.12*	acaitta	pey. p.a. 155.14 301.22	acaii	abs. 188.12*
				acaitta	pey. p.a. 155.14 301.22

Fig. 5. Merging of word indexes (Index I: blue, Index II: yellow, and I & II merged: blue and yellow as green) (Color figure online)

Databases are ideal for storing a large amount of data and, depending on the database query language, the available data format and the tool, allow simple to more complex search queries to be performed. In addition, database management systems such as Heurist also support the creation of websites, including a search mask, display of search results, and presentation of data, in short time.

To make the creation of project-specific, autonomous databases as easy as possible, we have developed a databasing on demand approach so that users may create customized databases automatically. Therefore we use Heurist to create database instances.[2]

The precondition for the databasing on demand approach is that the poems, dictionaries, and comments are stored in TEI or CSV format. Otherwise, the transformation from Word to TEI, as described in Sect. 5, must be performed first. Then, a TEI-specific database scheme must be created (either manually or by creating a template, which is then reusable). The databasing on demand process encompasses the three steps 1. transforming from TEI to CSV based on the extended EpiDoc XSLT Stylesheets to parse the TEI files and create the particular data view, 2. importing data into the database, and then 3. publishing data, respectively.

1. **Transform from TEI to CSV:** Input data are TEI (poems, commentaries) files. Since there are very different possibilities to structure such a scheme. It is recommended to follow the TEI project example. We have written java program to process the transformation from all TEI files (poems, commentaries) to one CSV file.

[2] https://heurist.fdm.uni-hamburg.de/html/heurist/?db=CSMC_UWA_NETamil2 (internal access).

2. **Import data into the database:** The created CSV file from the previous step is the input file. The CSV file was imported into the database. When importing, the user must make the mapping from the columns in the table and the fields in the database manually.
3. **Publishing data:** During or after the importing process, a web page, see Fig. 6, can be created easily and in short time.

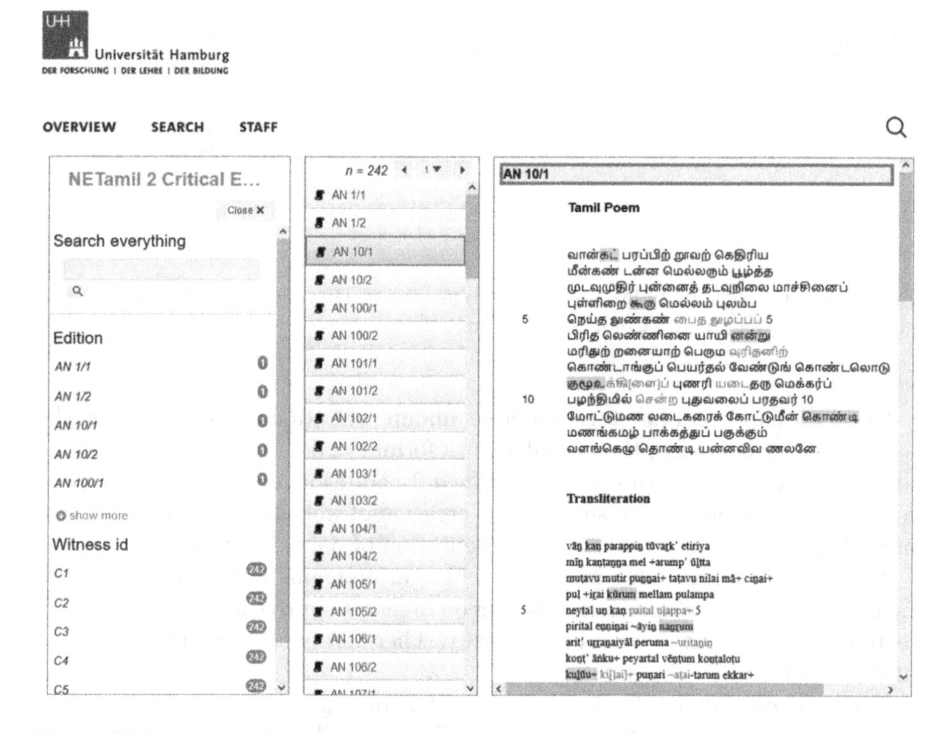

Fig. 6. Website created with Heurist; left: search area, middle: result set, right: data representation

Even though Heurist offers a kind of drag and drop functionality for creating a webpage, the user needs knowledge of the automatically generated source code (PHP and HTML) so that project-specific adjustments can be made. The usage of some database functionalities requires a certain know-how, so that a close cooperation between humanities scholars and computer scientists is very advantageous and recommended.

We used the above three steps for poems and comments. For creating links to other dictionaries, such as the second Dravidian Etymological Dictionary, they were automatically created in the PHP script for the users' view. However, creating links automatically to the poems in the database require another approach. For this, a database query was made to obtain the database IDs of the poems.

A Java program was then written which automatically creates a link to the database entry for the reference to a poem (see Fig. 7). As a result, users can use the created information system[3] as an interactive version of poems, dictionaries, and commentaries.

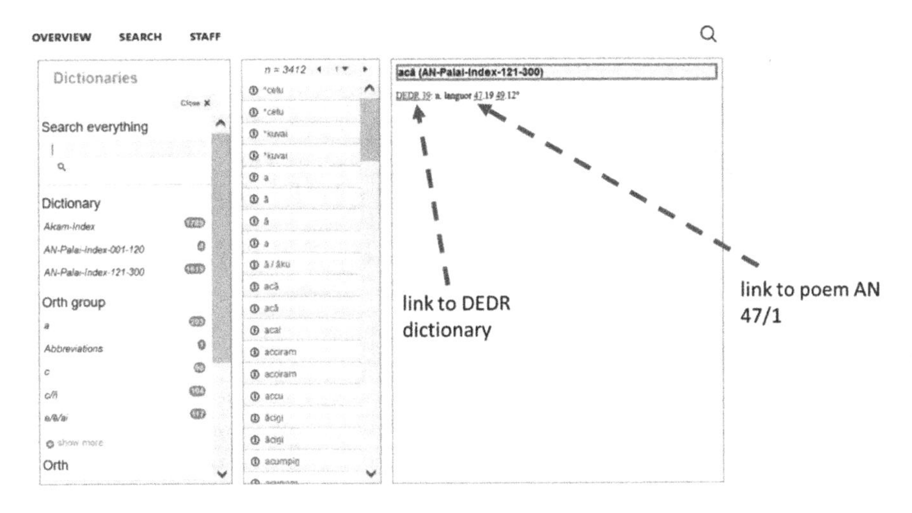

Fig. 7. Dictionary with links to the DEDR dictionary and poems; left: search area, middle: result set, right: data representation with links to other dictionaries and poems

8 An Annotation System for the Humanities

As depicted in Fig. 2 and presented in Sect. 6 we created a viewer for TEI documents, that contain for instance poems written in Tamil, that stem from the transformation of DOCX documents into TEI. The poems where transcribed from palm-leaf manuscripts and editions that were photographed and stored as images in a database. We add the images to the viewer for those who are authorized to access them. Images of manuscripts and editions are associated with the images, as the DOCX documents, at where the poems stem from, contain the names of the images. A humanities scholar may want to note in his or her texts that a specific region in an image is damaged by a tape worm. One way of doing that is to add a description to the text, that contains the name of the image and a description at where the damaged region is located at. A description could contain for instance the coordinates of the region in pixel units. If another scholar reads the text, then he or she needs to search for the image and if the image is found, for the correct position in the image. We follow another approach, at where the humanities scholars can interactively add annotations to the images,

[3] https://heurist.fdm.uni-hamburg.de/html/heurist/?db=CSMC_UWA_NETamil2&website&id=981&pageid=976 (internal access).

as depicted in Fig. 8. For the annotation of the images, we use Annotorious[4], which is open-source and implemented in JavaScript. As Annotorious is written in JavaScript, it runs only on client side. However, annotations are serializable into JSON, we send to the server at where the viewer and Heurist are hosted at. Annotations are stored in the Heurist database and were exported together with the images and contents of the TEI documents into any format.

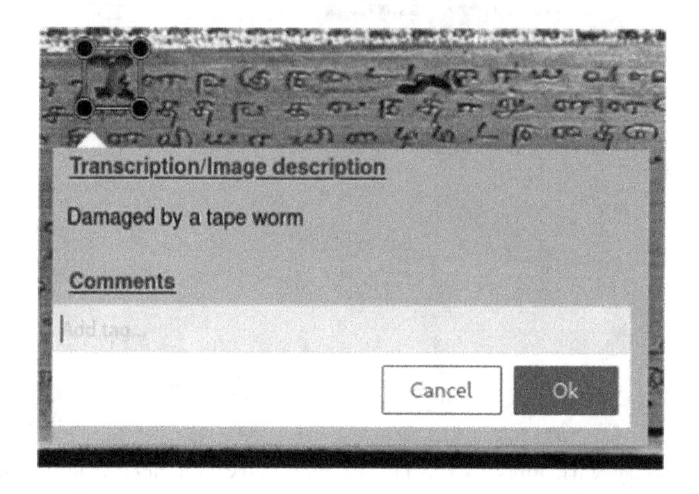

Fig. 8. Annotation of a palm-leave manuscript

9 Application and Results

From the point of view of a scholar in the humanities, the approach described here promises a number of advantages, both with respect to ongoing as with respect to already accomplished work. Since nowadays manuscripts are no longer documents that are stored in a library, that can be consulted, often with great effort, and exploited for text edition, but are available or can be made available online, it is very tempting to complement a printed critical edition by an electronic one. This way it is possible to accompany the critical apparatus by direct access to all the (digitized) sources, a direct view of the manuscript images and the variants they attest. The advantages for ongoing text-critical work and continued improvements on the hands of the research community, no longer a single scholar, are obvious. Similarly, the direct interlinking of edition, word indexes and existing dictionaries simplifies the necessary lexical work and allows an efficient use of work already done in the same area. An additional benefit is the possible use of such a construct in teaching: direct access to all the relevant material for students, and even the possibility to make their own contributions. Attractive is also the user-friendliness of the conversion procedure which does

[4] https://github.com/recogito/annotorious.

not require advanced technical skills in IT. Many scholars prefer to continue working with a simple word-processing program, all the more when they have huge amounts of data which have already been produced in such formats. The same holds good for cases of collaboration with colleagues in technically less advanced areas.

10 Conclusion and Future Work

The presentation of data and the results of analysis plays an important role in the humanities. Tools such as Word that support the WYSIWYG (What You See Is What You Get) principle are often used. For this purpose, humanities scholars mainly use Word to present their research data such as critical editions. In addition, printed versions of critical editions are mainly required as a project result. Individually structured documentation and print versions of research data complicates automatic data exchange and linkage to other (similar) research data. Increasingly, however, humanities scholars' research data are being archived in widely used standards such as TEI. Since TEI is a machine-readable format and we want to support humanities scholars to continue using their established processes and tools, we have shown in this paper how to create a transformation from Word to TEI. Furthermore, we have shown how to use this standard to perform a databasing on demand to support the analysis process by performing a more advanced search over a large amount of data in a short time. We also presented an annotation system for adding annotations to a palm leaf, for example. The annotations were stored in a database and used for further analysis. All our approaches satisfy the FAIR principles that are elementary to good scientific practice in particular.

As future work, researchers from other fields are also supported by these approaches to represent critical editions and work with the new services. We are also working on merging the databases from different projects into one overarching information system (see [11]) so that, for example, federated searching is supported. The database of Tamil poems, dictionaries, and commentaries generated by databasing on demand could be integrated as part of the federated database system in a next step. From a data linking point of view, it is possible to include dynamic links in addition to the static links in order to enrich the existing information with further information.

References

1. Al-Shboul, M.K., Abrizah, A.: Information needs: developing personas of humanities scholars. J. Acad. Librariansh. **40**(5), 500–509 (2014)
2. Bender, M., Braun, T., Gehrke, M., Kuhr, F., Möller, R., Schiff, S.: Identifying and translating subjective content descriptions among texts. Int. J. Semant. Comput. **15**(4), 461–485 (2021). https://doi.org/10.1142/S1793351X21400122
3. DFG: Guidelines for Safeguarding Good Research Practice (2022). https://www.dfg.de/download/pdf/foerderung/rechtliche_rahmenbedingungen/gute_wissenschaftliche_praxis/kodex_gwp_en.pdf. Accessed 17 Jan 2022

4. Elliott, T., et al.: EpiDoc reference stylesheets (version 9) (2008–2017). https://sourceforge.net/p/epidoc/wiki/Stylesheets/. Accessed 22 Jan 2022
5. Elliott, T., et al.: EpiDoc guidelines: ancient documents in TEI XML (Version 9) (2007–2022). https://epidoc.stoa.org/gl/latest/. Accessed 22 Jan 2022
6. Everipedia: Leiden Conventions. https://everipedia.org/Leiden_Conventions. Accessed 17 Jan 2022
7. Friedrich, M., et al.: Ethical and responsible research at the CSMC (2022). https://www.csmc.uni-hamburg.de/files/uwa-csmc-ethics-code.pdf. Accessed 17 Jan 2022
8. GitHub repository: EFES: EpiDoc Front-End Services (2022). https://github.com/EpiDoc/EFES. Accessed 22 Jan 2021
9. HEURIST: A unique solution to the data management needs of Humanities researchers (2022). https://heuristnetwork.org/. Accessed 17 Jan 2022
10. Melzer, S., Peukert, H., Wang, H., Thiemann, S.: Model-based development of a federated database infrastructure to support the usability of cross-domain information systems. In: IEEE International Systems Conference (SysCon 2022), Montreal, Canada, 25–28 April 2022. IEEE (2022, accepted)
11. Melzer, S., Thiemann, S., Möller, R.: Modeling and simulating federated databases for early validation of federated searches using the broker-based SysML toolbox. In: IEEE International Systems Conference (SysCon 2021), Vancouver, BC, Canada, 15 April–15 May 2021, pp. 1–6. IEEE (2021). https://doi.org/10.1109/SysCon48628.2021.9447055
12. Parr, T.: The definitive ANTLR 4 reference. Pragmatic Bookshelf (2013)
13. Reisinger, F., Martens, L.: Database on demand - an online tool for the custom generation of FASTA-formatted sequence databases. PROTEOMICS **9**(18), 4421–4424 (2009). https://doi.org/10.1002/pmic.200900254, https://analyticalsciencejournals.onlinelibrary.wiley.com/doi/abs/10.1002/pmic.200900254
14. Vizcaíno, J.A., Reisinger, F., Côté, R., Martens, L.: PRIDE and "database on demand" as valuable tools for computational proteomics. In: Hamacher, M., Eisenacher, M., Stephan, C. (eds.) Data Mining in Proteomics. Methods in Molecular Biology (Methods and Protocols), vol. 696, pp. 93–105. Humana Press, Totowa (2011). https://doi.org/10.1007/978-1-60761-987-1_6
15. Wilden, E.: A critical edition and an annotated translation of the Akanāṉūṟu: Part 1, Kaliṟṟiyāṉainirai. Old commentary on Kaliṟṟiyāṉainirai KV - 90, word index of Akanāṉūṟu KV - 120. École Française d'Extrême-Orient (2018)

Handwriting Text Recognition

Best Practices for a Handwritten Text Recognition System

George Retsinas[1(✉)], Giorgos Sfikas[2,3,4], Basilis Gatos[2], and Christophoros Nikou[3]

[1] School of Electrical and Computer Engineering, National Technical University of Athens, Athens, Greece
gretsinas@central.ntua.gr

[2] Computational Intelligence Laboratory, Institute of Informatics and Telecommunications, National Center for Scientific Research "Demokritos", Athens, Greece
bgat@iit.demokritos.gr

[3] Department of Computer Science and Engineering, University of Ioannina, Ioannina, Greece
cnikou@cse.uoi.gr

[4] Department of Surveying and Geoinformatics Engineering, University of West Attica, Athens, Greece
gsfikas@uniwa.gr

Abstract. Handwritten text recognition has been developed rapidly in the recent years, following the rise of deep learning and its applications. Though deep learning methods provide notable boost in performance concerning text recognition, non-trivial deviation in performance can be detected even when small pre-processing or architectural/optimization elements are changed. This work follows a "best practice" rationale; highlight simple yet effective empirical practices that can further help training and provide well-performing handwritten text recognition systems. Specifically, we considered three basic aspects of a deep HTR system and we proposed simple yet effective solutions: 1) retain the aspect ratio of the images in the preprocessing step, 2) use max-pooling for converting the 3D feature map of CNN output into a sequence of features and 3) assist the training procedure via an additional CTC loss which acts as a shortcut on the max-pooled sequential features. Using these proposed simple modifications, one can attain close to state-of-the-art results, while considering a basic convolutional-recurrent (CNN+LSTM) architecture, for both IAM and RIMES datasets. Code is available at https://github.com/georgeretsi/HTR-best-practices/.

Keywords: Handwritten text recognition · Convolution - recurrent neural network · Best practices

1 Introduction

Handwritten Text Recognition (HTR) is an active area of research, combining ideas from both computer vision and natural language processing. Unlike recognition of machine-printed text, handwriting is related to a number of unique

© Springer Nature Switzerland AG 2022
S. Uchida et al. (Eds.): DAS 2022, LNCS 13237, pp. 247–259, 2022.
https://doi.org/10.1007/978-3-031-06555-2_17

characteristics that make the task much more challenging than traditional optical character recognition (OCR). The challenging nature of handwriting recognition stems mostly from the potentially high writing variability between individuals. To this end, along with visually decoding an image into sequence of characters, several HTR works adopt language models to reduce this innate ambiguity of handwritten characters, making use of contextual and semantic information.

In general, designing an effective and generalizable learning system is a ongoing challenge, with transferability between different learned writing styles more being not a given in most cases [22]. Neural Networks (NNs), among a variety of other learning systems, have been used for the recognition of handwriting from early on, with a span ranging between simpler subtasks such as single digit recognition [1] up to full, unconstrained offline HTR [7,21]. Following the rise of deep learning and its applications, recent developments in HTR are monopolized by Deep Neural Networks (DNNs). The seminal work of Graves et al. [9] played a pivotal role in the rise of deep learning for HTR applications by enabling the training of recurrent nets without assuming any prior character segmentation. A plethora of subsequent works for HTR relied on Graves et al. in order to train modern and notably effective DNNs [14,16,21,24].

This work focuses on finding best practices for building modern HTR systems. We explore a set of guidelines for training HTR DNNs, re-examining and extending ideas from several previous works of ours [23–25]. We start with a fairly common deep network architecture for HTR, consisting of a CNN backbone and a BiLSTM head, and we make simple yet effective architectural and training choices. These best practice suggestions can be categorized and summarized as follows:

1. **pre-processing:** retain aspect ratio of images and use batches of padded images in order to effectively use mini-batch Stochastic Gradient Descent (SGD)
2. **architectural:** replace the column-wise concatenation step between the CNN backbone and the recurrent head with a max-pooling step. Such a choice not only reduces the required parameters but has an intuitive motivation: we care only about the existence of a character and not its vertical position.
3. **training:** add an extra shortcut branch, consisting of a single 1D convolution layer, at the output of the CNN backbone. This branch results to an extra character sequence estimation, trained in parallel to the recurrent branch. Both branches use a CTC loss. The motivation behind such a choice comes from the increased difficulty of training recurrent layers. However, if such a straightforward shortcut exists, the output of the CNN backbone should converge to more discriminative features, ideal for fully harnessing the power of recurrent layers compared to an end-to-end training scheme.

The contribution of this paper is best highlighted through the experimental section, where we achieve state-of-the-art results with the aforementioned choices, despite the simplicity of the employed network. Furthermore, other state-of-the-art existing methods propose complex architectures and augmentation schemes which are orthogonal to our approach, highlighting the importance of the suggested best practices.

2 Related Work

As is the case with most, if not all, tasks in computer vision, modern HTR literature is dominated by neural network-based methods. Recurrent neural networks have become the baseline [15,21], as they naturally fit to the sequential nature of handwriting.

Recurrent-based approaches have thus practically overshadowed the previous state-of-the-art, which was based mostly on Hidden Markov Model (HMM)-based approaches. Since the introduction of the standard recurrent network paradigm [6,7], many key advances have emerged paving the way for very efficient HTR systems. A characteristic example is the integration of the Long Short-Term Memory models (LSTMs) into HTR systems [10], that effectively dealt with the vanishing gradient problem. More importantly, Graves et al. [9] introduced a very effective algorithm for training such HTR systems with sequence-based loss using dynamic programming. Specifically, this Connectionist Temporal Classification (CTC) method and corresponding output layer [8], a differentiable output layer that maps a sequential input into per-time unit softmax outputs, allows simultaneous sequence alignment and recognition with a suitable decoding scheme. Multi-dimensional recurrent networks have been considered for HTR [15], however there has been criticism that the extra computational overhead may not translate to an analogous increase in efficiency [21].

Even though in this work we will focus only on greedy decoding of a CTC-trained network, research on decoding schemes is also active [4], with the beam search algorithm being a popular approach, capable of exploiting an external lexicon as an implicit language model.

Sequence-to-Sequence approaches, involving translating an input sequence to an output sequence of a different length in general, became very popular when achieved state-of-the-art results in Natural Language Processing and gradually evolved to Transformer networks with attention mechanisms [29]. Such approaches were later adopted successfully by the HTR community [3,19,25,27,30].

Recent research directions include complex augmentation schemes [14,16,31], novel network architectures/modules (e.g. Seq2Seq/Transformers, Spatial Transformer Networks [5], deformable convolutions [24]) and multi-task losses with auxiliary training feeds (e.g. n-gram training [28]).

3 Proposed HTR System

In what follows, we will describe in detail the proposed HTR system with emphasis given on the suggested best practice modifications. The described system takes as input either a word or a line image and then returns the predicted sequence of characters based on an unconstrained greedy CTC decoding algorithm [8].

3.1 Preprocessing

The pre-processing steps, applied to every image, are:

1. All images are resized to a resolution of 128×1024 pixels for line images or 64×256 pixels. Initial images are padded (using the image median value, usually zero) in order to attain the aforementioned fixed size. If the initial image is greater than the predefined size, the image is rescaled. The padding option aims to preserve the existing aspect ratio of the text images.
2. During training, image augmentation is performed. A simple global affine augmentation is applied at every image, considering only rotation and skew of small magnitude in order to generate valid images. Additionally, gaussian noise is added to the images.
3. Each word/line transcription has spaces added before and after, i.e. "He rose from" is changed to "He rose from". This operation aims to assist the system to adapt to the marginal spaces that exist in the majority of the images during the training phase. For the testing phase, these additional spaces are discarded.

Augmentation operations are part of every modern deep learning system and can consistently provide increased performance, allowing better generalization [21]. The used augmentation scheme is very basic, trying to have minimal overhead from this step.

The addition of extra spaces in the transcription is not explicitly referred to recent existing works, but it is intuitive given the pad operation of step 1 which creates large empty margins. It has a minor yet positive impact to our system and thus is added as a step. Due to the reduced significance of this step in the overall performance, it is not explored in the experimental section.

On the other hand, we found the padding operation critical in many settings. A typical trade-off, met in many recent text recognition/spotting works, concerns the definition of the input size: using a predefined fixed size can assist the architectural design of CNNs and training time requirements, while retaining the initial image size by processing individually each image (e.g. [26]) may lead to better performance at the cost of discarding the mini-batch option.

Modern DNN training relies on creating batches of several images, since batch manipulation of images can notably affect the training time by fully utilizing the GPU resources. Thus image resizing is a widely-used first step for any vision problem when DNNs are involved. On the contrary, when using different sized images by processing each image individually and update the network's weights after a predefined number of images, as if a batch was processed, leads to an impractical time-consuming training procedure to otherwise lightweight DNNs, where the existing hardware is under-utilized.

In this work, contrary to the majority of existing approaches, we propose a simple, yet elegant, solution: we aim to retain the aspect ratio of the images and simultaneously organize them into batches. The images are transformed into the same, predefined, shape without resizing, if possible. Specifically, if the image size is smaller than the predefined size, we pad the image accordingly. The padding operation is performed equally at each direction, positioning the initial image at the center of the new one, with a fixed value, the median value of the initial image. If the image is larger than the predefined size, it is resized, affecting the aspect ratio. To assist the proposed approach, we can compute the average height and width over the whole set of the initial images and select an appropriate size in order to perform the aforementioned resize operation scarcely (only for very large words/sentences) and thus avoid deformations that are generated by frequently violating the aspect ratio.

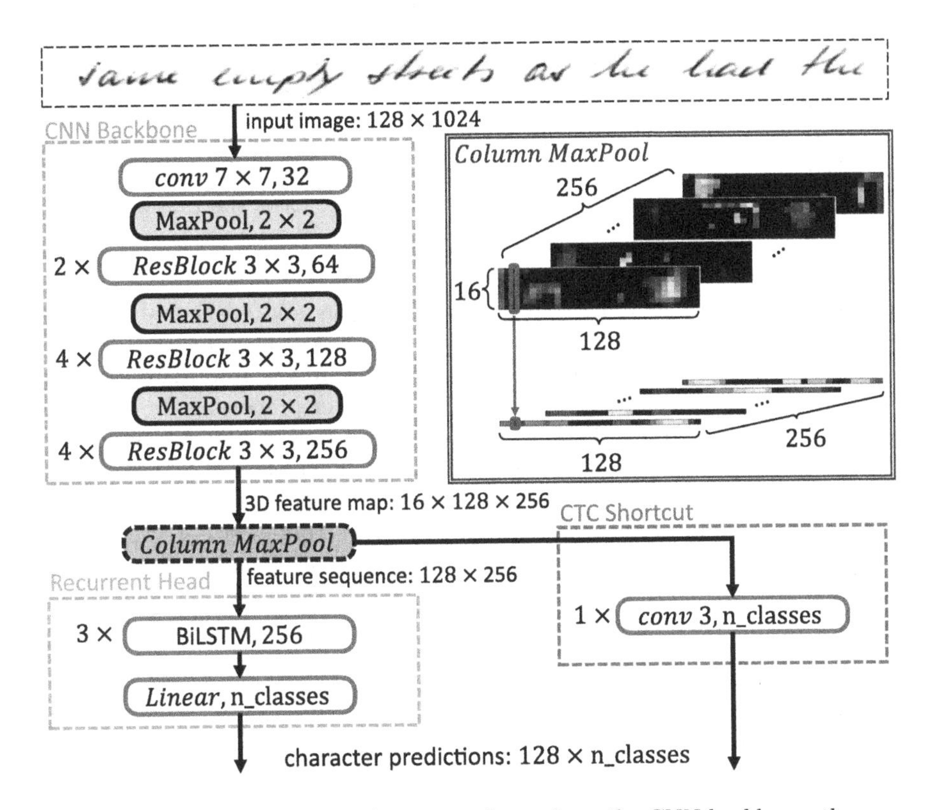

Fig. 1. Overview of the DNN architecture. Apart from the CNN backbone, the recurrent head, we also depict the auxiliary CTC shortcut branch which will be the core component of the proposed training modification. Furthermore, we visualize the column-wise max-pooling operation that is performed between the CNN backbone and the recurrent head.

3.2 Network Architecture

The model that we will use to test the proposed technique can be characterized as a convolutional-recurrent architecture (an architecture overview is depicted in Fig. 1). The convolutional-recurrent architecture can be broadly defined as a convolutional backbone being followed by a recurrent head, typically connected to a CTC loss. Convolutional-recurrent variants have given routinely very good results for HTR [5,21].

Convolutional Backbone: In our model, the convolutional backbone is made up of standard convolutional layers and ResNet blocks [12], interspersed with max-pooling and dropout layers. In particular, the first layer is a 7×7 convolution with 32 output channels, followed by cascades of 3×3 ResNet blocks [12]: a series of 2 ResNet blocks with 64 output channels, 4 ResNet blocks with 128 output channels and 4 ResNet blocks with 256 output channels. The standard convolution and the ResNet blocks are all followed by ReLU activations, Batch Normalization and dropout. Between cascades of blocks we downscale the produced feature map with 2×2 max-pooling operations of stride 2, as shown in Fig. 1. Overall, the convolutional backbone accepts a line image and outputs a tensor of size $h \times w \times d$ (e.g. assuming the line image case, the tensor is of size $16 \times 128 \times 256$).

Flattening Operation: The convolutional backbone output should be transformed into a sequence of features in order to processed by recurrent networks. Typical HTR approaches, assume a column-wise approach (towards the writing direction) to ideally simulate a character by character processing. In our work, the CNN output is flattened by a max-pooling operation in a column-wise manner. Flattening of the extracted feature maps by the widely-used concatenation operation would result into a sequence of length w with feature vectors of size hd, while max-pooling results to reduced feature vectors of size d. Apart from the apparent computational advantage, *column-wise max-pooling* achieves model translation invariance in the vertical direction. In fact, the reasoning behind max-pooling is that we care only about the existence of features related to a character and not their spatial position. This has been the major motivation for *column-wise max-pooling*, as successfully employed in our previous works [24,25,28].

Recurrent Head: The recurrent component consists of 3 stacked Bidirectional Long Short-Term Memory (BiLSTM) units of hidden size 256. These are followed by a linear projection layer, which converts the sequence to a size equal to the number of possible character tokens, $n_{classes}$ (including the blank character, required by CTC). The final output of the recurrent part can be translated into a sequence of probability distributions by applying a softmax operation. During evaluation, the aforementioned greedy decoding is performed by selecting the character with the highest probability at each step and then removing the blank characters from the resulting sequence [8].

3.3 Training Scheme

The training of the HTR system is performed via an Adam [13] optimizer using an initial learning rate of 0.001 which gradually decreases using a multistep scheduler. The overall training epochs are 240 and the scheduler decreases the learning rate by a factor of 0.1 at 120 and 180 epochs.

This optimizing scheme, with minor modifications, is commonly used for HTR systems. Nonetheless, we assume an end-to-end training approach where both the convolutional and the recurrent parts of the system are optimized through the final CTC loss. Even though this typical approach produces well-performing solutions, the LSTM head may encumbers the overall training procedure, since recurrent modules are known to exhibit convergence difficulties.

To circumvent this training hindrance, we introduce an auxiliary branch as shown in Fig. 1. We dub this extra module as a "CTC shortcut". In what follows, we describe this module and its functionality in detail.

CTC Shortcut: Architecture-wise, the CTC shortcut module consists only of a single 1D convolutional layer, with kernel size 3. Its output channels equal to the number of the possible character tokens ($n_{classes}$). Therefore, the 1D convolutional layer is responsible for straightforwardly encoding context-wise information and providing an alternative decoding path. Note that we strive for simplicity for this auxiliary component, since its aim is to assist the training of the main branch and thus a shallow convolutional part of only one layer is ideal for this task. We do not expect from the CTC branch to result to precise decodings.

The CTC shortcut is trained along with the main architecture using a multi-task loss by adding the corresponding CTC losses of the two branches with the appropriate weights. Specifically, if f_{cnn} represents the convolutional backbone, f_{rec} represents the recurrent part and $f_{shortcut}$ represents the proposed shortcut branch, while I is an input image and s its corresponding transcription, the multi-task loss is written as:

$$L_{CTC}(f_{rec}(f_{cnn}(I)); s) + 0.1\, L_{CTC}(f_{shortcut}(f_{cnn}(I)); s) \tag{1}$$

Since CTC shortcut acts only as an auxiliary training path, it is weighted by 0.1 to reduce its relative contribution to the overall loss.

The motivation behind this extra branch is rather simple: overall convergence is assisted by quickly generating discriminative features at the top of the CNN backbone through the straightforward 1D convolutional path, simplifying the training task for the recurrent part. Due to its training-oriented assisting nature, CTC shortcut is used only during training and omitted during evaluation. Therefore, this proposed shortcut does not introduce any overhead during inference.

4 Experimental Evaluation

Evaluation of the proposed system is performed on two widely used datasets, IAM [18] and Rimes [11]. The ablation study, considering different settings of the proposed methodology, is performed on the challenging IAM dataset, consisting of handwritten text from 657 different writers and partitioned into writer-independent train/validation/test sets (we use the same set partition as in [21]). All experiments follow the same setting: line-level or word-level recognition using a lexicon-free unconstrained greedy CTC decoding scheme. Character Error Rate (CER) and Word Error Rate (WER) metrics are reported in all cases (lower values are better).

4.1 Ablation Study

First, we explore the impact of the proposed modifications over both the validation and the test set of IAM dataset. Moreover, both line-level recognition (Table 1) and word-level recognition (Table 2) are considered. Specifically, we investigate the difference in performance when we: 1) use *resized* or *padded* (retain aspect-ratio case) input images, 2) use *concatenation* of *max-pooling* flattening operation between the convolutional backbone and the recurrent head and 3) use or not the *CTC shortcut during* the training process.

Table 1. Line-level recognition results for IAM dataset: exploring the impact of the proposed modifications.

Preprocessing	Flattening	CTC shortcut	Validation		Test	
			CER (%)	WER (%)	CER (%)	WER (%)
Resized	Concatenation	No	4.28	15.29	5.93	19.57
		Yes	3.72	13.18	5.11	16.96
Resized	Max-pooling	No	3.73	13.54	5.28	17.77
		Yes	3.47	12.77	4.85	16.19
Padded	Concatenation	No	4.06	14.40	5.54	18.60
		Yes	3.37	12.22	4.71	15.94
Padded	Max-pooling	No	3.46	12.55	4.93	16.81
		Yes	**3.21**	**11.89**	**4.62**	**15.89**

The following observations can be made:

- Retaining the aspect-ratio of the images (padded option) achieves improved results for the majority of cases.
- Performing the flattening operation via max-pooling not only is more cost-effective, but it has a positive impact on performance. This is more evident in line-level recognition setting.
- Training with a CTC shortcut module provides notable boost over all cases. For example, in line-level recognition the significant difference in performance when considering different flattening operations is considerably decreased when the CTC shortcut approach is adopted (e.g. for padded line-level recognition the WER performance difference drops from 1.79% to only 0.05%). This hints that the initial difference in performance is mainly attributed to difficulties in training (concatenated version has a much larger feature vector to manage). Note that evaluating the CTC shortcut branch yields poor

Table 2. Word-level recognition results for IAM dataset: exploring the impact of the proposed modifications.

Preprocessing	Flattening	CTC shortcut	Validation		Test	
			CER (%)	WER (%)	CER (%)	WER (%)
Resized	Concatenation	No	4.35	12.55	5.58	15.46
		Yes	4.27	12.02	5.46	15.13
Resized	Max-pooling	No	4.25	12.17	5.69	15.87
		Yes	4.09	11.65	5.23	14.40
Padded	Concatenation	No	4.17	11.99	5.66	15.66
		Yes	3.98	11.50	5.37	14.98
Padded	Max-pooling	No	4.00	11.25	5.43	15.06
		Yes	**3.76**	**10.76**	**5.14**	**14.33**

decodings, despite the notable performance increase of the main network. For example, assuming line-level recognition and the padded/max-pooling setting, we report 5.26% CER/19.76% WER for the validation set and 7.36% CER/25.66% WER for the test set.
- Applying all three modifications together achieves the best results across all setting and metrics.
- Word recognition reports improved results compared to line-level recognition with respect to the WER metric. This was expected, since word-level setting assumes perfect word segmentation. Interestingly enough, this is not the case for the CER metric. This can be explained by the lack of sufficient context (i.e. find a capital letter or a punctuation from the whole line information).

We also explore in more depth the CTC shortcut option, which seems to provide the best boost in performance. Specifically, we report the progress of both the loss and the CER/WER metrics (over the validation set) during the training procedure for the line-level recognition setting. The loss curves are depicted in Fig. 2, while the validation set evaluation metrics are reported in Fig. 3. As we can see, loss curves are similar, but the case of CTC shortcut consistently has slightly better behavior. The impact of the CTC shortcut is more clearly shown in CER/WER curves and thus solutions with greater generalization properties are expected when a model is trained along with the CTC shortcut.

4.2 Comparison to State-of-the-Art Systems

Finally, we compare our method to several existing state-of-the-art methods, as shown in Table 3. The reported methods follow the same setting: line-level lexicon-free recognition. The proposed HTR system along with the suggested modifications achieves results comparable to the best performing methods. Notably, it outperforms the majority of existing works for both datasets and metrics despite many of the reported methods propose novel elements to further increase performance that are in general orthogonal to our approach. For example, the work of Chowdhury et al. [3] presents better WER for the RIMES dataset while using a sequence-to-sequence approach (such models can produce

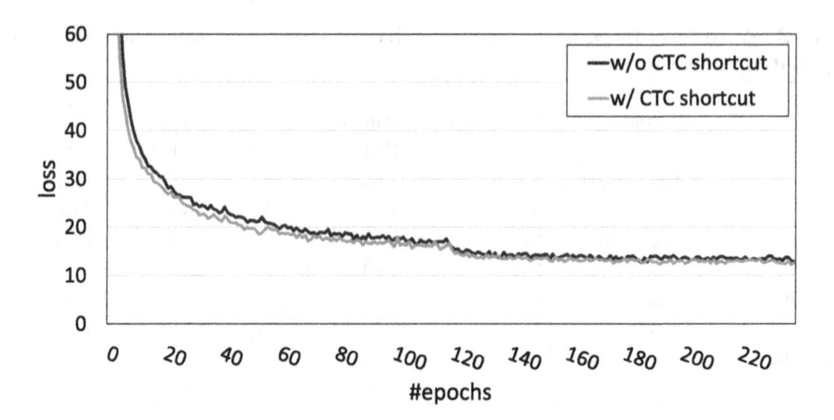

Fig. 2. Behavior of HTR performance in terms of loss value with and without the extra CTC shortcut branch during the training phase. Reported curves correspond to the proposed line-level HTR system trained on the IAM dataset.

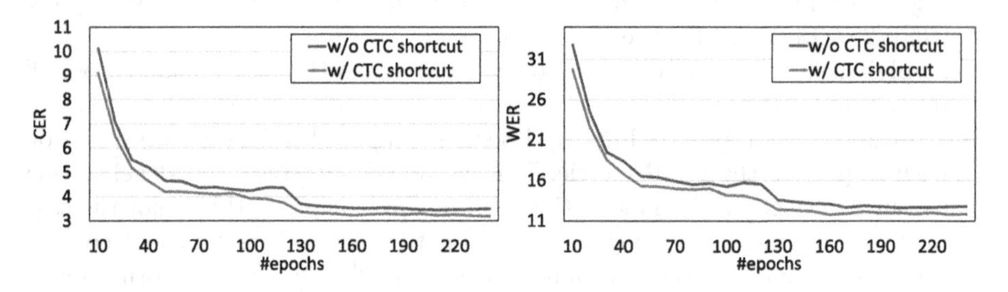

Fig. 3. Behavior of HTR performance in terms of CER (left) and WER (right) metrics with and without the extra CTC shortcut branch during the training phase. Reported curves correspond to the proposed line-level HTR system, trained on IAM dataset and evaluated on the validation set.

increased WER as implicit language models can be learnt [25]), while our previous work [24] achieves better CER for the IAM dataset while using similar network (max-pooling flattening and padded input images) along with deformable convolutions and a post processing uncertainty reduction algorithm.

Moreover, the very recent work of Luo et al. [16] manages to outperform our method for the word-level recognition setting on the IAM dataset by using a STN component and a complex augmentation method, where "optimal" augmentations are learnt. Specifically, our method achieves 5.14% CER/14.33%, while Luo et al. achieve 5.13% CER/13.35% WER for the exact same setting. Nonetheless, their initial baseline network, stripped of all the extra modules (which could be added to the proposed architecture without any problem), performs poorly: 7.39% CER and 19.12% WER.

Overall, we achieve very competitive results (outperforming other existing lexicon-free methods for line-level recognition on IAM) by only using a typical convolutional-recurrent architecture along with a set of simple, yet intuitive and effective modifications, forming an effective set of best practice suggestions which can be applied to the majority of HTR systems.

Table 3. Performance comparison for IAM/RIMES datasets (line-level recognition)

Method	IAM		RIMES	
	CER (%)	WER (%)	CER (%)	WER (%)
Chen et al. [2]	11.15	34.55	8.29	30.5
Pham et al. [20]	10.8	35.1	6.8	28.5
Khrishnan et al. [14]	9.78	32.89	–	–
Chowdhury et al. [3]	8.10	16.70	3.59	9.60
Puigcerver [21]	6.2	20.2	2.60	10.7
Khrishnan et al. [14]	9.78	32.89	–	–
Markou et al. [17]	6.14	20.04	3.34	11.23
Dutta et al. [5]	5.8	17.8	5.07	14.7
Wick et al. [30]	5.67	–	–	–
Michael et al. [19]	5.24	–	–	–
Tassopoulou et al. [28]	5.18	17.68	–	–
Yousef et al. [32]	4.9	–	–	–
Retsinas et al. [24]	4.55	16.08	3.04	10.56
Proposed	4.62	15.89	2.75	9.93

5 Conclusions

In this paper, we proposed a series of best practice modifications over typical convolutional-recurrent networks trained with CTC loss. Apart from presenting a fairly compact architecture based on residual blocks, we present three impactful modifications: 1) retain aspect-ratio of input images gathered in batches through a padding operation, 2) apply a column-wise max-pooling operation between the convolutional backbone and the recurrent head of a typical HTR architecture for reduced computational effort and increased performance and 3) enhance performance through a CTC shortcut during training in order to circumvent an end-to-end training over recurrent networks, which have been proven "difficult" to train in various settings. All proposed modifications have proven to be very helpful, considerably increasing the performance of the vanilla network. Overall, the proposed system achieves results in the ballpark of state-of-the-art, while being orthogonal to the majority of modern deep learning modules and approaches.

Acknowledgement. This research has been partially co - financed by the EU and Greek national funds through the Operational Program Competitiveness, Entrepreneurship and Innovation, under the calls: "RESEARCH - CREATE - INNOVATE", project *Culdile* (code T1EΔK - 03785) and "OPEN INNOVATION IN CULTURE", project *Bessarion* (T6YBΠ - 00214).

References

1. Bishop, C.M.: Pattern Recognition and Machine Learning. Springer, Heidelberg (2006)
2. Chen, Z., Wu, Y., Yin, F., Liu, C.L.: Simultaneous script identification and handwriting recognition via multi-task learning of recurrent neural networks. In: 14th IAPR International Conference on Document Analysis and Recognition (ICDAR), vol. 1, pp. 525–530. IEEE (2017)
3. Chowdhury, A., Vig, L.: An efficient end-to-end neural model for handwritten text recognition (2018)
4. Collobert, R., Hannun, A., Synnaeve, G.: A fully differentiable beam search decoder. In: International Conference on Machine Learning, pp. 1341–1350. PMLR (2019)
5. Dutta, K., Krishnan, P., Mathew, M., Jawahar, C.: Improving CNN-RNN hybrid networks for handwriting recognition. In: 2018 16th International Conference on Frontiers in Handwriting Recognition (ICFHR), pp. 80–85. IEEE (2018)
6. Fischer, A., Keller, A., Frinken, V., Bunke, H.: Lexicon-free handwritten word spotting using character HMMs. Pattern Recogn. Lett. **33**(7), 934–942 (2012)
7. Fischer, A.: Handwriting recognition in historical documents. Ph.D. thesis, Verlag nicht ermittelbar (2012)
8. Graves, A.: Connectionist temporal classification. In: Graves, A. (ed.) Supervised Sequence Labelling with Recurrent Neural Networks. Studies in Computational Intelligence, vol. 385, pp. 61–93. Springer, Heidelberg (2012). https://doi.org/10.1007/978-3-642-24797-2_7
9. Graves, A., Fernández, S., Gomez, F., Schmidhuber, J.: Connectionist temporal classification: labelling unsegmented sequence data with recurrent neural networks. In: Proceedings of the 23rd International Conference on Machine Learning, pp. 369–376 (2006)
10. Greff, K., Srivastava, R.K., Koutník, J., Steunebrink, B.R., Schmidhuber, J.: LSTM: a search space odyssey. IEEE Trans. Neural Netw. Learn. Syst. **28**(10), 2222–2232 (2016)
11. Grosicki, E., Carre, M., Brodin, J.M., Geoffrois, E.: Rimes evaluation campaign for handwritten mail processing (2008)
12. He, K., Zhang, X., Ren, S., Sun, J.: Deep residual learning for image recognition. In: Proceedings of the IEEE Conference on Computer Vision and Pattern Recognition, pp. 770–778 (2016)
13. Kingma, D.P., Ba, J.: Adam: a method for stochastic optimization. In: Proceedings of the International Conference on Learning Representations (ICLR) (2015)
14. Krishnan, P., Dutta, K., Jawahar, C.: Word spotting and recognition using deep embedding. In: 2018 13th IAPR International Workshop on Document Analysis Systems (DAS), pp. 1–6. IEEE (2018)
15. Leifert, G., Strau, T., Gr, T., Wustlich, W., Labahn, R., et al.: Cells in multidimensional recurrent neural networks. J. Mach. Learn. Res. **17**(97), 1–37 (2016)
16. Luo, C., Zhu, Y., Jin, L., Wang, Y.: Learn to augment: joint data augmentation and network optimization for text recognition. In: Proceedings of the IEEE/CVF Conference on Computer Vision and Pattern Recognition, pp. 13746–13755 (2020)
17. Markou, K., et al.: A convolutional recurrent neural network for the handwritten text recognition of historical Greek manuscripts. In: Del Bimbo, A., et al. (eds.) ICPR 2021. LNCS, vol. 12667, pp. 249–262. Springer, Cham (2021). https://doi.org/10.1007/978-3-030-68787-8_18

18. Marti, U.V., Bunke, H.: The IAM-database: an English sentence database for offline handwriting recognition. Int. J. Doc. Anal. Recogn. **5**(1), 39–46 (2002)
19. Michael, J., Labahn, R., Grüning, T., Zöllner, J.: Evaluating sequence-to-sequence models for handwritten text recognition. In: 2019 International Conference on Document Analysis and Recognition (ICDAR), pp. 1286–1293. IEEE (2019)
20. Pham, V., Bluche, T., Kermorvant, C., Louradour, J.: Dropout improves recurrent neural networks for handwriting recognition. In: 2014 14th International Conference on Frontiers in Handwriting Recognition, pp. 285–290. IEEE (2014)
21. Puigcerver, J.: Are multidimensional recurrent layers really necessary for handwritten text recognition? In: 2017 14th IAPR International Conference on Document Analysis and Recognition (ICDAR), vol. 1, pp. 67–72. IEEE (2017)
22. Retsinas, G., Sfikas, G., Gatos, B.: Transferable deep features for keyword spotting. In: Multidisciplinary Digital Publishing Institute Proceedings, vol. 2, p. 89 (2018)
23. Retsinas, G., Sfikas, G., Nikou, C.: Iterative weighted transductive learning for handwriting recognition. In: Lladós, J., Lopresti, D., Uchida, S. (eds.) ICDAR 2021. LNCS, vol. 12824, pp. 587–601. Springer, Cham (2021). https://doi.org/10.1007/978-3-030-86337-1_39
24. Retsinas, G., Sfikas, G., Nikou, C., Maragos, P.: Deformation-invariant networks for handwritten text recognition. In: 2021 IEEE International Conference on Image Processing (ICIP), pp. 949–953. IEEE (2021)
25. Retsinas, G., Sfikas, G., Nikou, C., Maragos, P.: From Seq2Seq recognition to handwritten word embeddings. In: Proceedings of the British Machine Vision Conference (BMVC) (2021)
26. Sudholt, S., Fink, G.A.: PHOCNet: a deep convolutional neural network for word spotting in handwritten documents. In: Proceedings of the 15th International Conference on Frontiers in Handwriting Recognition (ICFHR), pp. 277–282 (2016)
27. Sueiras, J., Ruiz, V., Sanchez, A., Velez, J.F.: Offline continuous handwriting recognition using sequence to sequence neural networks. Neurocomputing **289**, 119–128 (2018)
28. Tassopoulou, V., Retsinas, G., Maragos, P.: Enhancing handwritten text recognition with n-gram sequence decomposition and multitask learning. In: 2020 25th International Conference on Pattern Recognition (ICPR), pp. 10555–10560. IEEE (2021)
29. Vaswani, A., et al.: Attention is all you need. In: Advances in Neural Information Processing Systems, pp. 5998–6008 (2017)
30. Wick, C., Zöllner, J., Grüning, T.: Transformer for handwritten text recognition using bidirectional post-decoding. In: Lladós, J., Lopresti, D., Uchida, S. (eds.) ICDAR 2021. LNCS, vol. 12823, pp. 112–126. Springer, Cham (2021). https://doi.org/10.1007/978-3-030-86334-0_8
31. Wigington, C., Stewart, S., Davis, B., Barrett, B., Price, B., Cohen, S.: Data augmentation for recognition of handwritten words and lines using a CNN-LSTM network. In: 2017 14th IAPR International Conference on Document Analysis and Recognition (ICDAR), vol. 1, pp. 639–645. IEEE (2017)
32. Yousef, M., Hussain, K.F., Mohammed, U.S.: Accurate, data-efficient, unconstrained text recognition with convolutional neural networks. Pattern Recogn. **108**, 107482 (2020)

Rescoring Sequence-to-Sequence Models for Text Line Recognition with CTC-Prefixes

Christoph Wick[1], Jochen Zöllner[1,2](✉), and Tobias Grüning[1]

[1] Planet AI GmbH, Warnowufer 60, 18057 Rostock, Germany
{christoph.wick,tobias.gruening}@planet-ai.de
[2] Computational Intelligence Technology Lab, Department of Mathematics, University of Rostock, 18051 Rostock, Germany
jochen.zoellner@uni-rostock.de

Abstract. In contrast to Connectionist Temporal Classification (CTC) approaches, Sequence-To-Sequence (S2S) models for Handwritten Text Recognition (HTR) suffer from errors such as skipped or repeated words which often occur at the end of a sequence. In this paper, to combine the best of both approaches, we propose to use the CTC-Prefix-Score during S2S decoding. Hereby, during beam search, paths that are invalid according to the CTC confidence matrix are penalised. Our network architecture is composed of a Convolutional Neural Network (CNN) as visual backbone, bidirectional Long-Short-Term-Memory-Cells (LSTMs) as encoder, and a decoder which is a Transformer with inserted mutual attention layers. The CTC confidences are computed on the encoder while the Transformer is only used for character-wise S2S decoding. We evaluate this setup on three HTR data sets: IAM, Rimes, and StAZH. On IAM, we achieve a competitive Character Error Rate (CER) of 2.95% when pretraining our model on synthetic data and including a character-based language model for contemporary English. Compared to other state-of-the-art approaches, our model requires about 10–20 times less parameters. Access our shared implementations via this link to GitHub.

Keywords: Text Line Recognition · Handwritten text recognition · Document analysis · Sequence-To-Sequence · CTC

1 Introduction

Optical Character Recognition (OCR), the transcription of digital images into machine-actionable text, is still a challenging problem even though there were great advancements in recent years. A typical OCR pipeline consists of two steps: text detection and text recognition. Currently, most text recognition systems act on lines of text, and thus convert a sequence of arbitrary length (here the width of the text line image) into another sequence of different (typically shorter) length. There are two established approaches to tackle this problem: S2S [11] and CTC [4].

© Springer Nature Switzerland AG 2022
S. Uchida et al. (Eds.): DAS 2022, LNCS 13237, pp. 260–274, 2022.
https://doi.org/10.1007/978-3-031-06555-2_18

S2S approaches are a very generic approach as they do not constrain the lengths of the input or output sequence, nor their ordering[1]. An S2S system comprises two modules: the *encoder* encodes the line images into features. The *decoder*, starting from a Start-of-Sequence (⟨sos⟩)-token, decodes the encoded line character-(or token)-wise until an End-of-Sequence (⟨eos⟩)-token is emitted.

Since written visual and decoded characters are ordered in the task of Text Line Recognition (TLR), the CTC algorithm can be applied. Here, analogous to S2S, the line image is first encoded, then CTC decodes the complete sequence in one step: at each position of the encoded feature sequence, CTC predicts either the desired character or a special *blank* character, meaning there is no output. This results in a so-called *confidence matrix* that stores a probability distribution for each character plus the blank for each position in the line. Best path decoding determines the most likely characters, then repeated character predictions are fused and blanks are erased to obtain the final transcription.

S2S has some advantages which are eminent in better accuracies compared to CTC (see, e.g., [10] or [7]), but there are also fundamental problems when it comes to TLR (see, e.g., [14]): a crucial problem is that S2S approaches tend to predict ⟨eos⟩-tokens prematurely or delayed which is why parts of the original sequence are cropped or repeated several times. Similarly, repetition or skipping of characters, digits, or whole words can also occur within the sequence. In TLR, this can result in severe errors if, for example, repeating digits in a large number are "skipped". The reason for this behaviour is that the mutual-attention modules are not always certain about the snippet that shall be decoded next. There are attempts to tackle this problem by modifying the attention. A comparison of six different attention versions, e.g., monotonic (force left to right transcription) or penalised (penalising features that were already attended to), was performed by [10] for HTR. In contrast, CTC approaches do not suffer from any of these problems by design. Hence, combining the best of both worlds is reasonable but non-trivial since CTC and S2S have very different concepts for decoding.

Watanabe et al. [12] proposed an algorithm to use CTC scores during decoding with an S2S approach. Their network architecture is constructed similar to S2S approaches using attention, however, a second CTC-based decoder is added to the shared encoder. Upon decoding, the CTC-Prefix-Score is computed to weigh the next character probabilities of the decoder. Since [12] only applied their approach to speech recognition, in this paper, we examine the applicability on TLR which, to the best of our knowledge, has not been performed, yet. Thereto, we setup a traditional CNN/LSTM encoder, and a Transformer-based decoder. Optionally, we add a character-based Language Model (LM).

We evaluate our models on two contemporary and one historic handwritten datasets each in a different language: IAM [8] for English, Rimes [1] for French, and StAZH (not published, yet) for historic Swiss-German.

To summarise, in this paper, we contribute the following:

- We propose a hybrid CTC/Transformer decoder architecture for TLR using both the CTC confidence matrix and S2S based on [12].

[1] Arbitrary ordering is required, e.g., for translation tasks.

- Additionally, we apply pretraining on synthetic data and a LM to achieve state-of-the-art results on IAM.
- We share our Tensorflow-based implementation[2] of the network architectures, the CTC-Prefix-Scorer, and our beam search.

The remainder of this paper is structured as follows: first, we discuss related work in the area of TLR and HTR. Next, we introduce our methodology and describe the computation of CTC-Prefix-Scores for joint CTC/Transformer-decoding. Afterwards, we present the three datasets and evaluate our methods. We conclude with a discussion of our results and future work.

2 Related Work

In recent years, TLR and HTR have been studied quite excessively. A thorough literature review of handwritten OCR is provided by Memon et al. [9].

Bluche et al. [2] proposed gated convolutional layers in a CNN/LSTM/CTC-based approach achieving a CER of 3.2% on IAM when using a LM with a limited vocabulary size of 50K words. In [10], Michael et al. compare different attention mechanisms in the decoder for S2S-based HTR. Their best model on IAM yielded a CER of 4.87% without the usage of external data or a LM.

Yousef et al. [16] applied Fully Convolutional Networks (FCNs) without any recurrent connections trained with the CTC loss function. On IAM, they reached a challenging CER of 4.9% without the use of additional data or a LM.

Transformers for TLR were first introduced by Kang et al. [5]. They proposed a CNN/Transformer encoder and a Transformer decoder yielding a CER of 4.67% on IAM when pretraining their model on synthetic data, setting a new best value for open vocabulary HTR. This approach was extended by Wick et al. [14] who proposed a bidirectional decoding scheme: one Transformer reads the line forwards, another one backwards. The final results is obtained by voting.

Diaz et al. [3] compared different network architectures and decoding schemes for TLR. They varied Recurrent Neural Networks (RNNs), Gated Recurrent Convolution Layers, and self-attention layers for the encoder, and CTC or Transformer-based S2S for the decoder. Choosing a LM and a pretrained model on an internal real-world dataset, they reached a new state-of-the-art CER of 2.75% using CTC and self-attention on the IAM downstream task.

Li et al. [7] proposed a pure S2S-Transformer-based approach, i.e., they replaced every CNN from the network architecture. Their large model with 558 million parameters achieved a CER of 2.89% on IAM if pretrained on several millions of synthetic lines. A separate LM was not applied since the decoder was already capable to learn an intrinsic LM.

[2] https://github.com/Planet-AI-GmbH/tfaip-hybrid-ctc-s2s.

Table 1. Overview of the three used datasets showing their language, alphabet size $|A|$, and the number of training, validation, and test lines.

Dataset	Language	$\|A\|$	# Train	# Val	# Test
IAM	English (en)	79	6,161	966	2,915
StAZH	Swiss-German (de-ch)	109	12,628	1,624	1,650
Rimes	French (fr)	100	10,171	1,162	778

3 Data

To evaluate our proposed methods, we use three different datasets for HTR: IAM [8], StAZH, and Rimes [1]. Table 1 summarises the language, alphabet size $|A|$, and number of lines for training, validation, and testing.

For the popular IAM-dataset we use Aachen's partition[3]. StAZH is an internal dataset of the European Union's Horizon 2020 READ project and thus not openly available, yet. The documents contain resolutions and enactments of the cabinet as well as the parliament of the canton of Zurich from 1803 to 1882. We included it for additional results on historical handwritings.

To train our English LM, we collected over 16 million text lines from news, webpages, and wikipedia. For the French LM, we used a wikipedia dump with about 30 million text lines. To train a Swiss-German LM only about 1 million text lines of contemporary Swiss-German language were collected.

4 Methods

In this section we present our proposed approach for line-based HTR. First, we describe preprocessing steps and our network architecture which comprises the CTC- and Transformer-based branches for decoding. Afterwards, we present the training of the model and how inference including the CTC-Prefix-Score and an optional LM is performed. Finally, we describe the training of the character-based LM and our line synthesiser for artificial lines used for pretraining the HTR models.

4.1 Preprocessing

To preprocess a line, we apply contrast normalisation without binarisation, normalisation of slant and height, then scale the line images to a fixed height of 64 pixels while maintaining their aspect ratio. To artificially increase the amount of training data, we augment the preprocessed images by applying minor disturbances to the statistics relevant for the normalisation algorithms. Furthermore, we combine dilation, erosion, and grid-like distortions [15] to simulate naturally occurring variations in handwritten text line images. These methods are applied to the preprocessed images randomly each with an independent probability of 50%.

[3] https://github.com/jpuigcerver/Laia/tree/master/egs/iam.

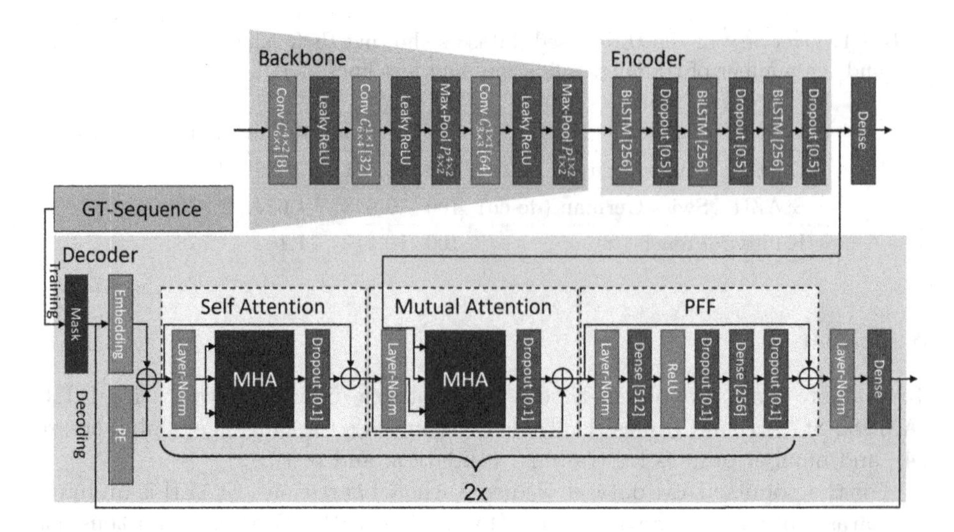

Fig. 1. Our proposed network architecture. The visual backbone is a CNN with a subsampling factor of 8, the encoder comprises three LSTM-layers, and the decoder is a Transformer build up by stacked self-attention, mutual-attention layers, and absolute positional encoding (PE).

4.2 Network-Architecture

Similar to current approaches, we propose a network architecture that is comprised of three basic components (see Fig. 1): a visual backbone that extracts high-level features with a limited receptive field, an encoder that can learn context, and a decoder that transcribes the features into characters. The backbone comprises three convolutional layers and two max-pooling layers which subsample the (rescaled) original line image with a dimension of $64 \times W \times 1$ to a feature map of $4 \times \frac{W}{8} \times 64$. The final output is height-concatenated to a dimension of $\frac{W}{8} \times 256$. The encoder consists of three stacked bidirectional LSTM-layers, each with 256 hidden nodes and a dropout rate of 0.5.

The Transformer decoder first applies a trainable embedding layer on the history of characters, and then adds sinusoidal absolute positional encoding to enable the attention modules to learn the order of characters. Note, that no additional positional encoding is applied to the outputs of the encoder since it is learned autonomously. Afterwards, to learn long range dependencies between the decoder features, a self-attention module, and to couple the textual with the visual features, a mutual-attention module are applied, each with 8 heads and 512 nodes. Last, a Pointwise Feed-Forward (PFF) layer with an output dimension of 256 is applied. Two blocks each consisting of one self-attention, one mutual-attention, and one PFF are stacked to form the final output. Ultimately, a dense layer maps the features to the alphabet size $|A| + 1$ (including the $\langle$eos$\rangle$-token).

Each Multi-Head-Attention (MHA) requires three inputs, a query Q, key K, and value V, whereby the output Y is computed as

$$Y_i = \text{Softmax}\left(\frac{Q'_i \cdot K'}{\sqrt{f}}\right) \cdot V' ,$$

where f is the number of features, i is the i-th entry of each matrix, and Q', K', V' are obtained by three independent linear transformations using the matrices W_Q, W_K, W_V. Several so-called attention heads using different transformations are computed in parallel and concatenated afterwards. In a self-attention module, all inputs are identical: $Q \equiv K \equiv V$.

The backbone and encoder form a sub-network that can be applied for HTR autonomously if decoded with CTC after appending an additional dense layer to map the encoder outputs to $|A| + 1$ (including the *blank* token). These outputs, the CTC confidences, are later used during training and decoding (see next Sections).

4.3 Training

The total loss L_{tot} for training comprises two components: the CTC-loss L_{CTC} computed on the CTC confidences, and the character-wise Cross-Entropy-loss L_{CE} applied to the decoder outputs. Both losses are weighted by the factor $\lambda_{CTC} \in [0, 1]$ which denotes the influence of L_{CTC}:

$$L_{\text{tot}} = \lambda_{CTC} \cdot L_{\text{CTC}} + (1 - \lambda_{CTC}) \cdot L_{\text{CE}}$$

Throughout this paper, we set $\lambda_{CTC} = 0.3$ which is a compromise of having a greater influence of the decoder but also a non-negligible encoder contribution.

Teacher forcing is applied to speed-up training by presenting the Ground Truth (GT)-sequence, prefixed with a $\langle\text{sos}\rangle$ token, to the decoder during training (see Fig. 1). The characters in the "future" are masked so that the network has no access for predicting the "next" character. Note that we do not apply label smoothing.

Our setup computes the exponential mean average with a decay of 0.99 of all weights during training to obtain the final model weights for evaluation. We use an ADAM-optimiser with global gradient clipping of 5.0. The learning rate is increased linearly from 0.0 to a maximum of 0.001 after five epochs and then reduced by a factor $\sqrt{5/i_{\text{epoch}}}$, where i_{epoch} is the current epoch. Training is finished if the model saturates on the validation dataset. The best model on the validation dataset is stored as final model.

4.4 Inference

During inference (see Fig. 2), the decoder transcribes the input character by character starting with a $\langle\text{sos}\rangle$ token using beam search with a beam width of n_{beams}. The confidences for the next character are a combination of three parts that each contributes costs $C \in \mathbb{R}^{|A|+1}$ (alphabet size plus $\langle\text{eos}\rangle$), the negative logarithms of the confidences:

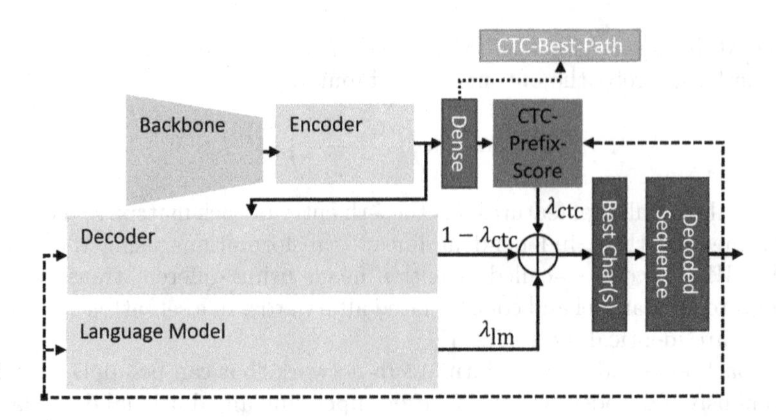

Fig. 2. The decoding scheme using beam search. The character costs of the CTC-Prefix-Score, the S2S decoder, and LM are weighted ($\lambda_{\mathrm{ctc}}, \lambda_{\mathrm{lm}}$) and summed up to obtain the next best characters for decoding. The history of the decoded sequence (dashed arrows) is required in all three modules to predict the next character. For comparison, best path decoding can be applied directly on the CTC confidences.

- The **decoder** directly outputs confidences for the next character C_{CE}.
- Based on the CTC confidences which are computed by the additional dense layer after the **encoder** (orange), the CTC-Prefix-Score C_{CTC} of the next sequence is computed (see next Section).
- A LM which computes confidences for the next character C_{LM} solely based on the previous sequence (see Sect. 4.5).

The total costs C_{tot} are computed as

$$C_{\mathrm{tot}} = \lambda_{CTC} \cdot C_{\mathrm{CTC}} + (1 - \lambda_{CTC}) \cdot C_{\mathrm{CE}} + \lambda_{LM} \cdot C_{\mathrm{LM}}$$

where $\lambda_{\mathrm{LM}} \in [0, 1]$ is the weight of the LM.

CTC Prefix Score. The CTC-Prefix-Score decoding of Watanabe et al. [12] is an addition to the S2S decoding scheme by penalising paths in the beam search that contradict the CTC confidences. Here, we provide only a coarse description of their algorithm and refer to the original paper for the algorithms and the mathematical details. For a deeper description of the CTC algorithm we refer to the original paper of Graves et al. [4].

The forward variables of CTC allow to compute the probability of a sequence of characters by summing up the confidences of all possible paths that result in this sequence. The CTC-Prefix-Score is the total score of all paths that start with a given but end with an arbitrary sequence. Watanabe et al. propose to add this score to rescale each path during beam search decoding of S2S. If a path emits an ⟨eos⟩-token the full CTC-score is used for rescaling. This overall setup prevents paths that prematurely end or skip characters since those receive a bad CTC-Prefix-Score. Unfortunately, the computation is slow since for each beam all possible path endings must be summed up for each decoding step.

Beam Search. The applied beam search tracks the best n_{beams} character sequences that have not yet finished, and all (or only the best) completed sequences, i.e., those that predicted an $\langle\text{eos}\rangle$-token. By default, for each unfinished sequence, C_{tot} and its parts are computed. Hereby, as already stated, the computation of the CTC-Prefix-Score is costly if evaluated for every possible next character. To speed up the computation, we determine so-called "pre-beams" (see, e.g., the usage in [6]) which splits the computation of C_{tot} in

$$C_{\text{pre}} = (1 - \lambda_{CTC}) \cdot C_{\text{CE}} + \lambda_{LM} \cdot C_{\text{LM}}$$

and

$$C_{\text{tot}} = \lambda_{CTC} \cdot C_{\text{CTC}} + C_{\text{pre}} \, .$$

After C_{pre} is computed, the search space is narrowed down by reducing the possible beams to a maximum of $1.5 \cdot n_{\text{beams}}$. Only the CTC-Prefix-Scores for the remaining characters $C_{\text{CE}} \in \mathbb{R}^{1.5 \cdot n_{\text{beams}}+1}$ are evaluated. Then the actual n_{beams} best beams are selected. This simplification (if $1.5 \cdot n_{\text{beams}} < |A|$) assumes that the best characters of the decoder are similar to those of the encoder and will lead to only minor differences in the result. To the best of our knowledge, there is no open-source Tensorflow implementation for beam search supporting pre-beams which is why we share our code on GitHub.

4.5 Language Model

Our character-based LM is a traditional Transformer (equivalent to the decoder of Fig. 1 without mutual-attention) which stacks six self-attention and subsequent PFF layers with 512 output nodes, and 2,048 nodes in the first PFF dense layer. We use 8 attention heads and do not apply dropout. The embedding layer has a dimension of 128. The training hyper-parameters are identical to the ones of training the HTR model.

Text samples are transformed and augmented by applying random transformations on the text:

1. The text is randomly cropped (start and end) to a length between 768 and 1,024 characters.
2. If the first character is upper case, its case is switched with a probability of 0.5, while if it is lower, it is switched with a probability of 0.1.
3. The IAM dataset does not follow traditional spacing rules, e.g., by inserting spaces before punctuation. Therefore, we add an additional rule for IAM only that randomly inserts spaces before, after, or on both sides of `;,-:'"?!.[](){}` with a probability of 0.1. This results in a LM that will have uncertainty on the exact location of spaces and special characters, but high ones on characters.

4.6 Synthetic Line Generation

Synthetic handwritten lines are used to pre-train the visual features and also to train the "intrinsic" LM of the decoder. To render synthetic lines (see Fig. 3 for

Drinking and driving can absolutely not be tolerated by anyone.
To cook, bring a large pot of water to a boil.
were 1,472 households with 92.0% of housing units

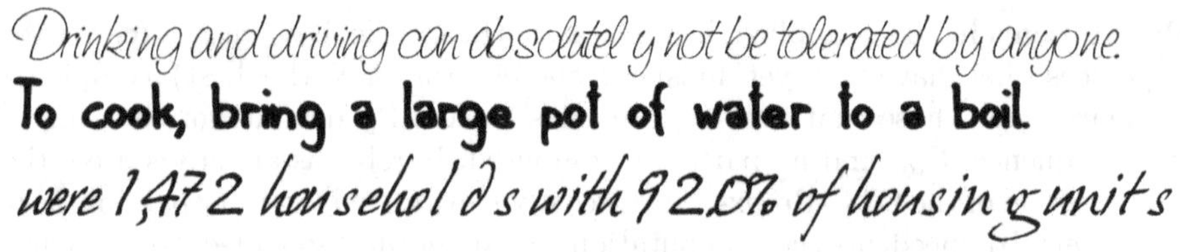

Fig. 3. Three synthetically rendered lines based on the English text corpus.

Table 2. Top-1 and top-10 accuracies of the trained LMs for correctly predicting the next character on the listed dataset. The model was trained on an additional corpus for the respective language.

Dataset	Language	Top-1	Top-10
IAM	en	58.1%	90.4%
StAZH	de-ch	52.5%	87.3%
Rimes	fr	58.5%	92.8%

some examples), we collected several cursive computer fonts. The sentences are generated analogously to the cropping rules used for the LM (see Sect. 4.5). Data augmentation during preprocessing (see Sect. 4.1) then varies the renderings with clean synthetic computer fonts to make the network more robust against real-world data.

5 Results

In this section, we present our evaluation results. The network architecture and decoding are implemented in the open-source framework *tfaip* [13] which is based on Tensorflow. The encoder plus backbone, decoder, and LM comprise 3.2, 1.6, and 19.0 million parameters, respectively. The combined model has about 24 million parameters in total.

First, we show the accuracies of our trained LMs and the performances of the pretrained models applied to the three datasets. Then, we measure the CERs when training on real data, optionally starting from the pretrained models. Next, the LM is added and n_{beams} is varied. We conclude our experiments with an ablation study and a comparison to related work.

5.1 Language Models

Table 2 shows the top-1 and top-10 accuracies, to correctly predict the next character, of the three LMs. The accuracies are stated for the validation datasets of IAM, StAZH, and Rimes, including the augmentations applied in Sect. 4.5. The independent training data was chosen from the respective datasets (see Sect. 3).

Table 3. CER given in percent of the models pretrained on artificial lines synthesised for the respective language. We list the CER for using solely the encoder with CTC best path decoding and our proposed CTC/Transformer combination.

Dataset	CTC		CTC/Trafo	
	Val.	Test	Val.	Test
IAM	17.4	19.5	15.3	17.3
StAZH	64.7	65.1	64.6	64.3
Rimes	24.3	25.1	22.4	23.0

Table 4. Influence of using pretrained models. The CER is given in percent. We list the CER for using solely the encoder with CTC best path decoding and our proposed CTC/Transformer combination.

		Validation		Test	
	Pretr.	CTC	CTC/Tr	CTC	CTC/Tr
IAM	No	3.63	3.38	5.47	5.10
IAM	Yes	3.17	2.60	4.99	3.96
StAZH	No	3.22	2.93	3.05	2.81
StAZH	Yes	3.17	2.64	3.06	2.66
Rimes	No	4.84	4.47	4.31	3.88
Rimes	Yes	4.87	4.14	4.25	3.49

All LMs yield a top-1 accuracy of over 50% and a top-10 accuracy of about 90%. Note that, as expected, the LM for StAZH is the worst because the data of the training corpus is contemporary Swiss-German (de-ch) while StAZH comprises historic documents where spellings are different. Since the French (fr) corpus is larger than the English (en) one, the accuracy is higher. Other reasons for the lower accuracy on English are the different whitespace rules and the additional augmentation to map this.

5.2 Pretraining on Synthetic Data

Table 3 lists the CERs for the pretrained models which are trained exclusively on synthetic data generated for each of the three languages. Naturally, since only synthetic data was used, the models perform very poorly on real world data, especially on StAZH which is the most difficult dataset. Combining CTC and the Transformer for decoding yields slightly better results than the standalone encoder with CTC which is explained by the fact that the decoder learned an intrinsic LM.

Table 5. The upper half shows the influence of using an additional LM during decoding by varying its weight λ_{lm}. The bottom half investigates the influence of varying the number of beams n_{beams} during beam search decoding. All CERs and WERs are given in percent.

	CER [%]				WER [%]			
LM	**0**	**0.1**	**0.5**	**1**	**0**	**0.1**	**0.5**	**1**
IAM	3.96	3.69	3.19	3.64	12.20	11.12	9.17	10.34
StAZH	2.66	2.66	2.79	3.81	11.87	11.75	12.48	16.72
Rimes	3.49	3.40	3.39	3.57	9.03	8.65	8.28	8.49
Beams	**1**	**5**	**10**	**20**	**1**	**5**	**10**	**20**
IAM	5.81	3.19	3.17	3.13	14.09	9.17	9.00	8.93
StAZH	4.37	2.79	2.79	2.79	15.54	12.48	12.48	12.50
Rimes	4.10	3.39	3.19	3.19	9.19	8.28	7.61	7.61

5.3 Training on Real Data

Table 4 shows the CER when training on the actual real-world data, optionally using the pretrained model from Sect. 5.2. On both the validation and test sets of the respective datasets, the CER shrinks when using the pretrained model and CTC/Transformer for decoding. However, the benefit is clearly higher for the IAM dataset. If only using CTC for decoding, pretraining has a smaller impact, and even leads to slightly worse results for StAZH on the test set and for Rimes on the validation set. Since the CTC branch only includes visual information, we explain this by the handwritten computer fonts which are more similar to IAM but differ considerably from the writings in the other two datasets. This however shows that the pretrained decoder actually learned an intrinsic LM because using the pretrained model always shows an improved CER when using CTC/Tr.

5.4 Enabling the Language Model

Even though the decoder of the HTR model is already capable to learn a LM as shown in the previous section, we examine if an additional specialised and larger LM (see Sect. 4.5) further improves the results. The upper half of Table 5 summarises the results by setting λ_{lm} to 0, 0.1, 0.5, and 1 ($n_{\mathrm{beams}} = 5$). $\lambda_{lm} = 0$ corresponds to using no LM and thus to the values of Table 4 (last column). Here, we only show the results when combining all three decoding paths (CTC/Tr/LM), see Fig. 2).

On IAM, the LM with $\lambda_{\mathrm{lm}} = 0.5$ yields clearly improved results (CER = 3.2%) which further reduces the CER by 20% (relative). This shows that the intrinsic LM of our HTR model is not yet powerful enough to "fully" learn the English language.

In contrast, using the LM trained on a contemporary Swiss-German corpus does not further improve the results on the historic StAZH dataset, instead the results worsen. The reason is that the historic and contemporary writings are too different and therefore induce errors instead of correcting misspellings.

Similar to IAM, the French LM fits well. However, since the French alphabet size is larger than the English one, a higher beam count is required to obtain the best results as shown in the next Section.

5.5 Varying the Beam Count

The bottom half of Table 5 lists the CERs and WERs when varying n_{beams} if using a LM with $\lambda_{\text{LM}} = 0.5$. Setting $n_{\text{beams}} = 1$ corresponds to best path decoding. The results show that increasing n_{beams} from one to five has the highest impact on the results. On IAM, further increases only led to small improvements. More beams on StAZH had no impact because the LM does not fit well. Setting $n_{\text{beams}} = 10$ further reduced the error rates on Rimes which we explain by the bigger alphabet size due to characters with diacritics such as "e, è, é, ê".

5.6 Ablation Study

Furthermore, we performed an ablation study to measure the impact of pretraining, the inclusion of a LM, and the decoder. We chose IAM, set $\lambda_{\text{LM}} = 0.5$ (if a LM is used), and $n_{\text{beams}} = 5$. The results of the CERs and WERs are listed in Table 6 (upper half).

As expected, the CER always benefits from starting with a model pretrained on synthetic data. In experiments A and B we only use the encoder combined with CTC-best-path-decoding which therefore primarily shows the influence of pretraining on the visual backbone and the BiLSTM-encoder: using pretraining improves the CER by about 9%, relatively. The Transformer decoder benefits by a larger margin (see, e.g., G and I where the CER is reduced by about 23%). The reason is that the Transformer can learn an intrinsic LM in addition to the improved visual features from the encoder.

Comparison of C–F to G–J, respectively, shows the impact of the CTC-Prefix-Score during decoding which always leads to improved results, independent of using pretraining or a LM. Furthermore, it is astonishing that combining a LM with a non-pretrained model results in very bad CERs (D). Visual inspection of the transcribed lines shows that there are many lines that are too short, i.e., words were skipped in between but mainly at the end. Here, the CTC-Prefix-Score (H) has a high impact since it penalises beams that do not cover all characters that are detected by the encoder.

Our best model (without additional training data, K) is obtained by combining all proposed improvements (pretraining on synthetic data, a LM, and adding the CTC-Prefix-Score) and a bigger beam width of $n_{\text{beams}} = 20$, resulting in a CER of 3.13% and a WER of 8.94%.

The last column of Table 6 lists the number of lines that can be processed per second, using a batch size of 1 and CPU decoding only. The numbers show that using S2S instead of CTC results in a slowdown of about 5–6. The combination

Table 6. Ablation study and comparison with related work. All CERs and WERs are given in percent. Transformers (Tr) in the encoder are comprised of stacked self-attention layers, Transformers in the decoder consist of mixed self- and mutual-attention layers. Our decoding is performed with 5 beams, exclusions with 20 beams are marked. "+ Data" denotes whether during training additional synthetic (syn) or real data is used, or if the IAM validation set is included (val). The LM column lists if a LM (trained on external corpora) with open vocabulary (Open) or with a limited vocabulary size (50K works) was used during inference. The number of parameters (#P) are given in millions. The last column lists the number of samples that can be processed per second using a batch size of 1 and a CPU only.

	Authors	Enc.	Dec.	+ Data	LM	CER	WER	#P	#/s
A	Ours	LSTM	CTC	No	No	5.47	17.93	3.2	10.77
B	Ours	LSTM	CTC	Syn	No	4.99	16.85	3.2	11.57
C	Ours	LSTM	Tr	No	No	5.61	16.24	4.8	1.90
D	Ours	LSTM	Tr	No	Open	14.38	18.25	24	0.50
E	Ours	LSTM	Tr	Syn	No	4.15	12.22	4.8	2.50
F	Ours	LSTM	Tr	Syn	Open	6.46	13.38	24	0.86
G	Ours	LSTM	CTC/Tr	No	No	5.09	15.88	4.8	0.69
H	Ours	LSTM	CTC/Tr	No	Open	4.33	12.69	24	0.37
I	Ours	LSTM	CTC/Tr	Syn	No	3.96	12.20	4.8	0.70
J	Ours	LSTM	CTC/Tr	Syn	Open	3.20	9.19	24	0.40
K	Ours (20)	LSTM	CTC/Tr	Syn	Open	**3.13**	**8.94**	24	0.18
L	Ours	LSTM	CTC/Tr	Syn/Val	Open	3.01	8.81	24	0.42
M	Ours (20)	LSTM	CTC/Tr	Syn/Val	Open	**2.95**	**8.66**	24	0.18
N	Bluche [2]	LSTM	CTC	No	50K	3.2	–	0.75	–
O	Michael [10]	LSTM	S2S	Val	No	4.87	–	–	–
P	Yousef [16]	FCN	CTC	No	No	4.9	–	3.4	–
Q	Kang [5]	Tr	Tr	Syn	No	4.67	15.45	–	–
R	Wick [14]	Tr	Bi-Tr	No	No	5.67	–	–	–
S	Diaz [3]	Tr	CTC	Syn/Real	Open	**2.75**	–	≈12	–
T	Li [7]	Tr	Tr	Syn	No	3.42	–	334	–
U	Li [7]	Tr	Tr	Syn	No	2.89	–	558	–

of S2S with the CTC-Prefix-Score further reduces the speed by a factor of about 3. Adding a LM reduces the decoding speed by another factor of about 2. Thus, with $n_{\text{beams}} = 5$, this results in a speed reduction of about 26.

Increasing n_{beams} by a factor of 4 does reduce the decoding speed by another factor of 2 (only). The dependency for low beam counts is non-linear since all beams can be processed in parallel.

5.7 Comparison with the State of the Art

Table 6 also includes performances of previous publications. Our best model yields a competitive CER of 3.13% which outperforms most previous models that, similar to our approach, do not include additional real-world data for training (N–R). We even outperform N who use, in contrast to our open vocabulary LM, a LM with a limited vocabulary size of 50,000 words. Comparing Q to E, which are comparable setups, in terms of including synthetic data and only using the Transformer for decoding, shows that our overall method is superior. The reasons might be our preprocessing or the different network architecture. R can be compared to C since both use the same preprocessing pipeline but a different network architecture which explains the similar results.

Setup S which achieves the current best value with a CER of 2.75% also includes a large amount of additional real-world data which is why a direct comparison must be considered as biased. Nevertheless, this verifies our observations that including additional data (here the validation set of IAM) further improves the results up to a CER of 3.01% and 2.95%, for $n_{\mathrm{beams}} = 5$ or 20 (L and M), respectively.

The large model of the recent publication of Li et al. [7] (U) also outperforms our best model. We expect that the increased model size in cooperation with an improved text synthesising method, as used by [7], explains their further improvements. In contrast, their base model (T) is significantly worse than our best model even though their model is significantly larger (334M vs 24M parameters). While our model employs a LM, Li et al. use a large decoder which should actually have more capacity for language modelling. The benefit of our approach is that our LM can be trained separately requiring only text data, which the transformer decoder as part of the full network has to be trained with fully annotated image and text data.

6 Future Work

In this paper, we proposed the combination of a CNN/LSTM-encoder and Transformer-decoder network for TLR. To solve intrinsic problems of the S2S-approach, we added the CTC-Prefix-Score [12] which was, to the best of our knowledge, not yet applied in the context of TLR. Furthermore, we added a separately trained Transformer as a LM. On the well-established IAM-dataset, we achieved a competitive CER of 2.95% significantly outperforming a current state-of-the-art model with ten times more parameters [7]. Our approach is outperformed solely by a significantly larger model ([7], 20 times more parameters), or by including an additional dataset with other real-world handwritten data [3].

A crucial disadvantage of our approach compared to CTC-based ones is the very slow decoding time. In practice, this must outweigh the gain in performance for the actual use-case. Even though we already incorporated some methods to speed up decoding, e.g., by reducing the search space of the CTC-Prefix-Score, other possibilities must be explored. A straightforward approach is to use tokens instead of single characters. Since a token comprises several characters, and can thus form syllables or even complete words, the number of iterated decoding steps is massively reduced. For TLR, this approach has already proven to be successful by [7].

Acknowledgments. This work was partially funded by the European Social Fund (ESF) and the Ministry of Education, Science and Culture of Mecklenburg-Western Pomerania (Germany) within the project Neural Extraction of Information, Structure and Symmetry in Images (NEISS) under grant no ESF/14-BM-A55-0006/19.

References

1. Augustin, E., Carré, M., Grosicki, E., Brodin, J.M., Geoffrois, E., Prêteux, F.: Rimes evaluation campaign for handwritten mail processing. In: International Workshop on Frontiers in Handwriting Recognition (IWFHR 2006), pp. 231–235 (2006)
2. Bluche, T., Messina, R.: Gated convolutional recurrent neural networks for multilingual handwriting recognition. In: 2017 14th IAPR International Conference on Document Analysis and Recognition (ICDAR), vol. 1, pp. 646–651. IEEE (2017)
3. Diaz, D.H., Qin, S., Ingle, R., Fujii, Y., Bissacco, A.: Rethinking text line recognition models. arXiv preprint arXiv:2104.07787 (2021)
4. Graves, A., Fernández, S., Gomez, F., Schmidhuber, J.: Connectionist temporal classification: labelling unsegmented sequence data with recurrent neural networks. In: Proceedings of the 23rd International Conference on Machine Learning, pp. 369–376. ACM (2006)
5. Kang, L., Riba, P., Rusiñol, M., Fornés, A., Villegas, M.: Pay attention to what you read: non-recurrent handwritten text-line recognition. arXiv preprint arXiv:2005.13044 (2020)
6. Li, C., et al.: ESPnet-SE: end-to-end speech enhancement and separation toolkit designed for ASR integration. In: Proceedings of Spoken Language Technology Workshop, pp. 785–792. IEEE (2021)
7. Li, M., et al.: TrOCR: transformer-based optical character recognition with pre-trained models. arXiv preprint arXiv:2109.10282 (2021)
8. Marti, U.V., Bunke, H.: The IAM-database: an English sentence database for offline handwriting recognition. Int. J. Doc. Anal. Recogn. **5**(1), 39–46 (2002)
9. Memon, J., Sami, M., Khan, R.A., Uddin, M.: Handwritten OCR: a comprehensive systematic literature review (SLR). IEEE Access **8**, 142642–142668 (2020)
10. Michael, J., Labahn, R., Grüning, T., Zöllner, J.: Evaluating sequence-to-sequence models for handwritten text recognition. In: 2019 International Conference on Document Analysis and Recognition (ICDAR), pp. 1286–1293. IEEE (2019)
11. Sutskever, I., Vinyals, O., Le, Q.V.: Sequence to sequence learning with neural networks. In: Advances in NIPS, pp. 3104–3112 (2014)
12. Watanabe, S., Hori, T., Kim, S., Hershey, J.R., Hayashi, T.: Hybrid CTC/attention architecture for end-to-end speech recognition. IEEE J. Sel. Top. Signal Process. **11**(8), 1240–1253 (2017)
13. Wick, C., et al.: tfaip-a generic and powerful research framework for deep learning based on Tensorflow. J. Open Sour. Softw. **6**(62), 3297 (2021)
14. Wick, C., Zöllner, J., Grüning, T.: Transformer for handwritten text recognition using bidirectional post-decoding. In: Lladós, J., Lopresti, D., Uchida, S. (eds.) ICDAR 2021. LNCS, vol. 12823, pp. 112–126. Springer, Cham (2021). https://doi.org/10.1007/978-3-030-86334-0_8
15. Wigington, C., Stewart, S., Davis, B., Barrett, B., Price, B., Cohen, S.: Data augmentation for recognition of handwritten words and lines using a CNN-LSTM network. In: ICDAR, pp. 639–645 (2017)
16. Yousef, M., Hussain, K.F., Mohammed, U.S.: Accurate, data-efficient, unconstrained text recognition with CNNs. Pattern Recognit. **108**, 107482 (2020)

A Light Transformer-Based Architecture for Handwritten Text Recognition

Killian Barrere(✉), Yann Soullard, Aurélie Lemaitre, and Bertrand Coüasnon

Univ. Rennes, CNRS, IRISA, Rennes, France
killian.barrere@irisa.fr

Abstract. Transformer models have been showing ground-breaking results in the domain of natural language processing. More recently, they started to gain interest in many others fields as in computer vision. Traditional Transformer models typically require a significant amount of training data to achieve satisfactory results. However, in the domain of handwritten text recognition, annotated data acquisition remains costly resulting in small datasets compared to those commonly used to train a Transformer-based model. Hence, training Transformer models able to transcribe handwritten text from images remains challenging. We propose a light encoder-decoder Transformer-based architecture for handwriting text recognition, containing a small number of parameters compared to traditional Transformer architectures. We trained our architecture using a hybrid loss, combining the well-known connectionist temporal classification with the cross-entropy. Experiments are conducted on the well-known IAM dataset with and without the use of additional synthetic data. We show that our network reaches state-of-the-art results in both cases, compared with other larger Transformer-based models.

Keywords: Light network · Hybrid loss · Transformer · Handwritten text recognition · Neural networks

1 Introduction

Handwritten Text Recognition (HTR) refers to the process of automatically recognizing the text written inside an image of a text-line, a paragraph or even whole pages, after a first step of document layout analysis. This task is valuable nowadays for a growing number of people as it enables the text to be available in a digitized format and resist in time. There is still a tremendous amount of document collections waiting to be processed, and HTR models have shown acceptable error rates on specific documents [16]. This is mostly thanks to the recent advances and growing interest in deep learning approaches. However, HTR remains challenging for a variety of reasons. The variability of writing styles, degraded documents, the need for data matching documents and the lack of annotated records limit the abilities of current deep learning approaches.

Popular network architectures are based on both convolutional layers and long short-term memory layers, trained using the Connectionist Temporal

© Springer Nature Switzerland AG 2022
S. Uchida et al. (Eds.): DAS 2022, LNCS 13237, pp. 275–290, 2022.
https://doi.org/10.1007/978-3-031-06555-2_19

Classification (CTC) loss function [8]. These layers result in an optical model, typically followed by a language model which aims to correct recognition errors. Fully Convolutional Networks have been investigated recently with the goal to significantly reduce training time by removing recurrent layers. Lately, few works based on Transformers [17] have been proposed for HTR [10,14,18]. Based on multi-head attention layers, a Transformer network proves to be an efficient alternative to recurrent layers. It enables a sequential analysis of the sequence thanks to positional encoding and efficient parallelism. Transformer-based models offer a great potential, but those architectures generally require a tremendous amount of annotated training data to be robust. This limits the efficiency of such networks in HTR tasks where annotated data are expensive.

In this paper, we propose a light encoder-decoder Transformer-based architecture for handwriting recognition. The network presented in this work is significantly smaller than traditional Transformer networks, which makes it lighter in the number of weights and easier to train on small datasets. Our architecture is trained using a hybrid loss combining both the CTC loss and the Cross-Entropy (CE) loss, which seems crucial.

The article is organized as follows. After reviewing related works in Sect. 2, our network architecture is presented with details in Sect. 3. Afterward, we present results obtained on the popular IAM dataset. As opposed to traditional Transformer approaches, we show that our approach reaches state-of-art results without any additional data nor transfer learning. Nevertheless, it is capable to outperform the other types of network architectures by using synthetic data.

2 Related Works

2.1 Standard Approaches for HTR

Popular network architectures used for Handwriting Text Recognition combine both convolutional layers and recurrent layers. A number of convolutional layers are stacked at the beginning of the network to extract local features from text-line images. Then, recurrent layers, and more specifically Bi-directional Long Short-Term Memory (BLSTM) layers are stacked to process the features sequentially and output character probabilities based on contextual dependencies. Such an architecture is frequently called a Convolutional Recurrent Neural Networks (CRNN) [3,13,15,16]. Models are generally trained using the well-known Connectionist Temporal Classification (CTC) loss [7]. It enables to deal with label sequences of shortest length than predicted sequences, without any knowledge about character segmentation.

Encoder-decoder based architectures have also been investigated for HTR. They typically rely on an attention mechanism and a decoder based on LSTM to sequentially predict the characters [2,6,12]. Michael et al. [12] propose to use a hybrid loss combining both a CTC loss applied to the encoder and a cross-entropy loss for the decoder. Such models can obtain low error rates on common datasets. However, they suffer from the lack of computation parallelization inherently due to recurrent layers, which impacts both training and inference time.

Recently, Fully Convolutional Networks (FCN) have been proposed for HTR. They refer to deep architectures composed of many convolutional layers and no recurrent layer. They often include the most recent innovations in deep learning, like gating mechanism or residual connections to obtain state of the art results [4,5,9,19]. These architectures benefit from the computational parallelism offered by convolutional layers, and hence can be trained faster than traditional architectures based on recurrent layers. However, they may require tremendous work to be optimized well and require data augmentation to attain state-of-the-art results. By removing recurrent layers, Fully Convolutional Networks might struggle to learn long-range contextual dependencies, which can be useful in HTR.

CRNN architectures dominate in the field of HTR, thanks to their ability to learn local and long-range features. However, they highly suffer from the notable training time of recurrent layers. The last few years, Fully Convolutional Networks obtained state-of-the-art results but they might experience difficulties related to long-range contextual dependencies.

2.2 Transformer-Based Architectures

In natural language processing, multi-head attention have been proposed by Vaswani et al. [17] inside the so-called Transformer model. They propose an effective alternative to recurrent layers, capable of handling broad context in a constant amount of operation while enabling efficient parallelism. Transformer models also started to gain interest in the field of HTR. Kang et al. [10] propose an end-to-end Transformer that aims for both recognizing handwritten text and modeling the language. They obtain very low error rates by implementing a big network architecture that requires synthetic data to be trained efficiently. Singh et al. [14] propose a Transformer-based architecture to address the problem of full-page handwriting recognition. They obtain promising results on full-page recognition thanks to the use of synthetic data again. Wick et al. [18] use a bi-directional Transformer architecture coupled with a voting mechanism and show that their architecture outperforms a standard Transformer-based architecture.

Transformer-based layers propose an efficient trade-off between CRNN and Fully Convolutional Networks. Multi-head attention layers offer indeed both parallelism and the ability to learn long-range contextual dependencies. However, to perform well, a Transformer-based architecture, methods from the state of the art rely on synthetic data. Such additional data require to be designed as a complement to the training data which might remain a challenging task.

3 Our Light Encoder-Decoder Transformer-Based Model

Most Transformer-based architectures are based on large and deep models using many parameters and requiring a large amount of training data to perform well. Handwritten datasets typically contains too few examples for a Transformer-based model to perform well [10]. Synthetic data seem to be an efficient solution

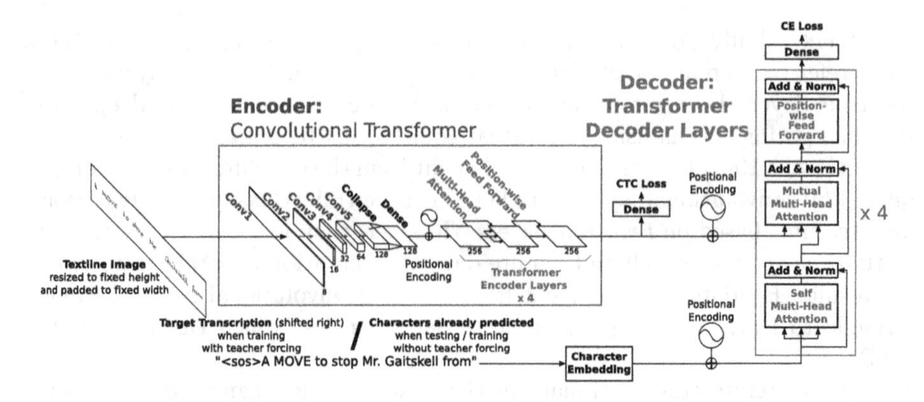

Fig. 1. Our encoder-decoder Transformer-based architecture. Our architecture is composed of an encoder combining convolutional layers and Transformer layers (Sect. 3.2), and of a Transformer-based decoder (Sect. 3.3).

but designing valuable synthetic data might prove to be a complex task. While it takes efforts to design synthetic data, they may result in deteriorated performance, especially for difficult datasets like historical documents.

In this work, by contrast, we aim for a Transformer-based architecture capable of obtaining state-of-the-art results without the need of additional data, while still benefiting from the usage of synthetic data. We propose a light encoder-decoder Transformer-based architecture that can be trained efficiently on datasets of limited size, without the need for additional data[1]. This section provides details about our network architecture.

3.1 Summary of the Architecture

We propose an end-to-end trainable Transformer-based architecture for HTR. Our network architecture is illustrated in Fig. 1. Our Transformer-based architecture remains low in the number of parameters, with 6.9M parameters, compared with traditional Transformer networks that might use up to 100M parameters. The architecture follows the principle of encoder-decoder models as it is composed of two key parts: an encoder and a decoder.

The encoder takes as input text-line images and aims to extract and process visual features. Our encoder is principally based on convolutional layers and Transformer encoder layers (Fig. 2). More details about the encoder are disclosed in Sect. 3.2.

The decoder subsequently uses the output of the encoder to sequentially predict the character sequence written inside the image (character by character). In addition, it also has access to the sequence of previously predicted characters.

[1] By additional data, we mean data from another dataset than the one which is studied and synthetic data. We nevertheless use data augmentation techniques to improve our models.

Thus the decoder might act as a language model at character level together with optical features from the input image. Section 3.3 provides more specific explanations about the decoder.

We train our models with a hybrid loss combining both the Connectionist Temporal Classification (CTC) loss [7] and the Cross-Entropy (CE) loss. This is discussed further in Sect. 3.4.

3.2 Network Encoder

The encoder is inspired by traditional CRNN and by Transformer layers [17]. We propose replacing the recurrent layers from the CRNN architecture by Transformer encoder layers. These layers refer to the encoder block of the Transformer architecture [17]. They are based on multi-head attention followed by a position-wise feed forward layer. Transformer encoder layers have the advantage of being more parallelizable than recurrent layers on GPU while being able to handle long-range contextual dependencies in a constant number of operations. Our convolutional Transformer encoder is illustrated in Fig. 2.

The first part of the encoder is composed of 5 convolutional blocks used to extract visual features from the image. Except the last one, each convolutional block is composed of a 2D convolutional layer with a kernel of size 3×3, a stride of 1 and no padding. The last convolutional block uses a kernel size of 4×2 to better match the shape of a character [3,13]. The number of filters in the convolutional layers are respectively equal to 8, 16, 32, 64 and 128. Each convolution layer is then followed by a LeakyReLU activation function. Following the activation function, we apply a layer normalization to ease the network training capabilities and increase the regularization capacities of the network. A 2×2 max pooling is used inside the first three convolutional blocks to decrease the size of intermediate feature maps. It also focuses the training process on the most impacting features and reduces the number of network parameters. Lastly, a dropout is applied with a probability of 0.2 at the end of each block.

Following the last convolutional block, a collapse layer is used to flatten the vertical dimension of the feature maps, therefore enabling us to easily work with Transformer layers. It is composed of a convolution layer with a kernel

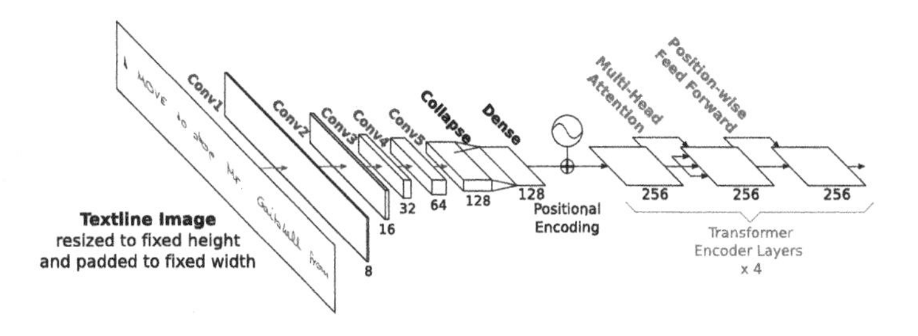

Fig. 2. Our Convolutional Transformer Encoder. This encoder is composed of a first stack of convolutional layers followed by Transformer Encoder layers.

size of width 1 and height similar to the height of the input feature maps. We subsequently apply a LeakyReLU activation function followed by a layer normalization.

Following that layer, we use a dense layer to increase the hidden size from 128 to 256. Before the Transformer encoder layers, sinusoidal positional encoding [17] is added to the output of the dense layer. We then use 4 stacked Transformer encoder layers. Each Transformer layer uses a hidden dimension of 256 and is composed of self-attention layers based on multi-head attention with 4 heads. They are then followed by a position-wise feed-forward layer with an intermediate feature size of 1024. Residual connections are used between each sub-layer. A dropout with a probability of 0.2 is applied to each sub-layer output.

The encoder of our architecture remains light in its number thanks to a small convolutional backbone, while additionally using a few numbers of parameters inside of Transformer layers. Compared to other Transformer-based architectures that use heavier convolutional neural networks like ResNets [10,14], our architecture only uses 5 convolutional layers. It results in a total of only 237k parameters for the convolutional backbone, which is lower than most of the convolutional backbone used in other Transformer-based architectures [10,14]. In addition, we maintain the number of intermediate neurons small inside Transformer layers, with a hidden size of 256. Heavier Transformer-based architectures might use up to 1,024 neurons inside Transformer layers.

3.3 Transformer Decoder

Our encoder-decoder Transformer-based model then processes the output from the encoder along with the sequence of previously predicted characters (or the target sequence shifted right when training with teacher forcing). An illustration of the decoder is shown in Fig. 1.

Our Transformer-based decoder uses both the output from the encoder part for mutual (or encoder-decoder) attention and the sequence of predicted characters. Sinusoidal positional encoding [17] is added to the output of the encoder. Despite the fact that our encoder already includes a positional encoding inside, we nevertheless find it beneficial to add a second positional encoding to the output of the encoder. Similarly to including an additional loss in the middle of a dense network, we believe that re-adding the information of the position in the middle of the architecture might help our architecture to converge. In a similar way, we apply positional encoding to the sequence of predicted characters, after applying a character-level embedding and before feeding it to the decoder part.

The Decoder is composed of a stack of 4 Transformer decoder layers [17]. Each layer takes as input the output from the previous layer or the embedded character sequence for the first layer. It is composed of a self-attention layer, using multi-head attention. Following the first sub-layer, a multi-headed encoder-decoder attention is applied. It uses the sequence coming from the encoder and apply weighted attention based on the encoder output and the output from the previous decoder sub-layer. Each attention sub-layer uses a hidden size of 256 and 4 heads. Lastly, we apply a position-wise feed forward layer composed of two dense layers with an intermediate feature size of 1024. As used in the encoder, a residual connection is applied to each sub-layer and dropout of probability 0.2 is used.

3.4 Hybrid Loss

To train the entire model, we use a hybrid loss combining both the Connectionist Temporal Classification (CTC) loss function [7] and a Cross-Entropy (CE) loss function. Such a hybrid loss has been introduced fist in the domain of HTR by Michael et al. [12] to train an encoder-decoder model. In this work, both losses are linearly combined as follow:

$$\mathcal{L} = \lambda \cdot \mathcal{L}_{CTC} + (1 - \lambda) \cdot \mathcal{L}_{CE} \tag{1}$$

where λ balances between the CTC loss and CE loss. For a given input observation $\mathbf{x} = (x_1, ..., x_T)$ of length T with label $\mathbf{y} = (y_1, ..., y_L)$ of length L, we have:

$$\mathcal{L}_{CTC} = -ln(\sum_{\pi \in \mathcal{B}^{-1}(\mathbf{y})} p(\pi|\mathbf{x})) \text{ and } \mathcal{L}_{CE} = -\sum_{i=1}^{L} y_i \cdot log(\hat{y}_i^{dec}) \tag{2}$$

with $\hat{y}_i^{dec}$ the probability to predict y_i in output of the decoder and $\pi \in \mathcal{B}^{-1}(\mathbf{y})$ a path from the encoder output that produces the label sequence $\mathbf{y}$ by applying the function $\mathcal{B}$ that maps the output sequence[2] of length T' to a sequence of length L with $T' > L$. Note that the CTC loss requires the probability of each character per frame in input with an additional blank label that refers to predict no character. The function $\mathcal{B}$ removes all repeated labels in the path and then removes blank labels (see [7] for more details).

Thus, the CE loss is applied using the output from the decoder, whereas the CTC is used with the output of the encoder. The decoder is therefore only trained using gradients coming from the CE loss, while the network encoder is trained using gradients coming from both the CTC and CE losses after backpropagation through the decoder.

Both losses require the probabilities for each class which are the characters and the *CTC Blank* label for the CTC loss and the characters and an *end-of-sequence* token for CE loss. Therefore, we apply a dense layer followed by a softmax at the end of the encoder and the decoder. Unlike the architecture proposed by Michael et al. [12], the character probabilities in output of the encoder are unused in input of the decoder. As illustrated in Fig. 1, the character probabilities are only used for the CTC loss and the output of the last hidden layer of the encoder is given in input of the decoder.

[2] Note that T' is equal to T if no reduction of the input sequence length is applied inside the network.

4 Experiments and Results

Handwritten datasets are generally too small to get traditional Transformer-based architectures that perform well. As stated before, additional data are generally used to improve the performance of such architectures. By contrast, we aimed for a relatively light Transformer-based architecture to perform well both with and without additional data.

In this section, we demonstrate that, compared to other Transformer-based architectures, our light Transformer architecture reaches state-of-the-art results without requiring any additional data. To prove the efficiency of our network, we conduct experiments on the well-known IAM dataset (1) without using additional data; (2) using synthetic data as often done in other works.

We start by introducing experimental settings as well as the process used to generate synthetic data. We investigate the impact on performance of the main components from our network. Later, we compare the results obtained with a larger version of our light architecture and show that a big architecture is unrequired for a task of HTR. We then show the interest of using a hybrid loss that include a CTC loss at the end of the encoder part. We conclude the section by comparing the performance of our architecture with the ones of other methods from state of the art.

4.1 Handwritten Text-line Data

IAM. The IAM offline handwriting database [11] is probably the most popular dataset in the domain. It is composed of text-line images of modern English handwriting, produced by several writers that have been asked to write a specific text. It has been extensively used in the literature.

To conduct our ablation study and compare our network with state of the art methods, we use the *aachen* split. While this is not the original split, the *aachen* split, is commonly used in the field to compare results. This split provides 6,482 text-line images in training, 976 for validation and 2,915 images to evaluate the performance of the model.

Synthetic Data. To further investigate our model, we use generated synthetic data in addition to the training dataset. Recent Transformer architectures in the field use synthetic data, and report significant gains by using such data [10,14].

We use textual data extracted from a collection of English Wikipedia articles. We then generate text-line images by using handwritten fonts[3]. Generated text-line images could be more realistic thanks to the recent advances on generative models [1]. However, this approach would also require training data, while we aim at performing well with limited amount of real training data. We generate various handwritings by setting different stroke widths and slant angles. Lastly, to simulate real handwritten variations, we apply some image transformations close to a usual data augmentation pipeline. We apply elastic distortions, vary

[3] The fonts are available on https://fonts.google.com and https://www.dafont.com.

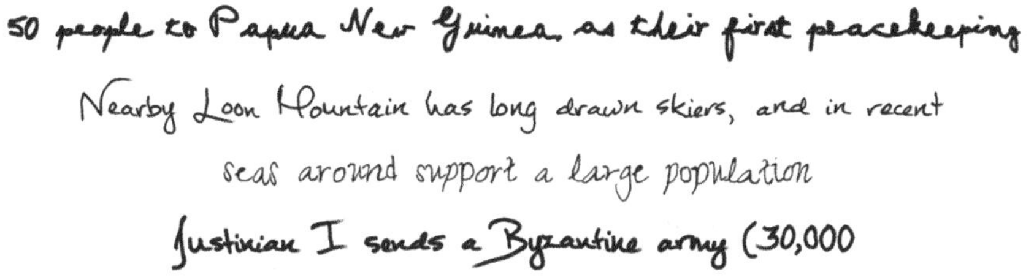

50 people to Papua New Guinea as their first peacekeeping

Nearby Loon Mountain has long drawn skiers, and in recent

seas around support a large population

Justinian I sends a Byzantine army (30,000

Fig. 3. Examples of generated English synthetic data, trying to match the IAM dataset.

the perspective of the image and add some noise to the generated image. Some examples of generated synthetic data are displayed in Fig. 3.

Synthetic text-line images are generated on the fly and are combined with the training dataset. In our experiments using synthetic data, we use 10,000 synthetic text-line images for each training epoch in addition to every real example from the IAM training dataset.

4.2 Experimental Settings

In this section, we describe the experimental settings we use for each of the proposed experiments, unless specified otherwise.

We train our model using teacher forcing with a probability of 1 by feeding the target sequence shifted right to the decoder. We employ masking inside the layers of self-attention in the decoder, therefore ensuring that each prediction of a character only depends on the sequence of characters before it. It follows the identical principle used by Vaswani et al. [17]. In addition to the usual gain in convergence speed, teacher forcing also allows our model to be trained in a parallel fashion, using a single decoder pass. Hence, teacher forcing helps in reducing training times. At test time, characters are predicted step by step, by applying one decoder pass for each character to be produced. The sequence of characters that have been previously predicted is provided in input of the decoder to predict the following character. We decode until an *end-of-sequence* token is predicted or up to 128 characters. Meanwhile, at training time, we use the length of the target sequence to reduce training times.

We use a custom learning rate policy as proposed by Vaswani et al. [17] with a first linear ramp for a warm-up phase of 4,000 steps. We then decay the learning rate following an invert square root function. When training with a hybrid loss, we use a value of $\lambda = 0.5$ to train our models as the two losses are of equivalent orders of magnitude.

Input images are resized to a fixed height of 128 pixels while keeping the aspect ratio. Following that, we apply data augmentation as usually done in the field to virtually increase the number of training examples. We then randomly apply elastic transform, random erosion or dilatation, random perspective and random padding both on the left and right size of the image. Images are then standardized and we add a gaussian noise. Each transformation is applied with a probability of 0.2. Images are then padded according to the largest image in each batch, and a similar process is applied to the target sequence.

To compare results with others, we measure our models performance with both Character Error Rate (CER) and Word Error Rate (WER). It is computed using the Levenshtein distance (also called edit distance), by measuring the number of inserted, replaced and deleted characters (respectively words) between the predicted text and the ground-truth. That distance is then normalized by the length of the ground-truth.

4.3 Ablation Study of the Main Components of Our Network

In this section, we investigate the impact on error rates of the main components of our architecture (the impact of the hybrid loss will be discussed later). We compare the following networks; each trained and evaluated on the IAM dataset with and without synthetic data:

- **Light Transformer** corresponds to our proposed architecture as detailed in Sect. 3.
- **CTNN** is our network encoder, trained and performing as a standalone. It is trained with CTC loss.
- **CRNN** is a common and popular network (it refers to a baseline architecture) composed of convolutional and recurrent layers. It uses the same convolutional layers than our Light Transformer (presented in Sect. 3.2) followed by a stack of four BLSTMs layers with a hidden size of 256. It is similar to CTNN, but uses recurrent layers instead of Transformer layers and is also trained using CTC loss.
- **CRNN + Decoder** combines both the CRNN and our network decoder. Compared to Light Transformer, we replaced transformer layers in the encoder by recurrent layers.

Results are available in Table 1. First, we discuss results without synthetic data. Our Light Transformer architecture is able to obtain better results than a baseline CRNN architecture on the IAM dataset. Compared to this baseline,

Table 1. Comparison between the results obtained by our Transformer-based architecture and a CRNN architecture on the test set of the IAM dataset (*aachen* split), with and without synthetic data added.

Architecture	# params.	IAM		IAM + Synth. data	
		CER (%)	WER (%)	CER (%)	WER (%)
CRNN (baseline)	1.7M	6.14	23.26	5.66	21.62
CRNN + Decoder	5.5M	6.92	21.16	5.36	18.01
CTNN (encoder)	3.2M	5.93	22.82	6.15	24.02
Light transformer	6.9M	**5.70**	**18.86**	**4.76**	**16.31**

our architecture is able to obtain a CER 22% better relatively without using any additional data, but with more parameters. When we remove the decoder from our Light Transformer, we observe that the resulting architecture (CTNN) performs worse. However, it is nevertheless able to obtain a better CER and WER than the baseline architecture. Besides, using a decoder in addition to a CRNN architecture seems to bring worse CER, even if the WER is improved. We believe that when using a Transformer-based decoder, the context from an architecture based on recurrent layers might not be good enough. Lastly, when we compare recurrent layers and Transformer layers in the encoder (CRNN + Decoder versus Light Transformer and CRNN versus CTNN), we observe a gain in the network performance.

Using synthetic data with our Light Transformer architecture, we observe a major improvement of the CER and WER. This gain seems to be far more beneficial for the decoder part, while it might even be unbeneficial for the encoder part performing as a standalone. On the one hand, synthetic data may not fit the real data well and this may affect more specifically the encoder part for which the images are given in input. On the other hand, the decoder trained to predict one character at a time might be able to act as a language model, benefiting from the amount of synthetic training data to learn the language. In addition, we also observe that the WER highly benefits from the usage of a decoder with, and without synthetic data.

As expected regarding a Transformer architecture, synthetic data bring a major improvement for our proposed architecture. Nevertheless, our architecture is able to perform well without any additional data, and we highlight the relevance of the different parts of our network.

4.4 Benefits of Using a Light Architecture

Traditional Transformer-based architecture are relatively heavier (up to 100M parameters) than our network (6,9M of parameters) and they generally perform poorly without additional data. To highlight the advantages of our light architecture, we propose a variant of it, resulting in a similar architecture with more parameters. Our scaled variant of the light architecture is mostly based on the same parameters described in Sect. 3. We use twice as many neurons inside Transformer layers for both the encoder and the decoder. While our light architecture uses a hidden size of 256, our scaled version uses 512 neurons instead. We also increase the number of attention heads to 8, therefore resulting in the identical number of neurons per head as in our light architecture. Lastly, intermediate feature size of position-wise feed-forward layers is also doubled, with 2,048 features instead of 1,024. It results in a scaled version of our architecture using 28M parameters instead of 6.9M parameters. We refer to this scaled version as Large Transformer. Table 2 shows the results obtained by both our light and large Transformer architectures.

We obtain slightly better results with our Light Transformer architecture, therefore indicating the heavy architecture might not prevail over our light architecture. This might be explained by the fact that the number of parameters is

Table 2. Comparison of the results obtained by our light Transformer-based architecture (Light Transformer) and a heavy version of it (Large Transformer) on the test set of the IAM dataset (*aachen* split).

Architecture	# params.	IAM		IAM + Synth. Data	
		CER (%)	WER (%)	CER (%)	WER (%)
Light transformer	6.9M	**5.70**	**18.86**	**4.76**	**16.31**
Large transformer	28M	5.79	19.67	4.87	17.67

lower. Our light architecture might be easier to train on a scenario in which we have a relatively low number of annotated training data for training traditional Transformer-based models. Even so, by adding synthetic data, we find that our light Transformer-based architecture performs better than Large Transformer version. Hence, we believe a big Transformer architecture is unnecessary to obtain good results on a specific type of documents.

In addition, a light architecture would require less resource to be trained efficiently. The training cost could then be reduced both in training time and in the number of training examples.

4.5 Interest of the Hybrid Loss

We evaluate how important it is to train our model with a hybrid loss composed of both the Connectionist Temporal Classification (CTC) loss and a Cross-Entropy (CE) loss. To do so, we compare the results obtained with our light architecture trained with CE only (without the CTC loss at the end of the encoder) and with a hybrid loss. Results are available in Table 3.

Using a hybrid loss seems to be crucial in our architecture, as our light Transformer model trained with a hybrid loss is able to attain lower error rates when compared to a training achieved with CE loss only. Despite the fact that our Transformer architecture relies on residual connections (that help gradients to flow), this may not be enough to efficiently train our architecture. Using a hybrid loss improves this. This is especially accurate when the model is only trained with the given original dataset, as the amount of data is too small for traditional Transformer-based models. However, even by using additional synthetic data, we find it beneficial to include a hybrid loss inside our training procedure.

Table 3. Comparison of the results obtained with our model trained with or without a hybrid loss on the test set of the *aachen* split of the IAM dataset. We present results with and without synthetic data.

Loss function(s)	IAM		IAM + Synth. data	
	CER (%)	WER (%)	CER (%)	WER (%)
CE only	10.29	26.36	6.76	19.62
Hybrid (CTC + CE)	**5.70**	**18.86**	**4.76**	**16.31**

No matter the amount of training data, we believe a hybrid loss is essential four our light architecture to converge quickly and efficiently. Adding an intermediate loss at the end of the encoder seems to assist the model in efficiently training the first layers in a deep architecture. In addition, the CTC loss is dedicated to recognize characters from input sequence of longer length which seems to be useful in our application. From our point of view, using a hybrid loss might help training an encoder-decoder network and combining the CTC and CE losses seems to be efficient for HTR.

4.6 Comparison with the State of the Art

The performance of our architecture is compared with the main results from the state of the art on the IAM dataset (Table 4). To allow a proper comparison, we do not include methods that use an explicit Language Model to correct outputs from the optical model. This is a general tendency, as most results from the state of the art use neither language model nor lexicon. We only report the CER as most of the works do not present WER results. However, we have shown WER results obtained with our light Transformer in the previous tables.

Without using synthetic data, we obtain a CER on the IAM test set of 5.70%. Compared to models based on FCN proposed by Yousef et al. [19] and Coquenet et al. [5], our architecture obtains worse results without synthetic data, as other transformer-based models. To the best of our knowledge, there are no published results on FCN with synthetic data included to which we can compare fairly. An FCN model might obtain better results than our approach when dealing with limited real samples while benefiting from the addition of synthetic data. However, we believe a transformer-based model will benefit more than an FCN from the addition of synthetic data due to the fact that it includes a decoder, capable of learning to model the language to some extent.

Compared to other Transformer-based networks, our architecture is able to obtain state-of-the-art results. As opposed to the Transformer model proposed

Table 4. Results on the IAM dataset (*aachen* split). We compare our architecture to methods based on CRNN [12], FCN [5,19] and Transformer [10,14,18].

Model	# params.	IAM	IAM + Synth. data
		CER (%)	CER (%)
CRNN + LSTM [12]		5.24	
FCN [19]	3.4M	**4.9**	
VAN (line level) [5]	1.7M	4.95	
Transformer [10]	100M	7.62	**4.67**
FPHR transformer [14]	28M		6.5
Forward transformer [18]	13M	6.03	
Bidi. transformer [18]	27M	5.67	
Our light transformer	6.9M	5.70	4.76

by Kang et al. [10], our light architecture performs well even without additional data, while Singh et al. [14] do not present result without synthetic data. Still, by including synthetic data to the training set, our model obtains even lower error rates, reaching a CER of 4.76%. Our proposed architecture reaches results close to the best performing architecture, while using 14 times less parameters.

In a general manner, our light architecture remains low in its number of parameters. Our model only uses 6.9M parameters which is far lower than the Transformer architectures from the state of the art [10,14,18]. Hence, we expect our model to be trained much faster in comparison, while requiring less resource.

Wick et al. [18] propose a medium-sized Transformer (Forward Transformer) which has 13M parameters. In their work, they propose to duplicate the architecture to perform both a forward and backward scan of the text-line image, and reduce the error rates. The resulting bidirectional Transformer attains similar CER to the one obtained by our architecture. Furthermore, our contribution is compatible with their work by adding a backward Transformer which might improve the results obtained by our network.

To conclude, compared to other Transformer-based networks, our light Transformer architecture reaches state of the art results, no matter the amount of training data with significantly less parameters.

5 Conclusion and Future Works

We present a light encoder-decoder network based on Transformer-like architecture for handwritten text recognition. We highlight the relevance to use a hybrid loss combining linearly the connectionist temporal classification loss and the cross-entropy loss, which seems crucial for encoder-decoder architecture to be trained efficiently. Compared to other Transformer-based models, our architecture remains light in the number of parameters and does not require any additional data to be trained efficiently. Our network architecture reaches results at the level of state-of-the-art Transformer-based models, with a 5.70% CER on the IAM test set. Using synthetic data, our architecture is able to attain a 4.76% CER, close to the best performing network.

As future works, we would like to apply our architecture to historical documents in which the number of annotated data is even more critical. Using a light architecture might be beneficial considering the moderate amount of training data, while a Transformer-like architecture based on multi-head attention might be useful for even more difficult writings compared to modern texts. Synthetic data prove to be an efficient solution to the limited number of training data, but the context of historical documents will require a thorough design of the generation process of synthetic data.

Acknowledgments. This work was granted access to the HPC resources of IDRIS under the allocation 2021-AD011012550 made by GENCI.

References

1. Bhunia, A.K., Khan, S., Cholakkal, H., Anwer, R.M., Khan, F.S., Shah, M.: Handwriting transformers. In: Proceedings of the IEEE/CVF International Conference on Computer Vision, pp. 1086–1094 (2021)
2. Bluche, T., Louradour, J., Messina, R.: Scan, attend and read: end-to-end handwritten paragraph recognition with MDLSTM attention. In: 14th IAPR ICDAR, vol. 1, pp. 1050–1055. IEEE (2017)
3. Bluche, T., Messina, R.: Gated convolutional recurrent neural networks for multilingual handwriting recognition. In: 14th IAPR ICDAR, pp. 646–651. IEEE (2017)
4. Coquenet, D., Chatelain, C., Paquet, T.: SPAN: a simple predict & align network for handwritten paragraph recognition. In: Lladós, J., Lopresti, D., Uchida, S. (eds.) ICDAR 2021. LNCS, vol. 12823, pp. 70–84. Springer, Cham (2021). https://doi.org/10.1007/978-3-030-86334-0_5
5. Coquenet, D., Chatelain, C., Paquet, T.: End-to-end handwritten paragraph text recognition using a vertical attention network. IEEE Trans. Pattern Anal. Mach. Intell. (2022)
6. Doetsch, P., Zeyer, A., Ney, H.: Bidirectional decoder networks for attention-based end-to-end offline handwriting recognition. In: 15th ICFHR, pp. 361–366 (2016)
7. Graves, A., Fernández, S., Gomez, F., Schmidhuber, J.: Connectionist temporal classification: labelling unsegmented sequence data with recurrent neural networks. In: 23rd International Conference on Machine Learning, pp. 369–376. ACM (2006)
8. Graves, A., Liwicki, M., Fernández, S., Bertolami, R., Bunke, H., Schmidhuber, J.: A novel connectionist system for unconstrained handwriting recognition. IEEE Trans. Pattern Anal. Mach. Intell. **31**(5), 855–868 (2008)
9. Ingle, R.R., Fujii, Y., Deselaers, T., Baccash, J., Popat, A.C.: A scalable handwritten text recognition system. In: 15th IAPR ICDAR. IEEE (2019)
10. Kang, L., Riba, P., Rusiñol, M., Fornés, A., Villegas, M.: Pay attention to what you read: non-recurrent handwritten text-line recognition. arXiv preprint arXiv:2005.13044 (2020)
11. Marti, U.V., Bunke, H.: The IAM-database: an English sentence database for offline handwriting recognition. IJDAR **5**(1), 39–46 (2002)
12. Michael, J., Labahn, R., Grüning, T., Zöllner, J.: Evaluating sequence-to-sequence models for handwritten text recognition. In: 2019 International Conference on Document Analysis and Recognition (ICDAR), pp. 1286–1293. IEEE (2019)
13. Puigcerver, J.: Are multidimensional recurrent layers really necessary for handwritten text recognition? In: 14th IAPR ICDAR, vol. 1, pp. 67–72. IEEE (2017)
14. Singh, S.S., Karayev, S.: Full page handwriting recognition via image to sequence extraction. In: Lladós, J., Lopresti, D., Uchida, S. (eds.) ICDAR 2021. LNCS, vol. 12823, pp. 55–69. Springer, Cham (2021). https://doi.org/10.1007/978-3-030-86334-0_4
15. Soullard, Y., Swaileh, W., Tranouez, P., Paquet, T., Chatelain, C.: Improving text recognition using optical and language model writer adaptation. In: International Conference on Document Analysis and Recognition, pp. 1175–1180 (2019)
16. Strauß, T., Leifert, G., Labahn, R., Hodel, T., Mühlberger, G.: ICFHR 2018 competition on automated text recognition on a read dataset. In: 216th ICFHR, pp. 477–482. IEEE (2018)
17. Vaswani, A., et al.: Attention is all you need. In: NIPS, pp. 5998–6008 (2017)

18. Wick, C., Zöllner, J., Grüning, T.: Transformer for handwritten text recognition using bidirectional post-decoding. In: Lladós, J., Lopresti, D., Uchida, S. (eds.) ICDAR 2021. LNCS, vol. 12823, pp. 112–126. Springer, Cham (2021). https://doi.org/10.1007/978-3-030-86334-0_8
19. Yousef, M., Hussain, K.F., Mohammed, U.S.: Accurate, data-efficient, unconstrained text recognition with convolutional neural networks. Pattern Recogn. **108**, 107482 (2020)

Effective Crowdsourcing in the EDT Project with Probabilistic Indexes

Joan Andreu Sánchez(✉), Enrique Vidal, and Vicente Bosch

tranSkriptorium AI, Valencia, Spain
{jandreu,evidal,vbosch}@transkriptorium.com
http://www.transkriptorium.com

Abstract. Many massive handwritten document images collections are available in archives and libraries worldwide with their textual contents being practically inaccessible. Automatic transcription results generally lack the level of accuracy needed for reliable text indexing and search purposes if the recognition systems are not trained with enough training data. Creating training data is expensive and time-consuming. The European Digital Treasures project intended to explore crowdsourcing techniques for producing accurate training data. This paper explores crowdsourcing techniques based on Probabilistic Indexes. A crowdsourcing tool was developed in which volunteers could amend incorrectly transcribed words. Confidence measures were used to guide and help the users in the correction process. In further steps, this new corrected data will be used to re-train the Probabilistic Indexing system.

Keywords: Crowdsourcing · Handwritten text recognition · Probabilistic indexing

1 Introduction

In recent years, massive quantities of historical handwritten documents are being scanned into digital images which are then made available through web sites of libraries and archives all over the world. As a result of these efforts, many massive text *image* collections are available through Internet. The interest of these efforts not withstanding, unfortunately these document images are largely useless for their primary purpose; namely, exploiting the wealth of information conveyed by the text captured in the document images. Therefore, there is a fast growing interest in automated methods which allow the users to search for the relevant textual information contained in these images which is required for their needs.

The "European Digital Treasures (EDT): Management of centennial archives in the 21st century" project[1] aimed at bringing joint European heritage, especially its digital versions, major visibility, outreach and use. Three main goals were defined in the EDT project:

[1] https://www.digitaltreasures.eu.

© Springer Nature Switzerland AG 2022
S. Uchida et al. (Eds.): DAS 2022, LNCS 13237, pp. 291–305, 2022.
https://doi.org/10.1007/978-3-031-06555-2_20

- To conceptualize and generate new business models that seek the profitability and economic sustainability of the digitized heritage of archives.
- To foster the development of new audiences especially focused on two groups: the young and the elderly - the latter so-called "golden-agers" or the "silver generation" made up of retirees and citizens aged 60+.
- To promote the transnational mobility of managers, historians, experts, graphic artists, industrial designers and archivists, working on the production of new technological products and interactive exhibitions that give support and visibility to major European cultural areas.

In order to use classical text information retrieval approaches, a first step is to convert the text images into digital text. Then, image textual content can be straightforwardly indexed for plain-text textual access. However, state-of-the-art *handwritten text recognition* (HTR) techniques lack the level of accuracy needed for reliable text indexing and search purposes [3,8,13].

Given the current situation of the HTR technology that have previously been described, in the last decade the Probabilistic Indexing (PrIx) technique [2,4–6,10,11] has emerged as a solid technique for making the searching in document images a reality. This technique provides a nice *trade-off* between the *recall* and the *precision* that allows the user to locate most of the relevant information that s/he is looking for in large image collections [12].

The approaches proposed here are training-based and therefore need some amount (tens to hundreds) of manually transcribed images to train the required optical and language models. Producing this training material is expensive and requires a lot of time. Crowdsourcing techniques have been explored in the past to overcome this problem and currently there exist successful platforms that rely on this approach.[2] Unfortunately, the crowd that participate in the projects deployed in these platforms do not receive help from automatic HTR systems, and consequently, the effort from the people can be expensive and tedious. The EDT project had an activity to explore assisted crowdsourcing alternatives that help the crowd to create and validate GT for improving HTR results and PrIx results. This paper explores the use of PrIx to create a comfortable and effective crowdsourcing platform in which users could produce large amounts of high-quality GT data. This platform was used in the EDT project with five collections from five different European archives.

2 Preparing the PrIx System for the Crowdsourcing Platform

The crowdsourcing platform developed by tranSkriptorium AI for the EDT project was a web-based tool in which the basic technologies that were used are based on HTR and PrIx techniques. The following steps were followed to build this web platform for each individual collection:

[2] http://www.zooniverse.org
http://www.fromthepage.org.

1. An initial set of images was defined to create an initial GT dataset. The number of images was different for each collection and it depended on the archive resources. The selected images were annotated with regions and baselines inside these regions. Archives were interested just on the transcripts of some parts of the images, and not in the full images, because of budget restrictions. A specific layout analysis system was trained for each collection that focused in detecting lines in the adequate regions [7].
2. The characteristics of the GT for training an HTR system were agreed with each archives individually. Some collections were forms with printed parts. Annotating these parts was considered relevant for some of the archives in order to make easier the location of information according to this printed text (for example, locating a person with a given name).
3. An HTR system was trained for each collection with current technology based on Deep Convolutional and Recurrent Neural Networks.
4. A PrIx system with new capabilities for changing the transcripts was created for each archive.

3 Description of the Collections

Five collections were processed in this research in different languages: Hungarian, Norwegian, Portuguese, Spanish, and English. We now describe the characteristic of each one.

3.1 EDT Hungary

This collection is composed of table images that contain Hungarian proper names in a left-most image region. Figure 1 shows examples of these images. This region was automatically detected by layout analysis techniques and only text lines in this region were detected and extracted. Significant difficulties of this collection include:

- Layout: these tables have a specific but fairly regular layout. They contain both printed and handwritten text, but most cells are typically empty. Only the proper names located in the left-most image region were interesting in the EDT project. Each proper name is preceded by a handwritten number that has to be avoided in the line detection process to prevent recognition problems.
- Optical modeling: words contain many diacritics and the text has been written by very many hands. This leads to a very high writing style variability.
- Language modeling: the handwritten text are mainly proper names and most of the them are abbreviated in a not consistent way.

For the GT preparation, the left columns were annotated with a rectangular region. The main figures of the dataset that was used for preparing the HTR and PrIx systems are reported in Table 1.

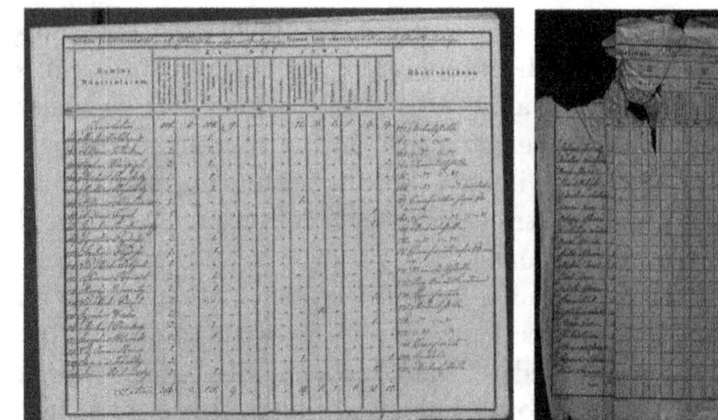
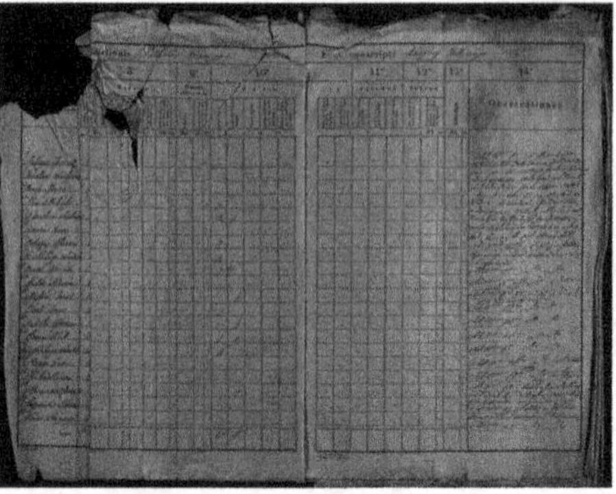

Fig. 1. Examples of text images from the EDT Hungary collection.

Table 1. Main values of the EDT Hungary dataset.

	Train	Validation	Test	Total GT
Images				410
Lines	7 000	800	672	8 472
Vocabulary	4 168	770	657	4 768
Character set				74
Running words	14 000	1 684	1 403	17 687
Running chars.	119 667	13 804	11 396	144 867

3.2 EDT Norway

This dataset is composed of images of register cards that contain mainly Norwegian proper names. Figure 2 shows examples of these images. Significant difficulties of this collection include:

- Layout: these cards have a fairly regular layout, but it is fairly complex. Both printed and handwritten text is considered for recognition. Each card contains record space for two persons, which appear in separate (left and right) image regions. However, handwritten information can be given for just one person or for both. Most card fields are typically empty and some stamps appear in many cards.
- Optical modeling: some special characters and diacritics are used and the cards have been filled by very many hands. This leads to very high writing style variability, and more so for the size of the initial capitals, which tends to be much larger than the main text body.
- Language modeling: the handwritten text contains many proper names and dates. Many date formats are used, but the archive wished to handle dates in a standard format. Only selected information items of the cards were interesting for the archive.

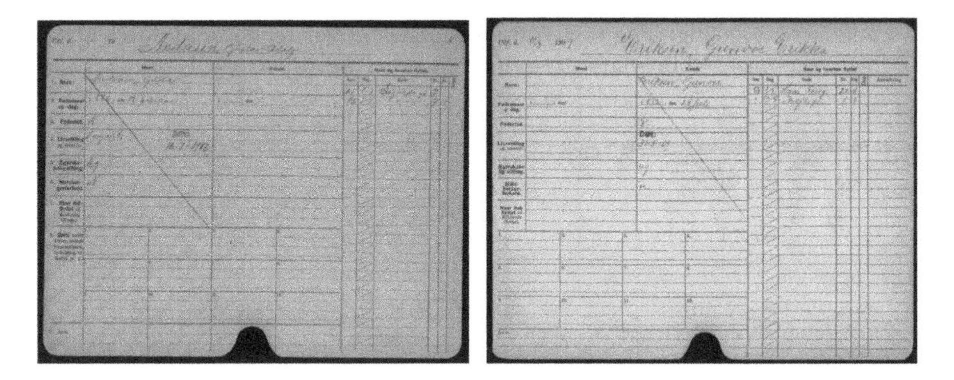

Fig. 2. Examples of text images from the EDT Norway collection.

For the GT preparation, the left column of each card was annotated with a rectangular region. Then, the space for two persons were also annotated with rectangular regions along with the proper name in the header. This made easier the training of a specific layout analysis system. Lines were detected only inside these regions. The main figures of this dataset are reported in Table 2. Printed and handwritten characters were modeled without distinction.

Table 2. Main values of the EDT Norway dataset.

	Train	Validation	Test	Total GT
Images				36
Lines	5 000	243	200	5 443
Vocabulary	1 588	180	131	1 676
Character set				74
Running words	12 069	581	328	13 100
Running chars.	65 405	3 195	2 479	71 079

3.3 EDT Portugal

This dataset is composed of grayscale images with running text written in Portuguese. Figure 3 shows examples of these images. Significant difficulties of this collection include:

- Layout: baseline detection is difficult because text lines generally exhibit a great amount of warping, along with very variable slopes and slant. Layout becomes often complex because of plenty marginalia and other more complex layout structures. Many images include parts of adjacent pages.
- Optical modeling: Text include many diacritics and has been written by several hands leading to a significant amount of writing style variability.

- Language modeling: The text contains many proper names and dates and a great amount of abbreviations and hyphenated words.

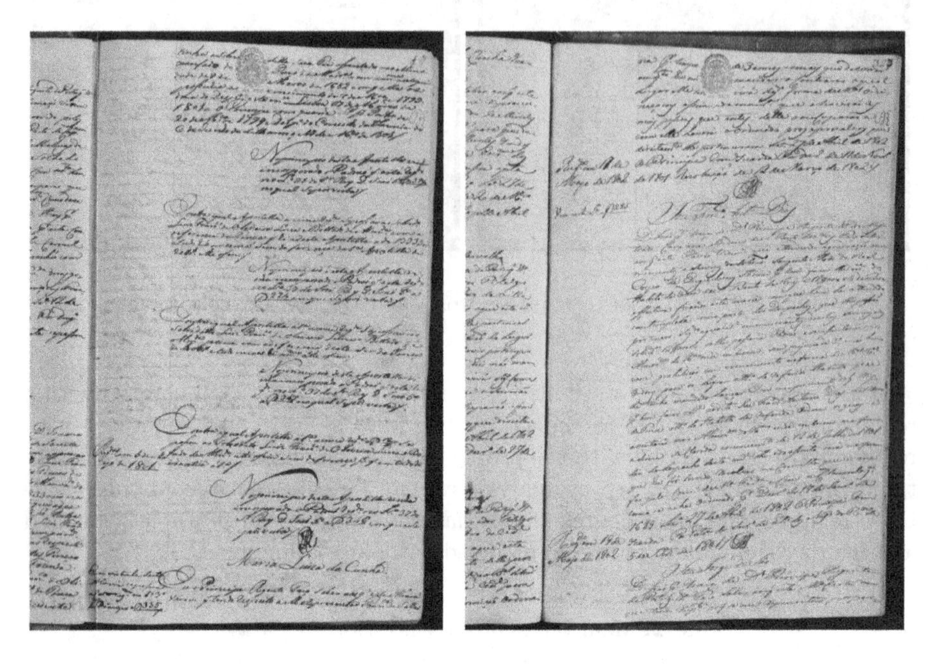

Fig. 3. Examples of text images from the EDT Portugal collection.

The main figures of this dataset that are used for preparing the HTR system are reported in Table 3.

Table 3. Main features of the EDT Portugal dataset.

	Train	Validation	Test	Total GT
Images				36
Lines	1 200	122	92	1 414
Vocabulary	2 034	526	424	2 281
Character set				78
Running words	11 338	1 131	816	13 285
Running chars.	62 785	6 359	4 685	73 829

3.4 EDT Spain

This dataset is composed of images that contain visa records of Spanish citizens for traveling worldwide. They were issued between years 1936 and 1939 in a

Spanish consulate based in Buenos Aires (Argentina). Figure 4 shows examples of these images. Significant difficulties of this collection include:

- Layout: Each image contains four visa forms and each form includes one (or several) picture(s) associated to each visa. Some dates are stamped and others are handwritten. Text lines often exhibit extreme slope.
- Optical modeling: Forms include printed text and are filled with handwritten text written by many hands, leading to a very high writing style variability.
- Language modeling: the handwritten text contains many proper names (given names, surnames, town and state names, countries, etc.) and dates. A large amount of words are heavily abbreviated. All the textual information in each visa was relevant for the archive.

Fig. 4. Examples of text images from the EDT Spain collection.

A small set of 99 images of the whole collection were selected for GT annotation. For layout analysis, each image was annotated with four rectangular regions to isolate each visa, and then the geometric data of each photograph region and all the baselines were annotated. Then each line was manually transcribed and annotated word by word with "semantic" tags.

The tags that have been used in this dataset and their meanings are:

- `<print>`: printed word of the form
- `<date>`: date, both stamped or handwritten
- `<gname>`: given name
- `<surname>`: surname (two surnames are used in Spanish)
- `<state>`: province
- `<country>`: country
- `<civilstate>`: civil state (single, married, etc.)
- `<residence>`: place of residence
- `<place>`: location (city, village, etc.)
- `<job>`: occupation
- `<age>`: years old

The main values of this dataset are reported in Table 4. Since the amount of handwritten text and printed text can significantly affect the recognition results, this table provides the amount of both types of text. Experiments will be reported without taking into account this difference. Since the tags are considered special characters, the same word with two different tags was considered two different words. This fact explains the large vocabulary that can be observed in the "Hand" columns.

Table 4. Main features of the EDT Spanish dataset. The vocabulary is shown both with and without numbers

	Train		Validation		Test		Total GT	
	Print	Hand.	Print	Hand.	Print	Hand.	Print	Hand.
Images							99	
Lines	6 500		250		261		7 011	
Vocabulary w numbers	317	2 900	48	254	49	276	333	3 057
Vocabulary w/o numbers	53	2 040	36	203	37	202	53	2 149
Character set							85	
Running words	14 662	10 530	575	390	602	455	15 839	11 375
Running chars.	108 226	72 019	4 329	2 585	4 503	3 046	117 058	77 650

3.5 EDT Malta

This collection is composed of grayscale images with lists of proper names and the name of the flight or the ship in which the people arrived to Malta. Figure 5 shows examples of these images. Significant difficulties of this collection include:

- Layout: the lines have quite slope, slant, skew and warping. Many images are physically degraded with many holes because of woodworm and other insects. Many images include parts of adjacent pages.
- Optical modeling: there are fainted text and several hands, and many flights and ship names are replaced with quotes (").
- Language modeling: the text contains mainly proper names and dates.

The main figures of this dataset used for preparing the HTR system are reported in Table 5.

4 HTR and PrIx Experiments and Results

An HTR system was trained for each collection using the prevalent technology based on Convolutional and Recurrent Neural Networks for optical modeling and N-gram for language modeling [1]. The datasets introduced in Sect. 3 were used to prepare an HTR system for each collection. These HTR systems were evaluated using the partitions described for each collection. CER and WER were computed and the results are shown in Table 6. Several values of N were

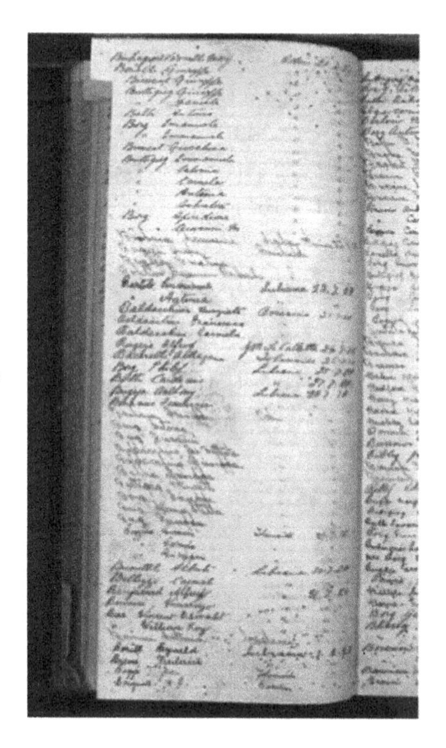
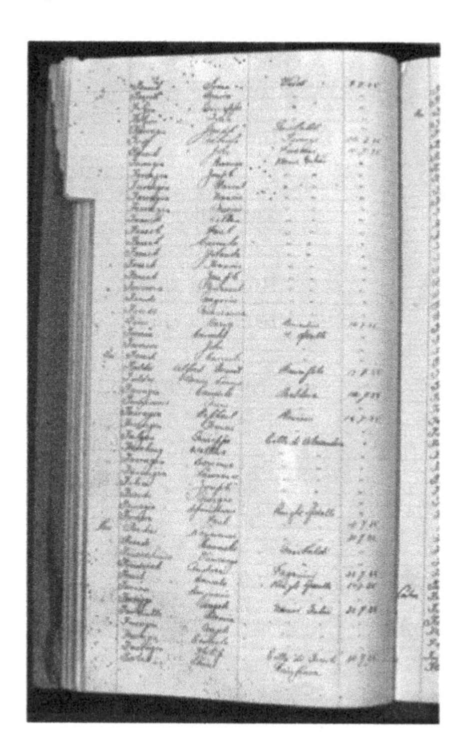

Fig. 5. Examples of text images from the EDT Malta collection.

used for the N-gram LM but only the results obtained with $N = 5$ are shown since this was the best value or other values of $N > 5$ did not get a significant improvement.

We observe that both CER and WER have large difference among collections. The worst results were obtained for the Portugal and Malta collections. In the case of the Portugal collection, the bad WER results were due mainly to the large amount of abbreviated forms. This WER could be decreased by using additional GT. In the case of the Malta collection, the problem was again with the amount of training data. Additional detail can be consulted in [9]. It is important to remark that if a searching system was used with these results, the obtained results could be disappointing for the user, since many results would lost. The PrIx try to overcome this situation.

In "*Probabilistic Indexing*" (PrIx) [2,4–6,11,12]. For each text image, a kind of "word heat map" is created. Each pixel in this map indicates the greater or lesser probability that this pixel is part of one or, generally, many possible words or character sequences with lexical sense. From these maps a PrIx is created as a list of hypotheses of character sequences (words) probably written in the image, along with their probability and location in the image.

PrIx makes possible to detect precise textual information for a given query and confidence threshold, even in the worst conditions of quality of the

Table 5. Main features of the EDT Malta dataset.

	Train	Validation	Test	Total GT
Images				49
Lines	2 200	230	101	2 531
Vocabulary	3 249	551	274	3 614
Character set				77
Running words	9 996	1 037	448	11 481
Running chars.	65 376	6 829	2 971	75 176

Table 6. CER and WER obtained for each collections.

Dataset	CER	WER
Hungary	8.8	25.7
Norway	5.6	13.4
Portugal	14.5	35.4
Spain	9.4	20.8
Malta	12.8	36.6

documentation or its digitization. This allows to carry out simple or complex searches based on expanded abbreviations and modernized word versions. By varying the confidence threshold, different related values of recall and precision can be obtained. Results are also reported in terms of overall average precision (AP), which are obtained by computing the area under Recall-Precision curves.

The average precision for the previous datasets and the corresponding R-P curves are shown in Fig. 6. The ideal situation would be a curve similar to the curve obtained in the Norway collection where the are under the curve is large.

The PrIx can provide more or less alternatives for a given word image. It depends on some parameters (specially the word graph density), but in general the more confusing is the word image, the more alternatives it has. If the parameters are adjusted to have lots of hypotheses then the recall can drop faster (see details in [5]).

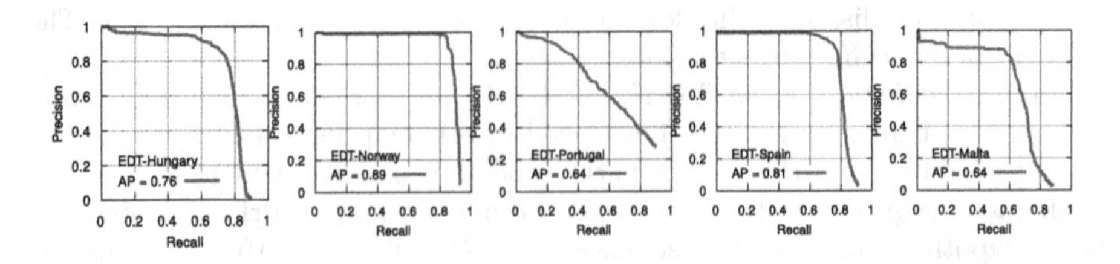

Fig. 6. R-P curves for the EDT datasets.

5 Crowdsourcing Platform

The main goal in this research was to produce new GT to retrain the HTR and PrIx systems from new data for each collection in order to improve the search results as much as possible. A crowdsourcing platform was prepared for each collection of the EDT project. It is important to remark that the full collection for each archive contained thousands of images. The archives that participate in the EDT project were in charge of choosing the pages to be reviewed by the crowd. They were also in charge of all recruitment process and in assisting the participants. This is noticiable because we did not have direct contact with final users and were not able to evaluate the platform performance by asking directly to final volunteers. We explain just the Spanish platform because all platforms are similar.

Before starting the crowdsourcing process, the whole collection of each archive was processed: first, automatic line detection was applied to each image with very good results [7]; then, PrIx were obtained for the detected lines. Once the platform was ready for each archive the reviewing process started.

The reviewing process was thought up in the following steps:

1. A volunteer chooses an image and ask for all spots to the PrIx system for which the confidence has a high score (e.g. 95%). The system provides all spots and the corresponding transcripts. (see Fig. 7 top). Printed words are not shown because they are all perfectly recognized and too much bounding boxes disturbs the users. But this is optional.
2. If the user is not satisfied with the resulting transcripts, then s/he can amend the errors by choosing a different alternative or writing the correct word in the edit box. (see Fig. 7 bottom).

When the user is satisfied with all shown spots, then s/he can accept the showed transcripts and they are marked as GT. The accepted transcripts change to a different color. The two previous steps are iteratively repeated by decreasing the threshold. Figure 8 shows the following step when the threshold is decreased to (e.g. 50%).

This process is repeated until the whole image is transcribed or until the user is satisfied with the generated GT. One problem with this platform was the situation in which a line was not correctly detected or the confidence was too low and some word was not included in the PrIx process. For such situation the user could click on the desired word image and then a bounding box was automatically created and the user could write the correct transcript associated to that position. Initially, they were able to adjust the bounding box but we removed this option because it was time-consuming.

At the time of writing this paper, the crowdsourcing process was just finished. In the case of the Spanish collection, it was composed by about 1 000 and the volunteers transcribed the whole collection in few weeks. Once this reviewing process was closed, the following step was to collect the GT prepared by participants and to analyze it to re-train the system.

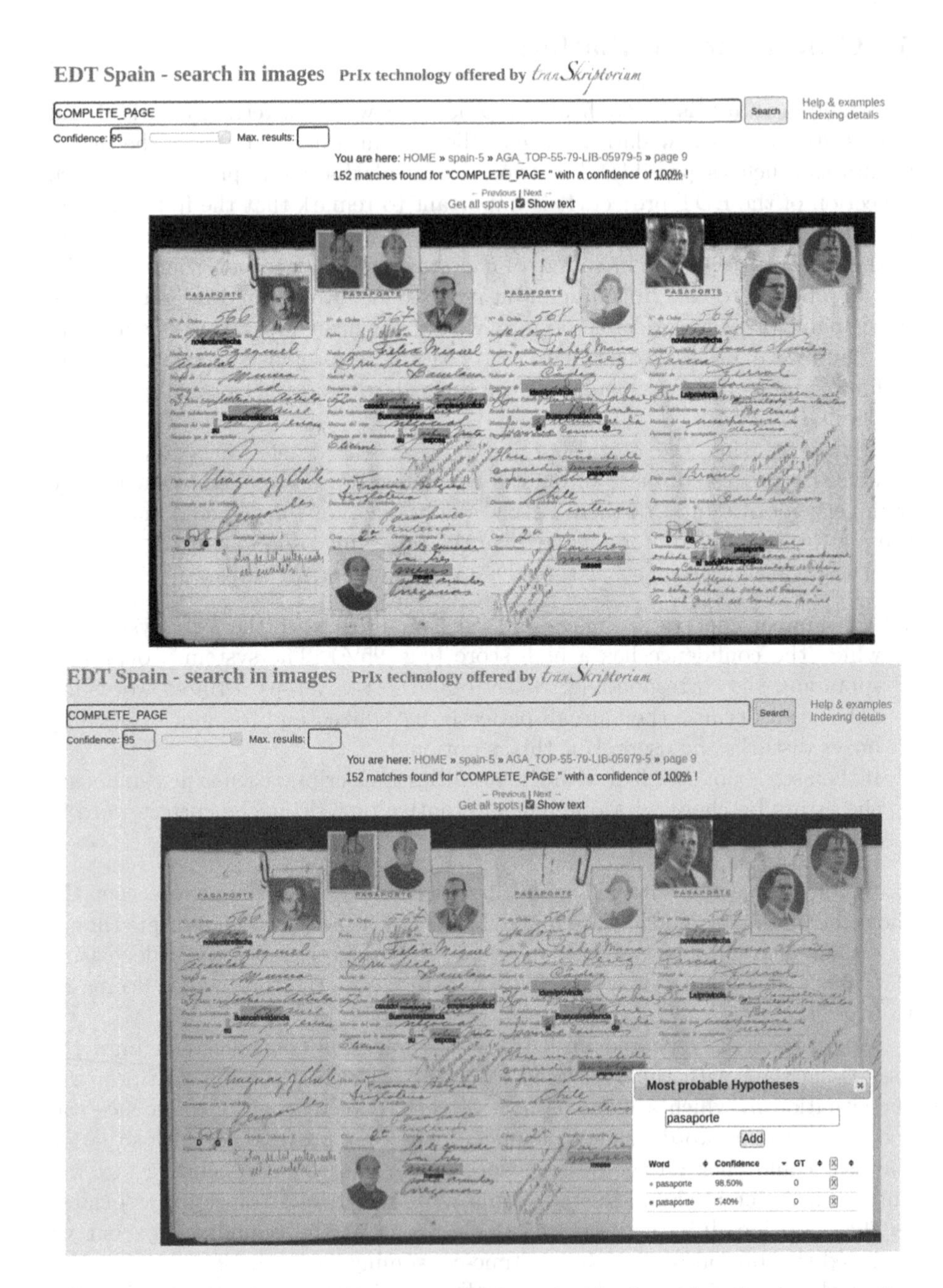

Fig. 7. Screenshot of the crowdsourcing platform for the Spanish collection when the confidence threshold was 95%.

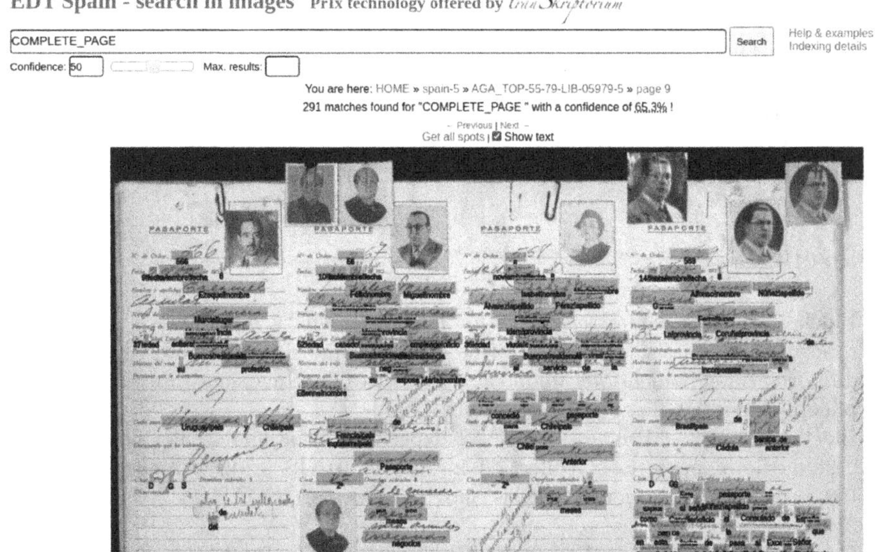

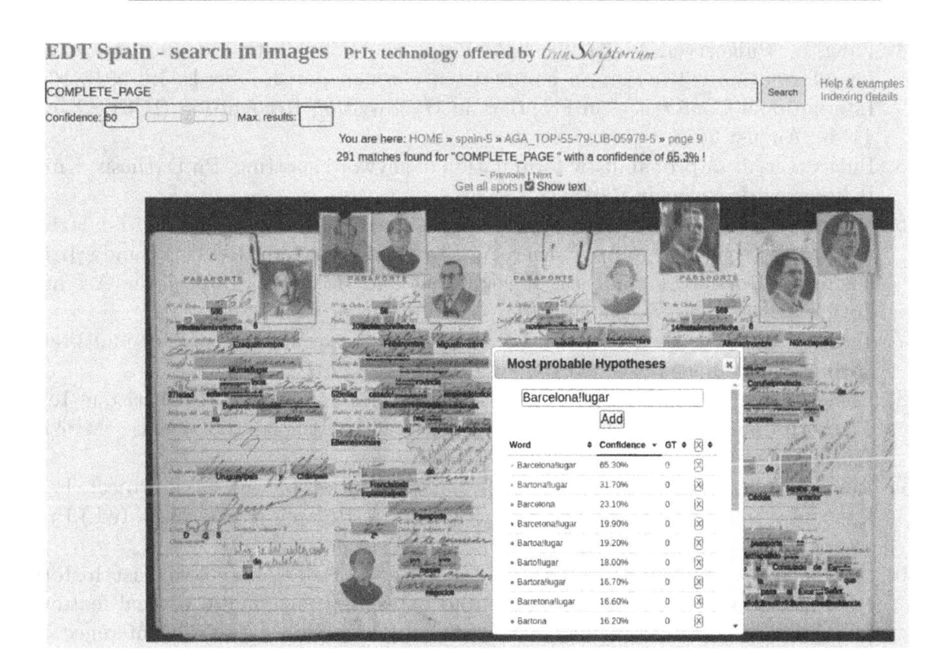

Fig. 8. Screenshot of the crowdsourcing platform for the Spanish collection when the confidence threshold was 50%.

6 Conclusions

A crowdsourcing platform has been introduced for HTR that is being used in the European EDT project. The platform is currently being used by final users. The transcripts provided by final users will be used to retrain the HTR system. As future work we intend to analyze the transcripts generated by final users and research the benefits of this transcripts to retrain the HTR system.

References

1. Bluche, T.: Deep neural networks for large vocabulary handwritten text recognition. Ph.D. thesis, Ecole Doctorale Informatique de Paris-Sud - Laboratoire d'Informatique pour la Mécanique et les Sciences de l'Ingénieur, May 2015. Discipline: Informatique
2. Bluche, T., et al.: Preparatory KWS experiments for large-scale indexing of a vast medieval manuscript collection in the HIMANIS project. In: International Conference on Document Analysis and Recognition (ICDAR), vol. 01, pp. 311–316, November 2017
3. Graves, A., Liwicki, M., Fernández, S., Bertolami, R., Bunke, H., Schmidhuber, J.: A novel connectionist system for unconstrained handwriting recognition. IEEE Trans. Pattern Anal. Mach. Intell. **31**(5), 855–868 (2009)
4. Lang, E., Puigcerver, J., Toselli, A.H., Vidal, E.: Probabilistic indexing and search for information extraction on handwritten German parish records. In: 2018 16th International Conference on Frontiers in Handwriting Recognition (ICFHR), pp. 44–49, August 2018
5. Puigcerver, J.: A probabilistic formulation of keyword spotting. Ph.D. thesis, Univ. Politècnica de València (2018)
6. Puigcerver, J., Toselli, A.H., Vidal, E.: Advances in handwritten keyword indexing and search technologies. In: Fischer, A., Liwicki, M., Ingold, R. (eds.) Handwritten Historical Document Analysis, Recognition, and Retrieval-State of the Art and Future Trends, vol. 89, pp. 175–193. World Scientific (2020)
7. Quirós, L., Vidal, E.: Evaluation of a region proposal architecture for multi-task document layout analysis. CoRR, abs/2106.11797 (2021)
8. Romero, V., Toselli, A.H., Vidal, E.: Multimodal Interactive Handwritten Text Transcription. Series in Machine Perception and Artificial Intelligence (MPAI). World Scientific Publishing (2012)
9. Sánchez, J.A., Vidal, E.: Handwritten text recognition for the EDT project. Part I: model training and automatic transcription. In: Bermejo, M.A., et al. (ed.) Proceedings of the EDT Alicante Workshop (2021, to appear)
10. Toselli, A.H., Romero, V., Vidal, E., Sánchez, J.A.: Making two vast historical manuscript collections searchable and extracting meaningful textual features through large-scale probabilistic indexing. In: 15th International Conference on Document Analysis and Recognition (ICDAR) (2019)
11. Vidal, E., et al.: The Carabela project and manuscript collection: large-scale probabilistic indexing and content-based classification. In: 17th International Conference on Frontiers in Handwriting Recognition (ICFHR), pp. 85–90 (2020)

12. Vidal, E., Sánchez, J.A.: Handwritten text recognition for the EDT project. Part II: textual information search in untranscribed manuscripts. In: Bermejo, M.A., et al. (ed.) Proceedings of the EDT Alicante Workshop (2021, to appear)
13. Vinciarelli, A., Bengio, S., Bunke, H.: Off-line recognition of unconstrained handwritten texts using HMMs and statistical language models. IEEE Trans. Pattern Anal. Mach. Intell. **26**(6), 709–720 (2004)

Applications in Handwriting

Paired Image to Image Translation for Strikethrough Removal from Handwritten Words

Raphaela Heil[1(✉)], Ekta Vats[2], and Anders Hast[1]

[1] Division for Visual Information and Interaction, Department of Information Technology, Uppsala University, Uppsala, Sweden
{raphaela.heil,anders.hast}@it.uu.se

[2] Centre for Digital Humanities Uppsala, Department of Archives, Libraries and Museums, Uppsala Universtiy, Uppsala, Sweden
ekta.vats@abm.uu.se

Abstract. Transcribing struck-through, handwritten words, for example for the purpose of genetic criticism, can pose a challenge to both humans and machines, due to the obstructive properties of the superimposed strokes. This paper investigates the use of paired image to image translation approaches to remove strikethrough strokes from handwritten words. Four different neural network architectures are examined, ranging from a few simple convolutional layers to deeper ones, employing Dense blocks. Experimental results, obtained from one synthetic and one genuine paired strikethrough dataset, confirm that the proposed paired models outperform the CycleGAN-based state of the art, while using less than a sixth of the trainable parameters.

Keywords: Strikethrough removal · Paired image to image translation · Handwritten words · Document image processing

1 Introduction

Struck-through words generally appear at a comparably low frequency in different kinds of handwritten documents. Regardless of this, reading what was once written and subsequently struck through can be of interest to scholars from the humanities, such as literature, history and genealogy [11]. In order to facilitate strikethrough-related research questions from such fields, strikethrough removal approaches have been proposed in the area of document image analysis [4,9,24].

In this work, we approach the problem from a new perspective, employing a paired image to image translation method. The underlying idea of this family of models consists of the use of two corresponding images, one from the source and one from the target domain [12]. Using these paired images, deep neural networks are trained to learn the transformation from the source to the target domain. As indicated by the name, this approach is generally limited by the availability of paired data. This poses a particular challenge when applying paired approaches

© Springer Nature Switzerland AG 2022
S. Uchida et al. (Eds.): DAS 2022, LNCS 13237, pp. 309–322, 2022.
https://doi.org/10.1007/978-3-031-06555-2_21

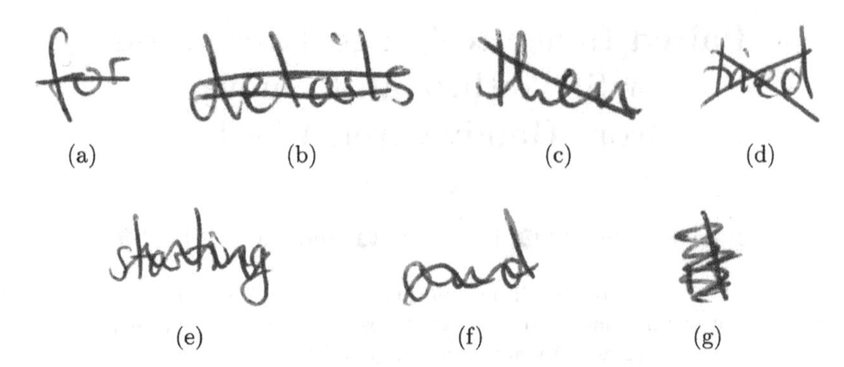

Fig. 1. Examples for the types of strikethrough strokes considered in this work. a) Single, b) Double, c) Diagonal, d) Cross, e) Zig Zag, f) Wave, g) Scratch. All samples are taken from the test set of $Dracula_{real}$, i.e. displaying genuine strikethrough stokes.

to the task of strikethrough removal. Once a word has been struck through, the original, clean word is permanently altered and effectively becomes unobtainable from the manuscript itself. In order to mitigate this problem, we examine the paired image to image translation setting using a combination of synthetic and genuine data. The general approach is implemented and examined via a selection of four deep neural networks of varying size and architectural complexity. All models are evaluated using a synthetic and a genuine test dataset [9].

The main contributions of this work are as follows. This work advances the state of the art in strikethrough removal by using a paired image to image translation approach. To the best of the authors' knowledge, this is the first attempt at using such an approach towards strikethrough removal. In order to overcome the issue of insufficient paired data, this work uses a combination of synthetic and genuine datasets, and also introduces a new dataset ($Dracula_{synth}$) to further supplement the training data. Four different models are evaluated and compared with the CycleGAN-based state of the art [9], using two existing datasets [7,8] and the new $Dracula_{synth}$ dataset. The $Dracula_{synth}$ dataset, the source code and the pre-trained models will be made publicly available.

2 Related Works

2.1 Strikethrough Processing

At the time of writing, three approaches for removing strikethrough strokes from handwritten words have been proposed. Firstly, Chaudhuri et al. introduce a graph-based approach, identifying edges belonging to the strikethrough stroke and removing them, using inpainting [4]. They report F_1 scores of 0.9116 on a custom, unbalanced database, containing strikethrough strokes of the types *single*, *multiple*, *slanted*, *crossed*, *zig zag* and *wavy* (cf. Fig. 1). Chaudhuri et al. do not consider the stroke type that we denote *scratch*.

In addition to this, Poddar et al. [24] employ a semi-supervised, generative adversarial network (GAN), which is trained using a combination of GAN-loss, Structural Similarity Index and L_1-norm. They report an average F_1 score of 0.9676 on their own synthetic IAM database, which contains strokes of types *straight*, *slanted*, *crossed*, *multiple straight* and *partial straight* and *partial slanted* strokes. In contrast to Poddar et al., we also consider strikethrough strokes of the types *wave*, *zig zag* and *scratch* in this work. As can be seen in Fig. 1, the latter strokes are considerably more challenging that the former ones.

Lastly, Heil et al. investigate the use of attribute-guided cycle consistent GANs (CycleGANs) [9]. They report F_1 scores of up to 0.8172, respectively 0.7376, for their synthetic and genuine test sets. In contrast to the two former approaches, the datasets and code from Heil et al. are publicly available [6–8]. We reuse their datasets in this work as the basis for our experiments as well as to compare our results to that of [9].

Besides the removal of strikethrough, a number of works have concerned themselves with other aspects related to the processing of struck-through words. Various approaches [4,9,23,25], have been proposed to classify whether a given image depicts a struck-through or clean word. In addition to this, a number of works have examined the extent to which strikethrough impacts the performance of writer identification [1] and text and character recognition approaches [2,15,18]. Some of the aforementioned works are briefly summarised in a survey from early 2021 by Dheemanth Urs and Chethan [5].

2.2 Paired Image to Image Translation

A variety of approaches for paired image to image translation have been proposed over the years. One prominent example, employing a conditional GAN, is *Pix2Pix* [12]. In the field of document image analysis, paired image to image translation approaches have for example been used in the form of auto-encoders to remove various types of noise [17,26], as well as to binarise manuscript images [3,21]. To the best of our knowledge, paired image to image translation approaches have not yet been used for the task of strikethrough removal.

2.3 Strikethrough Datasets

While strikethrough occurs naturally in a variety of real-world datasets, such as the IAM database [16], the amount and diversity are generally too small to be used to efficiently train larger neural networks. At the time of writing, only two datasets, which focus specifically on strikethrough-related research questions, are publicly available. The first one, which we will refer to as IAM_{synth} [7] in this work, is an IAM-based [16] dataset consisting of genuine handwritten words that have been altered with synthetic strikethrough strokes [9]. In addition to this, a smaller, but genuine, dataset of struck-through, handwritten words exists. This single-writer dataset, which we will refer to as $Dracula_{real}$ [8], contains handwritten word images and their struck-through counterparts from Bram Stoker's

Dracula. Heil et al. [9] collected the genuine dataset by scanning and aligning handwritten words, via [27], before and after the strikethrough was applied.

3 Image to Image Translation Models for Strikethrough Removal

In contrast to earlier works [9,24], we approach the task of strikethrough removal using paired image to image translation. For this, a model receives an image from the source domain as input and is trained to reproduce the image as it would appear in the target domain. In this work, we propose to consider struck-through images as the source and cleaned images, i.e. without strikethrough strokes, as the target domain.

We examine four different deep neural network architectures in the paired image to image translation setting and compare their performance to the attribute-guided CycleGANs proposed by Heil et al. [9]. The chosen architectures were selected with the aim of exploring models with a range of layer arrangements and varying amounts of trainable parameters. A brief overview of the architectures, and the names by which we will refer to them for the remainder of this work, are given below. Furthermore, Fig. 2 presents a schematic summary. For a more detailed description of the architectures, such as number of convolutional filters per layer, stride, padding, etc., the interested reader is referred to the code, accompanying this paper (cf. section A - Dataset and Code Availability).

SimpleCNN. As indicated by the name, this model constitutes a simple convolutional neural network, consisting of three up- and down-sampling layers with few (16, respectively 32) filters.

Shallow. This network consists of two convolutional down-, respectively up-sampling layers. It does not contain any intermediate bottleneck layers. The arrangement of layers in this architecture is the same as the outer layers in the *Generator* one.

UNet. This dense UNet [13] consists of one down- and up-sampling, as well as one bottleneck block, each of which consisting of a Dense block with four dense layers. The two outer dense blocks are furthermore connected via a skip connection. We use the implementation provided by [22].

Generator. This network uses the same architecture as the generator that was used in the attribute-guided CycleGAN by Heil et al. [9]. It consists of three convolutional up-, respectively down-sampling layers, with a Dense [10] bottleneck in between. Each of the convolutional layers is followed by a batch normalisation step, as well as a rectified linear unit (ReLU).

With the exception of *UNet*, all of the models above use a sigmoid as the final activation function. For *UNet*, we follow the original implementation, provided by *OctoPyTorch* [22], which uses the identity function as activation and combines the network with a *binary cross entropy with logits loss.*

Table 1 summarises the trainable parameter counts for the described models, as well as the attribute-guided CycleGAN from [9]. As can be seen from the table the chosen models cover a considerable range of trainable parameter amounts.

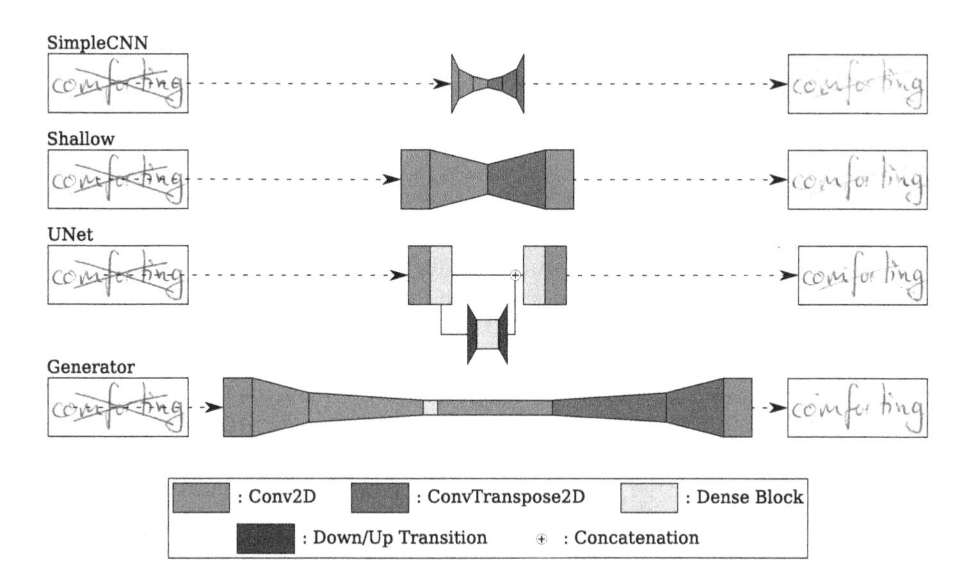

Fig. 2. Schematic overview over the four paired architectures examined in this work. Red boxes are regular 2D convolutional layers, blue transposed convolutions, green up-, respectively down-sampling layers and yellow ones dense blocks. Height of the boxes represents an approximate measure of the input, respectively output size, while the width indicates the number of convolutional filters (Conv2D and ConvTranspose2D), respectively number of dense layers (Dense Block). (Color figure online)

Table 1. Number of trainable parameters per model. It should be noted that the CycleGAN parameter count includes the generators *and* discriminators but not the pre-trained auxiliary discriminator.

Model name	Parameter count
SimpleCNN	28 065
Shallow	154 241
UNet	181 585
Generator	1 345 217
Attribute-guided CycleGAN [9]	8 217 604

Table 2. Summary of the three datasets used in this work. Numbers indicate the amount of words contained in each dataset split.

Dataset	Train	Validation	Test	Multi-writer
IAM_{synth}	3066	273	819	Yes
$Dracula_{real}$	126	126	378	No
$Dracula_{synth}$	5×126	N/A	N/A	No

4 Experiment Setup

4.1 Datasets

As mentioned above, at the time of writing, two strikethrough-related datasets, IAM_{synth} and $Dracula_{real}$, are publicly available. We base our experiments on these and introduce an additional synthetic dataset. The latter one, which we will refer to as $Dracula_{synth}$, consists of the clean images from the $Dracula_{real}$ training split, to which synthetic strikethrough was applied via the method proposed in [9]. We repeat the generation process five times, using different seed values for the random number generator, resulting in five different strikethrough strokes for each individual word image. In addition to combining all of these generated images into one large training set, we also consider the separate partitions, numbered 0 to 4, each of which contains only one instance of any given word image. It should be noted that $Dracula_{synth}$ is used exclusively during the training phase. For validation and testing, the original images from $Dracula_{real}$ are used. However, if necessary for future investigations, synthetic strikethrough can also be generated for the original validation and test words, using the approach outlined above. Table 2 summarises the three described datasets with respect to the number of images per split and the number of writers.

4.2 Neural Network Training Protocol

All models are implemented in PyTorch 1.7 [20] and are trained for a total of 30 epochs with a batch size of four, using the Adam [14] optimiser with default parameters. We use a regular binary cross entropy loss for all models, except *UNet*, for which the binary cross entropy with logits loss is used, as mentioned above. Each model is trained from scratch using each of the following datasets:

- IAM_{synth} training split
- $Dracula_{synth}$ training partitions 0–4, each individually
- a combination of all $Dracula_{synth}$ training partitions, i.e. 630 images

For all of the models, we monitor the performance, measured via the F_1 score, using the validation split of the respective dataset (IAM_{synth} or $Dracula_{real}$). We retain each model's weights from the best performing epoch for the evaluation on the respective test split. Each model is separately retrained 30 times, yielding 30 sets of weights per model and dataset combination.

In addition to the models described above, we also train a number of CycleGANs, following the procedure described in [9]. As above, each model is retrained 30 times.

Regardless of architecture, all images are, where necessary, converted to greyscale, inverted so that ink pixels have the highest intensities in the image, and are scaled to a height of 128 and padded with the background colour (black) to a fixed width of 512.

4.3 Evaluation Protocol

Each of the trained models is applied to the struck-through images from the test split of $Dracula_{real}$. The resulting image is inverted and rescaled to its original dimensions and additional padding is removed, where applicable. Subsequently, the cleaned image is compared with the corresponding ground-truth, calculating the *Root Mean Square Error* (RMSE) of the greyscale images, as well as the F_1 score of the Otsu-binarised [19] ones. We calculate the F_1 score following the formula provided by [4], which is repeated below:

$$DetectionRate(DR) = \frac{O2O}{N} \tag{1}$$

$$RecognitionAccuracy(RA) = \frac{O2O}{M} \tag{2}$$

$$F_1 = \frac{2 * DR * RA}{DR + RA} \tag{3}$$

where M is the number of ink pixels in the image, cleaned by the respective model; N the number of ink pixels in the ground-truth image, and $O2O$ the number of matching pixels between the two images.

Unless otherwise stated, we report the average and standard deviation, summarised over 30 repeated training runs for each of the model and dataset combinations.

5 Results and Analysis

In the following sections, the performances of the four paired image to image translation models are presented, compared with the *CycleGAN*'s results and analysed. Besides the quantitative evaluation in Subsects. 5.1 to 5.3, a number of qualitative results are also presented in Subsect. 5.4.

5.1 Models Trained on IAM_{synth}

In a first step, we compare the four paired models with the unpaired CycleGAN approach, when trained on the IAM_{synth} dataset and evaluated on the IAM_{synth} and $Dracula_{real}$ test splits. Tables 3 and 4 present the respective mean F_1 and *RMSEs* scores for the five architectures, summarising the 30 repeated training runs. Values reported for [9] stem from our own experiments, using the code by Heil et al. [6].

As can be seen from Table 3, there is some degree of variation between the four paired models, when evaluating them on the IAM_{synth} test split. However, all of them outperform the attribute-guided CycleGAN by a considerable margin. For the F_1 score, improvements range from 7.4% points (pp) for the *SimpleCNN*, up to roughly 17 pp for the *Generator*. Similarly, the former model outperforms the *CycleGAN* by approximately 4 pp with respect to the *RMSE*, while the latter achieves a performance improvement of 9 pp. Comparing the four paired approaches with each other, the two larger models *Generator* and *UNet* prominently outperform the two smaller ones.

Considering the model performances on the $Dracula_{real}$ test split, shown in Table 4, it can be noted that the four paired architectures do not drastically differ from each other. For the F_1 score, the *SimpleCNN* and *Generator* constitute the lower and upper bounds, respectively. Interestingly, with regard to the *RMSE*, *SimpleCNN* performs slightly better than *Generator*, which ranks second in the overall comparison of values. A possible explanation for this could be that *Generator* leaves behind traces of strikethrough which are removed by the binarisation, required for the F_1 score, but remain visible and make an impact in the *RMSE* calculation.

In the overall ranking, the *CycleGAN* reaches the fifth (F_1), respectively third position (*RMSE*), consistently being outperformed by the *Generator* model with a margin of 3.8 pp for the F_1 score and roughly 0.5 pp for the *RMSE*. Although this gain in cleaning performance is relatively small, the *Generator* includes the additional benefit of having less than a sixth of the *CycleGAN*'s trainable parameters. Taking the amount of trainable parameters further into account, the *Shallow* and *UNet* models provide a good trade-off between performance and size in this evaluation setting.

Table 3. Mean F_1 scores (higher better, range [0,1]) and *RMSE*s (lower better, range [0,1]) for the five architectures, trained and evaluated on the train and test splits, respectively, of IAM_{synth}. Standard deviation over thirty training runs given in parentheses. Best model marked in bold.

Model	F_1	RMSE
SimpleCNN	0.8727 (± 0.0042)	0.0753 (± 0.0025)
Shallow	0.9163 (± 0.0045)	0.0558 (± 0.0025)
UNet	0.9599 (± 0.0015)	0.0301 (± 0.0012)
Generator	**0.9697 (± 0.0012)**	**0.0237 (± 0.0016)**
Attribute-guided CycleGAN [9]	0.7981 (± 0.0284)	0.1172 (± 0.0286)

Table 4. Mean F_1 scores (higher better, range [0,1]) and *RMSEs* (lower better, range [0,1]) for the five architectures, trained on IAM_{synth} and evaluated on the test split of $Dracula_{real}$. Standard deviation over thirty training runs given in parentheses. Best model marked in bold.

Model	F_1	RMSE
SimpleCNN	0.7204 (± 0.0303)	**0.0827 (± 0.0038)**
Shallow	0.7450 (± 0.0028)	0.0932 (± 0.0044)
UNet	0.7451 (± 0.0013)	0.1005 (± 0.0033)
Generator	**0.7577 (± 0.0035)**	0.0868 (± 0.0021)
Attribute-guided CycleGAN [9]	0.7189 (± 0.0243)	0.0927 (± 0.0212)

Table 5. Mean F_1 scores (higher better, range [0,1]) and *RMSEs* (lower better, range [0,1]) for the five architectures, trained on individual partitions of $Dracula_{synth}$ and evaluated on the test split of $Dracula_{real}$. Standard deviations over five partitions with thirty training runs each, given in parentheses. Best model marked in bold.

Model	F_1	RMSE
SimpleCNN	0.7327 (± 0.0046)	0.0757 (± 0.0008)
Shallow	0.7648 (± 0.0052)	0.0709 (± 0.0023)
UNet	0.7482 (± 0.0031)	0.0761 (± 0.0047)
Generator	**0.7872 (± 0.0059)**	**0.0655 (± 0.0026)**
Attribute-guided CycleGAN [9]	0.5073 (± 0.1484)	0.1317 (± 0.0312)

Table 6. Mean F_1 scores (higher better, range [0,1] and RMSEs (lower better, range [0,1]) for the five architectures, trained on the aggregated partitions of $Dracula_{synth}$ and evaluated on the test split of $Dracula_{real}$. Standard deviation over thirty training runs given in parentheses. Best model marked in bold.

Model	F_1	RMSE
SimpleCNN	0.7543 (± 0.0034)	0.0718 (± 0.0011)
Shallow	0.7825 (± 0.0049)	0.0681 (± 0.0029)
UNet	0.7662 (± 0.0064)	0.0734 (± 0.0038)
Generator	**0.8122 (± 0.0031)**	**0.0592 (± 0.0015)**
Attribute-guided CycleGAN [9]	0.6788 (± 0.0516)	0.1148 (± 0.0373)

5.2 Models Trained on Individual Partitions of $Dracula_{synth}$

Table 5 summarises the performances for the five architectures trained on the five individual partitions of synthetic data from $Dracula_{synth}$, that were generated based on the train split of $Dracula_{real}$. As can be seen from the table, there is a considerable difference in performance between the paired models and the attribute-guided *CycleGAN*. Again, the *Generator* models perform best in comparison to the other paired approaches. It can also be noted that, with the exception of the *CycleGAN*, on average all models trained on the $Dracula_{synth}$ training partitions outperform their counterparts trained on IAM_{synth}(cf. Table 4).

5.3 Models Trained on the Aggregation of Partitions from $Dracula_{synth}$

Following the large improvements gained by training on the individual partitions of $Dracula_{synth}$, all models were retrained from scratch on the aggregation of the five partitions. This aggregating step was taken in order to investigate the impact of a more diverse set of strikethrough strokes, applied to handwriting from the target domain. Table 6 shows the resulting F_1 and *RMSE* scores for this experiment. The accumulated dataset yields moderate, yet consistent, increases for all models. Although a more substantial improvement of 17 pp for the F_1 score can be observed for the attribute-guided *CycleGAN*, it still performs considerably worse than the paired approaches. Future experiments may investigate the impact of further increasing the size and diversity of synthetic datasets based on clean images from the target handwriting domain.

5.4 Qualitative Results

In order to demonstrate the range of strikethrough removal capabilities of the evaluated models, we present a number of hand-picked positive ('cherry-picked') and negative ('lemon-picked') cases from the $Dracula_{real}$ test split. These images

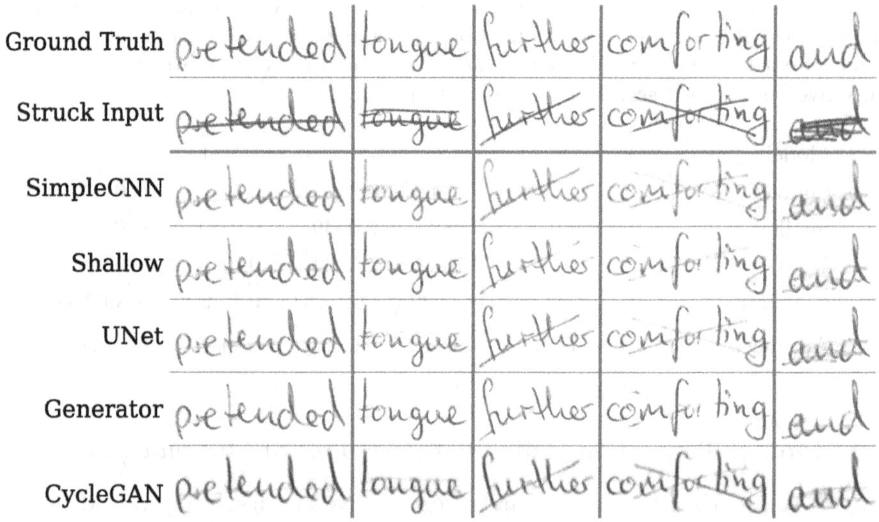

Fig. 3. Cherry-picked examples for the five models. All images are taken from the $Dracula_{real}$ test split and were processed by the respective model. Results are shown as the mean greyscale images, averaged over 30 model repetitions.

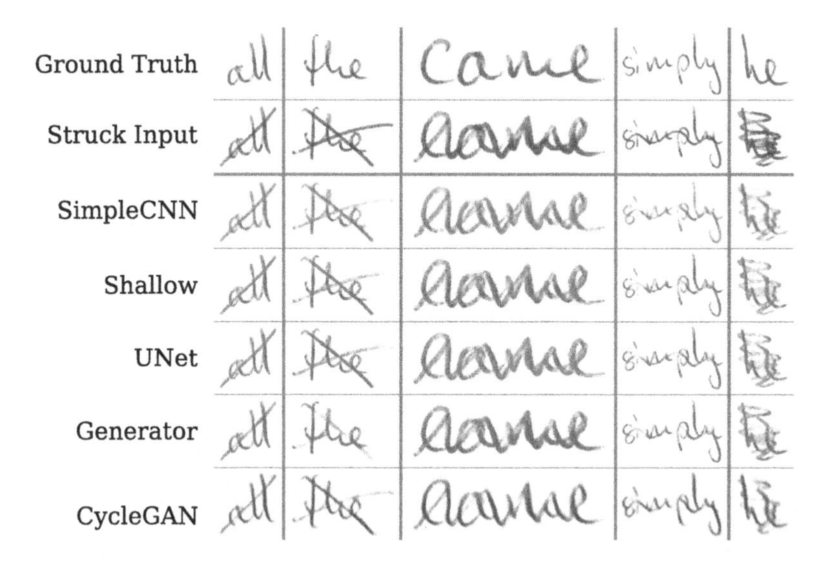

Fig. 4. Lemon-picked examples for the five models. All images are taken from the $Dracula_{real}$ test split and were processed by the respective model. Results are shown as the mean greyscale images, averaged over 30 model repetitions.

are shown in Fig. 3 and Fig. 4, respectively, and were obtained by calculating the mean of the 30 greyscale model outputs for each architecture, trained on the aggregated partitions of $Dracula_{synth}$. As can be seen from Fig. 3, most of the models manage to remove a fair portion of the genuine strikethrough, despite being only trained on synthetic strikethrough strokes. Additionally, it can be noted that the mean images for the paired approaches are generally more crisp than those obtained from *CycleGAN*, which appear more blurry, indicating less agreement between individual model checkpoints.

In contrast to the figure above, Fig. 4 depicts mean images for lemon-picked examples. Some of the models remove portions of the strokes, for example, the majority of one of the diagonal strokes in the cross sample cleaned by the *Generator* (second to last row, second image column, word the).

Overall, inspecting the rest of the cleaned images, not pictured here for brevity, a general trend can be noted for different types of strikethrough strokes. Generally, most of the *single*, *double* and *diagonal* strokes are removed convincingly. Shorter, *scratched* out words are often cleaned less than their longer counterparts. Strokes of types *cross*, *zig zag* and *wave* are cleaned considerably less often than the other stroke types.

6 Conclusions

In this work we have examined four paired image to image translation models and compared them with a state of the art unpaired strikethrough removal approach [9]. Based on the presented results and analyses, we draw the following conclusions:

1. Paired image to image translation approaches outperform the attribute-guided CycleGAN, proposed by [9], in all of the experiments presented in this work. The examined models not only outperform the state of the art in terms of strikethrough removal performance but also contain considerably fewer trainable parameters, making them cheaper and faster to train per epoch. Although the best results are obtained from the largest paired approach, the smaller evaluated models still display a considerable cleaning performance and may therefore still be of interest in scenarios with limited computing resources.
2. Using the custom synthetic strikethrough dataset, based on clean words from the same writer as the target domain (i.e. here $Dracula_{synth}$ for $Dracula_{real}$), yields better results in a paired image to image translation setting than the much more diverse IAM_{synth} dataset under the same experiment conditions.
3. Upon inspection of the qualitative results, stroke types can be separated into two groups, based on level of difficulty. *Single*, *double* and *diagonal* lines are generally cleaned more easily than *crosses*, *zig zag* and *waves*. For *scratched* out words, a trend can be noted, indicating that longer words are cleaned more easily than shorter ones.

Overall, considering the use case of strikethrough removal in an archival context, for example as preprocessing step for genetic criticism, where one or few handwriting styles are present in the data, we recommend to explore options to create an in-domain, synthetic dataset, similar to the approach that was taken for $Dracula_{synth}$. In the presented case, slightly more than 100 clean words from the target handwriting style, combined with synthetic strikethrough, yielded models with considerable strikethrough removal abilities, generating convincingly cleaned words.

In the future, we aim to expand the scope of genuine strikethrough removal datasets in order to further investigate a variety of approaches and the impact of more diverse handwriting styles.

Acknowledgements. The computations were enabled by resources provided by the Swedish National Infrastructure for Computing (SNIC) at Chalmers Centre for Computational Science and Engineering (C3SE) partially funded by the Swedish Research Council through grant agreement no. 2018-05973. This work is partially supported by Riksbankens Jubileumsfond (RJ) (Dnr P19-0103:1).

A Dataset and Code Availability

$Dracula_{synth}$: https://doi.org/10.5281/zenodo.6406538
Code: https://doi.org/10.5281/zenodo.6406284.

References

1. Adak, C., Chaudhuri, B.B., Blumenstein, M.: Impact of struck-out text on writer identification. In: 2017 International Joint Conference on Neural Networks (IJCNN), pp. 1465–1471 (2017). https://doi.org/10.1109/IJCNN.2017.7966025
2. Brink, A., van der Klauw, H., Schomaker, L.: Automatic removal of crossed-out handwritten text and the effect on writer verification and identification. In: Document Recognition and Retrieval XV, vol. 6815, pp. 79–88. SPIE (2008). https://doi.org/10.1117/12.766466
3. Calvo-Zaragoza, J., Gallego, A.J.: A selectional auto-encoder approach for document image binarization. Pattern Recogn. **86**, 37–47 (2019). https://doi.org/10.1016/j.patcog.2018.08.011
4. Chaudhuri, B.B., Adak, C.: An approach for detecting and cleaning of struck-out handwritten text. Pattern Recogn. **61**, 282–294 (2017). https://doi.org/10.1016/j.patcog.2016.07.032
5. Dheemanth Urs, R., Chethan, H.K.: A study on identification and cleaning of struck-out words in handwritten documents. In: Jeena Jacob, I., Kolandapalayam Shanmugam, S., Piramuthu, S., Falkowski-Gilski, P. (eds.) Data Intelligence and Cognitive Informatics. AIS, pp. 87–95. Springer, Singapore (2021). https://doi.org/10.1007/978-981-15-8530-2_6
6. Heil, R.: RaphaelaHeil/strikethrough-removal-cyclegans: release for publication. Zenodo (2021). https://doi.org/10.5281/zenodo.4767169, version 1.0
7. Heil, R., Vats, E., Hast, A.: IAM Strikethrough Database. Zenodo (2021). https://doi.org/10.5281/zenodo.4767095, version 1.0.0
8. Heil, R., Vats, E., Hast, A.: Single-Writer Strikethrough Dataset (1.0.0). Zenodo (May 2021). https://doi.org/10.5281/zenodo.4765063
9. Heil, R., Vats, E., Hast, A.: Strikethrough removal from handwritten words using CycleGANs. In: Lladós, J., Lopresti, D., Uchida, S. (eds.) ICDAR 2021. LNCS, vol. 12824, pp. 572–586. Springer, Cham (2021). https://doi.org/10.1007/978-3-030-86337-1_38
10. Huang, G., Liu, Z., Van Der Maaten, L., Weinberger, K.Q.: Densely connected convolutional networks. In: 2017 IEEE Conference on Computer Vision and Pattern Recognition (CVPR), pp. 2261–2269 (2017). https://doi.org/10.1109/CVPR.2017.243
11. Hulle, D.V.: The stuff of fiction: digital editing, multiple drafts and the extended mind. Textual Cultures, **8**(1), 23–37 (2013). http://www.jstor.org/stable/10.2979/textcult.8.1.23
12. Isola, P., Zhu, J.Y., Zhou, T., Efros, A.A.: Image-to-image translation with conditional adversarial networks. In: 2017 IEEE Conference on Computer Vision and Pattern Recognition (CVPR), pp. 5967–5976 (2017). https://doi.org/10.1109/CVPR.2017.632
13. Jégou, S., Drozdzal, M., Vazquez, D., Romero, A., Bengio, Y.: The one hundred layers tiramisu: fully convolutional densenets for semantic segmentation. In: 2017 IEEE Conference on Computer Vision and Pattern Recognition Workshops (CVPRW), pp. 1175–1183 (2017). https://doi.org/10.1109/CVPRW.2017.156
14. Kingma, D.P., Ba, J.: Adam: a method for stochastic optimization. In: Bengio, Y., LeCun, Y. (eds.) 3rd International Conference on Learning Representations, ICLR 2015, San Diego, CA, 7–9 May 2015, Conference Track Proceedings (2015). http://arxiv.org/abs/1412.6980

15. Likforman-Sulem, L., Vinciarelli, A.: Hmm-based offline recognition of handwritten words crossed out with different kind of strokes. In: Proceedings of the 11th International Conference on Frontiers in Handwriting Recognition, vol. 11, pp. 70–75 (2008). http://eprints.gla.ac.uk/59027/
16. Marti, U.V., Bunke, H.: The iam-database: an english sentence database for offline handwriting recognition. Int. J. Doc. Anal. Recogn. **5**(1), 39–46 (2002). https://doi.org/10.1007/s100320200071
17. Neji, H., Nogueras-Iso, J., Lacasta, J., Ben Halima, M., Alimi, A.M.: Adversarial autoencoders for denoising digitized historical documents: the use case of incunabula. In: 2019 International Conference on Document Analysis and Recognition Workshops (ICDARW), vol. 6, pp. 31–34 (2019). https://doi.org/10.1109/ICDARW.2019.50112
18. Nisa, H., Thom, J.A., Ciesielski, V., Tennakoon, R.: A deep learning approach to handwritten text recognition in the presence of struck-out text. In: 2019 International Conference on Image and Vision Computing New Zealand (IVCNZ), pp. 1–6 (2019). https://doi.org/10.1109/IVCNZ48456.2019.8961024
19. Otsu, N.: A threshold selection method from gray-level histograms. IEEE Trans. Syst. Man Cybern. **9**(1), 62–66 (1979). https://doi.org/10.1109/TSMC.1979.4310076
20. Paszke, A., et al.: PyTorch: an imperative style, high-performance deep learning library. In: Wallach, H., Larochelle, H., Beygelzimer, A., d'Alché-Buc, F., Fox, E., Garnett, R. (eds.) Advances in Neural Information Processing Systems 32, pp. 8024–8035. Curran Associates, Inc. (2019). http://papers.neurips.cc/paper/9015-pytorch-an-imperative-style-high-performance-deep-learning-library.pdf
21. Paulus, E., Burie, J.-C., Verbeek, F.J.: Binarization strategy using multiple convolutional autoencoder network for old sundanese manuscript images. In: Barney Smith, E.H., Pal, U. (eds.) ICDAR 2021 Workshops. LNCS, vol. 12917, pp. 142–157. Springer, Cham (2021). https://doi.org/10.1007/978-3-030-86159-9_10
22. Pielawski, N.: OctoPyTorch: segmentation Neural Networks (2021). https://github.com/npielawski/octopytorch, commit: 6e65f23
23. Poddar, A., Chakraborty, A., Mukhopadhyay, J., Biswas, P.K.: Detection and localisation of struck-out-strokes in handwritten manuscripts. In: Barney Smith, E.H., Pal, U. (eds.) ICDAR 2021 Workshops. LNCS, vol. 12917, pp. 98–112. Springer, Cham (2021). https://doi.org/10.1007/978-3-030-86159-9_7
24. Poddar, A., Chakraborty, A., Mukhopadhyay, J., Biswas, P.K.: Texrgan: a deep adversarial framework for text restoration from deformed handwritten documents. In: Proceedings of the Twelfth Indian Conference on Computer Vision, Graphics and Image Processing. ICVGIP 2021, Association for Computing Machinery, New York (2021). https://doi.org/10.1145/3490035.3490306
25. Shivakumara, P., et al.: A connected component-based deep learning model for multi-type struck-out component classification. In: Barney Smith, E.H., Pal, U. (eds.) ICDAR 2021 Workshops. LNCS, vol. 12917, pp. 158–173. Springer, Cham (2021). https://doi.org/10.1007/978-3-030-86159-9_11
26. Zhao, G., Liu, J., Jiang, J., Guan, H., Wen, J.R.: Skip-connected deep convolutional autoencoder for restoration of document images. In: 2018 24th International Conference on Pattern Recognition (ICPR), pp. 2935–2940 (2018). https://doi.org/10.1109/ICPR.2018.8546199
27. Öfverstedt, J., Lindblad, J., Sladoje, N.: Fast and robust symmetric image registration based on distances combining intensity and spatial information. IEEE Trans. Image Process. **28**(7), 3584–3597 (2019). https://doi.org/10.1109/TIP.2019.2899947

Revealing Reliable Signatures by Learning Top-Rank Pairs

Xiaotong Ji[1(✉)], Yan Zheng[1], Daiki Suehiro[1,2], and Seiichi Uchida[1]

[1] Kyushu University, Fukuoka, Japan
{xiaotong.ji,yan.zheng}@human.ait.kyushu-u.ac.jp,
{suehiro,uchida}@ait.kyushu-u.ac.jp
[2] RIKEN Center for Advanced Intelligence Project, Tokyo, Japan

Abstract. Signature verification, as a crucial practical documentation analysis task, has been continuously studied by researchers in machine learning and pattern recognition fields. In specific scenarios like confirming financial documents and legal instruments, ensuring the absolute reliability of signatures is of top priority. In this work, we proposed a new method to learn "top-rank pairs" for writer-independent offline signature verification tasks. By this scheme, it is possible to maximize the number of absolutely reliable signatures. More precisely, our method to learn top-rank pairs aims at pushing positive samples beyond negative samples, after pairing each of them with a genuine reference signature. In the experiment, BHSig-B and BHSig-H datasets are used for evaluation, on which the proposed model achieves overwhelming better pos@top (the ratio of absolute top positive samples to all of the positive samples) while showing encouraging performance on both Area Under the Curve (AUC) and accuracy.

Keywords: Writer-independent signature verification · Top-rank learning · Absolute top

1 Introduction

As one of the most important topics in document processing systems, signature verification has become an indispensable issue in modern society [1,2]. Precisely, it plays important role in enhancing security and privacy in various fields, such as finance, medical, forensic agreements, etc. As innumerable significant documents are signed almost every moment throughout the world, automatically examining the genuineness of the signed signatures has become a crucial subject. Since misjudgment is hardly allowed especially in serious and formal situations like in forensic usages, obtaining "highly reliable" signatures is of great importance.

Figure 1 illustrates two scenarios of signature verification. In the writer-dependent scenario (a), it is possible to prepare the verifiers specialized for individuals. In contrast, in the writer-independent scenario (b), we can prepare

X. Ji and Y. Zheng—The author contribute equally to this paper.

© Springer Nature Switzerland AG 2022
S. Uchida et al. (Eds.): DAS 2022, LNCS 13237, pp. 323–337, 2022.
https://doi.org/10.1007/978-3-031-06555-2_22

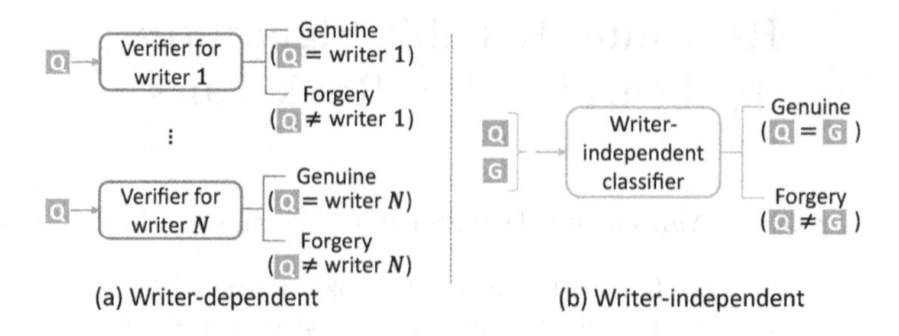

Fig. 1. Two scenarios of signature verification. 'Q' and 'G' are the query signature and a genuine reference signature, respectively.

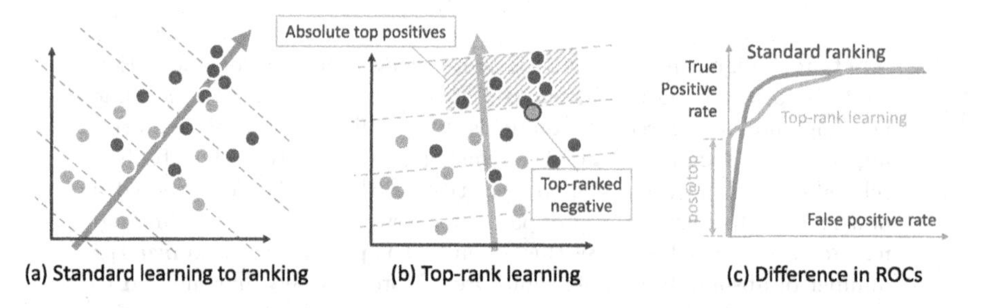

Fig. 2. (a), (b) Two ranking methods and (c) their corresponding ROCs. In (a) and (b), a thick arrow is a ranking function that gives higher rank scores to positive samples (purple circles) than negative samples (pink circles). Each dotted line is an "equi-distance" line. (Color figure online)

only a single and universal classifier that judges whether a pair of a questioned signature (i.e., a query) and a genuine reference signature are written by the same person or not. Consequently, for a reliable writer-independent verification, the classifier needs (i) to deal with various signatures of various individuals and (ii) not to accept unreliable pairs with some confidence.

In this paper, we propose a new method to learn *top-rank pairs* for highly-reliable writer-independent signature verification. The proposed method is inspired by top-rank learning [3–5]. Top-rank learning is one of the ranking tasks but is different from the standard ranking task. Figure 2 shows the difference of top-rank learning from standard learning to rank. The objective of the standard ranking task (a) is to determine the ranking function that evaluates positive samples much higher than negative samples as possible. This objective is equivalent to maximize AUC. In contrast, top-rank learning (b) has a different objective to maximize *absolute top positives*, which are highly-reliable positive samples in the sense that no negative sample has a higher rank than them. In (c), the very beginning of the ROC of top-rank learning is a vertical part; this means that there are several positive samples that have no negative samples ranked higher than them. Consequently, Top-rank learning can derive absolute top positives as highly reliable positive samples. The ratio of absolute top positives over all positives is called "*pos@top.*"

The proposed method to learn top-rank pairs accepts a paired feature vector for utilizing the promising property of top-rank learning for writer-independent signature verification. As shown in Fig. 1(b), the writer-independent scenario is based on the pairwise evaluation between a query and a genuine reference. To integrate this pairwise evaluation into top-rank learning, we first concatenate q and g into a single paired feature vector $\boldsymbol{x} = q \oplus g$, where q and g denote the feature vector of a query and a genuine reference, respectively. The paired vector $\boldsymbol{x}$ is treated as positive when q is written by the genuine writer and denoted as $\boldsymbol{x}^+$; similarly, when q is written by a forgery, $\boldsymbol{x}$ is negative and denoted as $\boldsymbol{x}^-$.

We train the ranking function r with the positive paired vectors $\{\boldsymbol{x}^+\}$ and the negative paired vectors $\{\boldsymbol{x}^-\}$, by top-rank learning. As the result, we could have highly reliable positives as absolute top positives; more specifically, a paired signature in the absolute top positives is "a more genuine pair" than the most genuine-like negative pair (i.e., the most-skilled forgery), called *top-ranked negative*, therefore highly reliable. Although being an absolute top positive is much harder than just being ranked higher, we could make a highly reliable verification by using the absolute top positives and the trained ranking function r.

To prove the reliability of the proposed method in terms of pos@top, we conduct writer-independent offline signature verification experiments with two publicly-available datasets: BHSig-B and BHSig-H. We use SigNet [1] as not only the extractor of deep feature vectors (q and g) from individual signatures but also a comparative method. SigNet is the current state-of-the-art model of offline signature verification. The experimental results prove that the proposed method outperforms SigNet not only pos@top but also other conventional evaluation metrics, such as accuracy, AUC, FAR, and FRR.

Our contributions are arranged as follows:

- We propose a novel method to learn top-rank pairs. This method is the first application of top-rank learning to conduct a writer-independent signature verification task to the best knowledge of the authors, notwithstanding that the concepts of top-rank learning and absolute top positives are particularly appropriate to highly reliable signature verification tasks.
- Experiments on two signature datasets have been done to evaluate the effect of the proposed method, including the comparison with the SigNet. Especially, the fact that the proposed method achieves higher pos@top proves that the trained ranking function gives a more reliable score that guarantees "absolutely genuine" signatures.

2 Related Work

The signature verification task has attracted great attention from researchers since it has been proposed [6,7]. Generally, signature verification is divided into online [8] as well as offline [9,10] fashions. Online signatures offer pressure and stroke order information that is favorable to time series analysis methods [11] like Dynamic Time Warping (DTW) [12]. On the other hand, offline signature verification should be carried out only by making full use of image feature information [13]. As a result, acquiring efficacious features from offline signatures [7,14,15] has become a highly anticipated challenge.

In recent years, CNNs have been widely used in signature verification tasks thanks to their excellent representation learning ability [16,17]. Among CNN-based models, Siamese network [18,19] is one of the common choices when it comes to signature verification tasks. Specifically, a Siamese network is composed of two identically structured CNNs with shared weights, particularly powerful for similarity learning, which is a preferable learning objective in signature verification. For example, Dey et al. proposed a Siamese network-based model that optimizes the distances between two inputted signatures, shows outstanding performance on several famous signature datasets [1]. Wei et al. also employed the Siamese network, and by utilizing inverse gray information with multiple attention modules, their work showed encouraging results as well [20]. However, none of the those approaches target revealing highly reliable genuine signatures.

To acquire the highly reliable genuine signatures, learning to rank [21,22] is a more reasonable approach than learning to classify. This is because learning to rank allows us to rank the signatures in order of genuineness. Bipartite ranking [23] is one of the most standard learning-to rank-approach. The goal of the bipartite ranking is to find a scoring function that gives a higher value to positive samples than negative samples. This goal corresponds to the maximization of AUC (Area under the ROC curve), and thus bipartite ranking has been used in various domains [24–26].

As a special form of bipartite ranking, the top-rank learning strategy [3–5] possesses characteristics that are more suitable for absolute top positive hunting. In contrast to the standard bipartite ranking, top-rank learning aims at maximizing the absolute top positives, that is, maximizing the number of positive samples ranked higher than any negative sample. Therefore, top-rank learning is suitable for some tough tasks that require possibly highly reliable (e.g., medical diagnosis [27]). To the best of our knowledge, this is the first application of top-rank learning to the signature verification task.

TopRank CNN [27] is a representation learning version of the conventional top-rank learning scheme, which combines the favorable characteristics of both CNN and the top-rank learning scheme. To be more specific, when encountered with entangled features in which positive is chaotically tied with negative, conventional top-rank learning methods without representation learning capability like TopPush [3] can hardly achieve a high pos@top. That is to say, the representation learning ability of CNN structure makes the TopRank CNN a more powerful top-rank learning method. Moreover, for avoiding the easily-happen over-fitting phenomenon, TopRank CNN considerately attached the max operation with the p-norm relaxation.

Despite the superiority of ranking schemes, studies that apply ranking strategies on signature verification tasks are still in great demand. Chhabra et al. in [28] proposed a Deep Triplet Ranking CNN, aiming at ranking the input sig-

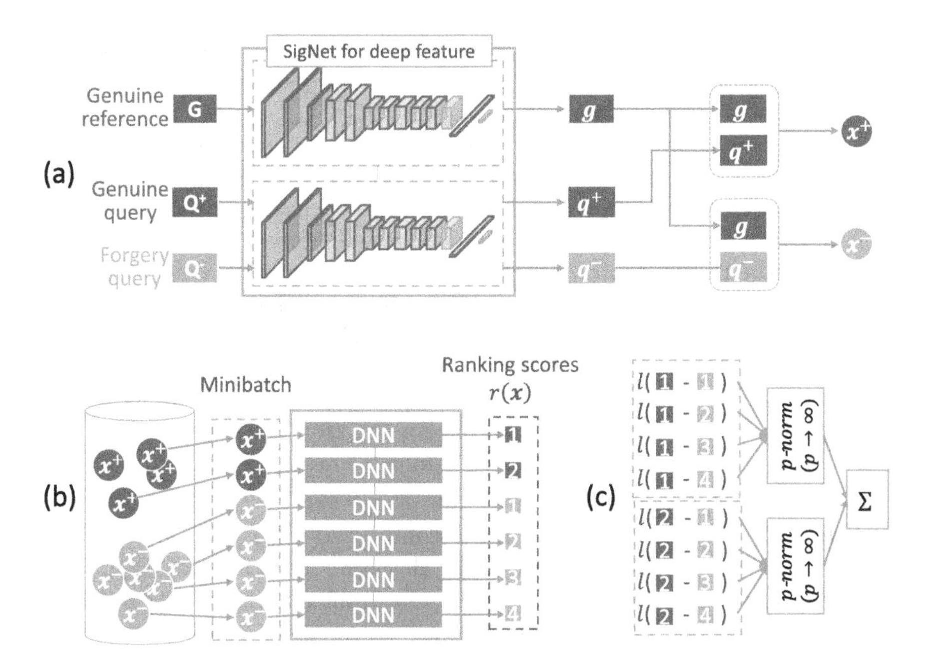

Fig. 3. The overall structure of the proposed method to learn *top-rank pairs* for writer-independent signature verification. (a) Feature representation of paired samples. (b) Learning top-rank pairs with their representation. (c) Top-rank loss function $\mathcal{L}_{\text{TopRank}}$.

natures in genuine-like order by minimizing the distance between genuine and anchors. In the same year, Zheng et al. proposed to utilize RankSVM for writer-dependent classification, to ensure the generalization performance on imbalanced data [2]. However, even if they care about ranking results to some extent, no existing studies have been dedicated to the absolute top genuine signatures yet. To address this issue, this work is mainly designed to focus on obtaining the absolute top genuine signatures, implemented by learning top-rank pairs for writer-independent offline signature verification tasks.

3 Learning Top-Rank Pairs

Figure 3 shows the overview of the proposed method to learn top-rank pairs for writer-independent offline signature verification. The proposed method consists of two steps: a representation learning step and a top-rank learning step. Figure 3(a) shows the former step and (b) and (c) show the latter step.

3.1 Feature Representation of Paired Samples

As shown in Fig. 1(b), each input is a pair of a genuine reference sample g and a query sample q for writer-independent signature verification. Then the paired samples (g, q) are fed to some function to evaluate their discrepancy. If the evaluation result shows a large discrepancy, the query is supposed to be a forgery; otherwise, the query is genuine.

Now we concatenate the two d-dimensional feature vectors (g and q) into a $2d$-dimensional single vector as shown in Fig. 3(a). Although the concatenation doubles the feature dimensionality, it allows us to treat the paired samples in a simple way. Specifically, we consider a (Genuine g, Genuine q^+)-pair as a positive sample with the feature vector $\boldsymbol{x}^+ = g \oplus q^+$ and a (Genuine g, Forgery q^-)-pair as a negative sample with $\boldsymbol{x}^- = g \oplus q^-$. If we have m (Genuine, Genuine)-pairs and n (Genuine, Forgery)-pairs, we have two sets $\boldsymbol{\Omega}^+ = \{\boldsymbol{x}_i^+ \mid i = 1, \ldots, m\}$ and $\boldsymbol{\Omega}^- = \{\boldsymbol{x}_j^+ \mid j = 1, \ldots, n\}$.

Under this representation, the writer-independent signature verification task is simply formulated as a problem to have a function $r(\boldsymbol{x})$ that gives a large value for $\boldsymbol{x}_i^+$ and a small value for $\boldsymbol{x}_j^-$. Ideally, we want to have $r(\boldsymbol{x})$ that satisfies $r(\boldsymbol{x}_i^+) > r(\boldsymbol{x}_j^-)$ for arbitrary $\boldsymbol{x}_i^+$ and $\boldsymbol{x}_j^-$. In this case, we have a constant threshold θ that satisfies $\max_j r(\boldsymbol{x}_j^-) < \theta < \min_i r(\boldsymbol{x}_i^+)$. If $r(\boldsymbol{x}) > \theta$, $\boldsymbol{x}$ is simply determined as a (Genuine, Genuine)-pair. However, in reality, we do not have the ideal r in advance; therefore we need to optimize (i.e., train) r so that it becomes closer to the ideal case under some criterion. In Sect.3.2, pos@top is used as the criterion so that trained r gives more absolute tops.

As indicated in Fig. 3(a), each signature is initially represented as a d-dimensional vector g (or q) by SigNet [1], which is still a state-of-the-art signature verification model realized by metric learning with the contrastive loss. Although it is possible to use another model to have the initial feature vector, we use SigNet throughout this paper. The details of SigNet will be described in Sect. 3.4.

3.2 Optimization to Learn Top-Rank Pairs

We then use a top-rank learning model for optimizing the ranking function $r(\boldsymbol{x})$. As noted in Sect. 1, top-rank learning aims to maximize pos@top, which is formulated as:

$$\text{pos@top} = \frac{1}{m} \sum_{i=1}^{m} I\left(r(\boldsymbol{x}_i^+) > \max_{1 \leq j \leq n} r(\boldsymbol{x}_j^-)\right), \tag{1}$$

where $I(z)$ is the indicator function. pos@top evaluates the number of positive samples with a higher value than any negative samples. The positive samples that satisfy the condition in Eq. 1 are called absolute top positives or just simply absolute tops. Absolute tops are very "reliable" positive samples because they are more positive than the top-ranked negative, that is, the "hardest" negative $\max_{1 \leq j \leq n} r(\boldsymbol{x}_j^-)$.

Among various optimization criteria, pos@top has promising properties for the writer-independent signature verification task. Maximization of pos@top is equivalent to the maximization of absolute tops—this means we can have reliable positive samples to the utmost extent. In a very strict signature verification task, the query sample q is verified as genuine only when the concatenated vector $\boldsymbol{x} = g \oplus q$ becomes one of the absolute tops. Therefore, having more absolute tops by maximizing pos@top will give more chance that the query sample is completely trusted as genuine.

Note that we call $r(\boldsymbol{x})$ as a "ranking" function, instead of just a scoring function. In Eq. (1), the value of the function r is used just for the comparison of samples. This suggests that the value of r has no absolute meaning. In fact, if we have a maximum pos@top by a certain $r(\boldsymbol{x})$, $\phi(r(\boldsymbol{x}))$ also achieves the same pos@top, where ϕ is a monotonically-increasing function. Consequently, the ranking function r specifies the order (i.e., the rank) among samples.

We will optimize r to maximize $\boldsymbol{x}^+$ in pos@top. Top-Rank Learning is the existing problem to maximize pos@top for a training set whose samples are individual (i.e., unpaired) vectors. In contrast, our problem to learn top-rank pairs is a new ranking problem for the paired samples, and applicable to various ranking problems where the relative relations between two vectors are important[1].

As noted in Sect. 3.3 and shown in Fig. 3(b), we train r along with a deep neural network (DNN) to have a reasonable feature space to have more pos@top. However, there are some risks when we maximize Eq. (1) directly using a DNN, if it has a high representation flexibility. The most realistic case is that the DNN overfits some outliers or noise. For example, if a negative outlier is distributed over the positive training samples, achieving the perfect pos@top is not a reasonable goal.

To avoid such risks, we employ the p-norm relaxation technique [27,29]. More specifically, we convert the maximization of pos@top into the minimization of the following loss:

$$\mathcal{L}_{\text{TopRank}}\left(\boldsymbol{\Omega}^+, \boldsymbol{\Omega}^-\right) = \frac{1}{m}\sum_{i=1}^{m}\left(\sum_{j=1}^{n}\left(l(r(\boldsymbol{x}_i^+) - r(\boldsymbol{x}_j^-))\right)^p\right)^{\frac{1}{p}}, \tag{2}$$

where $l(z) = \log(1 + e^{-z})$ is a surrogate loss. Figure 3(c) illustrates $\mathcal{L}_{\text{TopRank}}$. When $p = \infty$, Eq. (2) is reduced to $\mathcal{L}_{\text{TopRank}} = \frac{1}{m}\sum_{i=1}^{m}\max_{1\le j\le n} l(r(\boldsymbol{x}_i^+) - r(\boldsymbol{x}_j^-))$, which is equivalent to the original pos@top loss of Eq. (1). If we set p at a large value (e.g., 32), the Eq. (2) approaches the original loss of pos@top. In [27], it is noted that it is better not to select a too large p, because it has a risk of over-fitting and the overflow error in the implementation.

3.3 Learning Top-Rank Pairs with Their Representation

In order to have a final feature representation to have a more pos@top, we apply a DNN to convert $\boldsymbol{x}$ non-linearly during the training process of r. Figure 3(b) shows the process with the DNN. Each of the paired feature vectors in a minibatch is

[1] Theoretically, learning top-rank pairs can be extended to handle vectors obtained by concatenating three or more individual vectors. With this extension, our method can rank the mutual relationship among multiple vectors.

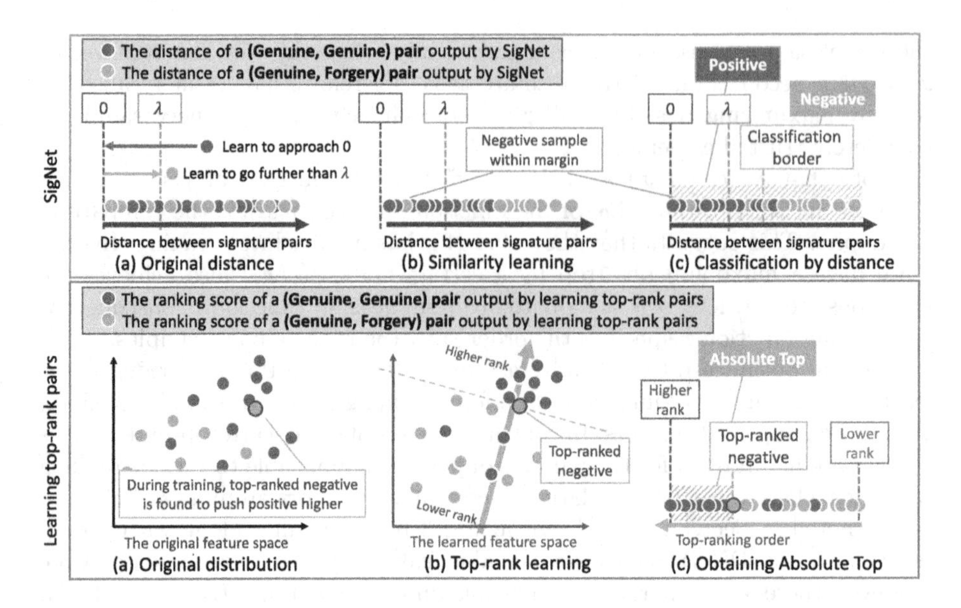

Fig. 4. The learning mechanisms of SigNet (top) and the learning top-rank pairs (bottom).

fed to a DNN. In the DNN, the vectors are converted to another feature space and then their ranking score r is calculated. The parameters of DNN are trained by the loss function $\mathcal{L}_{\text{TopRank}}$.

3.4 Initial Features by SigNet

As noted in Sect. 3.1, we need to have initial vectors (g and q) for individual signatures by an arbitrary signature image representation method. SigNet [1] is the current state-of-the-art signature verification method and achieved high performance in standard accuracy measures. As shown in Fig. 3(a), SigNet is based on metric learning with a contrastive loss and takes a pair of a reference signature and a query signature as its input images. For all pairs of reference and query signatures, SigNet is optimized to decrease the distance between (Genuine, Genuine)-pairs and increase the distance between (Genuine, Forgery)-pairs. The trained network can convert a reference image into g and a query into q. Then a positive sample $\boldsymbol{x}_i^+$ or a negative sample $\boldsymbol{x}_j^-$ is obtained by the concatenation $g \oplus q$, as described in Sect. 3.1.

Theoretically, we can conduct the end-to-end training of SigNet in Fig. 3(a) and DNN in (b). In this paper, however, we fix the SigNet model after its independent training with the contrastive loss. This is simply to make the comparison between the proposed method and SigNet as fair as possible. (In other words, we want to purely observe the effect of pos@top maximization and thus minimize the extra effect of further representation learning in SigNet.)

One might misunderstand that the metric learning result by SigNet and the ranking result by top-rank learning in the proposed method are almost identical; however, as shown in Fig. 4, they are very different. As we emphasized so far, the proposed method aims to have more pos@top; this means we have a clear boundary between the absolute tops and the others. In contrast, SigNet has no such function. Consequently, SigNet might have a risk that a forgery has a very small distance with a genuine. Finally, this forgery will be wrongly considered as one of the reliable positives, which are determined by applying a threshold λ to the distance by SigNet.

4 Experiments

In this section, we demonstrate the effectiveness of the proposed method on signature verification tasks. Specifically, we consider a comparative experiment with SigNet which is known as the outstanding method for the signature verification tasks.

4.1 Datasets

In this work, the BHSig260 offline signature dataset[2] is used for the experiments[3], which composes two subsets where one set of signatures are signed in Bengali (named BHSig-B) and the other in Hindi (named BHSig-H). The BHSig-B dataset includes 100 writers in total, each of them possesses 24 genuine signatures and 30 skillfully forged signatures. On the other hand, BHSig-H dataset contains 160 writers, each of them own genuine and forged signatures same as BHSig-B. In the experiments, both of the datasets are divided into training, validation, and test set to the ratio of 8: 1: 1.

Following the writer-independent setting, we evaluate the verification performance using the pair of signatures. That is, the task is to verify that the given pair is (Genuine, Genuine) or (Genuine, Forgery). We prepare a total of 276 (Genuine, Genuine)-pairs for each writer and a total of 720 (Genuine, Forgery)-pairs for each writer.

4.2 Experimental Settings

Setting of the SigNet. SigNet is also based on a Siamese network architecture, whose optimization objective is similarity measurement. In the experiment, we followed the training setting noted in [1], except for the modifications of the data division.

[2] Available at http://www.gpds.ulpgc.es/download.

[3] CEDAR dataset that is also used in [1] was not used in this work because it has achieved 100% accuracy in the test set. Besides, GPDS 300 and GPDS Synthetic Signature Corpus datasets are restricted from obtaining.

Table 1. The comparison between the proposed method and SigNet on BHSig-B and BHSig-H datasets.

Dataset	Approaches	**pos@top** (↑)	Accuracy (↑)	AUC (↑)	FAR (↓)	FRR (↓)
BHSig-B	**proposed**	**0.283**	**0.806**	**0.889**	**0.222**	**0.222**
	SigNet	0.000	0.756	0.847	0.246	0.247
BHSig-H	**proposed**	**0.114**	**0.836**	**0.908**	**0.179**	**0.178**
	SigNet	0.000	0.817	0.891	0.192	0.192

Setting of the Proposed Method. As introduced in Sect. 3.4, we used the extracted features from the trained SigNet (124G Floating point operations (FLOPs)) as the initial features. For the learning top-rank pairs with their representation, we used a simple architecture, 4 fully-connected layers (2048, 1024, 512 and 128 nodes respectively) with ReLU function (1G FLOPs). The hyper-parameter p of the loss function Sect. 2 is chosen from $\{2, 4, 8, 16, 32\}$ based on the validation pos@top. As a result, we obtained p=4 for BHSig-B and p=16 for BHSig-H. The following results and the visualization are obtained by using these hyper-parameters.

4.3 Evaluation Metrics

In the experiment, pos@top, accuracy, AUC, False Rejection Rate (FRR), and False Acceptance Rate (FAR) are used to comprehensively evaluate the proposed method and SigNet.

- **pos@top**: The ratio of the absolute top (Genuine, Genuine) signature pairs to the number of all of the (Genuine, Genuine) signature pairs (see also Eq. (1)).
- Accuracy: The maximum result of the average between True Positive Rate (TPR) and True Negative Rate (TNR), following the definition in [1].
- AUC: Area under the ROC curve.
- FAR: The ratio of the number of falsely accepted (Genuine, Forgery) signature pairs divided by the number of all (Genuine, Forgery) signature pairs.
- FRR: The ratio of the number of falsely rejected (Genuine, Genuine) signature pairs divided by the number of all (Genuine, Genuine) signature pairs.

4.4 Quantitative and Qualitative Evaluations

The quantitative evaluations of the proposed method and SigNet on two datasets are shown in Table 1. Remarkably, the proposed method achieved an overwhelming better performance on pos@top, while pos@top of SigNet is 0 for both datasets. This proves that the proposed method can reveal absolute top positive signature pairs (i.e., highly reliable signature pairs). Furthermore, the proposed method also outperformed the comparison method on all other evaluation criteria, accuracy and AUC, lower FAR, and FRR.

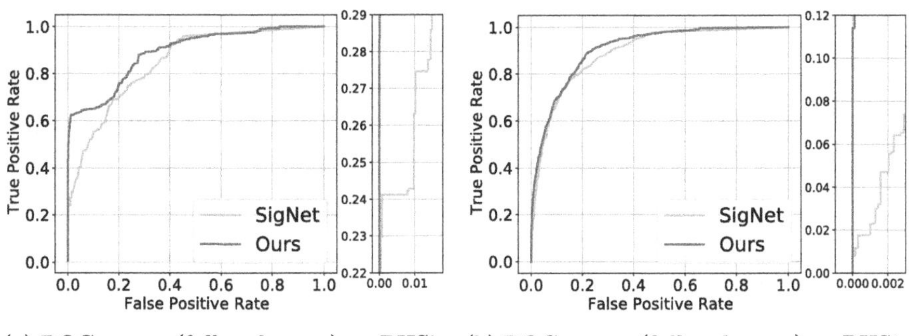

(a) ROC curves (full and zoom) on BHSig-B

(b) ROC curves (full and zoom) on BHSig-H

Fig. 5. The comparison of ROC curves between the proposed method and SigNet on BHSig-B and BHSig-H datasets.

The ROC curves of the proposed method and SigNet on two datasets are shown in Fig. 5 respectively, each followed by a corresponding zoom view of the beginning part of ROC curves. As a more intuitive demonstration of Table 1, other than the larger AUC, it is extremely obvious that the proposed method achieved higher pos@top, which is the True Positive Rate (TPR) for $x = 0$.

Figures 6(a) and (b) show the distributions of features from the proposed method mapped by Principal Component Analysis (PCA) on two datasets, colored by ranking scores (normalized within the range [0,1]). From the visualizations, we can easily notice the absolute top positive sample pairs distinguished by the top-ranked negative.

Following the feature distributions, Figs. 6(c) and (d) give a more intuitive representation of the ranking order for the learning top-rank pairs. These graphs include information of (1) where did the top-ranked negative appear and (2) how many (Genuine, Genuine) and (Genuine, Forgery)-pairs are scored to the same rank. It could be observed that the first negative appeared after a portion of positive from these two graphs. On the other hand, Figs. 6(e) and (f) show the ranking conditions of SigNet. Since the first negative pair shows on the top of the ranking, this causes 0 pos@top for both datasets.

As shown in Fig. 7, the absolute top (Genuine, Genuine)-pairs in (a) and (b) show great similarity to their counterparts. Especially, the consistency in their strokes and preference of oblique could be easily noticed even with the naked eye. On the other hand, both the non-absolute top (Genuine, Genuine)-pairs in (c) and (d) and (Genuine, Forgery)-pairs in (e) and (f) show less similarity to their corresponding signatures, no matter whether they are written by the same writer or not. Notwithstanding that they are all assigned with low ranking scores by learning top-rank pairs, such resemblance between two different classes could easily incur misclassification in conventional methods. Thus, our results shed light on the validity of the claimed effectiveness of the proposed method to maximize the pos@top.

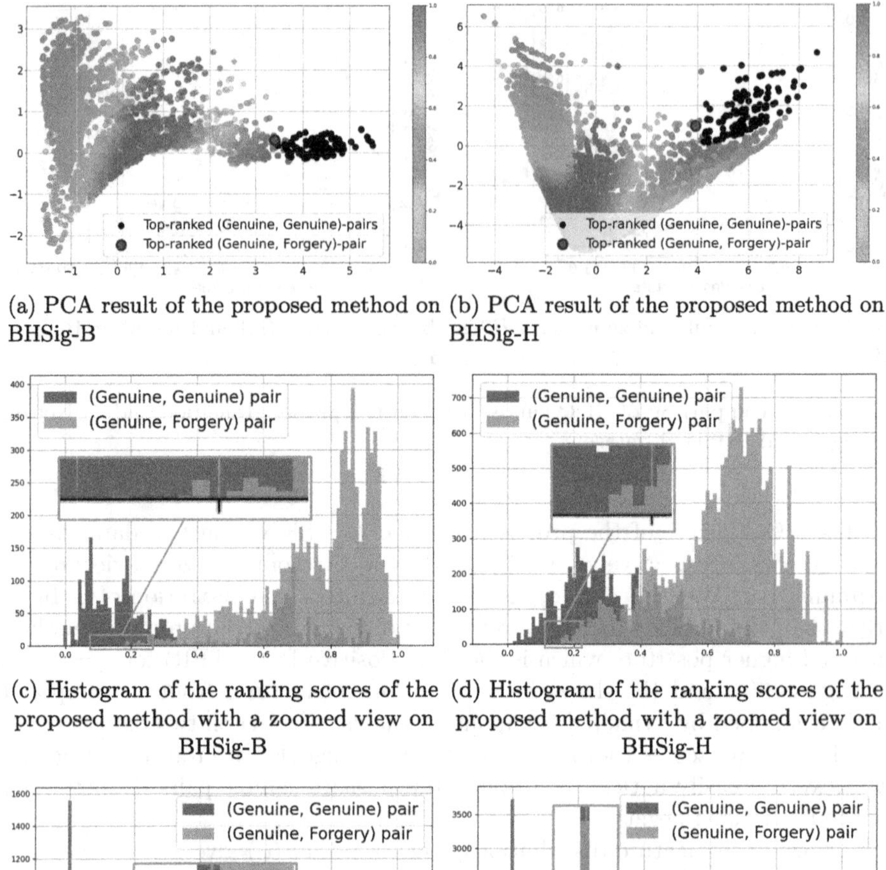

(a) PCA result of the proposed method on BHSig-B

(b) PCA result of the proposed method on BHSig-H

(c) Histogram of the ranking scores of the proposed method with a zoomed view on BHSig-B

(d) Histogram of the ranking scores of the proposed method with a zoomed view on BHSig-H

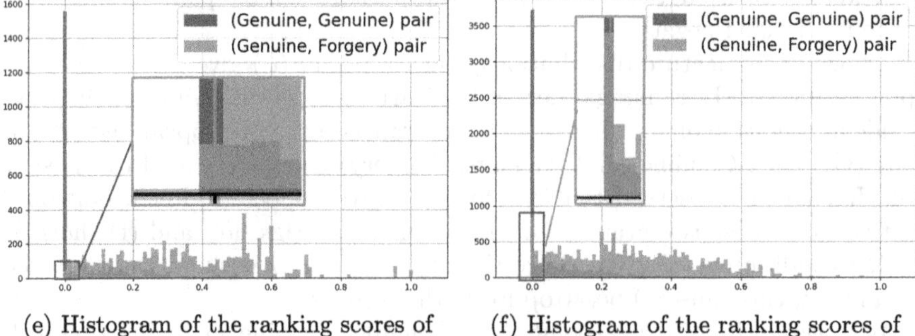

(e) Histogram of the ranking scores of SigNet with a zoomed view on BHSig-B

(f) Histogram of the ranking scores of SigNet with a zoomed view on BHSig-H

Fig. 6. (a) and (b) PCA visualizations of feature distribution for the proposed method. The top-ranked negative and the absolute top positives are highlighted. (c) – (f) Distributions of the ranking scores as histograms. The horizontal and vertical axes represent the ranking score and #samples, respectively.

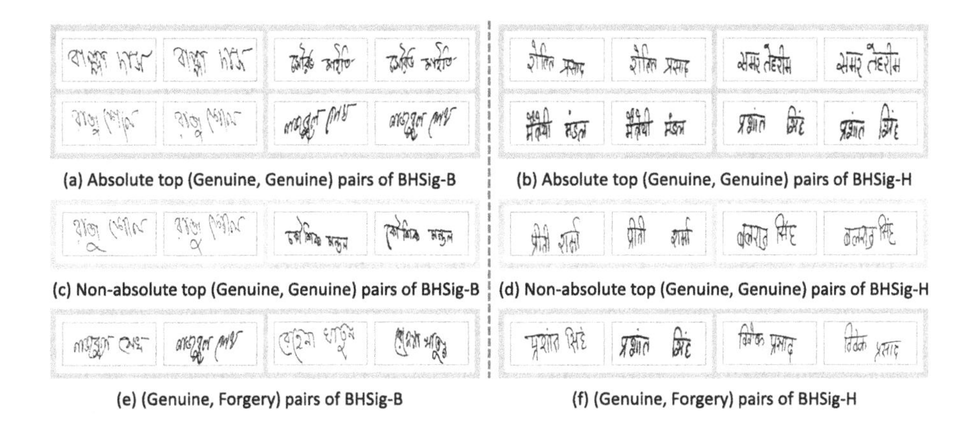

(a) Absolute top (Genuine, Genuine) pairs of BHSig-B

(b) Absolute top (Genuine, Genuine) pairs of BHSig-H

(c) Non-absolute top (Genuine, Genuine) pairs of BHSig-B

(d) Non-absolute top (Genuine, Genuine) pairs of BHSig-H

(e) (Genuine, Forgery) pairs of BHSig-B

(f) (Genuine, Forgery) pairs of BHSig-H

Fig. 7. Examples of (a) (b) absolute top (Genuine, Genuine)-pairs, (c) (d) non-absolute top (Genuine, Genuine)-pairs, and (e) (f) (Genuine, Forgery)-pairs from BHSig-B and BHSig-H respectively.

5 Conclusion

As a critical application especially for formal scenarios like forensic handwriting analysis, signature verification played an important role since it has been proposed. In this work, we proposed a writer-independent signature verification model for learning top-rank pairs. What is novel and interesting for this model is that the optimization objective of top-rank learning is to maximize the pos@top, to say the highly reliable signature pairs in this case. This optimization goal has fulfilled the requirement of the intuitive need of signature verification tasks to acquire reliable genuine signatures, not only to naively classify positive from negative. Through two experiments on data set BHSig-B and BHSig-H, the effectiveness of pos@top maximization has been proved compared with a metric learning-based network, the SigNet. Besides, the performance of the proposed model on the AUC, accuracy, and other common evaluation criteria frequently used in signature verification shown encouraging results as well.

Acknowledgement. This work was supported by JSPS KAKENHI Grant Number JP21J21934, Grant-in-Aid for JSPS Fellows and JST, ACT-X Grant Number JPMJAX200G, Japan.

References

1. Dey, S., Dutta, A., Toledo, J.I., Ghosh, S.K., Lladós, J., Pal, U.: Signet: convolutional siamese network for writer independent offline signature verification, CoRR (2017). http://arxiv.org/abs/1707.02131
2. Zheng, Y., Zheng, Y., Ohyama, W., Suehiro, D., Uchida, S.: Ranksvm for offline signature verification. In: Proceedings of the ICDAR, pp. 928–933 (2019)
3. Li, N., Jin, R., Zhou. Z.: Top rank optimization in linear time. In: Proceedings of the NIPS, pp. 1502–1510 (2014)

4. Frery, J., Habrard, A., Sebban, M., Caelen, O., He-Guelton, L.: Efficient top rank optimization with gradient boosting for supervised anomaly detection. In: Ceci, M., Hollmén, J., Todorovski, L., Vens, C., Džeroski, S. (eds.) ECML PKDD 2017. LNCS (LNAI), vol. 10534, pp. 20–35. Springer, Cham (2017). https://doi.org/10.1007/978-3-319-71249-9_2
5. Boyd, S.P., Cortes, C., Mohri, M., Radovanovic, A.: Accuracy at the top. In: Proceedings of the NIPS, pp. 962–970 (2012)
6. Impedovo, D., Pirlo, G.: Automatic signature verification: the state of the art. IEEE Trans. Syst. Man Cybern. Part C **38**, 609–635 (2008)
7. Hafemann, L.G., Sabourin, R., Oliveira, L.S.: Offline handwritten signature verification - literature review. In: Proceedings of the IPTA, pp. 1–8 (2017)
8. Lee, L.L., Berger, T., Aviczer, E.: Reliable online human signature verification systems. IEEE Trans. Pattern Anal. Mach. Intell. **18**, 643–647 (1996)
9. Kalera, M.K., Srihari, S.N., Xu, A.: Offline signature verification and identification using distance statistics. Int. J. Pattern Recognit. Artif. Intell. **18**, 1339–1360 (2004)
10. Ferrer, M.A., Vargas-Bonilla, J.F., Morales, A., Ordonez, A.: Robustness of offline signature verification based on gray level features. IEEE Trans. Inf. Forensics Secur. **7**, 966–977 (2012)
11. Lai, S., Jin, L.: Recurrent adaptation networks for online signature verification. IEEE Trans. Inf. Forensics Secur. **14**, 1624–1637 (2019)
12. Okawa, M.: Template matching using time-series averaging and DTW with dependent warping for online signature verification. IEEE Access **7**, 81010–81019 (2019)
13. Banerjee, D., Chatterjee, B., Bhowal, P., Bhattacharyya, T., Malakar, S., Sarkar, R.: A new wrapper feature selection method for language-invariant offline signature verification. Expert Syst. Appl. **186**, 115756 (2021)
14. Okawa, M.: From bovw to VLAD with KAZE features: offline signature verification considering cognitive processes of forensic experts. Pattern Recognit. Lett. **113**, 75–82 (2018)
15. Ruiz-del-Solar, J., Devia, C., Loncomilla, P., Concha, F.: Offline signature verification using local interest points and descriptors. In: Ruiz-Shulcloper, J., Kropatsch, W.G. (eds.) CIARP 2008. LNCS, vol. 5197, pp. 22–29. Springer, Heidelberg (2008). https://doi.org/10.1007/978-3-540-85920-8_3
16. Hafemann, L.G., Sabourin, R., Oliveira, L.S.: Learning features for offline handwritten signature verification using deep convolutional neural networks. Pattern Recognit. **70**, 163–176 (2017)
17. Souza, V.L.F., Oliveira, A.L.I., Sabourin, R.: A writer-independent approach for offline signature verification using deep convolutional neural networks features. In: Proceedings of the BRACIS, pp. 212–217 (2018)
18. Melekhov, I., Kannala, J., Rahtu, E .: Siamese network features for image matching. In: Proceedings of the ICPR, pp. 378–383 (2016)
19. Guo, Q., Feng, W., Zhou, C., Huang, R., Wan, L., Wang, S.: Learning dynamic siamese network for visual object tracking. In: Proceedings of the ICCV, pp. 1781–1789 (2017)
20. Wei, P., Li, H., Hu, P.: Inverse discriminative networks for handwritten signature verification. In: Proceedings of the CVPR, pp. 5764–5772 (2019)
21. Trotman, A.: Learning to rank. Inf. Retr. **8**, 359–381 (2005)
22. Burges, C.J.C., et al.: Learning to rank using gradient descent. In: Proceedings of the ICML, vol. 119, pp. 89–96 (2005)
23. Agarwal, S., Graepel, T., Herbrich, R., Har-Peled, S., Roth, D.: Generalization bounds for the area under the ROC curve. J. Mach. Learn. Res. **6**, 393–425 (2005)

24. Usunier, N., Amini, M.-R., Goutte, C.: Multiview semi-supervised learning for ranking multilingual documents. In: Gunopulos, D., Hofmann, T., Malerba, D., Vazirgiannis, M. (eds.) ECML PKDD 2011. LNCS (LNAI), vol. 6913, pp. 443–458. Springer, Heidelberg (2011). https://doi.org/10.1007/978-3-642-23808-6_29
25. Charoenphakdee, N., Lee, J., Jin, Y., Wanvarie, D., Sugiyama, M.: Learning only from relevant keywords and unlabeled documents. In: Proceedings of the EMNLP-IJCNLP, pp. 3991–4000 (2019)
26. Mehta, S., Pimplikar, R., Singh, A., Varshney, L. R., Visweswariah, K.: Efficient multifaceted screening of job applicants. In: Proceedings of the EDBT/ICDT, pp. 661–671 (2013)
27. Zheng, Y., Zheng, Y., Suehiro, D., Uchida, S.: Top-rank convolutional neural network and its application to medical image-based diagnosis. Pattern Recognit. **120**, 108138 (2021)
28. Chhabra O., Chakraborty, S.: Siamese triple ranking convolution network in signature forgery detection (2019)
29. Rudin, C.: The p-norm push: a simple convex ranking algorithm that concentrates at the top of the list. J. Mach. Learn. Res. **10**, 2233–2271 (2009)

On-the-Fly Deformations for Keyword Spotting

George Retsinas[1], Giorgos Sfikas[4(✉)], Basilis Gatos[2], and Christophoros Nikou[3]

[1] School of Electrical and Computer Engineering, National Technical University of Athens, Athens, Greece
gretsinas@central.ntua.gr

[2] Computational Intelligence Laboratory, Institute of Informatics and Telecommunications, National Center for Scientific Research "Demokritos", Athens, Greece
bgat@iit.demokritos.gr

[3] Department of Computer Science and Engineering, University of Ioannina, Ioannina, Greece
cnikou@cse.uoi.gr

[4] Department of Surveying and Geoinformatics Engineering, University of West Attica, Athens, Greece
gsfikas@uniwa.gr

Abstract. Modern Keyword Spotting systems rely on deep learning approaches to build effective neural networks which provide state-of-the-art results. Despite their evident success, these deep models have proven to be sensitive with respect to the input images; a small deformation, almost indistinguishable to the human eye, may considerably alter the resulting retrieval list. To address this issue, we propose a novel "on-the-fly" approach which deforms an input image to better match the query image, aiming to stabilize the aforementioned sensitivity. Results on the IAM dataset verify the effectiveness of the proposed method, which outperforms existing Query-by-Example approaches. Code is available at https://github.com/georgeretsi/defKWS/.

Keywords: Keyword spotting · Query by example · Phoc descriptor · Non-rigid deformations · Image warping · Deep learning

1 Introduction

Distance measures are critical in any retrieval paradigm. Keyword Spotting (KWS), defined as the problem of retrieving relevant word instances over a digitized document, is no exception to this rule. A typical KWS pipeline consists of i) a way to produce compact word image descriptions ii) a distance measure that is used to compare query words against the target content, iii) returning to the user the words that are the closest to the query, in terms of the distance measure considered. We can consider this processing pipeline in terms of the characteristics of the space where the produced image descriptions reside. The choice of

© Springer Nature Switzerland AG 2022
S. Uchida et al. (Eds.): DAS 2022, LNCS 13237, pp. 338–351, 2022.
https://doi.org/10.1007/978-3-031-06555-2_23

this space is interrelated with the choice of certain topological properties; the selected distance measure induces a specific topology for the considered space, and it is in this regard that it is important in the current application context. Consequently, it is perhaps more useful to consider the problem of defining a descriptor *jointly* with the problem of defining a useful distance measure.

Depending on the input format of the query, we can distinguish two scenarios: Query by Example (QbE) and Query by String (QbS). In the QbE scenario the query is given in the form of an image, while in the QbS scenario the query corresponds to a text string. We consider only the QbE KWS scenario.

In this work, we propose focusing on determining a distance measure to accurately perform KWS. We cast the KWS problem as follows: Given a query image $\mathbf{I_q}$ we aim to find a valid spatial transformation $\mathcal{T}$ over an input image $\mathbf{I_w}$, such that the two images are as *visually close* as possible.

Implementation-wise, the distance of the images is to be minimized in the feature domain, by extracting feature vectors for both images using a deep neural network. The transformations are implemented via a deformation set of parameters $\mathbf{d}$ in an iterative manner. Overall, we seek appropriate parameters $\mathbf{d}$, which minimize the distance between the two images using a gradient-based optimization algorithm. This concept is depicted in Fig. 1, where the deformed image is generated by the proposed algorithm for a large number of iterations (100). In a sense, we seek a "least resistance" transformation path, where transformations do not disrupt the content of the image. In terms of the manifold hypothesis, this path can be understood as a geodesic connecting the two images along the data manifold; the required distance is defined to be the length of this path.

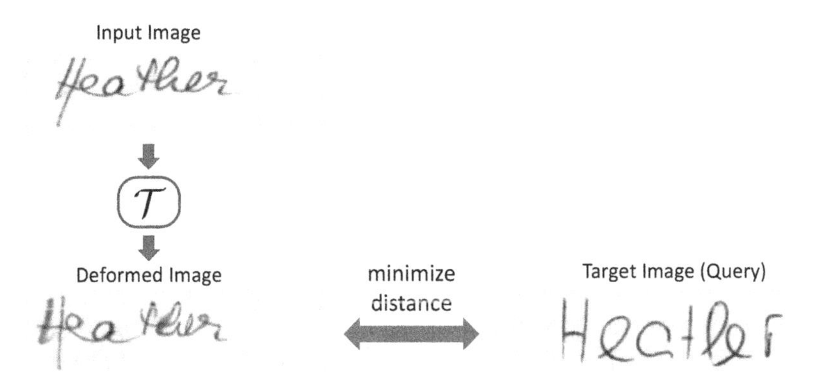

Fig. 1. Visualization of the paper motivation: transform an input image in such a manner that it becomes sufficiently similar to a given target image. The deformed version of the image was created by the proposed algorithm over an impractically large number of iterations (100).

Even though the core idea of computing a deformation may be attractive for improving performance (it is in this sense that we refer to it as "on-the-fly"), it comes at the cost of computational effort. An iterative process should be performed for each query-word pair. As the experimental section suggests, even when using a very small number of iterations (e.g., 3) we can achieve a considerable boost in performance. This is attributed to the sensitivity of modern deep learning systems, where small deformations, almost indistinguishable to the human eye, can notably alter the retrieval list of a query.

Furthermore, as a minor contribution, we propose a simple yet effective deep model based on residual blocks which is used to estimate Pyramidal-Histogram-of-Characters (PHOC) [1] representations. Using this fairly compact network of ~8 million parameters, we achieve performance in the ballpark of the current state-of-the-art methods. Using this model, along with the proposed on-the-fly deformation scheme we outperform existing methods for the case of QbE keyword spotting scenario for the challenging IAM dataset.

The remainder of this paper is structured as follows. In Sect. 2 we briefly review related work. In Sect. 3 we present the architecture of a reference KWS system that will serve as baseline as well as a backbone of the proposed "on-the-fly deformation" approach, which is presented in Sect. 4. Experiments are presented in Sect. 5. We close the paper with a short discussion of our conclusion and plans for future work in Sect. 6.

2 Related Work

Recent developments in computer vision and KWS in particular have redirected the interest of the community from learning-free feature extraction approaches [4,15] to learning-based/deep learning approaches [12,16,25]. A stepping stone towards this shift was the the seminal work of Almazán et al. [1], which first introduced the Pyramidal-Histogram-of-Characters (PHOC) representation. The main idea was to embed both word images and text strings into this common subspace of PHOC embedding, allowing both QbE and QbS scenarios. Since this method was introduced before the surge of deep learning, it was trained with Support Vector Machines (SVMs) but later inspired a large number of deep learning methods to tackle the KWS problem using the attribute-based rationale of PHOC embeddings. A characteristic example is the introduction of convolutional networks as PHOC estimators [9,12,21,25,27]. Apart from this direction of attribute-based approaches, Wilkinson et al. [28] used a triplet CNN, accepting pairs of positive word matches together with a negative word match, Retsinas et al. [20] extracted word image features from a Seq2Seq recognition system, while Krishnan et al. [13] adopted a hybrid approach which includes PHOC and one-hot word representations, along with a semantic-driven normalized word embedding, aiming to extract an effective joint image-text feature space.

Ideas related to shape deformation and deformation-induced distances have been proposed and used in different contexts. Rigid and non-rigid deformations (or "warps") are important in image and volume registration [14]. Other examples of context include proposing a distance between shapes and using it to learn a manifold [3,24], using non-rigid matching between shape and template to create a shape description [2], or using a deformation-based distance to cluster a dataset [23]. Geometric deformations have also garnered much interest in the context of deformable convolutions and spatial transformers [19].

Notably, the proposed work "flirts" with the idea of constructing an implicit manifold path, consisted of valid deformation steps between word representations. Despite the fact that manifold-based approaches have been successfully deployed for QbE KWS along with traditional feature extraction methods [22,26], we are not aware of recent deep learning approaches towards this direction.

3 Reference KWS System

First, we will describe the reference KWS system that we use, which will serve both as baseline and as the backbone of the proposed on-the-fly deformation approach. In what follows, we describe the preprocessing steps, the proposed architecture and the training process.

3.1 Preprocessing

The pre-processing steps, applied to every word image, are: 1. All images are resized to a resolution of 64×256 pixels. Initial images are padded (using the image median value, usually zero) in order to attain the aforementioned fixed size. If the initial image is greater than the predefined size, the image is rescaled. The padding option aims to preserve the existing aspect ratio of the text images. 2. A simple global affine augmentation is applied at every image, considering only rotation and skew of a small magnitude in order to generate valid images.

3.2 Proposed Architecture

The overall architecture of the proposed PHOC estimator is described in Fig. 2. We distinguish 4 components:

1. **Convolutional Backbone:** The convolutional backbone is made up of standard convolutional layers and ResNet blocks [6], interspersed with max-pooling and dropout layers. In particular, we have a 7×7 convolution with 32 output channels, a series of $2 3 \times 3$ ResNet blocks with 64 output channels, $4 3 \times 3$ ResNet blocks with 128 output channels and $4 3 \times 3$ ResNet blocks with 256 output channels. The standard convolution and the ResNet blocks are all followed by ReLU activations, Batch Normalization and dropout. Between the aforementioned series of ResNet blocks, a 2×2 max pooling of stride 2 is applied in order to spatially downscale the produced feature map. Overall, the convolutional backbone accepts a word image of size 64×256 and outputs a tensor of size $8 \times 32 \times 256$ (the last dimension corresponds to the feature space). The CNN backbone is followed by a *column-wise max pooling* operation, which transforms the output of the CNN (a feature map tensor of size $8 \times 32 \times 256$), into a sequence of feature vectors along the x-axis (size 32×256).

2. **1D Convolutional Part:** The sequence of features, as generated by the CNN backbone and the column-wise max-pooling operation, is then processed by a set of 1D convolutional layers in order to add contextual information to the extracted features. This contextual information is in line with the underlying representation of PHOC, where an attribute corresponds to both the class (character) and its relative position in the images (e.g. the rightmost 'a'). The 1D CNN part consists of three consecutive 1D convolutional layers with kernel size 5, stride 2 and 256 channels. Between the layers, ReLU activations, Batch Normalization and Dropout modules are used. Since the convolutional layers are strided, the output of this part would be a reduced feature sequence of size 4×256. This sequence is then concatenated into a single feature vector of size 1024.
3. **Linear Head:** The final part of the proposed architecture takes as input the concatenated feature vector and produces a PHOC estimation. Pyramidal representations up to 4 levels are used and thus the resulting PHOC embedding has a size of 390. The linear head consists only of two linear layers with a ReLU activation between them and a sigmoid activation on the output (as in [25]).

3.3 Training Process

As in the original PHOCNet paper [25], we train our system using the binary cross-entropy (BCE) loss. Training samples are augmented using a typical affine transformation approach. The training of the proposed system is performed via an Adam [8] optimizer using an initial learning rate of 0.001 which gradually decreases using a multistep scheduler. The overall training epochs are 240 and the scheduler decreases the learning rate by a factor of 0.1 at 120 and 180 epochs. Training samples were fed in batches of 64 images.

3.4 Retrieval Application

For the considered QbE setting, retrieval can be trivially performed by comparing image descriptors. A widely-used distance metric for KWS applications is the cosine distance [16,18,27] and thus we also perform comparisons by this metric. Even though initial works on neural net-based PHOC estimators (e.g. [25,27]) performed comparisons on the PHOC estimation space, as these estimations can be also used straightforwardly for the QbS scenario, it has been proven in practice that features drawn from intermediate layers are more effective [17] in QbE applications. Therefore, we select the concatenated output of the 1D CNN part as the default image descriptor.

4 On-the-Fly Deformation

In this section, we will describe in detail the proposed on-the-fly deformation approach, where a word image is spatially transformed in order to be as close as possible to the query image with respect to their corresponding feature vectors.

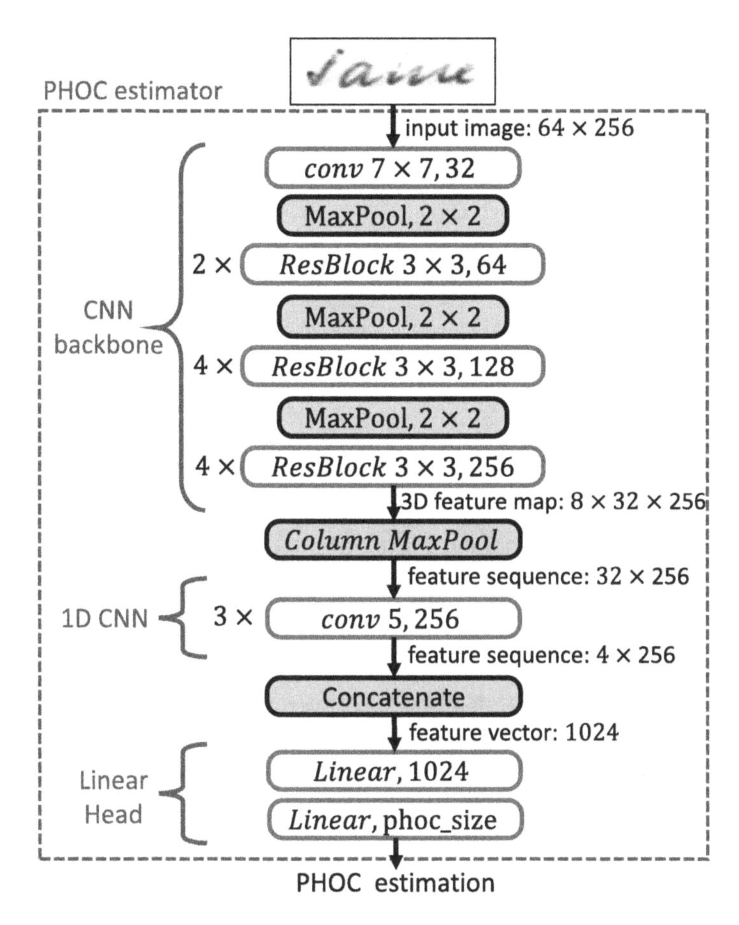

Fig. 2. Overview of the reference architecture consisted of three distinct architecture blocks: CNN backbone, 1D CNN part and Linear Head.

4.1 Considered Deformations

First, we describe the considered deformations which can be applied to the image during the proposed process. Specifically, we consider three categories of increasing deformation "freedom":

1. **Global Affine:** A typical affine transformation is applied on the image, defined by a 2×3 transformation matrix.
2. **Local Affine:** Seeking more refined transformation, we consider a local affine approach, where the image is split to overlapping segments along the x-axis and a typical affine transformation is applied in each segment. Consistency-wise, bilinear interpolation is performed in order to compute the per pixel translation of neighboring segments.

3. **Local Deformation:** Image patches are deformed according to independent x-y pixel translation vectors. Patches of 8×8 are considered and the deformation vectors correspond to the center of these patches.

Examples of aforementioned deformation categories are depicted in Fig. 3.

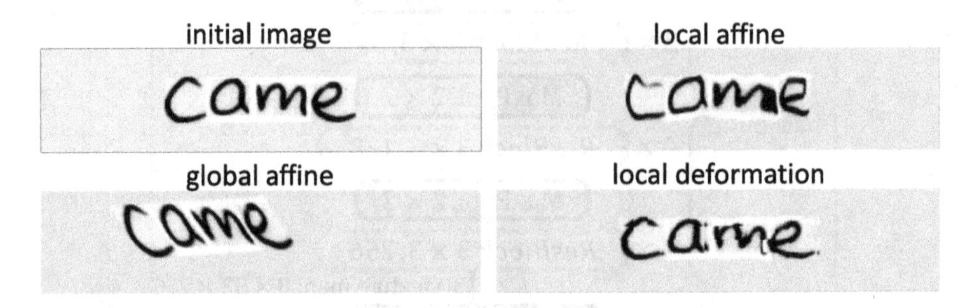

Fig. 3. Visualization of the possible deformation categories. Deformation parameters were selected in order to clearly show the effect of the deformations without significantly distorting the image.

The application of these deformations over the initial image is performed through a grid sampler which allows back-propagation, as defined in Spatial Transformer Networks [7], since the deformation parameters are to be iteratively defined.

Overall, the deformation parameters $\mathbf{d}$ consist of three distinct subsets, $\mathbf{d} = \{\mathbf{d}_{ga}, \mathbf{d}_{la}, \mathbf{d}_{ld}\}$ (ga: global affine, la: local affine and ld: local deformation) which form an interpolation grid. The interpolation grid is applied over the image via a grid sampling technique, generating a transformed image: $\mathbf{I}' = \mathcal{T}(\mathbf{I}; \mathbf{d})$.

4.2 Query-Based Deformation

As we have already stated, we aim to find an appropriate transformation of a word image with respect to a target query image. Given a feature extractor f, a bilinear grid sampler $\mathcal{T}$ controlled by deformation parameters d and a pair of input/query images $\mathbf{I_w}/\mathbf{I_q}$, we could opt to maximize the quantity

$$S_C(f(\mathcal{T}(\mathbf{I}_w; \mathbf{d})), f(\mathbf{I}_q)) \tag{1}$$

where S_C denotes the cosine similarity metric. Equation 1 hints towards the main objective function component, which we proceed to regularize by adding intuitive constraints. First, deformation parameters should be minimized in order to avoid inconsistencies and abrupt transformations ($\|\mathbf{d}\|_2$). As we have already mentioned, we want to create a "least resistance" path of consecutive transformations and thus we do not want to deviate from the solution of the previous step by allowing large changes in deformation parameters. Furthermore, a per-pixel comparison between the transformed image and the query ($\|\mathcal{T}(\mathbf{I}_w; \mathbf{d}) - \mathbf{I}_q\|_F$) may be helpful to adapt finer details.

Summing up the aforementioned loss components, we finalize our proposal for the objective function as:

$$\begin{aligned}\mathcal{L}(\mathbf{d}) =& 1 - S_C(f(\mathcal{T}(\mathbf{I}_w; \mathbf{d})), f(\mathbf{I}_q)) + \\ & a\|\mathcal{T}(\mathbf{I}_w; \mathbf{d}) - \mathbf{I}_q\|_2 + b\|\mathbf{d}\|_2 \end{aligned} \tag{2}$$

The optimization of Eq. 2 is performed via Adam [8]. A sketch of the proposed algorithm is presented in Algorithm 1. Loss hyperparameters are empirically set to $a = 10$ and $b = 1$ by visually observing the resulted transformations. The critical hyper-parameters to be set are the learning rate of Adam and the number of iterations. These hyperparameters are correlated and essentially define the trade-off of improvement vs cost of computation.

Algorithm 1. On-the-fly Deformation

Input: Adam hyperparameters, number of iterations K,
initial deformation $\mathbf{d}_0$, loss hyperparameters a, b
Output: optimized deformation parameters $\mathbf{d}_K$
1: Initialize $\mathbf{d}$ as $\mathbf{d}_0$
2: **for** $i = 0$ to $K - 1$ **do**
3: Forward Pass: Compute $\mathcal{L}(\mathbf{d}_i)$ according to Eq. 2
4: Backward Pass: Compute $\nabla\mathcal{L}(\mathbf{d}_i)$
5: Adam Update: $\mathbf{d}_{i+1}$
6: **end for**

4.3 Implementation Aspects

The extra computational effort introduced by the proposed iterative approach is not trivial. Transforming each and every word image with respect to a specific query is inefficient for large-scale applications. Nonetheless, direct comparison of the initial feature vectors yields notable retrieval performance and thus we expect all the relevant images to be brought up in the first places of the retrieval list. Therefore, instead of applying the proposed method to every word image, we can use only a subset of the N_w most relevant images, as done in [15]. This way, the computationally intense proposed algorithm is performed for only a dozens of words.

5 Experimental Evaluation

Evaluation of the proposed system is performed on the IAM dataset, consisted of handwritten text from 657 different writers and partitioned into writer - independent train/validation/test sets. As IAM is a large and multi-writer dataset, it is very challenging and typically used as the standard benchmark of comparison for Keyword Spotting methods. The setting under investigation is Query-by-Example (QbE) spotting. As evaluation metrics, we use the standard metrics that are used in the related literature: mean average precision (MAP). Following the majority of KWS works, images in test set with more than one occurrence that do not belong to the official stopword list comprise the query list [1,25]. Results are reported as average values over 3 separate runs.

5.1 Ablation Study

First, we report basic spotting performance when the proposed KWS system is used. Compared to applying cosine distance on the PHOC estimations which corresponds to 88.78% MAP, using the proposed intermediate feature vectors (i.e., the concatenated output of the 1D CNN part) results to the significantly increased performance of 91.88% MAP.

An interesting aspect of the problem at hand is the sensitivity of the deep learning model to small deformation on the input image. This observation is not restricted to the specific architecture, but pertains to deep neural nets in general. Specifically, it is closely related to the recent field of adversarial examples [5], where a network can be "fooled" and misclassify an input image by adding barely noticeable noise to the image.

This observation would be our motivation for dramatically reducing the number of required steps for the proposed iterative algorithm. If our objective would be to aim at well-performing image deformations of large magnitude, then we should cautiously perform many steps of the proposed algorithm with a small learning rate. Nonetheless, in practice we can greatly simplify the procedure by performing 2-3 steps with a larger learning rate, under the implicit assumption that a solution with considerably improved performance exists in the immediate neighborhood of the image.

To support this idea, we report the per-query difference in spotting performance when applying random transformations from the described categories. The magnitude of the deformation parameters are close to zero, essentially having no visible impact on the image. Specifically, the mean absolute difference in AP for all considered queries is $\sim$1.5%, even though the overall MAP lies in the ballpark of the initial performance ($91.88 \pm 0.18\%$). This means that applying visually trivial deformations could generate notable fluctuations in per-query performance. Therefore, we aim to harness this fluctuations and find the proper deformations that can have a positive impact on the overall performance through the proposed algorithm.

Relying on the idea of minor yet impactful deformations, we set the learning rate to 0.01, the default iterations to $K = 3$ and the considered subset to the $N_w = 50$ more relevant images.

In the following ablation experiments, we examine different aspects of the proposed iterative algorithm over the validation set of the IAM (first validation set of the official writer-independent partitions), In Table 1 we report the impact of the different categories of deformations. As we can see, the deformations of higher degree of freedom, i.e. local affine and local deformation, yield better results. Nonetheless, the best results are reported when all three types of deformation are used together. Therefore, in the upcoming experiments, the combination of the three deformation types is the default approach.

Table 1. Exploring the impact of the different categories of deformations and their combinations.

Deformation type	MAP (%)
Reference	95.59
gaffine	95.91
laffine	96.22
ldeform	96.19
gaffine + laffine	96.12
gaffine + ldeform	96.14
laffine + ldeform	96.32
gaffine + laffine + ldeform	96.40

Table 2 contains the results for different number N_w of retrieved words. Along the performance metric, we report the required extra time per query. We observe that in all considered cases, there is a noticeable performance increase. Performance is gradually increased up to 50 retrieved words. For 75 retrieved words, no further increase is observed and thus $N_w = 50$ is set as the default value for the rest of the experiments. As expected, the time requirements are almost linearly increased by the number of the retrieved words. In the proposed setting the per-query retrieval time is under 1 s, but this linear correlation hints that applying the proposed algorithm to the full retrieval list of thousand of words would lead to impractical time requirements.

Table 2. Exploring the impact of the N_w (size of considered subset for applying the deformation algorithm) to both performance and time requirements. The IAM validation set is used for evaluation.

N_w	MAP (%)	Time (sec/query)
Reference	95.59	–
10	96.21	0.14
25	96.33	0.30
50	96.40	0.57
75	96.39	0.83

Finally, we explore the impact of K, i.e. the number of iterations required by the proposed approach. In Table 3, we summarize both the MAP and the required time per query for different values of K. Under the proposed framework, even a single iteration can provide a boost in performance. Nonetheless, (too) many iterations seem to have a negative effect on performance. This can be attributed to the rather high learning rate (0.01), selected specifically for performing a small number of iterations. Again, the time requirements have a linear dependence to the number of iterations, as expected by the computational complexity of the algorithm. Even though we achieve significant boost by simply following a few steps towards the gradient direction, the case of visually intuitive transformations (cf. Fig. 1) still requires a large number of constrained steps. The reported time requirements raise an obstacle towards this direction, since even for only 50 images, several seconds of run-time are required.

Table 3. Exploring the impact of the number of iterations K to both performance and time requirements. The IAM validation set is used for evaluation.

K	MAP (%)	Time (sec/query)
Reference	95.59	–
1	95.97	0.24
2	96.34	0.40
3	96.40	0.57
4	96.26	0.74
5	96.23	0.91
10	96.11	1.75
15	96.07	2.61
20	95.98	3.45

5.2 Comparison to State-of-the-Art Systems

A comparison of our method versus state-of-the-art methods for KWS is presented in Table 4. Notably, the proposed reference system achieves performance in the ballpark of the state-of-the-art approaches, supporting the effectiveness of the extracted deep features. More importantly, the proposed iterative on-the-fly approach achieves a significant boost of over 1% MAP compared to the reference system, outperforming existing KWS methods for the QbE scenario.

Table 4. Comparison of state-of-the-art KWS approaches for the IAM dataset and the QbE setting.

Method	MAP (%)
PHOCNet [25]	72.51
HWNet [11]	80.61
Triplet-CNN [28]	81.58
PHOCNet-TPP [27]	82.74
DeepEmbed [9]	84.25
Deep Descriptors [18]	84.68
Zoning Ensemble PHOCNet [21]	87.48
End2End Embed [10]	89.07
DeepEmbed [10]	90.38
HWNetV2 [12]	92.41
NormSpot [13]	92.54
Seq2Emb [20]	92.04
Proposed Systems	
Reference system	91.88
On-the-fly deformations	93.07

6 Conclusions and Future Work

In this work, we proposed an iterative approach which provides on-the-fly deformations capable to minimize the distance between image descriptions. Considered deformations spatially transform the image according to a minimization loss, following a gradient-based optimization approach. The proposed algorithm is utilized in QbE KWS setting, where our aim is to transform an input word image to be as close as possible to a query image, with respect to an extracted feature space. The feature space in our case is created by a deep neural network, acting as PHOC estimator, while the features are drawn from an intermediate layer. An interesting aspect of this work is the observation that no visually intense deformations are required for achieving considerable boost in performance. This phenomenon can be attributed to the nature of neural networks, which are susceptible to adversarial attacks, a field that has been studied extensively in the recent years. This observation leads to a cost-effective algorithm, since a few steps of the algorithm are sufficient for attaining increased performance. In fact, the proposed approach achieves the best performance over existing state-of-the-art approaches, at the cost of the required on-the-fly estimation of an appropriate deformation over a reduced set of images.

Still, several interesting research questions need to be addressed as future work: is there an efficient way to generate larger yet "valid" deformations? can we connect this idea to manifold exploration and shortest path approaches through a transformation space? can we apply the deformation protocol only to the query image, as an efficient alternative, in order to be as close as possible to its neighbors?

Acknowledgment. This research has been partially co - financed by the EU and Greek national funds through the Operational Program Competitiveness, Entrepreneurship and Innovation, under the calls : "RESEARCH - CREATE - INNOVATE", project *Culdile* (code T1EΔK - 03785) and "OPEN INNOVATION IN CULTURE", project *Bessarion* (T6YBΠ - 00214).

References

1. Almazán, J., Gordo, A., Fornés, A., Valveny, E.: Word spotting and recognition with embedded attributes. IEEE Trans. Pattern Anal. Mach. Intell. **36**(12), 2552–2566 (2014)
2. Cootes, T.F., Twining, C.J., Babalola, K.O., Taylor, C.J.: Diffeomorphic statistical shape models. Image Vis. Comput. **26**(3), 326–332 (2008)
3. Gerber, S., Tasdizen, T., Joshi, S., Whitaker, R.: On the manifold structure of the space of brain images. In: Yang, G.-Z., Hawkes, D., Rueckert, D., Noble, A., Taylor, C. (eds.) MICCAI 2009. LNCS, vol. 5761, pp. 305–312. Springer, Heidelberg (2009). https://doi.org/10.1007/978-3-642-04268-3_38
4. Giotis, A.P., Sfikas, G., Nikou, C., Gatos, B.: Shape-based word spotting in handwritten document images. In: 13th International conference on document analysis and recognition (ICDAR), pp. 561–565. IEEE (2015)
5. Goodfellow, I.J., Shlens, J., Szegedy, C.: Explaining and harnessing adversarial examples. In: Proceedings of the International Conference on Learning Representations (ICLR) (2015)
6. He, K., Zhang, X., Ren, S., Sun, J.: Deep residual learning for image recognition. In: Proceedings of the IEEE Conference on Computer Vision and Pattern Recognition. pp. 770–778 (2016)
7. Jaderberg, M., Simonyan, K., Zisserman, A., et al.: Spatial transformer networks. Adv. Neural Inf. Process. Syst. **28**, 2017–2025 (2015)
8. Kingma, D.P., Ba, J.: Adam: a method for stochastic optimization. In: Proceedings of the International Conference on Learning Representations (ICLR) (2015)
9. Krishnan, P., Dutta, K., Jawahar, C.V.: Deep feature embedding for accurate recognition and retrieval of handwritten text. In: Proceedings of the 15th International Conference on Frontiers in Handwriting Recognition (ICFHR), pp. 289–294 (2016)
10. Krishnan, P., Dutta, K., Jawahar, C.: Word spotting and recognition using deep embedding. In: 2018 13th IAPR International Workshop on Document Analysis Systems (DAS), pp. 1–6. IEEE (2018)
11. Krishnan, P., Jawahar, C.V.: Matching handwritten document images. In: Leibe, B., Matas, J., Sebe, N., Welling, M. (eds.) ECCV 2016. LNCS, vol. 9905, pp. 766–782. Springer, Cham (2016). https://doi.org/10.1007/978-3-319-46448-0_46
12. Krishnan, P., Jawahar, C.V.: HWNet v2: an efficient word image representation for handwritten documents. IJDAR **22**(4), 387–405 (2019). https://doi.org/10.1007/s10032-019-00336-x
13. Krishnan, P., Jawahar, C.: Bringing semantics into word image representation. Pattern Recogn. **108**, 107542 (2020)

14. Noblet, V., Heinrich, C., Heitz, F., Armspach, J.P.: 3-D deformable image registration: a topology preservation scheme based on hierarchical deformation models and interval analysis optimization. IEEE Trans. Image Proc. **14**(5), 553–566 (2005)
15. Retsinas, G., Louloudis, G., Stamatopoulos, N., Gatos, B.: Efficient learning-free keyword spotting. IEEE Trans. Pattern Anal. Mach. Intell. **41**(7), 1587–1600 (2018)
16. Retsinas, G., Louloudis, G., Stamatopoulos, N., Sfikas, G., Gatos, B.: An alternative deep feature approach to line level keyword spotting. In: Proceedings of the IEEE/CVF Conference on Computer Vision and Pattern Recognition, pp. 12658–12666 (2019)
17. Retsinas, G., Sfikas, G., Gatos, B.: Transferable deep features for keyword spotting. In: International Workshop on Computational Intelligence for Multimedia Understanding (IWCIM), held in conjunction with EUSIPCO (2017)
18. Retsinas, G., Sfikas, G., Louloudis, G., Stamatopoulos, N., Gatos, B.: Compact deep descriptors for keyword spotting. In: 2018 16th International Conference on Frontiers in Handwriting Recognition (ICFHR), pp. 315–320. IEEE (2018)
19. Retsinas, G., Sfikas, G., Nikou, C., Maragos, P.: Deformation-invariant networks for handwritten text recognition. In: 2021 IEEE International Conference on Image Processing (ICIP), pp. 949–953. IEEE (2021)
20. Retsinas, G., Sfikas, G., Nikou, C., Maragos, P.: From Seq2Seq recognition to handwritten word embeddings. In: Proceedings of the British Machine Vision Conference (BMVC) (2021)
21. Retsinas, G., Sfikas, G., Stamatopoulos, N., Louloudis, G., Gatos, B.: Exploring critical aspects of cnn-based keyword spotting. a phocnet study. In: 13th IAPR International Workshop on Document Analysis Systems (DAS), pp. 13–18. IEEE (2018)
22. Retsinas, G., Stamatopoulos, N., Louloudis, G., Sfikas, G., Gatos, B.: Nonlinear manifold embedding on keyword spotting using t-sne. In: 2017 14th IAPR International Conference on Document Analysis and Recognition (ICDAR), vol. 1, pp. 487–492. IEEE (2017)
23. Sfikas, G., Heinrich, C., Nikou, C.: Multiple atlas inference and population analysis using spectral clustering. In: 2010 20th International Conference on Pattern Recognition, pp. 2500–2503. IEEE (2010)
24. Sfikas, Giorgos, Nikou, Christophoros: Bayesian multiview manifold learning applied to hippocampus shape and clinical score data. In: Müller, H., et al. (eds.) MCV/BAMBI -2016. LNCS, vol. 10081, pp. 160–171. Springer, Cham (2017). https://doi.org/10.1007/978-3-319-61188-4_15
25. Sudholt, S., Fink, G.A.: PHOCNet: a deep convolutional neural network for word spotting in handwritten documents. In: Proceedings of the 15th International Conference on Frontiers in Handwriting Recognition (ICFHR), pp. 277–282 (2016)
26. Sudholt, S., Fink, G.A.: A modified isomap approach to manifold learning in word spotting. In: Gall, J., Gehler, P., Leibe, B. (eds.) GCPR 2015. LNCS, vol. 9358, pp. 529–539. Springer, Cham (2015). https://doi.org/10.1007/978-3-319-24947-6_44
27. Sudholt, S., Fink, G.A.: Evaluating word string embeddings and loss functions for CNN-based word spotting. In: 2017 14th IAPR International Conference On Document Analysis And Recognition (ICDAR), vol. 1, pp. 493–498. IEEE (2017)
28. Wilkinson, T., Brun, A.: Semantic and verbatim word spotting using deep neural networks. In: 2016 15th International Conference on Frontiers in Handwriting Recognition (ICFHR), pp. 307–312. IEEE (2016)

Writer Identification and Writer Retrieval Using Vision Transformer for Forensic Documents

Michael Koepf, Florian Kleber(✉), and Robert Sablatnig

Computer Vision Lab, Institute of Visual Computing and Human-Centered Technology, TU Wien, Favoritenstraße 9/193-1, Vienna, Austria
{michael.koepf,florian.kleber,robert.sablatnig}@tuwien.ac.at
https://cvl.tuwien.ac.at/

Abstract. Writer identification and writer retrieval deal with the analysis of handwritten documents regarding the authorship and are used, for example, in forensic investigations. In this paper, we present a writer identification and writer retrieval method based on Vision Transformers. This is in contrast to the current state of the art, which mainly uses traditional Convolutional-Neural-Network-approaches. The evaluation of our self-attention-based and convolution-free method is done on two public datasets (CVL Database and dataset of the ICDAR 2013 Competition on Writer Identification) as well as a forensic dataset (WRITE dataset). The proposed system achieves a top-1 accuracy up to 99% (CVL) and 97% (ICDAR 2013). In addition, the impact of the used script (Latin and Greek) and the used writing style (cursive handwriting and block letters) on the recognition rate are analyzed and presented.

Keywords: Writer identification · Writer retrieval · Forensics · Vision Transformer

1 Introduction

The author of a handwritten document can be identified by analyzing specific characteristics, such as strokes, in the script and comparing these to a set of reference documents of potential writers. This procedure is carried out by a handwriting expert and accepted by law in modern legal systems (e.g., §314 ZPO, Austrian code of civil procedure). Thus, it can be used in forensic analysis to conclude if common authorship of a "questioned document" [36] and a handwriting specimen of a suspect is given.

Writer Identification (WI) deals with determining the authorship of a handwritten document. On the other hand, Writer Retrieval (WR) is the task of finding all documents written by the same person as a given reference document within a set of documents. If the handwritten documents are digitized, pattern recognition and statistical learning methods can be used as auxiliary technologies in forensics. In particular, the latter may be used to retrieve documents

© Springer Nature Switzerland AG 2022
S. Uchida et al. (Eds.): DAS 2022, LNCS 13237, pp. 352–366, 2022.
https://doi.org/10.1007/978-3-031-06555-2_24

with similar handwriting from a collection, such as a law enforcement agency archive. A handwriting expert then only needs to analyze the retrieved subset, which can lead to an identification of the writer (cf. [5]). The essential difference between WI and WR is in the conceptual output—the writer's identity vs. a list of documents [4]. Furthermore, WI/WR can also be used in non-forensic applications to allow libraries or archives to search non-indexed collections based on the script.

Variations of the writing of one individual challenge WI/WR methodologies. A person may write a letter in different forms, for example, depending on its position in a word. Other variations include the usage of different scripts or the writing style. Further, accidental modifications (e.g., caused by the used pen) and intentional modifications (e.g., attempted obfuscation) may occur [14].

Figure 1 illustrates the influence of different factors on the appearance and variability of handwriting.

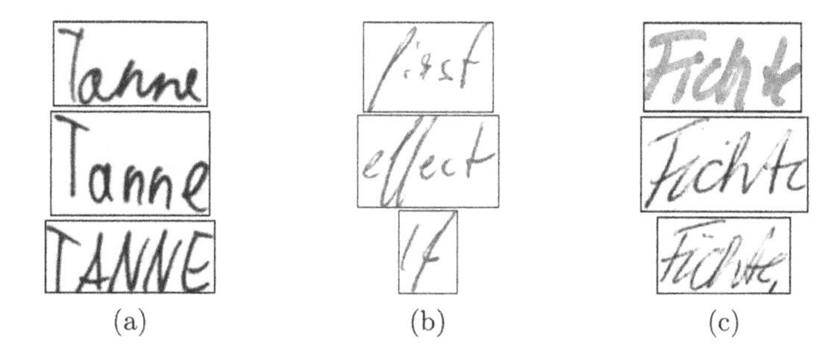

Fig. 1. Illustration of selected variations in handwriting: (a) The German word 'Tanne' (Engl. fir) written in different styles (cursive handwriting, block letters) by the writer with ID 13 from the WRITE dataset. (b) The English words 'first', 'effect' and 'if' written by the writer with ID 2 from the CVL Database [25,26]. The appearance of the letter 'f' differs depending on its position in a word. (c) The German word 'Fichte' (Engl. spruce) written with different pens by the writer with ID 16 from the WRITE dataset.

This paper presents an offline WI/WR method based on a purely self-attention-based Deep Neural Network (DNN) architecture, this is, a DNN without convolutional layers. In particular, we use a Vision Transformer (ViT)—an architecture that is based on the Transformer model [41] originating from the field of Natural Language Processing. Self-attention-based DNNs (e.g., [42]) are performance-wise the current state of the art for image classification on the ImageNet dataset [11]. In addition, they have also been showing their capabilities in other fields, for example, image retrieval [13]. Current state-of-the-art WI/WR methods mainly use traditional Convolutional Neural Networks (CNNs) (e.g., [10,17]) or hand-crafted features (e.g., [15]). To the best of our knowledge, this paper is the first to present ViT for offline WI/WR.

In detail, our contributions are as follows:

- We will show that Vision Transformers is a suitable DNN architecture for WI and WR and competes with state of the art.
- We thoroughly evaluate all steps of our proposed method using publicly available datasets dedicated to document WI/WR (CVL Database [25,26] and dataset of ICDAR2013 Competition on WI/WR [33]).
- Additionally, we analyze the impact of the script (Latin and Greek) and the writing style (cursive handwriting and block letters) on the accuracy of the proposed method.

This work is organized as follows: Sect. 2 gives a brief overview of the current state of the art. Section 3 describes the methodology used. The experimental setup, the results and a comparison to state of the art are presented in Sects. 4 and 5. Finally, a short conclusion is given in Sect. 6.

2 Related Work

In this work, we distinguish between systems *with enrollment* and *without enrollment*. The former has at least one sample for each potential writer in the training set, while in the latter the writers of the training set and test set are disjoint (cf. [6]).

2.1 Methods with Enrollment

Khan et al. [24] use Scale-Invariant Feature Transform (SIFT) [34] and RootSIFT [2] descriptors in combination with Gaussian Mixture Models (GMMs). Their pipeline operates on word-level images by training three different GMMs for each writer to represent the intra-class similarity (2 GMMs) and the inter-class similarity (1 GMM).

He and Schomaker [20] propose a CNN with two branches. The first branch is a feature pyramid that uses an entire word-level image as input, and the second one uses extracted image patches of the same image. The intermediate feature maps of the first branch are fused at different depth levels with the ones of the second branch. To identify the writer of a word, the predictions of all patches are averaged.

Kumar and Sharma [28] make use of a CNN that operates directly on page-level images, making an additional segmentation obsolete. Javidi and Jampour [22] use a combination of two different feature descriptors—one feature vector is obtained from a CNN, and an additional one represents the thickness of the handwriting.

Another work conducted by He and Schomaker [21] uses a series of convolutional blocks combined with a Recurrent Neural Network.

2.2 Methods Without Enrollment

Christlein et al. [9] use local features extracted from image patches. Their proposed CNN architecture is trained with patches centered at the handwriting contour. The embeddings obtained from the activation functions of the CNN's second-last layer are encoded with GMM supervectors [8] followed by an additional normalization step (ZCA-whitening).

Instead of relying on local features extracted from image patches, Tang and Wu [40] present a system using global features with the capability of processing images containing several lines of handwriting in a single pass, making an additional encoding obsolete.

Christlein et al. [7] utilize Exemplar-SVMs [35]. They encode PCA-whitened RootSIFT [2] descriptors detected at the script contour using GMM supervectors [8].

Keglevic et al. [23] propose a method that uses a Triplet CNN. Image patches centered at SIFT [34] key points are used to train a three-branch-network on a similarity measure. The last linear layer of the network serves as an embedding, and Vector of Locally Aggregated Descriptors (VLAD) [3] is used as encoding, followed by an additional whitening.

Opposed to other DL-based methods, which primarily train CNNs from scratch, Liang et al. [30] make use of transfer learning and utilize a pre-trained ResNet-50 [19] and also VLAD [3] as encoding.

3 WI/WR Using Vision Transformer

This paper analyzes the use of the ViT-Lite-7/4 model [18] trained from scratch for WI/WR. In the following, the preprocessing, the model's architecture and the aggregation/encoding is described. We propose two different methods: one that uses enrollment and one that does not rely on a previous enrollment of the writers. In the former, the ViT acts as a feature-extractor-classifier-combination; the output is a writer ranking. In the latter, the ViT only serves as a feature extractor that produces embeddings that are encoded to obtain a feature vector for each document. For this method, k-nearest neighbors (kNN) is used for classification and the result is a ranking of documents. Figure 2 illustrates the proposed pipeline on a conceptual level.

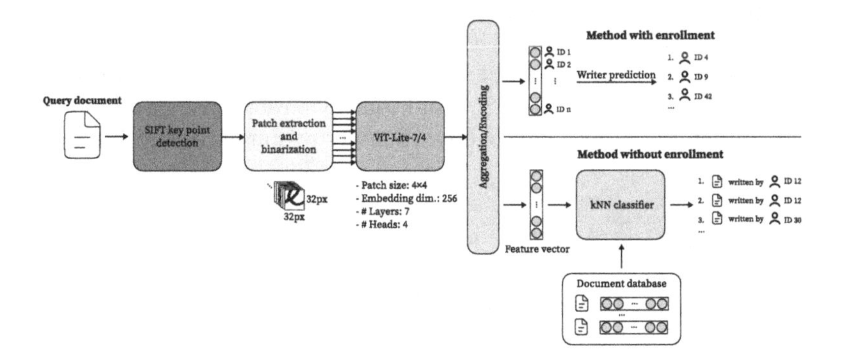

Fig. 2. Conceptual view of the proposed pipeline.

3.1 Preprocessing

Our proposed method uses segmented page-level images as input and relies on features extracted from image patches. For the patch extraction step, we adopt the approach of [15,16,23] and use SIFT [34] to detect key points in the handwriting. The prior smoothing value σ is set to an empirically defined value of 3.75. This reflects a trade-off between the number of key points and the amount of obtained patches. The patches are centered at the key points, and—as in [9,23]—a resolution of 32×32 pixels is used, followed by a subsequent binarization using Otsu [37] to account for variations of the used pen. An example demonstrating the result of the preprocessing is shown in Fig. 3.

(a) (b) (c)

Fig. 3. Illustration of the preprocessing using the first three words ('Imagine a vast') of text # 1 written by the writer with ID 13 from the CVL Database: (a) Original handwriting (b) Handwriting with detected SIFT [34] key points ($\sigma = 3.75$) (c) A subset of the extracted and binarized patches arranged as a reconstruction of the original handwriting.

3.2 ViT-Lite

State-of-the-art ViT model variants (e.g., ViT-B, ViT-L, ViT-H [12]) rely on pre-training with mid-size to large datasets (ImageNet [11], JFT-300M). The ViT-Lite models [18] are shallower networks that achieve reasonable accuracies on small datasets containing low-resolution images without the need of pre-training. The best performing ViT-Lite variant—ViT-Lite-7/4—yields an accuracy of 91.38% on CIFAR-10 [27] (with additional use of a dataset-specific augmentation strategy) and 99.29% on the MNIST dataset [29]. The ViT-Lite-7/4 uses a Transformer encoder with seven layers, an input image resolution of 32×32, and a patch size of 4×4 pixels. The flattened patches are projected into an embedding of dimension 256 and four attention heads are used. Based on these characteristics, we have chosen to apply this architecture to the task of WI/WR.

3.3 Aggregation/Encoding

Since our proposed method relies on local features extracted from image patches, an aggregation/encoding is necessary to either produce the ranking of writers (method with enrollment) or obtain a feature vector for an entire page (method without enrollment).

Method with Enrollment. We adopt the technique of [17] and use mean aggregation. For all patches of a page, we average the activation functions of the ViT's last layer ("multilayer perceptron (MLP) head" [12]). This results in a vector of a dimension equal to the number of writers in the respective training set. Each component of the MLP head corresponds to one writer. The ranking of writers can be obtained by sorting the components of the vector of aggregated values in descending order—the highest value corresponds to the most likely writer.

Method Without Enrollment. The vector is aggregated analogous to the method with enrollment and—as part of the encoding step—additionally L^2 normalized. However, in contrast to the method with enrollment, the vector cannot be used to predict the writer directly. Instead, it serves as a feature vector that acts as a surrogate of a document in a vector space model [38]. After calculating this feature vector for every document in the collection (i.e., the test set), a ranking for a given reference document ('query document') is obtained with kNN. The higher a document is ranked, the more similar it should be to the query document.

4 Experimental Setup

4.1 Datasets

We use the following datasets for training and evaluation: CVL [25,26], ICDAR 2013 Competition on Writer Identification [33] and WRITE.

The CVL Database [25,26] is a collection of 1,604 pages contributed by 310 writers with five to seven pages per writer. One reference text is German; six are in English. The included train-test split can be used to evaluate systems without enrollment, and the reference texts of the test set are a subset of the ones in the training set. Alternatively, He and Schomaker [20,21] propose a split for WI with enrollment: text # 1, # 2 and # 3 of each writer are used for training and all others for testing. Throughout this paper, we refer to the former as *CVL w/o enrollment* and the latter as *CVL w/enrollment.*

The ICDAR 2013 dataset [33] contains 1,400 page-level images written by 350 writers. Each writer contributed two English and two Greek texts. The training set and the test set use different reference texts. As with the CVL w/o enrollment dataset, the predefined split is suitable for WI/WR without enrollment.

The WRITE dataset consists of 202 documents written by 16 writers with two to 25 pages per writer. The reference texts reflect handwriting specimens as used in Austrian penal institutions. The forms vary slightly between institutions, with the standard-setting being a German reference text and the digits 0 to 9 in Arabic numerals. Additionally, a form may advise a person to write parts of the text or the entire text in a specific writing style (cursive handwriting or block letters) or to write the digits additionally in Roman numerals. However, not every writer has a specimen in both writing styles. The WRITE dataset cannot be published due to copyright constraints.

Table 1 gives a detailed listing of the distribution of writers and documents on the used training sets and test sets.

Table 1. Distribution of writers and pages on the considered training and test sets.

Dataset	# Writers	Training Set		Test Set	
		# Writers	# Pages	# Writers	# Pages
CVL w/enrollment [20,21]	310	310	929	310	675
CVL w/o enrollment [25,26]	310	27	189	283	1,415
ICDAR 2013 [33]	350	100	400	250	1,000
WRITE	16	n/a	n/a	16	202

In addition, we also evaluate our proposed method on the following subsets:

- CVL w/enrollment–English: 310 English texts from the test set (text # 4)
- ICDAR 2013–Greek: 500 Greek texts from the test set
- ICDAR 2013–English: 500 English texts from the test set
- WRITE–Cursive handwriting: 184 text regions in cursive handwriting
- WRITE–Block letters: 145 text regions in block letters

4.2 Training Details

For the training, we adopt the setup of [18]. We use AdamW [32] with its default beta values ($\beta_1 = 0.9$, $\beta_2 = 0.999$) and a weight decay of 0.03 as optimizer and an additional learning rate warm-up. As loss function, cross-entropy loss with label smoothing [39] (smoothing value $\epsilon = 0.1$) is used. Furthermore, we introduce two additional regularization techniques: First, data augmentation by rotating the image patches randomly at an angle between –25 and +25°, and second, early stopping in case the validation loss does not improve for 10 epochs.

Furthermore, we define a validation hold-out to estimate the generalization error during the training process. For the CVL w/enrollment training set, we use text # 2 of each writer for validation, for the CVL w/o enrollment training set text # 2 and text # 4. For ICDAR 2013, we use one text per writer for validation and balance the validation set with respect to the script.

We optimize the learning rate, the number of warm-up epochs, and the batch size on each training set with grid search. The entire training process results in three models (*CVL w/enrollment*, *CVL w/o enrollment*, *ICDAR 2013*).

4.3 Evaluation

Method with Enrollment. As an evaluation measure, the top-k criterion is used. A document is counted as correctly classified when the author is among the top-k predicted (ranked) ones. The top-k accuracy is calculated as the quotient of correctly classified and the total number of documents.

Method Without Enrollment. We consider the following setup for each test set:

- *Feature extractor*: we evaluate all applicable ViTs, this is, all models where none of the documents of the respective test were used for training.
- *Generation of the feature vector*: alternatively to the MLP head, we also use the embedding of the transformer encoder. For the chosen architecture, this results in a feature vector of dimension 256.
- *Metrics*: for kNN, we experiment with six different metrics (Canberra distance, Chebyshev distance, Cityblock ('Manhattan') distance, Cosine distance, standardized Euclidean distance and squared Euclidean distance).

For the performance evaluation, we follow the procedure of the ICDAR 2013 Competition on Writer Identification [33]. Every document is used once as a query and the pairwise distance between this document and all other documents is calculated. According to the distances, a ranking of the nearest documents for each document is created. As evaluation measures, the hard-top-k and soft-top-k criteria are used. With hard-top-k, a query document is only considered as classified correctly if *all* of the k highest-ranked documents were written by the same writer as the query document. For soft-top-k it suffices when *any* of the k highest-ranked documents originate from the same person as the query document. The respective accuracies are calculated analogous to the top-k accuracy.

Additionally, we also evaluate our method with the mean Average Precision (mAP) [31]—a measure for ranked retrieval results. The mAP is defined as the arithmetic mean of the Average Precision (AP), this is,

$$\frac{1}{N}\sum_{i=1}^{N} \mathrm{AP}_i \tag{1}$$

where N denotes the number of queries (i.e., the number of documents in the test set). The AP [31] is calculated as

$$\frac{\sum_r \frac{r'}{r}}{R}. \tag{2}$$

r are the ranks of documents that originate from the same writer as the query, r' corresponds to the number of correctly retrieved documents up to and including rank r and R denotes the remaining number of documents of the writer in the test set.

5 Results

This section provides an overview of the results obtained from our experiments and a comparison to state of the art. As they coincide, the hard-top-1 and soft-top-1 criteria are reported as 'top-1' throughout this section. In tables, the soft-top-k and hard-top-k criteria are abbreviated as 's-top-k' and 'h-top-k'.

For the proposed method with enrollment, the results are given in terms of the top-k criterion ($k \in \{1, 2, 3, 5, 10\}$). For the method without enrollment, we report the results with the best top-1 accuracy for each trained model that was evaluated as a feature extractor. The hard-top-k criterion is limited to $k = 2$ on the entire CVL w/o enrollment test set[1], $k = 3$ on the entire ICDAR 2013 test set and $k = 1$ on all other datasets used for evaluation. The values of k for the soft-top-k criterion are analogous to those of the top-k criterion. Additionally, we also report the mAP.

5.1 CVL Dataset

Our method achieves a top-1 accuracy of 99.0% on CVL w/enrollment. Considering the five highest-ranked writers of each document, the accuracy already reaches 99.9%. On CVL w/enrollment–English, the results are nearly identical to those on the entire test set (see Table 2a).

The best results on CVL w/o enrollment are obtained with the ViT trained on ICDAR 2013 in combination with the output of the transformer encoder for the generation of the feature vector and the standardized Euclidean distance as metric (top-1 accuracy: 97.4%). Furthermore, the ICDAR 2013 model performs better than the one trained on the dedicated training set (see Table 2b). This can be explained due to the small number of writers in the CVL w/o enrollment training set (see Table 1).

Table 2. Results on the CVL test sets.

Dataset	Top-1	Top-2	Top-3	Top-5	Top-10
CVL w/ enrollment	0.990	0.993	0.996	0.999	0.999
CVL w/ enrollment–English	0.990	0.994	0.997	0.997	0.997

(a) CVL w/ enrollment and CVL w/ enrollment–English

Method	Top-1	H-Top-2	S-Top-2	S-Top-3	S-Top-5	S-Top-10	mAP
CVL w/o enrollment + transformer encoder + std. Euclidean	0.966	0.925	0.974	0.977	0.979	0.984	0.906
ICDAR 2013 + transformer encoder + std. Euclidean	**0.974**	**0.950**	**0.979**	**0.982**	**0.984**	**0.987**	**0.928**

(b) CVL w/o enrollment

5.2 ICDAR 2013 WI/WR Competition Dataset

On the entire ICDAR 2013 dataset, our method achieves a top-1 accuracy of 97.0% (ICDAR 2013 model, output of the transformer encoder for the feature vector generation, Canberra distance).

On the language subsets, we also found that the ViT trained on ICDAR 2013 performs best. On ICDAR 2013–Greek, the top-1 accuracy is 98.0%. However, with the CVL w/enrollment model marginally better results for the soft-top-2 criterion (99.2% vs. 99.0%) and the soft-top-10 criterion (100.0% vs. 99.8%) can be achieved. On ICDAR 2013–English, the top-1 accuracy drops to 92.6%.

[1] The writer with ID 431 only copied three of his five assigned reference texts.

Table 3 lists the best results for each model. It can be observed that performance-wise, the model trained on CVL w/o enrollment is noticeably inferior compared to the other two. Furthermore, all our models perform better on the Greek language subset in general. Unexpectedly, this also applies to models trained with texts containing only Latin characters. When comparing the best setups of the ICDAR 2013 model and the CVL w/enrollment model—in proportion to the top-1 accuracy (97.0% vs. 96.1%)—a higher performance difference can be noticed for the hard-top-2 (74.7% vs. 65.4%) and hard-top-3 criterion (54.3% vs. 41.4%) on the entire test set. Thus, we conclude that it is beneficial to use a model trained on all different scripts in the test set for our proposed approach.

Table 3. Results on the ICDAR 2013 test set.

Method	Top-1	H-Top-2	H-Top-3	S-Top-2	S-Top-3	S-Top-5	S-Top-10	mAP
CVL w/ enrollment + transformer encoder + std. Euclidean	0.961	0.654	0.414	0.976	**0.983**	0.984	**0.991**	0.792
CVL w/o enrollment + transformer encoder + std. Euclidean	0.840	0.446	0.227	0.907	0.939	0.954	0.970	0.635
ICDAR 2013 + transformer encoder + Canberra	**0.970**	**0.747**	**0.543**	**0.980**	**0.983**	**0.986**	0.990	**0.844**

(a) Entire test set

Method	Top-1	S-Top-2	S-Top-3	S-Top-5	S-Top-10	mAP
CVL w/ enrollment + transformer encoder + std. Euclidean	0.976	**0.992**	0.992	0.994	**1.000**	0.989
CVL w/o enrollment + transformer encoder + std. Euclidean	0.866	0.904	0.936	0.968	0.980	0.909
ICDAR 2013 + transformer encoder + std. Euclidean	**0.980**	0.990	**0.994**	**0.998**	0.998	**0.991**

(b) ICDAR 2013–Greek

Method	Top-1	S-Top-2	S-Top-3	S-Top-5	S-Top-10	mAP
CVL w/ enrollment + transformer encoder + std. Euclidean	0.916	0.938	**0.958**	**0.964**	**0.968**	0.941
CVL w/o enrollment + transformer encoder + std. Euclidean	0.768	0.844	0.862	0.888	0.912	0.828
ICDAR 2013 + transformer encoder + Canberra	**0.926**	**0.946**	0.952	0.960	0.966	**0.946**

(c) ICDAR 2013–English

5.3 WRITE Dataset

The CVL w/enrollment model obtains the best results on the WRITE dataset and its evaluated subsets. The top-1 accuracy is 85.1% on the entire dataset (MLP head for generating the feature vector in combination with the standardized Euclidean distance). The same setup yields the highest accuracies on WRITE–Block letters (84.8%). In contrast, on WRITE–Cursive handwriting, using the output of the transformer encoder instead of the MLP head results in higher recognition rates. It can be observed that the accuracies on the WRITE dataset are far lower than on the other datasets. One reason for this might be that the distribution of the pages per writer is highly imbalanced, resulting in a more challenging retrieval task.

The highest top-1 accuracy on WRITE–Cursive handwriting is 1.1% better than on WRITE–Block letters. However, for the soft-top-3, soft-top-5 and soft-top-10 criteria and the mAP, the accuracies are higher on WRITE–Block letters. Regarding the top-1 accuracy, the difference between the ViT trained on CVL w/enrollment and the one trained on ICDAR 2013 for the respective best settings is 3.4% on the entire dataset and 2.7% on the cursive writing style subset. In contrast, on WRITE–Block letters, the ICDAR 2013 model is performance-wise more inferior (–4.8%). Furthermore, the difference in the top-1 accuracy between the writing style subsets of the best setups on the WRITE dataset is significantly lower than those on the language subsets of ICDAR 2013. Thus, for Latin script, we did not find any evidence that the writing style used for training significantly influences our proposed method's recognition rate. The detailed results are shown in Table 4.

Table 4. Results on the WRITE dataset.

Method	Top-1	S-Top-2	S-Top-3	S-Top-5	S-Top-10	mAP
CVL w/ enrollment + MLP head + std. Euclidean	**0.851**	**0.886**	**0.901**	**0.911**	**0.921**	**0.534**
CVL w/o enrollment + transformer encoder + std. Euclidean	0.777	0.856	0.881	0.881	0.906	0.514
ICDAR 2013 + transformer encoder + std. Euclidean	0.817	0.876	0.896	0.906	**0.921**	0.533

(a) Entire dataset

Method	Top-1	S-Top-2	S-Top-3	S-Top-5	S-Top-10	mAP
CVL w/ enrollment + transformer encoder + std. Euclidean	**0.859**	**0.886**	**0.908**	**0.908**	**0.918**	**0.570**
CVL w/o enrollment + MLP head + Cityblock	0.810	0.837	0.853	0.880	0.913	0.530
ICDAR 2013 + transformer encoder + Cosine	0.832	0.864	0.897	0.902	0.913	0.551

(b) WRITE–Cursive handwriting

Method	Top-1	S-Top-2	S-Top-3	S-Top-5	S-Top-10	mAP
CVL w/ enrollment + MLP head + std. Euclidean	**0.848**	**0.876**	**0.910**	**0.931**	**0.972**	**0.593**
CVL w/o enrollment + MLP head + std. Euclidean	0.779	0.848	0.862	0.917	0.959	0.534
ICDAR 2013 + transformer encoder + std. Euclidean	0.800	0.828	0.855	0.897	0.938	0.587

(c) WRITE–Block letters

5.4 Comparison to State of the Art

Table 5 lists the comparison to state of the art on the evaluated test sets.

While our method is 0.4% below the highest top-1 accuracy [21] on the entire CVL w/enrollment test set, we achieve a better result for the top-5 criterion. For CVL w/enrollment–English, our proposed method outperforms one of the three other systems [22] and achieves the same performance as [24] (top-1: 99.0%).

While we were not able to improve on any criterion on CVL w/o enrollment, our proposed method only performs slightly worse for the top-1 criteria compared to [40] (−2.3%).

On ICDAR 2013, we improved on the top-1 accuracy compared to [17] by 8.5% and compared to [30] by 2.0%.

Table 5. Comparison to state of the art.

	Top-1	Top-5
He and Schomaker [20]	0.991	0.994
He and Schomaker [21]	**0.994**	0.997
Ours	0.990	**0.999**

(a) CVL w/ enrollment

	Top-1	Top-5
Khan et al. [24]	0.990	n/a
Kumar and Sharma [28]	**0.994**	**1.000**
Javidi and Jampour [22]	0.962	n/a
Ours	0.990	0.997

(b) CVL w/ enrollment–English

	Top-1	H-Top-2	S-Top-5	mAP
Fiel and Sablatnig [17]	0.989	0.976	0.993	n/a
Christlein et al. [9]	0.994	0.988	n/a	0.978
Tang and Wu [40]	**0.997**	**0.990**	**0.998**	n/a
Christlein et al. [7]	0.992	0.984	0.996	**0.980**
Liang et al. [30]	0.990	0.982	0.993	0.970
Ours (ICDAR 2013 + transformer encoder + std. Euclidean)	0.974	0.950	0.984	0.928

(c) CVL w/o enrollment

	Top-1	H-Top-2	H-Top-3	S-Top-5	mAP
Fiel and Sablatnig [17]	0.885	0.405	0.158	0.960	n/a
Christlein et al. [9]	0.989	0.832	0.613	n/a	0.886
Tang and Wu [40]	0.990	0.844	**0.681**	0.992	n/a
Christlein et al. [7]	**0.997**	**0.848**	0.635	**0.998**	**0.894**
Keglevic et al. [23]	0.989	0.779	0.564	n/a	0.861
Liang et al. [30]	0.950	0.452	0.369	0.981	0.650
Ours (ICDAR 2013 + transformer encoder + Canberra)	0.970	0.747	0.543	0.986	0.844

(d) ICDAR 2013

6 Conclusion

In this work, we introduced Vision Transformers to the field of WI/WR. The choice of using ViTs was primarily based on their superior performance in image classification compared to traditional CNNs [42]. Our proposed method—a ViT trained from scratch with image patches extracted at SIFT key points—achieves results comparable to state of the art when using enrollment. For WI/WR without enrollment, future research is necessary to further improve the results, especially on the hard-top-k criterion. Additionally, we analyzed the influence of different scripts (Latin and Greek) and different writing styles (cursive handwriting and block letters) on the recognition rate. Our experiments have shown that training a model on all scripts present in the test set is beneficial. In contrast, the writing style seems to have less influence on the recognition rate.

Future research directions may include the usage of ViT architectures with convolutional layers, using pre-trained ViT models (transfer learning), or applying ViTs on word-level images. Especially the latter constitutes an interesting research topic when it comes to explainability. The attention could be visualized (e.g., Attention Rollout [1]) to find regions in the handwriting that are semantically relevant for the model.

Acknowledgements. The project has been funded by the Austrian security research programme KIRAS of the Federal Ministry of Agriculture, Regions and Tourism (BMLRT) under the Grant Agreement 879687. The computational results presented have been achieved in part using the Vienna Scientific Cluster (VSC).

References

1. Abnar, S., Zuidema, W.: Quantifying attention flow in transformers. In: Proceedings of the 58th Annual Meeting of the Association for Computational Linguistics, pp. 4190–4197. Association for Computational Linguistics (2020). https://doi.org/10.18653/v1/2020.acl-main.385
2. Arandjelovic, R., Zisserman, A.: Three things everyone should know to improve object retrieval. In: 2012 IEEE Conference on Computer Vision and Pattern Recognition, Providence, pp. 2911–2918. IEEE (2012). https://doi.org/10.1109/CVPR.2012.6248018
3. Arandjelovic, R., Zisserman, A.: All about VLAD. In: 2013 IEEE Conference on Computer Vision and Pattern Recognition, Portland, pp. 1578–1585. IEEE (2013). https://doi.org/10.1109/CVPR.2013.207
4. Atanasiu, V., Likforman-Sulem, L., Vincent, N.: Writer retrieval - exploration of a novel biometric scenario using perceptual features derived from script orientation. In: 2011 International Conference on Document Analysis and Recognition, Beijing, pp. 628–632. IEEE (2011). https://doi.org/10.1109/ICDAR.2011.132
5. Bensefia, A., Paquet, T., Heutte, L.: A writer identification and verification system. Pattern Recogn. Lett. **26**(13), 2080–2092 (2005). https://doi.org/10.1016/j.patrec.2005.03.024
6. Christlein, V.: Handwriting analysis with focus on writer identification and writer retrieval. Ph.D. thesis. Friedrich-Alexander-Universität Erlangen-Nürnberg (2018)
7. Christlein, V., Bernecker, D., Hönig, F., Maier, A., Angelopoulou, E.: Writer identification using GMM supervectors and exemplar-SVMs. Pattern Recogn. **63**, 258–267 (2017). https://doi.org/10.1016/j.patcog.2016.10.005
8. Christlein, V., Bernecker, D., Honig, F., Angelopoulou, E.: Writer identification and verification using GMM supervectors. In: IEEE Winter Conference on Applications of Computer Vision, Steamboat Springs, CO, pp. 998–1005. IEEE (2014). https://doi.org/10.1109/WACV.2014.6835995
9. Christlein, V., Bernecker, D., Maier, A., Angelopoulou, E.: Offline writer identification using convolutional neural network activation features. In: Gall, J., Gehler, P., Leibe, B. (eds.) GCPR 2015. LNCS, vol. 9358, pp. 540–552. Springer, Cham (2015). https://doi.org/10.1007/978-3-319-24947-6_45
10. Christlein, V., Maier, A.: Encoding CNN activations for writer recognition. In: 2018 13th IAPR International Workshop on Document Analysis Systems (DAS), Vienna, pp. 169–174. IEEE (2018). https://doi.org/10.1109/DAS.2018.9
11. Deng, J., Dong, W., Socher, R., Li, L.J., Li, K., Fei-Fei, L.: ImageNet: a large-scale hierarchical image database. In: 2009 IEEE Conference on Computer Vision and Pattern Recognition, Miami, pp. 248–255. IEEE (2009). https://doi.org/10.1109/CVPR.2009.5206848

12. Dosovitskiy, A., et al.: An image is worth 16x16 words: transformers for image recognition at scale. arXiv:2010.11929 [cs] (2020)
13. El-Nouby, A., Neverova, N., Laptev, I., Jégou, H.: Training vision transformers for image retrieval. arXiv:2102.05644 [cs] (2021)
14. Ellen, D., Day, S., Davies, C.: Scientific Examination of Documents: Methods and Techniques, 4th edn. CRC Press, Boca Raton (2018). https://doi.org/10.4324/9780429491917
15. Fiel, S., Sablatnig, R.: Writer retrieval and writer identification using local features. In: 2012 10th IAPR International Workshop on Document Analysis Systems, Gold Coast, Queenslands, pp. 145–149. IEEE (2012). https://doi.org/10.1109/DAS.2012.99
16. Fiel, S., Sablatnig, R.: Writer identification and writer retrieval using the Fisher vector on visual vocabularies. In: 2013 12th International Conference on Document Analysis and Recognition, Washington, pp. 545–549. IEEE (2013). https://doi.org/10.1109/ICDAR.2013.114
17. Fiel, S., Sablatnig, R.: Writer identification and retrieval using a convolutional neural network. In: Azzopardi, G., Petkov, N. (eds.) CAIP 2015. LNCS, vol. 9257, pp. 26–37. Springer, Cham (2015). https://doi.org/10.1007/978-3-319-23117-4_3
18. Hassani, A., Walton, S., Shah, N., Abuduweili, A., Li, J., Shi, H.: Escaping the big data paradigm with compact transformers. arXiv:2104.05704 [cs] (2021)
19. He, K., Zhang, X., Ren, S., Sun, J.: Deep residual learning for image recognition. In: 2016 IEEE Conference on Computer Vision and Pattern Recognition (CVPR), Las Vegas, pp. 770–778. IEEE (2016). https://doi.org/10.1109/CVPR.2016.90
20. He, S., Schomaker, L.: FragNet: writer identification using deep fragment networks. IEEE Trans. Inf. Forensics Secur. **15**, 3013–3022 (2020). https://doi.org/10.1109/TIFS.2020.2981236
21. He, S., Schomaker, L.: GR-RNN: global-context residual recurrent neural networks for writer identification. Pattern Recogn. **117**, 107975 (2021). https://doi.org/10.1016/j.patcog.2021.107975
22. Javidi, M., Jampour, M.: A deep learning framework for text-independent writer identification. Eng. Appl. Artif. Intell. **95**, 103912 (2020). https://doi.org/10.1016/j.engappai.2020.103912
23. Keglevic, M., Fiel, S., Sablatnig, R.: Learning features for writer retrieval and identification using triplet CNNs. In: 2018 16th International Conference on Frontiers in Handwriting Recognition (ICFHR), Niagara Falls, pp. 211–216. IEEE (2018). https://doi.org/10.1109/ICFHR-2018.2018.00045
24. Khan, F.A., Khelifi, F., Tahir, M.A., Bouridane, A.: Dissimilarity Gaussian mixture models for efficient offline handwritten text-independent identification using SIFT and RootSIFT descriptors. IEEE Trans. Inf. Forensics Secur. **14**(2), 289–303 (2019). https://doi.org/10.1109/TIFS.2018.2850011
25. Kleber, F., Fiel, S., Diem, M., Sablatnig, R.: CVL-database: an off-line database for writer retrieval, writer identification and word spotting. In: 2013 12th International Conference on Document Analysis and Recognition, Washington, pp. 560–564. IEEE (2013). https://doi.org/10.1109/ICDAR.2013.117
26. Kleber, F., Fiel, S., Diem, M., Sablatnig, R.: CVL database - an off-line database for writer retrieval. Writer Ident. Word Spotting (2018). https://doi.org/10.5281/ZENODO.1492267
27. Krizhevsky, A.: Learning multiple layers of features from tiny images. University of Toronto, Technical report (2009)

28. Kumar, P., Sharma, A.: Segmentation-free writer identification based on convolutional neural network. Comput. Electr. Eng. **85**, 106707 (2020). https://doi.org/10.1016/j.compeleceng.2020.106707
29. LeCun, Y., Cortes, C., Burges, C.: Mnist handwritten digit database. ATT Labs, 2 (2010). http://yann.lecun.com/exdb/mnist
30. Liang, D., Wu, M., Hu, Y.: Offline writer identification using convolutional neural network and VLAD descriptors. In: Sun, X., Zhang, X., Xia, Z., Bertino, E. (eds.) ICAIS 2021. LNCS, vol. 12736, pp. 253–264. Springer, Cham (2021). https://doi.org/10.1007/978-3-030-78609-0_22
31. Liu, L., Özsu, M.T. (eds.): Encyclopedia of Database Systems. Springer, Boston (2009). https://doi.org/10.1007/978-0-387-39940-9
32. Loshchilov, I., Hutter, F.: Decoupled weight decay regularization. In: ICLR (2019)
33. Louloudis, G., Gatos, B., Stamatopoulos, N., Papandreou, A.: ICDAR 2013 Competition on Writer Identification. In: 2013 12th International Conference on Document Analysis and Recognition, Washington, pp. 1397–1401. IEEE (2013). https://doi.org/10.1109/ICDAR.2013.282
34. Lowe, D.G.: Distinctive image features from scale-invariant keypoints. Int. J. Comput. Vis. **60**(2), 91–110 (2004). https://doi.org/10.1023/B:VISI.0000029664.99615.94
35. Malisiewicz, T., Gupta, A., Efros, A.A.: Ensemble of exemplar-SVMs for object detection and beyond. In: 2011 International Conference on Computer Vision, Barcelona, pp. 89–96. IEEE (2011). https://doi.org/10.1109/ICCV.2011.6126229
36. Osborn, A.S.: Questioned Documents; a Study of Questioned Documents with an Outline of Methods by Which the Facts May be Discovered and Shown. Lawyers Co-operative Publishing Company, Rochester (1910)
37. Otsu, N.: A threshold selection method from gray-level histograms. IEEE Trans. Syst. Man Cybern. **9**(1), 62–66 (1979). https://doi.org/10.1109/TSMC.1979.4310076
38. Salton, G., Wong, A., Yang, C.S.: A vector space model for automatic indexing. Commun. ACM **18**(11), 613–620 (1975). https://doi.org/10.1145/361219.361220
39. Szegedy, C., Vanhoucke, V., Ioffe, S., Shlens, J., Wojna, Z.: Rethinking the inception architecture for computer vision. In: 2016 IEEE Conference on Computer Vision and Pattern Recognition (CVPR), Las Vegas, pp. 2818–2826. IEEE (2016). https://doi.org/10.1109/CVPR.2016.308
40. Tang, Y., Wu, X.: Text-Independent writer identification via CNN features and joint Bayesian. In: 2016 15th International Conference on Frontiers in Handwriting Recognition (ICFHR), Shenzhen, pp. 566–571. IEEE (2016). https://doi.org/10.1109/ICFHR.2016.0109
41. Vaswani, A., et al.: Attention is all you need. In: Guyon, I., Luxburg, U.V., Bengio, S., Wallach, H., Fergus, R., Vishwanathan, S., Garnett, R. (eds.) Advances in Neural Information Processing Systems. vol. 30. Curran Associates, Inc. (2017)
42. Zhai, X., Kolesnikov, A., Houlsby, N., Beyer, L.: Scaling vision transformers. arXiv:2106.04560 [cs] (2021)

Approximate Search for Keywords in Handwritten Text Images

José Andrés[1(✉)], Alejandro H. Toselli[2], and Enrique Vidal[1]

[1] PRHLT Research Center, Universitat Politècnica de València, Valencia, Spain
{joanmo2,evidal}@prhlt.upv.es

[2] College of Computer and Information Science, Northeastern University, Boston, MA, USA
a.toselli@northeastern.edu

Abstract. Thanks to the ability to deal with the intrinsic uncertainty of handwritten text in historical documents, Probabilistic Indexing (PrIx) has emerged as an alternative to traditional automatic transcription to retrieve information from such documents. Using PrIx, adequate search techniques have been developed that not only allow for typical single-word queries, but also support complex multi-word boolean and word-sequence queries, which are commonly used in many free-text document search applications today. Here we focus on another type of text-image PrIx-based queries; namely approximate (or "fuzzy") word spelling, also commonly provided by many conventional plain-text search tools. When handwritten historical documents are considered, approximate spelling has proved to be a remarkably useful search asset in practice. However, its performance had not been formally assessed so far. We explain how approximate-spelling has been developed for large-scale PrIx's and provide an empirical analysis of precision-recall performance and computational efficiency. Experiments with the well-known "Bentham Papers" large manuscript collection show that the proposed approximate-spelling search techniques generally improve the already good search accuracy of exact-spelling queries, while computing performance gracefully scales up to deal with very large collections of handwritten text images.

Keywords: Handwritten text processing · Keyword spotting · Approximate search · Levenshtein distance · Information search and retrieval

1 Introduction

Recent advances in handwritten text recognition (HTR) [3,7,13,14] make it possible to achieve very precise transcripts of relatively modern and simple handwritten documents. However, for millions of historical manuscripts, diverse and contrived writing styles, complex and non-uniform page layouts and poor preservation quality, among others, prevent state of the art HTR techniques achieve usable transcription quality for most applications.

© Springer Nature Switzerland AG 2022
S. Uchida et al. (Eds.): DAS 2022, LNCS 13237, pp. 367–381, 2022.
https://doi.org/10.1007/978-3-031-06555-2_25

A major hurdle in this kind of manuscripts is the inherent ambiguity and uncertainty raised by archaic and inconsistent word spelling and often erratic use of abbreviations. These problems are very frequent, generally affecting to as many as one out of three running words in many historical collections. In contrast with HTR, where the target is to obtain a unique, single best interpretation of the text images, the Probabilistic Indexing (PrIx) framework [12,19,20] was proposed to explicitly deal with these intrinsic word-level uncertainties.

PrIx draws from ideas and concepts previously developed for keyword spotting (KWS), both in speech signals and text images. However, rather than caring for "key" words, any element in an image which is likely enough to be interpreted as a word is detected and stored, along with its relevance probability and its location in the image. Depending on the observed optical and/or linguistic uncertainty, PrIx retains just one, a few, or maybe many alternative textual interpretations of image regions which are stained, blurred and/or contain arcane word forms or abbreviations. Moreover, by mainly focusing on word (rather than word sequence) interpretation, PrIx is largely insensitive to page layout and reading order problems which typically hinder severely the capabilities of HTR to achieve good transcripts.

On the other hand, the PrIx framework goes far beyond the scope of KWS, because PrIx is not limited to "keyword" search applications, nor is it constrained by possibly prohibitive computational demands at query time. Instead, probabilistic indices (PrIx's) are computed "off-line" for all images of a collection, thereby allowing fast and light search computing at query time. Moreover, PrIx's can be used for many other applications, such as text analytics [6,16], including textual content based document classification [6,11].

Focusing on search applications, PrIx easily allows complex queries that go far beyond the typical single keyword queries of traditional KWS. In particular, full support for standard multi-word boolean and word-sequence queries have been developed in [18] and used in many PrIx search applications for large and huge collections of handwritten text documents[1].

Most of these applications also provide another classical set of free-text handy search tools; namely, wildcard and approximate (or "fuzzy") spelling. While these tools are generally considered remarkably useful search assets in practice, their performance had not been formally assessed so far. This paper reports a first study in this direction; specifically, it explains how approximate-spelling has been developed for large-scale PrIx's and provides an empirical analysis of precision-recall performance and computational efficiency.

The experiments have been carried out on handwritten text images of a large collection which is well known in the document processing community: The Bentham Papers.[2] A standard ground-truthed dataset from this collection has been used to evaluate search accuracy and computational efficiency, and the full collection with almost 90 000 page images has also been used to assess computational scalability. This collection was probabilistically indexed some time ago and it is

---

[1] See http://prhlt-carabela.prhlt.upv.es/PrIxDemos for a list of PrIx live demonstrators.

[2] https://www.benthampapers.ucl.ac.uk.

now publicly available through a search demonstrator[3] which includes, among many other advanced search tools, the approximate word spelling techniques presented in this paper.

2 Probabilistic Indexing and Search

In the PrIx framework, KWS is seen as the binary classification problem of deciding whether a particular image region x is *relevant* for a given query word v, i.e. try to answer the following question: "Is v actually written in x?". As in [19] and [12], we denote the image-region word *relevance probability* (RP) as $P(R{=}1 \mid X{=}x, V{=}v)$, but for the sake of conciseness, we will omit the random variable names, and for $R = 1$, we will simply write R. As discussed in [20], this RP can be simply approximated as:

$$P(R \mid x, v) \;=\; \sum_{b \sqsubseteq x} P(R, b \mid x, v) \;\approx\; \max_{b \sqsubseteq x} P(v \mid x, b) \tag{1}$$

where b is any small, word-sized image sub-region or bounding box (BB), and with $b \sqsubseteq x$ we mean the set of all BBs contained in x. $P(v \mid x, b)$ is just the posterior probability needed to "recognize" the BB image (x, b). Therefore, assuming the computational complexity entailed by the maximization in (1) is algorithmically managed, any sufficiently accurate isolated word classifier can be used to obtain $P(R \mid x, v)$. Image region word RPs do not explicitly take into account where the considered words may appear in the region x, but the precise positions of the words within x are easily obtained as a by-product [20].

An alternative to Eq. (1) to compute $P(R \mid x, v)$ is to use a suitable segmentation-free *word-sequence* recognizer [12,19,20]:

$$P(R \mid v, x) \;=\; \sum_{w} P(R, w \mid v, x) \;=\; \sum_{w : v \in w} P(w \mid x) \tag{2}$$

where w is the sequence of words of any possible transcript of x and with $v \in w$ we mean that v is one of the words of w. So the RP can be computed using state-of-the-art optical and language models and processing steps similar to those employed in handwritten text recognition (cf. Sect. 4.4), even though no actual text transcripts are explicitly produced in PrIx. It is worth noting that this computing approach allows very precise modelling of the words in their natural linguistic context, which is an important advantage with respect to other approaches which just try to (optically) model words isolatedly.

In this PrIx approach, character-level optical and language models are generally adopted, but good word-level performance is achieved by determining RPs for *"pseudo-words"*, which are arbitrary character sequences that are likely-enough to correspond to actual words. Pseudo-words are automatically "discovered" in the very test images being processed and stored along with their relevance probabilities and image locations [12,17]. The resulting indexed elements are referred to as *"pseudo-word spots"*.

[3] http://prhlt-kws.prhlt.upv.es/bentham.

This word-level indexing approach has proved to be very robust and flexible. So far it has been used to very successfully index several large iconic manuscript collections, such as the medieval French CHANCERY collection [2], the BENTHAM PAPERS [16], the Spanish CARABELA collection [6], and the massive collection of FINNISH COURT RECORDS, among others.[4]

2.1 Multi-word Boolean Queries

The PrIx framework can be naturally extended to deal with queries, q, formulated as boolean combinations of individual-word queries, $v_1, ..., v_K$, using the three basic boolean operators: *OR*, *AND* and *NOT*, respectively denoted as "$\vee$", "$\wedge$" and "$\neg$" [18].

For the sake of clarity, we denote the relevance probability of a single-word query $q = v_i$ as $P(\mathcal{R}_{v_i}) \stackrel{\text{def}}{=} P(\mathcal{R} \mid v_i, x)$. Thus, the relevance probability of x for a K-fold *AND* query $q = v_1 \wedge v_2 \wedge \cdots v_K$ is then written as $P(\mathcal{R}_{v_1} \wedge \mathcal{R}_{v_2} \cdots \wedge \mathcal{R}_{v_K})$. Similarly, the probability for an *OR* query is denoted as $P(\mathcal{R}_{v_1} \vee \mathcal{R}_{v_2} \cdots \vee \mathcal{R}_{v_K})$.

In [18] good approximations were proposed and assessed for efficiently computing the RP of arbitrarily complex combinations of these boolean operators:

$$P(\mathcal{R}_{v_1} \wedge \mathcal{R}_{v_2} \cdots \wedge \mathcal{R}_{v_K}) \approx \min\left(P(\mathcal{R}_{v_1}), P(\mathcal{R}_{v_2}), \ldots, P(\mathcal{R}_{v_K})\right)$$
$$P(\mathcal{R}_{v_1} \vee \mathcal{R}_{v_2} \cdots \vee \mathcal{R}_{v_K}) \approx \max\left(P(\mathcal{R}_{v_1}), P(\mathcal{R}_{v_2}), \ldots, P(\mathcal{R}_{v_K})\right)$$

In addition, the relevance probability of the *NOT* operator applied to a boolean query combination, B, is computed as:

$$P(\neg B) = 1 - P(B)$$

Using these approximations, relevance probability of any arbitrary Boolean combination of single-keyword queries can be easily and very efficiently computed. For example, to search for image regions containing both the words "cat" and "dog" but none of the words "mouse" or "rabbit" we can issue the query $q = \mathsf{cat} \wedge \mathsf{dog} \wedge \neg(\mathsf{mouse} \vee \mathsf{rabbit}))$ and the relevance probability of a region x is computed as:

$$P(R \mid x, q) \equiv P(R \mid x, \mathsf{cat} \wedge \mathsf{dog} \wedge \neg(\mathsf{mouse} \vee \mathsf{rabbit})) \approx$$
$$\min\Big(P(\mathcal{R}_{\mathsf{cat}}),\ P(\mathcal{R}_{\mathsf{dog}}),\ \big(1 - \max(P(\mathcal{R}_{\mathsf{mouse}}), P(\mathcal{R}_{\mathsf{rabbit}}))\big)\Big)$$

3 Approximate-Spelling Queries

Every entry of the PrIx of a given document is a pseudo-word, accompanied by its RP and its location within the document. Therefore, all conventional search tools available for querying plain text documents are potentially available to search for information in the PrIx's of a collection of text images. In this section

[4] See a list of demonstrators at: http://transcriptorium.eu/demots/KWSdemos.

we are interested in queries which include approximate (also called "fuzzy" or "elastic") word spelling. Since PrIx's are generally very large (as compared with plain text), computing efficiency becomes a major concern in our case.

In what follows, without loss of generality, we restrict ourselves to single-word approximate-spelling queries. Of course, these queries can then be straightforwardly combined with other approximate- and/or exact-word queries into arbitrarily complex boolean expressions as outlined in Sect. 2.1.

A single-word approximate-spelling query is given by a *base word*, along with an indication that some *flexibility* is allowed as to how the word is expected to be spelled in the documents considered. We model this flexibility in terms of the Levenshtein or edit distance [10], $d(v, v')$, which is the number of edit operations (i.e., single character insertions, substitutions or deletions) that need to be applied to a word v to produce another word v'. By specifying a maximum edit distance, the degree of allowed flexibility can be easily controlled.

This flexibility is particularly useful to deal with historical text, where the evolution of the language through time generally leads to important uncertainties about how a word we are interested in might have been spelled in the documents of a large collection. In addition, it is also useful when script, writing style, and/or optical image quality problems make it difficult (for machines and humans alike) to tell which are the exact characters actually written. In these cases, as discussed in Sect. 2, PrIx's do generally offer a large number of spelling alternatives. However, in many of these occasions, the word form used in a conventional query may happen not to be among any of the indexed alternatives, or it is one with very low RP. For any of these reasons, or for both, approximate-spelling queries constitute a powerful tool that do help users to retrieve textual information from PrIx's which would be difficult to find otherwise.

An approximate-spelling query with edit distance threshold d_0, can be seen as a boolean *OR* query composed of all the spelling variations of the given base word allowed by applying up to d_0 edit operations. However, the number of possible variations is generally huge. For instance, assuming a 26 character alphabet, more than 75 000 spelling variations of the word "aptitude" are within an edit distance of $d_0 = 2$ and more than 750 000 are within $d_0 = 3$. Fortunately enough, we do not need to consider all the variations, but only those which actually appear in the given PrIx (clearly, the RP of all the others is *null* and would never be retrieved!). That is, given a base word v we can first search for all the indexed pseudo-words v' such that $d(v, v') \leq d_0$ and then issue a regular multi-word *OR* query with only these pseudo-words.

A Levenshtein automaton [15] can be used for this purpose. Let $q = v \sim d_0$ denote an approximate-spelling query with base word v and threshold d_0, and let S be the set of pseudo-words in the PrIx. The language accepted by the Levenshtein automaton for this query is $L(v, d_0) = \{v' : d(v, v') \leq d_0\}$ and let $L_S(v, d_0) \stackrel{\text{def}}{=} L(v, d_0) \cap S$ be the set of PrIx pseudo-words in $L(v, d_0)$.

Then $v \sim d_0$ is equivalent to the *OR* query "$v_1 \vee \cdots \vee v_n$", where $v_i \in L_S(v, d_0), 1 \leq i \leq n$ and, as discussed in Sect. 2.1, the RP of any image region x for this query is computed as:

$$P(R \mid x, v{\sim}d_0) \;\approx \max_{v' \in L_S(v,d_0)} P(R \mid x, v') \tag{3}$$

For example, consider the query "aptitude$\sim$1" and let S be the set of pseudo-words in the PrIx computed for the experiments in Sect. 5.1. Then,

$$L_S(\mathsf{aptitude}, 1) = \{\mathsf{aptitude},\ \mathsf{apptitude},\ \mathsf{aptitute},\ \mathsf{aptitudes},\ \mathsf{apttude}, \mathsf{aptitutde}\}.$$

And using the simplified notation and approximations of Sect. 2.1, the RP of a certain image region for this query is calculated as:

$$P(\mathcal{R}_{\mathsf{aptitude}\sim 1}) = P(\mathcal{R}_{\mathsf{aptitude}\vee\mathsf{apptitude}\vee\mathsf{aptitute}\vee\mathsf{aptitudes}\vee\mathsf{apttude}\vee\mathsf{aptitutde}}) \approx \max(P(\mathcal{R}_{\mathsf{aptitude}}), P(\mathcal{R}_{\mathsf{apptitude}}), P(\mathcal{R}_{\mathsf{aptitute}}), P(\mathcal{R}_{\mathsf{aptitudes}}), P(\mathcal{R}_{\mathsf{apttude}}), P(\mathcal{R}_{\mathsf{aptitutde}}))$$

3.1 Algorithmics

The Levenshtein distance calculation is a well known problem which has been solved with different approaches over the years. Typically, it is computed by dynamic programming [21]. If we want to compute $L_S(v, d_0)$ employing this algorithm, this leads to a time complexity of $O(KNM)$, where $K = |S|$ (the number of unique pseudo-words in the PrIx), N is the length of v and M is the average length of a pseudo-word. This is clearly prohibitive in this context, as the number of unique pseudo-words is typically huge.

Another popular solution is to employ the Levenshtein automaton as discussed before. However, in order to make it work efficiently, specialized automata for all the allowed distance thresholds need to be used, which leads to several implementation difficulties.

Finally, as a compromise between computational efficiency and development simplicity, we have followed the ideas presented in [9] but employing, rather than a *trie*, a *Directed Acyclic Word Graph* (DAWG) [5]. It is a memory-efficient data structure that avoids prefix, suffix and infix redundancies of the strings to be stored. This approach entails two main phases: the creation of the DAWG and the search for the pseudo-words in $L_S(v, d_0)$.

On the one hand, the DAWG is created off-line. In this phase, the vocabulary S of unique PrIx pseudo-words is obtained and, following the procedure detailed in [5], the DAWG containing all the pseudo-words in S is created.

On the other hand, the search for the pseudo-words in $L_S(v, d_0)$ takes place on-line. For a query of the type $v \sim d_0$, a traversal of the DAWG is performed, where the Levenshtein distance from each prefix of a pseudo-word to all the possible prefixes of v is calculated. By doing this, only the distance between the few prefixes of the query and all the prefixes of the pseudo-words in S need to be calculated, thereby avoiding to recompute prefixes that would otherwise be calculated multiple times. Moreover, as we are only interested in pseudo-words in $L_S(v, d_0)$ (i.e., those within a distance d_0 of v), if the Levenshtein distance between any prefix of a pseudo-word and all the prefixes of v is larger than d_0, we can stop the search since no longer prefix of this one can be within the allowed distance.

This leads to an algorithm with a time complexity of $O(KN)$, where N is the length of v (in the worst case we would have to compute the Levenshtein distance between all the prefixes of v and all the prefixes of S). The space complexity is $O(E)$, where E is the number of edges of the DAWG, which is never greater than the number of characters in S and we have empirically observed it is in fact sensibly smaller in practice (cf. Sect. 5.2).

4 Dataset, Assessment, Queries, and Empirical Settings

To evaluate empirically the approximate-spelling search performance, the following sections will describe the dataset, evaluation measures, query sets, and experimental setup used to obtain the PrIx and to run the search process.

4.1 Dataset

The Bentham Papers collection [4] is a large set of documents written by the renowned English philosopher and reformer Jeremy Bentham (1748–1832) and his secretarial staff about different topics. It contains about 90 000 handwritten page images, which are challenging since they were written by several hands and have many crossed-out words, difficult to read pages, etc. Moreover, portions of the collection are written in French and Spanish and occasionally in Latin and ancient Greek. The Bentham papers images are available from the Bentham Project and the British Library.[5]

In this work we have employed the whole collection in experiments aimed to assess computing efficiency, along with the same relatively smaller experimental dataset used in [16], which comprises 1 213 transcribed page images. Table 1 shows only information relevant to this work about the test set partition. Details of the full dataset and the training partitions can be seen in [16].

Table 1. Statistics of the test set defined for the Bentham experimental dataset and for the query sets Q1, Q2 and Q3, used in the evaluation experiments.

Test set		Q1	Q2	Q3
Pages	357	357	253	352
Lines	12 363	12 080	896	3 070
Running words	89 870	87 070	965	3 629
Unique words (Lexicon)	6 988	6 953	861	650

4.2 Query Selection

Three query sets, also posted in Table. 1, were defined to assess the effectiveness of single-word approximate-spelling queries. First we adopt the same query

[5] https://www.ucl.ac.uk/bentham-project and https://www.bl.uk.

selection criterion of [16], where all the test-set words longer than 1 character are used (6 953 transliterated keywords). This criterion ensures that all these keywords are relevant (i.e., each appears in at least one test-set image), thereby allowing mAP to be properly computed. This set is referred to as Q1. The results of [16] for this query set are here considered as a baseline.

In addition we consider a more specific query set, explicitly aimed at assessing the capability of approximate-spelling search to find information which would be impossible to find using exact-spelling. In this set, referred to as Q2, 861 words from Q1 have been chosen such that their individual mAP is zero using exact-word spelling, as in [16]; that is, they are in the GT but not in any spot of the PrIx associated to the line where they appear in the GT.

Q1 and Q2 are aimed at fully automatic, objective evaluation based on the given GT transcripts. However, when using approximate-spelling queries, not all retrieved results which are objective true positives are actually useful, or desired by the user. For instance, the query "able∼1" would be expanded into "able ∨ table ∨ dable ∨ cable ∨ sable ∨ ... ∨ abole", which would retrieve many spots that most probably were not intended by the user. Clearly, for typical information retrieval tasks, only words "semantically related" with the query (base) word are expected to be of user interest.

Measuring semantic kinship or the degree of user liking of retrieval results is elusive and, moreover, it requires important human effort for each system evaluation. Nevertheless, as a first attempt to (at least roughly) qualify the usefulness of approximate-spelling search, we have performed an additional experiment that takes into account the "semantic similitude" between the query (base) word and its approximate-spelling expansions. To this end, a small set of 650 words from Q1, referred to as Q3, was manually selected by a user. This selection mainly includes typical or "likely" query words and excludes words that are not expected to be used in approximate-spelling queries, such as prepositions, conjunctions, numerals, etc.

4.3 Evaluation Protocols and Measures

Two main evaluation protocols are used in this paper to asses accuracy and usefulness of approximate-spelling search results. The first one is called "objective" and will be used with the three previously defined query sets for fully automatic evaluation based only on GT transcripts. The second protocol, called "subjective", will be used with the Q3 query set to take into account the semantics of the query (base) words and their approximate-spelling expansions. Both of them are carried out at line level, as in [16].

In the "objective" protocol, a relevant detected entry, called *"hit"*, or (hereafter) *true positive* (TP), is defined as an event where a query that matches the pseudo-word of a PrIx spot, also matches one of the words of the GT transcript of the text line geometrically associated to that spot. Otherwise. it is called a *false positive* (FP).

On the other hand, in the "subjective" protocol, a TP event is defined when a query that matches the pseudo-word of a PrIx spot, also matches one of the words of the GT transcript of the text line associated to that spot and this word is "semantically related" with the query (base) word, according to (subjective) user judgment. Otherwise. it is called *false positive* (FP).

In any case, a spot is *detected* when a query matches a pseudo-word, regardless it also matches or not any word of the GT transcript of the line associated with that spot. The number of detected spots is the sum of TPs plus FPs.

The total number of *relevant* entries for a query depends on the evaluation protocol adopted. In the "objective" protocol, the total number of *relevant* entries for a query is the total number of matches of the query with the line reference transcripts (assuming a match when a query matches one of the words of that transcript). In addition, for an entry to be counted as *relevant* in the "subjective" protocol, at least one of the matched words must be "semantically related" with the query (base) word.

Finally, the number of relevant spots undetected, also known "misses" or *false negatives* (FNs), is the total number of *"relevant"* spots minus the TPs.

For a set of queries $\mathcal{Q}$ and a relevance threshold τ, the recall, $\rho_q(\tau)$, and the precision, $\pi_q(\tau)$, for a given query $q \in \mathcal{Q}$ are defined as:

$$\rho_q(\tau) \;=\; \frac{h(q,\tau)}{r(q)}, \qquad \pi_q(\tau) \;=\; \frac{h(q,\tau)}{d(q,\tau)} \tag{4}$$

where $r(q)$ is the number of spots which are *relevant* for q, $d(q,\tau)$ is the number of spots *detected* with relevance threshold τ and $h(q,\tau)$ is the number of TPs; i.e., detected spots which are relevant. The interrelated trade-off between recall and precision can be displayed as the so-called *recall-precision* (R-P) curve, $\pi_q(\rho_q)$.

The area under this curve for a query q is called the *average precision* (AP) of q, and the mean of AP values over all $q \in \mathcal{Q}$ is called the *mean AP* (mAP) of $\mathcal{Q}$. See [19,20] for additional details. In our results, we provide mAP values, with confidence intervals (at 95%, $\alpha = 0.025$), computed using the bootstrap method [1] with 10 000 repetitions.

These metrics aim at assessing the usefulness of search results. In addition, to assess computational performance of the proposed search technique, conventional query response time and required memory are used.

4.4 Experimental Settings

The present experiments are based on the very same PrIx's as in [16], produced using Convolutional-Recurrent Neural Network (CRNN) optical models and a n-gram character language model, as outlined in Sect. 2. Image regions were assumed to be text-lines detected an extracted using the system described in [8]. More details about the architecture and meta-parameters adopted for the different models and processing steps are provided in [16].

The basic search engine for PrIx exact-spelling queries is the same used in many previous works (see footnote 4). Moreover, to create the DAWG and compute edit distances, as discussed in Sect. 3.1, the Brmson's library "dawg-levenshtein"has been used.

Finally, the experiments for assessing computational performance were conducted on a single core of an Intel i9-10900X 3.70GHz CPU.

5 Experiments and Results

To evaluate the performance of the proposed approximate-spelling search approach, in the following sub-sections we will report the retrieval and computational performance, followed by a set of illustrative real search examples, aimed to help better understand the results.

5.1 Retrieval Performance

Table 2 reports mAP results according to the "objective" and "subjective" protocols established in Sect. 4.3) for the Bentham Papers query sets Q1, Q2 and Q3 as described in Sect. 4.2.

Table 2. Exact and approximate-spelling retrieval performance (mAP) for "objective" and "subjective" evaluation protocols, using the three query sets defined in Sect. 4.2. 95% confidence intervals are never larger than 0.03.

Query set	Q1 (6 953 words)	Q2 (861 words)	Q3 (650 words)	
Evaluation criteria	Objective	Objective	Objective	Subjective
Exact (baseline)	0.76	0.00	0.78	0.78
Approx. $d_0 = 1$	0.81	0.40	0.83	0.81
Approx. $d_0 = 2$	0.85	0.67	0.84	0.65

As expected, for Q1, approximate-spelling "objective" mAP improves significantly with respect to exact-spelling. The improvement is huge for Q2, for which the exact-spelling baseline mAP was 0. The great success of Q2 queries mainly stems from the frequent PrIx pseudo-word hypotheses which are not exactly correct but are lexicographically close to the correct (GT) transcripts.

We recall that the results for Q3 are preliminary and only aim to roughly measure the *usefulness* of approximate-spelling search, taking into account the "semantic similitude" between the query (base) word and its approximate-spelling expansions, as discussed in Sect. 4.3.

Firstly, since the "subjective" TPs are always a subset of the TPs obtained according to the "objective" protocol, "objective" mAP values are always greater than or equal than the corresponding "subjective" results.

Secondly, the approximate-spelling results with $d_0 = 1$ are similar for both protocols, which means that most of the retrieved spots with $d_0 = 1$ are *useful*; i.e., semantically related to the (base) query word. Nevertheless, it is worth recalling that Q3 was manually selected by a user and consequently, the results might be optimistic. In any case, approximate-spelling search with $d_0 = 1$ can be considered a useful asset that would allow enhancing the user search experience – more so when the amount of retrieved spots provided by exact-spelling search is lower than expected.

Finally, for approximate-spelling with $d_O = 2$, the "subjective" mAP falls significantly with respect to corresponding "objective" result. Clearly, by allowing larger spelling tolerance, it becomes increasingly more likely to retrieve spots that are semantically unrelated to the query (base) word. Yet, approximate-spelling search with $d_O = 2$ may still be a useful asset in some cases, as illustrated in Sect. 5.3. Therefore, the use of approximate-spelling search with $d_O = 2$ is mainly recommended in cases where the amounts of spots retrieved by exact spelling and $d_O = 1$ are much smaller than expected.

5.2 Computational Performance

During the experiments performed in Sect. 5.1, we measured the memory and time requirements of the approximate-spelling queries. Moreover, in order to show how the proposed data structures and algorithms behave on a real, large-scale dataset, we also measured the computational performance over the full Bentham Papers collection, which encompasses almost 90 000 images, with more than 25 million estimated running words [16].[6]

Memory and time performance of approximate-spelling queries are reported in Table 3. While the full collection is more than 250 times larger than the experimental ground-truthed dataset, the number of unique pseudo-words is only 6.2 times larger. The proposed DAWG data structure properly takes advantage of this fact and allows dealing with the large full collection with quite reasonable memory requirements.

On the other hand, query computing time requirements only grow sublinearly, not only with the dataset size, but also with the size of the DAWG; but, as expected, it grows fast with the edit distance threshold. Nevertheless, results show that the proposed approach is very efficient and adequately scales up to support very large manuscript collections.

Table 3. Memory usage (MB) and single-query response time (milliseconds) of approximate-spelling search. (*) Estimated as discussed in [16] for Bentham Full.

Dataset	Bentham GT	Bentham Full	Scale factor
Number of images	357	89 911	251.9
Running words (*)	89 870	25 487 932	283.6
Number of unique pseudo-words ($\|S\|$)	5 951 009	37 172 635	6.2
Number of characters in S	55 654 799	355 514 400	6.4
DAWG size (edges)	3 158 261	24 818 936	7.8
Memory usage (MB)	192	1 602	8.3
Query time (ms) for $d_O = 1$	0.7	1.3	1.9
Query time (ms) for $d_O = 2$	10.5	24.7	2.4

[6] See the public search interface here: http://prhlt-kws.prhlt.upv.es/bentham.

It is important to mention that, even for the full collection and the larger edit distance threshold, the reported computing times are much smaller than the corresponding times required to deal with the Boolean OR combination of the (often very large) lists of expanded pseudo-words. Therefore, future works should aim at reducing the OR combination computing time, rather than at further improving the proposed methods to obtain the lists of pseudo-words.

Finally, we would like to remark that these results are only approximate because of uncertainties associated with the complexity of the data structures and software in general.

5.3 Illustrative Retrieval Examples

Subjective true positives. Consider the query "government∼1", which is expanded to many alternatives, such as: government ∨ governments ∨ goverment ∨ governent ∨ sovernment ∨ guvernement ∨ overnment ∨ gevernment ∨ ...". Thanks to the approximate-spelling search, we will retrieve spots of two words that users would usually expect, "government" and "governments", but we also get spots of the words "goverment", "gevernment", "sovernment" and "governent", which clearly correspond to the word "government", slightly misspelled either by the writer, or by the PrIx process. These matches would be considered as *subjective true positives*, as they are semantically related with the query (base) word. Another example of useful approximate-spelling query is "judicial∼2" which would expand to "judicially ∨ judiciary ∨ judicing ∨ judical" among others. Examples are shown in Fig. 1 and Fig. 2.

Objective true positives. As discussed in Sect. 4.3, when employing the "objective" evaluation protocol, the semantic similitude between the query (base) word and its expansions are not taken into account for determining a TP. For example, the query "sever∼2" would expand, among others, to "sever ∨ severe ∨ severs ∨ seven ∨ never ∨ fever ∨ server ∨ rever ∨ lever ∨ lever ∨ severed ∨ beer ∨ seek ∨ seed ∨ saver ∨ over". Two high RP examples can be seen in Fig. 3. In addition, we would like to remark that the examples shown previously as *subjective true positives* are also *objective true positives*, given that the set of *subjective true positives* is a subset of the *objective true positives* set.

False positives. Moreover, as in exact spelling search, approximate search can also retrieve usual FPs, generally due to the inherent uncertainties in the images. For example, the query "supposes∼1" would retrieve the PrIx spot shown in Fig. 4 (left), with the high RP hypothesis "suppose". This is a FP because the word written in the image is "supreme", with an edit distance from "supposes" greater than one.

Misses. Finally, despite improving search performance, approximate-spelling search is still also susceptible to missing keywords. For example, in Fig. 4 (right) the text written in the image is "kingdoms" but no hypothesis for words within a $d_0 = 1$ edit distance from "kingdom" are in the PrIx spots of this image region with sufficiently high RP. Instead, the PrIx provides two high RP spots with hypotheses for "produces" and "prudence". Therefore, the query "kingdom$\sim$1" would not retrieve this spot (unless a very low confidence threshold is set – which would surely provide many false alarms).

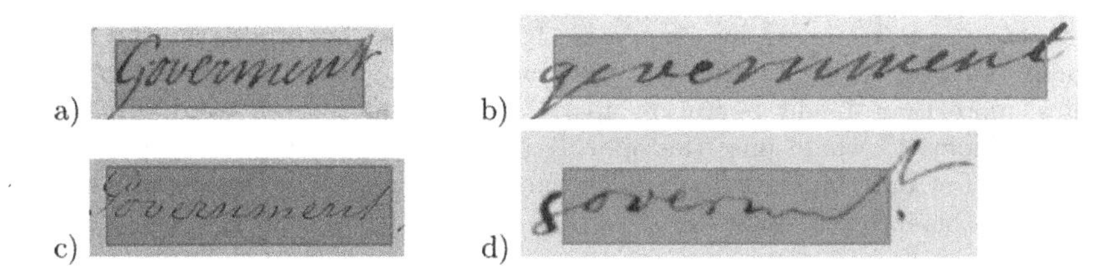

Fig. 1. *Objective* and *subjective* TPs retrieved by "government$\sim$1". The GT for all these spots is "government". In the case a) the PrIx provides a high RP hypothesis for "goverment", in b) for "gevernment", in c) for "sovernment", and in d) for "governent". All these pseudo-words are within edit distance 1 of the query base word and semantically related to it. Therefore, they are considered both *objective* and *subjective* TPs. In cases a) and b) the word was misspelled by the writer, while in c) the PrIx process misspelled the initial "s" for capital "g" and in d) the writing was too sloppy, leading the PrIx to delete the "m".

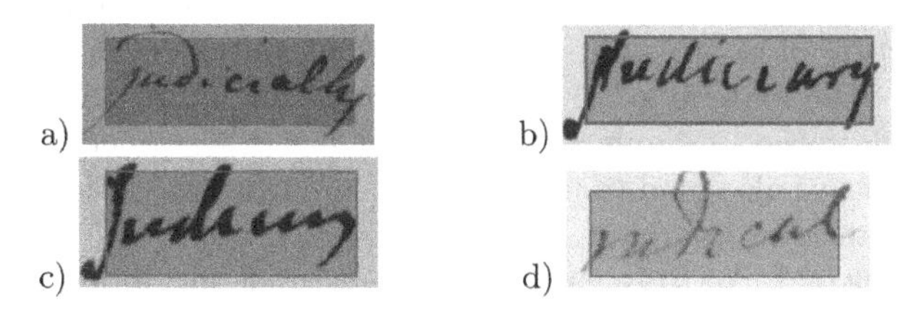

Fig. 2. *Objective* and *subjective* TPs retrieved by "judicial$\sim$2". In the case a) the PrIx provides a high RP hypothesis for "judicially", in b) for "judiciary", in c) for "judicing", and in d) for "judical". The GT in all these cases is the same as the highest RP hypothesis for each BB. All these pseudo-words are within edit distance 2 of the query base word and they are semantically related to it. Therefore, they are both *objective* and *subjective* TPs.

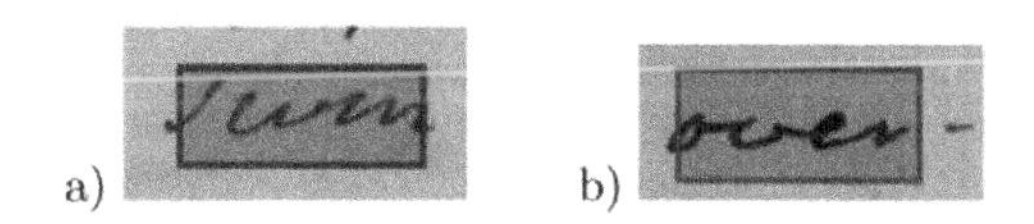

Fig. 3. *Objective* TPs but *subjective* FPs retrieved by the query "sever$\sim$2". The hypotheses with highest RP are "seven" for a) and "over" for b). Both are *objective* TPs because they are within d_0 = 2 edit distance from "sever" and match the corresponding GT words. However, both spots are *subjective* FPs, because they are not only different form "sever", but also semantically unrelated with it.

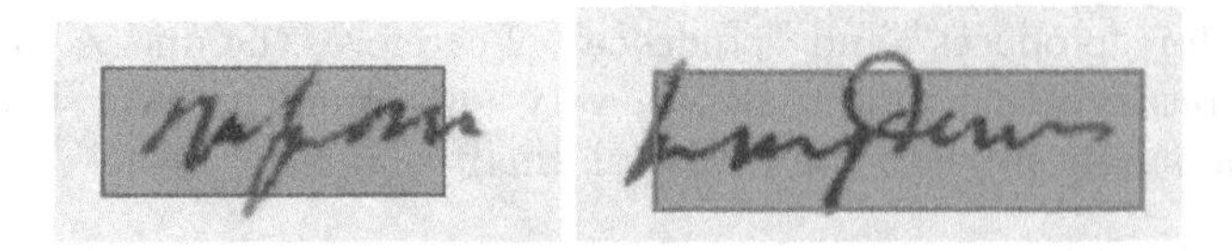

Fig. 4. *Left:* FP retrieved by "supposes∼1". The best hypothesis in the PrIx is "suppose" but, according to the GT, the word written in the image is "supreme", which is more than 1 edit operation apart from the query base word. *Right:* The query "kingdom∼1" would miss this spot. According to the GT, the written word is "kingdoms", but only high RP hypotheses such as "produces" or "prudence", exist in the PrIx for this image region.

6 Conclusion

Our study shows the ability of approximate-spelling queries to help the users dealing with the intrinsic uncertainty of handwritten text images. It also shows how these queries affect positively the retrieval performance of the system. However, this type of queries may also retrieve words that were not intended to be retrieved when the user made the query.

As future work, we plan to develop an adequate framework to compute relevance probability of approximate-spelling queries not only using the relevance probabilities of the matching PrIx pseudo-words, but also taking into account priors that reflect the dissimilarity from each matching pseudo-word and the query. Moreover, we also plan to improve the *subjective* evaluation of the usefulness of the expansions of a query.

Finally, we would like to remark that the methods for obtaining the lists of pseudo-words that are within a given edit distance d_0 are already more than satisfactory and therefore, future efforts should be focused only on reducing the OR combination computing time.

Acknowledgments. Work partially supported by the research grants: Generalitat Valenciana under project DeepPattern (PROMETEO/2019/121), grant PID2020-116813RB-I00a funded by MCIN/AEI/ 10.13039/501100011033, grant RTI2018-095645-B-C22 funded by MCIN/AEI/ 10.13039/501100011033 and by "ERDF A way of making Europe". Computing infrastructure was provided by the EU-FEDER Comunitat Valenciana 2014–2020 grant IDIFEDER/2018/025.

References

1. Bisani, M., Ney, H.: Bootstrap estimates for confidence intervals in ASR performance evaluation. In: 2004 IEEE International Conference on Acoustics, Speech, and Signal Processing, vol. 1, pp. I-409. IEEE (2004)
2. Bluche, T., et al.: Preparatory KWS experiments for large-scale indexing of a vast medieval manuscript collection in the HIMANIS Project. In: 2017 14th IAPR International Conference on Document Analysis and Recognition (ICDAR), vol. 01, pp. 311–316 (2017)

3. Bluche, T.: Deep neural networks for large vocabulary handwritten text recognition. Ph.D. thesis, Paris 11 (2015)
4. Causer, T., Wallace, V.: Building a volunteer community: results and findings from Transcribe Bentham. Digital Humanit. Q. **6**(2) (2012)
5. Daciuk, J., Mihov, S., Watson, B.W., Watson, R.E.: Incremental construction of minimal acyclic finite-state automata. Comput. Linguist. **26**(1), 3–16 (2000)
6. Vidal, E., et al.: The Carabela project and manuscript collection: large-scale probabilistic indexing and content-based classification. In: 16th ICFHR (2020)
7. Graves, A., Liwicki, M., Fernández, S., Bertolami, R., Bunke, H., Schmidhuber, J.: A novel connectionist system for unconstrained handwriting recognition. IEEE Trans. Pattern Anal. Mach. Intell. **31**(5), 855–868 (2008)
8. Grüning, T., Leifert, G., Strauß, T., Michael, J., Labahn, R.: A two-stage method for text line detection in historical documents. Int. J. Doc. Anal. Recogn. (IJDAR) **22**(3), 285–302 (2019). https://doi.org/10.1007/s10032-019-00332-1
9. Hanov, S.: Fast and easy Levenshtein distance using a Trie. Steve Hanov's Blog, **30**, 4–30 (2013). http://stevehanov.ca/blog/index.php?id=114
10. Levenshtein, V.I.: Binary codes capable of correcting deletions, insertions and reversals. Sov. Phys. Dokl. **10**(8), 707–710 (1966). doklady Akademii Nauk SSSR, V163 No4 845–848 1965
11. Prieto, J.R., Vidal, E.: Textual-content-based classification of bundles of untranscribed of manuscript images. In: Proceedings of the International Conference on Pattern Recognition, ICPR (2020)
12. Puigcerver, J.: A probabilistic formulation of keyword spotting. Ph.D. thesis. Univ. Politècnica de València (2018)
13. Romero, V., Toselli, A., Vidal, E.: Multimodal interactive handwritten text recognition, machine perception and artificial intelligence, vol. 80. Word Scientific (2012)
14. Sánchez, J.A., Romero, V., Toselli, A.H., Villegas, M., Vidal, E.: A set of benchmarks for handwritten text recognition on historical documents. Pattern Recogn. **94**, 122–134 (2019)
15. Schulz, K.U., Mihov, S.: Fast string correction with Levenshtein automata. Int. J. Doc. Anal. Recogn. **5**(1), 67–85 (2002). https://doi.org/10.1007/s10032-002-0082-8
16. Toselli, A., Romero, V., Vidal, E., Sánchez, J.: Making two vast historical manuscript collections searchable and extracting meaningful textual features through large-scale probabilistic indexing. In: 2019 15th IAPR International Conference on Document Analysis and Recognition (ICDAR) (2019)
17. Toselli, A.H., Puigcerver, J., Vidal, E.: Two methods to improve confidence scores for lexicon-free word spotting in handwritten text. In: Proceedings of 15th ICFHR, pp. 349–354 (2016)
18. Toselli, A.H., Vidal, E., Puigcerver, J., Noya-García, E.: Probabilistic multi-word spotting in handwritten text images. Pattern Anal. Appl. **22**(1), 23–32 (2018). https://doi.org/10.1007/s10044-018-0742-z
19. Toselli, A.H., Vidal, E., Romero, V., Frinken, V.: HMM word graph based keyword spotting in handwritten document images. Inform. Sci. **370–371**, 497–518 (2016)
20. Vidal, E., Toselli, A.H., Puigcerver, J.: A probabilistic framework for lexicon-based keyword spotting in handwritten text images. arXiv preprint arXiv:2104.04556 (2021)
21. Wagner, R.A., Fischer, M.J.: The string-to-string correction problem. J. ACM (JACM) **21**(1), 168–173 (1974)

Keyword Spotting with Quaternionic ResNet: Application to Spotting in Greek Manuscripts

Giorgos Sfikas[1,3,4](✉), George Retsinas[2], Angelos P. Giotis[3], Basilis Gatos[1], and Christophoros Nikou[3]

[1] Computational Intelligence Laboratory, Institute of Informatics and Telecommunications, National Center for Scientific Research "Demokritos", Athens, Greece
bgat@iit.demokritos.gr

[2] School of Electrical and Computer Engineering, National Technical University of Athens, Athens, Greece
gretsinas@central.ntua.gr

[3] Department of Computer Science and Engineering, University of Ioannina, Ioannina, Greece
{agiotis,cnikou}@cse.uoi.gr

[4] Department of Surveying and Geoinformatics Engineering, University of West Attica, Athens, Greece
gsfikas@uniwa.gr

Abstract. Quaternionized versions of standard (real-valued) neural network layers have shown to lead to networks that are sparse and as effective as their real-valued counterparts. In this work, we explore their usefulness in the context of the Keyword Spotting task. Tests on a collection of manuscripts written in modern Greek show that the proposed quaternionic ResNet achieves excellent performance using only a small fraction of the memory footprint of its real-valued counterpart. Code is available at https://github.com/sfikas/quaternion-resnet-kws.

Keywords: Quaternions · Keyword spotting · Document image processing · Modern greek

1 Introduction

Keyword spotting (KWS) in document images is the go-to application when a search for a specific word or words and their instances inside a digitized document is required, but full recognition of the document may not be the optimal option. Handwritten Text Recognition (HTR) is in general not a trivial task, especially when the collection includes many different writing styles, is of bad digitization quality or is heavily degraded. Furthermore, while learning-based HTR has had many spectacular successes owing to the use of deep learning, accurate recognition relies on a well-trained system, which in turn requires a large and diverse annotated training set. Transferability of a model, i.e. training

© Springer Nature Switzerland AG 2022
S. Uchida et al. (Eds.): DAS 2022, LNCS 13237, pp. 382–396, 2022.
https://doi.org/10.1007/978-3-031-06555-2_26

for a specific style or styles of handwriting and applying to a different target is usually not to be considered as granted. For these reasons, KWS is still a very much viable alternative. Modern KWS methods also rely heavily on deep learning models [21], however in comparison to HTR methods and the training data requirements, they tend to be less resource-hungry as the task is objectively simpler.

Since the advent of convolutional neural networks for KWS [20,23], the literature on KWS methods largely follows and builds on the developments in deep learning for signal or vision applications in general. A recent trend is the use of layers that challenge the status of the two-dimensional canonical grid of input images and feature maps as an immutable parameter; graph convolutional networks operate on word graph inputs, and [35] use a neural network to map word image graph representations to Pyramidal Histogram of Character (PHOC) descriptions. With respect to a wider application scope, a recent trend in machine learning is the use of self-attention layers, popularized in the transformer model [32] initially proposed for a natural language processing setting. The mechanism of self-attention aspires to completely replace two other "pillars" of neural networks, namely fully-connected layers and convolution layers, and indeed has found success in a diverse array of application contexts. In computer vision it has quickly been employed in many different applications [9].

Creating sparse, light-weight neural networks is another important trend [19]. Many different techniques have been proposed in this regard, in an effort to create networks that are as efficient as networks that are larger and/or slower. One such technique is the use of quaternion network layers, which replace standard (real-valued) inputs, parameters and outputs with quaternion-valued components [1,38]. Quaternions form an algebra of intrinsically four-dimensional objects, equipped with its special version of the multiplication operation. Except a very well-known application in representing rotations in space, quaternion algebra has found uses in digital image processing and vision [3,10,26]. Starting in the 90s, a parallel line of research has given formulations that described neural networks with non-real values for inputs, features and layers [1,11,13,14]. Complex and hypercomplex neural networks have been rediscovered recently, with applications that further extended the scope of the first formulations. Following the deep learning "revolution" of the latter half of the 2010s,s, deep complex neural networks have been proposed in 2017 [31], and they were very soon to be followed by papers on quaternion neural networks [6,16,38], published almost simultaneously. These works covered quaternion versions for dense, convolution and recurrent layers. Later works [15] have explored quaternion versions of other layers, models and diverse applications (e.g. Generative Adversarial Networks [25] or CNN-RNN networks for speech recognition [16]). Also, theoretical extensions of the quaternionic layer were considered, which included exploring higher hypercomplex dimensions [2] or parameterizing a generic hypercomplex operation [30,36].

In the current work, we propose employing quaternion layers combined with a ResNet architecture, in order to create a light-weight KWS system. In a nutshell, compared to related prior work, our contribution with this paper is:

- the introduction of a Quaternionic version of the ResNet block, and the Quaternionic ResNet as a network that is comprised of this type of blocks.
- testing the proposed model for the problem of KWS, and showing that it can be as effective as its real-valued counterpart. The important advantage here (present also in quaternion networks in general) is that the network requires only one quarter of the parameters of a real-valued model that uses the same architecture (same number of layers, and Quaternionic layer blocks in place of standard layer blocks).

The remainder of this paper is structured as follows. In Sect. 2 we present elementary notions concerning the algebra of quaternions, and discuss how it is applied to create quaternionic variants of neural networks. We explain why quaternionic layers inherently lead to lightweight/"less-costly" networks in Sect. 3. In Sect. 4 we review the proposed quaternionic network for keyword spotting. Section 5 numerically confirms our approach compared to non-quaternionic architectures. We close the paper with Sect. 6 where we draw our conclusions and discuss future work.

2 Quaternions in Neural Networks

2.1 Elementary Notions

Quaternions have been introduced in the 19th century by Hamilton [34]. It was presented as a special kind of algebra over quadruples of real numbers, after having being realized that a triplet-based algebraic structure would necessarily be inherently constrained in terms of its properties. The definition of a quaternion is:

$$q = \alpha + \beta \boldsymbol{i} + \gamma \boldsymbol{j} + \delta \boldsymbol{k}, \tag{1}$$

where $\alpha, \beta, \gamma, \delta$ are real numbers and $\boldsymbol{i}, \boldsymbol{j}, \boldsymbol{k}$ are imaginary units. This definition is reminiscent of the definition of complex numbers, and indeed one may interpret quaternions $\mathbb{H}$ as a generalization of $\mathbb{C}$. The important difference is that while we have a single "real" part (α) in both $\mathbb{C}$ and $\mathbb{H}$, in quaternions we have three *independent* imaginary parts (β, γ, δ) and an equal number of imaginary units/axes ($\boldsymbol{i}, \boldsymbol{j}, \boldsymbol{k}$). This relation is made more evident with the Cayley-Dickson form of quaternions:

$$q = \chi + \psi \boldsymbol{j}, \tag{2}$$

where now χ and ψ are *complex* coefficients, $\chi = \alpha + \beta \boldsymbol{i}, \psi = \gamma + \delta \boldsymbol{i}$. The form Eq. 2 is equivalent to Eq. 1. One may arrive from one to the other after considering the following multiplication rule for imaginary units:

$$\boldsymbol{i}^2 = \boldsymbol{j}^2 = \boldsymbol{k}^2 = \boldsymbol{ijk} = -1,$$

$$\boldsymbol{ij} = -\boldsymbol{ji} = \boldsymbol{k}, \boldsymbol{jk} = -\boldsymbol{kj} = \boldsymbol{i}, \boldsymbol{ki} = -\boldsymbol{ik} = \boldsymbol{j}. \tag{3}$$

As illustrated in Eq. 3, multiplication for quaternions is not commutative, and for imaginary units in particular, changing the order of multiplications leads

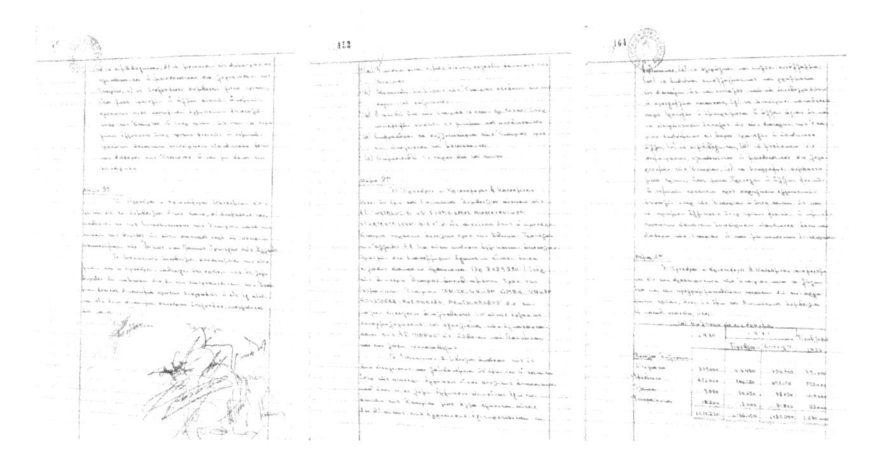

Fig. 1. Sample pages from the two datasets used in this work, "Memoirs" (top row) and "PIOP-DAS" (bottom row).

to the opposite of the initial result. In general, we can write that $pq \neq qp$ for $p, q \in \mathbb{H}$.

Addition is defined simply by adding up respective coefficients, and retains the "usual" properties with no big surprises (commutative law, associative law, and the distributive property combined with the aforementioned multiplication rule). These considerations lead to the following multiplication rule:

$$\begin{aligned} pq = &(\alpha_p\alpha_q - \beta_p\beta_q - \gamma_p\gamma_q - \delta_p\delta_q) + \\ &(\alpha_p\beta_q + \beta_p\alpha_q + \gamma_p\delta_q - \delta_p\gamma_q)\boldsymbol{i} + \\ &(\alpha_p\gamma_q - \beta_p\delta_q + \gamma_p\alpha_q + \delta_p\beta_q)\boldsymbol{j} + \\ &(\alpha_p\delta_q + \beta_p\gamma_q - \gamma_p\beta_q + \delta_p\alpha_q)\boldsymbol{k}, \end{aligned} \tag{4}$$

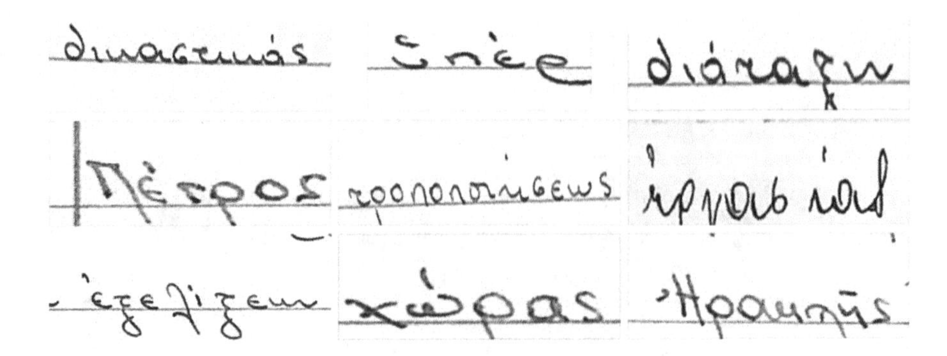

Fig. 2. Word image samples from the "PIOP-DAS" dataset.

where $p = \alpha_p + \beta_p \boldsymbol{i} + \gamma_p \boldsymbol{j} + \delta_p \boldsymbol{k}$ and $q = \alpha_q + \beta_q \boldsymbol{i} + \gamma_q \boldsymbol{j} + \delta_q \boldsymbol{k}$. This rule can also be shorthanded as:

$$pq = S(p)S(q) - V(p) \cdot V(q) + S(p)V(q) + S(q)V(p) + V(p) \times V(q), \tag{5}$$

where $\cdot$ and $\times$ denote the dot and cross product respectively, $S(\cdot)$ is the "scalar" part of the quaternion and $V(\cdot)$ is the "vector" part of the quaternion (i.e. $S(p) = \alpha \in \mathbb{R}$ and $V(p) = [\beta\ \gamma\ \delta]^T \in \mathbb{R}^3$).

Other useful relations include the definition of a quaternion conjugate, and a quaternion magnitude. These are defined as follows:

$$\bar{q} = \alpha - \beta \boldsymbol{i} - \gamma \boldsymbol{j} - \delta \boldsymbol{k}, \tag{6}$$

and

$$|q| = \sqrt{q\bar{q}} = \sqrt{\bar{q}q} = \sqrt{\alpha^2 + \beta^2 + \gamma^2 + \delta^2}. \tag{7}$$

Again, analogies to complex numbers can be straightforwardly drawn, considering $\gamma = 0, \delta = 0$ to obtain the well-known relations for complex numbers. Finally, we note that matrix calculus can be extended to matrices with quaternionic elements, $\mathbb{H}^{m \times n}$ [37].

2.2 Quaternionized Versions of Standard NN Layers

Quaternion Neural Networks (QNNs) are defined as neural networks that have quaternion-valued inputs, outputs and parameters. In order to deal with providing quaternion-valued inputs where real-valued scalars and matrices are available, the respective structures are padded with zero channel values when the dimensionality of the input is $d < 4$. In the present use-case, grayscale images are inherently one-dimensional -valued in each pixel, and colour images would be three-dimensional -valued, accounting for the Red, Green and Blue channels (using a different colorspace would lead to an analogous consideration). Hence in particular, a scalar value a is mapped to quaternion $a + 0\boldsymbol{i} + 0\boldsymbol{j} + \boldsymbol{k}$, and a colour value $[r, g, b]^T$ is mapped to quaternion $0 + r\boldsymbol{i} + g\boldsymbol{j} + b\boldsymbol{k}$ by convention.

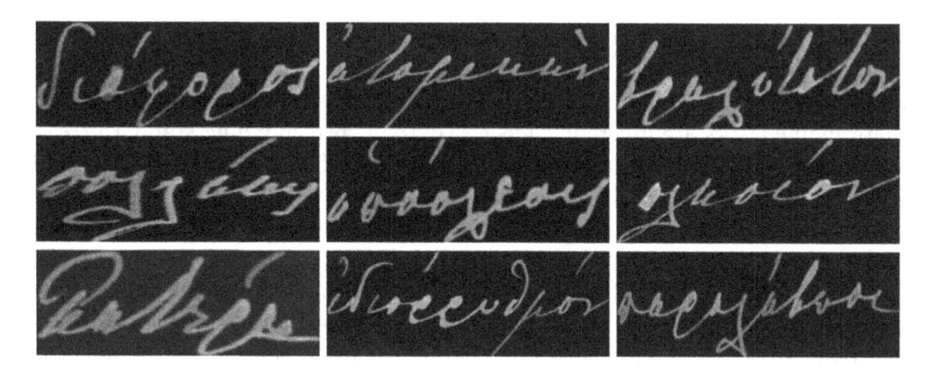

Fig. 3. Word image samples from the "Memoirs" dataset.

Dense Layer. A QNN comprises a cascade of layers as is the case with real-valued neural networks. The dense or fully-connected layer for quaternions can be written as

$$f_{dense}(x; W, h) = Wx + h$$

where $x \in \mathbb{H}^N$ is a quaternion-valued input vector, $W^{M \times N}$ is a quaternion-valued matrix of weights and $h \in \mathbb{H}^N$ denotes a quaternion-valued bias term. We assume that the input is of dimensionality equal to N, and the output dimensionality is equal to M.

Convolution Layer. Regarding convolution, we must note that a number of different options here are possible regarding the exact form of the convolution operation. In particular for two-dimensional convolution, which is of interest in an image processing network, there is a left-sided convolution, a right-sided convolution, and also a two-sided convolution (or "bi-convolution") [4], with the difference being in whether the convolution kernel multiplies the signal from the left or the right. The two-sided option corresponds to the case where an horizontal kernel multiplies the signal from the left, and a vertical kernel multiplies the signal from the right. In this work, we choose to use the left-multiplying convention, so formally we have:

$$f_{conv}(m; K) = K * m,$$

where $m \in \mathbb{H}^{M \times N}$ is the input feature map, and $K \in \mathbb{H}^{d \times d}$ is the convolution kernel.

Activation Functions. We use "split - activation" functions to handle non-linearities in the quaternion domain, which means that each quaternion channel is treated by the activation function separately, as if it were part of a tensor valued in $\mathbb{R}^4$. For our architecture, we use split-activation versions of standard real-valued functions (Rectified Linear Unit, the leaky Rectified Linear Unit, sigmoid). The same rationale is followed for other network components: the dropout layer and the batch normalization layer are applied as if a quaternionic tensor $H \times W \times D$ were its real isomorphic image $H \times W \times 4D$.

3 Why Are Quaternionic Layers Less Costly?

The main motivation behind using a quaternionic network for a given task is that quaternionic layers are inherently less costly than corresponding standard (real-valued) layers (And more importantly, without significant sacrifices in terms of performance). "Cost" here is to be understood in terms of total required *independent* parameters. The reason for this useful trait is that the definition of quaternionic layers comes with extensive parameter sharing. This is due to i) the definition of quaternionic ("Hamilton") multiplication itself, and ii) that every four real-valued tensor channels are grouped together and mapped to quaternion real/imaginary components. For example, an input color image, comprising of 3 channels plus 1 zero-padded channel would be mapped as a single-channel quaternionic 2D signal.

Let us illustrate this with a simple example. Assume a real-valued linear layer, without bias or activation, that transforms an input comprised of 4 neurons to an output of 4 neurons. In other words, an input $x \in \mathbb{R}^4$ is mapped to an output $y \in \mathbb{R}^4$, and the operation can be written as the matrix-vector multiplication:

$$\begin{bmatrix} y_1 \\ y_2 \\ y_3 \\ y_4 \end{bmatrix} = \begin{bmatrix} \alpha \; \beta \; \gamma \; \delta \\ \epsilon \; \zeta \; \eta \; \theta \\ \iota \; \kappa \; \lambda \; \mu \\ \nu \; \xi \; o \; \pi \end{bmatrix} \begin{bmatrix} x_1 \\ x_2 \\ x_3 \\ x_4 \end{bmatrix}, \tag{8}$$

where greek letters $\alpha, \beta, \cdots, \pi$ denote the $4 \times 4 = 16$ operation parameters. Assuming that we want to define a quaternionic linear layer on the same input and a same-sized output, the input and output vectors would be mapped to a single quaternion each; we shall again denote these quaternions using the notation x_1, x_2, x_3, x_4 and y_1, y_2, y_3, y_4, where components 1–4 correspond to the real and the the three imaginary quaternion components. As our input and output, under these terms, is a single quaternion and a single quaternion respectively, a quaternionic linear layer is composed of a single multiplication operation. This operation can be written as:

$$\begin{bmatrix} y_1 \\ y_2 \\ y_3 \\ y_4 \end{bmatrix} = \begin{bmatrix} \alpha & -\beta & -\gamma & -\delta \\ \beta & \alpha & -\delta & \gamma \\ \gamma & \delta & \alpha & -\beta \\ \delta & -\gamma & \beta & \alpha \end{bmatrix} \begin{bmatrix} x_1 \\ x_2 \\ x_3 \\ x_4 \end{bmatrix}, \tag{9}$$

where we have re-written Eq. 4 as a matrix-vector multiplication and changed notation accordingly. The parameters of the operation are again 16 as in Eq. 8, but they are grouped into 4 groups of 4, or in other words we have 4 *independent* parameters.

This paradigm is extended to cover any-size inputs and outputs, as long as they are multiples of 4. If they are not multiples of 4, they can easily be padded using zero-valued channels to the next closest multiple of 4 (The network will easily learn to ignore the paddings, hence there is no real overhead involved). Alternatively, an operation such as 1×1 convolution can be used to map a real-valued input to the desired channel multiple [30]. In general, an operation that is written as $y = Wx$, where $y \in \mathbb{R}^{4K}$, $x \in \mathbb{R}^{4L}$ and $W \in \mathbb{R}^{4K \times 4L}$ is thus mapped to a quaternionic operation $\hat{y} = \hat{W}\hat{x}$, where $\hat{y} \in \mathbb{H}^K$, $\hat{x} \in \mathbb{H}^L$ and $\hat{W} \in \mathbb{H}^{K \times L}$. Parameter vector $\hat{W}$ only contains $4 \times K \times L$ parameters compared to $4 \times 4 \times K \times L$ of W, hence we have a $4\times$ saving.

The above considerations hold for any operation, as long as it can be written as a linear transformation, or a composition that includes linear transformations. Hence, not only linear layers can be quaternionized, but convolutions [8,38] and deconvolutions [30], or resnet blocks as in the current work.

4 Proposed Model

The proposed model is structured as a feed-forward convolutional network, accepting a batch of word images as input, and processing them into a batch of fixed-size PHOC descriptor targets [24]. The input is set to the real part of the quaternion input map, and imaginary components are set to zero.

In our architecture, we group QNN processing layers in ResNet blocks. Each block groups a cascade of quaternion convolutions, structured as follows: We assume that a feature map x is the input of the block. A quaternion convolution of stride equal to 1 and kernel size equal to 3×3 is followed by a batch normalization layer and a ReLU activation. The result of this step is convolved by a second quaternion convolution with the same operation characteristics, followed by a second batch normalization layer. The output $\phi(x)$ is connected with the input x via a skip connection and followed by a final ReLU activation, so the block output can be written formally as $ReLU(\phi(x) + x)$.

On a high level, the network is composed as a sequence of a convolutional backbone, followed by a pooling and flattening layer, which in turn is topped by a fully-connected head. The backbone is made up of 7 ResNet blocks. These are parametrized, in terms of the number of input and output channels respectively, as follows: $(1, 16)$, $(16, 32)$, $(32, 64)$, $(64, 64)$, $(64, 64)$, $(64, 128)$, $(128, 32)$, where the nth tuple is equal to (#input channels, #output channels). Pooling is performed using pyramidal spatial pooling with 3 layers and an output size of $2,688$. Then, two quaternionic dense layers follow, mapping first to 256 neurons before passing to a dropout layer and the output logits. The logits are transformed into per-attribute (unigrams, bigrams) probabilities with a split-activation sigmoid function. The quantity to be optimized by the network is a binary cross-entropy (BCE) loss function, measuring divergence of the attribute estimate probabilities against the true PHOC values.

5 Experiments

5.1 Datasets

For our experiments, we have used two different collections of handwritten pages. The manuscripts are written in the modern greek language. The very few non-greek words which exist in the set have been manually removed for the tests,

Table 1. Model size comparison, in terms of total number of trainable parameters. For the quaternionic variants, the equivalent number of real-valued parameters is reported, to ease comparisons.

Model/Network type	"Small" size	"Standard" size
Quaternionic Resnet	1,317,428	5,946,164
Real-valued Resnet	5,233,840	23,730,864

as training would be too difficult for them due to their small sample size. Both datasets have been manually segmented into a number of word images, which we use as queries and retrieval targets for our Query-by-Example keyword spotting trials. Also, both datasets use the polytonic Greek script [24]. However, for the PIOP-DAS dataset we have only *monotonic* annotations, meaning that only the "acute" accent and the "diairesis" are used, and the actually existing text diacritics are either represented by an acute accent (this is the case for the grave accent and the circumflex) or not mapped at all (this is the case for the smooth and rough breatings, subscript). Samples of the pages contained in the set are shown in Fig. 1.

PIOP-DAS Dataset. The PIOP-DAS dataset consists of 22 manuscripts, that were scanned from the archives of the Greek "Πειραϊκή-Πατραϊκή " (*"Peiraiki-Patraiki"*) bank. The documents record 4 sessions of the bank governing commitee, held between December 1971–April 1972. The manuscript digitizations have been segmented into a total of 12,362 words. 80% of the dataset was assigned as the training set, and the rest is used as the test set. In absolute figures, 9,341 words and 3,021 words are assigned to the training and test set respectively. We use as queries all words in the test set, except those that appear only once. Examples of segmented words from this set are shown in Fig. 2. The dataset is publicly available at https://github.com/sfikas/piop-das-dataset.

Memoirs Dataset. The Memoirs dataset [5,7] comprises 46 manuscripts, written in the late 19th century in the form of a personal diary. The text is written by a single author, Sophia Trikoupi, who was the sister of the Greek prime minister Charilaos Trikoupis. A total of 4,941 is available in this dataset. We use the training and test partitions defined in [24], which comprise 2,000 words each. There are 941 remaining words that correspond to the validation set, which we do not use in this work. To select queries, we "initially" use the words selected in [5]. [5] uses a learning-free KWS baseline, so this query list is trimmed down subsequently here as some words do not exist in the test set. Examples of segmented words from this set are shown in Fig. 3. The dataset is publicly available at https://github.com/sfikas/sophia-trikoupi-handwritten-dataset/.

5.2 Hyperparameters and Other Training Considerations

Concerning the encoding of the Greek script into unigram descriptor bins, we choose to use separate bins for unaccented and accented characters, following [24] (for example, ω and $\acute{\omega}$ are treated as different letters. A shared bin is used for upper-case and lower-case versions of the same letter. We set the learning rate to 10^{-3}, and the batch size to 40. All images are resized to 32×128 pixels and cast as grayscale images before entering the network, either for training or inference. We use the Adam optimizer with weight decay set to 5×10^{-5}, and a cosine annealing scheduler with restarts every 300 epochs, for a maximum of 900 epochs of training. Data augmentation is employed during training, applying a small affine transform on each drawn training sample [22].

5.3 Results

We have run trials testing the proposed Quaternionic model against models that differ in terms of the domain of their structural components (layers, parameters, inputs and outputs) and in terms of their size. With regard to the size of the models, we name the two sizes "standard" and "small". The "standard" model corresponds exactly to the architecture presented in Sect. 4. The "small" model comprises only 3 resnet blocks, instead of 7 blocks for the "standard" model. Experiments were run over both datasets PIOP-DAS and Memoirs. Model training results may be examined in Figs. 4, 5 and Table 2. We report loss and accuracy in terms of the progression of model training, as well as the best accuracy figures by the end of training and best attained overall. We show training and test loss, both computed as a binary cross-entropy average over data on the training and test sets respectively. Over the same plots, model accuracy is measured in terms of mean average precision (MAP).

We can see that in all cases the quaternion-valued and the real-valued models, regardless of size fare equally well, with almost excellent results. In one case –in particular, Memoirs/"small" network – the quaternionic network achieves better accuracy figures that its real counterpart. We believe however that the most important trait of the quaternionic models is related to the size of the respective architectures, as reported in Table 1. Therein, we see that quaternionic

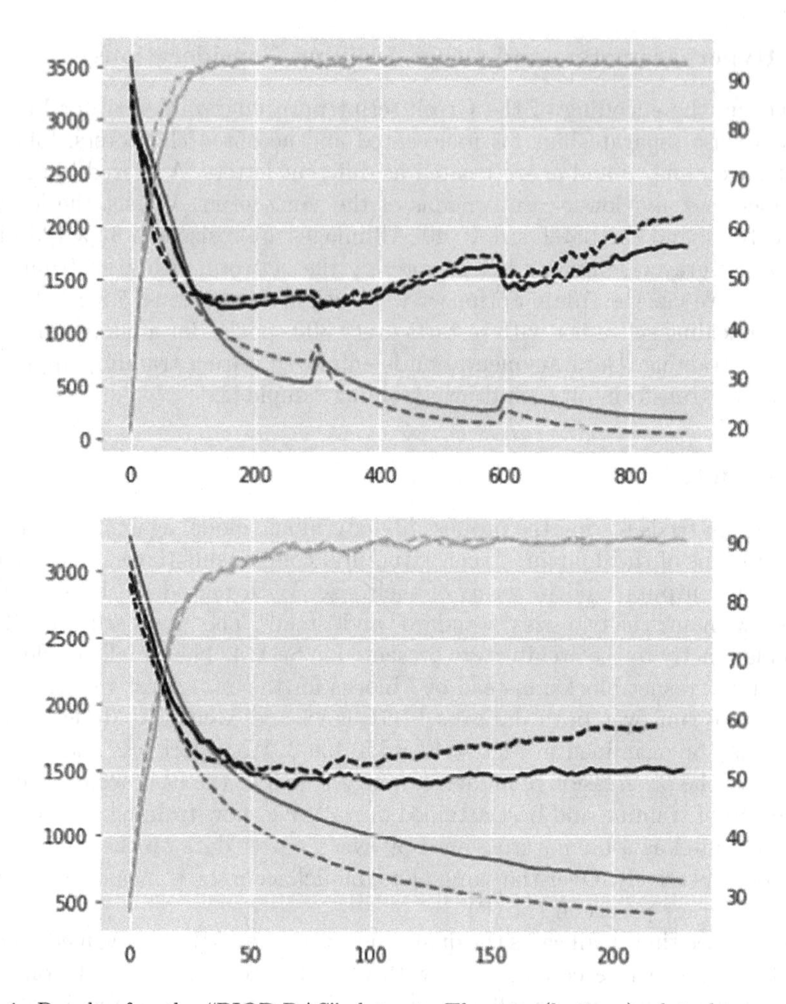

Fig. 4. Results for the "PIOP-DAS" dataset: The top (bottom) plot shows results for the "standard" ("small") network. Training set BCE loss, test set BCE loss and MAP are shown using red, blue and green colors respectively. Solid lines correspond to the proposed Quaternionic models. Dashed lines correspond to real-valued networks with the same architecture as the proposed networks, but using standard real-valued components in place of the quaternionic ones. The horizontal axis corresponds to the number of training epochs. The vertical axis corresponds to loss value (shown on the left, lower is better) and map accuracy percentage (shown on the right, higher is better). The plots were smoothed with a uniform kernel of size 10 to ease visualization.

models enjoy a much smaller size in terms of number of trainable parameters (see discussion in Sect. 3 for a theoretical justification of this result). The eventual gain is an operation of significantly lower complexity.

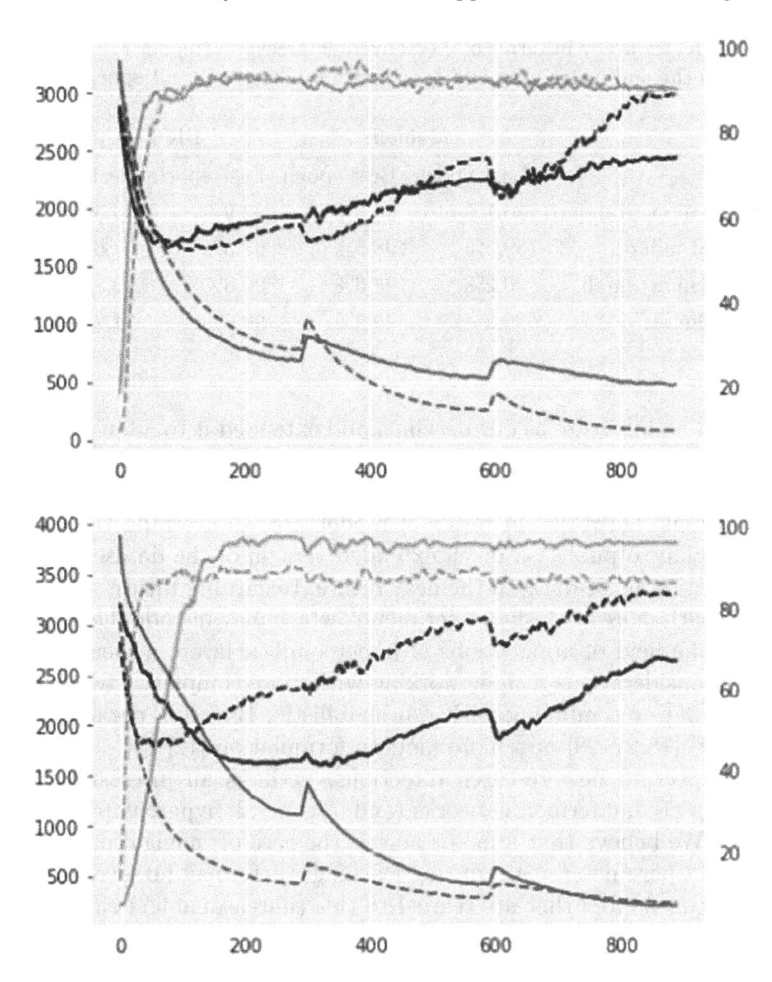

Fig. 5. Results for the "Memoirs" dataset: The top (bottom) plot shows results for the "large" ("small") network. Training set BCE loss, test set BCE loss and MAP are shown using red, blue and green colors respectively. Solid lines correspond to the proposed Quaternionic models. Dashed lines correspond to real-valued networks with the same architecture as the proposed networks, but using standard real-valued components in place of the quaternionic ones. The horizontal axis corresponds to the number of training epochs. The vertical axis corresponds to loss value (shown on the left, lower is better) and map accuracy percentage (shown on the right, higher is better). The plots were smoothed with a uniform kernel of size 10 to ease visualization (Color figure online).

6 Conclusion and Future Work

With this work we have introduced a novel Quaternionic ResNet block and validated the value of using Quaternionized versions of standard (i.e., real-valued) layers in the context of the document keyword spotting task. We have used a

Table 2. Model accuracy in terms of mean average precision (higher is better). Results correspond to the end of training and best figure reported over all epochs.

MAP%	Memoirs		PIOP-DAS	
	Last epoch	Best epoch	Last epoch	Best epoch
Quaternion/Standard	96.1%	98.6%	92.9%	94.3%
Real/Standard	89.3%	98.6%	93.3%	94.6%
Quaternion/Small	90.2%	94.6%	90.3%	92.8%
Real/Small	86.3	94.5%	90.0%	92.2%

PHOC-based architecture as our baseline, and extended it to a Quaternion Neural Network. Our results show that the proposed QNN has achieved excellent retrieval performance while being much less resource-demanding compared to non-quaternionic networks, in terms of model size. Concerning the new PIOP dataset, we plan to publish a much extended version of the dataset, along with richer annotation meta-data, in the near future. Regarding future work, we plan to continue with more extensive experiments, which may include the newer developments in the field of applications of hypercomplex layers in neural networks [36]. Other considerations include working with more complex network architectures [17,33,35] or combining with a probabilistic, Bayesian paradigm using a more "classic" [18,27–29] or a more modern formulation [12].

With respect to new research directions, perhaps an interesting question would be *why* are quaternion networks (and in general, hypercomplex networks) so effective. We believe that a factor here is the ease of "navigation" on a much more compact parameter space during learning; there may however exist other, more important factors that are related to this (unreasonable?) effectiveness.

Acknowledgments. This research has been partially co-financed by the EU and Greek national funds through the Operational Program Competitiveness, Entrepreneurship and Innovation, under the calls "RESEARCH - CREATE - INNOVATE" (project *Culdile* - code T1EΔK-03785, project *Impala* - code T1EΔK-04517) and "OPEN INNOVATION IN CULTURE" (project *Bessarion* - T6YBΠ-00214).

References

1. Arena, P., Fortuna, L., Occhipinti, L., Xibilia, M.G.: Neural networks for quaternion-valued function approximation. In: Proceedings of IEEE International Symposium on Circuits and Systems-ISCAS 1994, vol. 6, pp. 307–310. IEEE (1994)
2. Bojesomo, A., Liatsis, P., Marzouqi, H.A.: Traffic flow prediction using deep sedenion networks. arXiv preprint arXiv:2012.03874 (2020)
3. Ell, T.A., Le Bihan, N., Sangwine, S.J.: Quaternion Fourier Transforms for Signal and Image Processing. Wiley, Hoboken (2014)
4. Ell, T.A., Sangwine, S.J.: Hypercomplex Fourier transforms of color images. IEEE Trans. Image Process. **16**(1), 22–35 (2007)

5. Gatos, B., et al.: GRPOLY-DB: An old Greek polytonic document image database. In: Proceedings of the International Conference on Document Analysis and Recognition (ICDAR), pp. 646–650. IEEE (2015)
6. Gaudet, C.J., Maida, A.S.: Deep quaternion networks. In: 2018 International Joint Conference on Neural Networks (IJCNN), pp. 1–8. IEEE (2018)
7. Giotis, A.P., Sfikas, G., Nikou, C., Gatos, B.: Shape-based word spotting in handwritten document images. In: 13th International conference on document analysis and recognition (ICDAR), pp. 561–565. IEEE (2015)
8. Grassucci, E., Zhang, A., Comminiello, D.: Lightweight convolutional neural networks by hypercomplex parameterization. arXiv preprint arXiv:2110.04176 (2021)
9. Han, K., et al.: A survey on visual transformer. CoRR abs/2012.12556 (2020). https://arxiv.org/abs/2012.12556
10. Hui, W., Xiao-Hui, W., Yue, Z., Jie, Y.: Color texture segmentation using quaternion-Gabor filters. In: 2006 International Conference on Image Processing, pp. 745–748. IEEE (2006)
11. Isokawa, T., Kusakabe, T., Matsui, N., Peper, F.: Quaternion neural network and its application. In: Palade, V., Howlett, R.J., Jain, L. (eds.) KES 2003. LNCS (LNAI), vol. 2774, pp. 318–324. Springer, Heidelberg (2003). https://doi.org/10.1007/978-3-540-45226-3_44
12. Kobyzev, I., Prince, S.J., Brubaker, M.A.: Normalizing flows: an introduction and review of current methods. IEEE Trans. Pattern Anal. Mach. Intell. **43**(11), 3964–3979 (2020)
13. Leung, H., Haykin, S.: The complex backpropagation algorithm. IEEE Trans. Signal Process. **39**(9), 2101–2104 (1991)
14. Nitta, T.: A quaternary version of the back-propagation algorithm. In: Proceedings of ICNN'95-International Conference on Neural Networks. vol. 5, pp. 2753–2756. IEEE (1995)
15. Parcollet, T., Morchid, M., Linarès, G.: A survey of quaternion neural networks. Artif. Intell. Rev. **53**(4), 2957–2982 (2019). https://doi.org/10.1007/s10462-019-09752-1
16. Parcollet, T., et al.: Quaternion convolutional neural networks for end-to-end automatic speech recognition. arXiv preprint arXiv:1806.07789 (2018)
17. Prieto, J.R., Vidal, E.: Improved graph methods for table layout understanding. In: Lladós, J., Lopresti, D., Uchida, S. (eds.) ICDAR 2021. LNCS, vol. 12822, pp. 507–522. Springer, Cham (2021). https://doi.org/10.1007/978-3-030-86331-9_33
18. Prince, S.J.: Computer Vision: Models, Learning, and Inference. Cambridge University Press, Cambridge (2012)
19. Retsinas, G., Elafrou, A., Goumas, G., Maragos, P.: Weight pruning via adaptive sparsity loss. arXiv preprint arXiv:2006.02768 (2020)
20. Retsinas, G., Louloudis, G., Stamatopoulos, N., Gatos, B.: Efficient learning-free keyword spotting. IEEE Trans. Pattern Anal. Mach. Intell. **41**(7), 1587–1600 (2018)
21. Retsinas, G., Sfikas, G., Nikou, C., Maragos, P.: From Seq2Seq recognition to handwritten word embeddings. In: Proceedings of the British Machine Vision Conference (BMVC) (2021)
22. Retsinas, G., Sfikas, G., Stamatopoulos, N., Louloudis, G., Gatos, B.: Exploring critical aspects of CNN-based keyword spotting. a phocnet study. In: Proceedings of the International Workshop on Document Analysis Systems (DAS), pp. 13–18. IEEE (2018)
23. Rusakov, E., Sudholt, S., Wolf, F., Fink, G.A.: Exploring architectures for CNN-based word spotting. arXiv preprint arXiv:1806.10866 (2018)

24. Sfikas, G., Giotis, A.P., Louloudis, G., Gatos, B.: Using attributes for word spotting and recognition in polytonic greek documents. In: Proceedings of the International Conference on Document Analysis and Recognition (ICDAR), pp. 686–690. IEEE (2015)
25. Sfikas, G., Giotis, A.P., Retsinas, G., Nikou, C.: Quaternion generative adversarial networks for inscription detection in byzantine monuments. In: Del Bimbo, A., Cucchiara, R., Sclaroff, S., Farinella, G.M., Mei, T., Bertini, M., Escalante, H.J., Vezzani, R. (eds.) ICPR 2021. LNCS, vol. 12667, pp. 171–184. Springer, Cham (2021). https://doi.org/10.1007/978-3-030-68787-8_12
26. Sfikas, G., Ioannidis, D., Tzovaras, D.: Quaternion Harris for multispectral keypoint detection. In: Proceedings of the International Conference on Image Processing (ICIP), pp. 11–15. IEEE (2020)
27. Sfikas, G., Nikou, C., Galatsanos, N., Heinrich, C.: MR brain tissue classification using an edge-preserving spatially variant bayesian mixture model. In: Metaxas, D., Axel, L., Fichtinger, G., Székely, G. (eds.) MICCAI 2008. LNCS, vol. 5241, pp. 43–50. Springer, Heidelberg (2008). https://doi.org/10.1007/978-3-540-85988-8_6
28. Sfikas, G., Nikou, C., Galatsanos, N., Heinrich, C.: Majorization-minimization mixture model determination in image segmentation. In: Proceedings of the IEEE International Conference on Computer Vision and Pattern Recognition (CVPR), pp. 2169–2176. IEEE (2011)
29. Sfikas, G., Retsinas, G., Gatos, B.: A PHOC decoder for lexicon-free handwritten word recognition. In: Proceedings of the International Conference on Document Analysis and Recognition (ICDAR), vol. 1, pp. 513–518. IEEE (2017)
30. Sfikas, G., Retsinas, G., Gatos, B.: Hypercomplex generative adversarial networks for lightweight semantic labeling. In: International Conference on Pattern Recognition and Artificial Intelligence (2022)
31. Trabelsi, C., et al.: Deep complex networks. arXiv preprint arXiv:1705.09792 (2017)
32. Vaswani, A., et al.: Attention is all you need. In: Advances in neural information processing systems (NIPS), pp. 5998–6008 (2017)
33. Vidal, E., Toselli, A.H.: Probabilistic indexing and search for hyphenated words. In: Lladós, J., Lopresti, D., Uchida, S. (eds.) ICDAR 2021. LNCS, vol. 12822, pp. 426–442. Springer, Cham (2021). https://doi.org/10.1007/978-3-030-86331-9_28
34. Vince, J.: Quaternions for Computer Graphics. Springer, London (2021). https://doi.org/10.1007/978-1-4471-7509-4
35. Wolf, F., Fischer, A., Fink, G.A.: Graph convolutional neural networks for learning attribute representations for word spotting. In: Lladós, J., Lopresti, D., Uchida, S. (eds.) ICDAR 2021. LNCS, vol. 12821, pp. 50–64. Springer, Cham (2021). https://doi.org/10.1007/978-3-030-86549-8_4
36. Zhang, A., et al.: Beyond fully-connected layers with quaternions: parameterization of hypercomplex multiplications with $1/n$ parameters. In: Proceedings of the International Conference on Learning Representations (ICLR) (2021)
37. Zhang, F.: Quaternions and matrices of quaternions. Linear Algebra Appl. **251**, 21–57 (1997)
38. Zhu, X., Xu, Y., Xu, H., Chen, C.: Quaternion convolutional neural networks. In: Proceedings of the European Conference on Computer Vision (ECCV), pp. 631–647 (2018)

Open-Source Software and Benchmarking

A Comprehensive Comparison of Open-Source Libraries for Handwritten Text Recognition in Norwegian

Martin Maarand[1](✉), Yngvil Beyer[2], Andre Kåsen[2], Knut T. Fosseide[3], and Christopher Kermorvant[1]

[1] TEKLIA, Paris, France
maarand@teklia.com
[2] National Library of Norway, Oslo, Norway
[3] Lumex, Oslo, Norway

Abstract. In this paper, we introduce an open database of historical handwritten documents fully annotated in Norwegian, the first of its kind, allowing the development of handwritten text recognition models (HTR) in Norwegian. In order to evaluate the performance of state-of-the-art HTR models on this new base, we conducted a systematic survey of open-source HTR libraries published between 2019 and 2021, identified ten libraries and selected four of them to train HTR models. We trained twelve models in different configurations and compared their performance on both random and scripter-based data splitting. The best recognition results were obtained by the PyLaia and Kaldi libraries which have different and complementary characteristics, suggesting that they should be combined to further improve the results.

Keywords: Handwriting recognition · Norwegian language · Open-source

1 Introduction

Thanks to the recent progress of handwritten text recognition (HTR) systems based on deep learning, the automatic transcription of handwritten documents has become a realistic objective for an increasing number of cultural heritage institutions. Archives and libraries are increasingly willing to include HTR systems in their digitization workflow similarly to their long standing use of OCR (optical character recognition) in the digitization of printed documents. Their goal is to index and make searchable all digitized documents, whether printed or handwritten. Several research projects such as READ/Transkribus [18] and eScriptorium [13] have shown that HTR can now be used at a large scale for automatically transcribing historical handwritten documents. They have demonstrated that high-level accuracy can be reached when HTR models are specifically trained on a representative sample of the target data that has been manually transcribed. However, if the goal is to automatically transcribe the complete

© Springer Nature Switzerland AG 2022
S. Uchida et al. (Eds.): DAS 2022, LNCS 13237, pp. 399–413, 2022.
https://doi.org/10.1007/978-3-031-06555-2_27

Fig. 1. A sample of pages from the dataset: letters from Henrik Ibsen (1872), Camilla Collett (1877) and Harriet Backer (1919).

collections of large libraries or archival institutions consisting of handwritten documents with a huge stylistic variety, manual annotation of a large representative sample rapidly becomes prohibitively costly. In such cases, generic recognition systems, independent from the writer, the period of time and even the language, are needed, as noted by [8].

However, cultural heritage institutions did not wait for the arrival of HTR engines to proceed with the transcription of their documents. Many collections have been researched and transcribed manually and sometimes even edited and published. These research or publishing projects are usually focused on one author, one historical period or one type of document, which means that the transcribed corpus is usually very homogeneous: all the documents are from the same author or from a limited number of authors or of the same type. These sets of transcribed documents can be used as a basis for the constitution of training corpora for HTR engines, but they contain an important bias: their selection was not carried out randomly. They are therefore not representative of all the documents in the collections of the institution, and the engines trained on these documents will consequently have a low generalization capacity. However, should we proceed with a new sampling and a new transcription of documents and put aside all these already transcribed documents?

Once the training corpus has been identified, the implementation of HTR processing requires the choice of a library and the training of the models. Today the technologies used by the state-of-the-art models are relatively homogeneous and based on the same Deep Learning algorithms. Several open-source libraries are available to train these models. The choice of one library over another is difficult for a non-expert because very few direct comparisons are published. Scientific articles generally present results for models obtained after an advanced expert optimization of parameters and hyper-parameters, the full details of which are not always available. The databases used for these comparisons are reference databases, prepared and normalized, which have been used for many years and on which the systems are over-optimized IAM [16], RIMES [2]. The complexities of the implementations to obtain satisfactory results are generally not evaluated.

We propose in this paper to study these different aspects using a new database of Norwegian handwritten documents. Our contributions are the following:

- we introduce and describe a new database of handwritten documents in Norwegian, manually transcribed and freely accessible;
- we present the result of a comprehensive survey of the libraries used in recent publications on handwriting recognition and a comparison of these libraries according to criteria allowing to guide the choice of one or several of them for a HTR project;
- we present a comparison of the error rates obtained on this new database by models trained with a selection of libraries;
- we study the generalization capabilities of these models to new writers on the Norwegian database.

2 Related Work

Spoken by about five million people, mainly in Norway, Norwegian is considered a low-resource language. Even though some speech recognition or machine translation services are available online, the linguistic resources, spoken or written corpora, needed to develop automatic language processing tools are very limited. To the best of our knowledge, there are no digitized corpora of handwritten documents available in Norwegian [9]. An important work has been done within the Norwegian National Library to collect a corpus of electronic texts of sufficient size to train a BERT-type language model [14]. These language models can be used in other applications such as speech recognition in Norwegian [22]. The constitution of corpora allowing the training of named entity extraction models is also recent [10].

Regarding the comparison of handwriting recognition systems, the most common practice in the scientific literature is to compare a new system to the state-of-the-art on reference bases, reporting results published in other papers [3–5,11,12,24,28]. It is quite rare to see authors of new algorithms re-implementing and re-training state-of-the-art models for their evaluation [20,27], due to the significant work involved in re-implementing several models. This approach would not be very reliable for a comparison anyway, because authors tend to optimize their own system more than their competitors do, in order to maximize their chance to have their results published. Benchmarks of handwriting recognition services have been published for printed text recognition (OCR) [7], but only with generic models and services, without training. Another approach to promote the comparison of handwriting recognition systems is to organize competitions [25] or to publish datasets that can be used as benchmarks [26].

3 The Hugin-Munin Dataset for HTR in Norwegian

3.1 Overview

Since the beginning of the millennium, the National Library of Norway (NLN) has been a heavy user of Optical Character Recognition (OCR) i.e. automatic recognition of characters in printed books. In the last couple of years, neural networks have successfully improved recognition technology beyond printed text. This improvement has enabled the application of such technology to handwriting at a large scale. However, the heterogeneity of handwritten documents is a bigger challenge in handwritten text recognition (HTR) than in OCR. There are considerable variations between both writers and writing styles.

The digital instance of NLN is nb.no/search, a digital search engine. With this, users can discover and explore the wide range of texts that constitute the textual cultural heritage of Norway. Yet, for handwritten documents in the collection such a search function has not been available. One of the long-term goals of our present work is *searchability* in nb.no/search or more precisely a fulltext index. Searchability will streamline private archives and make this kind of material available in the same way as printed books. Another long-term goal is to provide readability or a reading support function, since it is not a given that users of today can read old handwriting.

The present work has profited greatly from earlier transcription efforts in different institutions in Norway. Both Collett and Kielland are part of an editorial philological series at the NLN[1], while for Ibsen and Munch the transcriptions have been generously shared with us by the University of Oslo[2] and the Munch Museum[3], respectively.

3.2 Dataset

The current dataset consists of private correspondences and diaries of 12 Norwegian writers with a significant representation in the collection at the NLN. All the documents were written between 1820 and 1950, and they are owned and digitized by the NLN. The various collected transcriptions have been converted to PAGE XML. The selected writers represent a variation in styles of handwriting and orthography.

Some of the writers were selected because they had already been transcribed in other projects, whereas others were selected due to on-going editorial philological projects or forthcoming dissemination activities, but also due to requests from the Norwegian research community[4].

The dataset consists of 164,922 words or tokens in 23,732 lines. It will be published open-source on Zenodo. The images will be distributed in a suitable format and the transcriptions in PAGE XML format.

We have defined two different splits for the experiments:

Random Split: we randomly split all the pages of the dataset in 80% for training, 10% for validation and 10% for testing. This is the standard protocol in machine leaning but it assumes that the sampling has been uniformly performed on the whole corpus, which is not the case.

---

1 https://www.nb.no/forskning/nb-kilder/.
2 https://www.ibsen.uio.no/.
3 https://emunch.no/english.xhtml.
4 An example of this is Anker & Schjønsby's *Lyset i flatene: i arkivet etter Harriet Backer* (2021).

Table 1. Number of pages by writers in train, validation and test sets for the random split and the writer split.

Writer	Lifespan	Random split			Writer split		
		Train	Val	Test	Train	Val	Test
Backer, Harriet	1845–1932	58	9	10	58	9	0
Bonnevie, Kristine	1872–1948	43	5	5	43	5	0
Broch, Lagertha	1864–1952	43			43		
Collett, Camilla	1813–1895	68	10	10	68	10	0
Garborg, Hulda	1862–1934	166	30	16	166	30	0
Hertzberg, Ebbe	1847–1912	48	6	6	48	6	0
Ibsen, Henrik	1828–1906	42	4	5	42	4	0
Kielland, Kitty	1843–1914	34	5	5	0	0	44
Munch, Edvard	1863–1944	33	5	5	0	0	43
Nielsen, Petronelle	1797–1886	58			58		
Thiis, Jens	1870–1942	41	4	4	41	4	0
Undset, Sigrid	1882–1949	40	5	5	0	0	50
Total		674	83	71	567	68	137

Table 2. Count of pages, lines, words and characters in the dataset. (vertical text lines were ignored).

	Pages	Lines	Words	Chars
Train set	674	19,653	139,205	637,689
Validation set	83	2,286	13,916	61,560
Test set	71	1,793	11,801	52,831
Total	828	23,732	164,922	752,080

Writer Split: we chose three writers that had the lowest number of pages in the train set and moved all of their pages to the test set (Kielland, Munch and Undset). In addition, we removed all the other writers from the test set. In the end, as it can be seen in Table 1 there are 16% fewer pages in the train set compared to the random split. This split allows estimating the generalization capacity of the models to unseen writers.

3.3 Transcription Process

The transcriptions in the dataset were initially produced by matching existing transcriptions to the images (text-to-image), or by using pre-existing or self-trained HTR models in Transkribus. This process was followed by one round of proofreading. The proofreading was mostly done by students, with little or no prior experience with transcription. In the next phase of the project the transcriptions will be controlled by the project leaders in order to avoid inconsistencies and to further improve the “ground truth”.

As mentioned above, the output of earlier efforts and projects has been an important source of transcriptions. Such output has made it possible to use a functionality called text-to-image. This functionality makes it possible to align text lines and image lines.

3.4 Language

Today Norway has two written languages, Bokmål and Nynorsk. Danish was the de facto official language in Norway until at least 1814. Due to standardization efforts in the early 1700 s Dano-Norwegian (Danish used by Norwegians) and Danish remained almost identical throughout the the 19th century [19]. During the 19th century, several orthographic shifts took place e.g. shift from aa to å, and different ways of writing the letter ø (ö, ó, ø). Dano-Norwegian subsequently developed into Bokmål. Nynorsk, on the other hand, was conceived by the linguist Ivar Aasen in the mid-1850s with strong emphasis on the spoken dialects of Norway as well with the Old Norse language in mind. The existence of two parallel languages has resulted in a lively debate both before and after Norway gained full independence in 1905.

4 HTR Libraries and Models

4.1 Selection of the Libraries

We conducted a survey of the HTR libraries used in recent scientific peer-reviewed articles published by the document processing community. We screened the papers published in conferences where most of the work on HTR are published: the International Conference on Document Analysis and Recognition (ICDAR), the International Conference on Frontiers of Handwriting Recognition (ICFHR), the International Workshop on Document Analysis Systems (DAS) and the International Conference on Pattern Recognition (ICPR) between 2019 and 2021. Our inclusion criteria in our study were as follows:

- the code of the system must be open source;
- the system must be compared to state-of-the-art systems on publicly available databases of handwritten documents in European languages.

Based on the open-source criteria, we identified the ten libraries that are presented in Table 3. We also included the HTR+, the system available in Transkribus even though it is not open-source because it is currently a standard tool in the community. We collected the following information on the libraries based on their source code repository:

- the type of deep learning framework used by the library;
- the number of commits and the number of different contributors to the source code;
- the date of the last commit.

Table 3. Survey of open-source HTR libraries used in publications in major document processing conferences between 2019 and 2022.

Name	Framework	Last commit	Commits	Contrib.	Last version
Kaldi [1]	Kaldi	18/12/2021	9223	100	–
Kraken [13]	PyTorch	19/12/2021	1486	18	11/2021
PyLaia [24]	PyTorch	08/02/2021	860	4	12/2020
HTR-Flor++ [20]	TensorFlow 2	8/12/2021	280	4	10/2020
PyTorchOCR [4]	PyTorch	10/09/2021	24	1	–
VerticalAttentionOCR [5]	PyTorch	3/12/2021	21	1	–
Convolve, Attend & Spell [12]	PyTorch	24/06/2019	20	2	–
HRS[3]	TensorFlow	19/03/2021	20	2	–
ContentDistillation [11]	PyTorch	13/06/2020	3	1	–
Origaminet [28]	PyTorch	13/06/2020	2	2	–
HTR+ [17]	–	–	NA	NA	–

We have selected the libraries to be evaluated according to the following criteria:

- number of commits: a large number of contributions to the library indicates regular updates, added features and active development;
- number of contributors: the code associated with publications is often the work of a single person, the main author of the publication. A low number of contributors can indicate a difficulty in handling the code and puts the maintenance and future development of the library at risk;
- date of the last commit: the quality and security of a software requires a follow-up of the dependencies updates and the application of security patches. The last commit must be recent.
- date of the last version: good software development practices recommend to index the stable and validated states of a software by numbered releases. The presence of a release is an indication of software quality.

Based on these criteria, we have selected the Kaldi, Kraken, PyLaia and HTR-Flor++ libraries. The other libraries have been discarded mainly for lack of contributors or updates. As previously mentioned, HTR+ has also been added to the selection because it is a *de facto* reference due to the success of the Transkribus platform.

4.2 Description of the Selected Libraries

Kaldi [1] is a library developed for speech recognition and adapted to HTR.

Kraken [13] is a turn-key OCR system optimized for historical and non-Latin script material. Since it was developed for the recognition of connected scripts such as Arabic, it is also suited to the recognition of handwritten cursive text.

PyLaia [24] is a deep learning toolkit for handwritten document analysis based on PyTorch. It is one of the HTR engines available in Transkribus.

HTR-Flor++ [20] is a framework for HTR that implements different state-of-the-art architectures, based on TensorFlow.

HTR+ [17] is the HTR system developed in the framework of the READ project and available in Transkribus.

4.3 Training of the Models

In order to make a fair comparison, we have trained handwriting recognition models with each of the libraries following the provided documentation. A first model was trained with the default parameters, without optimization, which corresponds to a training performed by a non-expert (basic model). We then contacted the creators of libraries, when possible, and asked them for advice to improve the model obtained. We trained several models following their advice and selected the best one (expert model). We also trained PyLaia and HTR+ models using the Transkribus platform, which provides features for PyLaia that are not available in the open-source version. As HTR+ is not open-source, it is not possible to train or use it outside the Transkribus platform.

The details of the different models are as follows:

Kaldi Basic: we trained a model according to the Bentham recipe provided in Kaldi source code. The text is modelled at BPE level with a ngram (n=3). The separate language model is trained only on the line transcriptions that come from the train set. As shown in [1] the Kaldi model training has two steps. At first a model is trained from the transcription and line image pairs ("flat start"). Then it is used to align the transcriptions on the line images. Finally another model is trained on these alignments. When the training is finished, only the last model is needed for inference - the "flat start" model can be discarded. The neural network is composed of 10 layers of SDNN. It has 6 convolution layers and 3 TDNN layers, with batch normalizations and ReLUs in between and an output layer with softmax. The lines input images are resized to a fixed height of 40 pixels while keeping the aspect ratio.

Kaldi Expert: we increased the number of SDNN layers to 15. (4 extra convolution layers and one extra TDNN layer were added).

Kraken Basic: the model was trained with `ketos` default parameters (input height 48). The model has 3 convolution and 3 LSTM layers and it uses group normalization.

Kraken Expert: we were provided with a better model (by the authors of the library) that had 120 as input height. In addition, there was an issue with image preprocessing that hindered the performance. By using the binary format, this unnecessary preprocessing was turned off and the results got better. This model has 4 convolution and 3 LSTM layers. Group normalization layers have been replaced with dropout and max pooling.

PyLaia Basic: the model was trained with default parameters, except for the input height that was fixed to 128, because without fixing the line height the model was not capable to learn much. The model has 4 convolution and 3 LSTM layers.

PyLaia Expert: we tried to emulate the model that was used in Transkribus as closely as possible, but not all the parameters used were available in the open source version. It has the same number of layers, but the number of features in convolution layers is different, max pooling is different and it uses batch normalization. Also, this one uses a different learning rate and has longer patience for early stopping.

HTR-Flor++ Basic: we used the default model (Flor). It has 6 convolution and 2 GRU layers.

HTR-Flor++ Expert: we tried other already implemented models (Bluche: HTR-Flor++ expert-a, Puigcerver: HTR-Flor++ expert-b, Puigcerver Octave CNN: HTR-Flor++ expert-c)[5]

HTR+ Basic: we trained a HTR+ model in Transkribus from scratch. The parameters were not disclosed in the Transkribus interface.

HTR+ Expert: we trained a model using a pre-trained model in Danish. The parameters were not disclosed in the Transkribus interface.

Training Data. The automatically created line polygons were very noisy, sometimes cutting too much of the text, making them almost unreadable. To deal with that, we decided to use bounding box extraction, which gave better results than polygon extraction in preliminary tests. Now the lines images can be noisy as well, sometimes containing (part of) the line above and below, but at least the text is visible.

Also, we ignore vertical lines in this evaluation. The performance on them could be measured in a future work.

The libraries need a transcriptions file that contains a link to the line image and the transcriptions or something similar. PyLaia requires the user to create a file with all the available symbols and transform the data before training. Other libraries do it automatically from the training data, which avoids issues from possibly malformed symbol files.

Connectionist temporal classification (CTC) loss [6] is the objective function that is used by the models to learn how to recognize the handwriting. Except for Kaldi model, that uses a lattice free maximum mutual information [23] as the objective function.

The models use early stopping, meaning that the training will be stopped if the model stops improving. Kaldi, however, uses a preset number of epochs and will complete them and then combine the model checkpoints of different epochs to produce a final model.

To compare the results of different models we used two metrics - character error rate (CER) and word error rate (WER). CER is the edit distance on character level between the predicted transcription and ground truth divided by the length of the ground truth transcription. WER is calculated in a similar way, but on word level.

[5] The model architectures can be seen here: https://github.com/arthurflor23/handwritten-text-recognition/tree/master/doc/arch.

Table 4. Comparison of the performance of the different models configurations (basic and expert) measured with Character Error Rates (CER) and Word Error Rates (WER) on the train, validation and test sets with random data split.

Model	Height	Augm.	Train		Val		Test	
			CER	WER	CER	WER	CER	WER
Kaldi basic	40	No	5.30	12.05	11.61	26.19	10.76	24.85
Kaldi expert	40	No	4.71	11.10	10.29	24.17	9.18	**22.19**
Kraken basic	48	No	51.95	76.52	64.60	89.72	64.44	89.49
Kraken expert	120	Yes	0.40	1.31	12.05	30.29	12.20	31.28
PyLaia basic	128	No	1.37	4.45	11.02	28.09	10.87	27.62
PyLaia basic	128	Yes	3.08	9.39	10.44	26.50	10.10	26.30
PyLaia expert	64	Yes	3.73	10.66	11.70	28.90	12.75	31.12
PyLaia expert	128	Yes	1.68	5.30	9.15	24.28	**8.86**	23.79
HTR-Flor++ basic	128	Yes	–	–	–	–	11.49	31.59
HTR-Flor++ expert-a	128	Yes	–	–	–	–	56.10	82.21
HTR-Flor++ expert-b	128	Yes	–	–	–	–	12.62	32.33
HTR-Flor++ expert-c	128	Yes	–	–	–	–	11.04	29.70
HTR+ basic	N/A	N/A	2.98	–	7.17	–	9.14	21.81
HTR+ expert	N/A	N/A	2.58	–	6.34	–	8.31	20.30

Table 5. Detailed analysis of the CER for different classes of characters on the different sets for the best HTR model (PyLaia expert) on the random split and the writer split.

Random split							
		Lowercase	Uppercase	Digits	Special	Accents	Punctuation
Train	number	569,687	28,908	3,205	12,269	125	23,506
	CER	1.4	1.6	2.7	2.1	22.4	8.4
Validation	number	55,457	2,314	324	1,236	15	2,215
	CER	7.7	18.6	32.4	14.0	73.3	30.0
Test	number	47,516	2,085	319	1,040	20	1,851
	CER	7.7	13.4	24.8	14.5	75.0	28.2

5 Results

5.1 Random Split

The first set of experiments was conducted on the Random Split. Table 4 reports the CER and WER for all the models on the train, validation and test sets.

As a first general remark, we found that the discussion with the creators of the libraries was often very beneficial to know how to properly configure the

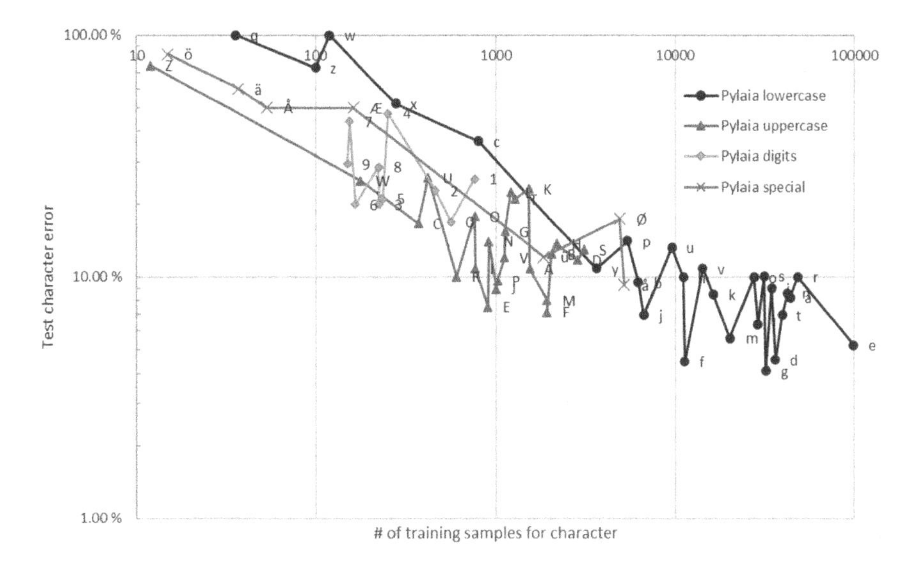

Fig. 2. Character Error Rate (CER) on the test set with respect to the number of training samples for each character, for the best HTR model (PyLaia) on the random split.

models or test parameters that could improve the results. In the case of Kraken, the results were very bad with the default architecture and the advice of the library experts allowed us to obtain competitive results. In the case of the other libraries, the discussion with the experts allowed us to validate that we had the optimal configuration.

The best CER on the test set was obtained with PyLaia using the optimized architecture with 128 pixel line height and data augmentation. The second best CER on test set was obtained by Kaldi using the optimized architecture, which also yielded the best WER. The better results of Kaldi in terms of WER can be explained by the fact that it uses an explicit language model (ngram of BPE), while the other systems model the dependencies between characters with recurrent neural networks.

It can also be noted that other systems perform better on line images with higher resolution (line height). This is especially true for Kraken which performs very poorly with the default height of 48 pixels.

The impact of data augmentation can be observed on the results of the PyLaia basic model. With data augmentation, over-training is reduced, the error rate in learning increases from 1.37% to 3.08% but the error rates in validation and testing decrease.

It should be noted that the Kaldi systems seem to suffer less from over-training than the other systems: their error rate on the training set is always higher but their error rates on the validation and test sets are among the best. This can be attributed to the use of language models by Kaldi system.

Table 6. Detailed analysis of the confusion between characters for the best HTR model (PyLaia expert) on the Random split.

Char	# Confusions	Relative confusion	Conf. 1		Conf. 2		Conf. 3		Others
a	271	7.38%	o	2.9%	e	1.93%	æ	0.79%	1.77%
b	42	8.08%	l	2.9%	t	1.54%	h	1.35%	2.31%
e	207	2.60%	a	0.5%	o	0.39%	i	0.29%	1.46%
h	86	8.13%	s	2.5%	t	1.13%	k	0.85%	3.69%
m	74	4.49%	n	2.61%	v	0.61%	i	0.24%	1.03%
n	189	5.59%	r	1.72%	m	1.18%	v	0.68%	2.01%
o	162	7.98%	a	3.20%	e	1.87%	ø	1.04%	1.87%
r	198	5.18%	s	0.89%	n	0.89%	v	0.55%	2.85%
s	188	7.25%	r	1.74%	h	1.04%	e	0.81%	3.66%
F	5	5.21%	T	2.1%	f	1.04%	d	1.04%	1.04%
L	13	20.00%	t	9.2%	l	3.08%	d	3.08%	4.62%
æ	34	7.93%	e	2.3%	a	2.10%	d	0.93%	2.56%
ø	56	14.74%	o	6.1%	å	2.37%	e	1.58%	4.74%
å	21	11.60%	ø	4.4%	a	3.32%	u	1.11%	2.76%

Finally, the HTR+ expert model, based on a Danish model, outperformed all the other systems. However its results are not directly comparable since this model has not been trained on exactly in the same conditions: both the line extraction and the evaluation were done in Transkribus and are not open-source, so they may be different from our line extraction and CER/WER metrics.

We conducted a detailed analysis of the CER for each characters with respect to their number of training samples, presented on Fig. 2. Lower case letters are by far the majority and therefore the best recognized, except for some rare letters (*qzwx*). Numbers and special characters are very poorly represented (except å and ø) and therefore very poorly recognized. Capital letters are in a intermediate situation, with a relatively small number of samples but relatively low error rates. A summary of the CER for different classes of characters is presented on Table 5. One can note that punctuation is particularly difficult to recognize: even with almost as many examples as capital letters, their error rate is twice as high.

Finally, we analyzed the most frequent confusions between characters on the test set for the best HTR model (PyLaia expert). An extract of the confusion table is presented in Table 6. Compared with results from printed OCR [21], the HTR confusions are more spread across confusion alternatives and the typical OCR single character substitutions like e-o-c, h-b, n-u, m-n, i-l-I while present in the HTR results, have less relative weight. The same is true for the typical OCR character substitutions for the Norwegian special characters: ø-o, æ- a or e, å - a, which are also present in the HTR with less relative weight while some confusion not often seen in OCR such as å-i, ø-, æ-o and å-r are relatively important.

Table 7. Comparison of the performance of the different configuration (basic and expert) of Kaldi and PyLaia models measured with Character Error Rates (CER) and Word Error Rates (WER) on the train, validation and test sets with the writer split.

Model	Height	Augm.	Train		Val		Test	
			CER	WER	CER	WER	CER	WER
Kaldi basic	40	No	4.90	11.34	12.57	28.10	24.24	44.49
Kaldi expert	40	No	4.37	10.48	11.03	25.79	**21.79**	**42.13**
PyLaia basic	128	Yes	2.70	8.25	10.64	27.58	24.36	49.42
PyLaia expert	128	Yes	1.64	5.40	9.53	25.90	22.74	47.95

OCR errors are generally caused by noise and low contrast where the basic characters are in essence very similar, but HTR errors are generally caused by different graphical representations of different characters most prominent when comparing different writers, but there might also be large differences of character representations for a single writer. Some common characters, e.g. 'a','g' and 'r' have generally different topology in cursive handwriting compared to print which also affects the confusion alternatives.

5.2 Random Split by Writer with Unseen Writers

We chose the best models from the previous experiments (PyLaia and Kaldi) and trained them on the Writer split. This experiment allows us to evaluate the generalization capabilities of the different models to new writers. As it might be expected, the models perform a lot worse on unseen writers. PyLaia and Kaldi obtain very similar results, with a slight advantage to Kaldi. Again, this advantage can be attributed to the use of a language model by Kaldi but also to the lower resolution of the input images, which may reduce over-training.

6 Conclusion

In this article we have introduced a database of handwritten historical documents in Norwegian. This database is the first of its kind and constitutes a valuable resource for the development of handwritten text recognition models (HTR) in Norwegian. In order to evaluate the performance of state-of-the-art HTR models on this new database, we conducted a systematic survey of open-source HTR libraries published between 2019 and 2021. We selected four libraries, amongst ten, according to criteria of quality and sustainability of their source code based on software development metrics. We trained twelve models in different configurations and compared their performance on both random data splitting and writer-based data splitting to evaluate their generalization capabilities to writers not seen during training. Finally, we studied the most frequent confusions between characters. The best recognition results were obtained by the Kaldi library which uses a language model and PyLaia which uses higher resolution images and data augmentation during training. A combination of these different techniques, in a single model or by voting, should further increase the performance of HTR models. Recently proposed models based on transformers[15] should also be added to the benchmark.

Acknowledgements. We thank Daniel Stoekl, Benjamin Kiessling, Joan Andreu Sanchez, Arthur Flôr for their advices during the training of the HTR models with their respecting libraries. This work was supported by the Research Council of Norway through the 328598 IKTPLUSS HuginMunin project.

References

1. Arora, A., et al.: Using ASR methods for OCR. In: International Conference on Document Analysis and Recognition (2019)
2. Augustin, E., Brodin, J.M., Carré, M., Geoffrois, E., Grosicki, E., Prêteux, F.: RIMES evaluation campaign for handwritten mail processing. In: International Conference on Document Analysis and Recognition, p. 5 (2006)
3. Chammas, E., Mokbel, C., Likforman-Sulem, L.: Handwriting recognition of historical documents with few labeled data. In: International Workshop on Document Analysis Systems, pp. 43–48. IEEE (2018)
4. Coquenet, D., Chatelain, C., Paquet, T.: Recurrence-free unconstrained handwritten text recognition using gated fully convolutional network. In: International Conference on Frontiers in Handwriting Recognition, pp. 19–24 (2020)
5. Coquenet, D., Chatelain, C., Paquet, T.: End-to-end handwritten paragraph text recognition using a vertical attention network. IEEE Trans. Pattern Anal. Mach. Intell. (2022)
6. Graves, A., Fernández, S., Gomez, F., Schmidhuber, J.: Connectionist temporal classification: labelling unsegmented sequence data with recurrent neural networks. In: International Conference on Machine Learning, pp. 369–376 (2006)
7. Hegghammer, T.: OCR with tesseract, Amazon textract, and Google document AI: a benchmarking experiment. J. Comput. Soc. Sci. (2021)
8. Hodel, T., Schoch, D., Schneider, C., Purcell, J.: General models for handwritten text recognition: feasibility and state-of-the art. German kurrent as an example. J. Open Humanit. Data **7**(13), 1–10 (2021)
9. Hussain, R., Raza, A., Siddiqi, I., et al.: A comprehensive survey of handwritten document benchmarks: structure, usage and evaluation. J. Image Video Proc. **2015**, 46 (2015). https://doi.org/10.1186/s13640-015-0102-5
10. Jørgensen, F., Aasmoe, T., Ruud Husevåg, A.S., Øvrelid, L., Velldal, E.: NorNE: annotating named entities for Norwegian. In: Language Resources and Evaluation Conference (2020)
11. Kang, L., Riba, P., Rusiñol, M., Fornés, A., Villegas, M.: Distilling content from style for handwritten word recognition. In: International Conference on Frontiers in Handwriting Recognition (2020)
12. Kang, L., Toledo, J.I., Riba, P., Villegas, M., Fornés, A., Rusiñol, M.: Convolve, attend and spell: an attention-based sequence-to-sequence model for handwritten word recognition. In: Brox, T., Bruhn, A., Fritz, M. (eds.) GCPR 2018. LNCS, vol. 11269, pp. 459–472. Springer, Cham (2019). https://doi.org/10.1007/978-3-030-12939-2_32

13. Kiessling, B., Tissot, R., Stokes, P., Stökl Ben Ezra, D.: eScriptorium: an open source platform for historical document analysis. In: International Conference on Document Analysis and Recognition Workshops, vol. 2, p. 19 (2019)
14. Kummervold, P.E., de la Rosa, J., Wetjen, F., Brygfjeld, S.A.: Operationalizing a national digital library: the case for a norwegian transformer model. In: Nordic Conference on Computational Linguistics (2021)
15. Li, M., et al.: Trocr: transformer-based optical character recognition with pre-trained models (2021). https://arxiv.org/abs/2109.10282
16. Marti, U.V., Bunke, H.: The IAM-database: an English sentence database for offline handwriting recognition. IJDAR **5**, 39–46 (2002). https://doi.org/10.1007/s100320200071
17. Michael, J., Weidemann, M., Labahn, R.: Htr engine based on nns p 3 optimizing speed and performance-htr +. Technical report, READ-H2020 Project 674943 (2018)
18. Muehlberger, G., et al.: Transforming scholarship in the archives through handwritten text recognition: transkribus as a case study. J. Doc. (2019)
19. Nesse, A., Sandøy, H.: Norsk Språkhistorie IV: Tidslinjer. Novus, Oslo (2018)
20. Neto, A.F.S., Bezerra, B.L.D., Toselli, A.H., Lima, E.B.: HTR-flor++: a handwritten text recognition system based on a pipeline of optical and language models. In: ACM Symposium on Document Engineering (2020)
21. Nguyen, T.T.H., Jatowt, A., Coustaty, M., Nguyen, N.V., Doucet, A.: Deep statistical analysis of OCR rrrors for effective post-OCR processing. In: Joint Conference on Digital Libraries (2019)
22. Ortiz, P., Burud, S.: Bert attends the conversation: improving low-resource conversational asr (2021). https://arxiv.org/abs/2110.02267
23. Povey, D., et al.: Purely sequence-trained neural networks for asr based on lattice-free mmi. In: Interspeech, pp. 2751–2755 (2016)
24. Puigcerver, J., Mocholí, C.: PyLaia (2018). https://github.com/jpuigcerver/PyLaia
25. Strauß, T., Leifert, G., Labahn, R., Mühlberger, G.: Competition on automated text recognition on a read dataset. In: International Conference on Frontiers in Handwriting Recognition (2018)
26. Sánchez, J.A., Romero, V., Toselli, A., Villegas, M., Vidal, E.: A set of benchmarks for handwritten text recognition on historical documents. Pattern Recogn. **94**, 122–134 (2019)
27. Toiganbayeva, N., et al.: KOHTD: kazakh offline handwritten text dataset (2021). https://arxiv.org/abs/2110.04075
28. Yousef, M., Bishop, T.E.: Origaminet: weakly-supervised, segmentation-free, one-step, full page textrecognition by learning to unfold. In: Conference on Computer Vision and Pattern Recognition (2020)

Open Source Handwritten Text Recognition on Medieval Manuscripts Using Mixed Models and Document-Specific Finetuning

Christian Reul[1(✉)], Stefan Tomasek[1], Florian Langhanki[1], and Uwe Springmann[2]

[1] University of Würzburg, Würzburg, Germany
{christian.reul,florian.langhanki}@uni-wuerzburg.de,
stefan.tomasek@germanistik.uni-wuerzburg.de
[2] CIS, LMU Munich, Munich, Germany
springmann@cis.uni-muenchen.de

Abstract. This paper deals with the task of practical and open source Handwritten Text Recognition (HTR) on German medieval manuscripts. We report on our efforts to construct mixed recognition models which can be applied out-of-the-box without any further document-specific training but also serve as a starting point for finetuning by training a new model on a few pages of transcribed text (ground truth). To train the mixed models we collected a corpus of 35 manuscripts and ca. 12.5k text lines for two widely used handwriting styles, Gothic and Bastarda cursives. Evaluating the mixed models out-of-the-box on four unseen manuscripts resulted in an average Character Error Rate (CER) of 6.22%. After training on 2, 4 and eventually 32 pages the CER dropped to 3.27%, 2.58%, and 1.65%, respectively. While the in-domain recognition and training of models (Bastarda model to Bastarda material, Gothic to Gothic) unsurprisingly yielded the best results, finetuning out-of-domain models to unseen scripts was still shown to be superior to training from scratch. Our new mixed models have been made openly available to the community.

Keywords: Handwritten text recognition · Medieval manuscripts · Mixed models · Document-specific finetuning

1 Introduction

Efficient methods of Automatic Text Recognition (ATR) of images of either printed (OCR) or handwritten (HTR) material rely on the availability of pretrained recognition models that have been trained on a wide variety of various glyphs (the specific incarnations of an alphabet of characters on paper) that one wants to recognize. Appropriately trained *mixed models* have shown to result

© Springer Nature Switzerland AG 2022
S. Uchida et al. (Eds.): DAS 2022, LNCS 13237, pp. 414–428, 2022.
https://doi.org/10.1007/978-3-031-06555-2_28

in Character Error Rates (CERs) below 2% even for historical printed material previously unaccessible to automated recognition [15]. Automatically transcribed text resulting from the out-of-the-box application of available mixed models will already enable use cases such as searching and can enable corpus-analytic studies if large text collections are available.

For other downstream tasks such as determining that certain words are not contained in a text or preparing a critical edition, much lower error rates have to be achieved. Before embarking on a time-consuming manual correction project, automatic methods can again come to the rescue to a certain extent. Following the lead of Breuel et al. [1] it was shown that printing-specific neural-network-based models can be trained that deliver better results than mixed models [17] which in turn have been shown to provide an exellent starting point for the training procedure [13].

In the end, even for the result of these fine-tuned models it may still be necessary to manually correct the outcome. However, the best way to achieve one's goal of a certain acceptable residual error rate (no printed text can ever be 100% error-free) with minimal human effort is to combine out-of-the-box recognition with a so-called Iterative Training Approach (ITA) which has already been shown to be highly effective when dealing with early printed books [12]. The idea is to continously retrain specialized models on manually transcribed lines (Ground Truth, GT), apply them to new data, and use the ever-increasing accuracy to keep the error rates and therefore the required correction effort as low as possible at all times.

While the above method has previously been shown to work well with printed material, we here report on experiments with medieval German manuscripts. Because handwritten material is much less regular than printed text, resulting from variation of glyphs among different writers and even the same writer, we expect considerably higher error rates. Just how low an error rate can be achieved by the application of a pretrained mixed model that gets finetuned to a specific document with moderate effort (a few pages of GT) is the topic of the current paper.

A second goal is to explore what effort it takes to train a finetuned model based on a mixed model. As we make our models openly available[1], the question is: What increase in recognition quality can be achieved by continuously training a finetuned model on an ever larger amount of generated GT?

The third goal explores the possibility of adapting a mixed model out of the domain of the training material it is based on. Would it still be possible to finetune a mixed model with a few pages of GT to an out-of-domain manuscript successfully?

The remainder of the paper is structured as follows: After an overview of related work in Sect. 2 we introduce the data required for training and evaluation in Sect. 3 and explain the methodology of our experiments in Sect. 4. The experiments are described in Sect. 5 and discussed in Sect. 6 before Sect. 7 concludes the paper.

[1] https://github.com/Calamari-OCR/calamari_models_experimental.

2 Related Work

For a thorough literature review of HTR methods we refer to a recent survey by Memon et al. [8] and to the description of a set of benchmarks by Sanchez et al. [16].

The Transkribus platform [6] offers a comprehensive selection of publicly available models[2] for a variety of languages and different epochs. However, of the 85 models available at the time of writing, only 14 can be used with the open source[3] PyLaia engine [10] while the vast majority (71 models) are only available for the proprietary HTR+ engine [9] which can exclusively be used via Transkribus. The vast majority of the public models were trained using texts from the 16th century or later.

Hawk et al. (see [4]) reported on experiments with mixed models dealing with Caroline minuscle scripts using the OCRopus OCR engine[4]. Several experiments with varying numbers of different manuscripts in the training set showed that when applying a mixed model to a manuscript which has not been part of the training pool, models trained on a wider variety of manuscripts perform better. However, when a model is applied to material that the model has already seen during training (same manuscript but different pages), the trend is mostly the reverse.

In [18] Stökl Ben Ezra et al. present an open annotated dataset[5] and pre-trained script-specific as well as a mixed model for recognition and page segmentation on Medieval Hebrew manuscripts.

Hodel et al. [5] deal with mixed models for German *Kurrent* script. They present an open test set[6] comprising 2,426 lines collected from minutes of the meetings of the Swiss Federal Council between 1848 and 1903. Evaluating three HTR+ and one PyLaia Kurrent model on the test set resulted in median CER values from 2.76% to 13.30%.

3 Data Sets

To perform our experiments we first had to collect and produce data, both as page images of medieval manuscripts as well as the corresponding GT in the form of diplomatic[7] (i.e. faithful to the written image) transcriptions. The training and evaluation sets we compiled are described in this section.

To ensure maximum flexibility and connectivity we always collected the original color images and used the PAGE XML format [11] to store any further information like region and line coordinates, transcriptions, etc.

---

[2] https://readcoop.eu/transkribus/public-models.
[3] https://github.com/jpuigcerver/PyLaia.
[4] https://github.com/ocropus/ocropy.
[5] https://zenodo.org/record/5167263.
[6] https://zenodo.org/record/4746342.
[7] https://en.wikipedia.org/wiki/Diplomatics#Diplomatic_editions_and_transcription.

Table 1. Training data used for our experiments. A part from the *subcorpus* we list the number of *manuscripts* as well as the respective number of *pages* and *lines* (all and selected for training).

subcorpus	# works	all		selected		centuries			
		# pages	# lines	# pages	# lines	13	14	15	16
Kindheit Jesu	5	128	6,244	18	576	3	2		
Parzival	6	36	1,685	19	771			6	
Faithful Transcriptions	12	73	2,483	44	1,381		2	9	1
Marienleben	6	28	1,232	19	791		4	2	
Medical Tracts	6	26	891	21	699	1	1	4	
Sum	35	291	12,535	121	4,218	4	9	21	1

Our training corpus was collected and produced from various sources and within several projects as summarized in Table 1: The *Kindheit Jesu (Childhood of Jesus)* editorial project at University of Würzburg, the *Parzival* (*Percival*)[8] digital editorial project at University of Bern, some theological manuscripts transcribed during the *Faithful Transcriptions* transcribathon[9] [3], several manuscripts containing the *Marienleben (Life of Mary)* and some medieval medical tracts.

These manuscripts have been chosen to cover both a span of several centuries (13th to 16th) of origin as well as a certain variety of writing styles typical for this time: Gothic and Bastarda cursive.[10]

Manuscript pages from the first two projects were manually transcribed while for the last two projects the segmentation of the images and the subsequent transcriptions was done via the open source OCR4all[11] framework. All pre-existing transcriptions had to be adapted to our transcription guidelines to have a uniform representation of glyphs to characters.

In total, the training stock consists of 35 manuscripts written by at least 30 different hands and comprising ca. 12,5k lines which can be broken down into close to 8,5k lines of Gothic and ca. 4k lines of Bastarda cursives. Figure 1 shows some representative example lines and their corresponding transcriptions.

To evaluate our models we collected 212 pages comprising close to 9k lines from five additional manuscripts, three written in Gothic and two in Bastarda cursives (cf. Fig. 2 for some example lines): One manuscript containing the *Kindheit Jesu* (Handschrift-B) written by Konrad von Fußesbrunnen, two additional manuscripts about the life of Mary (*Driu liet von der maget*) by Brother Wernher (Wernher-Krakau and Wernher-Wien), and as Bastarda examples we chose two manuscripts of the moral doctrine *Der Welsche Gast* by Thomasin von

[8] https://www.parzival.unibe.ch/englishpresentation.html.
[9] https://lab.sbb.berlin/events/faithful-transcriptions-2/?lang=en.
[10] https://www.adfontes.uzh.ch/tutorium/schriften-lesen/schriftgeschichte/bastarda-und-gotische-kursive.
[11] https://github.com/ocr4all.

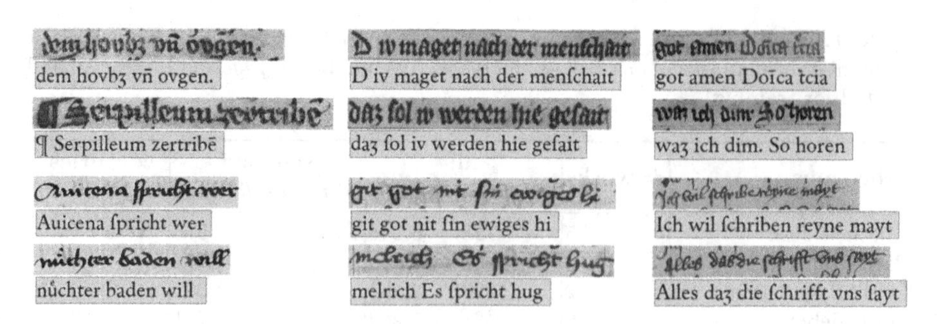

Fig. 1. Two line images for each of six representative manuscripts and their corresponding transcription for the training corpus (top: Gothic, bottom: Bastarda).

Zirclaere[12] (Gast-1 and Gast-2) from the digital editorial project of the University of Heidelberg. Wernher-Wien was written by a hand already present in the training data and added for comparison.

For our experiments we randomly selected 32 pages as our maximum training set and then repeatedly cut it in half to obtain further sets comprising 16, 8, 4, and 2 pages. The rest of the data we used as a fixed evaluation set. Table 2 lists the details.

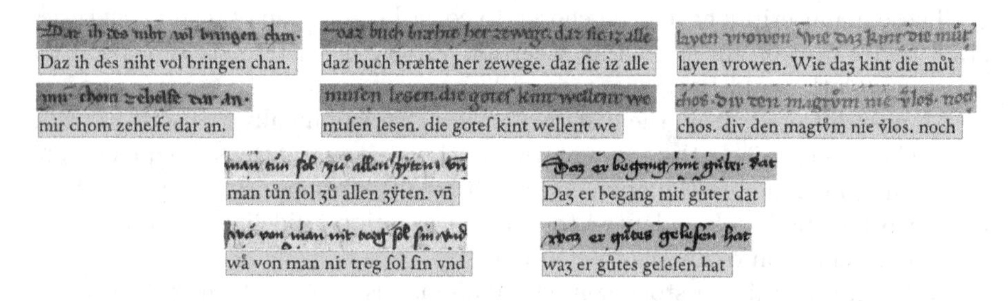

Fig. 2. Two line images and the corresponding transcription for the five manuscripts used for evaluation. top: Handschrift-B, Wernher-Krakau, Wernher-Wien; bottom: Gast-1, Gast-2

4 Methods

Training mixed models which can be applied out-of-the-box as well as serve as a starting point for document-specific training is a challenging task: On the one hand, the models need to be very robust and generalize well on as many documents as possible. On the other hand, the models should still be quite

---

[12] https://digi.ub.uni-heidelberg.de/wgd.

Table 2. Number of available *pages* and *lines* for the evaluation manuscripts, devided into a fixed evaluation set (*Eval*) and five training batches for the ITA (*Train*).

manuscript	date	eval # pages	eval # lines	train # lines for # pages 2	4	8	16	32
Handschrift-B	1250-1275	8	592	152	304	607	1.215	2.430
Wernher-Krakau	1200-1225	18	363	49	96	197	349	687
Wernher-Wien	1250-1275	10	241	49	98	196	391	752
Gast-1	1450-1475	8	546	141	261	521	1.070	2.160
Gast-2	ca. 1300	8	222	58	114	217	442	881

specifically geared towards a certain type of material in order to produce the best possible results.

To deal with this challenge, we devised the following pipeline where each training starts from the model resulting from the previous step:

1. A strong mixed model for printed types based on [15] serves as a foundation for all training processes. Despite the apparent difference to manuscript material we expect this to provide a better starting point compared to starting from scratch, i.e. a random parameter allocation.
2. Both handwriting styles (Gothic and Bastarda) are trained together to show the model as much data as possible in order to learn general features, adapt to noise, etc. The result is a combined mixed model covering both styles.
3. Finally, the existing models are refined by exclusively training on Gothic script or Bastarda data, respectively. This results in a mixed model for each handwriting style.

An important factor when dealing with ATR or deep learning in general is the selection and optimization of hyperparameters which define the structure of the network. We refrain from evaluating a wide variety of configurations, but stick to the following networks which have been shown to be very useful in the past and are predefined in the open source[13] Calamari ATR engine [19] which we used for our experiments:

- *def*: The original and comparatively shallow Calamari default network structure consisting of two convolutional neural networks (CNNs) (40 and 60 3×3 filters respectively), each followed by a 2×2 max pooling layer, and a concluding LSTM network layer with 200 cells using dropout (0.5)[14].
- *htr+*: An adaptation of the standard network structure of the Transkribus platform with more complex variations regarding filter sizes, strides, etc. (see [9] for details).

[13] https://github.com/Calamari-OCR/calamari.

[14] In Calamari short notation:
conv=40:3 × 3, pool=2 × 2, conv=60:3 × 3, pool=2 × 2, lstm=200, dropout=0.5.

- *deep3*: An alternative deeper network structure which extends the default network by another convolutional layer and two additional LSTMs. This network yielded good results during our previous experiments with mixed models for printed material and is becoming the new Calamari default[15].

Analogously to the method described in [15] we apply a two-stage training approach to reduce the influence of a few overrepresented manuscripts while utilizing all of the available training data: For each individual step in the training workflow we first train on all available data, to show the model as much material as possible. Then, during the so-called refinement stage, we use the resulting model from stage one as a starting point and run another full training process while only using selected pages for each manuscript. The selected pages have been determined beforehand by simply drawing single pages at random until a predetermined number of lines is surpassed or all pages have been drawn for a single manuscript. Since almost all works comprised more than or at least close to 150 lines of GT we chose this number as our cutoff value. In this way a balanced training corpus is constructed which gives equal weight to all manuscripts.

Second, we varied the input data by using different preprocessing results, i.e. mainly different binarization techniques, which can also be regarded as a form of data augmentation. We used two methods from the *ocrd-olena* package[16] (Wolf, Sauvola MS Split), the binary and normalized grayscale output produced by OCRopus' *ocropus-nlbin* script, as well as the SBB binarization technique[17].

As mentioned in the introduction, the most cost-effective way to get to a low-error output consists in avoiding manual correction work as much as possible by building better recognition models and thus replacing human effort by a higher but much cheaper computational load. This issue of getting to a reasonable recognition accuracy as fast as possible is addressed by the so-called *Iterative Training Approach* [12] which consists of the following steps:

1. Transcribe a small number of lines from scratch or correct the output of a suitable mixed model, if available
2. Train a document-specific model using all available GT (including the GT from earlier iteratons)
3. Apply the model to further lines which have not yet been transcribed
4. Correct the output
5. Repeat steps 2–4

In this paper the ITA is utilized two times: First, when transcribing manuscripts to produce initial GT for the evaluation set. Second, we simulate an ITA during the second experiment by iteratively doubling the training pages.

[15] In Calamari short notation:
conv=40:3 $\times$ 3, pool=2 $\times$ 2, conv=60:3 $\times$ 3, pool=2 $\times$ 2, conv=120:3 $\times$ 3, pool=2 $\times$ 2, lstm=200, lstm=200, lstm=200, dropout=0.5.

[16] https://github.com/OCR-D/ocrd_olena.

[17] https://github.com/qurator-spk/sbb_binarization.

5 Experiments

For our experiments we used Calamari version 2.1. All training runs followed the cross-fold training methodology of [14] producing a voting ensemble of five individual voters and combining their respective outputs via the confidence values of each individual recognized character. Apart from better results this also reduces the variance among the results considerably. In addition, we used the same random seed for the experiments to standardize all processes involving randomness (data shuffling, augmentation etc.).

To determine the end of each training process we utilized the default early stopping criterion which evaluates the current model against the held out validation data after each epoch. The training stops if the validation CER did not improve for five consecutive times but at the latest after 100 epochs. The number of samples after which an evaluation was performed was set according to the size of the respective training set but always to at least 1,000 steps. Regarding further (hyper-) parameters we stuck to the Calamari defaults and well established best practices. Most notably this includes augmenting each sample five times by using image degradation transformations for each written line as well as utilizing a general weight decay of 10^{-5} for all layers and an EMA weight decay of 0.99.

The model recognition output is evaluated against the GT by the Levenshtein edit distance which measures the CER. As for preprocessing, we exclusively use the sbb binarization output for recognition and training in our experiments since it has been proven to provide good results for a variety of image conditions.

5.1 Determining the Best Starting Model

We first performed only a few selected experiments to determine the best general approach regarding the network structure and the degree of generalization of the mixed models.

1. Start off with a small amount of GT to obtain a first book-specific model which hopefully already considerably outperforms the initial mixed model.
2. Build a strong specialized model by applying the ITA procedure of iteratively adding more training data to the training pool. This model's output may already be sufficient for many use-cases and allows to efficiently weed out the remaining errors manually.

All training processes are performed for each of the three network structures introduced above (default, htr+, and deep3). To examine the influence of the different handwriting styles we compare the output of the combined, Bastarda and Gothic models trained according to Sect. 4. Despite these variations, all trainings follow the two-stage training procedure: 1) train on all available pages, 2) use the output from 1) and refine it on the more balanced set of selected pages.

We evaluate the resulting models on two selected works, i.e. Wernher-Krakau (Gothic script) and Gast-1 (Bastarda). For each of the works we define a subset

Table 3. The upper part of the table lists the CERs (in %) for the two manuscripts *Wernher-Krakau* (Gothic) and *Gast-1* (Bastarda) recognized with the three pretrained mixed models (*combined*, *Bastarda*, and *Gothic*). The CERs are averaged results for out-of-the-box and after document-specific training on 4 and 16 pages and given for three different *network* structures (*default*, *htr+*, and *deep3*). Finally, the CERs are averaged again for each network (last column) and each pretrained model (last row). The respective best values are marked in bold.

	Wernher-Krakau			**Gast-1**			
network	combined	Bastarda	Gothic	combined	Bastarda	Gothic	avg.
default	4.62	9.16	**4.44**	6.23	**5.85**	16.51	7.80
htr+	**2.98**	6.79	3.23	4.42	**4.24**	12.25	5.65
deep3	3.15	6.88	**3.05**	**3.82**	3.97	10.05	**5.15**
avg.	3.58	7.61	**3.57**	4.82	**4.69**	12.93	-

Table 4. Detailed results for the deep3 network showing the individual CERs for the two manuscripts *Wernher-Krakau* (Gothic) and *Gast-1* (Bastarda) after training from scratch (*no PT* = no pretraining) and with pretrained models (out-of-the-box and after training on 4/16 pages).

	Wernher-Krakau				**Gast-1**			
# pages	no PT	combined	Bastarda	Gothic	no PT	combined	Bastarda	Gothic
ootb	-	6.53	16.56	6.21	-	6.92	7.24	24.88
4	5.27	1.62	2.51	1.59	8.33	2.71	2.80	3.28
16	2.29	1.31	1.56	1.35	2.89	1.82	1.88	1.99

of evaluation pages which remain unchanged for all upcoming experiments to ensure comparability. Table 3 sums up the overall results while Table 4 lists the results for the best performing deep3 network in detail.

As expected, the two deeper networks perform considerably better than the shallow one with deep3 achieving the lowest average CER (5.15% compared to 5.65%/7.80% for def/htr+). Applying Bastarda models to Bastarda material and Gothic to Gothic achieves the best results, closely followed by the combined model. Based on these outcomes we will perform the upcoming experiments using deep3 as our network structure and always apply the most-suitable (like-for-like) model.

5.2 Iterative Document-Specific Training

After determining the best starting model we want to take a closer look at the behaviour of the models produced during the ITA. This experiment is carried out on all five available evaluation manuscripts. To simulate the ITA we define fixed training pages for each iteration of the training process. For practical reasons we always stick to full pages independent of their number of lines or tokens. We start out with a very managable amount of GT consisting of only two pages and always double that number during the following iterations (2, 4, 8, 16, and 32 pages) where each set of pages completely subsumes the previous set. Note that each iteration starts from the initial mixed model and not from the model produced during the previous iteration. This is important, since we expect the learned *knowledge* of the mixed model to gradually diminish during more specific training. By beginning the training from the initial model each time we hope to counteract this forgetting effect. Each book-specific model as well as the initial mixed model is applied to a fixed evaluation set, i.e. all pages not used for training. Table 5 sums up the results.

Table 5. Each row lists the CERs and improvement rates (%) obtained by training on the given number of pages (*# pages*) with row *0* showing the out-of-the-box results. Each training was performed by starting from scratch (column *FS*) and by utilizing pretrained model as a starting point (*PT*). Column *Impr.* shows the improvement of PT over FS and of PT over the previous iteration. Wernher-Wien is special since it was produced by a hand already present in the training data and thus represents a true in-domain application of our pretrained model. It is therefore not included in the average values given in the lower last column *Avg. All*.

	Wernher-Krakau			**Handschrift B**			**Avg. Gothic**			**Wernher-Wien**		
# pages	FS	PT	Impr.	FS	PT	Impr.	FS	PT	Impr.	FS	PT	Impr.
0	-	6.21	-	-	4.90	-	-	5.55	-	-	2.99	-
2	13.67	1.95	86/69	10.73	2.61	76/47	12.20	2.28	81/59	22.42	2.39	89/20
4	5.27	1.59	70/19	7.68	2.30	70/12	6.48	1.95	70/15	7.95	2.11	74/12
8	2.57	1.45	44/9	4.07	1.89	54/18	3.32	1.67	50/14	4.24	1.99	53/6
16	2.29	1.35	41/7	3.81	1.64	57/13	3.05	1.50	51/11	3.10	1.84	41/8
32	1.56	1.31	16/3	3.30	1.38	58/16	2.43	1.35	45/10	2.33	1.57	33/15

	Gast-1			**Gast-2**			**Avg. Bastarda**			**Avg. All**		
# pages	FS	PT	Impr.	FS	PT	Impr.	FS	PT	Impr.	FS	PT	Impr.
0	-	7.24	-	-	6.54	-	-	6.89	-	-	6.22	-
2	16.69	3.50	79/52	43.39	5.00	89/24	30.04	4.25	86/38	21.12	3.27	85/48
4	8.33	2.80	66/20	21.68	3.61	83/28	15.01	3.21	79/25	10.74	2.58	76/21
8	4.65	2.26	51/19	11.98	3.09	74/14	8.32	2.68	68/17	5.82	2.17	63/16
16	2.89	1.88	35/17	6.12	2.87	53/7	4.51	2.38	47/11	3.78	1.94	49/11
32	2.26	1.64	27/13	3.73	2.25	40/22	3.00	1.95	35/18	2.71	1.65	39/15

6 Discussion

Inspecting column *Avg. All* of Table 5 shows that applying the pretrained mixed models out-of-the-box achieves an average CER of 6.22% which then quickly improves during finetuning following the ITA. Just two pages of GT are enough to achieve a CER of 3.27% improving the out-of-the-box output by 48% on average. As expected, further iterations lead to further improvements, resulting in an average CER of 1.65% when utilizing 32 pages of GT. Using a pretrained model as a starting point for the document-specific trainings leads to significantly lower error rates compared to starting from scratch, with the improvement factors diminishing with more pages of GT being added, ranging from 85% (2 pages) to 39% (32 pages). Finally, the CERs for the manuscripts written using Gothic handwriting (excluding Wernher-Wien) are considerably lower than the ones of the Bastarda manuscripts, both for the out-of-the-box recognition (5.56% for Gothic fonts, 6.89% for Bastarda ones) as well as for document-specific training (on average the CERs for Gothic is about 35% lower compared to Bastarda).

Overall, these results are very promising. The CER achieved by applying the pretrained mixed models out-of-the-box (on average 6.22% for both scripts combined) is already good enough for some downstream tasks like (error-tolerant) full-text-search and certainly allows for a much quicker GT production compared to transcribing from scratch. The subsequent document-specific training utilizing the mixed models as a starting point quickly led to vast improvements even when using just a few pages of GT. Two pages were enough to instantly improve the out-of-the-box output considerably (48% on average), resulting in a CER of 3.27%. These small amounts of GT can easily be produced by a single researcher, especially when starting from an already quite low-error recognition output of a mixed model or a document-specific model trained during an earlier iteration.

A thorough document-specific training using 32 pages of GT yielded an excellent average CER of 1.65% with four out of the five evaluation manuscript reaching CERs of considerably below 2%. The only exception is Gast-2 (CER of 2.25%) which was somewhat expected since it represents the most challenging work and each page only comprises less than thirty rather short lines. For comparison, this is about 2.5 times less than the other bastard manuscript Gast-1. The high quality of HTR output not only enables a wide variety of possibilities for subsequent use of the generated texts but can also serve for further very efficient GT production, as the ITA does not have to stop here. In fact, increasing the number of pages used for training from 16 to 32 still yielded a very notable improvement factor of 15%, indicating further room for improvement. However, the overall return on investment, i.e. the gains in CER in relation to the necessary amount of human effort, is diminishing when more and more material-specific GT is produced and trained.

Since the ITA can eventually be continued forever, it is up to the users to choose the strategy best suited to their material, use-case, and quality requirements. For example, when the goal is to transcribe an entire manuscript to prepare a critical editions, which requires (basically) error-free text, it naturally makes sense to continue material-specific training until the transcription is finished. On the contrary, when the goal is to reach a certain target CER required to perform specific downstream tasks the ITA should of course only be used until this target is reached to then process the remaining pages fully automatically.

Either way, the results show that utilizing a pretrained model as a starting point for the ITA is almost mandatory, as the effect compared to starting from scratch is immense. When using just two pages of training material the CERs improves by 85% on average when incorporating a pretrained model. As expected, this effect decreases with a rising number of training pages. Yet, even when using a considerable amount of GT, i.e. 32 pages, the CER still improves by a very notable 39%.

Most trainings can be completed within a couple of hours even when using a standard desktop PC without a GPU, making this approach highly feasible for the practicing humanist. This answers the first and second question from the introduction: Just training an available pretrained model on a few pages of GT that can be transcribed within a few hours, may provide recognized text with an error level in the low single digits. If the amount of training material is enlarged, the errors can further be diminished, albeit at a lower improvement rate.

This leaves the third question of what to do if only an out-of-domain pretrained model is available: Should one train from scratch on newly transcribed GT alone or would it still be better to start training with an existing out-of-domain model? A first indication that this might indeed be the case is apparent from comparing the results on Wernher-Wien whose Gothic hand was already present in the pretrained Gothic model (Table 5): While the out-of-the-box recognition is the best among all Gothic manuscripts, this advantage quickly vanishes with increasing document-specific training. This impression is reinforced by the out-of-domain experiments in Table 3 where, as expected, applying out-of-domain models (Bastarda to Gothic and vice versa) leads to considerably worse results. Using out-of-domain models as a starting point for document-specific training still yields far superior results compared to starting from scratch, especially when not a lot of document-specific GT is available: Training the out-of-domain Bastarda model (with an out-of-the-box CER of 16.56%) resulted in vast improvements (2.51%/1.56%) even compared to training from scratch (5.27%/2.29%). The same tendencies can be observed for the Bastarda manuscript Gast-1 with a Gothic out-of-the-box result of 24.88% and trained results of 3.28%/1.99%, compared to training from scratch of 8.33%/2.89%. The takeway from these observations is that users should not shy away from working with material for which no perfectly suited mixed model is available, but simply use the closest match and work from there.

Finally, we take a look at the change of the error distribution during the ITA exemplified with the Wernher-Krakau manuscript (Table 6). The biggest source of error in the out-of-the-box output are the dots at the end of each verse which are very small and often merge with the preceeding letter. Further dominant errors are the confusion of a *w* for *vv* as well as the deletion of whitespaces and diacritical marks like the superscript *e*, *v*, and the abbreviation mark for *er* (represented as a hook or a zig-zag sign above; marked as @ in the confusion table). Document-specific training on 4 pages considerably improves the recognition of whitespaces and pushes almost all *w*/*vv* confusions and most diacritical errors beyond the ten most frequent errors, and close to 80% of the dot-related errors have vanished. After training on 16 pages, dots and whitespace-related errors are responsible for the vast majority (close to 35%) of remaining errors.

Table 6. The ten most common confusions for 18 pages of the Wernher-Krakau manuscript showing the *GT*, prediction (*PRED*), the absolute number of occurrences of an error (*CNT*) and its fraction of total errors in percent (*%*), resulting from the *out-of-the-box* application of a mixed model and document-specific training using *4/16 pages*.

out-of-the-box				4 page training				16 page training			
GT	PRED	CNT	%	GT	PRED	CNT	%	GT	PRED	CNT	%
.		121	13.2	.		25	10.7	.		24	12.1
ͤ		62	6.7		.	14	6.0		␣	17	8.6
w	vv	44	9.6	␣		14	6.0		.	15	7.6
␣		35	3.8		␣	12	5.2	␣		12	6.1
ͮ		33	3.6	i		9	3.9	i		10	5.1
@		28	3.0	z	l	7	3.0	in	m	3	3.0
z	l	26	2.8	ͤ		6	2.6	æ	a	3	1.5
d	c	24	2.6	t		5	2.1	n	u	3	1.5
u	i	15	1.6	i	u	4	1.7	ͮ		3	1.5
	.	14	1.5	n	u	4	1.7		ͤ	3	1.5

7 Conclusion and Future Work

After creating a training corpus comprising 35 Medieval German manuscripts and close to 13k lines for two widely used German medieval handwritten styles, Gothic and Bastarda, we were able to train several highly performant mixed models. Evaluation on four previously unseen manuscripts yielded very low error rates, both for the out-of-the-box application of mixed models (average CER below 6%) as well as for document-specific training. For the latter, the quality of the result strongly depended on the amount of training material used, ranging from average CERs of 3.28% for just two pages of GT to 1.68% for thoroughly trained models (32 pages). A large share of this efficiency can be attributed to using the mixed models as a starting point for each individual material-specific training. Pretraining showed to be highly effective with average improvement rates ranging from 86% to 38% depending of the number of pages used for training. This not only held true for the in-domain application of models (Bastarda model to Bastarda material, Gothic to Gothic) but also when using out-of-domain models as a foundation for finetuning.

Better and more widely applicable pretrained models can be constructed the more GT is made openly available by individual researchers and groups. To foster this spirit of open collaboration we made our own pretrained models openly available.

Regarding future work, we want to utilize the intrinsic confidence values of the ATR engine. For example, these confidence values could be used to identify individual lines the existing model struggled with the most and then transcribe these lines in a targeted way to maximize the training effect in an active learning-like approach. A first implementation for this, highlighting uncertain characters, has already been made available via OCR4all [12]. In addition, aggregated confidence information could also provide an indicator for the current text quality and therefore serve as a stopping criterium for the ITA.

Current research strongly pushes towards recurrent-free approaches based on attention and transformer networks. Kang et al. [7] first introduced a non-recurrent architecture using multi-head self-attention layers not only on a visual but also on a textual level. Many refinements and combinations with existing approaches, e.g. [2], have followed since. While a broad application of these developments in the practical area is still hindered by the lack of stable and resource-efficient open source implementations, it seems clear that this is the direction into which ATR is heading in the future.

Acknowledgement. The authors would like to thank our student research assistants Lisa Gugel, Kiara Hart, Ursula Heß, Annika Müller, and Anne Schmid for their extensive segmentation and transcription work as well as Maximilian Nöth and Maximilian Wehner for supporting the data preparation.

This work was partially funded by the German Research Foundation (DFG) under project no. 460665940.

References

1. Breuel, T.M., Ul-Hasan, A., Al-Azawi, M.A., Shafait, F.: High-performance OCR for printed English and Fraktur using LSTM networks. In: 12th International Conference on Document Analysis and Recognition (ICDAR), pp. 683–687. IEEE (2013). https://doi.org/10.1109/ICDAR.2013.140
2. Diaz, D.H., Qin, S., Ingle, R., Fujii, Y., Bissacco, A.: Rethinking text line recognition models. arXiv preprint (2021). https://arxiv.org/abs/2104.07787
3. Eichenberger, N., Suwelack, H., Schröer, A.: Faithful transcriptions. 027.7 J. Libr. Cult. (2021). https://doi.org/10.21428/1bfadeb6.d3bdbcd2
4. Hawk, B.W., Karaisl, A., White, N.: Modelling medieval hands: practical OCR for caroline minuscule. Digit. Humaniti. Q. **13**(1) (2019). http://www.digitalhumanities.org/dhq/vol/13/1/000412/000412.html
5. Hodel, T., Schoch, D., Schneider, C., Purcell, J.: General models for handwritten text recognition: feasibility and state-of-the art. German kurrent as an example. J. Open Humanit. Data **7**(13), 1–10 (2021). https://doi.org/10.5334/johd.46
6. Kahle, P., Colutto, S., Hackl, G., Mühlberger, G.: Transkribus-a service platform for transcription, recognition and retrieval of historical documents. In: 14th IAPR International Conference on Document Analysis and Recognition (ICDAR), vol. 4, pp. 19–24. IEEE (2017). https://doi.org/10.1109/ICDAR.2017.307

7. Kang, L., Riba, P., Rusiñol, M., Fornés, A., Villegas, M.: Pay attention to what you read: non-recurrent handwritten text-line recognition. arXiv preprint (2020). arXiv:2005.13044, https://arxiv.org/abs/2005.13044
8. Memon, J., Sami, M., Khan, R.A., Uddin, M.: Handwritten optical character recognition (OCR): a comprehensive systematic literature review (SLR). IEEE Access **8**, 142642–142668 (2020). https://doi.org/10.1109/ACCESS.2020.3012542
9. Michael, J., Weidemann, M., Labahn, R.: HTR engine based on NNs P3. Horizon 2020 Technical report (2018). https://readcoop.eu/wp-content/uploads/2018/12/Del_D7_9.pdf
10. Mocholí Calvo, C., et al.: Development and experimentation of a deep learning system for convolutional and recurrent neural networks. Ph.D. thesis. Universitat Politècnica de València (2018)
11. Pletschacher, S., Antonacopoulos, A.: The PAGE (page analysis and ground-truth elements) format framework. In: 20th International Conference on Pattern Recognition, pp. 257–260. IEEE (2010). https://doi.org/10.1109/ICPR.2010.72
12. Reul, C., et al.: OCR4all-an open-source tool providing a (semi-)automatic OCR workflow for historical printings. Appl. Sci. **9**(22), 4853 (2019). https://doi.org/10.3390/app9224853
13. Reul, C., Springmann, U., Wick, C., Puppe, F.: Improving OCR accuracy on early printed books by combining pretraining, voting, and active learning. JLCL: Spec. Issue Autom. Text Layout Recognit. **33**(1), 3–24 (2018). https://jlcl.org/content/2-allissues/2-heft1-2018/jlcl_2018-1_1.pdf
14. Reul, C., Springmann, U., Wick, C., Puppe, F.: Improving OCR accuracy on early printed books by utilizing cross fold training and voting. In: 2018 13th IAPR International Workshop on Document Analysis Systems (DAS), pp. 423–428. IEEE (2018). https://doi.org/10.1109/DAS.2018.30
15. Reul, C., Wick, C., Noeth, M., Wehner, M., Springmann, U.: Mixed model OCR training on historical Latin script for Out-of-the-box recognition and finetuning. In: The 6th International Workshop on Historical Document Imaging and Processing, pp. 7–12 (2021). https://doi.org/10.1145/3476887.3476910
16. Sánchez, J.A., Romero, V., Toselli, A.H., Villegas, M., Vidal, E.: A set of benchmarks for handwritten text recognition on historical documents. Pattern Recognit. **94**, 122–134 (2019). https://doi.org/10.1016/j.patcog.2019.05.025
17. Springmann, U., Lüdeling, A.: OCR of historical printings with an application to building diachronic corpora: a case study using the RIDGES herbal corpus. Digit. Humanit. Q. **11**(2) (2017), http://www.digitalhumanities.org/dhq/vol/11/2/000288/000288.html
18. Stökl Ben Ezra, D., Brown-DeVost, B., Jablonski, P., Lapin, H., Kiessling, B., Lolli, E.: BiblIA-a general model for medieval hebrew manuscripts and an open annotated dataset. In: The 6th International Workshop on Historical Document Imaging and Processing, pp. 61–66 (2021). https://doi.org/10.1145/3476887.3476896
19. Wick, C., Reul, C., Puppe, F.: Calamari-a high-performance tensorflow-based deep learning package for optical character recognition. Digit. Humanit. Q. **14**(2) (2020). http://www.digitalhumanities.org/dhq/vol/14/2/000451/000451.html

A Comprehensive Study of Open-Source Libraries for Named Entity Recognition on Handwritten Historical Documents

Claire Bizon Monroc[1,2](✉), Blanche Miret[1], Marie-Laurence Bonhomme[1], and Christopher Kermorvant[1,3]

[1] TEKLIA, Paris, France
claire.bizonm@gmail.com
[2] DYOGENE-INRIA, Paris, France
[3] Normandie Université, LITIS, Rouen, France

Abstract. In this paper, we propose an evaluation of several state-of-the-art open-source natural language processing (NLP) libraries for named entity recognition (NER) on handwritten historical documents: spaCy, Stanza and Flair. The comparison is carried out on three low-resource multilingual datasets of handwritten historical documents: HOME (a multilingual corpus of medieval charters), Balsac (a corpus of parish records from Quebec), and Esposalles (a corpus of marriage records in Catalan). We study the impact of the document recognition processes (text line detection and handwriting recognition) on the performance of the NER. We show that current off-the-shelf NER libraries yield state-of-the-art results, even on low-resource languages or multilingual documents using multilingual models. We show, in an end-to-end evaluation, that text line detection errors have a greater impact than handwriting recognition errors. Finally, we also report state-of-the-art results on the public Esposalles dataset.

Keywords: Named entity recognition · Text line detection · Handwritten historical documents

1 Introduction

As more and more historical documents are digitized, large collections of document images are constituted in patrimonial institutions. Automatic natural language processing (NLP) can be a valuable tool for researchers in the humanities who analyze and interpret large masses of unstructured documents [19]. For historical document images, in particular, the development of efficient document layout analysis (DLA) and handwritten text recognition (HTR) systems make their textual content available for downstream information extraction tasks with classical NLP tools. Named entity recognition (NER), in particular, has a potential to improve the user experience of humanities researchers by providing search and exploration facilities in large corpora. Moreover, recent years have

© Springer Nature Switzerland AG 2022
S. Uchida et al. (Eds.): DAS 2022, LNCS 13237, pp. 429–444, 2022.
https://doi.org/10.1007/978-3-031-06555-2_29

seen the rise of NLP as a major element of computational humanities, leading to the development of efficient neural models for NER tasks, to the point that NER on standard contemporary large-resource languages has been called a solved problem [19]. These developments triggered the birth of several libraries, making state-of-the-art neural architectures easily available for both industrial development and research. Yet, the application of these state-of-the-art tools to historical documents still gives rise to specific challenges.

NER systems performance are highly dependent on the target language and domain. As previously noted [5], applying standard tools on textual data from a different field without adaptation data can lead to a dramatic drop of performance. Yet, out-of-the-box information extraction models in libraries are generally trained and evaluated on contemporary language data extracted from Wikipedia or modern newspapers. Although notable efforts have been made to develop corpora and language models for historical German [18] for example, textual data is still mostly extracted from XIXth century documents.

Medieval documents, such as charters, are generally written in *dead languages*, such as Latin or old versions of European languages, for which there is no native speaker and no production of new documents. Very few of the massive annotated language corpora that drove the success of neural NLP tools in English and German are available for historical languages [18], and model performance on historical versions of their languages can suffer due to diachronic changes and lack of standardization in naming entities in old texts. Moreover, classical NLP tools applied to historical documents can only be the last step in a pipeline composed of at least a line recognition step followed by HTR. The quality of the HTR plays a large part in the decreased performances of NER models on automatically processed historical documents, and designing meaningful metrics to evaluate those results has proved challenging [14]. Another challenge is the existence of nested entities. Examples in the literature have mostly focused on flat entities, but medieval name identification often embed location names (e.g., *Fridericus, Dei gracia dux Austrie et Styrie*, a *name* entity containing the *location* entities *Austrie* and *Styrie*). Furthermore, nested entities raise challenges in terms of modeling and evaluation.

In this paper, we explore these different aspects through our experiments using three state-of-the art natural language processing libraries for named entity recognition (NER) on handwritten historical documents (spaCy [16], Stanza [22], and Flair [3]) using three different datasets of handwritten historical documents: a multilingual corpus of medieval charters (HOME), a corpus of Quebec parish records in French (Balsac), and a public corpus of marriage records in Catalan (Esposalles). The contributions of this paper are the following: **(1)** We provide a comprehensive study of open-source NER tools and metrics on low-resource, multilingual handwritten documents , highlight the different impact levels on this task considering the preceding steps in real use case scenarios (e.g., text line detection and HTR). and show that they can be considered as a sustainable option for this particular type of documents. **(3)** We propose a different use of nested entities compared to [9], with a better exploitation of the data. **(4)** We obtain state-of-the-art performance on the Esposalles dataset outperforming the best published results. These results in particular point out how current off-the-shelf NER libraries offer state-of-the art tools that are able to reach and even surpass tailor-made models.

Next, we continue with the presentation of related work in Sect. 2. Section 3 describes the three historical corpora and the challenges they raise. The open-source libraries are presented in Sect. 4. The hyperparameters and the different metrics used to evaluate our models are detailed in Sect. 5. Finally, we draw conclusions and highlight some future perspectives in Sect. 6.

2 Related Work

The amount of previous research on NER for handwritten historical document is relatively limited in comparison to the vast amount of published work on contemporary documents. Some authors proposed to extract named entities without explicit HTR [1], by extracting specific features from the images and predicting the named entity (NE) with a bidirectional long short-term memory (BLSTM) neural network. Their method was tested on relatively simple documents (Georges Washington letters, Queensland State Archives dataset, IAM dataset) but they claimed it can be used on any language in Latin script, in which an NE usually starts with a capital letter.

Several types of HTR and NER pipelines on the Esposalles dataset were explored [21], using different types of graphical and neural-based models for NER. This research showed that when the HTR error rate is as low as 5%, the NER performance is as good as on recognized text as on the groundtruth. On the same dataset, another study proposed to train a single convolutional-sequential model that jointly performs handwriting recognition and named entity extraction [10], with the objective to reduce the error propagation between the two steps.

Other researchers trained CRF-based NER models using linguistic features on a large corpus of medieval cartularies and charters in Medio Latin [2]. The authors obtained a score greater than 90% for persons and locations, but using manually transcribed text. A comparison of two different approaches to NER with a newly produced fine-grained multilingual dataset of medieval charters, was recently proposed [9]. One approach was a traditional sequential approach (HTR and NER) whereas the second one was an integrated approach in which the model transcribes and predicts the NEs simultaneously. The integrated approach yielded the best results, but the experiments were carried on manually segmented lines. Two-stage (HTR and afterward NER) and end-to-end architectures (NER directly at image level) were also compared [23]. Manually annotated the known IAM and George Washington datasets with NE labels and their experiments showed, contrary to previous research [9], that a two-stage model can achieve higher scores than an end-to-end one.

Another comparison of five different types of NER systems (statistical, rule-based and neural) was explored [25], on two collections of letters in English from the 17th and 18th centuries, but without retraining the models, and showed that ensemble methods can improve NER results.

Table 1. Number of pages, lines, words and entities in the different corpora.

	HOME				Esposalles	Balsac
No. of	Czech	German	Latin	All	Catalan	French
Page	202	173	126	501	125	896
Line	3,591	3,199	1,971	8,761	3,827	45,479
Word	66,257	77,086	35,759	179,102	39,527	205,165
Entity	4,117	4,419	3,315	11,851	16,782	25,564

After massive digitization campaign of historical newspapers in most western libraries, many studies have been conducted on the performance of NER on digitized texts. A previous study tackled NER on old Finnish newspapers [17], showing that with an OCR error rate around 25% to 30%, the NER performance is highly degraded for systems based on linguistic methods. Moreover, the performance of a NER was proved [14] that it can drop with at least 30% points, when the character error rate (CER) of the OCR process increases from 1% up to 7% by adding synthetic noise to the documents. To this scope, the CLEF HIPE 2020 evaluation [11,12] gathered 13 teams on shared tasks dedicated to the evaluation of named entity processing on historical newspapers in French, German and English. This evaluation showed that NER on historical newspapers are capable of dealing with historical inputs and reach performance comparable to those obtained on contemporary texts when enough training data is available but confirm an important degradation of the performance in presence of OCR noise.

3 Handwritten Historical Document Corpora

We conducted experiments on three different corpora of historical documents annotated with named entities: HOME, Balsac, and Esposalles. Detailed statistics on the corpora can be found in Table 1. Also, Table 2 presents for each corpus the following statistics per language: type of entity, the average number of tokens and characters in an entity, the total number of entities and the number of nested entities[1].

HOME is a corpus of annotated images of handwritten medieval charters, 499 charters from the Archives of the Bohemian Crown, as well as archives of several monasteries, chosen for their historic and linguistic significance to the history of Central Europe. A corrected and augmented version of the corpus has been made available by the National Archives of the Czech Republic and served as the basis for the work presented in this paper. A HOME Czech charter example is presented in Fig. 1 (1). Contrary to previous research [9], we take into account the fact that entities can be split across two lines, and therefore consider such groups as constituting one single entity.

[1] Note that Table 2 presents a count for *person* entities, which do not appear as such in the Esposalles corpus. Therefore, the number of *person* entities in this table is actually the sum of *name* and *surname* entities in the dataset.

Table 2. For each type of entity, average entity lengths in number of tokens (words), characters (chars), and entities counts for each corpus and language. **PER** = Person, **LOC** = Location, **DAT** = Date, **OCC** = Occupation, **ST** = Civil state.

			HOME				Esposalles	Balsac
			Czech	German	Latin	All	Catalan	French
PER	Length	Token	3.0	5.1	2.9	3.6	1.61	2.33
		Char	20.4	32.9	21.3	24.2	9.62	14.27
	Count	Entity	1,997	1,356	1,374	4,727	9,167	15,810
		Nested	3	1	5	9	–	–
LOC	Length	Token	1.2	1.0	1.1	1.1	1.89	2.08
		Char	9.4	7.2	8.3	8.1	9.35	12.33
	Count	Entity	1,956	2,909	1,826	6,691	2,959	2,823
		Nested	1,054	1,035	840	2,929	–	–
DAT	Length	Token	8.5	10.7	7.4	9.0	–	5.21
		Char	58.0	66.5	56.0	60.4	–	23.28
	Count	Entity	164	154	115	433	–	4,551
		Nested	1	0	0	1	–	–
OCC	Length	Token	–	–	–	–	1.25	1.31
		Char	–	–	–	–	7.15	10.44
	Count	Entity	–	–	–	–	3,207	2,380
		Nested	–	–	–	–	–	–
ST	Length	Token	–	–	–	–	1.01	–
		Char	–	–	–	–	6.82	–
	Count	Entity	–	–	–	–	1,449	–
		Nested	–	–	–	–	–	–

Balsac corpus contains 896 images of parish records from Quebec regions, from 1850 to 1916. The records contain baptism, marriage and death records, and are an important data source for social, historical and genealogical research. A Balsac parish record example is presented in Fig. 1 (2). The pages are organized in two columns, with a small margin paragraph for each act, specifying a reference number and the name of the act's main subject, and a main column with the proper content of the act. Both margins and main paragraphs were annotated and used for training.

Esposalles dataset [13] is a subset of 125 pages from the Esposalles database, which contains marriage license records from the Archives of the Cathedral of Barcelona, written in Catalan by a single writer from the XVIIth century. An Esposalles record example is presented in Fig. 1 (3). This dataset was used in the 2017 ICDAR Information Extraction in Historical Handwritten Records campaign, and it is still possible to submit results for its tasks online[2]. For the purpose of comparing results on this dataset to those we got on HOME and Balsac,

[2] https://rrc.cvc.uab.es/?ch=10&com=introduction.

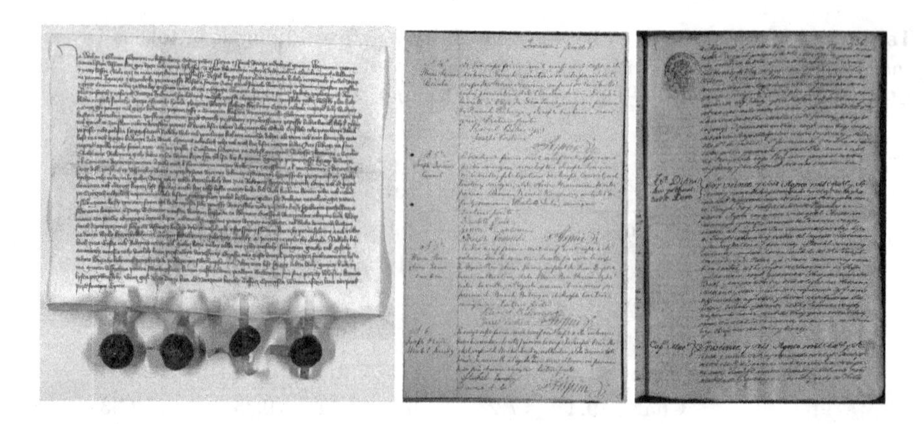

Fig. 1. Page examples from the three datasets: (1) HOME Czech charter (2) Balsac parish record (3) Esposalles record.

we performed the basic track NER task of the competition (where the entities to identify are name, surname, occupation, location and state).

3.1 Nested Entities in HOME Corpus

Previous research [9] ignored nested entities and continuations (situations where an entity overlaps two lines in a charter). Their solution was useful to reduce the complexity of the NER task and create a benchmark to compare sequential and combined approaches. Nevertheless, disregarding nested entities is suboptimal. First, nested entities actually are a large part of the annotated entities, up to half of them for entities of type *location*. Ignoring them, therefore, could mean disregarding an important volume of valuable information. Secondly, removing nested annotations increases ambiguity in the labels, as nested entities are then annotated with the labels of their parent entities. For example, the following entity extracted from a Latin charter refers to a person:

Nos Fridericus, Dei gracia dux Austrie et Styrie [...] *profitemur et recognoscimus*
PER: Fridericus, Dei gracia dux Austrie et Styrie

We refer to this PER entity as *parent*. When ignoring nested entities, the tokens *Austrie* and *Styrie* are labelled as *person*, but when met outside a nested entity, the same tokens are annotated as *location*. We refer to these entities as *nested*.

regni Hungarie et ducatus Austrie
LOC: Hungarie - Austrie

To avoid these ambiguities, we decided to incorporate the present nested entities. There are several ways to perform this: (1) giving such entities two labels (their own and the label of their parent). This would create a multilabel classification problem, which is not supported by most of the out-of-the-box NER tools that our work aims to compare. (2) training two models, one with the labels of the parent entities and the other with the labels of the nested entities, to predict the overlapping entities separately. A lighter way to solve the issue is to "*flatten*" nested entities, that is to systematically use nested labels when they exist, and reduce parent entities to their core to avoid double labelling. The following example illustrates this method.

Nos Fridericus, Dei gracia dux Austrie et Styrie [...] *profitemur et recognoscimus*
PER: Fridericus, Dei gracia dux Austrie et Styrie

becomes

Nos Fridericus, Dei gracia dux Austrie et Styrie [...] *profitemur et recognoscimus*
PER: Fridericus, Dei gracia dux
LOC: Austrie - Styrie

This solution is also relevant to one of the main purposes of applying entity extraction to historical documents: to link those entities to indexes of people or locations, or use the recognized entities to fill databases, which also requires entity linking (EL), so that if one record concerns a person already in the database, for example, it gets correctly linked to that individual. The loss of specific information concerning the relation between parent and nested entities can be compensated by providing neighborhood context as an input to an EL model. The identification of the core part of an entity was done following a simple rule: after extracting nested entities, all parts of the entity with more than a defined threshold of characters are kept. To avoid the potential creation of meaningless entities from various linking and stopwords, we filtered out for each language a list of non-semantically significant words when they were located at the edge of entities, i.e., "et cetera" in Latin, "und" in German or "z", "od" in Czech.

4 Named Entity Recognition Libraries

We compared the performance of three open-source and widely used NLP Python libraries on the NER task: Stanza, Flair and spaCy.

Stanza (v1.1.1)[3] [22] was created and maintained by the Stanford NLP group. It provides different NLP tools that can be used in a pipeline (e.g., tokenization, NER) and pre-trained neural models supporting 66 different languages, including German, Czech, Latin, Catalan and French.

Flair (v0.7)[4] [3] provides, besides state-of-the-art NLP models, contextual word and character embeddings, a feature that is now also supported by Stanza.

[3] https://stanfordnlp.github.io/stanza.
[4] https://github.com/flairNLP/flair.

Moreover, Flair allows stacking embeddings, a technique that allowed the authors to exceed state-of-the-art NER results on German, architecture that we consider in this study.

spaCy (v2.3.5)[5] [16] is explicitly designed to ease deploying models into production. The library comes with out-of-the-box support of multiple languages, including Czech, German, Catalan and French. We built on pre-trained models to train a NER pipeline whenever they were made available by the library. When they were not, we trained the NER on the available basic language defaults.

5 Experiments

In this section, we describe our experimental setup: the data, the metrics and the different model configurations and training procedures. First, for the HOME and Balsac datasets, the documents were randomly split in three training, validation and test sets with an 80/10/10 ratio. NER models were then evaluated on the test corpus of their language or on all languages for multilingual models. The files in IOB (short for inside, outside, beginning) required by the NER tools and that were used for training, validation and test, were created from the XML documents provided by Transkribus[6], which was the tool used for transcription and annotation. For Balsac, there was originally no *person* annotation, but two distinct *name* and *surname* annotations. As the end-goal was to identify individuals, they were reunited in a unique tag during pre-processing. Regarding Esposalles, for training and testing, files in the IOB format were created from the text files containing the text line transcriptions, and the text files containing the entity categories for each word of the text line, available after downloading the Esposalles dataset[7]. We used the train/validation/test sets defined for the ICDAR 2017 competition [13], both for training the HTR and the NER models.

5.1 Evaluation Metrics

Basic Page-Level Metric. This first metric compares the entities found in the prediction with the ones from the groundtruth, at page level, disregarding the positions and number of occurrences of the entities, and ignoring punctuation. If an entity from the groundtruth is predicted with the true label and text, then it is considered as a true positive, regardless of where in the page and how many times it appeared. This metric returns the precision, recall and F1 scores.

Nerval for Automatic Transcription. The Nerval[8] metric is used for the evaluation of the NER models on the automatic transcriptions. These are first aligned with the groundtruth at character level, by minimizing the Levenshtein distance

[5] https://spacy.io.
[6] https://readcoop.eu/transkribus/.
[7] https://rrc.cvc.uab.es/?ch=10&com=tasks.
[8] https://gitlab.com/teklia/nerval.

between them. Each entity in the groundtruth is then matched with a corresponding entity in the aligned transcription, with the same entity label, or an empty character string if no match is found. If the distance between the two entities is less than threshold (by default 30%) of the groundtruth entity length, the predicted entity is considered as recognized. For the purpose of matching detected entities to existing datasets, we estimated that a 70% match between the entity texts was an acceptable threshold.

Esposalles Metric. This metric is described in [13]. For each marriage record, the text and the label of the entity are compared to the entities from the groundtruth transcriptions and annotations. The two transcriptions are aligned and the score for each entity can be: 0 if the entity label is wrong, 1 if both the transcription and the label are exact, and $1 - CER$, between the automatic transcription and the groundtruth for that entity if the label is correct, but the transcription is different. This metric does not return precision, recall and F1 scores but a global accuracy score for each entity category and all entities.

Text line detection metric. The metric used to evaluate the line detection is the average precision at a 0.75 intersection-over-union (IoU) threshold (AP@0.75), which means that a detected line is matched to an annotated text line if they intersect by at least 75% of their surfaces [6]. The AP@0.75 is the proportion of correctly predicted text lines based on this IoU threshold.

Text Recognition Metric. Character error rate (CER) and word error rate (WER) are used to evaluate the text recognition models. CER corresponds to the Levenshtein distance between two strings, that is, the minimum number of character editions to be made to transform one string to the other. WER uses the same principle at word level.

5.2 Hyperparameters and Model Training

After an optimization using grid search, the chosen hyperparameters were (hidden layer dimension, learning rate, dropout): Flair (256, 0.1, -), spaCy (64, 0.001, -), and for Stanza (256, 0.01, 0.3).

We trained two Stanza models for each corpus: *Stanza Basic*, the default Stanza configuration with word2vec embeddings [20], pre-trained for CoNLL 2018, default configuration[9]. We also explored *Stanza Fasttext*, the default Stanza configuration, but with FastText embeddings [7].

For Flair, the models were trained with a combination of word2vec word embeddings and contextualized character embeddings [3]. For Czech, French, and German, we used pre-trained Flair embeddings (contextual character embeddings provided by the library). For Latin, we used Fasttext pre-trained embeddings and we also trained our own Flair embeddings on a large corpus of

[9] https://github.com/stanfordnlp/stanza-train.

Table 3. Micro-averaged page-level percent of precision, recall and F1 scores by language and (monolingual) NER model, on manual transcriptions, with the *Basic page-level metric.*

	HOME			Esposalles	Balsac	
	Czech	German	Latin	Catalan	French	All
ICFHR 2020 [9] (different strategy for nested entities and different metric)						
Precision	70.8	85.2	59.7	–	–	–
Recall	50.7	67.9	44.4	–	–	–
F1	59.1	75.6	50.9	–	–	–
Stanza Basic						
Precision	83.5	92.8	87.3	97.5	77.3	87.7
Recall	83.5	92.6	85.5	98.2	86.3	89.2
F1	83.5	92.7	**86.4**	97.8	81.6	88.4
Stanza Fasttext						
Precision	84.3	93.8	84.3	96.8	78.7	87.6
Recall	86.0	92.5	85.2	97.8	86.6	89.6
F1	**85.1**	**93.1**	84.7	97.3	82.5	**88.5**
Flair						
Precision	80.8	90.0	82.6	96.1	94.1	88.7
Recall	77.3	87.0	76.5	96.5	93.5	82.3
F1	79.0	88.5	79.4	96.3	**93.8**	85.1
spaCy v2						
Precision	81.1	90.2	n/a	97.5	84.3	88.3
Recall	76.3	89.8	n/a	98.3	84.3	87.2
F1	78.6	90.0	n/a	**97.9**	84.3	87.7

10 millions tokens available in the public domain and collected by The Latin Library[10]. There were no Flair embeddings for Catalan either, thus we chose French embeddings for the Esposalles dataset. The multilingual Flair model for HOME was trained by feeding all the charter transcriptions in the train set to Flair pre-trained (on 300+ languages) multilingual character embeddings. Regarding spaCy, for German and French, for which three different models are available, we used the medium-sized model `de_core_news_md`, and `fr_core_news_sm`, respectively. For Czech and Catalan, there are no pre-trained models. Thus, we trained our NER models with the default linguistic features support. Latin is, also, not supported by this library, thus, we used the multilingual spaCy model. Finally, the text line detection were performed with Doc-UFCN [6] and the automatic text recognition by Kaldi [4].

[10] https://github.com/cltk/lat_text_latin_library.

Table 4. Performance comparison of multilingual vs monolingual NER on the HOME corpus (manual transcriptions) with the *basic page-level metric.*

	HOME					
	Czech	German	Latin	Czech	German	Latin
	Flair			spaCy v2		
Precision	80.8	90.0	82.6	81.1	90.2	n/a
Recall	77.3	87.0	76.5	76.3	89.8	n/a
F1	*79.0*	*88.5*	*79.4*	78.6	90.0	n/a
	Flair multilingual			spaCy v2 multilingual		
Precision	75.1	88.8	84.1	81.6	91.0	84.5
Recall	73.7	84.8	79.5	80.6	90.6	82.1
F1	74.4	86.8	81.8	**81.0**	**90.8**	**83.3**

Table 5. Performance comparison when applying NER to manually annotated text (Ann.) and automatically recognized text (Rec.) with nerval metric. "–": no multilingual models for Balsac and Esposalles, "n/a": for spaCy in Latin, since there is no Latin base model.

	HOME						Esposalles		Balsac	
	Czech		German		Latin		Catalan		French	
	Ann	Rec	Ann	Rec	Ann	Rec	Ann	Rec	Ann	Rec
Stanza Basic										
Precision	79.1	60.8	93.8	84.0	86.9	77.8	97.6	96.5	77.9	69.7
Recall	86.8	40.0	93.3	67.5	88.9	67.5	98.0	96.6	85.5	78.3
F1	82.8	48.3	93.5	74.8	**87.9**	**72.3**	97.8	96.5	81.5	73.7
Stanza Fasttext										
Precision	79.1	63.1	95.2	85.1	84.7	75.6	97.1	96.3	79.2	71.9
Recall	88.4	42.4	93.7	67.5	87.9	67.0	97.5	96.2	86.3	78.8
F1	**83.5**	**50.7**	**94.4**	75.3	86.3	71.0	97.3	96.2	82.6	75.2
Flair										
Precision	77.7	57.2	92.1	86.5	84.3	76.6	97.9	96.8	93.7	86.0
Recall	81.8	34.5	88.1	65.6	82.1	67.3	98.2	97.1	93.1	85.6
F1	79.7	43.0	90.1	**75.7**	83.2	71.6	**98.0**	**96.9**	**93.4**	**85.8**
Flair multilingual										
Precision	73.9	53.4	89.9	79.7	83.0	73.0	–	–	–	–
Recall	77.6	33.4	85.9	62.8	81.0	64.1	–	–	–	–
F1	75.7	41.1	87.9	70.3	82.0	68.3	–	–	–	–
spaCy										
Precision	76.4	61.5	91.9	83.8	n/a	n/a	97.1	95.8	83.1	74.5
Recall	80.3	37.9	91.0	67.5	n/a	n/a	97.8	96.3	83.6	76.7
F1	78.3	46.9	91.5	74.7	n/a	n/a	97.5	96.1	83.4	75.6
spaCy multilingual										
Precision	77.5	58.2	91.5	83.5	85.6	75.3	–	–	–	–
Recall	83.4	38.4	90.6	68.3	86.3	66.0	–	–	–	–
F1	80.4	46.3	91.1	75.1	85.9	70.3	–	–	–	–

Table 6. Evaluation of the text line detection using the *average precision @0.75* for Doc-UFCN model and evaluation of the HTR process using character error rate (CER) and word error rate (WER) on the test sets of the different corpora.

	HOME				Balsac	Esposalles
AP@0.75	48.57				91.13	–
	Kaldi HOME (multilingual)				Kaldi Balsac	Kaldi Esposalles
	Czech	German	Latin	all	French	Catalan
CER	8.70	7.48	10.37	8.93	6.41	1.32
WER	29.71	26.40	35.59	29.26	17.41	3.51

6 Results

Results on the Manual Transcriptions. Table 3 presents the results of our experiments using the different NER models on the manually transcribed HOME, Balsac, and Esposalles datasets. Table 3 does not allow us to pick a clear winner out of the three NER libraries. Averaging the scores over all the datasets, *Stanza FastText* performs better on two of the HOME languages. However, spaCy and Flair obtain the best results on the Esposalles and Balsac datasets. Therefore, it is unclear which library will perform best on a given dataset, thus, practically, testing the three libraries is our recommendation. Table 4 offers a comparison of the results obtained by a multilingual NER model and a monolingual one, on the manually transcribed HOME dataset. This table shows that Flair performs better when it is specialized for one language, but spaCy benefits from the larger amount of training data of the multilingual model, outperforming both a monolingual one and Flair, on all the HOME languages. These results show that using multilingual NER models is a sustainable option for low-resource languages or when the documents contain multiple languages. Both tables reported the results with the *Basic page-level metric*.

Results on the Manual and Automatic Transcriptions. Table 5 presents how NER results are impacted by text line detection and handwriting text recognition (HTR), both on manual and HTR transcriptions, with the *Metric for aligned automatic transcriptions*. We observe that there is only a marginal decrease in accuracy on the Esposalles dataset between the NER on the manual transcription and the one on the HTR text, and that can be explained by how there is no text line detection step for Esposalles as the line images are given, and the HTR model for Esposalles has a 3.51% WER as shown in Table 6, which is very low. We can draw a similar (and expected) conclusion looking at the results on Balsac, where the accuracy always drops by less than ten percentage points between manual and transcribed text, and where the text line detection works well and the HTR model has a WER of 17.41%, which is moderate. On the other end of the spectrum, we see that the text line detection does not perform well on the HOME dataset (AP@0.75 of 48.57), and the HTR model has

Table 7. Results from various NER libraries on the Esposalles dataset, manual and automatic (Kaldi) transcription, using the *Esposalles metric.*

	PER (name)	PER (surname)	LOC	OCC	ST	all
Naver Labs [21]						
auto						0.95
CITlab-ARGUS-2 [13]						
auto						0.92
spaCy v2						
manual	0.98	0.97	0.98	0.98	0.98	0.98
auto	0.97	0.95	0.96	0.97	0.98	0.96
spaCy v3						
manual	0.97	0.96	0.95	0.97	0.98	0.97
auto	0.96	0.94	0.93	0.97	0.98	0.95
Flair						
manual	0.99	0.99	0.98	0.98	0.98	0.98
auto	0.98	0.95	0.96	0.98	0.98	**0.97**
Stanza Basic						
manual	0.96	0.98	0.94	0.97	0.97	0.97
auto	0.96	0.95	0.92	0.97	0.97	0.96
Stanza + FastText						
manual	0.96	0.97	0.94	0.97	0.97	0.96
auto	0.96	0.95	0.92	0.97	0.97	0.95

a WER between 25% and 35%, depending on the language. It should be noted that the test sets for HOME are rather small: 17 pages for the Czech language, 14 pages for German, and 11 pages for Latin. Therefore, one page on which the lines are badly detected can heavily impact the final results. Even on the manual text, the Czech language is the one on which the NER models perform the most poorly. However, the dramatic drop in recall on the recognized text may be in part explained by the quality of the line detection. The average difference in number of lines between the groundtruth and the detected lines in Czech are three lines, with one page with only 8.9% of lines detected, one page with only 75% of lines detected. This is worse for German, for which the average difference in line count is of 1.3 lines. For Latin, the number of detected lines is consistent with the groundtruth. Thus, the recall drops less between manual and automatic transcription in Latin than it does for German, although the WER of the HTR engine is almost ten points high for Latin. It seems that the quality of the text line detection heavily impacts the recall, while the performance of the HTR models have a more effected precision. Regarding the performance of the multilingual models, they offer a viable solution on the HOME dataset, with performance scores close to the monolingual models. It is worth noting that the results presented for multilingual Flair or multilingual spaCy are based on a multilingual HTR, which opens perspectives for a fully multilingual processing workflow.

Results on the Esposalles Dataset. Table 7 compares the performance scores on the manual and HTR transcriptions of the Esposalles dataset, with the previous best results on this task, using the *Esposalles metric*. Using the dedicated Esposalles scorer, Flair outperforms the best published result on this dataset [21]. All the tested models yield similar and competitive results.

7 Conclusions and Future Work

In this study, we presented a comparison of three off-the-shelf open-source NER libraries on three historical datasets which include five different languages: Czech, German, Latin, Catalan and French. The NER models were evaluated on both manual and automatic transcriptions of the handwritten documents. We showed that, in the case of low or moderate error rates, while the drop in performance of the NER on the recognized text is limited, the quality of text line detection has a large impact on the results. We also showed that multilingual models, for both HTR and NER, demonstrated competitive performance and represent a viable option in case of scarce training data. None of the three compared library outperforms the other, thus, we recommend testing the three of them (with default hyperparameters). Finally, using Flair NER, we established state-of-the-art results on the Esposalles dataset. As future work, we envision a study of other off-the-shelf models for NER, Transformer-based [24], recently provided by spaCy and Stanza. We are also considering in exploring some multitask methods, especially curated for nested entities in digitized documents [8,15].

Acknowledgement. This work is part of the *HOME History of Medieval Europe* research project, supported by the French National Research Agency (Grant agreements No. ANR-17-JPCH-0006) through the European JPI Cultural Heritage and Global Change. The authors wish to thank Emanuela Boroş for the fruitful discussions and her careful reading of the manuscript.

References

1. Adak, C., Chaudhuri, B.B., Blumenstein, M.: Named entity recognition from unstructured handwritten document images. In: Workshop on Document Analysis Systems, pp. 375–380 (2016)
2. Aguilar, S.T., Tannier, X., Chastang, P.: Named entity recognition applied on a data base of medieval latin charters. The case of chartae burgundiae. In: International Workshop on Computational History (2016)
3. Akbik, A., Blythe, D., Vollgraf, R.: Contextual string embeddings for sequence labeling. In: International Conference on Computational Linguistics (2018)
4. Arora, A., et al.: Using ASR methods for OCR. In: International Conference on Document Analysis and Recognition, pp. 663–668 (2019)

5. Bamman, D.: Natural language processing for the long tail. In: Digital Humanities (2017)
6. Boillet, M., Kermorvant, C., Paquet, T.: Robust text line detection in historical documents: learning and evaluation methods. IJDAR (2022). https://doi.org/10.1007/s10032-022-00395-7
7. Bojanowski, P., Grave, E., Joulin, A., Mikolov, T.: Enriching word vectors with subword information. Trans. Assoc. Comput. Linguist. **5**, 135–146 (2017)
8. Boroş, E., et al.: Alleviating digitization errors in named entity recognition for historical documents. In: Conference on Computational Natural Language Learning, pp. 431–441 (2020)
9. Boros, E., et al.: A comparison of sequential and combined approaches for named entity recognition in a corpus of handwritten medieval charters. In: International Conference on Frontiers in Handwriting Recognition, pp. 79–84 (2020)
10. Carbonell, M., Villegas, M., Fornés, A., Lladós, J.: Joint recognition of handwritten text and named entities with a neural end-to-end model. In: International Workshop on Document Analysis Systems (2018)
11. Ehrmann, M., Romanello, M., Flückiger, A., Clematide, S.: Extended overview of clef hipe 2020: named entity processing on historical newspapers. In: CEUR Workshop Proceedings. No. 2696, CEUR-WS (2020)
12. Ehrmann, M., Romanello, M., Flückiger, A., Clematide, S.: Overview of CLEF HIPE 2020: named entity recognition and linking on historical newspapers. In: Arampatzis, A., et al. (eds.) CLEF 2020. LNCS, vol. 12260, pp. 288–310. Springer, Cham (2020). https://doi.org/10.1007/978-3-030-58219-7_21
13. Fornés, A., et al.: Icdar 2017 competition on information extraction in historical handwritten records. In: International Conference on Document Analysis and Recognition (2017)
14. Hamdi, A., Jean-Caurant, A., Sidere, N., Coustaty, M., Doucet, A.: An analysis of the performance of named entity recognition over ocred documents. In: Joint Conference on Digital Libraries (2019)
15. Hamdi, A., Carel, E., Joseph, A., Coustaty, M., Doucet, A.: Information extraction from invoices. In: Lladós, J., Lopresti, D., Uchida, S. (eds.) ICDAR 2021. LNCS, vol. 12822, pp. 699–714. Springer, Cham (2021). https://doi.org/10.1007/978-3-030-86331-9_45
16. Honnibal, M., Montani, I., Van Landeghem, S., Boyd, A.: spaCy: Industrial-strength Natural Language Processing in Python (2020)
17. Kettunen, K., Ruokolainen, T.: Names, right or wrong: named entities in an ocred historical finnish newspaper collection. In: International Conference on Digital Access to Textual Cultural Heritage (2017)
18. Labusch, K., Zu, S., Kulturbesitz, B., Neudecker, C., Zellhöfer, D.: Bert for named entity recognition in contemporary and historical german. In: Conference on Natural Language Processing (2019)
19. McGillivray, B., Poibeau, T., Ruiz, P.: Digital humanities and natural language processing: "Je t'aime... Moi non plus". Digit. Humanit. Q. **14**(2) (2020)
20. Mikolov, T., Chen, K., Corrado, G., Dean, J.: Efficient estimation of word representations in vector space. In: International Conference on Learning Representations (2013)
21. Prasad, A., Déjean, H., Meunier, J., Weidemann, M., Michael, J., Leifert, G.: Bench-marking information extraction in semi-structured historical handwritten records. CoRR (2018)

22. Qi, P., Zhang, Y., Zhang, Y., Bolton, J., Manning, C.D.: Stanza: A Python natural language processing toolkit for many human languages. In: Annual Meeting of the Association for Computational Linguistics: System Demonstrations (2020)
23. Tüselmann, O., Wolf, F., Fink, G.A.: Are end-to-end systems really necessary for NER on handwritten document images? In: Lladós, J., Lopresti, D., Uchida, S. (eds.) ICDAR 2021. LNCS, vol. 12822, pp. 808–822. Springer, Cham (2021). https://doi.org/10.1007/978-3-030-86331-9_52
24. Vaswani, A., et al.: Attention is all you need. In: Advances in Neural Information Processing Systems, pp. 5998–6008 (2017)
25. Won, M., Murrieta-Flores, P., Martins, B.: Frontiers in Digital Humanities **5** (2018)

A Benchmark of Named Entity Recognition Approaches in Historical Documents Application to 19th Century French Directories

N. Abadie[1], E. Carlinet[2], J. Chazalon[2](✉), and B. Duménieu[3]

[1] LASTIG, Univ. Gustave Eiffel, IGN-ENSG, 94160 Saint-Mandé, France
nathalie-f.abadie@ign.fr

[2] EPITA Research & Development Laboratory (LRDE), Le Kremlin-Bicêtre, France
{edwin.carlinet,joseph.chazalon}@lrde.epita.fr

[3] CRH-EHESS, Paris, France
bertrand.dumenieu@ehess.fr

Abstract. Named entity recognition (NER) is a necessary step in many pipelines targeting historical documents. Indeed, such natural language processing techniques identify which class each text token belongs to, e.g. "person name", "location", "number". Introducing a new public dataset built from 19th century French directories, we first assess how noisy modern, off-the-shelf OCR are. Then, we compare modern CNN- and Transformer-based NER techniques which can be reasonably used in the context of historical document analysis. We measure their requirements in terms of training data, the effects of OCR noise on their performance, and show how Transformer-based NER can benefit from unsupervised pre-training and supervised fine-tuning on noisy data. Results can be reproduced using resources available at https://github.com/soduco/paper-ner-bench-das22 and https://zenodo.org/record/6394464.

Keywords: Historical documents · Natural language processing · Named entity recognition · OCR noise · Annotation cost

1 Introduction

OCRed texts are generally not sufficient to build a high level semantic view of a collection of historical documents. A subsequent stage is often needed to extract the pieces of information most likely to be searched for by users, such as named entities: persons, organisations, dates, places, etc. Indeed, being able to properly tag text tokens unlocks the ability to relate entities and provide colleagues from other fields with databases ready for exploitation.

Being active research topics, OCR and named entity recognition (NER) are still difficult tasks when applied to historical text documents. OCR approaches

N. Abadie, E. Carlinet, J. Chazalon and B. Duménieu—All authors contributed equally.

© Springer Nature Switzerland AG 2022
S. Uchida et al. (Eds.): DAS 2022, LNCS 13237, pp. 445–460, 2022.
https://doi.org/10.1007/978-3-031-06555-2_30

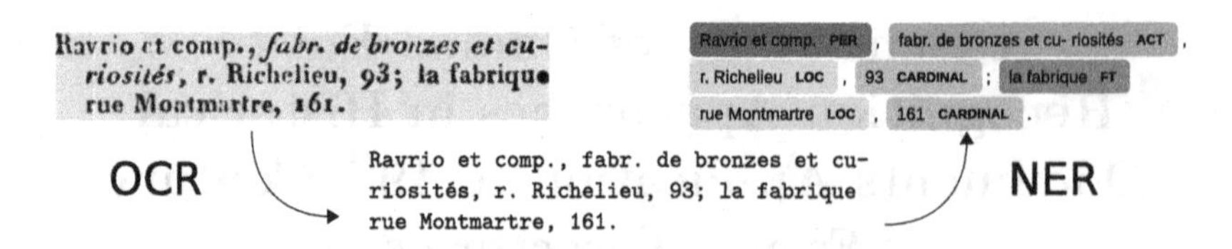

Fig. 1. Overview of the pipeline under study. From previously-extracted images of directory entries, we perform OCR and named entity recognition (NER) using different techniques. We aim at answering the following questions: *How noisy are modern, out-of-the-box OCR systems? What is the behaviour of NER when OCR is noisy? Can NER be made more robust to OCR noise?*

used for modern documents are likely to struggle even on printed historical documents due to multiple causes related to text readability (low resolution scans, inconsistent printing rules, artefacts, show-through), document complexity (intricate and versatile page layout, use of ancient fonts and special glyphs) and the variability inherent to the great diversity of historical sources. On the other hand, the semantics of entities in NER approaches developed for modern texts may be different from those in ancient texts.

In this article, we focus on a corpus of printed trade directories of Paris from the XIXth century, containing hundreds of pages long lists of people with their activity and address. They provide fine-grained knowledge to study the social dynamics of the city over time. As they originate from different publishers, they show a diversity in layout, information organisation and printing quality, which adds to the poor digitising quality to make OCR and NER challenging tasks.

Trade directories have been leveraged in recent work to identify polluted urban soils [2] and locate all gas stations in the city of Providence over the last century. In an ongoing research project, we aim at producing structured spatio-temporal data from the entries of the Paris trade directories to study the dynamics of the fraction of the XIXth century Parisian society reachable through these sources. Therefore, we investigate several state-of-the-art OCR and NER approaches to assess their usability to process the corpus (Fig. 1).

The contributions of this article are as follows: (i) We review state-of-the-art OCR and NER systems for historical documents (Sect. 2). (ii) We introduce a new dataset suitable for OCR and NER evaluation (Sect. 3). (iii) We measure the performance of three modern OCR systems on real data (Sect. 4). (iv) We evaluate modern NER approaches: their requirements in terms of training data, and the effects of pre-training (Sect. 5). (v) We show that Transformer-based NER can benefit from pre-training and fine-tuning to improve its performance on noisy OCR (Sect. 6).

2 OCR and NER on Historical Texts

The directory processing pipeline presented in [2] and illustrated in Fig. 1, includes an OCR step done with Tesseract, and a NER step to identify company names and addresses, performed using regular expressions. This section reviews existing OCR and NER approaches on historical texts and presents some works assessing the effects of OCR quality on the NER performance and the proposed solutions.

2.1 Optical Character Recognition of Historical Texts

Among the large number of OCR solutions, being either open, free, or paid software, available as libraries, Python packages, binaries, or cloud API, not all options seem suitable for historical document processing. We chose to avoid in our study paid and closed-source solutions. This notably discards Transkribus [25], which relies on the commercial system ABBYY's Finereader as well as on two handwritten transcription engines, to process text.

Most of the current state-of-the-art open-source OCR systems, like Tesseract [22], OCRopus [3], and the recent Kraken [8], Calamari [27] and PERO OCR [10] are based on a pipeline of convolutional neural networks (CNNs) and long short-term memory networks (LSTM). Although this model produces good results with modern texts, it still faces challenges with ancient texts, such as the lack of annotated data for learning, or different transcription styles for training data.

To overcome the limitations due to different transcription styles in training data, PERO OCR adds a Transcription Style Block layer to a classical model based on a CNN and a Recurrent Neural Network components [10]. This block takes the image of the text and a Transcription Style Identifier as inputs and helps the network decide what kind of transcription style to use as output.

2.2 Named Entity Recognition in Historical Texts

Many approaches have been designed to recognise named entities, ranging from handcrafted rules to supervised approaches [18]. Rule based approaches look for portions of the text that match patterns like in [2,20] or dictionary (gazetteers, author lists, etc.) entries like in [14,17]. Such kind of approaches achieve very good results when applied to a specialised domain corpus and when an exhaustive lexicon is available, but at high system engineering cost [18].

Supervised approaches include both approaches implementing supervised learning algorithms with careful text feature engineering, and deep learning based approaches which automatically build their own features to classify tokens into named entity categories. In recent years, the latter have grown dramatically, yielding state-of-the-art performances as shown in the recent survey proposed by [13]. This survey concludes that fine-tuning general-purpose contextualised language models with domain-specific data is very likely to give good performance for use cases with domain-specific texts and few training data. This strategy has been adopted by [11] to extract named entities in OCRed historical texts in German, French, and English. However, the NER performance drops significantly as the quality of the OCR decreases and is correlated with its decrease.

Several recent studies have focused on the impact of OCR quality on NER results. Most of the time, they have evaluated NER approaches based on deep learning architectures as they seem to adapt more easily to OCR errors than rule-based or more classical supervised approaches. [24] used the English model *en-core-web-lg* provided by the SpaCy [23] library to perform NER on a corpus made of many journal articles with different levels of OCR errors. For each OCRed article, a ground truth text is available so that the Word Error Rate (WER) can be computed. The performance of the NER model with respect to OCR quality is eventually assessed by computing the F1-score for each NER class and each article, i.e., each WER value. [6] performed a similar but more extensive evaluation on four supervised NER models: CoreNLP using Conditional Random Fields and three deep neural models, BLSTM-CNN, BLSTM-CRF, and BLSTM-CNN-CRF. They tested them on CoNLL-02 and CoNLL-03 NER benchmark corpora, degraded by applying four different types of OCR noise. Overall, NER F-measure drops from 90% to 50% when the Word Error Rate increases from 8% to 50%. However, models based on deep neural networks seem less sensitive to OCR errors. Two approaches have been proposed by [7] and [16] to reduce the negative impact of OCR errors on NER performance on historical texts. The former applies a spelling correction tool to several corpora with variable OCR error rates. As long as OCR errors remain low ($CER < 2\%$ and $WER < 10\%$), this strategy makes it possible to maintain good NER results. However, the F1-score starts to decrease significantly when OCR errors exceed these thresholds. The latter work focuses on adapting the training data to facilitate the generalisation of an off-the-shelf NER model from modern texts to historical texts. Finally, reusing a model trained on clean modern data, including embeddings computed on a historical corpus, and fine-tuned on a noisy historical ground truth has proven to be the most effective strategy.

2.3 Pipeline Summary

Based on those works, we chose to test three OCR systems, namely, Tesseract, PERO OCR, and Kraken. We also adopted two deep-learning-based French language models, available in packaged software libraries, already trained for the NER task and that can be adapted to the domain of historical directories: SpaCy NLP pipelines and CamemBERT. In Sect. 3, we will explain the evaluation protocol used to assess the combined performance of these OCR and NER systems.

Tesseract is a long-living project, born as a closed-sourced OCR at Hewlett-Packard in the eighties; it was progressively modernized, then open-sourced in 2005. From 2006 until November 2018, it was developed by Google and is still very active. We used in our tests the version 4.1.1, released Dec. 26, 2019. Version 5, released on Nov. 30, 2021, has not been integrated in our tests yet.

Kraken is a project created by Benjamin Kiessling several years ago (development can be traced back to 2015), and is actively used in the open-source eScriptorium project [9]. As no pre-trained model for modern French was easily available, we used the default English text recognition model trained on modern printed English by Benjamin Kiessling on 2019. Models can be easily found and downloaded thanks to their hosting on Zenodo.

PERO OCR is a very recent project (started in 2020) from the Brno University of Technology in the Czech Republic. Their authors used many state-of-the-art techniques to train it very efficiently. We used the version from the master branch of their GitHub repository, updated on Sep 15, 2021. We used the pre-trained weights provided by the authors on the same repository, created on Oct. 9, 2020 from European texts with Latin, Greek, and Cyrillic scripts.

SpaCy is a software library that offers NLP components assembled in modular pipelines specialised by language. Although BERT is available in the latest version of SpaCy (v3), the pipeline for French does not provide a NER layer at the time of our experiments (as of January 2022). Hence, we rely on SpaCy's ad hoc pipeline trained on French corpora and capable of Named Entity Recognition. The global architecture of these pipelines have not been published yet, but are explained by the developers on their website. Words are first encoded into local context-aware embeddings using a window-based CNN similar to [4]. The decision layer is an adaptation of the transition-based model presented in [12]. As words are processed sequentially, their vectors are concatenated with those of the last known entities to encode the nearby predicted semantics. The classification layer relies on a finite-state machine whose transition probabilities are learnt using a multi-layer perceptron.

CamemBERT is well-known adaptation of the BERT Transformer-based model for the French language [5,15,26]. Such language models have become a new paradigm for NER [13]. The learned contextual embeddings can be used as distributed representations for input instead of traditional embeddings like Google Word2Vec, and they can be further fine-tuned for NER by adding an additional output layer, usually referred to as a "head". They can also be pre-trained in an unsupervised way on large amount of unlabelled texts for domain adaptation.

3 Dataset

This section presents the historical sources that we selected and their contents. It also details the construction of the groundtruth dataset leveraged in our experiments and the metrics used to evaluate OCR and NER results. The resulting dataset is publicly available [1] under the permissive *Creative Commons Attribution 4.0 International* licence.

3.1 A Selection of Paris Trade Directories from 1798 to 1854

The directories are available from different libraries in Paris and have been digitised independently in various levels of quality. They cover a large period of time and originate from several publishers. Therefore, their contents, indexes, layouts, methods of printing, etc. may vary a lot from one directory to another (see Fig. 2). We want our groundtruth dataset to be representative of the diversity of directories available in the period.

Directories contain lists of people with various information attached. For instance, the directory published by *Didot* in 1854 contains three redundant lists of people sorted by name, by activity and by street name. A typical example entry from this directory is "*Batton (D.-A.)* ✱, *professeur au Conservatoire*

Non-Commerçans. (Paris). 269
Chardin, R. Pavée, 16. — B. C.
Chardin, R. Michel Lepelletier, 21.
Chardon, (Ve.) R. S. Marc, 15.
Cheviron, R. Chapon, 13.
Chimay, (Mme.) R. de Varennes, 31.
Choart-Duplessis, R. de Turenne, 31.

AMADOU ET ALLUMETTES. — Pour les ALLUMETTES OXIGÉNÉES. Voyez BRIQUETS PHYSIQUES.
DARRAS (Thomas), r. de la Vieille-Monnaie, 10.
Briquets et veilleuses, mèches à quinquets, mèches soufrées, pierres, agaric de chêne, liége en planches, bouchons.
GALLIENNE je., r. de la Heaumerie, 3.
Brûle-tout, boîtes à briquet, mèches à vin et à quinquet, veilleuses mèches, soufrées; pierres, agaric, bouchons, liége.
LESOY, r. Aubry-le-Boucher, 43.

BAUDOYER (place).
IX Arr. Hôtel-de-Ville.) ←
Rue Tixeranderie, pourtour St-Gervais, Saint-Antoine et Renaud-Lefèvre.
1 Lissoty (Vve), vins.
2* Privé, distillateur.
Lemoine-Cluzel et Leroy, nouveautés.
Chantrier, court.-gourm.
26* Longpré aîné, bijoutier en or et argent.
Saint-Omer, émailleur.
Cellier (A.), graveur-ciseleur.*
Rousseau (J.), bijoutier en or.*
Benoît, orfèvre-fabr.
Lérésy, doreur.
30 Bouton, fab. de cuir verni.*
31 Pardon, vins.
Bourguille, fabr. de presses.
Vaudain, passementier.
Finino jne, bronze doré.
Rabé aîné, fabr. de ballons.*
Gaulin, chapelier.
Moisy, tabletier.
49* Cendrier aîné, prop.
Desmarests, fab. bottes d'emballage.
Ferrand, lapidaire.
7 Ecole communale de jeunes filles.
Berthelot, vins.
6 Verstaen, serrurier-mécanicien.
8 Michel, brossier.
9 Labottière, serrurier.
10 Sacrez, vins.
12 Baudoin, épic.
13 Lejard, clouteries et crépins.
14 Baduel (Vve), fab. de et tapisseries.
10 Lainé jeune, vins.
Jumelles omnibus et entreprise générale des Omnibus.
11 Melouzay, vins en gros, et à Bercy, Port, 31.
12 Combaud, coiffeur.
Monmain (P.), vins en gros.
13 Dufailly, sculpt. fabr. de carton-pierre.

Fig. 2. Examples of directory layouts and contents. *Top*: Duverneuil et La Tynna 1806 - index by name; *Middle*: Deflandre 1828 - index by activity; *Bottom*: Bottin 1851 - index by street name.

de musique et de déclamation, Saint-Georges, 47.". It begins with the person's name and surname, here inside parentheses. The glyph denotes that this person was awarded the *Légion d'Honneur*. Then comes a description of the person's activity, here his profession (professor at the Conservatory of music and declamation), but it can also be a social status. Such descriptions range from a single word to paragraphs describing the occupation in full detail. The street name and the number where the person lives or carries out their activities end the entry.

These are the pieces of information we want to extract, deduplicate and structure to build a spatio-temporal database. Except for some potentially wordy activity descriptions, they correspond to named entities. However, while most entries contain the same types of named entities, their order and the way they are written vary from one directory/index to another. To provide examples of each entry structure, pages from each type have thus been annotated.

3.2 A Dataset for OCR and NER Evaluation

The simplified data extraction pipeline depicted in Fig. 3 processes the documents presented in Sect. 3.1. First, the page layout extraction and entry segmentation are performed with a semi-automated system and checked by a human. The resulting images, representing 8765 entries, are the inputs of a customizable two-stage pipeline.

OCR Stage. The first stage of the pipeline aims at extracting the raw text from the images. Individual image fragments are created from each segmented entry and fed to an OCR system for text extraction. An entry might span over multiple text lines but is always a single block. Thus, the most adapted mode is chosen when the OCR system allows for the detection mode (e.g. the *block* mode

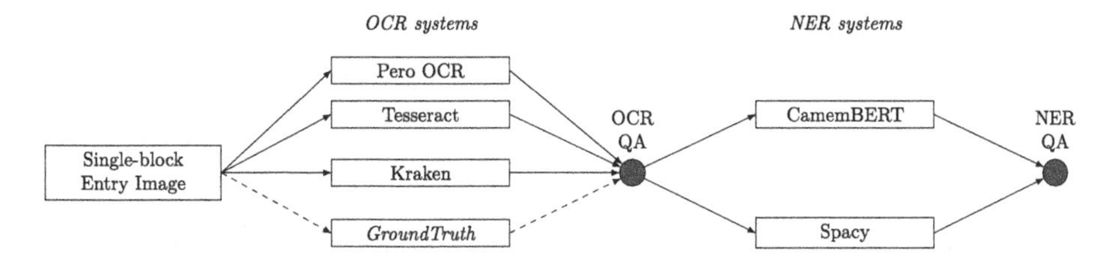

Fig. 3. Data extraction pipeline with two quality control checkpoints. The NER Q.A. checkpoint may either assess the NER system in isolation (using the dashed *groundtruth* path), or may evaluate the joint work of a NER system with an OCR system.

for *Tesseract*). Some glyphs used in this dataset might be unknown to the OCR, and some like ❋, which do not even have a Unicode codepoint, were annotated using Unicode Private User Area 1. Furthermore, as some annotations guidelines where unclear to human annotators, some projections rules were applied: whitespace, dash, dots and a couple of commonly confused characters where projected to well-defined codepoints. The same normalisation was applied to OCR predictions. At the end of this first stage, an OCR Quality Assessment (Q.A.) between the *normalised* groundtruth text and the OCR output is performed on the basis of the 8, 765 entries which were manually controlled, totalling 424, 764 characters (including 54, 387 spacing characters). Entries are 49.0-characters long on average.

NER Stage. The second stage of the pipeline aims at extracting the named entities from a text with a NER system. This text can either originate from the OCR outputs in a real-word scenario, or from groundtruth text in order to evaluate the NER performance independently. There are 5 types of entity to detect (see. Table 1). The NER system has to classify non-overlapping parts of the text into one of these entities (or none of them). A second Q.A. (namely, NER Q.A.) is performed at the end of this stage between the groundtruth and the NER output.

Note that the dataset used to assess the NER stage is a subset of the entries. Indeed, we need to ensure that the datasets contain the same entries whichever the OCR used in the previous stage. Entries where the OCR produced an empty string and those for which no entity could be projected from the groundtruth have to be ignored. We filter the set of entries by keeping only the entries that are always valid at the end of the stage 1. Therefore, the 8, 765 reference entries were manually annotated with 34, 242 entities; entries contain 3.9 entities on average. Projecting reference tagged entities on OCR predictions resulted in a variable loss of entries. For PERO OCR, 8, 392 valid entries were generated, for Tesseract 8, 700 and for Kraken 7, 990. The resulting intersection of the sets of valid entries contained 7, 725 entries for the tree OCR systems (and the reference), or 8, 341 entries if we consider PERO OCR and Tesseract only.

Table 1. Entities to recognise in the dataset.

Entity	Count	Description
PER	8788	a person name or a business name, usually referred to as several person names. First names, initials, or civility mentions are included. E.g.: *Alibert (Prosper)*, *Allamand frères et Hersent,Heurtemotte (Vve).*
TITLE	483	an honorary title, either text, glyphs or a combination of the two. E.g: O. ✱ for *Officier de la Légion d'Honneur* or [GM] for the Great Medal at the London exhibition.
ACT	6472	the profession or social status of a person summarised in the single concept of activity. E.g.: *horlogerie*, *export en tous genres*, *Conseiller d'État*, *propriétaire.*
LOC	9709	mostly street names (*r. de la Jussienne*), but may also be neighbourhoods (*Marais du Temple*) or any indirect spatial references (*au Palais Royal*).
CARDINAL	8747	a street number, as part of an address (16, 5 bis), or a range of numbers (e.g. 23-25 or 5 *à* 9).
FT	43	a geographic entity type, used to give more details on a location, e.g. *boutique*, *atelier*, *fab.* or *dépôt.*

3.3 Metrics for OCR and NER Quality Assessment

OCR Q.A. The predicted text by the OCR system is aligned with the groundtruth text using standard tools from Stephen V. Rice's thesis [19,21]. The Character Error Rate (CER) is computed at the entry level and at the global level, defined as the ratio between the number of errors (insertions/deletions/substitutions) over the reference text length. Word Error Rate is hard to define for our tokens and was not considered.

NER Q.A. The NER system outputs a text with tags that enclose the entities. To assess the quality of the entity extraction, we rely on a technique similar as for the OCR evaluation to build the *NER-target*. The *NER-target* is different from the groundtruth because it should not involve the errors committed during the previous stages. The OCR text is first aligned with the groundtruth text to form the NER-*input* (where *input* is a placeholder for *pero* if the input text is from PERO, NER-*tesseract*, NER-*reference*...). The tags of the groundtruth are then projected in the alignment on NER-*input* to provide the *NER-target.* The NER system then runs on NER-*input* and outputs the *NER-prediction.* The precision, recall, and f-measure (or f-score) are computed considering only the exact matches between entities of the *NER-target* and those from the *NER-prediction*, i.e. pairs of entries for which the type, start and end positions are exactly the same. Precision is the ratio of exact matches divided by the number

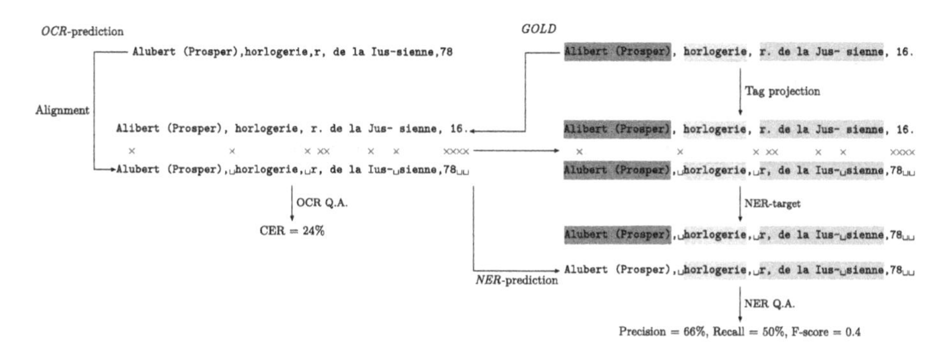

Fig. 4. OCR and NER evaluation protocol example. The OCR prediction is first aligned on the GOLD text to get the OCR quality. The same alignment is used to project the GOLD tags on the OCR prediction. Projected tags enable the NER system run on the OCR predicted text to be assessed.

of predicted entries, and recall is defined as the ratio of exact matches divided by the number of expected entries; the f-measure is their harmonic mean.

The evaluation process is illustrated on Fig. 4. The OCR and the groundtruth texts are first aligned to evaluate the OCR accuracy. As there are 11 mismatches over 56 aligned characters, the CER is thus 24%. This alignment is then used to back-project the groundtruth tags to build the tagged NER-*target*. Finally, the NER system runs on the OCR text; its output is compared to the NER groundtruth. There is only 2 over 3 tags matching in the prediction (precision), while only 2 over 4 entities are matched in the reference (recall). It follows an overall f-score of 0.4.

4 OCR Benchmark

This section focuses on the evaluation of the performance of the three open-source OCR systems we selected, as described in Sect. 2.3: Tesseract v4, Kraken and PERO OCR. The OCRs are used "out-of-the-box" with their default pre-trained models. No fine-tuning was performed as we did not get enough annotated data for such a task. The dataset used to perform this evaluation is composed of the 8,765 entries (containing 424,764 characters) from the dataset we previously introduced. The single-column, cropped images of entries are used as input of each OCR system. As the pages were previously deskewed, the text is mostly horizontal except for a few cases. The expected output is the human transcription of these images provided in the dataset. Before computing the Character Error Rate (CER) for each entry, each text prediction is normalised with the same basic rules as the ones used to post-process human transcription: dashes, quotes and character codes for glyphs like stars or hands are normalised.

Figure 5 compares the performance of the OCR systems on our dataset. We can see Kraken's performance are not as good as the two first OCR. This is partially due to the fact that the closest available model is for English text

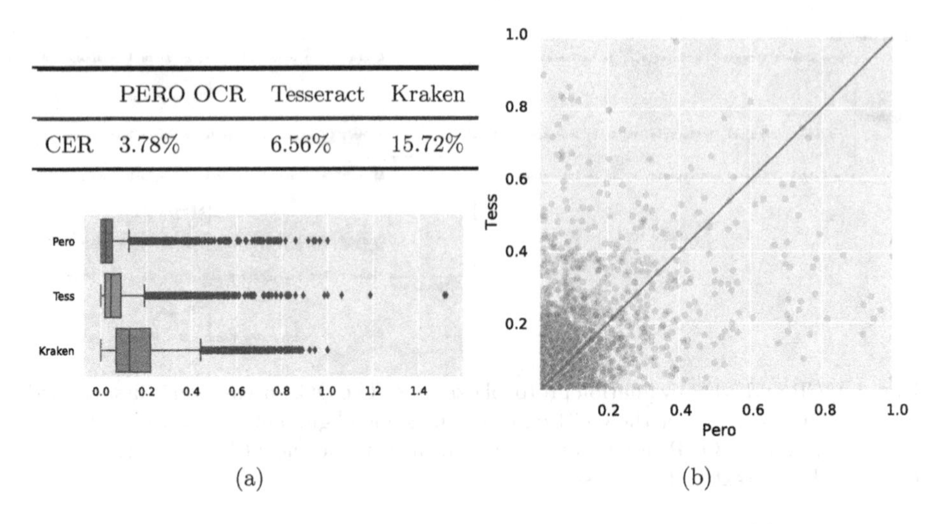

	PERO OCR	Tesseract	Kraken
CER	3.78%	6.56%	15.72%

(a) (b)

Fig. 5. CER at entry-level for PERO OCR, Kraken and Tesseract. (a) Global CER and distribution of the CER per entry. (b) joint plot of the per-entry error rate showing that PERO OCR and Tesseract do not fail on the same entries.

and so it misses French specific symbols. On the other hand, even when using a French model trained on French XIXth century documents, the performance does not increase (and relaxing the character matching rules does not help either). Tesseract and PERO OCR are performing better on this dataset "out-of-the-box". With no fine-tuning, PERO OCR gets the best accuracy with less than 4% character errors. Many of them are even due to a bad line detection in case of multi-lines entries and are not related to the OCR system itself. Figure 5 (b) shows that errors from the two best OCR are not committed on the same entries (if so, all points would be on the diagonal line) and that combining the outputs of PERO OCR and Tesseract could improve the overall recognition quality.

5 NER Sensibility to the Number of Training Examples

The constitution of annotated datasets to train a NER model is a critical preliminary step. Often done manually, possibly with bootstrapped annotations, this task is tedious, time-consuming, and error-prone. The ability of a model to perform well even with a few training examples is a practical criterion to consider. In this first experiment, we investigate the NER performance of SpaCy and CamemBERT when fine-tuned with an increasing number of training examples.

5.1 Training and Evaluation Protocol

The following models form our baseline for both NER experiments. Their short names written in square brackets will be used to reference them from now on.

- **SpaCy NER pipeline for French [SpaCy NER]**: We use the pipeline *fr_core_news_lg* provided by SpaCy v.3.2.1 [23], already trained for NER on the French corpora deep-sequoia and wikiner-fr. We stress again that we use the CNN version of this pipeline, not the transformer-based available in SpaCy v3.
- **Huggingface CamemBERT [CmBERT]**: We rely on the implementation of BERT models provided by the software library Huggingface (transformers v.4.15.0, datasets v.1.17.0). We chose to reuse a CamemBERT model published on the Huggingface repository[1] and trained for NER on wikiner-fr.
- **CamemBERT pre-trained on Paris directories [CmBERT+ptrn]**: To evaluate whether adapting CamemBERT to the domain increases its performance, we do an unsupervised pre-training of CmBERT for next sentence prediction and masked language modelling, using approx. 845, 000 entries randomly sampled and OCRed with PERO. The model is trained for 3 epochs and is available online[2].

Each model is then fine-tuned on subsets of the ground truth of increasing size. The NER metrics are eventually measured against a common test set. The procedure for creating these sets is as follows.

As the structure of entries varies across directories, the models may learn to overfit on a subset of directories with specific features. To reduce the evaluation bias, we start by leaving out 3 directories (1, 690 entries, ≈19%) from the ground truth as a test set containing entries from unseen directories.

Then, a stratified sampling based on the source directory of each entry is run on the remainder to create a training set (6, 373 entries, ≈73% of the gold reference) and a development set (709 entries, ≈8%). The development set is used to evaluate the model during the training phase. This resampling procedure is a convenient way to shape both sets, so they reflect the diversity of directories within the ground truth.

To generate smaller training sets, we start from the initial training set and iteratively split it in half using the same stratified sampling strategy as for the train/dev split to maintain the relative frequency of directories. We stop if a directory has only one entry left, or if the current training subset contains less than 30 entries, maintaining the relative frequency of directories within it. Applying this procedure to the initial training set produces 8 training subsets containing 49, 99, 199, 398, 796, 1593, 3186, and 6373 entries.

The three models are fine-tuned on the NER task 5 times using each of the 8 training subsets, with an early stopping criterion based on the number of training steps without improvement of the F1-score. This patience threshold is set to 1, 600 steps for SpaCy NER and 3 evaluations (1 evaluation every 100 steps) for CmBERT and CmBERT+ptrn. The metrics are measured for the 24 resulting NER models on the common test set and averaged over the runs.

[1] https://huggingface.co/Jean-Baptiste/camembert-ner.
[2] https://huggingface.co/HueyNemud/das22-10-camembert_pretrained.

5.2 Results and Discussion

Figure 6 displays the averaged precision, recall, and F1-score for all models on the 8 subsets created from the groundtruth. CmBERT, CmBERT+ptrn and SpaCy NER display the same behaviour: the performances increase dramatically with the number of training examples and rapidly reach an area of slower progress around 1000 examples. The F1 score increases by 4.6 points between 49 and 796 examples for CmBERT (resp. 1.6 for CmBERT+ptrn and 5.1 for SpaCy NER) but only by 1 point between 796 and 6373 examples (resp. 0.6 and 1.4). The models derived from CamemBERT always outperform the SpaCy model.

It appears that pre-training the CamemBERT model on OCR text seems worth it only when the training set used to fine-tune the NER layer is small. This effect might be due to the differences in nature between the training subsets, whose texts are manually corrected, and the noisy OCR texts used to pretrain CamemBERT. Indeed, the learned embeddings from pre-training are specialised to noisy texts and therefore less adapted to clean text. The pre-training aims at adapting the model to the vocabulary of the domain and to the errors caused by the OCR, which reveals not helpful and even counterproductive when the texts do not contain these types of errors.

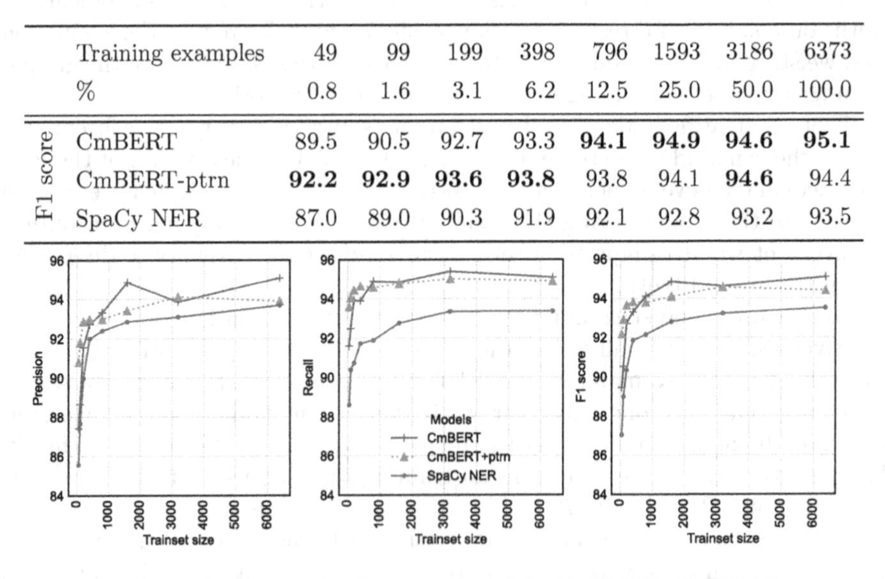

	Training examples	49	99	199	398	796	1593	3186	6373
	%	0.8	1.6	3.1	6.2	12.5	25.0	50.0	100.0
F1 score	CmBERT	89.5	90.5	92.7	93.3	**94.1**	**94.9**	**94.6**	**95.1**
	CmBERT-ptrn	**92.2**	**92.9**	**93.6**	**93.8**	93.8	94.1	**94.6**	94.4
	SpaCy NER	87.0	89.0	90.3	91.9	92.1	92.8	93.2	93.5

Fig. 6. Metrics measured on the fine-tuned models CmBERT, CmBERT+ptrn and SpaCy NER for 8 training sets of increasing sizes.

6 Impact of OCR Noise on Named Entity Recognition

Noise introduced by OCR is known to have a negative impact on NER, because it alters the structure and lexicon of the input texts, moving them away from the language model known to the NER process. In real-life situations, the models are often trained on texts without such noise, even though the texts to be annotated are extracted with OCR. In this second experiment, we aim at assessing the most appropriate strategy to build a NER model tolerant to OCR noise.

6.1 Training and Evaluation Protocol

Only CmBERT and CmBERT+ptrn are considered since the first experiment shows that SpaCy NER is outperformed for all sizes of training sets. We leave Kraken aside as its performance hinders the projection of NER labels from the ground truth, thus dramatically reducing the size of labelled sets. Because the NER metrics are entity-wise, they may underestimate the NER performances on very noisy texts with fragmented entities. We leverage the labelled sets of entries NER-*reference*, NER-*pero* and NER-*tesseract* created as explained in Sect. 3. Each dataset is split into training development, and test sets following the same protocol as described in Sect. 5.1, except this time we do not need to create smaller training sets. As the NER sets contain 8, 341 entries (see Sect. 3.2), the produced train sets (resp. development and test) count 6, 004 entries - 72% of the total (resp. 668 - 8% and 1, 669 - 20%). CmBERT and CmBERT+ptrn are fine-tuned on the training sets built from NER-*reference* and NER-*pero*. The training parameters are mostly the same as in Sect. 5.1, only this time the patience threshold is set to 5 evaluations. Finally the metrics are measured against the three tests sets.

6.2 Results and Discussion

The measured F1-score are given in Fig. 7. Results clearly show that models perform best when both the pre-training and the NER fine-tuning are similar to the processed texts in term of OCR noise.

In our tests, pre-training the model brings a slight gain in performance (≈0.5%). We did not pre-train or fine-tune with texts extracted with Tesseract. However, despite a loss of performance, the model pre-trained and fine-tuned on NER-pero still gives the best results. This is probably due to the fact that the texts produced by PERO OCR feature characteristics intermediary between human transcriptions and texts produced by Tesseract. This OCR tool removes the characters recognised with low confidence, which is probably a great help to the NER.

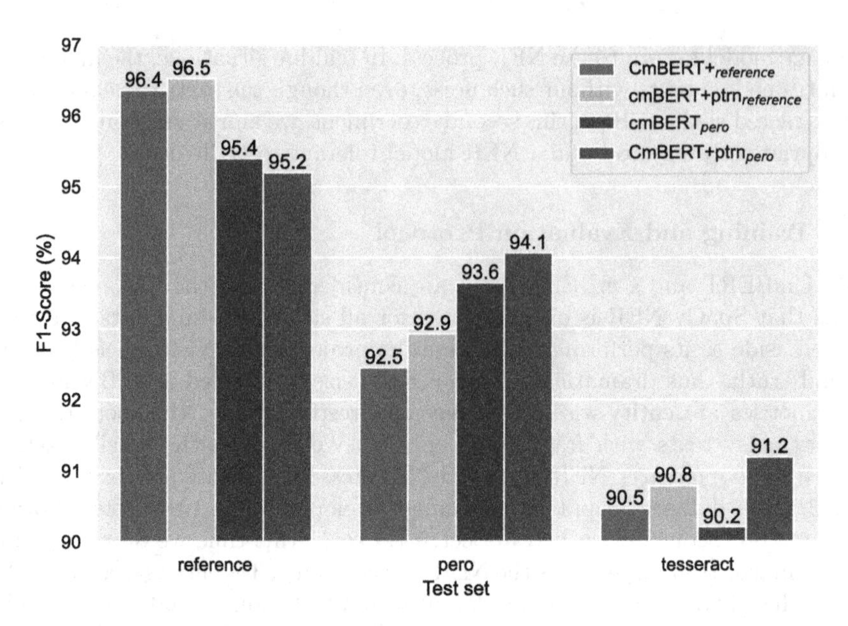

Fig. 7. NER F1-scores in presence of OCR noise in the training and testing sets, grouped by test set. The dataset used for training is noted in indice after the model name (e.g. CmBERT_{pero} for CmBERT fine-tuned on NER-*pero*).

7 Conclusion and Future Works

We assessed the performance of three modern OCR systems on a set of historical sources of great interest in social history. Although PERO OCR clearly outperforms its competitors, the qualitative analysis of OCR errors shows that its failure cases are not the same as Tesseract. This calls for leveraging both OCR systems in a complementary way to get the best of the two worlds. The evaluation of SpaCy NER and CamemBERT (with and without pre-training) showed that BERT-based NER can benefit from pre-training and fine-tuning on a corpus produced with the same process as the texts to annotate. Furthermore, it seems that all three models achieve good performance with relatively few training examples. With a F1-score of 92% with only 49 training examples, the pre-trained CamemBERT model is a good choice to serve as a bootstrapping model to quickly produce large training sets and therefore lower the burden of creating a ground truth from scratch. Besides, as directory entries always have the same structure - at least within a given index - we could take advantage of NER results and some simple rules to identify entries within pages instead of relying on the page layout only, or even interactively generate per-index NER models to take advantage of the low amount of training samples required. We plan to further explore the robustness of the considered NER models by introducing realistic OCR noise in order to identify possible critical points, in terms of noise level or in terms of entities or structural elements affected.

Acknowledgments. This work is supported by the French National Research Agency (ANR), as part of the SODUCO project, under Grant ANR-18-CE38-0013. The authors want to thank S. Bacciochi, P. Cristofoli and J. Perret for helping to create the reference dataset, L. Morice for annotating data, as well as G. Thomas, P. Abi Saad, R. Lelièvre, D. Mignon, T. Cavaciuti and P. Sadki for contributing to the annotation platform.

References

1. Abadie, N., et al.: A dataset of french trade directories from the 19th century (FTD), March 2022. https://doi.org/10.5281/zenodo.6394464
2. Bell, S., et al.: Automated data extraction from historical city directories: the rise and fall of mid-century gas stations in providence. RI. PLoS One **15**(8), 1–12 (2020)
3. Breuel, T.M.: The OCRopus open source OCR system. In: Document Recognition and Retrieval XV, vol. 6815, p. 68150F. International Society for Optics and Photonics (2008)
4. Collobert, R., Weston, J., Bottou, L., Karlen, M., Kavukcuoglu, K., Kuksa, P.: Natural language processing (almost) from scratch. J. Mach. Learn. Res. **12**, 2493–2537 (2011)
5. Devlin, J., Chang, M.W., Lee, K., Toutanova, K.: BERT: Pre-training of deep bidirectional transformers for language understanding. In: Proceedings of NAACL-HLT, pp. 4171–4186 (2019)
6. Hamdi, A., Jean-Caurant, A., Sidère, N., Coustaty, M., Doucet, A.: Assessing and minimizing the impact of OCR quality on named entity recognition. In: Hall, M., Merčun, T., Risse, T., Duchateau, F. (eds.) TPDL 2020. LNCS, vol. 12246, pp. 87–101. Springer, Cham (2020). https://doi.org/10.1007/978-3-030-54956-5_7
7. Huynh, V.-N., Hamdi, A., Doucet, A.: When to use OCR post-correction for named entity recognition? In: Ishita, E., Pang, N.L.S., Zhou, L. (eds.) ICADL 2020. LNCS, vol. 12504, pp. 33–42. Springer, Cham (2020). https://doi.org/10.1007/978-3-030-64452-9_3
8. Kiessling, B.: Kraken contributors. http://kraken.re
9. Kiessling, B., Tissot, R., Stokes, P., Stokl Ben Ezra, D.: eScriptorium: An open source platform for historical document analysis. In: International Conference on Document Analysis and Recognition Workshops, p. 19. IEEE (2019)
10. Kohút, J., Hradiš, M.: TS-Net: OCR trained to switch between text transcription styles. In: Lladós, J., Lopresti, D., Uchida, S. (eds.) ICDAR 2021. LNCS, vol. 12824, pp. 478–493. Springer, Cham (2021). https://doi.org/10.1007/978-3-030-86337-1_32
11. Labusch, K., Neudecker, C.: Named entity disambiguation and linking historic newspaper OCR with bert. In: CLEF (2020)
12. Lample, G., Ballesteros, M., Subramanian, S., Kawakami, K., Dyer, C.: Neural architectures for named entity recognition. In: Proceedings of NAACL-HLT. pp. 260–270 (2016)
13. Li, J., Sun, A., Han, J., Li, C.: A survey on deep learning for named entity recognition. IEEE Trans. Knowl. Data Eng. **34**(1), 50–70 (2020)
14. Mansouri, A., Affendey, L.S., Mamat, A.: Named entity recognition approaches. TAL **52**(1), 339–344 (2008)
15. Martin, L., et al.: CamemBERT: a tasty French language model. In: ProProceedings of the 58th Annual Meeting of the Association for Computational Linguistics, pp. 7203–7219 (2020)

16. März, L., Schweter, S., Poerner, N., Roth, B., Schütze, H.: Data centric domain adaptation for historical text with OCR rrrors. In: Lladós, J., Lopresti, D., Uchida, S. (eds.) ICDAR 2021. LNCS, vol. 12822, pp. 748–761. Springer, Cham (2021). https://doi.org/10.1007/978-3-030-86331-9_48
17. Maurel, D., Friburger, N., Antoine, J.Y., Eshkol-Taravella, I., Nouvel, D.: Casen: a transducer cascade to recognize french named entities. TAL **52**(1), 69–96 (2011)
18. Nadeau, D., Sekine, S.: A survey of named entity recognition and classification. Lingvisticae Investigationes **30**(1), 3–26 (2007)
19. Neudecker, C., Baierer, K., Gerber, M., Christian, C., Apostolos, A., Stefan, P.: A survey of OCR evaluation tools and metrics. In: The 6th International Workshop on Historical Document Imaging and Processing, pp. 13–18 (2021)
20. Nouvel, D., Antoine, J.Y., Friburger, N., Soulet, A.: Recognizing named entities using automatically extracted transduction rules. In: 5th Language and Technology Conference, pp. 136–140. Poznan, Poland (2011)
21. Santos, E.A.: Ocr evaluation tools for the 21st century. In: Proceedings of the Workshop on Computational Methods for Endangered Languages, vol. 1 (2019)
22. Smith, R.: An overview of the tesseract OCR engine. In: International Conference on Document Analysis and Recognition, vol. 2, pp. 629–633. IEEE (2007)
23. Spacy authors. https://spacy.io/
24. van Strien, D., Beelen, K., Ardanuy, M.C., Hosseini, K., McGillivray, B., Colavizza, G.: Assessing the impact of OCR quality on downstream NLP tasks (2020)
25. Transkribus contributors. https://readcoop.eu/transkribus
26. Vaswani, A., et al.: Attention is all you need. In: Advances in Neural Information Processing Systems, pp. 5998–6008 (2017)
27. Wick, C., Reul, C., Puppe, F.: Calamari-a high-performance tensorflow-based deep learning package for optical character recognition. Digit. Humanit. Q. **14**(1) (2020)

NCERT5K-IITRPR: A Benchmark Dataset for Non-textual Component Detection in School Books

Hadia Showkat Kawoosa, Mandhatya Singh, Manoj Manikrao Joshi, and Puneet Goyal(✉)

Indian Institute of Technology Ropar, 140001 Rupnagar, Punjab, India
{hadia.21csz0012,2017csz0003,2020csm1013,puneet}@iitrpr.ac.in

Abstract. The STEM subjects books heavily rely on Non-textual Components (NTCs) such as charts, geometric figures, and equations to demonstrate the underlying complex concepts. However, the accessibility of STEM subjects for Blind and Visually Impaired (BVIP) students is a primary concern, especially in developing countries such as India. BVIP uses assistive technologies (ATs) like optical character recognition (OCR) and screen readers for reading/writing purposes. While parsing, such ATs skip NTCs and mainly rely on alternative texts to describe these visualization components. Integration of effective and automated document layout parsing frameworks for extracting data from non-textual components of digital documents are required with existing ATs for making these NTCs accessible. Although, the primary concern is the absence of an adequately annotated textbook dataset on which layout recognition and other vision-based frameworks can be trained. To improve the accessibility and automated parsing of such books, we introduce a new NCERT5K-IITRPR dataset of National Council of Educational Research and Training (NCERT) school books. Twenty-three annotated books covering more than 5000 pages from the eighth to twelve standards have been considered. The NCERT label objects are structurally different from the existing document layout analysis (DLA) dataset objects and contain diverse label categories. We benchmark the NCERT5K-IITRPR dataset with multiple object detection methods. A systematic analysis of detectors shows the label complexity and fine-tuning necessity of the NCERT5K-IITRPR dataset. We hope that our dataset helps in improving the accessibility of NCERT Books for BVIP students.

Keywords: Graphical object detection · NCERT books · Assistive reading · NCERT dataset · Document layout analysis

1 Introduction

Approximately 2.2 billion people across the world are having some level of vision impairment[1] and nearly 62 million people are in India [1]. The onset of the

[1] https://www.who.int/news-room/fact-sheets/detail/blindness-and-visual-impairment.

© Springer Nature Switzerland AG 2022
S. Uchida et al. (Eds.): DAS 2022, LNCS 13237, pp. 461–475, 2022.
https://doi.org/10.1007/978-3-031-06555-2_31

COVID-19 era and its post effect led to the virtualization of schools as a new norm. This has severely impacted the educational scenario for BVIP students, pressing more on the need for inclusive education [2]. Inclusive education for BVIPs and especially in the STEM domain is one of the major challenges in developing countries such as India [3,4]. Inclusive education provides a collaborative and respectful environment for BVIP students among their normal sighted peers and develops a "sense of belonging" [5–7]. BVIP students face challenges in accessing regular school books or study materials, particularly STEM subjects that steadily employ NTCs/visualizations such as charts, graphs, tables, and equations. The widely used Assistive Technologies (ATs) such as speech-to-text software, OCRs or screen readers could not effectively interpret or detect these NTCs [8]. Such systems mainly rely on alternative text if it has been provided along with the NTCs. The accessibility scenario for accessing the textual content is comparatively better. Texts can be converted easily into braille or audio format through braille embossers and speech-to-text software, but for NTCs, it requires effective and complex vision-based methods.

The NTCs play an essential role in understanding document images as it compactly contains the underlying critical information. Detection and localization of these components is a primary step in the document parsing and understanding process. This work here is towards this initial and crucial step. Recently, with advancement in the deep learning domain, several works have been proposed for graphical object detection in document images [9–16]. Prior to discussing the details of the proposed dataset, we make some preliminary observations shown below:

1. The existing assistive schemes and solutions related to inclusive education, including government initiatives[2], majorly focus on the accessibility of the textual part. However, such solutions only provide partial inclusiveness as the majority of school books (India), including the widely used NCERT books, are designed in such a way to portray complex concepts via pictorial or non-textual representations. Such books use intuitive and non-standard visualization schemes to help students understand the underlying concepts. The NTCs used in these books do not follow the standard layouts settings.
2. The tactile-based solutions for the NTCs have scalability issues, which include hardware devices' requirements and production constraints. Instant production of the tactile solution is a challenging task, thus affecting making these solutions less scalable; in lieu, the audio-based solutions are relatively more scalable, affordable, and efficient.
3. General DLA datasets [9,17–20] label categories are structurally different from NCERT textbooks label categories. The school books are articulated extensively in a pictorial intuition form and are non-identical with the NTCs used in scientific, technical, or business documents. This makes the object detection methods, trained on general DLA datasets, perform marginal on the NCERT or related school book datasets.

[2] https://ncert.nic.in/pdf/accessibility/6.pdf.

One of the primary reasons for the inaccessibility of such books despite rapid advancement in deep learning and computer vision-based approaches is the absence of publically available annotated datasets of related school books. Due to the lack of ground truth containing NTCs, existing detectors and document parsing models can not be trained. Publicly available datasets further allow the scientific community to train and evaluate different models and build automated mechanisms to understand those books. The proposed NCERT5K-IITRPR dataset provides annotation for non-textual objects (table, chart, figures, equation, image, circuit diagram, and logo) and contains school-level books specific instead of general document layout annotations. The NCERT books are heavily loaded with such label categories. Especially, the labels such as charts, circuit diagrams, and equations (chemical equations) are different from the other document layout analysis datasets. Additionally, fine-grained label classifications are absent i.e., only the figure class represents all the graphical objects in the existing datasets. However, there are parallel fine-grained labels in the proposed dataset, such as figures, images, logos, and charts. To the best of our knowledge, it is the largest manually annotated dataset (PASCAL VOC format) of NCERT textbooks. We will release the dataset to support the development of effective and advanced BVIP accessible solutions for NCERT books. The main contributions are given below:

- We introduce a novel dataset for NTCs detection in NCERT school books. This dataset is among the first to provide manual annotation details in NCERT school books to the best of our knowledge. The label categories are structurally dissimilar to the existing DLA labels.
- Extensive bench-marking of one stage and two-stage object detectors have been performed.
- Though smaller than other document layout recognition datasets, NCERT5K-IITRPR offers more variability due to inherent variations in label categories. This has been empirically demonstrated in the experiments.

The remainder of this paper is organized as follows. Section 2 provides details of related datasets. Section 3 describes the NCERT5K-IITRPR dataset. Section 4 explains the benchmarking information and results. Section 5 provides the conclusion and future scope.

2 Related Datasets

The majority of the digital documents are in Portable Document Format (PDF), with more than 2.5 trillion documents [17]. Existing DLA datasets are mainly sourced from research papers, magazines, and financial records. The datasets are primarily prepared in manual, automatic, and hybrid ways. The larger datasets, in general, are prepared through an automated way to mitigate the lack of training data. However, manual annotation is costly but effective in terms of annotation quality. A comparison of our dataset with the existing datasets is present in Table 1.

Table 1. Statistics of existing datasets and our newly generated NCERT5K-IITRPR dataset.

Dataset	Label categories	Images
ICDAR-2013 [21]	Table	238
ICDAR-POD-2017 [9]	Table, Figure, Equation	2417
IIIT-AR-13K [18]	Table, Figure, Natural Images, Logo, Signature	13K
PubLayNet [17]	Table, Figure, Title, Text, List	360K
DeepFigures [11]	Table, Figure	5.5M
TableBank [20]	Table	417K
FinTabNet [22]	Table	90K
PubTabNet [23]	Table	568K
UNLV [24]	Table	3K
DocBank [19]	Table	500K
NCERT5K-IITRPR	**Table, Figure, Equation, Image, Logo, Circuit Diagram, Chart**	**5K**

The popular ICDAR-POD-2017 [9] dataset for page object detection (tables, figures and mathematical equations) consists of 2,417 document images (train-1600 images, test-817 images). The documents are selected from a corpus of 1,500 research articles from the CiteSeer platform. It comprises single-column, double-column as well as multi-column layout pages. DeepFigures [11] is prepared through automatic annotation of tables and figures in 5.5 million document pages. PublayNet [17] datasets annotations are also generated using the automated method with 360k images. This dataset also contains detailed annotations for textual parts such as title, list, text, etc. DocBank [19] datasets provides token-level annotations for supporting both NLP and computer vision approaches. The documents are acquired from arXiV.com, including the latex source files for semantic annotations. It contains a total of 500k pages with 400k pages for training and 50k-50k pages for validation and testing. IIIT-AR-13K [18] dataset contains 13K document pages over five label categories. The train-val-test distribution is set to 70%, 15%, and 15%.

Among all the label categories the "table" label is mostly explored due to its ubiquitous and representation utility. Several datasets such as ICDAR-2013 [21], UNLV [24], Marmot [25], PubTabNet [23], TableBank [20] have been proposed in literature for table detection and recognition task. The UNLV [24] dataset contains scanned document images with only table objects. The UNLV (2,889 page images), MARMOT (2,000 page images) and ICDAR-2013 (238 page images) datasets are comparatively smaller in size. FinTabNet [22] dataset contains multi-level and complex tables sourced from the financial reports of corporate organisation. It contains 89,646 document pages with a total of 112,887 tables. TableBank [20] and PubTabNet [23] are among the recent and largest annotated datasets for table detection and recognition with 417k and 568k+

images. The NCERT school books labels have significantly complex and varied structural properties than these document layout datasets.

3 NCERT5K-IITRPR Dataset

This section discusses the dataset aspects and related rationales in detail including the statistics, the diversity of the label categories, and annotation methods used for preparing the dataset. This annotated dataset will be made available on request.

Table 2. Statistics of our newly generated NCERT5K-IITRPR dataset. (Here, AS-Arts & Science, C-Commerce, P1-Part1, P2-Part2.)

	Subject	Standard	Total pages	Chapters
1	Mathematics	Eight	132	17
2	General Science	Eight	146	19
3	Mathematics P1	Nine	146	7
4	Mathematics P2	Nine	138	9
5	Science and Technology	Nine	226	18
6	Science and Technology P1	Ten	154	10
7	Science and Technology P2	Ten	130	10
8	Mathematics Part 1	Ten	186	6
9	Mathematics Part 2	Ten	178	7
10	Chemistry	Eleven	282	16
11	Physics	Eleven	266	14
12	Biology	Eleven	228	16
13	Mathematics & Statistics (AS) P1	Eleven	242	9
14	Mathematics & Statistics (AS) P2	Eleven	222	9
15	Mathematics Statistics (C) P1	Eleven	146	10
16	Mathematics & Statistics (C) P2	Eleven	176	10
17	Chemistry	Twelve	364	16
18	Physics	Twelve	376	14
19	Biology	Twelve	356	15
20	Mathematics & Statistics (AS) P1	Twelve	280	8
21	Mathematics & Statistics (AS) P2	Twelve	288	7
22	Mathematics & Statistics (C) P2	Twelve	182	9
23	Mathematics & Statistics (C) P1	Twelve	202	8
Total			5,046	264

3.1 Source and Statistics

The NCERT books have been widely used, in India, for more than five decades now. From the official website of NCERT[3], we have collected a total of 23 books of Science and Mathematics subjects of standard eight to twelve in the English language to prepare our dataset. The details of this dataset are shown in Table 2. Apart from core streams, mathematics subjects from arts & science, and commerce have also been considered. A total of thirteen mathematics and ten science books has been annotated. The number of pages vary from 130 to 376 in these books.

We have annotated seven categories in the current dataset version (table, chart, figure, image, equation, circuit diagram, logo). A total of 5,046 (5K) page images have been considered. A total of around 10K bounding boxes are generated as shown in Table 3. A total of 1.6K tables, 0.5K charts, 2K figures, 1.5K images, 1.3K equations, 0.2K circuit diagrams, and 3.3K logos are considered. We have considered an approximately 70-20-10 split for training, validation, and test sets. Table 3 shows the annotation wise statistics of proposed dataset.

Table 3. Annotation wise statistics of NCERT5K-IITRPR dataset. The percentage denotes the split proportion within categories.

Page objects	NCERT5K-IITRPR			
	Training	Validation	Test	Total
Table	1,169 (72%)	287 (17.7%)	167 (10.3%)	1,623
Chart	341 (66.9%)	101 (19.8%)	68 (13.3%)	510
Figure	1,455 (70.7%)	439 (21.3%)	163 (7.9%)	2,057
Image	1,060 (67.9%)	337 (21.6%)	165 (10.6%)	1,562
Equation	968 (73.7%)	252 (19.2%)	93 (7.1%)	1,313
Circuit diagram	159 (75%)	32 (15.1%)	21 (9.9%)	212
Logo	2,377 (70.9%)	660 (19.7%)	315 (9.4%)	3,352
Total	7,529	2,108	992	10,629

3.2 Label Categories

The NCERT books contain several categories of NTCs objects. The NTCs in NCERT5K-IITRPR dataset are comparatively more complex and dissimilar than the parallel categories present in general DLA datasets. For example, unlike in the generic DLA datasets, where the table category mainly consists of textual content, in the NCERT5K-IITRPR dataset, this category inscribes complex NTCs such as images, figures, chemical and mathematical formulae, etc. (as shown in Fig. 1). The labels have been categorized based on the following conditions:

[3] https://ncert.nic.in/textbook.php.

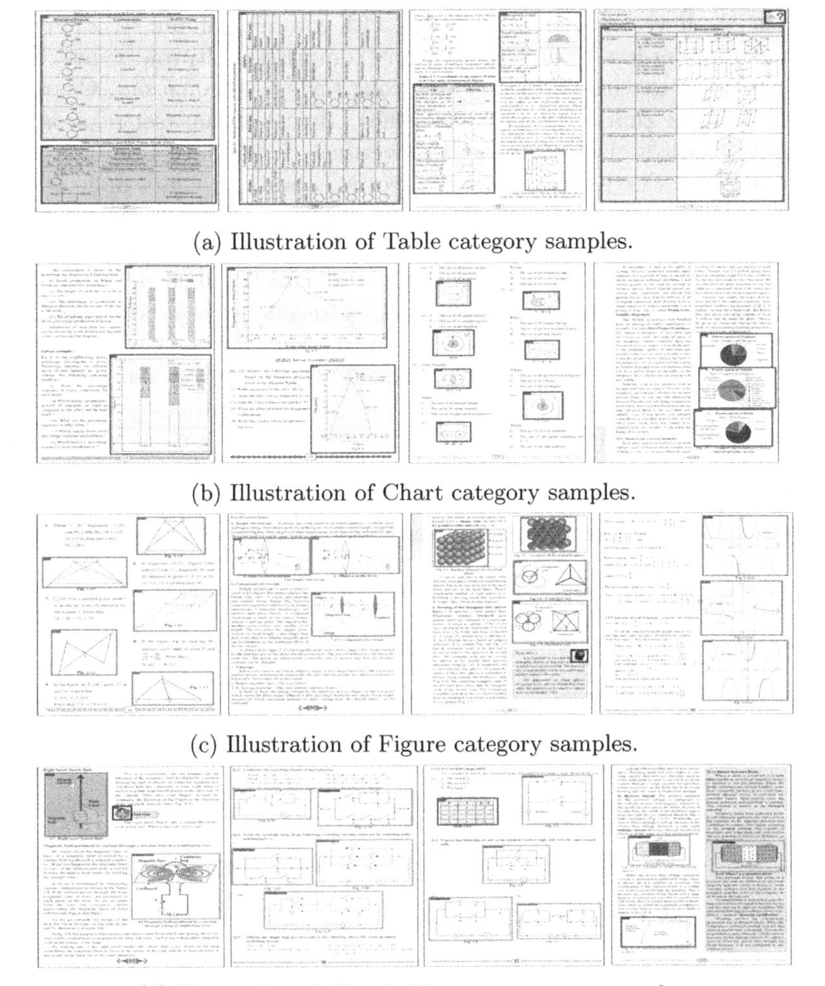

(a) Illustration of Table category samples.

(b) Illustration of Chart category samples.

(c) Illustration of Figure category samples.

(d) Illustration of Circuit Diagram category samples.

Fig. 1. Illustration of label diversification in proposed dataset.

- The label categories cover almost all elements so that reliable and dedicated parsing methods [26,27] can be applied to extract the content or summarize the incised information.
- The dissimilarity between the categories is distinctive so that a visual model can effectively discriminate between the categories and smoothly capture distinctive patterns.
- The categories are abundantly present in such school books.

A total of seven categories have been chosen: Table (T), Charts (C), Equation (E), Image (I), Figure (F), Circuit Diagram (CD), Logo (L). Sample images from some categories are shown in Fig. 1 demonstrating the diversity in label categories. We have considered the NTCs with row and column structure (both borders and border-less). The table category comprises multi-row, multi-column, and oriented tables with images, figures, long texts, and equations. Chart category contains line chart, bar chart, pie chart, venn diagram (set relation), and flow charts. Both horizontal, as well as vertical charts have been considered. The chart categories images can be directly parsed via chart understanding systems. Figures contain geometrical figures with cartesian coordinates-based graphs spanned in single or multiple quadrants, geometrical diagrams involving circles, triangles, rectangles, and other trigonometric figures. Ray diagram, which is popularly used in physics subjects, is considered in this category. Only chemical equations have been considered in the equation category as these are prevalent in NCERT science school books, especially eleven and twelve standards. Semantically, these equations are different than mathematical equations. Some chemical equations consist of chemical structure as well as alphanumeric tokens. Natural images, human images, or any general object images are considered in the image category. We have considered the Logo category, and this is the most prevalent label category present on almost every page image in the corpus.

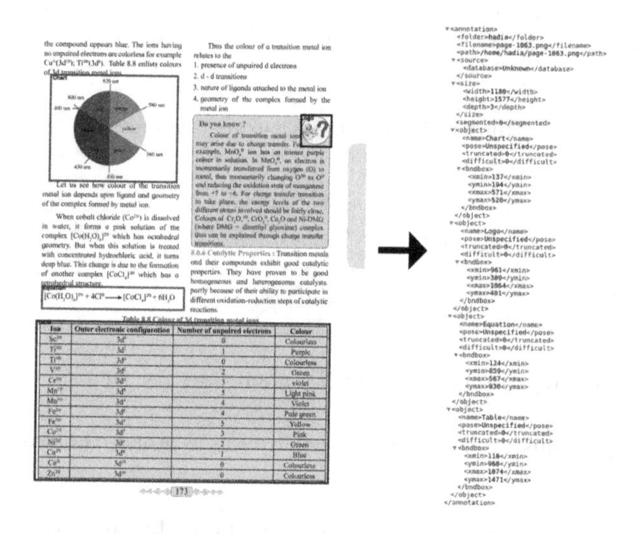

Fig. 2. A sample of NTCs representation and corresponding XML annotations.

3.3 Annotation Method

We took a total of 23 NCERT books. The books include Science and Mathematics subjects from eighth to twelve standards and are in English. The books are originally available in pdf format. So, we have converted these pdf pages into respective image files. We have experimented with off-the-shelf pdf to image converter libraries such as PyMuPDF and pdf2image[4]. Upon evaluation, the results obtained on PyMuPDF showed better results. Also, we observed that the resolution of images is affected (reduced) after the conversion process. So, to improve the overall resolution, we have increased the image resolution by four times by applying a zooming factor of 2. The final images are saved in PNG format. A total of around 5K images are generated. We then manually annotated the bounding boxes around each object type using the Labelimg[5] annotation tool. Annotations are saved in Pascal VOC format as XML files (a sample is shown in Fig. 2). A total of around 10K bounding boxes around various NTCs are generated, with annotation of 1.6k tables, 3.3k logos, 1.5k Images, 2.5k figures, 0.5k charts, and 0.3k circuit diagrams. The dataset is divided as follows: 70% as training, 20% as validation, and 10% as testing.

4 Benchmarking

The performance is evaluated on a total of 5,046 images extracted. For our baselines, we choose the most popular object detection methods. Models are evaluated in terms of mAP (0.5), and mAP (0.5:0.95), where mAP (0.5) is the average precision with an Intersection Over Union (IoU) of 0.5 and mAP (0.5:0.95) is the mean average precision with an IoU between 0.5 to 0.95.

The objective of a DLA task is to extract the pre-defined NTCs units in a multi-unit page images. Consider a document page P is composed of discrete unit set $U = \{U_0, U_1, U_2..U_n\}$, where n is the total number of NTC units, and set $L = \{L_1, L_2, ..., L_7\}$ is the set of 7 labels considered. Each unit U_i is annotated as $(L_j, (x_0, y_0, x_1, y_1))$, where the true label of U_i is L_j (where $j = 1:7$) and its bounding box coordinates are denoted as (x_0, y_0, x_1, y_1). We intend to find an automated estimation function F: (P,L) $\rightarrow$ S such that given the document page image P and the label set L, the prediction set S, providing the predicted labels $\tilde{L}$ and bounding box coordinates for the NTCs identified in P, is computed.

4.1 Models

All models are trained and evaluated on NVIDIA Tesla P100 GPU. We have used the default values of the parameters across all experiments.

- Scaled YOLOv4 [28]: We have experimented with the scaling model of YOLOv4 based on the Cross Stage Partial Network (CSP) approach. The scaling is done on depth, width, resolution, and structure. We have used a fully CSP-sized model YOLOv4-P5 version for our experiments, which is essentially designed for real-time object detection with the best speed and accuracy trade-off. We used publicly available PyTorch-based implementations. Images are preprocessed to a size of 640*640, and a batch size of 8 is maintained. The model was trained for 200 epochs. We use SGD as an

[4] https://pypi.org/project/pdf2image/.
[5] https://github.com/tzutalin/labelImg.

optimizer with a learning rate of 0.01 and momentum of 0.937 with a weight decay of 0.0005.
- YOLOv5[6]: We have used the YOLOv5x model with the CSPDarknet module as a feature extractor for our experiment. We have used PyTorch-based implementations of YOLOv5x. Images are resized to a size of 640*640, and a batch size of 4 is maintained. The model was trained for 200 epochs. As the training time increases, the model gradually converges. We use an SGD optimizer, a learning rate of 0.01, and a weight decay of 0.0005.
- Faster-RCNN [29]: We have experimented with two variants of Faster-RCNN model from detectron 2 (PyTorch) based framework. The two variants FRCNN-A and FRCNN-B utilizes ResNeXt101-FPN and ResNeXt101-C4 backbone architectures as feature extractors. The models were trained for 5000 iterations with batch size of 32. We use SGD optimizer with a learning rate of 0.001. A momentum of 0.9 and a weight decay of 0.001 has been considered. For the FRCNN-A model the anchor scales are taken as 32, 64, 128, 256, 512 and anchor ratios as 0.5, 1.0, 2.0.

4.2 Experimental Setup

The NCERT5K-IITRPR dataset consists of dissimilar and some different NTC label categories, which are not similar to the parallel label distributions in other larger datasets such as COCO, DocBank, or PubLayNet. So, we have utilized layered benchmarking experiments to demonstrate the efficacy and complexity of our dataset. A set of two experiments are devised to perform the benchmarking process based on the following preset: 1) how well the existing single-stage detectors (SSD) and double stage detectors (DSD), which are trained on larger object detection datasets such as the COCO dataset when fine-tuned on NCERT5K-IITRPR dataset, can recognize the label categories; 2) how well the existing detectors (SSD and DSD) can recognize the label categories when trained from scratch on the NCERT5K-IITRPR dataset. The experiments are described below:

- Benchmarking Experiment 1 (BE-1): In this set, mAP(0.5) and mAP(0.5: 0.95) metric is computed of COCO pre-trained models with fine-tuning on NCERT5K-IITRPR dataset.

    • BE-1A: SSD (COCO Pre-trained - Scaled YOLOv4-P5 and YOLOV5x)
    • BE-1B: DSD (COCO Pre-trained - FRCNN-A & FRCNN-B)

- Benchmarking Experiment 2 (BE-2): In this set, mAP(0.5) and mAP(0.5: 0.95) is computed of trained models trained from scratch on our train set and tested on the NCERT5K-IITRPR Test set.

    • BE-2A: SSD (Scaled YOLOv4-P5 and YOLOV5x)
    • BE-2B: DSD (FRCNN-A & FRCNN-B)

[6] https://github.com/ultralytics/yolov5.

Table 4. Quantitative results of BE-1.

Category	BE-1A				BE-1B			
	Scaled-YOLOv4		YOLOv5x		FRCNN-A		FRCNN-B	
	mAP (0.5)	mAP (0.5:0.95)	mAP (0.5)	mAP (0.5:0.95)	mAP (0.5)	mAP (0.5:0.95)	mAP (0.5)	mAP (0.5:0.95)
Table	0.991	0.889	0.988	0.879	0.983	0.831	0.982	0.857
Chart	0.904	0.712	0.899	0.691	0.600	0.4577	0.569	0.440
Figure	0.888	0.689	0.855	0.672	0.887	0.625	0.886	0.622
Image	0.875	0.690	0.858	0.676	0.805	0.587	0.807	0.595
Equation	0.906	0.629	0.899	0.629	0.645	0.346	0.566	0.286
Circuit diagram	0.943	0.796	0.924	0.815	0.517	0.404	0.726	0.575
Logo	0.983	0.758	0.987	0.770	0.984	0.726	0.982	0.710
Average	**0.927**	**0.738**	0.916	0.733	0.774	0.568	0.788	0.583

4.3 Results and Analysis

The quantitative results are shown in Table 4 and Table 5. Overall the performance of single-stage detectors in the BE-1 setting is better, specifically, the SSDs performance (BE-1A) is better than the DSDs (BE-1B). The YOLOv4-P5 model achieves the best overall mAP(0.5) and mAP(0.5:0.95) scores. The model generates tighter bounding boxes around smaller objects with higher confidences, as shown in Fig. 3(b). The smaller, larger, and complex structured tables are also accurately detected. Although the model skips over some equation categories, it classifies better than the other models. The fine-tuned YOLOv5x model also shows fair accuracy but lags in inference speed compared to YOLOv4-P5. The model can also detect small objects but fails in classifying objects with close similarities such as circuit diagrams and figures, figures and charts, etc., which share common structural elements. It also fails to detect chart objects, as shown in Fig. 3(a). However, in the BE-2 setting, i.e., without using the pre-trained weights, both SSDs and DSDs models performance is lowered as they fail to detect smaller and complex objects, especially the labels with inter-class similarities such as chart and figure as shown in Fig. 3(c, d).

In the BE-2 setting, the YOLOv4-P5 displays low mAP scores for detecting equations and charts categories. The major miss-classifications are in fine-grained categories with the highest inter-class similarities, such as figures, images, and circuit diagrams. For tables and logos, the classification and detection accuracy are comparatively higher. The model obtains tighter bounding boxes for correctly classified objects and displays higher confidence scores. In BE-2A the YOLOv5x, model shows considerable improvement in classifying Circuit Diagrams and demonstrates comparable performance on tables and logos. Compared to YOLOv4-P5 (without pre-trained), this model detects more equations. Although this model detects more objects, it miss-classifies charts and figures besides skipping some objects.

Table 5. Quantitative results of BE-2.

Category	BE-2A				BE-2B			
	Scaled-YOLOv4		YOLOv5x		FRCNN-A		FRCNN-B	
	mAP (0.5)	mAP (0.5:0.95)	mAP (0.5)	mAP (0.5:0.95)	mAP (0.5)	mAP (0.5:0.95)	mAP (0.5)	mAP (0.5:0.95)
Table	0.992	0.885	0.988	0.873	0.945	0.728	0.916	0.687
Chart	0.792	0.599	0.818	0.592	0.572	0.389	0.522	0.286
Figure	0.861	0.667	0.884	0.683	0.884	0.563	0.853	0.517
Image	0.902	0.692	0.861	0.660	0.819	0.521	0.719	0.408
Equation	0.789	0.506	0.926	0.595	0.559	0.211	0.484	0.179
Circuit diagram	0.923	0.768	0.927	0.805	0.544	0.352	0.646	0.418
Logo	0.987	0.753	0.985	0.755	0.981	0.704	0.977	0.668
Average	0.892	0.695	**0.913**	**0.709**	0.758	0.495	0.731	0.452

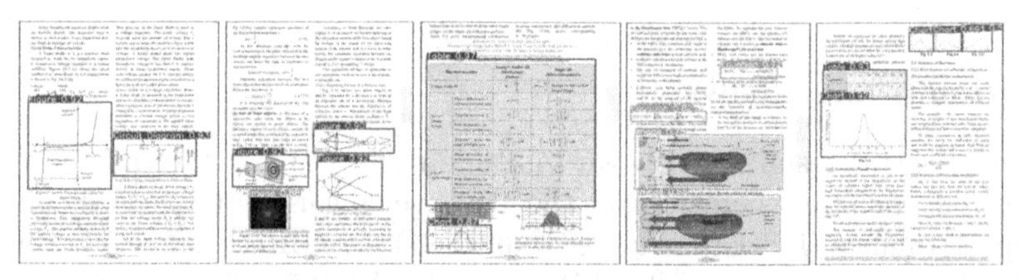

(a) Illustration of BE-1 (YOLOv5x) sample results.

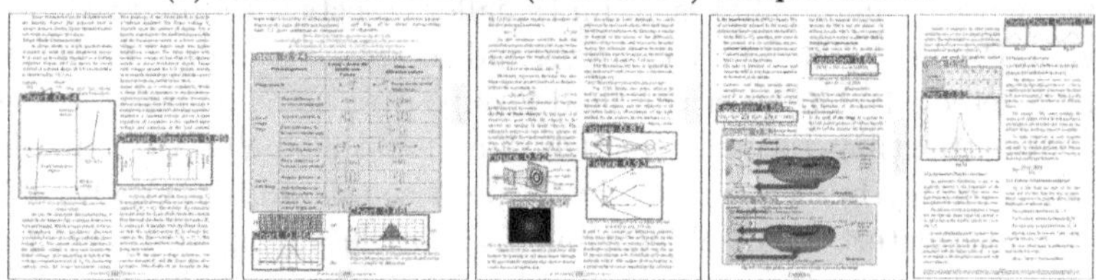

(b) Illustration of BE-1 (YOLOv4-P5) sample results.

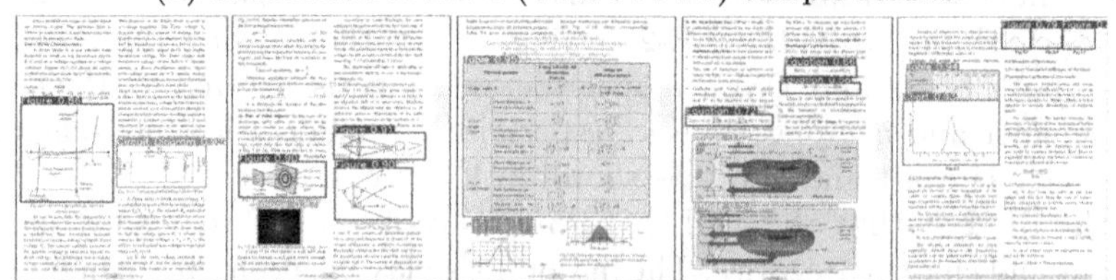

(c) Illustration of BE-2 (YOLOv5x) sample results.

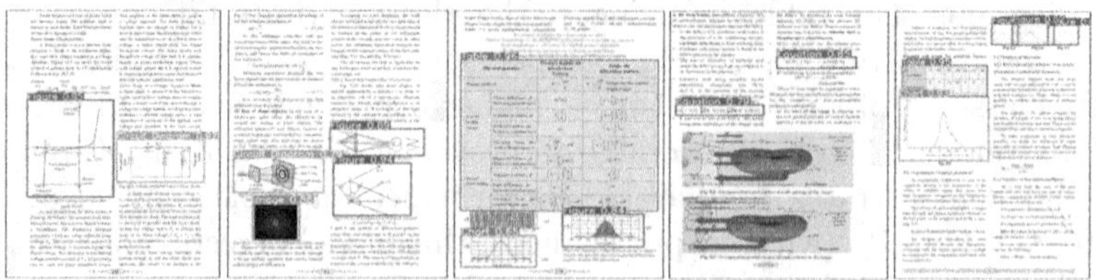

(d) Illustration of BE-2 (YOLOv4-P5) sample results.

Fig. 3. Sample prediction results of top performing models.

The experiments highlight the usability of this dataset in fine-tuning scenario. It highlights that the fine-tuning with the training set is adequate. Also, it highlights the complexity within label categories.

5 Conclusion

To improve the accessibility of and favor the development of effective parsing methods for NCERT school books, we present NCERT5K-IITRPR, with more than 5000 multi-object document pages, built using manual annotation with diverse label categories. The dataset contains fine-grained and diverse label categories. We have performed the benchmarking of different detectors and analyzed the effectiveness of NCERT5K-IITRPR. Experimental results demonstrate the complexity of the proposed dataset and its fine-tuning necessity. NCERT5K-IITRPR will allow the development of book parsers and improve accessibility scenarios for BVIP students.

In the future, we would like to extend the dataset with additional labels (such as mathematical equations) and we would like to analyze the performance of domain-related pre-trained models i.e. models pre-trained on larger DLA datasets. Also, we plan to develop a holistic and end-to-end model for data extraction and summarization of NCERT and similar books.

Acknowledgement. This research is supported by the DST under CSRI grant DST/CSRI/2018/234.

References

1. Vignesh, D., Gupta, N., Kalaivani, M., Goswami, A.K., Nongkynrih, B., Gupta, S.K.: Prevalence of visual impairment and its association with vision-related quality of life among elderly persons in a resettlement colony of Delhi. J. Fam. Med. Prim. Care **8**(4), 1432 (2019)
2. Gothwal, V.K., Kodavati, K., Subramanian, A.: Life in lockdown: impact of COVID-19 lockdown measures on the lives of visually impaired school-age children and their families in India. Ophthalmic Physiol. Optics. **42**(2), 301–310 (2021)
3. Tomar, G., Garg, V.: Making steam accessible for inclusive classroom. Glob. J. Enterp. Inf. Syst. **12**(4), 94–101 (2020)
4. Dey, S., Vidhya, Y., Bhushan, S., Neerukonda, M.: Creating an accessible technology ecosystem for learning science and math: a case of visually impaired children in Indian schools (2019)
5. Jelas, Z.M., Ali, M.M.: Inclusive education in Malaysia: policy and practice. Int. J. Inclusive Educ. **18**(10), 991–1003 (2014)
6. Lamichhane, K.: Teaching students with visual impairments in an inclusive educational setting: a case from Nepal. Int. J. Incl. Educ. **21**(1), 1–13 (2017)
7. Asamoah, E., Ofori-Dua, K., Cudjoe, E., Abdullah, A., Nyarko, J.A.: Inclusive education: perception of visually impaired students, students without disability, and teachers in Ghana. SAGE Open **8**(4), 2158244018807791 (2018)
8. Bansal, A., Garg, H., Jadhav, N., Kumar, S., Balakrishnan, M.: RAVI: reading assistant for visually impaired

9. Gao, L., Yi, X., Jiang, Z., Hao, L., Tang, Z.: ICDAR2017 competition on page object detection. In: 2017 14th IAPR International Conference on Document Analysis and Recognition (ICDAR), vol. 1, pp. 1417–1422. IEEE (2017)
10. Yi, X., Gao, L., Liao, Y., Zhang, X., Liu, R., Jiang, Z.: CNN based page object detection in document images. In: 2017 14th IAPR International Conference on Document Analysis and Recognition (ICDAR), vol. 1, pp. 230–235. IEEE (2017)
11. Siegel, N., Lourie, N., Power, R., Ammar, W.: Extracting scientific figures with distantly supervised neural networks. In: Proceedings of the 18th ACM/IEEE on Joint Conference on Digital Libraries, pp. 223–232 (2018)
12. Kavasidis, I., Pino, C., Palazzo, S., Rundo, F., Giordano, D., Messina, P., Spampinato, C.: A saliency-based convolutional neural network for table and chart detection in digitized documents. In: Ricci, E., Rota Bulò, S., Snoek, C., Lanz, O., Messelodi, S., Sebe, N. (eds.) ICIAP 2019. LNCS, vol. 11752, pp. 292–302. Springer, Cham (2019). https://doi.org/10.1007/978-3-030-30645-8_27
13. Saha, R., Mondal, A., Jawahar, C.V.: Graphical object detection in document images. In: 2019 International Conference on Document Analysis and Recognition (ICDAR), pp. 51–58. IEEE (2019)
14. Singh, M., Goyal, P.: DeepDoT: deep framework for detection of tables in document images. In: Singh, S.K., Roy, P., Raman, B., Nagabhushan, P. (eds.) CVIP 2020. CCIS, vol. 1377, pp. 421–432. Springer, Singapore (2021). https://doi.org/10.1007/978-981-16-1092-9_35
15. Singh, M., Goyal, P.: ChartSight: an automated scheme for assisting visually impaired in understanding scientific charts. (2021)
16. Yepes, A.J., Zhong, X., Burdick, D.: ICDAR 2021 competition on scientific literature parsing. arXiv preprint arXiv:2106.14616 (2021)
17. Zhong, X., Tang, J., Yepes, A.J.: PubLayNet: largest dataset ever for document layout analysis. In: 2019 International Conference on Document Analysis and Recognition (ICDAR), pp. 1015–1022. IEEE (2019)
18. Mondal, A., Lipps, P., Jawahar, C.V.: IIIT-AR-13K: a new dataset for graphical object detection in documents. In: Bai, X., Karatzas, D., Lopresti, D. (eds.) DAS 2020. LNCS, vol. 12116, pp. 216–230. Springer, Cham (2020). https://doi.org/10.1007/978-3-030-57058-3_16
19. Li, M., et al.: DocBank: a benchmark dataset for document layout analysis. arXiv preprint arXiv:2006.01038 (2020)
20. Li, M., Cui, L., Huang, S., Wei, F., Zhou, M., Li, Z.: TableBank: table benchmark for image-based table detection and recognition. In: Proceedings of the 12th Language Resources and Evaluation Conference, pp. 1918–1925. (2020)
21. Göbel, M., Hassan, T., Oro, E., Orsi, G.: ICDAR 2013 table competition. In 2013 12th International Conference on Document Analysis and Recognition, pp. 1449–1453. IEEE (2013)
22. Zheng, X., Burdick, D., Popa, L., Zhong, P., Wang, N.X.R.: Global table extractor (GTE): a framework for joint table identification and cell structure recognition using visual context. Winter Conference for Applications in Computer Vision (WACV) (2021)
23. Zhong, X., ShafieiBavani, E., Jimeno Yepes, A.: Image-based table recognition: data, model, and evaluation. In: Vedaldi, A., Bischof, H., Brox, T., Frahm, J.-M. (eds.) ECCV 2020. LNCS, vol. 12366, pp. 564–580. Springer, Cham (2020). https://doi.org/10.1007/978-3-030-58589-1_34
24. Shahab, A., Shafait, F., Kieninger, T., Dengel, A.: An open approach towards the benchmarking of table structure recognition systems. In: Proceedings of the 9th IAPR International Workshop on Document Analysis Systems, pp. 113–120 (2010)

25. Fang, J., Tao, X., Tang, Z., Qiu, R., Liu, Y.: Dataset, ground-truth and performance metrics for table detection evaluation. In: 2012 10th IAPR International Workshop on Document Analysis Systems, pp. 445–449. IEEE (2012)
26. Morris, D., Tang, P., Ewerth, R.: A neural approach for text extraction from scholarly figures. In: 2019 International Conference on Document Analysis and Recognition (ICDAR), pp. 1438–1443. IEEE (2019)
27. Luo, J., Li, Z., Wang, J., Lin, C.Y.: ChartOCR: data extraction from charts images via a deep hybrid framework. In: 2021 IEEE Winter Conference on Applications of Computer Vision (WACV). The Computer Vision Foundation (2021)
28. Wang, C.Y., Bochkovskiy, A., Liao, H.Y.M.: Scaled-YOLOv4: scaling cross stage partial network. In: Proceedings of the IEEE/CVF Conference on Computer Vision and Pattern Recognition (CVPR), pp. 13029–13038 (2021)
29. Ren, S., He, K., Girshick, R., Sun, J.: Faster R-CNN: towards real-time object detection with region proposal networks. Adv. Neural. Inf. Process. Syst. **28**, 91–99 (2015)

Poster Session 1

ReadOCR: A Novel Dataset and Readability Assessment of OCRed Texts

Hai Thi Tuyet Nguyen[1(✉)], Adam Jatowt[2], Mickaël Coustaty[3], and Antoine Doucet[3]

[1] Posts and Telecommunications Institute of Technology, Ho Chi Minh, Vietnam
tuyethai@ptithcm.edu.vn

[2] Department of Computer Science, University of Innsbruck, Innsbruck, Austria
adam.jatowt@uibk.ac.at

[3] L3i, La Rochelle University, La Rochelle, France
{mickael.coustaty,antoine.doucet}@univ-lr.fr

Abstract. Results of digitisation projects sometimes suffer from the limitations of optical character recognition software which is mainly designed for modern texts. Prior work has examined the impact of OCR errors on information retrieval (IR) and downstream natural language processing (NLP) tasks. However, questions remain open regarding the actual readability of the OCRed text to the end users, especially, considering that traditional OCR quality metrics consider only syntactic or surface features and are quite limited. This paper proposes a novel dataset and conducts a pilot study to investigate these questions.

Keywords: Readability assessment · OCR errors · Hierarchical attention network · BERT

1 Introduction

Considerable efforts have been devoted to transforming historical documents into electronic form for better preservation and easier access. Applying modern OCR technologies on old documents often leads to noisy outputs which negatively affect reading, retrieving, and other processes in digitized collections [2,18]. Whereas there have been many studies conducted regarding the impact of OCR errors on IR and NLP tasks [4,14,17], the influence on the reading ease remains still an open question. Usually, common metrics such as word error rate (WER) and character error rate (CER) are used to validate the quality of digitised text, implicitly including its readability.

Text readability (aka. reading ease) is however affected by many factors, such as lexical sophistication, syntactic complexity, discourse cohesion, and background knowledge [6]. Several readability formulas and machine learning techniques have been suggested in the past to assess an input text's readability. However, all of these approaches are designed to work on clean text, i.e., one

© Springer Nature Switzerland AG 2022
S. Uchida et al. (Eds.): DAS 2022, LNCS 13237, pp. 479–491, 2022.
https://doi.org/10.1007/978-3-031-06555-2_32

without any OCR errors. In other words, studies on readability of digitized texts are largely missing.

In this paper, we propose a novel dataset and conduct several experiments on this data in order to answer question how OCR errors impact readability of texts. The contributions of our work are as follows:

1. We introduce a novel dataset for readability assessment of OCRed texts. Studies on this dataset can help to understand the impact of OCR errors on reading. Additionally, based on the dataset, one can have more clues to decide whether the quality of target OCRed texts is acceptable for reading or not. Future systems can train and test their readability assessment models based on our proposed corpus. The corpus is publicly available and freely accessible[1].
2. We investigate the following: (i) the relation between readability reduction (i.e., the reduction of an original text's readability score due to OCR process) and standard error rates like WER, CER; (ii) the relation between the readability reduction and the readability of an original text; (iii) the impact of corrupted lexical, grammatical words, and two typical OCR error types on the readability.
3. Finally, we apply state-of-the-art methods for the readability assessment of noisy texts in our dataset.

2 Proposed Dataset

2.1 Document Collection

Several corpora have been proposed for studying text readability including Weebit [20], OneStopEnglish [19], Newsela [21], and CommonLit[2]. Whereas Weebit [20] and Newsela [21] classify texts into specific classes according to the age group for which the text is designed, OneStopEnglish [19] includes texts of three reading levels: beginner, intermediate, and advanced. Instead of assigning age-dependent classes or broad reading levels like the three above datasets, CommonLit provides actual readability scores for each text.

Our objective is to observe the effects of OCR errors at different reading ease scores, hence, we investigate two relations: (1) one between the readability of OCRed texts and their error rate metrics, and (2) one between the readability of OCRed texts and the readability of their original (non-OCRed) versions. In order to do that, we need a corpus composed of both original texts labelled with their readability scores and of their OCRed versions which are corrupted to different degrees and are also labelled with readability scores. Since no such dataset exists, we have decided to create and share one.

According to our observation, a small error rate may not affect a coarse-grained reading level of a corrupted text (i.e., OCRed text). For example, changing WER from 10% to 15% of the same document may not affect the reading

[1] https://tinyurl.com/ReadOCR.
[2] https://www.kaggle.com/c/commonlitreadabilityprize/data.

label of that document (e.g., the document may be still judged as being of the intermediate difficulty level). By using specific fine-grained readability scores rather than coarse-grained broad labels, we can more reliably compute correlation levels for our analysis. Based on this requirement, we have used and adapted an existing readability dataset that provides the actual reading ease scores instead of the broad readability classes which the texts belong to. Among the popular readability datasets, only the CommonLit dataset meets our requirements, hence we utilized it in our experiments. This corpus contains literary passages from different time periods and their fine-grained reading ease scores. It includes 2,834 texts collected from several sources, such as Wikipedia[3], Africanstorybook[4], Commonlit[5], etc.

2.2 Proposed Text Corpus

Since we wanted to manually assess readability of OCRed texts with different error degrees, we have sampled a subset from the CommonLit dataset. We randomly chose 161 files and added noise to the data at the word and document levels to mimic varying quality of OCR under different word error rates. The original (non-corrupted) files are also included in the proposed corpus along with their 483 corrupted versions. In total, there are 644 files whose detailed information is indicated in Table 2. This corpus is split into two parts (84% for training and 16% for testing) for conducting experiments discussed in Sect. 3.

The way we corrupted the original texts at word-level and document-level is described below:

- Word-level: we first picked all words belonging to a common English dictionary, and we randomly replaced some of their characters with plausible characters mimicking OCR errors. The plausible characters have been deduced from a confusion matrix which was created based on the alignment data corpus which is publicly provided through the ICDAR 2017 competition for post-OCR correction [5]. This corpus has been compiled from nine sources, mainly from the National Library of France and the British Library. It also represents a relatively wide time range. Its documents have different degradation levels under independent preservation practices. All of these characteristics make this corpus representative for typical OCR errors. Examples of plausible characters are illustrated in Table 1.
- Document-level: According to [1], average WERs range from 1% to 10%. Yet, some datasets have higher WERs [12]. In order to study the readability on a wider range of WERs, we corrupted original data with six WER levels ranging from 7% to 32%, with a step of 5%. In particular, we randomly chose words from each document based on the required WER. Each chosen word was then replaced by its randomly selected corrupted version that was generated at the word level.

[3] https://en.wikipedia.org.
[4] https://www.africanstorybook.org.
[5] https://www.commonlit.org.

Table 1. Examples of plausible characters (i.e., OCRed chars) of the original characters (i.e., GT chars) [13].

GT chars	OCRed chars
a	u, n, e, i,
c	e, o,
e	o, c,
f	t, l, i,
h	b, i, n,
o	e, a, c,

Table 2. The numbers of files and tokens of the constructed corpus and its split parts.

Stats	Parts	Original	Corrupted	Total
Files	All	161	483	644
	Train	135	405	540
	Test	26	78	104
Tokens	All	27,809	83,670	111,479
	Train	23,320	70,170	93,490
	Test	4,489	13,500	17,989

After creating the noisy versions of the original texts, we asked three volunteers[6] to read all the texts and assign a score for each noisy text to indicate how understandable it is in comparison with its original text. Similar to classical WER values we wanted to obtain fine-grained scores on the scale ranging from 0 to 100, hence, the annotators were asked to assign their scores within this range.

In order to assess the agreement on continuous data (i.e. detailed scores given by the annotators) we applied intra-class correlation coefficient - ICC [15,16], which ranges from 0 to 1. Among 6 cases of ICC, the best suited one for our case is ICC2k. ICC2k is the reliability estimated for the means of k raters with each text target rated by a random sample of k raters. We have obtained an ICC2k value of 0.865 which indicates a good reliability in our corpus according to a guideline of selecting an intraclass correlation [10].

In addition to the assigned scores, the annotators made also notes to justify their scores. By closely examining the obtained comments, we found that the annotators attempted to understand the texts by guessing the meaning of errors or sometimes by simply ignoring them. In general, the scores were dependent on how hard it was to guess the meaning of the erroneous words, how these affected the annotators' overall understanding of the texts, the number of these errors, and how much the errors disturbed the reading flow.

[6] The three annotators are sophomores, two of them are law students, and one is an information technology student.

Table 3. Pearson correlation coefficients between *ComScores* and *ReadScores* according to WER levels that are illustrated in Fig. 2 (except level 0).

WER	0.07	0.12	0.17	0.22	0.27	0.32
Correlation	0.204	0.080	0.182	0.319	0.267	0.003

2.3 Dataset Analysis

To answer our leading research question on the relation of readability and error rates, we used both the given scores (denoted as *ReadScore*) and the reading difficulty or reduction (denoted as *DiffScore*) computed as the difference between 100 and *ReadScore*. Min-max normalization was applied on the scores given by each annotator to ensure score comparability among the annotators.

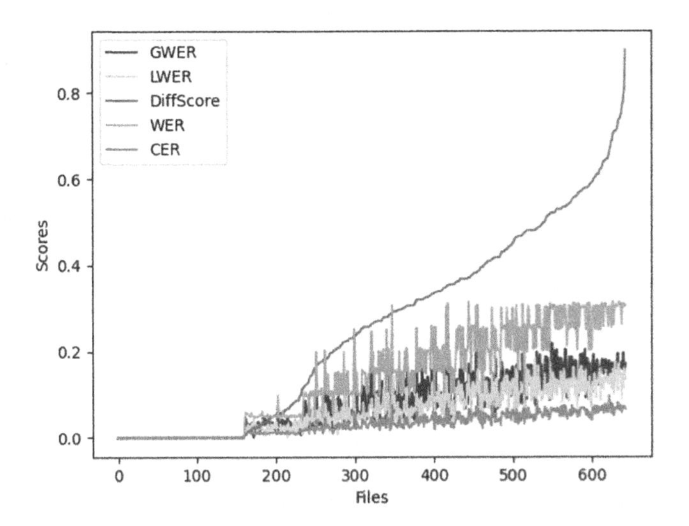

Fig. 1. Grammatical word error rate (GWER), lexical word error rate (LWER), WER, CER, and the *DiffScore* of the whole corpus whose documents are ordered on X-axis by their *DiffScores*. Pearson correlation coefficients between the other metrics and the *DiffScore* are 0.902, 0.910, 0.941, and 0.931, respectively.

Our analysis on the corrupted corpus shows that the *DiffScore* of noisy texts correlates well with the common error rates. However, this is not true for all WER values. For example, Fig. 1 shows that WER remains around 0.32 even when the *DiffScore* increases further.

Regarding the relation between the *ReadScore* and the original reading ease CommonLit scores (denoted as *ComScore*), we examined it with respect to six levels of WER: 0.07, 0.12, 0.17, 0.22, 0.27, and 0.32. Figure 2 illustrates the relations between *ComScores* and *ReadScores* which are grouped by all WER levels. The detailed correlation values of the examined relations according to six

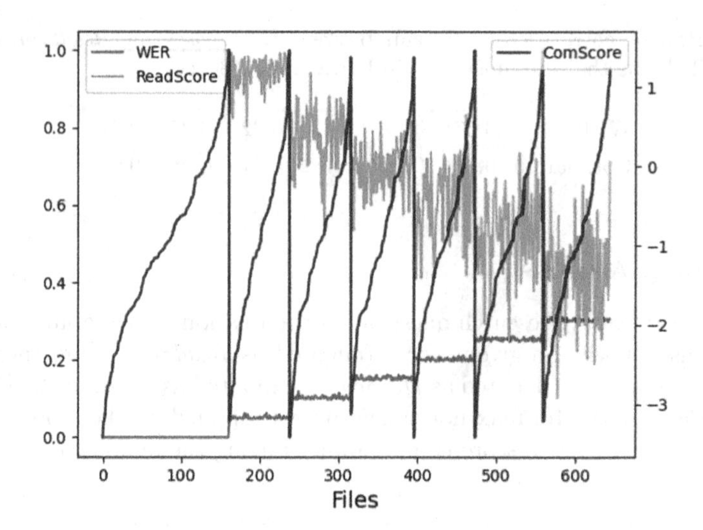

Fig. 2. *ComScores* and *ReadScores* of the whole corpus. The left Y axis shows *ReadScores* and WER, the right Y axis indicates *ComScores*. These scores are grouped according to all WER levels.

levels of WER are presented in Table 3. We expect that the more readable the original text is, the more readable its noisy version should be. In other words, the higher the *ComScore* is, the larger the *ReadScore* should be, at a given WER. Yet, in contrast to our expectation, the correlations between these scores fluctuate along with WER values. When texts are relatively noisy (e.g., WER of 0.32), it is probably difficult to understand them regardless of the reading ease scores of their non-corrupted (original) versions. The highest WER for our dataset is 0.32, however in reality OCRed texts (especially historical ones) could be even noisier. Further study is needed to investigate higher corruption levels.

Another analysis on the proposed corpus is to study the impact of the corrupted lexical and grammatical words on readability. Lexical words include nouns, verbs, and adjectives, while grammatical words (typically considered as stop words) consist of articles, pronouns, and conjunctions. We expect that corrupted lexical words have a high effect on readability than corrupted grammatical words. The correlation between the *DiffScore* and the error rate of the lexical words is a bit higher than the one for grammatical words, with 0.910 and 0.902, respectively. The difference between these two correlations is not so high since the readability of noisy texts relies not only on vocabulary but also on the reading flow and overall text understanding.

The next study on the ReadOCR corpus is to observe the effect of two common types of OCR errors, namely, *real-word* and *non-word* errors, on readability. Whereas *non-word* errors do not exist in a dictionary, *real-word* errors are dictionary entries but they are used in the incorrect context. With the above properties, *real-word* errors often mislead readers. However, according to our statistics, the rate of *real-word* errors correlates less with the *DiffScore* than that of *non-word* errors, with correlation values of 0.871 and 0.926, respectively. A possible reason for this lower correlation is that around a quarter of *real-word* errors in our corpus are stop words which are less important than lexical words.

3 Readability Assessment

In addition to the studies on relations between WER, CER, and the *ReadScore* or the *DiffScore*, we apply several readability assessment methods to predict the readability reduction and compare them against the results obtained so far.

3.1 Methods

Text readability scores can be assessed by traditional readability formulas and machine learning techniques. We utilize both of them to compute the readability reduction. The traditional readability formulas are used to assess the readability scores of corrupted and non-corrupted texts, and then to compute readability reduction. These formulas mainly focus on some lexical and syntactic information such as word length, sentence length, and word difficulty. As for machine learning techniques, we apply two state-of-the-art approaches to predict readability reduction of corrupted texts. One approach relies on a hierarchical attention mechanism to capture syntactic and structural information, while the other transfers knowledge of pre-trained models to predict the target scores.

Traditional Reading Formulas. We measured the readability of documents in our corpus using two popular reading formulas, Flesch-Kincaid grade level (FKGL) [9] and Dale-Chall readability formula (DCRF) [7]. FKGL and DCRF represent, respectively, the number of years of education or the educational grade levels generally required to understand a given text. Whereas FKGL depends on a sentence's length and a number of syllables per word, DCRF relies on a sentence's length and a list of 3,000 words that fourth-grade American students could reliably understand. A word that does not exist in the list is deemed as a difficult word. The formulas for these metrics are given in Eqs. (1) and (2). Using FKGL and DCRF, we first calculated the readability of each original text and its noisy versions. These scores were then used to compute the readability reduction as in Eq. (3).

$$FKGL = 0.39(\frac{totalWords}{totalSentences}) + 11.8(\frac{totalSyllables}{totalWords}) - 15.59 \tag{1}$$

$$DCRF = 15.79(\frac{difficultWords}{totalWords}) + 0.0496(\frac{totalWords}{totalSentences}) \tag{2}$$

$$reduction = \frac{originalRead - noisyRead}{originalRead} \tag{3}$$

Hierarchical Attention Network (HAN). HAN is one of the best performing approaches for readability assessment. In our experiment, we utilize it to predict readability reduction of texts.

HAN's architecture [22] contains 4 parts: a word encoder, a word-level attention layer, a sentence encoder, and a sentence-level attention layer. Words of each sentence are embedded into word vectors through an embedding matrix. Bidirectional GRU is then applied to encode contextual information from both directions of words. Since each word may contribute differently to the representation of the sentence, the word attention mechanism is used to extract informative words and aggregate their representation to form a sentence vector. Given the sentence vectors, they can be used to represent a document vector in a similar way. The encoded sentence is the bidirectional hidden state generated by bidirectional GRU. The attention mechanism is used again to compute the importance of the sentences in the document. The resulting vector becomes the final representation of the document and can be used as features for the target task. The hierarchical attention mechanism is expected to better capture the document structure and lead to accurate predictions of readability reduction.

For the experiments, we split documents into sentences and tokenize each sentence using NLTK library [3]. Each document contains up to 80 sentences, each of which has a maximum of 70 words. The model uses GoogleNews word2vec as word embeddings and is trained with a batch size of 16 and 50 epochs.

Transformer. Martinc *et al.* [11] reported positive results by transferring the knowledge of pre-trained BERT models [8] for predicting text readability of multiple corpora. This approach leverages the pretrained neural language model for the prediction task. In particular, a fully connected layer is put on the top of the pretrained model. The whole model is fine-tuned on a new data to predict readability reduction. An obvious shortcoming of fine-tuning BERT model is that the model cannot handle long documents as the maximum sequence length is limited to 512 tokens. Fortunately, the texts in our corpus are relatively short, therefore, they are not affected by this issue.

We utilized the pretrained BERT model with 12 layers of size 768 and 12 self-attention heads, i.e., bert-base-uncased model. The model was then fine-tuned with a linear layer, as well as with the same batch size, and the same number of epochs as the HAN model.

3.2 Experimental Results

We use our corpus to analyze the performance of the above methods. Since the corpus size is limited, as mentioned in Sect. 2.2, we split the dataset into two parts: train and test. Then k-fold cross validation is applied ($k = 5$ in our

Table 4. MSE and correlations between the *DiffScore* and DCRF reduction (i.e., DCRFRed), FKGL reduction (i.e., FKGLRed), BERT's prediction, HAN's prediction, CER, WER on the test data.

Scores	MSE	Pearson
DCRFRed	0.014	0.863
FKGLRed	0.129	−0.380
BERT prediction	**0.003**	0.960
HAN prediction	0.012	0.854
CER	0.085	0.945
WER	0.026	**0.967**

experiment) on the training set. The best model of each method is evaluated on the same test part. More details about the data splits are given in Table 2.

Mean Squared Error (MSE) is a common metric used for regression. We utilize it to validate our models by computing MSE between the *DiffScore* and the predicted values. Besides MSE, we plot the predicted values together with the reading difficulty scores (Figs. 3a and 3b) and compute the correlations shown in Table 4.

We notice that the BERT model is better than the HAN model in both MSE and correlation when comparing them on the same test data. The BERT model has a smaller MSE than the HAN model (0.003 vs. 0.012). The *DiffScore* has also a stronger correlation with the BERT model's prediction than the HAN model's (0.960 vs. 0.854). This is expected, as HAN is good at capturing document structure whereas BERT is a language model and focuses on text semantics. The sub-word segmentation of the BERT tokenizer also helps alleviating the out-of-vocabulary problem in corrupted texts. Moreover, BERT is more robust to noise when fine-tuning on noisy corpus.

MSE, correlation, and line plots are also used to compare performance of other traditional readability scores (i.e., FKGL, DCRF) as well as error rates (i.e., WER, CER) in readability assessment on corrupted texts. In particular, we computed MSE and correlation between the *DiffScore* and the other scores, which are reported in Table 4. All the computed correlation coefficients in this table are statistically significant with p-value under 0.05. Moreover, we plotted all scores according to HAN and BERT models, i.e., Figs. 3a and 3b, respectively. The results reveal that WER has the strongest correlation with the *DiffScore* whereas both MSE and line plots support the best performance of BERT model. In fact, MSE of WER and the *DiffScore* is approximately 9 times higher than that of BERT predictions and the *DiffScore*. Figures 3a and 3b also indicate that the resulting predictions of BERT model are much closer to the *DiffScore* than the error rate lines.

Regarding two traditional readability scores, FKGL reduction seems to be ineffective with a high MSE and negative correlation, DCRF reduction has comparable MSE to WER but its correlation is much lower than other scores.

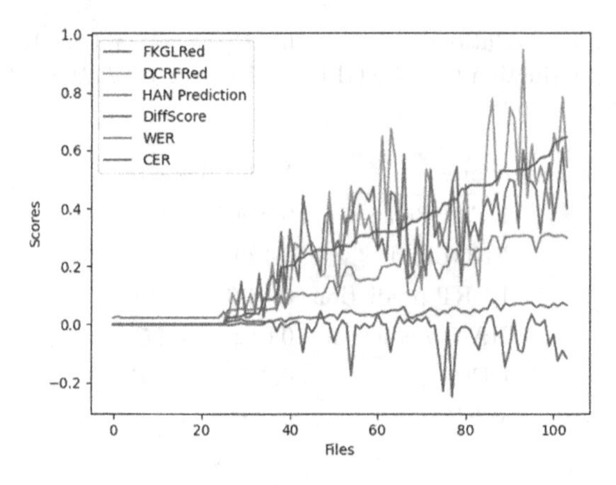

(a) Predictions of HAN model.

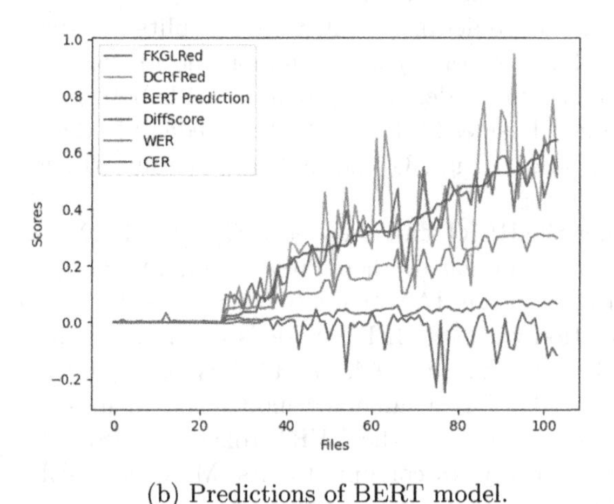

(b) Predictions of BERT model.

Fig. 3. Different scores in assessing readability reduction of the test data: traditional readability scores (FKGLRed as FKGL reduction, DCRFRed as DCRF reduction); error rates (WER, CER); reading difficulty or reduction as *DiffScore*; predictions of HAN and BERT models denoted as HAN prediction and BERT prediction, respectively.

In terms of DCRF, it depends on a list of easy words used (effectively assuming all other words as difficult). Since noisy texts contain many such words, their DCRFs are higher than DCRFs of their original texts. In terms of FKGL, corrupted texts often contain more words than their original version since OCR easily recognizes noise as punctuation, thus a tokenizer separates them into some words. When a number of syllables remains almost unchanged while the number of words increases, FKGL of a corrupted word becomes often lower than that of its original one. It should be noted that the increase of the total number of words is lower than that of difficult words.

Some examples of readability scores on different corruption levels (different error rates) of the same original text are illustrated in Table 5. Regarding two conventional scores, i.e., FKGL and DCRF, FKGL does not work effectively with the reduction as it obtains negative value due to the above-mentioned reasons (e.g., noisy texts have more words than their original one); DCRF exhibits some positive correlation with the *DiffScore* since it considers number of difficult words, and noisy words are often difficult ones. Similar to DCRF, error rates correlate well with the *DiffScore*; Nonetheless, readability is affected by not only the number of difficult words, the number of errors but also by other factors. BERT model implicitly takes into account many of such factors and hence gives good scores, which are closer to the *DiffScore*.

Table 5. Examples of different readability scores for the different corrupted versions of the same text.

Text	Radiosurgery is surgery using radiation, that is, the deitruction of precisely selected areas of lissue using ionizing radiation ra1her than excision with a blade	Radiosurgery uts sur ery using radiation, that is, the des1ruction of precisély selected areat of tissul using ionizing rndiation rather than excision with n blade	Radiosurgery is surgery using radiation, ihat is, • he destruction of precisely selectd areas ol tissue using ionizing radiation rather than excision with a blade
FKGL reduction	−0.004	−0.089	0.01
DCRF reduction	0.01	0.13	0.119
HAN prediction	0.023	0.289	0.246
BERT prediction	0.098	0.414	0.347
WER	0.048	0.259	0.2
CER	0.008	0.041	0.034
DiffScore	0.023	0.48	0.368

4 Conclusions

Our paper is the first work on the topic of readability assessment of OCRed texts. We provide a novel dataset and analyze the impact of OCR errors on readability as well as test two traditional measures and two SOTA baselines on our dataset. Among interesting findings, we observe that WER highly correlates with the reading difficulty. Nonetheless, the best BERT model has a smaller MSE and its prediction is much closer to the *DiffScore* than WER. By using this kind of model, one could potentially estimate the readability reduction of OCRed texts without having access to the original versions of those texts. Furthermore, the impact of the corrupted lexical words has been found to be not much higher than that of corrupted grammatical words since readability of noisy texts is additionally affected by other factors. Besides mis-recognized characters, layout errors may cause serious problems on text readability. In the next step of our future work, we will consider such errors.

Acknowledgements. This work has been supported by the "ANNA" and "Au-delà des Pyrénées" projects funded by the Nouvelle-Aquitaine region.

References

1. Abdulkader, A., Casey, M.R.: Low cost correction of OCR errors using learning in a multi-engine environment. In: 10th International Conference on Document Analysis and Recognition, ICDAR 2009, pp. 576–580. IEEE Computer Society (2009)
2. Bazzo, G.T., Lorentz, G.A., Suarez Vargas, D., Moreira, V.P.: Assessing the impact of OCR errors in information retrieval. In: Jose, J.M., et al. (eds.) ECIR 2020. LNCS, vol. 12036, pp. 102–109. Springer, Cham (2020). https://doi.org/10.1007/978-3-030-45442-5_13
3. Bird, S.: NLTK: the natural language toolkit. In: Proceedings of the COLING/ACL 2006 Interactive Presentation Sessions, pp. 69–72 (2006)
4. Boros, E., et al.: Alleviating digitization errors in named entity recognition for historical documents. In: Proceedings of the 24th Conference on Computational Natural Language Learning, CoNLL 2020, pp. 431–441. Association for Computational Linguistics (2020)
5. Chiron, G., Doucet, A., Coustaty, M., Moreux, J.P.: ICDAR 2017 competition on post-OCR text correction. In: 14th IAPR International Conference on Document Analysis and Recognition, pp. 1423–1428. IEEE (2017)
6. Crossley, S.A., Skalicky, S., Dascalu, M., McNamara, D.S., Kyle, K.: Predicting text comprehension, processing, and familiarity in adult readers: new approaches to readability formulas. Discourse Process. **54**(5–6), 340–359 (2017)
7. Dale, E., Chall, J.S.: A formula for predicting readability: instructions. Educ. Res. Bull. **27**, 37–54 (1948)
8. Devlin, J., Chang, M., Lee, K., Toutanova, K.: BERT: pre-training of deep bidirectional transformers for language understanding. In: Proceedings of the 2019 Conference of the North American Chapter of the Association for Computational Linguistics: Human Language Technologies, NAACL-HLT 2019), pp. 4171–4186. Association for Computational Linguistics (2019)
9. Kincaid, J.P., Fishburne, R.P., Jr., Rogers, R.L., Chissom, B.S.: Derivation of new readability formulas (automated readability index, fog count and flesch reading ease formula) for Navy enlisted personnel. Tech. rep, Naval Technical Training Command Millington TN Research Branch (1975)
10. Koo, T., Li, M.: A guideline of selecting and reporting intraclass correlation coefficients for reliability research. J. Chiropr. Med. **15**(2), 155–163 (2016)
11. Martinc, M., Pollak, S., Robnik-Šikonja, M.: Supervised and unsupervised neural approaches to text readability. Comput. Linguist. **47**(1), 141–179 (2021)
12. Nguyen, T.T.H., Jatowt, A., Coustaty, M., Doucet, A.: Survey of post-OCR processing approaches. ACM Comput. Surv. **54**(6), 1–37 (2021)
13. Nguyen, T., Jatowt, A., Coustaty, M., Nguyen, N., Doucet, A.: Deep statistical analysis of OCR errors for effective post-OCR processing. In: 19th ACM/IEEE Joint Conference on Digital Libraries, pp. 29–38 (2019)

14. Linhares Pontes, E., Hamdi, A., Sidere, N., Doucet, A.: Impact of OCR quality on named entity linking. In: Jatowt, A., Maeda, A., Syn, S.Y. (eds.) ICADL 2019. LNCS, vol. 11853, pp. 102–115. Springer, Cham (2019). https://doi.org/10.1007/978-3-030-34058-2_11
15. Ranganathan, P., Pramesh, C., Aggarwal, R.: Common pitfalls in statistical analysis: measures of agreement. Perspect. Clin. Res. **8**, 187 (2017)
16. Shrout, P.E., Fleiss, J.L.: Intraclass correlations: uses in assessing rater reliability. Psychol. Bull. **86**(2), 420 (1979)
17. van Strien, D., Beelen, K., Ardanuy, M.C., Hosseini, K., McGillivray, B., Colavizza, G.: Assessing the impact of OCR quality on downstream NLP tasks. In: Proceedings of the 12th International Conference on Agents and Artificial Intelligence, ICAART 2020. pp. 484–496. SCITEPRESS (2020)
18. Traub, M.C., van Ossenbruggen, J., Hardman, L.: Impact analysis of OCR quality on research tasks in digital archives. In: Kapidakis, S., Mazurek, C., Werla, M. (eds.) TPDL 2015. LNCS, vol. 9316, pp. 252–263. Springer, Cham (2015). https://doi.org/10.1007/978-3-319-24592-8_19
19. Vajjala, S., Lučić, I.: OneStopEnglish corpus: a new corpus for automatic readability assessment and text simplification. In: Proceedings of the Thirteenth Workshop on Innovative Use of NLP for Building Educational Applications, pp. 297–304 (2018)
20. Vajjala, S., Meurers, D.: On improving the accuracy of readability classification using insights from second language acquisition. In: Proceedings of the Seventh Workshop on Building Educational Applications Using NLP, pp. 163–173 (2012)
21. Xu, W., Callison-Burch, C., Napoles, C.: Problems in current text simplification research: new data can help. Trans. Assoc. Comput. Linguist. **3**, 283–297 (2015)
22. Yang, Z., Yang, D., Dyer, C., He, X., Smola, A., Hovy, E.: Hierarchical attention networks for document classification, pp. 1480–1489. Association for Computational Linguistics, San Diego (2016)

Hard and Soft Labeling for Hebrew Paleography: A Case Study

Ahmad Droby[1(✉)], Daria Vasyutinsky Shapira[1], Irina Rabaev[2], Berat Kurar Barakat[1], and Jihad El-Sana[1]

[1] Ben-Gurion University of the Negev, Beer-Sheva, Israel
{drobya,dariavas,berat,el-sana}@post.bgu.ac.il
[2] Shamoon College of Engineering, Beer-Sheva, Israel
irinar@ac.sce.ac.il

Abstract. Paleography studies the writing styles of manuscripts and recognizes different styles and modes of scripts. We explore the applicability of hard and soft-labeling for training deep-learning models to classify Hebrew scripts. In contrast to the hard-labeling scheme, where each document image has one label representing its class, the soft-labeling approach labels an image by a label vector. Each element of the vector is the similarity of the document image to a certain regional writing style or graphical mode. In addition, we introduce a dataset of medieval Hebrew manuscripts that provides complete coverage of major Hebrew writing styles and modes. A Hebrew paleography expert manually annotated the ground truth for soft-labeling. We compare the applicability of soft and hard-labeling approaches on the presented dataset, analyze, and discuss the findings.

Keywords: Digital paleography · Medieval Hebrew manuscripts · Script type classification · Soft-labeling · Convolutional neural network

1 Introduction

Paleographic analysis of a historical document can determine the place and date when the manuscript was written. In some cases, it is even possible to identify the scribe, verify the authenticity of a manuscripts, or obtain other essential information. The continuing digitization of manuscript collections held by various libraries resulted in the availability of a large number of digital manuscripts. A professional paleographer can only process a limited number of manuscripts, and there are still manuscript collections lacking even a basic catalogue. Hence, the processing must be automated, and for the development, evaluation, and comparison of algorithms, benchmark datasets are required.

The survival rates of Hebrew manuscripts are much lower in comparison with that of Latin, Greek or Arabic ones. There are about one thousand fragments of manuscripts in Hebrew scripts that survived from the Middle ages (for this matter, beginning of the 10th century - 1540). At the current state of research it

© Springer Nature Switzerland AG 2022
S. Uchida et al. (Eds.): DAS 2022, LNCS 13237, pp. 492–506, 2022.
https://doi.org/10.1007/978-3-031-06555-2_33

is impossible to estimate their number more precisely. The ongoing digitization of Hebrew manuscripts is very advanced, because of the continuous efforts of most of the world's leading libraries. Alongside it, the Institute for Microfilmed Hebrew Manuscripts at the National Library of Israel has already assembled most of the known Hebrew manuscripts on microfilms and digital images. Thus, the automatic recognition and analysis of Hebrew manuscripts and historical documents is an urgent desideratum.

Among the surviving Hebrew manuscripts, about three thousand are dated, and these are included into the SfarData database[1] of Hebrew paleography and codicology, completed by Malachi Beit-Arié and his team.

In this paper, we present our research on automatic classification of Hebrew manuscripts into fourteen categories according to the script types and graphical modes. To train a deep neural network, we compiled a dataset of manuscripts where all of these categories are present. The margins between categories of writing styles are sometimes fuzzy and overlap on visual appearances level. To categorize the document, paleographers examine the visual appearance of the handwriting as well as the codicological data, e.g., the media on which the document was written. Since we are working with digital images only, we are unable to utilize the codicological data. We hypothesize that hard-labeling may not be the ideal way for training the deep-learning model to recognize the writing category. Therefore, for each page image we decided to add an additional level of labeling - a soft label. The soft label is a label vector, where each element indicates the similarity of the document's script to a specific script type or mode. An expert in Hebrew paleography manually annotated the soft label for each document.

The main contributions of this paper include: (1) We experiment with two different ground truth labeling schemes for training a deep-learning model and analyze the obtained results. We also discuss the issues of paleographic analysis of Hebrew writings, as well as their specificities in the context of automated processing. (2) We present a benchmark dataset of Hebrew manuscripts compiled especially for developing and evaluating machine learning algorithms. To the best of our knowledge, this is the first dataset in Hebrew that includes samples of major Hebrew writing types and modes to address digital paleography community. The dataset contains page images from 171 different manuscripts covering fourteen categories of writing, and accompanied by hard and soft labels. The dataset can be downloaded from Zenodo repository https://zenodo.org/record/6387471. We believe that this dataset will help to leverage automatic Hebrew historical documents processing, and the historical document processing in general.

2 Related Work

Throughout the last decade, various computer vision algorithms have been used for paleography analysis. Earlier techniques relied on hand-crafted features,

[1] http://sfardata.nli.org.il/.

which were often based on textural and grapheme-based descriptors, and their combination [13–15]. During the recent years, deep learning approaches have set new benchmarks in a variety of academic fields, and have been also adapted for paleographic analysis [5,6,11,16,27]. Keglevic *et al.* [18] propose to use a triplet CNN to measure the similarity of two image patches. Abdalhaleem *et al.* [1] investigate in-writer differences in manuscripts. Their methodology is built on Siamese convolutional neural networks, which are trained to recognize little differences in a person's writing style. Studer *et al.* [25] explored the effect of ImageNet pre-training for various historical document analysis tasks, including style categorization of Latin manuscripts. They experimented with VGG19 [22], Inception V3 [26], ResNet152 [12], DenseNET12 [17], and additional well-known architectures. The models trained from scratch achieved 39%–46% accuracy rate, whereas the pre-trained models achieved a 49%–55% accuracy rate.

Two major competitions on the categorization of medieval handwritings in Latin script [7,8] were organized in 2016 and 2017. The goal of the competitions was to classify medieval Latin scripts into 12 categories based on their writing styles. The findings reveal that deep learning models can accurately recognize Latin script types with more than 80% accuracy on homogeneous document collections and about 60% accuracy on heterogeneous document collections.

There have been few works on Hebrew document paleography. Wolf *et al.* [29] explored handwriting matching and paleographic classification, focusing on the documents from the Cairo Genizah collection. Dhali *et al.* [9] use textural and grapheme-based features with support vector regression to determine the date of ancient texts from the Dead Sea Scrolls collection. Ben Ezra *et al.* [24] train a model for establishing the reading order of the main text by detecting insertion markers that indicate marginal additions. They used a corpus of 17 manuscripts of Tannaitic Rabbinic compositions dated from the $10th$ to $15th$ centuries. The international Israeli and French team [21,28] work on a project that combines handwritten text recognition of Medieval Hebrew documents with a crowdsourcing-based process for training and correcting the HRT model. Their project focuses on a subset of rabbinic works dated to 1-500 CE. The aforementioned projects work on different datasets and different kind of manuscripts, and each project is solving a different part of the puzzle. These projects complement each other for the final goal of recognition of the handwritten text in historical documents. In this work, we train a deep-learning model to classify medieval Hebrew scripts into fourteen classes.

3 Hebrew Paleography

Manuscripts are studied by means of paleography and codicology, that explores the writing and the material on which manuscripts are written, respectively. The theoretical basis of Hebrew paleography and codicology are formulated in the works of Malachi Beit-Arié, Norman Golb, Benjamin Richler, Colette Sirat [2–4,19,20,23,30].

Hebrew manuscripts refers to manuscripts written in Hebrew characters, as the language was often adopted from the host societies (Ladino, Judeo-Arabic, Yiddish etc.). Geographically, the spread of the Hebrew manuscripts was larger than Latin, Greek or Arabic manuscripts. Hebrew scripts themselves were influenced by the local traditions and often resemble the manuscripts of the host societies in scribal manner, material and ways of production.

There are six main types of the Hebrew script: Oriental, Sefardic, Ashkenazi, Italian, Byzantine and Yemenite. The writing styles of Hebrew manuscripts could be classified into two branches based on their geographic origin. Oriental, Sefardic and Yemenite styles developed in Islamic regions and were influenced by the Arabic calligraphy, while Ashkenazi and Italian styles evolved in Europe and were somewhat influenced by Latin scripts. The Byzantine type displays hybrid influences and probably the influences of Greek scripts.

Our project aims to recognize the main types of the Hebrew script, and their modes (square, semi-square, cursive). Paleographically, the backbone of our research is the SfarData. Malachi Beit-Arié and his team met with our team, discussed our project, gave us their full support, and allowed us to use their database in its entirety. Our team's paleographer, who is herself a student of Malachi Beit-Arié, handpicked digitized pages from the manuscripts described in the SfarData as the raw material for our project.

4 VML-HP-ext Dataset Description

The Hebrew paleography dataset is a valuable resource both for creating a large-scale paleographic examination of Hebrew manuscripts, and assessing and benchmarking scripts classification methods. In this paper we present an extended VML-HP-ext (Visual Media Lab - Hebrew Paleography Extended) dataset. The initial version of the VML_HP dataset was presented in [10]. It consists of pages excerpted from about 60 manuscripts, their corresponding hard labels, and the official split intro training and two testing sets. Compared to the first version, the extended dataset includes sample pages from three times more manuscripts. Every manuscript was carefully selected by our team's paleographer. The majority of the manuscripts used in this dataset are kept in the National Library of Israel, the British Library, and the Bibliothèque nationale de France. Almost all manuscripts in the Oriental square script belong to the National library of Russia (we used b/w microfilms from the collection of the Institute for Microfilmed Hebrew Manuscripts at the National Library of Israel). We only included pages with one script type and one script mode per page. For example, Sephardic square only, and not main text in Sephardic square and comments in Sephardic cursive. The main challenge when compiling the dataset was the limited amount of available digitized manuscripts. For some script types (Italian, Byzantine) the shortage was more pronounced; for others (Ashkenazi, Sephardic) we had manuscripts in abundance. Keeping the dataset balanced was a challenge by itself.

Table 1. Summary of the extended VML-HP-ext dataset. Some scripts do not have semi-cursive or cursive modes. Mss = manuscripts, pp = pages.

Type	Mode					
	Square		Semi-square		Cursive	
	#Mss	#pp	#Mss	#pp	#Mss	#pp
Ashkenazi	14	56	12	48	12	48
Byzantine	7	49	12	48	–	–
Italian	5	50	11	44	5	50
Oriental	15	45	11	44	–	–
Sephardic	15	45	16	48	12	48
Yemenite	24	92	–	–	–	–

The enlarged VML-HP-ext collection contains 715 page images excerpted from 171 different manuscripts. We also provide the official split of the VML-HP-ext into training, typical test, and blind test sets. Typical test set includes unseen pages of the manuscripts from the training set. While training and typical test sets are disjoint on page level, they do share the same set of manuscripts. Therefore, we also provide the blind test set, which consist of manuscripts that do not appear in the training set. The blind test set imitates a real-life scenario, where scholar would like to obtain a classification for a previously unseen document. Tables 1 and 2 summarize the extended VML-HP-ext dataset.

5 Case Study

In the following section, we present and discuss our experiments on the extended VMP-HP dataset. We report the results using hard label classification model and compare it with our previous results. In addition, we explore the use of the newly introduced soft labels to train a regression model, which can be used for classification.

5.1 Hard-Label Classification

This experiment aims at evaluating classification models on the extended dataset. We trained and evaluated several architectures on the extended dataset.

Table 2. The VML-HP-ext dataset - official split. Mss = manuscripts, pp = pages

Set	# Mss	# pp
Train	130	400
Typical test	130	143
Blind test	41	172
Total	171	715

The models were trained until convergence using 50K patches extracted from pages in the train set. The model was trained using binary cross entropy loss function. The patches were extracted using the patch generation method proposed in our previous work [10], which extracts patches with uniform text scale and on average 5 lines in each patch.

Table 3 shows the precision, recall, F1, and accuracy measures of the models on the blind test set. As seen, ResNet50 outperforms all the other model on every metric, achieving an accuracy of 60% which is significantly higher than the accuracy we obtained on the old dataset, which was 42.1%. Obviously, showing the benefits of the extended dataset, which include more varying handwriting in each script type. Table 4 presents the precision, recall, and F1 measures of ResNet50 for each class. As seen, there are some classes such as Italian square, Byzantine semi-square, and Ashkenazi semi-square that are frequently classified incorrectly as Italian semi-square, Byzantine square, and Ashkenazi cursive respectively (see the confusion matrix in Fig. 1). Considering those classes share the same regional style, these results suggest that there may be ambiguity in the definition of the script types; i.e., we hypothesize that there is no clear-cut between such graphical modes, rather, they lie on a spectrum between square and cursive (Fig. 2).

Table 3. Evaluation results of several classification models on blind test set of the extended dataset.

Model	Avg. Precision	Avg. Recall	Avg. F1-score	Accuracy
DenseNet	58	53	53	53
AlexNet	55	52	52	51
VGG19	60	57	56	56
ResNet50	**63**	**60**	**59**	**60**
SqueezeNet	58	55	54	55

Table 4. Evaluation results of classification model with ResNet50 architecture on the extended dataset.

Label	Square			Semi-square			Cursive		
	P	R	F1	P	R	F1	P	R	F1
Ashkenazi	0.78	0.90	0.83	0.63	0.47	0.54	0.50	0.57	0.53
Byzantine	0.14	0.07	0.09	0.29	0.65	0.40		-	
Italian	0.51	0.14	0.21	0.33	0.42	0.37	0.68	0.97	0.80
Oriental	0.95	0.69	0.80	0.50	0.75	0.60		-	
Sephardic	0.86	0.88	0.87	0.87	0.45	0.59	0.93	0.73	0.82
Yemenite	0.90	0.67	0.77		-			-	

	P	R	F1		
Average	0.63	0.60	0.59	**Accuracy**	0.60

P: precision, R: recall, F1: F1-score

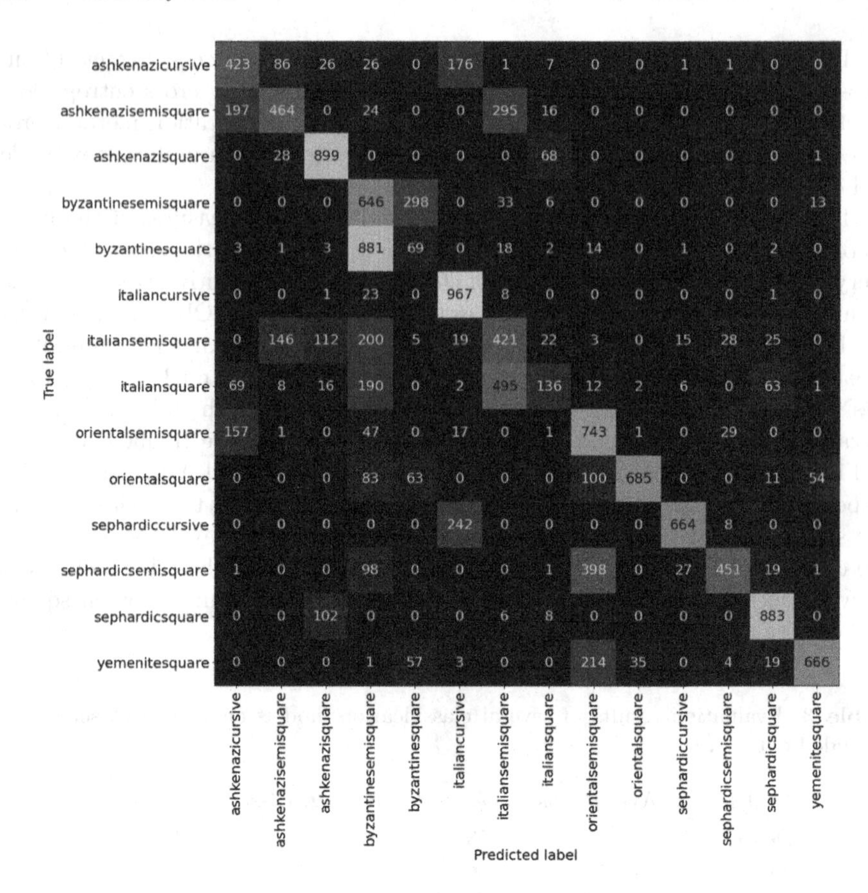

Fig. 1. Confusion matrix of classification model with ResNet50 architecture trained using the hard-labels.

5.2 Soft-Label Regression

As we have mentioned in Sect. 1, the margins between categories of writing styles are blurred, and there is an overlap between characteristics of different styles of writing. To categorize the document into writing category, paleographers rely both on visual appearance and codicological data (such as the media on which the document was written). However, we deal with only digital images and can not utilize codicological information. We hypothesize that hard labels may not be the best way to characterize the writing style of a document. Therefore, we added a second level of labeling - a soft label - for each page. The soft level is a label vector, where each element specifies the degree of similarity between the processed document and the certain script type or mode. The soft-labeling were done by an Expert Hebrew paleographer.

In a soft-labeling scheme, we label each manuscript using a vector of size eight. The first six elements of the vector express the degree of similarity of the manuscript to belong to certain regional type (Ashkenazi, Italian, Sephardic,

Correctly predicted patches

Input						
Prediction	Ashkenazi cursive	Byzantine semi-square	Italian semi-square	Oriental semi-square	Sephardic cursive	Ashkenazi semi-square

Incorrectly predicted patches

Input						
Prediction	Ashkenazi cursive	Ashkenazi semi-square	Italian square	Sephardic cursive	Sephardic square	Byzantine semi-square
GT	Ashkenazi semi-square	Italian semi-square	Italian semi-square	Italian cursive	Italian square	Byzantine square

Fig. 2. Sample results from the ResNet50 classification model

Oriental, Byzantine and Yemenite) and the last two elements are the degrees of similarity to certain graphical mode, square and cursive (similar values for both square and cursive indicate the semi-square mode). Similar to the previous experiment, we extracted $50K$ patches and assign each patch a vector of probability values corresponding to a regional and graphical types. We trained a regression model with a ResNet50 backbone on the mentioned $50K$ patches with mean squared error loss function. The model was trained until convergence, which happened after 10 epochs.

Figure 3 reports sample results. To evaluate the model numerically, we calculated the Root Mean Square Error (RMSE) on the blind test set. RMSE is calculated according to the following formula:

$$RMSE = \sqrt{\frac{\sum_{i=0}^{N} ||y(i) - \hat{y}(i)||^2}{N}},$$

where $y(i)$ is the predicted label for patch i, and $\hat{y}(i)$ is its actual label.

The trained model achieved RMSE of about 0.24. Although, this might give us an indication that the model give good results (as can be seen in Fig. 3), it is not very meaningful and does not show how this model compare against other classification methods. Therefore, arose a need to convert the predicted soft-label to hard-labels. Next, we will explore two different conversion methods and report corresponding results.

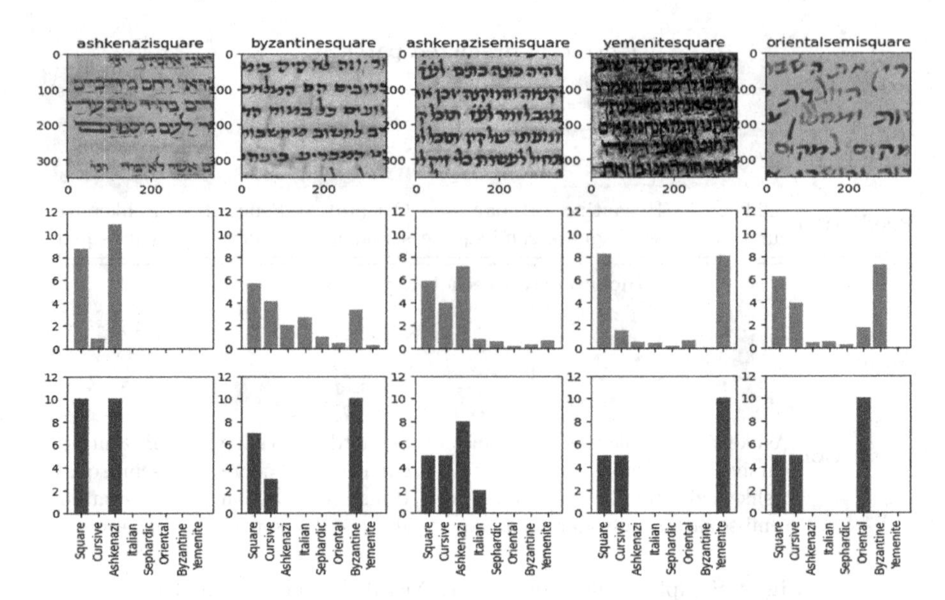

Fig. 3. Sample results of the regression model. Top row: the input patch with its ground-truth label. Second row: the predicted soft-label. Bottom row: the ground-truth soft-label.

5.3 Maximum Score Class Assignment

In this approach, a predicted soft-label s is converted to a hard label according to the following formula:

$$Regional(s) = argmax\{s(r)\};_{r \in \{\text{Ashkenazi, Byzantine, Italian, Oriental, Sephardic, Yemenite}\}}$$

$$Graphical(s) = \begin{cases} argmax_{g \in \{\text{square, cursive}\}}\{s(g)\}, & s(\text{square}), s(\text{cursive}) < T \\ \text{semi-square}, & \text{else} \end{cases}$$

In other words, the label is determined by taking the regional style and graphical mode with the maximum score unless both, the square and cursive, scores are under a predefined threshold T (we set $T = 0.3$), in which case the graphical mode is determined to be as semi-square.

Table 6 presents the evaluation results of the regression model after converting the soft-labels. It is important to note that this method introduces new labels that are not present in the dataset, such as Byzantine cursive, Oriental cursive, and Yemenite semi-square. The conversion method achieved an accuracy of 47%.

Table 5. The results of the regression model for the regional style classes only.

Label	Precision	Recall	F1-score
Ashkenazi	0.60	0.88	0.72
Byzantine	0.47	0.84	0.60
Italian	0.74	0.53	0.62
Oriental	0.92	0.37	0.53
Sephardic	0.85	0.68	0.76
Yemenite	0.83	0.64	0.72
Accuracy			0.66
Macro avg	0.74	0.66	0.66
Weighted avg	0.73	0.66	0.66

The regression model obtains an accuracy of 67% using the regional style labels only, as seen in Table 5. This indicates that the graphical style scores hinders the classification more than the regional ones. The model encounters the highest confusion between the Italian and Ashkenzi patches, as illustrated in the confusion matrix of the regional classification (Fig. 4). Most of the incorrect predictions are minor mistakes that a human paleographer could also have made. Neverthless, the confusion between Sephardic and Italian is a more serious error.

Table 6. Evaluation results of the regression model with maximum score class conversion method.

Label	Square			Semi-square			Cursive		
	P	R	F1	P	R	F1	P	R	F1
Ashkenazi	1.00	0.79	0.88	0.32	0.40	0.36	0.29	0.09	0.14
Byzantine	0.27	0.06	0.10	0.24	0.81	0.37		-	
Italian	0.00	0.00	0.00	0.22	0.66	0.33	0.18	0.18	0.18
Oriental	0.88	0.61	0.72	0.17	0.07	0.09		-	
Sephardic	0.98	0.36	0.52	0.32	0.64	0.43	0.99	0.15	0.25
Yemenite	0.83	0.31	0.45		-			-	

Average	P	R	F1	**Accuracy**
	0.50	0.37	0.34	0.37

P: precision, R: recall, F1: F1-score

5.4 Nearest Neighbor Label Conversion

This approach utilizes the soft and hard labels in the training set. It calculates the distances between the predicted labels and the soft-labels in the training

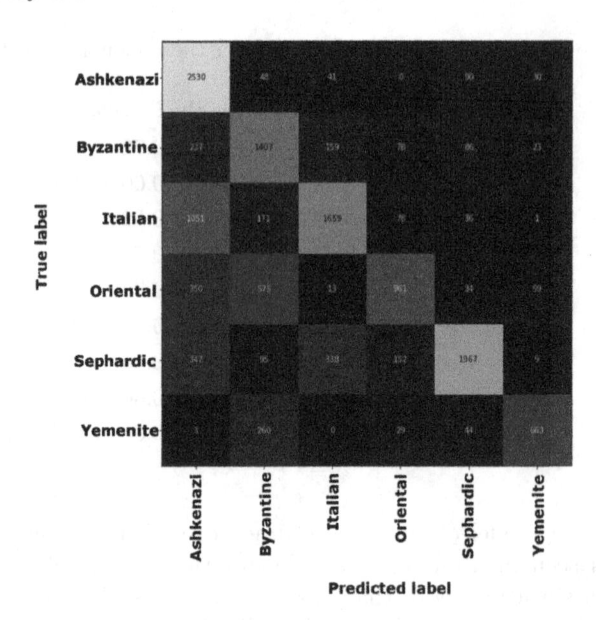

Fig. 4. The confusion matrix of the classification for the regional style classes.

set and converts each predicted soft-label to the nearest hard-label in the train set. Figure 5 presets sample results of this conversion, and Table 7 reports the classification accuracy using this method. The method obtains 46% accuracy, reaching results on par with the previous conversion method.

5.5 Comparison Between Soft and Hard-Label Classification

As we have reported earlier, the hard-label classification obtains higher accuracy in comparison with soft-labeling configuration. However, this does not tell the whole story as the soft-labeling regression model offers more insight on the script style. For examples, for a square graphical style text that has some cursive characteristics, using hard-label classification most probably will classify this text as square or semi-square, but a regression model will indicate "how

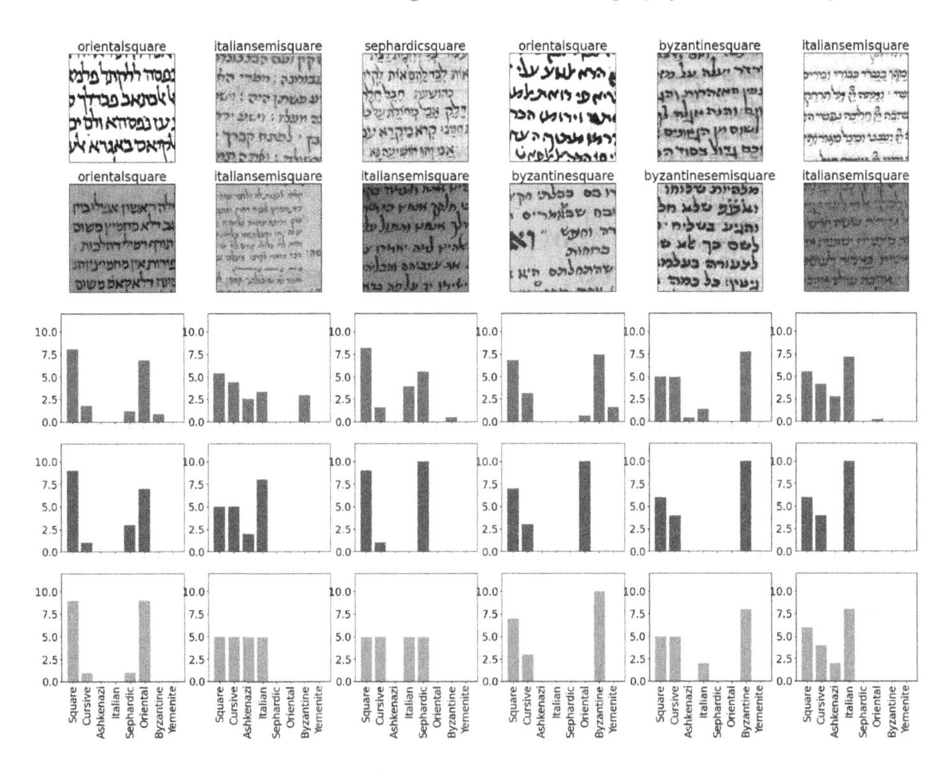

Fig. 5. Sample results of regression model with the nearest neighbor label conversion method. Top row: input patch with its ground-truth label. Second row: The nearest neighbor of the input patch. Third row: The predicted label of the input patch. Fourth row: The ground-truth soft-label of the input patch. Bottom row: The ground-truth soft-label of the nearest neighbor patch.

much cursive" this text is. Furthermore, such a regression model can offer paleography experts a tool to analyze the fluidity of the Hebrew script style, as a text can have multiple regional style characteristics while having varying level of squareness/cohesiveness.

Table 7. Evaluation results of the regression model with the nearest neighbor label conversion method.

Label	Square			Semi-square			Cursive		
	P	R	F1	P	R	F1	P	R	F1
Ashkenazi	0.98	0.57	0.72	0.43	0.67	0.52	0.50	0.01	0.03
Byzantine	0.03	0.01	0.01	0.25	0.87	0.38		-	
Italian	0.00	0.00	0.00	0.23	0.64	0.33	0.39	0.21	0.27
Oriental	0.99	0.65	0.79	0.29	0.06	0.10		-	
Sephardic	0.99	0.37	0.54	0.49	0.73	0.58	1.00	0.01	0.02
Yemenite	0.88	0.63	0.73		-			-	

	P	R	F1		
Average	0.53	0.40	0.37	**Accuracy**	0.40

P: precision, R: recall, F1: F1-score

6 Conclusion and Further Research

In this paper, we investigated the use of two types of labeling for Hebrew script types classification, hard and soft-labeling. Hard-labeling refer to the traditional labeling where each page is labeled with one script type. Soft-labeling assigns a vector of size eight to each page. The vector indicts how similar this page's writing style is to each geographical type and graphical mode. To perform the experiments, we compiled the VML-HP-ext dataset that covers major Hebrew script types. The dataset includes soft-labels for each page in addition to hard-labels.

We trained and evaluated several classification models on the hard-labeling configuration. ResNet50 topped the list with an accuracy of 60%. In addition, we experimented with soft-labeling, training a regression model to predict the similarity values of each image to each geographical and graphical type. Since such a model cannot be directly compared with regular hard-label classification, we proposed and evaluated two methods that convert soft labels to hard labels. We conclude that while the soft-labeling provides more information about the script style, e.g., how square or cursive it is, using the regression model with the conversion methods does not reach the accuracy of the models trained using hard-labeling.

In future work, we plan to experiment with additional ways to interpret the soft-labels and convert them to hard-labels. In addition, we want to experiment with unsupervised or semi-supervised classification, which may give us a more precise definition of the script type.

Acknowledgment. This research was partially supported by The Frankel Center for Computer Science at Ben-Gurion University.

References

1. Abdalhaleem, A., Barakat, B.K., El-Sana, J.: Case study: fine writing style classification using siamese neural network. In: 2nd International Workshop on Arabic and Derived Script Analysis and Recognition, pp. 62–66 (2018)
2. Beit-Arié, M.: Hebrew codicology. Tentative Typology of Technical Practices Employed in Hebrew Dated Medieval Manuscripts, Jerusalem (1981)
3. Beit-Arié, M.: Hebrew Codicology. ZFDM Repository (2021). https://doi.org/10.25592/uhhfdm.8849
4. Beit-Arié, M., Engel, E.: Specimens of mediaeval Hebrew scripts, vol. 3. Israel Academy of Sciences and Humanities (1987, 2002, 2017)
5. Christlein, V., Bernecker, D., Maier, A., Angelopoulou, E.: Offline writer identification using convolutional neural network activation Features. In: Gall, J., Gehler, P., Leibe, B. (eds.) Pattern Recognition, GCPR 2015. LNCS, vol. 9358, pp. 540–552. Springer, Cham (2015). https://doi.org/10.1007/978-3-319-24947-6_45
6. Christlein, V., Gropp, M., Fiel, S., Maier, A.: Unsupervised feature learning for writer identification and writer retrieval. In: 14th International Conference on Document Analysis and Recognition, vol. 1, pp. 991–997 (2017)
7. Cloppet, F., Eglin, V., Helias-Baron, M., Kieu, C., Vincent, N., Stutzmann, D.: ICDAR2017 competition on the classification of medieval handwritings in Latin script. In: 14th International Conference on Document Analysis and Recognition, vol. 1, pp. 1371–1376 (2017)
8. Cloppet, F., Eglin, V., Stutzmann, D., Vincent, N., et al.: ICFHR2016 competition on the classification of medieval handwritings in Latin script. In: 15th International Conference on Frontiers in Handwriting Recognition, pp. 590–595 (2016)
9. Dhali, M.A., Jansen, C.N., de Wit, J.W., Schomaker, L.: Feature-extraction methods for historical manuscript dating based on writing style development. Pattern Recogn. Lett. **131**, 413–420 (2020)
10. Droby, A., Kurar Barakat, B., Vasyutinsky Shapira, D., Rabaev, I., El-Sana, J.: VML-HP: Hebrew paleography dataset. In: Lladós, J., Lopresti, D., Uchida, S. (eds.) Document Analysis and Recognition – ICDAR 2021. LNCS, vol. 12824, pp. 205–220. Springer, Cham (2021). https://doi.org/10.1007/978-3-030-86337-1_14
11. Fiel, S., Sablatnig, R.: Writer identification and writer retrieval using the fisher vector on visual vocabularies. In: 12th International Conference on Document Analysis and Recognition, pp. 545–549 (2013)
12. He, K., Zhang, X., Ren, S., Sun, J.: Deep residual learning for image recognition. In: IEEE Conference on Computer Vision and Pattern Recognition, pp. 770–778 (2016)
13. He, S., Samara, P., Burgers, J., Schomaker, L.: Discovering visual element evolutions for historical document dating. In: 15th International Conference on Frontiers in Handwriting Recognition, pp. 7–12 (2016)
14. He, S., Samara, P., Burgers, J., Schomaker, L.: Historical manuscript dating based on temporal pattern codebook. Comput. Vis. Image Underst. **152**, 167–175 (2016)
15. He, S., Sammara, P., Burgers, J., Schomaker, L.: Towards style-based dating of historical documents. In: 14th International Conference on Frontiers in Handwriting Recognition, pp. 265–270 (2014)
16. Hosoe, M., Yamada, T., Kato, K., Yamamoto, K.: Offline text-independent writer identification based on writer-independent model using conditional autoencoder. In: 16th International Conference on Frontiers in Handwriting Recognition, pp. 441–446 (2018)

17. Huang, G., Liu, Z., Van Der Maaten, L., Weinberger, K.Q.: Densely connected convolutional networks. In: IEEE Conference on Computer Vision and Pattern Recognition, pp. 4700–4708 (2017)
18. Keglevic, M., Fiel, S., Sablatnig, R.: Learning features for writer retrieval and identification using triplet CNNs. In: 16th International Conference on Frontiers in Handwriting Recognition, pp. 211–216 (2018)
19. Richler, B.: Hebrew manuscripts in the Vatican library: catalogue, pp. 1–790 (2008)
20. Richler, B., Beit-Arié, M.: Hebrew manuscripts in the biblioteca palatina in parma: catalogue; palaeographical and codicological descriptions (2011)
21. Schor, U., Raziel-Kretzmer, V., Lavee, M., Kuflik, T.: Digital research library for multi-hierarchical interrelated texts: from 'Tikkoun Sofrim' text production to text modeling. In: Classics@18 (2021)
22. Simonyan, K., Zisserman, A.: Very deep convolutional networks for large-scale image recognition. arXiv preprint arXiv:1409.1556 (2014)
23. Sirat, C.: Hebrew Manuscripts of the Middle Ages. Cambridge University Press, Cambridge (2002)
24. Stökl Ben Ezra, D., Brown-DeVost, B., Jablonski, P.: Exploiting insertion symbols for marginal additions in the recognition process to establish reading order. In: Barney Smith, E.H., Pal, U. (eds.) Document Analysis and Recognition – ICDAR 2021 Workshops. LNCS, vol. 12917, pp. 317–324. Springer, Cham (2021). https://doi.org/10.1007/978-3-030-86159-9_22
25. Studer, L., et al.: A comprehensive study of imagenet pre-training for historical document image analysis. In: International Conference on Document Analysis and Recognition, pp. 720–725 (2019)
26. Szegedy, C., et al.: Going deeper with convolutions. In: IEEE Conference on Computer Vision and Pattern Recognition, pp. 1–9 (2015)
27. Vidal-Gorène, C., Decours-Perez, A.: A computational approach of Armenian paleography. In: Barney Smith, E.H., Pal, U. (eds.) Document Analysis and Recognition – ICDAR 2021 Workshops. LNCS, vol. 12917, pp. 295–305. Springer, Cham (2021). https://doi.org/10.1007/978-3-030-86159-9_20
28. Wecker, A.J., et al.: Tikkoun sofrim: a webapp for personalization and adaptation of crowdsourcing transcriptions. In: Adjunct Publication of the 27th Conference on User Modeling, Adaptation and Personalization, pp. 109–110 (2019)
29. Wolf, L., Potikha, L., Dershowitz, N., Shweka, R., Choueka, Y.: Computerized paleography: tools for historical manuscripts. In: 18th IEEE International Conference on Image Processing, pp. 3545–3548 (2011)
30. Yardeni, A., et al.: The Book of Hebrew Script: History, Palaeography, Script Styles, Calligraphy and Design. Carta Jerusalem, Jerusalem (1997)

AttentionHTR: Handwritten Text Recognition Based on Attention Encoder-Decoder Networks

Dmitrijs Kass[1] and Ekta Vats[2(✉)]

[1] Department of Information Technology, Uppsala University, Uppsala, Sweden
dmitrijs.kass@it.uu.se
[2] Centre for Digital Humanities Uppsala, Department of ALM, Uppsala University, Uppsala, Sweden
ekta.vats@abm.uu.se

Abstract. This work proposes an attention-based sequence-to-sequence model for handwritten word recognition and explores transfer learning for data-efficient training of HTR systems. To overcome training data scarcity, this work leverages models pre-trained on scene text images as a starting point towards tailoring the handwriting recognition models. ResNet feature extraction and bidirectional LSTM-based sequence modeling stages together form an encoder. The prediction stage consists of a decoder and a content-based attention mechanism. The effectiveness of the proposed end-to-end HTR system has been empirically evaluated on a novel multi-writer dataset Imgur5K and the IAM dataset. The experimental results evaluate the performance of the HTR framework, further supported by an in-depth analysis of the error cases. Source code and pre-trained models are available at GitHub (https://github.com/dmitrijsk/AttentionHTR).

Keywords: Handwritten text recognition · Attention encoder-decoder networks · Sequence-to-sequence model · Transfer learning · Multi-writer

1 Introduction

Historical archives and cultural institutions contain rich heritage collections from historical times that are to be digitized to prevent degradation over time. The importance of digitization has led to a strong research interest in designing methods for automatic handwritten text recognition (HTR). However, handwritten text possesses variability in handwriting styles, and the documents are often heavily degraded. Such issues render the digitization of handwritten material more challenging and suggest the need to have more sophisticated HTR systems.

The current state-of-the-art in HTR is dominated by deep learning-based methods that require a significant amount of training data. Further, to accurately model the variability in writing styles in a multi-writer scenario, it is important

© Springer Nature Switzerland AG 2022
S. Uchida et al. (Eds.): DAS 2022, LNCS 13237, pp. 507–522, 2022.
https://doi.org/10.1007/978-3-031-06555-2_34

to train the neural network on a variety of handwritten texts. However, only a limited amount of annotated data is available to train an end-to-end HTR model from scratch, and that affects the performance of the HTR system. This work proposes an end-to-end HTR system based on attention encoder-decoder architecture and presents a transfer learning-based approach for data-efficient training. In an effort toward designing a data-driven HTR pipeline, a Scene Text Recognition (STR) benchmark model [1] is studied, which is trained on nearly 14 million synthetic scene text word images. The idea is to fine-tune the STR benchmark model on different handwritten word datasets.

The novelty and technical contributions of this work are as follows. (a) The goal of this work is to leverage transfer learning to enable HTR in cases where training data is scarce. (b) Scene text images and a new dataset, Imgur5K [13], are used for transfer learning. The Imgur5K dataset contains a handwritten text by approximately 5000 writers, which allows our proposed model to generalize better on unseen examples. (c) Our transfer learning-based framework produces a model that is applicable in the real world as it was trained on word examples from thousands of authors, with varying imaging conditions. (d) The proposed attention-based architecture is simple, modular, and reproducible, more data can be easily added in the pipeline, further strengthening the model's accuracy. (e) A comprehensive error analysis using different techniques such as bias-variance analysis, character-level analysis, and visual analysis is presented. (f) An ablation study is conducted to demonstrate the importance of the proposed framework.

2 Related Work

In document analysis literature, popular approaches towards handwriting recognition include Hidden Markov Models [17], Recurrent Neural Networks (RNN) [7], CNN-RNN hybrid architectures [6], and attention-based sequence to sequence (seq2seq) models [3,12,20]. Recent developments [11,15] suggest that the current state-of-the-art in HTR is advancing towards using attention-based seq2seq models. RNNs are commonly used to model the temporal nature of the text, and the attention mechanism performs well when used with RNNs to focus on the useful features at each time step [12].

Seq2seq models follow an encoder-decoder architecture, where one network encodes an input sequence into a fixed-length vector representation, and the other decodes it into a target sequence. However, using a fixed-length vector affects the performance, and therefore attention mechanisms are gaining importance due to enabling an automatic search for the most relevant elements of an input sequence to predict a target sequence [2].

This work investigates attention-based encoder-decoder networks [1] for handwriting recognition, based on ResNet for feature extraction, bidirectional LSTMs for sequence modeling, a content-based attention mechanism for prediction [2], and transfer learning from STR domain to HTR to overcome lack of training data. To this end, the STR benchmark [1] has been studied, and the pre-trained models from STR are used in this work to fine-tune the HTR models. In [1], the models were trained end-to-end on the union of MJSynth [10] and SynthText [8] datasets.

After filtering out words containing non-alphanumeric characters, these datasets contain 8.9M and 5.5M word-level images for training, respectively.

The related methods include [3,11,12,15,20], where [3] is the first attempt towards using an attention-based model for HTR. The limitation of this work is that it requires pre-training of the features extracted from the encoding stage using the Connectionist Temporal Classification (CTC) loss to be relevant. The method proposed in [20] is based on a sliding window approach. Limitations of this work include an increase in overhead due to sliding window size initialization and parameter tuning. The model proposed in [12] consists of an encoder, an attention mechanism, and a decoder. The task of the encoder is to extract deep visual features and it is implemented as a combination of two neural networks: a Convolutional Neural Network (CNN) and a bidirectional Gated Recurrent Unit (BGRU). The task of the decoder is to predict one character at a time. Each time a character is being predicted an attention mechanism automatically defines a context as a combination of input features that are the most relevant for the current predicted character and improves the predictive performance of the decoder. Similar to the proposed method, [12] does not require pre-processing, a pre-defined lexicon, or a language model. The method proposed in [15] combines CNN and a deep bidirectional LSTM, and also investigates various attention mechanisms and positional encodings to address the problem of input-output text sequence alignment. For decoding the actual character sequence, it uses a separate RNN. Recently, [11] introduced Candidate Fusion as a novel approach toward integrating a language model into a seq2seq architecture. The proposed method performs well in comparison with the related methods using the IAM dataset under the same experimental settings.

3 Attention-Based Encoder-Decoder Network

The overall pipeline of the proposed HTR method is presented in Fig. 1. The model architecture consists of four stages: transformation, feature extraction, sequence modeling, and prediction, discussed as follows.

Transformation Stage. Since the handwritten words appear in irregular shapes (e.g. tilted, skewed, curved), the input word images are normalized using the thin-plate spline (TPS) transformation [4]. The localization network of TPS takes an input image and learns the coordinates of fiducial points that are used to capture the shape of the text. Coordinates of fiducial points are regressed by a CNN and the number of points is a hyper-parameter. TPS also consists of a grid generator that maps the fiducial points on the input image $\mathbf{X}$ to the normalized image $\tilde{\mathbf{X}}$, and then a sampler interpolates pixels from the input to the normalized image.

Feature Extraction Stage. A 32-layer residual neural network (ResNet) [9] is used to encode a normalized 100×32 greyscale input image into a 2D visual feature map $\mathbf{V} = \{\mathbf{v}_i\}, i = 1..I$, where I is the number of columns in the feature map. The output visual feature map has 512 channels $\times$ 26 columns. Each

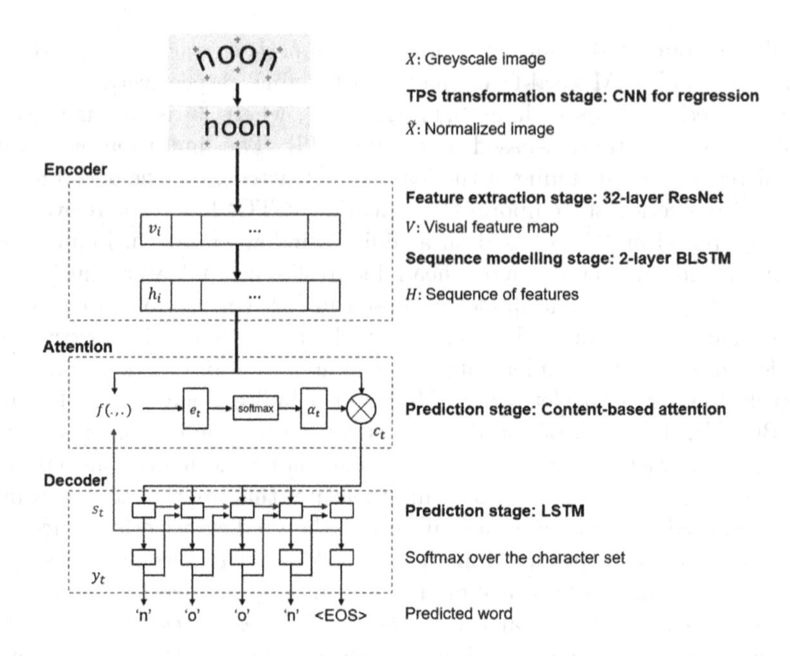

Fig. 1. Attention-based encoder-decoder architecture for HTR.

column of the feature map corresponds to a receptive field on the transformed input image. Columns in the feature map are in the same order as receptive fields on the image from left to right [18].

Sequence Modeling Stage. The features $\mathbf{V}$ from the feature extraction stage are reshaped into sequences of features $\mathbf{H}$, where each column in a feature map $\mathbf{v}_i \in \mathbf{V}$ is used as a sequence frame [18]. A bidirectional LSTM (BLSTM) is used for sequence modeling which allows contextual information within a sequence (from both sides) to be captured, and renders the recognition of character sequences simpler and more stable as compared to recognizing each character independently. For example, a contrast between the character heights in "il" can help recognize them correctly, as opposed to recognizing each character independently [18]. Finally, two BLSTMs are stacked to learn a higher level of abstractions. Dimensions of output feature sequence are 256×26.

Prediction Stage. An attention-based decoder is used to improve character sequence predictions. The decoder is a unidirectional LSTM and attention is content-based. The prediction loop has T time steps, which is the maximum length of a predicted word. At each time step $t = 1..T$ the decoder calculates a probability distribution over the character set

$$\mathbf{y}_t = \text{softmax}(\mathbf{W}_0 \mathbf{s}_t + \mathbf{b}_0)$$

where $\mathbf{W}_0$ and $\mathbf{b}_0$ are trainable parameters and $\mathbf{s}_t$ is a hidden state of decoder LSTM at time t. A character with the highest probability is used as a prediction.

The decoder can generate sequences of variable lengths, but the maximum length is fixed as a hyper-parameter. We used the maximum length of 26 characters, including an end-of-sequence (EOS) token that signals the decoder to stop making predictions after EOS is emitted. The character set is also fixed. A case-insensitive model used further in experiments uses an alphanumeric character set of length 37, which consists of 26 lower-case Latin letters, 10 digits, and an EOS token. A character set of a case-sensitive model also includes 26 upper-case Latin letters and 32 special characters (˜ˆ\!”#$%&′()∗+, −./ :; <=>?@[]·{|}), and has a total length of 95 characters.

A hidden state of the decoder LSTM $\mathbf{s}_t$ is conditioned on the previous prediction $\mathbf{y}_{t-1}$, a context vector $\mathbf{c}_t$ and a previous hidden state

$$\mathbf{s}_t = LSTM(\mathbf{y}_{t-1}, \mathbf{c}_t, \mathbf{s}_{t-1}).$$

A context $\mathbf{c}_t$ is calculated as a weighted sum of encoded feature vectors $\mathbf{H} = \mathbf{h}_1, \ldots, \mathbf{h}_I$ from the sequence modeling stage as

$$\mathbf{c}_t = \sum_{i=1}^{I} \alpha_{ti} \mathbf{h}_i,$$

where α_{ti} are normalized attention scores, also called attention weights or alignment factors, and are calculated as

$$\alpha_{ti} = \frac{\exp(e_{ti})}{\sum_{k=1}^{I} \exp(e_{tk})},$$

where e_{ti} are attention scores calculated using Bahdanau alignment function [2]

$$e_{ti} = f(\mathbf{s}_{t-1}, \mathbf{h}_i) = \mathbf{v}^\top \tanh(\mathbf{W}\mathbf{s}_{t-1} + \mathbf{V}\mathbf{h}_i + \mathbf{b}),$$

where $\mathbf{v}$, $\mathbf{W}$, $\mathbf{V}$ and $\mathbf{b}$ are trainable parameters.

Transfer Learning. Transfer learning is used in this work to resolve the problem of insufficient training data. In general, it relaxes the assumption that the training and test datasets must be identically distributed [21] and tries to transfer the learned features from the source domain to the target domain. The source domain here is STR, and the target domain is HTR. This paper uses the strategy of fine-tuning [21] by initializing the weights from a pre-trained model and continuing the learning process by updating all 49.6M parameters in it. The architecture of the fine-tuned model is the same as that of the pre-trained model. This, together with a significant overlap between the source and the target domains (synthetic scene text and handwritten text), allows the use of training datasets that are relatively small in size (up to 213K), compared to the training dataset used to obtain the pre-trained model (14.4M).

The learning rate is an important parameter to consider in fine-tuning, and its impact on the results is examined. A high learning rate can make the model perform poorly by completely changing the parameters of a pre-trained model.

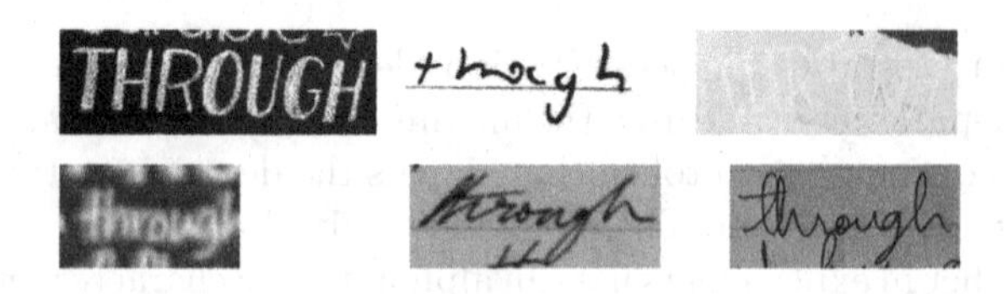

Fig. 2. Imgur5K dataset samples. Top row: training set. Bottom row: test set.

And a low learning rate can adjust a pre-trained model's parameters to a new dataset without making many adjustments to the original weights. A batch size of 32 has been selected based on the size of the training set and the memory requirements. Early stopping is used as a regularization method to avoid overfitting due to its effectiveness and simplicity.

4 Experimental Results

4.1 Datasets

Two word-level multi-writer datasets, IAM [14] and Imgur5K [13], are used. Raw image files were converted into a Lightning Memory-Mapped Database[1] (LMDB) before the experiments. The main benefit of LMDB is an extremely fast speed of read transactions, which is important for the training process.

The **IAM** handwriting database [14] contains 1,539 handwritten forms, written by 657 authors, and the total number of words is 115,320, with a lexicon of approximately 13,500 words. This work uses the most widespread RWTH Aachen partition into writer-exclusive training, validation, and test sets [12] to ensure a fair comparison with the related works.

The **Imgur5K** dataset [13] is a new multi-writer dataset that contains 8,094 images, with the handwriting of approximately 5,305 authors published on an online image-sharing and image-hosting service imgur.com. The total number of words is 135,375 with a lexicon of approximately 27,000 words. In comparison with the IAM dataset, the Imgur5K dataset contains 8 times more authors, 1.2 times more words, and 2 times larger lexicon. It is also a more challenging dataset in terms of background clutter and variability in pen types and styles. This work uses the same partitioning into document-exclusive training, validation, and test sets, as was provided from the dataset source. Figure 2 shows examples of a word *through* in both training and test sets.

The actual size of partitions differs from full partitions if an annotation is missing, the length of the annotation exceeds 25 characters, or if it contains characters that are not included in the model's character set. Furthermore, the IAM partitions contain only those words where the segmentation of corresponding lines was marked as "OK". The size of unfiltered and filtered partitions for both case-sensitive and case-insensitive models is shown in Table 1.

[1] https://lmdb.readthedocs.io/.

Table 1. Size of training, validation, and test partitions used in experiments.

Dataset	Character set	Training	Validation	Testing
IAM	All characters	47,981	7,554	20,305
	Case-sensitive (incl. special characters)	47,963	7,552	20,300
	Case-insensitive	41,228	6,225	17,381
Imgur5K	All characters	182,528	22,470	23,663
	Case-sensitive (incl. special characters)	164,857	19,932	21,509
	Case-insensitive	138,577	16,823	18,110
Total	All characters	230,509	30,024	43,968
	Case-sensitive (incl. special characters)	212,820	27,484	41,809
	Case-insensitive	179,805	23,048	35,491

Table 2. Validation set errors for the proposed method.

Char. set	Fine-tuned on	Val-CER			Val-WER		
		Imgur5K	IAM	Both	Imgur5K	IAM	Both
Case-insensitive	Benchmark	22.66	25.65	23.47	37.79	47.05	40.29
	IAM	27.58	3.65	21.12	40.83	11.79	32.98
	IAM→Imgur5K	**6.28**	7.73	6.67	**12.95**	20.06	14.87
	Imgur5K	6.70	8.48	7.18	13.48	22.43	15.90
	Imgur5K→IAM	13.69	**2.73**	10.73	23.43	**8.76**	19.47
	Imgur5K+IAM	7.17	3.24	**6.11**	13.79	9.82	**12.72**
Case-sensitive	Benchmark	29.77	42.81	33.35	49.73	59.07	52.30
	IAM	34.80	5.16	26.65	53.59	12.45	42.28
	IAM→Imgur5K	**9.32**	25.07	13.65	**19.61**	34.88	23.81
	Imgur5K	9.18	23.31	13.06	19.74	36.32	24.29
	Imgur5K→IAM	22.43	**4.88**	17.61	38.48	**11.80**	31.15
	Imgur5K+IAM	9.48	4.97	**8.24**	19.69	12.38	**17.68**

4.2 Hyper-parameters

AdaDelta optimizer is used with the hyper-parameters matching those in [1]: learning rate $\eta = 1$; decay constant $\rho = 0.95$; numerical stability term $\epsilon = 1\text{e}{-8}$; gradient clipping is 5; maximum number of iterations 300,000. Early stopping patience is set to 5 epochs, and training batch size is 32.

4.3 Results

Two standard metrics: character error rate (CER) and word error rate (WER) are used to evaluate the experimental results. CER is defined as a normalized Levenshtein distance:

$$CER = \frac{1}{N} \sum_{i=1}^{N} D(s_i, \hat{s}_i)/l_i,$$

Table 3. Validation WER on varying the learning rate η and decay constant ρ for the case-insensitive benchmark model, fine-tuned on the Imgur5K→IAM sequence. The lowest error is shown in **bold**. Cells with the lowest row-wise error have a grey fill.

Learning rate η	Decay constant ρ		
	0.9	0.95	0.99
1.5	9.69	11.07	11.47
1	9.59	**8.76**	9.98
0.1	9.27	8.96	8.96
0.01	9.30	9.49	9.03
0.001	11.33	10.86	10.67

where N is the number of annotated images in the set, $D(.)$ is the Levenshtein distance, s_i and $\hat{s}_i$ denote the ground truth text string and the corresponding predicted string, and l_i is the length of the ground truth string. Given the predictions are evaluated at word level, the WER corresponds to the misclassification error:

$$WER = \frac{1}{N}\sum_{i=1}^{N} \mathbb{1}\{s_i \neq \hat{s}_i\}.$$

Table 2 presents the validation set CER and WER. Models with a case-sensitive character set have 36 alphanumeric characters and a special EOS token. Models with a case-insensitive character set additionally have 26 upper-case letters and 32 special characters. In all experiments, training and validation sets are i.i.d. (independent and identically distributed). Sequential transfer learning using one and then the other dataset is denoted with a right arrow between the dataset names. Usage of two datasets as a union is denoted with a plus sign.

When transfer learning is performed sequentially, the lowest error on either Imgur5K or IAM is obtained when the corresponding dataset is used last in the fine-tuning sequence. The lowest error on the union of two datasets is obtained when both datasets are used for transfer learning at the same time. These models also have a competitive performance on stand-alone datasets. Furthermore, the results show that the model's domain of application determines the fine-tuning approach. If the unseen data is close to the IAM distribution then a fine-tuning sequence Imgur5K→IAM is expected to give the lowest error. If the unseen data is close to the Imgur5K distribution then two datasets should be used in a reversed sequence. And a more general-purpose handwritten text recognition model, as represented by a union of the two datasets, is more likely to be obtained by fine-tuning on the union of the datasets.

4.4 Hyper-parameter Tuning

AdaDelta optimizer's learning rate η and decay constant ρ are tuned for the case-insensitive benchmark model fine-tuned on the Imgur5K→IAM sequence. Table 3 summarizes the WER on the IAM validation set. The lowest error corresponds

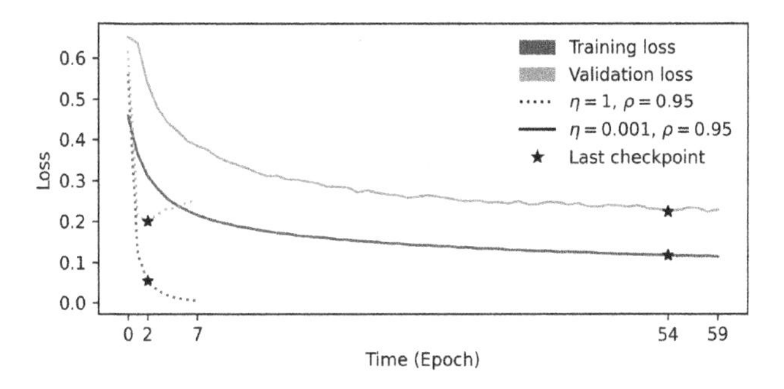

Fig. 3. Comparison of training and validation loss for the case-insensitive benchmark model fine-tuned on the Imgur5K→IAM sequence using two sets of hyper-parameters.

to the default $\eta = 1$ and $\rho = 0.95$. As the learning rate increases from 0.001 to 1.5, lower error values correspond to lower decay constants. This indicates that large steps in a gradient descent work better when less weight is given to the last gradient. This observation is consistent with [19] that suggests a large momentum (i.e., 0.9–0.99) with a constant learning rate can act as a pseudo increasing learning rate, and speed up the training. Whereas oversize momentum values can cause poor training results as the parameters of the pre-trained model get too distorted.

The IAM dataset is relatively small with only 41K training images, compared to 14.4M training images of the corresponding benchmark model, and it is therefore recommended to perform fine-tuning with low learning rates. However, a low learning rate $\eta = 0.001$ led to relatively high errors for all values of the decay constant. Figure 3 highlights that training with $\eta = 0.001$ was stopped when its training loss (solid blue line) was significantly above the training loss achieved with $\eta = 1$ (dotted blue line). This indicates an under-fitting problem caused by the early stopping mechanism. By switching the early stopping mechanism off and training for the full 300,000 iterations, we can achieve an error reduction from 10.86%→9.4% in validation WER. In future work, other regularization strategies will be investigated such as the norm penalty or dropout.

4.5 Ablation Study

An ablation study is conducted to demonstrate the effectiveness of transfer learning and attention mechanism for the given problem, and also to analyze the impact on the overall performance on the IAM dataset. This study is performed using a case-sensitive model trained and validated on the IAM dataset, and each model was trained five times using different random seeds. Table 4 presents the average validation set CER and WER. When transfer learning from the STR domain and the Imgur5K dataset is not used, the model's parameters are initialized randomly (indicated as "None" in the transfer learning column). When the attention mechanism is not used for prediction, it is replaced by the CTC.

The ablation study proves that transfer learning and attention mechanism help reduce both CER and WER. The lowest errors are achieved when both

Table 4. Ablation study highlighting the importance of transfer learning and attention mechanism. The lowest errors and corresponding components are shown in **bold**.

Transfer learning	Prediction	Val-CER, %	Val-WER, %
None	CTC	7.12	19.77
None	Attention	6.79	18.01
STR, Imgur5K	CTC	5.32	14.84
STR, Imgur5K	**Attention**	**4.84**	**11.97**

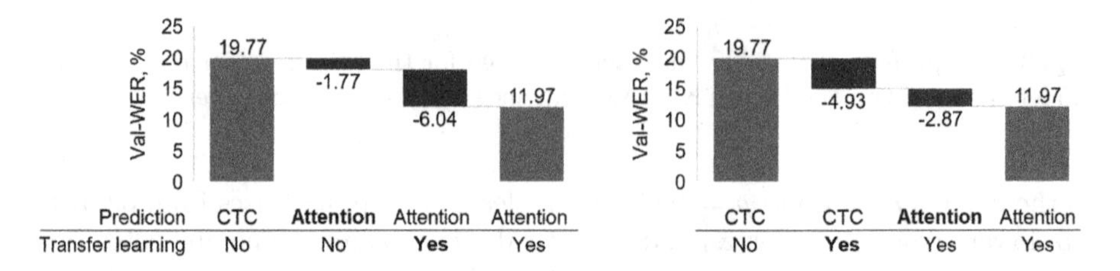

Fig. 4. The effect of transfer learning and attention mechanism on validation set WER for the IAM dataset when attention mechanism is introduced first (left), and when transfer learning is introduced first (right). Change is highlighted with **bold**.

transfer learning and attention mechanism are used. The effect of each component depends on which of them comes first, as illustrated in Fig. 4. Interestingly, in both cases, transfer learning is relatively more important among the two studied components in terms of error reduction.

4.6 Test Set Errors

Table 5 highlights the **test set CER and WER** for our best performing models on each dataset. A comparison is performed with related works on content attention-based seq2seq models, under the same experimental settings using the IAM dataset. The test set WER values suggest that the proposed method outperformed all the state-of-the-art methods [11,12,20]. The test set WER for the proposed method is 15.40%. Kang et al. [11] achieved a WER comparable to our work (15.91%) using a content-based attention mechanism. Kang et al. [11] also used location-based attention, achieving a WER of 15.15%, but since it explicitly includes the location information into the attention mechanism, it is not considered in this work. Furthermore, [11] incorporated a language model, unlike the proposed method. Though an RNN decoder can learn the relations between different characters in a word, it is worth investigating the use of language models to improve the performance of the proposed method as future work.

The test set CER results on the other hand suggest that Johannes et al. [15] outperformed all the state-of-the-art methods, with a test set CER of 5.24%. Similar to the proposed work, [15] did not use a language model. However, Kang et al. [11] improved upon their previous work [12] by integrating a language

Table 5. Results comparison with the state-of-the-art content attention-based seq2seq models for handwritten word recognition.

Character set	Test-CER			Test-WER		
	Imgur5K	IAM	Both	Imgur5K	IAM	Both
Ours, case-insensitive	6.46[a]	4.30[b]	5.96[c]	13.58[a]	12.82[b]	13.89[c]
Ours, case-sensitive	9.47[a]	6.50[b]	8.59[c]	20.45[a]	**15.40**[b]	18.97[c]
Kang *et al.* [12]	–	6.88	–	–	17.45	–
Bluche *et al.* [3]	–	12.60	–	–	–	–
Sueiras *et al.* [20]	–	8.80	–	–	23.80	–
Chowdhury *et al.* [5]	–	8.1	–	–	–	–
Johannes *et al.* [15]	–	**5.24**	–	–	–	–
Kang *et al.* [11][d]	–	5.79	–	–	15.91	–

Fine-tuning approaches: [a] IAM→Imgur5K; [b] Imgur5K→IAM; [c] Imgur5K+IAM.
[d]Also provides the results using location-based attention mechanism, which is not applicable in the considered content-based attention experimental settings.

model, and achieved a low error rate (5.79%). The proposed method achieved 6.50% CER and performs comparably with [15] and [11], outperforming the rest of the methods. Based on our error analysis (discussed in detail in the next section), the CER can be further reduced by using data augmentation, language modeling, and a different regularization method.

We also performed tests on the case-insensitive model that achieved 4.3% CER and 12.82% WER on IAM. This work is also the first attempt at using the Imgur5K dataset for handwritten word recognition (to the best of the authors' knowledge), and therefore no related work exists to compare the performance on the Imgur5K dataset. However, it can be observed from Table 5 that the proposed model has a relatively low test set CER and WER for Imgur5K, and can model a variety of handwritten words from this novel multi-writer dataset.

In summary, the proposed method has the lowest error rate when evaluated at the word level, and the third-lowest error rate when evaluate at the character level. The advantages of using the proposed approach in comparison with [15] and [11] include addressing the problem of training data scarcity, performing transfer learning from STR to HTR, and using Imgur5K dataset with word images from 5000 different writers. This allows the proposed HTR system to also handle examples that are rare due to insufficient annotated data. Our models are publicly available for further fine-tuning or predictions on unseen data.

5 Error Analysis

This section performs error analysis for our case-insensitive model performing best on the union of Imgur5K and IAM datasets.

5.1 Character-Level Error Analysis

To evaluate the model's predictive performance at the character level, precision and recall are computed from a confusion matrix. It is based on 33,844 images from 35,491 images in the test set where the length of the predicted word matches the length of the ground truth word. This is necessary to have 1-to-1 correspondence between true and predicted characters. Figure 5 presents a normalized confusion matrix with the ground truth characters in rows and predicted characters in columns. Only letters are shown for better readability. Values in each row are divided by the total number of characters in that row. Therefore, the main diagonal contains recall values, and each row adds up to 1, possibly with a rounding error. Off-diagonal values represent errors. The recall is also shown on the right margin and precision is shown on the bottom margin. All values are in percent, rounded to two decimal digits. Additionally, precision and recall are combined into a single metric, F1 score, and plotted against the probability mass (normalized frequency) of each character on Fig. 6. These two figures present several interesting and valuable insights:

- The model's errors pass a sanity check as the confusion matrix highlights that the most typical errors are between characters having similar ways of writing. For example, a is most often confused with o, and q with g.
- Letters z, q, j and x and all digits are significantly under-represented in the distribution of characters.
- In general, a lower probability mass of a character leads to a lower F1 score. However, the uniqueness of a character's way of writing seems to be an important factor. For example, digits *0* and *3* have approximately the same probability mass but hugely different F1 scores (77% and 97%). This difference is attributed to the fact that digit *3* is relatively unique and is, therefore, more difficult to confuse with other characters. The same conclusion applies to letter x, which achieves a relatively high F1 score despite being among the least frequent characters. Intuitively, characters that are somewhat unique in the way they are written need fewer examples in the training set to achieve high F1 scores. This also suggests that F1 scores can be used to prioritize characters for inclusion in the data augmentation, as a possible next step to improve the model's performance.

5.2 Bias-Variance Analysis

To gauge the model's performance, bias-variance analysis is performed according to [16]. It yields only approximate values of the test error components but is helpful and instructive from a practical point of view. Human-level error is used as a proxy for Bayes error, which is analogous to irreducible error and denotes the lowest error that can be achieved on the training set before the model's parameters start over-fitting. To estimate a human-level error, 22 human subjects from different backgrounds were asked to annotate the same set of randomly selected

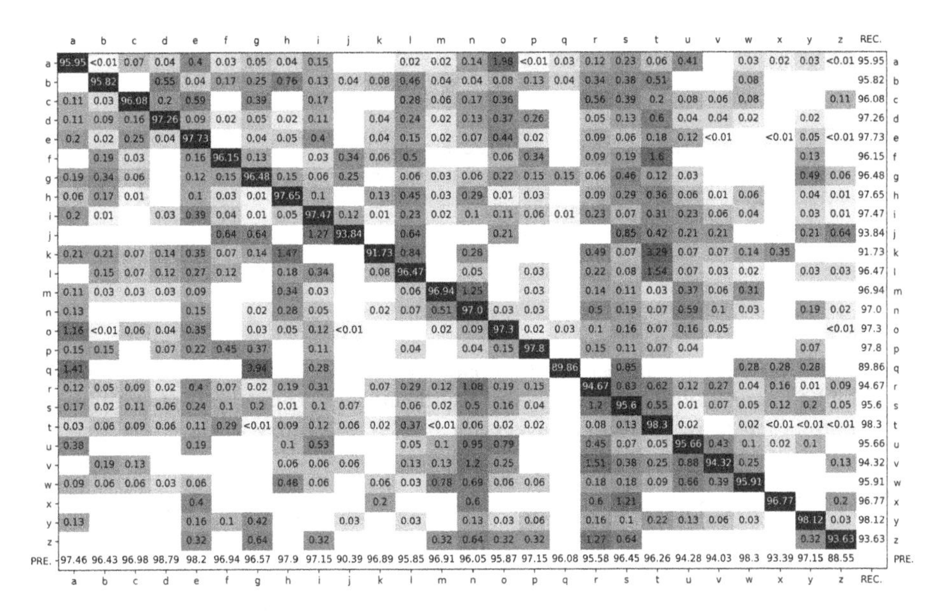

Fig. 5. Character-level error analysis using a confusion matrix with a row-wise normalization. The main diagonal represents recall. Each row adds up to 1, possibly with a rounding error. Abbreviations used: PRE. stands for precision, REC. stands for recall.

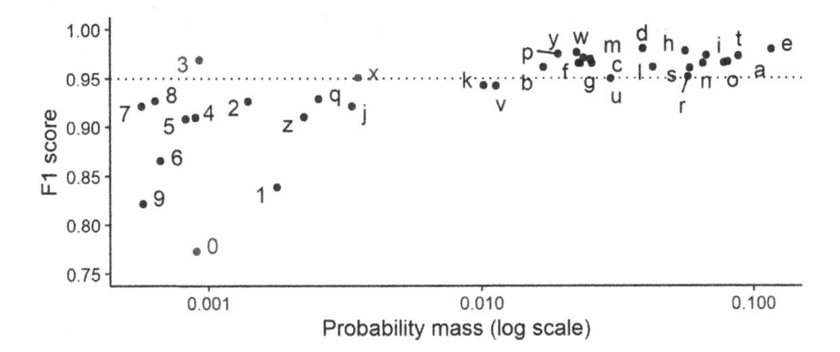

Fig. 6. F1 scores vs probability mass of lower-case alphanumeric characters. Characters discussed in the text are depicted in red. Figure best viewed in color.

100 misclassified images from the test set. The images that are misclassified by *all* reviewers contribute to the human-level error estimate.

In general, the bias of the model is the difference between Bayes error and the training set error. The variance is the difference between the training and validation set error. Additionally, the difference between the validation and test set error is denoted by the validation/test set mismatch error. Figure 7 summarizes the results of the bias-variance analysis. By definition, it is not feasible to reduce the Bayes error, which is 1.94% according to our estimate. It can be observed in Fig. 7 that the variance contributes the largest part of the error (7.53%) among other components and should be the primary focus of measures aimed at decreasing

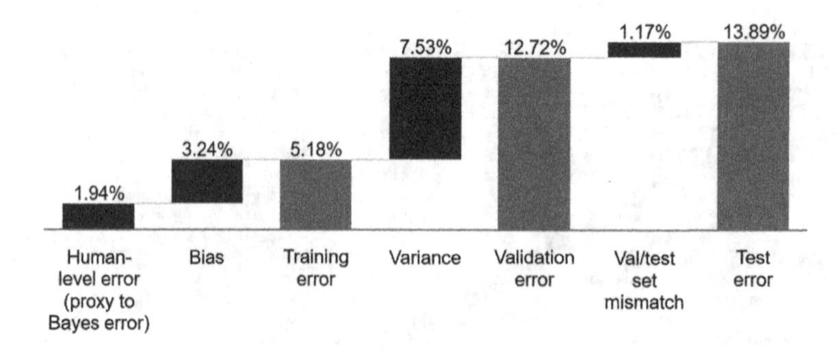

Fig. 7. Bias-variance decomposition. Red bars are components of the test set WER on the union of Imgur5K and IAM. Grey bars are the model's errors obtained by summing red bars from left to right. Human-level error is used as a proxy to Bayes error.

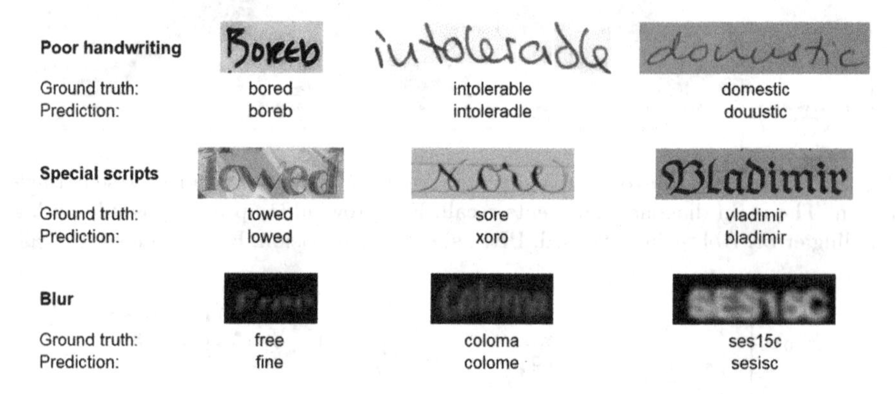

Fig. 8. Sample images from selected categories. Each category represents a possible main cause for misclassification, as identified through visual analysis.

the testing error of the model. Suitable strategies include using a larger training set, data augmentation, and a different regularization method (e.g. norm penalty, dropout) instead of early stopping. The second largest component is the bias (3.24%) that can be reduced by, for example, training a bigger neural network, using a different optimizer (e.g. Adam), and running the training longer. The smallest component (1.17%) is due to the validation/test set mismatch; this can be addressed by reviewing the partitioning strategy for the IAM dataset. However, this work follows the same splits as widely used in the HTR research for a fair comparison with other related work.

5.3 Visual Analysis of Images

To provide further insights into the possible causes for the model's errors, a visual inspection is performed on the same set of 100 misclassified images. The main problems contributing towards errors include poor handwriting, the use of special scripts (e.g. Gothic, curlicue, etc.), errors in some ground truth images, and other forms of issues such as low image resolution, blurry and rotated text, challenging backgrounds, strike-through words, or multiple words in an image. Figure 8 contains sample images representing some of the listed problems. The analysis suggests that the performance of the model can be further improved using data augmentation (with special scripts, blur, rotations, variety of backgrounds, etc.), and also fine-tuning on other multi-writer datasets. A language model is also foreseen to reduce the error rate on images with poor handwriting.

6 Conclusion

This work presented an end-to-end-HTR system based on attention encoder-decoder networks, which leverages pre-trained models trained on a large set of synthetic scene text images for handwritten word recognition. The problem of training data scarcity is addressed by performing transfer learning from STR to HTR, and using a new multi-writer dataset (Imgur5K) that contains word examples from 5000 different writers. This allows the proposed HTR system to also cope with rare examples due to insufficient annotated data. The experimental results on multi-writer datasets (IAM and Imgur5K) demonstrate the effectiveness of the proposed method. Under the given experimental settings, the proposed method outperformed the state-of-the art methods by achieving the lowest error rate when evaluated at the word level on the IAM dataset (test set WER 15.40%). At character level, the proposed method performed comparable with the state-of-the-art methods and achieved 6.50% test set CER. However, the character level error can be further reduced by using data augmentation, language modeling, and a different regularization method, which will be investigated as future work. Our source code and pre-trained models are publicly available for further fine-tuning or predictions on unseen data at GitHub[2].

Acknowledgment. The authors would like to thank the Centre for Digital Humanities Uppsala, SSBA, Anders Hast and Örjan Simonsson for their kind support and encouragement provided during the project. The computations were performed on resources provided by SNIC through Alvis @ C3SE under project SNIC 2021/7-47.

References

1. Baek, J., et al.: What is wrong with scene text recognition model comparisons? dataset and model analysis. In: IEEE International Conference on Computer Vision, pp. 4715–4723 (2019)
2. Bahdanau, D., Cho, K., Bengio, Y.: Neural machine translation by jointly learning to align and translate. arXiv preprint arXiv:1409.0473 (2014)
3. Bluche, T., Louradour, J., Messina, R.: Scan, attend and read: end-to-end handwritten paragraph recognition with mdlstm attention. In: 14th IAPR International Conference on Document Analysis and Recognition, vol. 1, pp. 1050–1055 (2017)
4. Bookstein, F.L.: Principal warps: thin-plate splines and the decomposition of deformations. IEEE Trans. Pattern Anal. Mach. Intell. **11**(6), 567–585 (1989)

[2] https://github.com/dmitrijsk/AttentionHTR.

5. Chowdhury, A., Vig, L.: An efficient end-to-end neural model for handwritten text recognition. arXiv preprint arXiv:1807.07965 (2018)
6. Dutta, K., Krishnan, P., Mathew, M., Jawahar, C.: Improving CNN-RNN hybrid networks for handwriting recognition. In: 2018 16th International Conference on Frontiers in Handwriting Recognition (ICFHR), pp. 80–85. IEEE (2018)
7. Frinken, V., Fischer, A., Manmatha, R., Bunke, H.: A novel word spotting method based on recurrent neural networks. IEEE Trans. Pattern Anal. Mach. Intell. **34**(2), 211–224 (2012)
8. Gupta, A., Vedaldi, A., Zisserman, A.: Synthetic data for text localisation in natural images. In: IEEE Conference on Computer Vision and Pattern Recognition, pp. 2315–2324 (2016)
9. He, K., Zhang, X., Ren, S., Sun, J.: Deep residual learning for image recognition. In: IEEE Conference on Computer Vision and Pattern Recognition, pp. 770–778 (2016)
10. Jaderberg, M., Simonyan, K., Vedaldi, A., Zisserman, A.: Synthetic data and artificial neural networks for natural scene text recognition. arXiv preprint arXiv:1406.2227 (2014)
11. Kang, L., Riba, P., Villegas, M., Fornés, A., Rusiñol, M.: Candidate fusion: integrating language modelling into a sequence-to-sequence handwritten word recognition architecture. Pattern Recogn. **112**, 107790 (2021)
12. Kang, L., Toledo, J.I., Riba, P., Villegas, M., Fornés, A., Rusiñol, M.: Convolve, attend and spell: an attention-based sequence-to-sequence model for handwritten word recognition. In: Brox, T., Bruhn, A., Fritz, M. (eds.) GCPR 2018. LNCS, vol. 11269, pp. 459–472. Springer, Cham (2019). https://doi.org/10.1007/978-3-030-12939-2_32
13. Krishnan, P., Kovvuri, R., Pang, G., Vassilev, B., Hassner, T.: Textstylebrush: transfer of text aesthetics from a single example. arXiv preprint arXiv:2106.08385 (2021)
14. Marti, U.V., Bunke, H.: The IAM-database: an English sentence database for offline handwriting recognition. Int. J. Doc. Anal. Recogn. **5**(1), 39–46 (2002)
15. Michael, J., Labahn, R., Grüning, T., Zöllner, J.: Evaluating sequence-to-sequence models for handwritten text recognition. In: 2019 International Conference on Document Analysis and Recognition (ICDAR), pp. 1286–1293. IEEE (2019)
16. Ng, A.: Lecture notes in cs230 deep learning (Stanford University, 2021 Fall) (2021)
17. Rodríguez-Serrano, J.A., Perronnin, F., et al.: A model-based sequence similarity with application to handwritten word spotting. IEEE Trans. Pattern Anal. Mach. Intell. **34**(11), 2108–2120 (2012)
18. Shi, B., Bai, X., Yao, C.: An end-to-end trainable neural network for image-based sequence recognition and its application to scene text recognition. IEEE Trans. Pattern Anal. Mach. Intell. **39**(11), 2298–2304 (2016)
19. Smith, L.N.: A disciplined approach to neural network hyper-parameters: Part 1-learning rate, batch size, momentum, and weight decay. arXiv preprint arXiv:1803.09820 (2018)
20. Sueiras, J., Ruiz, V., Sanchez, A., Velez, J.F.: Offline continuous handwriting recognition using sequence to sequence neural networks. Neurocomputing **289**, 119–128 (2018)
21. Tan, C., Sun, F., Kong, T., Zhang, W., Yang, C., Liu, C.: A survey on deep transfer learning. In: Kůrková, V., Manolopoulos, Y., Hammer, B., Iliadis, L., Maglogiannis, I. (eds.) ICANN 2018. LNCS, vol. 11141, pp. 270–279. Springer, Cham (2018). https://doi.org/10.1007/978-3-030-01424-7_27

HST-GAN: Historical Style Transfer GAN for Generating Historical Text Images

Boraq Madi(✉), Reem Alaasam, Ahmad Droby, and Jihad El-Sana

Ben-Gurion University of the Negev, Beersheba, Israel
{borak,rym,drobya}@post.bgu.ac.il, el-sana@cs.bgu.ac.il

Abstract. This paper presents Historical Style Transfer Generative Adversarial Networks (HST-GAN) for generating historical text images. Our model consists of three blocks: Encoder, Generator, and Discriminator. The Encoder encodes the style to be generated, and the Generator applies an encoded style, $\mathcal{S}$ to an input text image, $\mathcal{I}$, and generates a new image with the content of $\mathcal{I}$ and the style $\mathcal{S}$. The Discriminator encourages the Generator to enhance the quality of the generated images. Multiple loss functions are applied to ensure the generation of quality images. We evaluated our model against three challenging historical handwritten datasets of two different languages. In addition, we compare the performance of HST-GAN with the state of art approaches and show that HST-GAN provides the best generated images for the three tested datasets. We demonstrate the capability of HST-GAN to transfer multiple styles across domains by taking the style from one dataset and the content from another dataset and generate the content according to the desired style. We test the quality of the style domains transferring using a designated classifier and a human evaluation and show that the generated images are very similar to the original style.

Keywords: Historical handwritten styles transfer · Generative adversarial networks · Generating document images

1 Introduction

Over the recent years, deep learning models have gained popularity due to their effectiveness in solving image recognition and classification problems. Their ability to learn complex feature representation from data has been driving the development of sophisticated architectures and the preparation of training data. The quality and quantities of labeled datasets are essential for supervised machine learning algorithms. However, obtaining quality labeled data is time-consuming and expensive. In particular, historical documents datasets are inherently complicated to generate and difficult to annotate; they often require professional scholars.

Historical manuscripts datasets require proper preparation and pre-processing steps before applying learning stage of deep learning algorithms. Today, collecting datasets is no longer difficult, and many institutions provide access to most

© Springer Nature Switzerland AG 2022
S. Uchida et al. (Eds.): DAS 2022, LNCS 13237, pp. 523–537, 2022.
https://doi.org/10.1007/978-3-031-06555-2_35

of their digitized manuscripts. However, data preparation includes selecting the appropriate samples, annotating them, and splitting the data into training and testing. We often need to pre-process the dataset to extract main-text, remove side notes, and apply denoising and binarization. Challenging aspects, such as fading ink, ink stains, or missing parts, require advanced pre-processing steps such as restoration of the damaged regions of these manuscripts.

Manual restoration of historical manuscripts is time-consuming, requires highly trained experts, and may spoil the document's original writing style. Automatic approaches such as inpainting algorithms were developed to restore missing parts in natural images [14]. These approaches rely on surrounding pixels to extract clues for reconstructing a coherent result. It is challenging to extract these clues from historical documents, and the obtained results are often inaccurate, especially for missing parts. For example, applying inpainting to reconstruct missing parts of historical documents produces artifacts, such as inconsistent stroke width and blur level. This behavior is barely noticeable in natural images but quite striking in historical document images, mainly due to the rigid structure of the text and its sharp edges.

In recent years, Generative Adversarial Networks(GAN) [4] and their variations have led to impressive progress in image manipulation and generation. The learning process of GAN is guided by an image-to-image translation, which disentangles attributes from the input images and uses these attributions for generating new results. However, these models assume a complete mapping between two corresponding images in the dataset for different domains. This assumption is not always held, especially for historical manuscripts where the dataset is scarce and irreproducible, which makes establishing such mapping difficult. For these reasons, training these models for historical manuscript style transfer is unfeasible.

This paper presents HST-GAN, a GAN-Based novel Style Transfer approach for historical documents. Our approach transfers the input text to a style agnostic domain and then maps the style agnostic representation to the target style. A pixel-width skeleton representation of text carries little style features and can be perceived as a style agnostic domain. Obtaining the skeleton of a text image is practically disentangling the style from the text.

HST-GAN applies the style of reference image/s to an input document image. It transfers the input text image to a pixel width representation (skeleton), extracts style features, S_f, from reference images, and applies S_f to the skeleton representation to obtain the final result.

To perform these tasks, HST-GAN utilizes three networks: an *Encoder*, a *Generator* and a *Discriminator*. The Encoder learns the style features from the reference images. The Generator applies style features, S_f, to a text skeleton, T_s, and generates a text image of T_s, which have the style S_f. The Discriminator determines whether the obtained images belong to the target style class or not.

Previous works rely on training datasets that include content samples with different styles, which enables complete image-to-image mapping [12,24]. It is hard to locate corresponding patches with the same content on historical document images that have different text styles. Thus building a dataset that includes direct transfer between styles, such as in [12,24], is difficult. To overcome this limitation, we present a novel scheme (HST-GAN) that transfers text images via the skeleton domain.

Since HST-GAN synthesizes new historical documents with various writing styles, it could be perceived as a sophisticated augmentation procedure. This augmentation can expand datasets that are used to train models that address various historical document images processing tasks, such as writer style identification. In addition, HST-GAN can restore degraded and missing parts of manuscripts from manually generated sketches. Since the model inputs skeleton text similar to sketches, we can restore manuscripts interactively using those sketches. We have implemented HST-GAN, conducted various experimental studies, and obtained promising results.

2 Related Work

Style Transfer methods synthesize an output image based on style information extracted from a reference image and a content image. Gatys *et al.* [3] perceived the style of an image as a piece of information that is irrelevant to the position. They define the style and content of an image using features computed from a pre-trained VGG Network. The style is represented using a form of a Gram matrix with values of the second-order statistic of features, and the content is the high-level features extracted by the VGG Network. Their objective function matches the output style with the given style image and the output content with the content image. However, their objective function is slow since it relies on a heavy optimization process. Several approaches train feed-forward networks with a single forward pass [7,11]. However, these approaches are limited since their models apply to one or two specified styles, which means it cannot be generalized for any style. Other works accelerate the process by finding a general style representation to replace the Gram Matrix [5,13]. In particular, Huang *et al.* [5] introduced an adaptive instance normalization layer that aligns the content features variance and channel-wise mean with those of the style features. Inspired by this idea, many approaches adopted adaptive instance normalization layer for different tasks, such as high-quality generation [8], unsupervised translation [24], and text effect transfer [12].

Image-to-image Translation is performed by mapping an image from a source domain to a target domain using a generation function. One of the first works in this domain was proposed by Isola *et al.* [6], who introduced a unified translation framework using a supervised manner based on conditional GAN [19]. Their work combines two loss functions with paired samples from two different domains. Since acquiring paired image samples is challenging, several approaches [6,15] proposed frameworks that do not need pairs of images for the image-to-image translation. Zhu *et al.* [6] pair data by introducing a cycle loss which requires the translated image to be translated back to the input one, and Liu *et al.* [15] assume that any corresponding image pair from different domains can be mapped to the same latent representation in a shared-latent space. However, these approaches can learn the relations between only two domains. To generalize for multiple domains, Choi *et al.* [2] use a one-hot vector to specify the target domain, but extending the vector to a new domain is very expensive. To overcome this problem, Liu *et al.* [16] apply a few-shot algorithm, which is based on an unsupervised image-to-image translation framework.

Text-Effect transfer was first introduced by Yang *et al.* [23]. Their synthesizing idea of text effects is based on the analytic procedure of the regularity of the spatial distribution for text effects. Specifically, they characterize the stylized patches depending on the normalized position between the glyphs stokes and optimal scales. Although they have appealing results, their methods suffer from a heavy computational burden due to distance estimation and optimal scales detection. TET-GAN [24] is the first approach to use deep neural networks for text effect transfer. They train conditional GAN to accomplish style transfer and style removal. Their model, which consists of two generators, three encoders, and two discriminators, aims at disentangling and recombining the glyph(content) features and effect(style) features. However, its training is slow even for small datasets as a result of optimizing a vast amount of parameters and it is not robust to different glyph structures. These methods are limited to transfer only visual effect without font variations, i.e., the transferred image and the source image share the same font and text. Azadi *et al.* [1] address this problem by adopting two stacked conditional GAN models for font transfer and effect transfer. However, the model is limited to English big letters and low-resolution images of size 64×64. Li *et al.* [12] introduced a new model, FET-GAN, for text effect transfer with font variation for multiple text effects domains. They achieve the state-of-the-art in the text effect transfer task. However, their model is based on printed fonts which differ from the handwritten text. Handwritten text variation across the same writer and among different writers makes the task more challenging comparing to the printed font. In addition, their work assumes the text is centered in a clean image, which is not the case for historical handwritten documents.

To summarize, state-of-the-art text transfer methods are still unable to generate credible and diverse historical handwritten text automatically. Since all the mentioned approaches deal with clean font images with centered text, which is not the case for historical handwritten documents. Historical documents suffer from various degradation types such as bleed-through, stains, and faded ink. In addition, the handwriting text can vary across the same writer which is not the case for printed fonts. In this work, we focus on learning styles by considering calligraphic features and the background of document images.

2.1 Datasets

We experimented with the following three Arabic and Latin historical datasets.

DIVA-HisDB [21] is a historical Latin dataset that consists of three medieval manuscrip, each one with its style. We experimented with one sample manuscript,

CSG863, which has a challenging layout. It includes touching components, little overlapping letters, but no punctuation, as shown in Fig. 1(a).

AHTE dataset is a publicly available[1] historical Arabic handwritten dataset, which includes pages from different manuscripts. The dataset includes many degradation artifacts, such as bleed-through, stains, and faded ink. According to styles, we divide the dataset into two subsets: AHTE-A and AHTE-B. **AHTE-A** contains bold text with many ink stains but less punctuation and overlapping text compared to AHTE-B. It is more complex than DIVA-HisDB but less challenging than AHTE-B. An example image is shown in Fig. 1(b). **AHTE-B** includes complex layouts, crowded diacritics, and many overlapping words and sentences as shown in Fig. 1(c). Although it has sharp text, it includes background texture noise. It is the most complex among the three datasets due to the existence of many small details, such as punctuation.

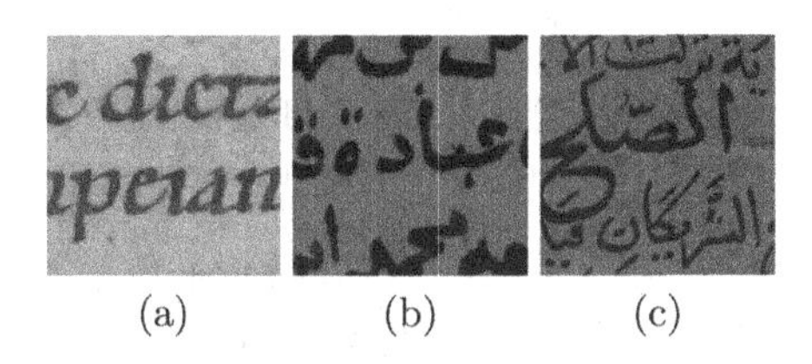

(a) (b) (c)

Fig. 1. Patch samples from the three datasets CSG803, AHTE-A and AHTE-B respectively

2.2 Data Preparation

The dataset consists of samples created from the three datasets. Each sample is a triplet that includes a patch P, its skeleton $\mathcal{S}(P)$, and a set of reference patches, which have styles similar to P. The P was cropped randomly from a document image and scaled to the size 256×256, as shown in Fig. 1. The skeleton of P, $\mathcal{S}(P)$, is generated following three steps: (1) converting P to gray-scale, (2) applying an image binarization, (3) using Medial Axis Transform (MAT) to get skeleton form. In this work we adopted the Otsu binarization. Following the steps mentioned above, we build a train and test sets of $9K$ and $3K$ triplets, respectively.

3 Method

We present Historical Style Transfer (HST-GAN), a model for transferring style via skeleton space among multiple visual styles of historical documents (Fig. 2).

[1] https://www.cs.bgu.ac.il/~berat/data/ahte_dataset.

(a) (b)

Fig. 2. (a) A historical patch from AHTE-B dataset. (b) The corresponding patch after applying otsu binarization and Medial Axis Transform.

3.1 Model Framework

The HST-GAN framework consists of three modules: an encoder, E, a generator, G, and a discriminator, D. Our goal is to train the model to learn several one-to-many mappings and transfers a text skeleton patch to multiple historical style domains according to reference patches.

The input to the encoder E is a set of reference patches R and its output is a latent vector, L as shown in Fig. 3. The reference patches R share the same style, and the latent vector L contains the encoded style of R. The encoder architecture includes five convolution blocks; each consists of a convolution layer followed by ReLU. The kernel size of the first block is 5×5, and the kernel size of the following four blocks is 3×3. Next, an adaptive average pooling is applied with 1×1 convolution layer for channel reduction, and its output is averaged to produce a fixed-size vector, 1×512. Finally, a sequence of fully connected layers with ReLU are appended to create the encoded style vector L of size 1×6144.

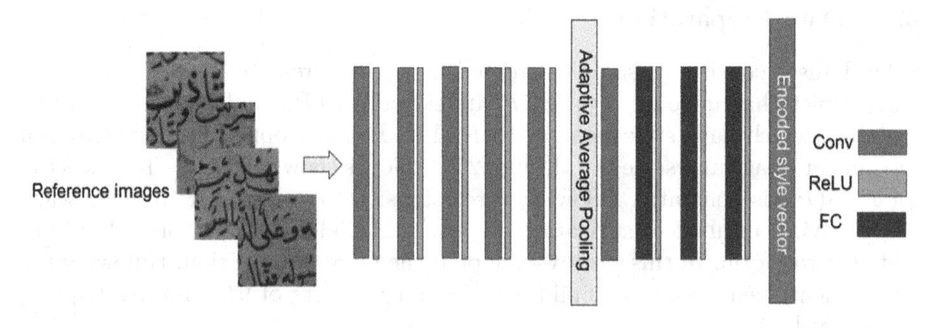

Fig. 3. The encoder composed of 5 convolution block followed by a sequence of FCs which input a vector of size 1×512 and outputs an encoded style vector L of size 1×6144

The Generator G inputs a skeleton text image x and outputs an image y^* conditioned with the encoded historical style L, as shown in Fig. 4. The output y^* shares the same style as the encoded reference images R. Moreover, the textual content of y^* is the same text in x.

The architecture of G consists of four convolution blocks, residual(Res) and ResAdaIN blocks, and three pairs of upsampling layers and convolution blocks, as shown in Fig. 4. Each convolution block includes convolution, ReLU, and instance normalization layers. The first block applies a 5×5 kernel and the remaining three apply 3×3 kernels. The residual(Res) and ResAdaIN blocks apply 3×3 kernels. The upsampling factor is two, and the kernel size is 5×5 in these three blocks.

Finally, the Refiner network consists of a sequence of Residual in Residual Dense Blocks(RRDBs) [22] and aims to increase sharpness and reduce noise (see Fig. 5), according to the finding of Wang *et al.* [22]. To preserve the dimension of the output image, we remove the upsampling layers from the RRDBs, as shown in Fig. 5.

The encoded style vector L is combined with the architecture of G via residual adaptive instance normalization blocks(ResAdaIN), similar to [12]. We segment L into three parts and pass them to the ResAdaIN layers.

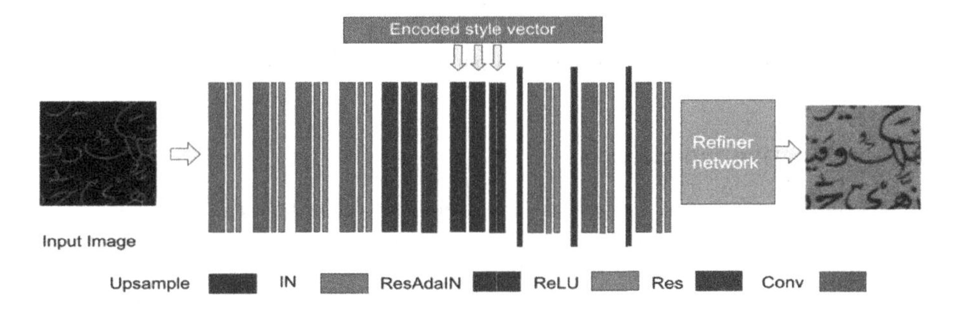

Fig. 4. The architecture of G inputs a skeleton text image x and outputs a historical image or transferred image y^* conditioned to the encoded historical style L.

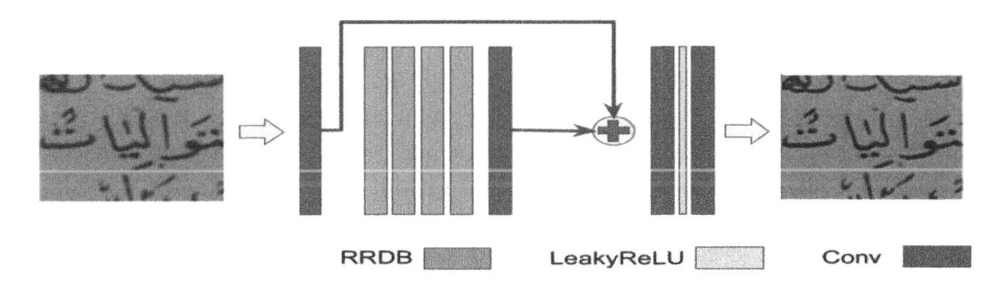

Fig. 5. Refiner network inputs an image and output an increased resolution, sharpness, and deblurring version.

The Discriminator D is patch-based discriminator [6] with few modifications, as illustrated in Fig. 6. The original version outputs a tensor with channel number equal to one. We modify the number of channels of the output tensor to C, which represents the number of historical style classes during the training phase, similar to [12]. The discriminator architecture includes one convolution block with a kernel of size 5×5 followed by three pairs of residual blocks using 3×3 kernel and average pooling. We append two layers, a sigmoid and an adaptive average pooling. The last convolution block builds a tensor of C classes as channel value and dimension of 32×32, as shown in Fig. 6.

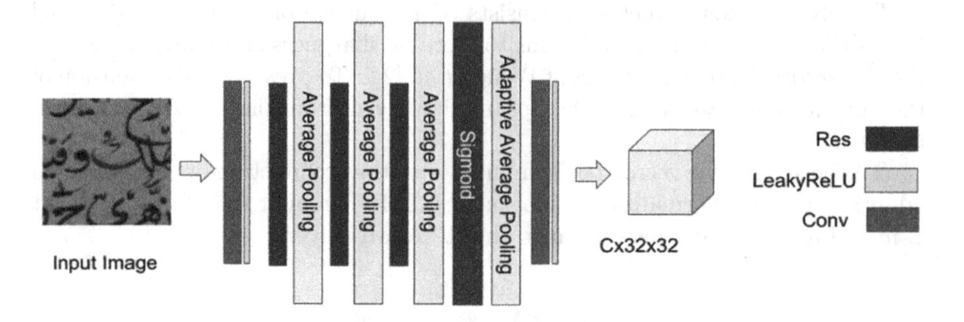

Fig. 6. The discriminator inputs a patch images and outputs a tensor C

3.2 Objective Function

The objective function combines Transfer Loss, Total Variation Loss, Style Consistency Loss, Feature Loss, and Adversarial loss. These loss functions are weighted by a hyperparameter, λ, that reflects the contribution of each loss.

Transfer Loss calculates the distance between the generated patch and the ground truth. Equation 1 presents the loss, where y is the ground truth and the generator input patch is x, and the encoded style of the reference patches is L.

$$L_{transfer} = E_{x,y}||y - G(x;L)||_1 \tag{1}$$

Reconstruction Loss measures the similarity between the generated patch and the respective ground truth. The generator inputs a patch, x, and its encoded style, L, and compare the output against x, as formulated in Eq. 2. The typical use of reconstruction loss assumes the input and the output are from the same domain [12,24]. In contrast, our input and output are from different domains; the input is a skeleton text patch, while the output is an RGB patch. Therefore, the reconstruction loss is not applicable for our model, as reported in Sect. 4.

$$L_{rec} = E_x||x - G(x;L)||_1 \tag{2}$$

Feature Loss compares the output of the generator G and the target image y on the feature spaces. We adopt a pre-trained VGG16 to extract high-level information and then apply L_1 metric, as formulated in Eq. 3, where ϕ_j is the activation function of the jth layer of the VGG16. We set $j = 4$ according to Liu's [17] experimental study.

$$L_f = E_{x,y}||\phi_j(y) - \phi_j(G(x; L))||_1 \tag{3}$$

Total Variation Loss is used to avoid color mutation, eliminate artifact noises, and encourage smooth regions. Equation 4 formulates this loss function, where y^* is the transferred image. The loss applies vertical and horizontal smoothing. We normalize the loss by width, height, and number of channel of y^*.

$$L_{tv} = \sum_{(i,j)\in y^*} (y^*_{i+1,j} - y^*_{i,j})^2 + \sum_{(i,j)\in y^*} (y^*_{i,j+1} - y^*_{i,j})^2 \tag{4}$$

Style Consistency Loss guides the encoder E to encode the reference images with the same style similarly. Since reference images sharing the same style, they should have similar latent vectors. This loss aims to minimize the distance between similar styles patches on the latent space. The loss function is applied to each encoded reference image before averaging, similar to Kulkarni *et al.* [10].

$$L_{consistent} = \sum_{k=1}^{R}\sum_{j=1}^{R} ||E(r_k) - E(r_j)||_1 \tag{5}$$

Adversarial loss combines the hinge version [20] with gradient penalty regularization [18] to ensure convergence and training stability, similar to [12]. For the style reference images, R, and the generated output $G(x; L)$, the adversarial loss is computed using the prediction score of discriminator D over the respective style channel corresponding to the style of R. We set the hyperparameter $\lambda_{GPR} = 10$ according to [20], to address gradient penalty regularization. The adversarial loss is expressed in Eq. 6.

$$\begin{aligned} L_{adv} = &\mathbb{E}_r[min(0, -1 + D(R)] + \\ &\mathbb{E}_x[min(0, -1 + D(G(x; L))] + \lambda_{GPR}\mathbb{E}_R|| \bigtriangledown D(R)||^2 \end{aligned} \tag{6}$$

The final objective function is the weighted sum of the previous loss functions, as shown in Eq. 7.

$$\begin{aligned} L = &\lambda_{transfer} * L_{transfer} + \lambda_f * L_f + \lambda_{tv} * L_{tv} + \\ &\lambda_{consistent} * L_{consistent} + \lambda_{adv} * L_{adv}. \end{aligned} \tag{7}$$

The components of our model G,D and E are optimized via the min-max criterion, $\min_{G,E} \max_D L$.

4 Experiments

In this section we present an experimental evaluation of our style transfer framework. We start by comparing the performance of our model with other approaches and then estimate the quality of the images produced by our framework using a deep learning model and human inspection.

4.1 Style Transfer Evaluation

As we have mentioned earlier, the reconstruction loss is not applicable for our model, i.e., including it in the objective loss produces unacceptable results. To demonstrate this, we train the FET-GAN [12] and TET-GAN [24] with their original configuration using the datasets we mentioned in Sect. 2.2. Recall that these configurations include the reconstruction loss. The resulting images are inaccurate and noisy, i.e., the model failed to obtain acceptable results. The second and third columns of Fig. 7 presents sample results.

We modify FET-GAN by removing the reconstruction loss from the objective function. The obtained results were better than the results with the reconstruction loss. However, these results include small black spots and gaps, as shown in Fig. 8. This kind of noise was encountered in the original paper but was ignored [12].

We managed to overcome these limitations by removing the reconstruction loss, and integrating the total variation loss in the objective loss of FET-GAN. Let us refer to the modified model as UFET-GAN. The total variation loss encourages the output to be smooth with coherent regions. The UFET-GAN manages to eliminate the black spots, as shown in Fig. 8.

Fig. 7. The first column shows skeleton inputs for AHTE-A, AHTE-B, and DIVA. The results of FET-GAN [12] and TET-GAN [24] are presented in second and third column respectively. The last column shows the target images.

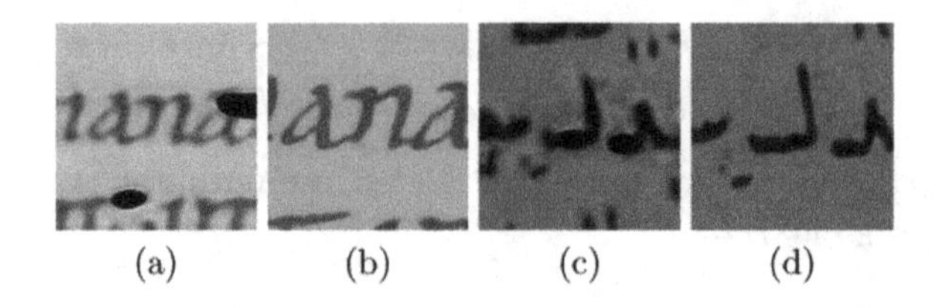

Fig. 8. (a) and (c) shows the black spots problem for AHTE-B and DIVA. (b) and (d) shows the results after adding total variation loss function

We compare our model, HST-GAN, with UFET-GAN, CycleGAN [25] and Pix2Pix [6]. The four models were trained and tested using the prepared dataset(Section 2.2). We visually compared the generated images and found that HST-GAN generates the best quality text images. It removes background noise and focuses on the text style, as illustrated in Fig. 9. As seen, HST-GAN and UFET-GAN generate smooth background and remove unnecessary background noise comparing to CycleGAN and Pix2Pix. For images with crowded diacritics and overlapping words as shown in the third row of Fig. 9, HST-GAN produced the best results and UFET-GAN comes second, while CycleGAN and Pix2Pix failed to generate the small details correctly.

We noticed that UFET-GAN generates images with higher blur level than the other approaches, while HST-GAN provides images with sharp edges. Overall, for the three test datasets, DIVA, AHTE-A and AHTE-B, HST-GAN generates images with the best quality.

For training HST-GAN, we set the hyperparameters for the adversarial loss as follow: $\lambda_{transfer} = 10$, $\lambda_f = 1$, $\lambda_{tv} = 0.001$, $\lambda_{consistent} = 1$, $\lambda_{adv} = 1$ and $\lambda_{GPR} = 10$. We set the hyperparameter, λ_{tv}, of the total variation loss to 0.001. In both HST-GAN and UFET-GAN, we use Adam optimizer [9] with a batch size of 4 over 20 epoch. We choose the number of reference images to be four in HTS-GAN, based on the ablation study of [12]. All the models were trained from scratch using a learning rate of 0.002.

4.2 Data Augmentation Using Style Transfer

We have shown that our model, HST-GAN, outperforms other approaches in terms of image quality. Since HST-GAN can project the style of a historical dataset to the content of another dataset, it can be used to augment historical dataset. To evaluate this, we project the style of AHTE-A to the content of AHTE-B and vice versa.

To transfer the style of AHTE-A to AHTE-B, we use a skeleton patch of size 256×256 from AHTE-A test set, and pass it to the model with random reference images from AHTE-B. This transfers the skeleton text patch to AHTE-B style as shown in Fig. 10. We perform the same procedure in the opposite direction from AHTE-B to AHTE-A. Let us refer to the generated sets, A_B and B_A, respectively.

Transferring the style from AHTE-A to AHTE-B and vice versa is evaluated using an authentic style classifier. The architecture of the *style classifier* is shown in Fig. 11. It includes four convolution blocks, max pooling, dropout, and sigmoid layers and outputs a vector with two entries that represents the probability of a patch being from AHTE-A or AHTE-B.

We generate a training and test sets from AHTE-A and AHTE-B by binarizing the patches using Otsu and labeling them using a hot-encoder scheme. The binarization guides the model to focus on learning to classify the text style based on the text itself and ignoring the background features. We train the model for 13 epochs using Adam optimizer. The model reaches style classification accuracy of 99.7%.

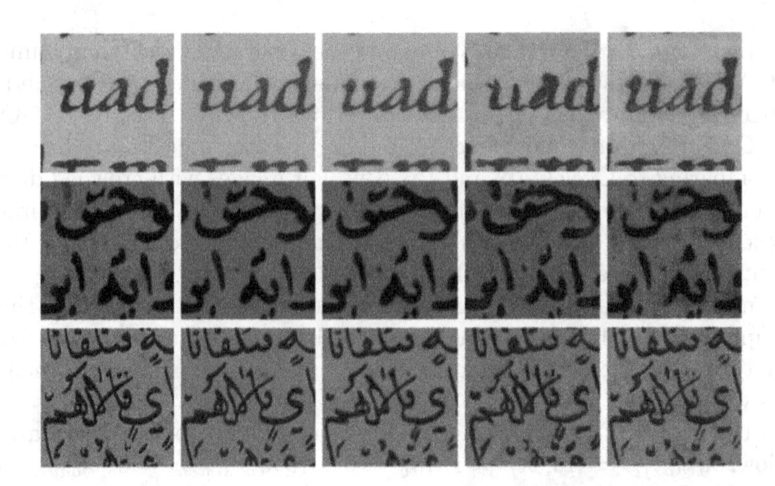

Fig. 9. The first column shows the original image from DIVA, AHTE-A and AHTE-B, followed by The results of our proposed model HST-GAN, UFET-GAN, CycleGAN and Pix2Pix respectively.

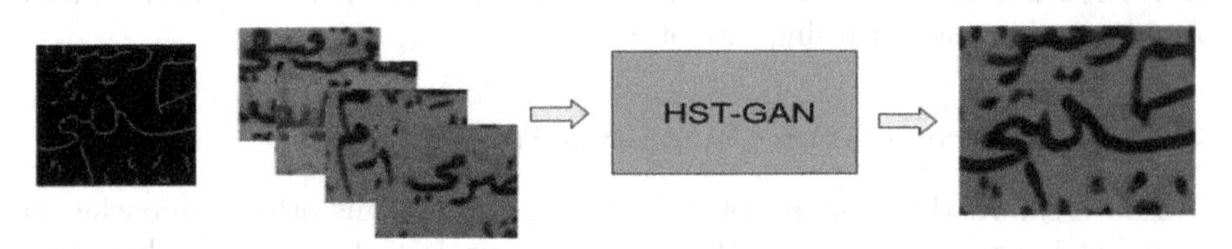

Fig. 10. Skeleton text patch from AHTE-A test set and reference images from AHTE-B. The transferred output have similar style to to AHTE-B style, while the content text is taken from the skeleton text patch.

We evaluate the performance of HST-GAN by applying the classifier to the two directions AHTE-A to AHTE-B and vice versa. The results of the classifier over the two test sets A_B and B_A are 91.9 and 92.5, respectively. These results indicate the quality of transferring; higher value means that HST-GAN generates high-quality patches or patches similar to the original ones to the degree that the classifier can't distinguish.

Fig. 11. The style classifier, which consists of four convolution blocks and two fully connected layer.

Qualitative Evaluation: To qualitatively evaluate the obtained results, we build a questionnaire consisting of 30 questions. Each question has an image, which is either a transferred image generated by our model or an original image from AHTE-A or AHTE-B. In addition, each question has two options containing a set of images from AHTE-A and AHTE-B as shown in Fig. 12.

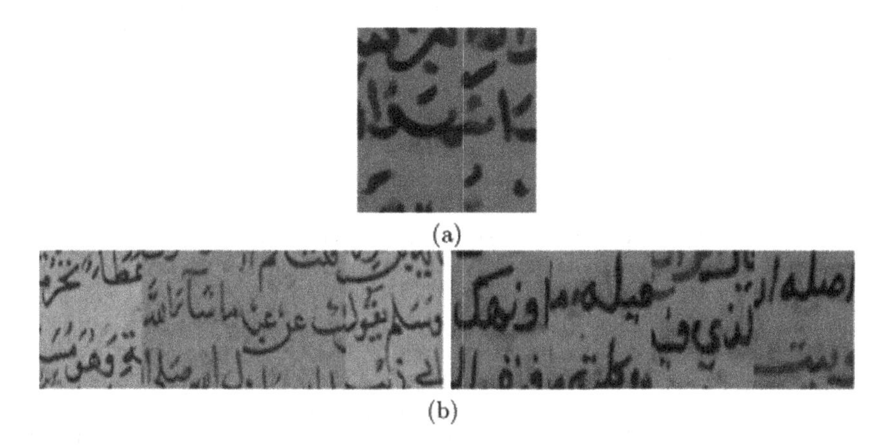

Fig. 12. An example of one of the 30 questions given in the questionnaire. The given image is a transfer image from AHTE-A to AHTE-B.

We have 25 respondents of different ages (13–60) who are native speakers of Arabic. The average score for the questions that involve AHTE-A or transferred image from B_A is 85% and for AHTE-B or transferred image from A_B is 82.1%. These results indicate the difficulty to distinguish between the transferred and real images. This questionnaire is also a sanity check for the style classifier. We get close results in both the questionnaire and the classifier on both datasets, demonstrating the classifier reliability.

5 Conclusion

In this paper, we present Historical Style Transfer GAN (HST-GAN) model for generating historical handwritten images. Our model inputs a reference image representing the desired style and a content image including the text content we aim to generate, and produces a high-quality historical text image. We test our model using challenging historical datasets, which include many degradation artifacts and show that our model manages to produce quality results.

We compare our model with other approaches and show that our model outpaced the state of art approaches in terms of the quality of historical text images.We evaluated the quality of the style transfer between two domains using a binary style classifier and a questionnaire.

References

1. Azadi, S., Fisher, M., Kim, V.G., Wang, Z., Shechtman, E., Darrell, T.: Multi-content gan for few-shot font style transfer. In: Proceedings of the IEEE Conference on Computer Vision and Pattern Recognition, pp. 7564–7573 (2018)
2. Choi, Y., Choi, M., Kim, M., Ha, J.W., Kim, S., Choo, J.: Stargan: unified generative adversarial networks for multi-domain image-to-image translation. In: Proceedings of the IEEE Conference on Computer Vision and Pattern Recognition, pp. 8789–8797 (2018)
3. Gatys, L.A., Ecker, A.S., Bethge, M.: Image style transfer using convolutional neural networks. In: Proceedings of the IEEE Conference on Computer Vision and Pattern Recognition, pp. 2414–2423 (2016)
4. Goodfellow, I.J., et al.: Generative adversarial networks. arXiv preprint arXiv:1406.2661 (2014)
5. Huang, X., Belongie, S.: Arbitrary style transfer in real-time with adaptive instance normalization. In: Proceedings of the IEEE International Conference on Computer Vision, pp. 1501–1510 (2017)
6. Isola, P., Zhu, J.Y., Zhou, T., Efros, A.A.: Image-to-image translation with conditional adversarial networks. In: Proceedings of the IEEE Conference on Computer Vision and Pattern Recognition, pp. 1125–1134 (2017)
7. Johnson, J., Alahi, A., Fei-Fei, L.: Perceptual losses for real-time style transfer and super-resolution. In: Leibe, B., Matas, J., Sebe, N., Welling, M. (eds.) ECCV 2016. LNCS, vol. 9906, pp. 694–711. Springer, Cham (2016). https://doi.org/10.1007/978-3-319-46475-6_43
8. Karras, T., Laine, S., Aila, T.: A style-based generator architecture for generative adversarial networks. In: Proceedings of the IEEE/CVF Conference on Computer Vision and Pattern Recognition, pp. 4401–4410 (2019)
9. Kingma, D.P., Ba, J.: Adam: a method for stochastic optimization. arXiv preprint arXiv:1412.6980 (2014)
10. Kulkarni, T.D., Whitney, W., Kohli, P., Tenenbaum, J.B.: Deep convolutional inverse graphics network. arXiv preprint arXiv:1503.03167 (2015)
11. Li, C., Wand, M.: Combining markov random fields and convolutional neural networks for image synthesis. In: Proceedings of the IEEE Conference on Computer Vision and Pattern Recognition, pp. 2479–2486 (2016)
12. Li, W., He, Y., Qi, Y., Li, Z., Tang, Y.: FET-GAN: font and effect transfer via k-shot adaptive instance normalization. In: Proceedings of the AAAI Conference on Artificial Intelligence, pp. 1717–1724 (2020)
13. Li, Y., Wang, N., Liu, J., Hou, X.: Demystifying neural style transfer. arXiv preprint arXiv:1701.01036 (2017)
14. Liu, G., Reda, F.A., Shih, K.J., Wang, T.C., Tao, A., Catanzaro, B.: Image inpainting for irregular holes using partial convolutions. In: Proceedings of the European Conference on Computer Vision (ECCV) (2018)
15. Liu, M.Y., Breuel, T., Kautz, J.: Unsupervised image-to-image translation networks. In: Advances in Neural Information Processing Systems, pp. 700–708 (2017)
16. Liu, X., He, P., Chen, W., Gao, J.: Multi-task deep neural networks for natural language understanding. arXiv preprint arXiv:1901.11504 (2019)
17. Liu, Y., Qin, Z., Luo, Z., Wang, H.: Auto-painter: cartoon image generation from sketch by using conditional generative adversarial networks. arXiv preprint arXiv:1705.01908 (2017)

18. Mescheder, L., Geiger, A., Nowozin, S.: Which training methods for gans do actually converge? In: International Conference on Machine Learning, pp. 3481–3490. PMLR (2018)
19. Mirza, M., Osindero, S.: Conditional generative adversarial nets. arXiv preprint arXiv:1411.1784 (2014)
20. Miyato, T., Kataoka, T., Koyama, M., Yoshida, Y.: Spectral normalization for generative adversarial networks. arXiv preprint arXiv:1802.05957 (2018)
21. Simistira, F., et al.: ICDAR 2017 competition on layout analysis for challenging medieval manuscripts. In: 2017 14th IAPR International Conference on Document Analysis and Recognition (ICDAR), vol. 1, pp. 1361–1370. IEEE (2017)
22. Wang, X., et al.: Esrgan: enhanced super-resolution generative adversarial networks. In: Proceedings of the European Conference on Computer Vision (ECCV) Workshops (2018)
23. Yang, S., Liu, J., Lian, Z., Guo, Z.: Awesome typography: statistics-based text effects transfer. In: Proceedings of the IEEE Conference on Computer Vision and Pattern Recognition, pp. 7464–7473 (2017)
24. Yang, S., Liu, J., Wang, W., Guo, Z.: Tet-gan: text effects transfer via stylization and destylization. In: Proceedings of the AAAI Conference on Artificial Intelligence, pp. 1238–1245 (2019)
25. Zhu, J.Y., Park, T., Isola, P., Efros, A.A.: Unpaired image-to-image translation using cycle-consistent adversarial networks. In: Proceedings of the IEEE International Conference on Computer Vision, pp. 2223–2232 (2017)

Challenging Children Handwriting Recognition Study Exploiting Synthetic, Mixed and Real Data

Sofiane Medjram(✉), Véronique Eglin(✉), and Stéphane Bres(✉)

Univ Lyon, INSA Lyon, CNRS, UCBL, LIRIS, UMR5205,
69621 Villeurbanne, France
medjram.sofiane@gmail.com, {veronique.eglin,stephane.bres}@insa-lyon.fr

Abstract. In this paper, we investigate the behavior of a MDLSTM-RNN architecture to recognize challenging children handwriting in French language. The system is trained across compositions of synthetic adult handwriting and small collections of real children dictations gathered from first classes elementary school. The paper presents the results of investigations concerning handwriting recognition in a context of weak annotated dataset and synthetic images generation for data augmentation.

Considering very poor databases of children handwriting, we propose series of experiments to show how the model can cope with small quantity of data. In the paper, we show that assuring a controlled variability of words of varying lengths composed by different instances of degraded or poorly-shaped characters, allows a better generalization of the Handwriting Text Recognition (HTR). We also investigate different choices and splitting propositions to compose both training and validation sets, with respect to children styles distributions. The paper presents conclusions of best suited strategies improving HTR accuracies. Compared to performances to train children real data only, the paper illustrates the impact of transfer learning from adults handwriting (from IAM dataset) and the impact of GAN handwriting styles augmentations to improve children handwriting recognition. We show in the paper also that data augmentation through scaling, rotation or even repeating same instances of words allows to enhance performances reaching sometimes human level.

Keywords: Children handwriting text recognition · Deep learning with MDLSTM-RNN · Synthetic image generation for data augmentation

1 Introduction

1.1 Children Handwriting Recognition Context and Problematic

Handwritten Text Recognition (HTR) domain is still a challenging research field in a context where training data are at least weakly annotated if not missing.

Supported by *the Study project* founded by the french Auvergne Rhône-Alpes Region.

© Springer Nature Switzerland AG 2022
S. Uchida et al. (Eds.): DAS 2022, LNCS 13237, pp. 538–552, 2022.
https://doi.org/10.1007/978-3-031-06555-2_36

Children's handwriting recognition represents an example that falls into this context of weakly annotated data, making deep learning based HTR systems ineffective. The high variance in handwriting styles (for the same child or different children) and the difficulty to read them pose significant hurdles in transcribing them to text [10].

Recently, in many deep learning models, a lot of annotated examples composed of word images and their corresponding transcriptions are required. The well-known IAM dataset (in English langugage, [9]) or the Rimes dataset (in French language, [1]) are two key references for Latin HTR. Unfortunately, due to the lack of annotated data and to high variances of handwriting styles, it has become really difficult to adapt an existing deep learning model to a new language especially when the necessary training data are missing. Transferring adult handwriting knowledge from these databases (IAM/Rimes) to Children handwriting recognition may be, to a certain extend, useful but the irregularity, complexity and individuality of children handwriting is so deep that it requires more investigation to be rigorously processed.

To tackle the challenging problem of children handwriting recognition, we chose a sub-optimal MDLSTM-RNN recognition system. Compared to recent architecture of this domain, like Transformers, MDLSTM-RNN architecture is more suitable to the problem addressed and to scarcity of data.

1.2 ScolEdit: A Small Real Children Handwriting Dataset

The data we have to deal with in our project, are dictations of elementary classes consisting of small sequences of words. In that context, we first created a small custom french children database named *Scoledit* thorough an annotation tool that we developed in open-source[1].

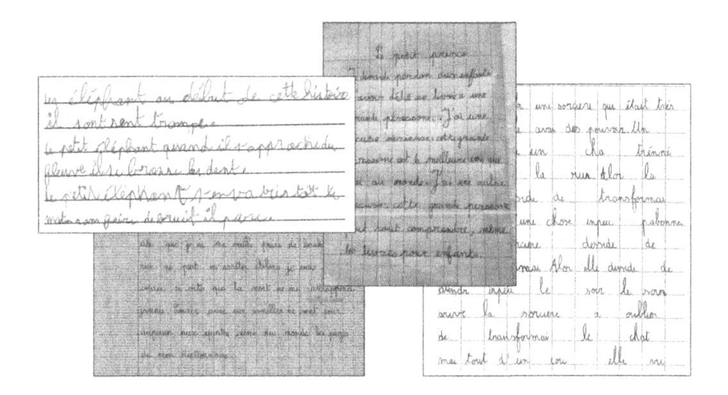

Fig. 1. Examples of dictation samples for different children from *ScolEdit* dataset.

[1] https://github.com/Sofiane23i/Study-Annotation-Tool/.

For each child handwriting loaded image[2], the annotation tool automatically detects words bounding boxes and display a new window for word annotation. The result of the annotation is presented in the conventional IAM Dataset format. The resulting *ScolEdit* dataset consists of 3832 word images of dictations written by 71 children, see Fig. 1.

1.3 Investigating Variable Training Datasets Composition

In the context of our study, the objective then is twofold: it consists in recognizing children's writings through short words sequences in a real-time dictation context (using a graphic tablet for an online acquisition), then identifying the error zones at the character level and finally displaying them back to the screen. Because data are real and acquired in a learning context for children, the *Scoledit* dataset is an essential part of the training data of the HTR system. It is obvious that data requirements are not satisfied for the complete training of the MDLSTM-RNN recognition network. So, we propose in the paper a complete study to measure the effects of introducing adults and children handwriting styles through synthetic and semi-synthetic data (composed by mixtures of real children handwriting and synthetic GAN-generated images). Each new configuration introduces a new amount of information leading to measurable improvements with the goal of providing a guideline for assisted dictation for children.

The paper is structured as follow: In section two, we present recent deep learning architectures from the literature for handwriting recognition; in section three, we will describe the annotation tool we developed to create children handwriting database ; in section four, we describe the HTR model used in our experiments to study the challenging problem of children handwriting recognition and in section five, we detail the protocols exploring the different scenarios and their results. We also present some guidelines required for an efficient use of the HTR system in a real data context where little annotated data is available. In section six, we present concluding remarks in regards with industrial concerns and exploitation.

2 Related Works for Latin Handwritten Text Recognition

Handwriting Recognition methods are divided into online and offline approaches. Online methods consist of tracking a digital pen and receiving information about position and stroke while text is written. Offline methods transcribe text contained in an image into a digital text. Offline methods are mostly used when the temporal dimension in the writing execution is not required, [11]. Recently, like in other application domains, deep learning approaches improved tremendously handwriting recognition accuracy [3,8,11]. In [11] off-line recognition is handled by Multi-Dimensional Recurrent Neural Networks, processing the image through several convolution layers CNN to produce deep spatial features of the image, and then passing them to a recurrent network RNN to produce deep contextual features. After being collapsed to a one dimensional sequence by summing

[2] http://scoledit.org/scoledition/corpus.php.

over the height axis, the sequence is processed through a softmax layer with a Connectionist Temporal Classification (CTC) loss handling the right alignment between the input and the output sequence.

In [3], an attention based model for end-to-end handwriting recognition has been proposed. Attention mechanisms are often used in seq2seq model in machine translation and speech recognition task, [5,7]. In handwriting recognition tasks, attention mechanism allows control of text transcription without segmenting the image into lines as a pre-processing step. It can process an entire page as a whole, while giving predictions for every single line and words in it [10].

Although models in [11] and [3] achieve pretty good results in offline handwriting recognition, they are slow in training due to use of LSTM layers. To overcome this drawback, Transformer have been introduced as efficient alternative solution to LSTM mechanism. In [8], the authors proposed a non-recurrent model for handwriting recognition that uses attention in spatial and contextual stages and provides character representation models enriched by training language model. Today, recent conclusions show that Transformers supply better recognition results compared with all other architectures because they totally avoid recursion, by processing sentences as a whole and by learning relationships between words through a multi-head attention. In our study, we have preferred to implement a classical recurrent engine, because the core of our problem is characterized by a scarcity of annotated children handwriting, also a character level recognition (we discarded the recognition of sequences of words and of complete sentences or paragraphs) and finally the need for soft learning.

3 Scoledit: A Real Children Handwriting Annotated Dataset

As we mention above, due the lack of children handwriting database, we created a custom one from elementary school copies of first classes. We selected 250 clean dictation copies of first classes[3] of 71 different students resulting in 3832 word images with corresponding annotation. The french annotation include lowercase letters only. The custom *Scoledit* dataset was created thanks to an annotation tool we developed in the project[4]. It consists in applying a succession of three steps

3.1 Line Cleaning

ScolEdit dataset contains a set of children copies of different classes. In our investigation, we are only interested in copies of children at elementary school. Considering there age, they learn to spell by writing on sheets of paper with guidelines without any restriction on the pen used as well as its color.

[3] http://scoledit.org/scoledition/corpus.php.
[4] https://github.com/Sofiane23i/Study-Annotation-Tool/.

To facilitate the creation of custom databases through our annotation tool, a preprocessing on children copies is made. It consists of cleaning copies from guidelines without losing children word strokes. To do that, we assume that pen color used by children is different from guidelines color. Using a simple k-means clustering based on HSV Color space is enough to separate children handwriting from background guidelines [4]. As the initialization of centroids and the number of clusters can be automatically estimated, the separation can be done even on multiple pen color (and remove handwritten overwriting, annotations or corrections by the teacher on the copy). Finally, some very small residual segments that might remain attached to the text after segmentation do not need to be removed because they do not affect the recognition of these irregular scripts (Fig. 2).

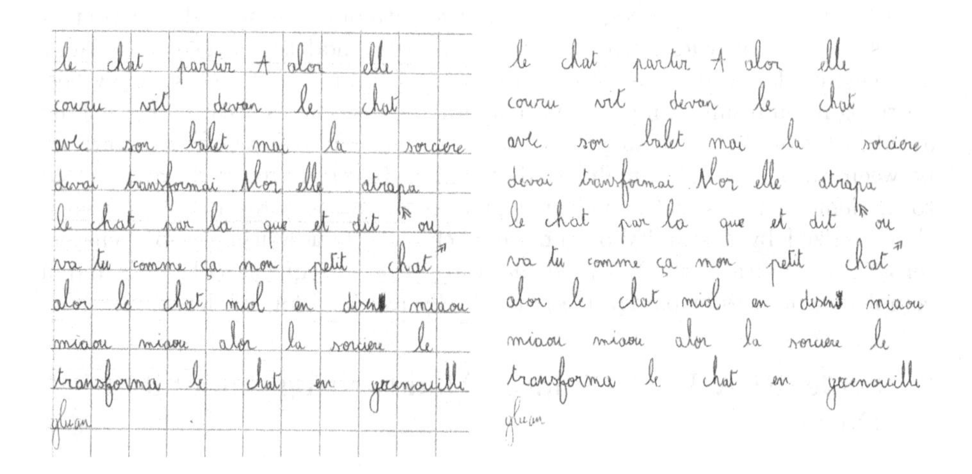

Fig. 2. Children copy cleaning using HSV color clustering.

3.2 Words Detection

The first step of the annotation process consists in words detection and delimitation with bounding boxes. To perform the word segmentation easily on short sequences of dictations, we used the dedicated neural network based word detector inspired from the ideas of Zhou [12] and Axler [2]. The model classifies each pixel as word (inner part or surrounding) or background pixel. For each pixel of the inner word class, an axis aligned bounding box around the word is predicted. Because usually multiple box candidates per word are predicted, a clustering algorithm is applied to them. The model is trained on the IAM dataset and then reuse on the ScolEdit dictation samples.

From the result of the model that detects words around bounding boxes, we adapted its script to extract detected words according to French language. The words will be extracted line by line from left to right, which makes the annotation easy to the user. Once words extraction is completed, a new window displaying words with an input field for ASCII annotation will be presented to the user.

3.3 Words Annotation on IAM Format

After filling input annotations of detected words, a XML file respecting IAM format will be generated and saved with the word images. These set of images with their annotated file can be directly used to train MDLSTM-RNN model on the lexicon and the corresponding handwriting style.

4 HTR Architecture and Recognition Scenarios

4.1 Standard MDLSTM-RNN Word Transcriber

The MDLSTM-RNN Encoder-Decoer model has been extensively used for handwriting recognition, alternating LSTM layers in four directions and subsampling convolutions for Encodeing and CTC operation as Decoding. It has been widely used to transcribe isolated text lines. In our context, we exploit his ability to recognize short sequences of individual words in a dictation context on quite challenging children's writings.

MDLSTM-RNN model receives as an input an image and returns as an output a transcription and a reliability score. The model transforms the input image, with respect to a time-step, into a features vector after a series of convolutions operations, which are then passed as an input to a bidirectional recurrent neural network. Operations series of the recurrent network are then decoded by a decoder in order to produce a clean transcription.

In formal way, MDLSTM-RNN presents a function that map an image (or matrix) Img of size ImgH x ImgW to a characters sequence (c1, c2, c3,....., cN). Text is recognized on character-level, which means words or texts not contained in the training data can be recognized correctly if the characters get correctly classified.

$$Img(HxW) -> (c1, c2, c3, ..., cN) \tag{1}$$

For each tested word, we measured the Character Error Rate (CER %) with the objective of requiring only few real words for a new writer.

In practice, we have designed different training scenarios for the recognition network, carefully mixing type and quantity of training and validation data.

In context of children handwriting, we have designed different training scenarios for the recognition network, carefully mixing type and quantity of training and validation data.

4.2 Mixing Real and Synthetic Data to Enhance the Recognition Rates

Data Augmentation. To compensate our limited training Scoledit dataset, we use some classical data augmentation techniques to artificially increase our training data: scaling and rotation (literally 10°C left and right) on words from the target dataset and synthetic data allowing to include new words from any size and lexicon. To perform this synthesis, we exploit Handwriting Synthesis GAN initially proposed by [6]. Generative models for handwriting synthesis are mostly used to produce realistic image samples based on online real examples. In situations where data are rare or difficult to collect, handwriting synthesis seem to be an efficient solution, adding new synthetic samples to the original training set, and so augmenting the set in a bootstrap manner.

Since the GAN is able to synthesize images from any handwriting styles learned, it is then possible to feed it with new children handwriting samples obtained from an on-line acquisition with tablet. This acquired images from real sources are useful to simulate character interactions depending on the adjacent characters.

Transfer Learning and Fine-Tuning. As the the Scoledit dataset only contains 3832 words (from two to ten characters per words), it is quite obvious to pretrain the HTR with other datasets. For that purspose, we explore two options during experiments:

- *Option 1*: we build the HTR model on a baseline dataset (IAM dataset, or GAN generated source domain) and then we transfer the trained weights to serve as initialization for the model to be trained on the target domain (our Scoledit dataset). It will serve as baselines in our experiments.
- *Option 2*: we build a model on combined data from 2 distinct sources : generic source (of a baseline dataset that can be IAM or synthetic images) and target source (a sub set of Scoledit dataset) into one general unique domain and we compare the performance of different compositions on different test datasets.

Considering one of the HTR trained models (presented in the Sect. 4.3), we can then experiment whether transfer learning helps to get better performing models on target Scoledit domain (Sect. 5).

4.3 Scenarios for HTR Training and Data Preparation

The experiments have been organized in different scenarios in an incremental improvement approach with different datasets mixing real and synthetic data. The Table 1 summarizes the different datasets employed in the experiments. They will be mentioned in the following sections. The test sets are composed either of real children samples (for scenario 1 et 2), which constitute the target of the study, i.e. a subset of the Scoledit dataset (named S1-D1), either of a set of images sharing the same style or lexicon as used in the training set (dataset called S2 and S3). The S1-D1 dataset is composed of 3640 real word images corresponding to 71 children for training, 192 words of 71 children for validation, see Fig. 1.

Supervised Validation and Domain Transfer. In the first serie of experiments, S1-D1 is used during training (alone or with pre-training). Here enumerated the three sub-scenarios including the learning from scratch on S1-D1, a transfer learning with IAM dataset and a transfer learning with synthetic data for three adult Latin handwriting styles. In any cases, this first scenario show the effect of transfering knowledge from pre-trained models with either IAM images with their own lexicon and styles, or synthetic images having their own style and their lexicon, or the ScolEdit lexicon.

1. Training the HTR with S1-D1 dataset from scratch and investigating two ways of selecting data for training and validation: random five cross-validation and manual validation data selection across all children handwriting styles.
2. Transfer learning from synthetic word images generated by GAN of three adult styles including 3032 words. The transfer operates then on real data of the S1-D1 dataset composed of 3640 word images.
3. Transfer learning from IAM dataset composed of 657 adult styles including 115320 words. Here again, the transfer operates on real data of the S1-D1 dataset composed of 3640 word images.

Configuration for Single Dictation Words. In the second serie of experiments, the system is trained to simulate the best behaviour for the dictation, selecting only six different words for training with appropriate repetitions (at least 320) for each writing style. Different combinations for the training datasets are studied here, suggesting different GAN writing styles (from 1 to 3). The experiment is presented to show effects of mislearning character interactions when the test lies on a real data of randomly chosen words. It also shows the efficiency of testing words of a dictation even produced by new children from the moment the order of characters is respected as learnt.

1. Training HTR with *S2* dataset composed of a combination of three adult synthetic styles of a limited vocabulary of six words repeated 320 times (S2-D1 for style 1, S2-D2 for style 2, S2-D3 for style 3 and S2-D123 for a mixture of the three styles). Those models are then tested on a subset of S1 (real children words having all the same character list). For each synthetic adult style, we collect 1856 images for training, 64 for validation.
2. Training HTR on S2-D123Aug dataset merging S2-D123 composed of a limited vocabulary of six words repeated 320 times for three adult handwriting styles with augmented data (including rotation and scaling changes) for a total of 28608 images for training, 1024 for validation.

Table 1. Datasets and test sets abreviation table.

Dataset name	Composition	Description	Train	Valid	Test
S1-D1		3832 Children Real Words collected from ScolEdit	3640	192	192
S2-D1	6 Unique Words	1920 Synthetic Words generated with Style1	1856	64	64
S2-D2	repeated 320 times	1920 Synthetic Words generated with Style2	1856	64	64
S2-D3		1920 Synthetic Words generated with Style3	1856	64	64
S2-D123		5952 Synthetic Words of Style1, Style2 and Style3	5696	256	256
S2-D123Aug		29632 Synthetic Words of Style1, Style2 and Style3 with Data Augmentation	28608	1024	1024
S3-D1	1736 Unique Words	1736 Synthetic Words generated with Style1	1672	64	64
S3-D123	no repetitions	5472 Synthetic Words of Style1, Style2 and Style3	5216	256	256
S3-D123Ch	words composed of one character to 11 characters	5472 Synthetic Words of Style1, Style2 and Style3 with Children Real data	8869	384	384

	Scenario one	Scenario two			Scenario three		
Test set name	S1-T1	S2-T1	S2-T2	S2-T3	S3-T1	S3-T2	S3-T3
Num images	192	64—256—1012	64—256—1012	64—256—1012	64—256—384	64—256—384	64—256—384
Description	Children real words	GAN adult : same styles as training set with different words composed of shuffled trained characters	GAN adult : different styles than training but same words	Same words than training but on real children samples	GAN adult: new styles and same six words of S2-T1	GAN adult : new styles and new words compared to training	Only on children real words

Large Lexicon Training with Transfer. We are interested here in exploiting the effect of training the HTR with a large lexicon covering a diversity of words of variant sizes, one character and B-grams compositions (from 2 to 11 characters). The experiments are composed of

1. Training HTR on S3-D1,S3-D2,S3-D3, and S3-D123 composed of a set of 1736 single words for each three synthetic adult styles (considered separately or merged). This experiment investigates the influence of a rich vocabulary composed of sequence of characters going from one single character to a maximum of eleven without any repetition. It also shares the dataset for each style as follows: 1672 images for training, 64 for validation.
2. Training the HTR on S3-D123Ch following the same initial configuration than previous but with the augmentation of 3640 real children images samples: 8896 images per style for training, 384 for validation.
3. Transfer learning form IAM dataset to S3-D123Ch dataset composed by three adult styles generated by GAN model of a rich vocabulary merged with real children data.

To sum up, in the first scenario called *Supervised validation and domain transfer*, we investigate the behavior of the model towards a wise choice of the validation set and only with S1-D1 dataset (alone or with pre-training). Then in the next one called *Automaton for dictation*, we investigated the HTR behaviour towards in its ability to reinforce recognition on a couple of selected words on four new synthetic datasets (S2-D1, S2-D2 ...etc.). Finally and according to conclusions deduced from both scenarios, we investigated a last scenario *Large lexicon with transfer* with larger vocabulary datasets (S3-D1, S3-D2 ...etc.).

5 Experiments and Results

Here we described the detailed results of the three scenarios investigating the impact of controlling the different datasets implied in the HTR training : data augmentation, transfer learning and supervision of the validation set. We use the classical recognition rates (CER in % and WER in %) and propose at least appropriate guidelines for a suitable use of the system for children handwriting recognition.

5.1 First Scenario: Supervised Selection of Validation Datasets and Domain Transfer

In this scenario, using S1-D1 dataset (as train, validation and test), we propose two experiments to evaluate the impact of choosing validation set randomly versus manually. We first propose to guide the system by supervising the choice of the validation set to ensure correct and robust performances. Then reinforced by this supervised selection strategy, the system will then be proposed to next experiments. The remaining experiments of this scenario evaluate the impact of transfer learning from adult handwriting styles to children ones, once with IAM database and a second with synthetic adult handwriting of three different styles.

Automatic 5 Cross-Validation vs Supervised Validation. We divide S1-D1 into 95% train, 2.5% validation and 2.5% test sets and carry out a 5 cross-validation learning procedure: first over the global dataset, all writing distribution combined and second by selecting balanced samples distribution over all children handwriting styles. For the second procedure, for each distribution (corresponding to child copy), we divide data into 95% training, 2.5% validation and 2.5% test, and we merge them along all distributions.

As we can see in the performances of Table 2 (*val* column for the lines *CV1* to *CV5*), there is a large gap between cross-validations (9% in CER (character error) and 11% in WA (word accuracy) (CV2 and CV3)) which reflects that data are different and not related.

With respect to the amount of available annotated data, although cross-validation is a fast process that allows several experiments, model performances are not as satisfying as hoped. Here, we observe that manual validation set selection improves performances to 9.5% Character Error Rate and 77.08% Word accuracy (Table 2 val columns line ManuS).

To confirms the impact of previous two experiments, we tested, in addition, current model on the residual dataset (S1-T1, see second table in Table 1) that contains new lexicon words and new children handwriting styles.

Transfer Learning from IAM and GAN Adult Styles. From previous experiences, we saw the positive effect of carefully choosing validation set among all children copies. Compared to the baseline of IAM results (Table 2 last line), performances obtained are not very satisfying (val columns, Character error rate: 9.50%. Word accuracy: 77.08%). For improving the performance, and considering the small amount of annotated data, we propose to study the impact of transferring learning from different adult styles (from IAM and synthetic dataset) to children styles.

We investigate two experiments of transfer learning, one from the huge database IAM (150320 word images) and second from small data of 3000 word images generated by GAN of three different styles. The aims of these experiences is to see if they improve children handwriting recognition and to see if synthetic data can be used to enhance performances and replace missing real data. Results in Table 2 shows that transfer learning from IAM dataset improves performances of children handwriting (Val column 6.38% CER). In other words, considering handwriting complexity and differences between children and adults, the HTR system handled easily both complexities with same architecture. Concerning transfer learning with synthetic samples, even with few amount of synthetic images per adult style (3000 words per style), the model shows satisfying performances on real child data (Val column 10.46% CER, see Table 2). It shows that synthetic handwriting data can be used to complete favourably the lack of real data for word recognition.

Table 2. HTR five cross-validation performances vs Manual selection comparison and effect of transfer learning.

S1-D1 Dataset	Transfer learning	Val set CER	Val set WA	S1-T1 CER	S1-T1 WA
CV1	No	17.96%	65.1%	37.81%	50.52%
CV2	No	14.62%	66.66%	39.17%	45.31%
CV3	No	23.89%	55.72%	37.56%	49.47%
CV4	No	19.28%	58.33%	35.19%	48.95%
CV5	No	17.52%	62.74%	35.07%	49.47%
ManuS	No	**09.50%**	**77.08%**	**31.09%**	**54.68%**
ManuS	GAN 3 styles	10.46%	74.47%	44.01%	41.66%
ManuS	IAM	**06.38%**	**82.81%**	**28.87%**	**56.25%**
IAM Dataset	No	09.71%	78.50%	–	–

5.2 Second Scenario: Training Focused on Dictation Words

Several observations can be deduced from the first experiment benches concerning the amount of real data to be trained and the impact of their variability. We intend here to see the HTR behavior on limited lexicon words repeated several time by the same person, especially its ability to generalize on same lexicon words of real children (when trained on synthetic data) or on new lexicon words. (words of a new vocabulary whose characters have been learned by the model.)

Once again, due of scarcity of real data, we created synthetic images into five datasets for the three different adult styles (S2-D1, S2-D2, S2-D3, S2-D123, S2-D123Aug as described in Table 1). For each experiment (concerning respectively each of the S2 dataset), we divided 95% of the data for training, 5% for validation. For test sets, we generate three different datasets named S2-T1, S2-T2, S2-T3 described in Table 1.

Table 3. HTR performances on a reduced set of six words of a dictation generated by synthetic GAN of three different adult styles.

Dataset	Val		S2-T1		S2-T2		S2-T3	
	CER	WA	CER	WA	CER	WA	CER	WA
S2-D1	0%	100%	83%	0%	3.25%	95.31%	-%	-%
S2-D2	0%	100%	83%	0%	8.04%	84.73%	-%	-%
S2-D3	0%	100%	83%	0%	11.50%	81.25%	-%	-%
S2-D123	0%	100%	75%	12.10%	2.48%	94.53%	34.93%	56.25%
S2-D123Aug	0%	100%	88%	0%	1.61%	97.99%	20.36%	74.57%

According to results given in Table 3 (val columns, S2-T2 and S-T3), we can see that the HTR system performs very well on test data of same words even on a different handwriting style and it can achieve human level performance (up to 95%). Unfortunately, it is not performing when evaluated on different words even composed of same characters that learned (Table 3 column S2-T1). We can also notice that the data augmentation positively impact the recognition rate of the HTR (when trained on S2-D123Aug).

5.3 Third Scenario: Large Lexicon Training with Transfer

We saw in the second scenario that the HTR performs pretty well on datasets composed of limited lexicon words when repeated several times. The system reaches most of times human level: it generalizes correctly on real words of same lexicon but not when characters of the lexicon are shifted or shuffled. Therefore, compared to generalization performances of the first scenario, we investigate in the third scenario the behavior of the HTR on a dataset of maximal variability, containing single words without repetition composed of one to eleven characters.

As previously, to overcome the data scarcity, we created synthetic samples into two datasets for the three different adult styles (S3-D1, S3-D123), and we also mixed the last one with real children world images (S3-D123 Ch) with and without applying IAM transfer learning. Once again here, for each experiment, we divided 95% for train, 5% for validation. Each experiment has been tested on several test sets S3-T1, S3-T2, S3-T3 as described in Table 1.

Table 4. HTR performances on large word lexicon generated by synthetic GAN of three adult styles alone or combined with real data, and the effect of the transfer learning.

Dataset	Transfer	Val		S3-T1		S3-T2		S3-T3	
S3		CER	WA	CER	WA	CER	WA	CER	WA
-D1	No	3.72%	89.06%	2.41%	92.18%	12.86%	81.25%	82.31%	3.12%
-D123	No	1.16%	95.70%	0%	100%	8.04%	7.39%	82.31%	3.12%
-D123Ch	No	7.18%	80.46%	0.72%	95.31%	4.50%	90.62%	8.22%	74.47%
-D123Ch	IAM	**5.47%**	**85.67%**	**0.24%**	**98.43%**	**21.86%**	**64.06%**	**5.52%**	**84.11%**

The experiment results (in Table 4) show the impact of using a database enriched by a large amount of never repeated words compared to the results obtained on a short lexicon dataset (second scenario). Even with this diversity, the performances against real children data remain quite low (Table 4, columns S3-T3, two first lines).

Finally, when we combine the S3-D123 dataset with real children data (to form S3-D123 Ch), the HTR increases in performance on new samples of children handwriting (Table 4, columns S3-T3), and performs still much better when we transfer the learning from IAM (last line of Table 4, S3-T3).

In this section, we presented how the model performs and generalize much better when data lexicon is rich and amount of data is large. More precisely, well selecting training, validation and test sets reflect correct and robust performances and draw clear ideas to improve performances of the model. In addition, when real data are scarse, we can favourably complete them using synthetic handwriting samples. We showed also how using synthetic data during training helps the model to generalize on real data. Moreover, we have also seen how transfer learning from huge datasets of adults handwriting improve model performances on real children data, and improve performances much better when real children data are combined with synthetic data of rich lexicon and different styles from those of IAM (CER:5.52%—WA:84.11%).

6 Conclusions

In this paper, we presented how MDLSTM-RNN network deals well with children handwriting complexities and how it can take advantage of training scenarios mixing data from real to synthetic, changing lexicons or writing styles as adjustment parameters of the training data. Our investigations show that amount and quality of data is another crucial factor : Synthetic data and augmentation have proved to be essential in achieving better recognition results but remain poor substitutes for diverse and real handwriting samples. Moreover, when annotated data remains limited (less than 5.000), we show how the HTR performs efficiently when it receives appropriate training and validation data. Random data selecting for training and validation may work for some application if the data volume is huge and the problem complexity is simple. When the amount of data is small and the complexity of writing is important, it is convenient to choose carefully training and validation sets for a better generalization and model convergence.

Finally, the combination of synthetic data and the pre-training of the HTR are also key factors in improving performances. By facilitating network training with synthetic data, each new child handwriting can be artificially imitated by generated data. The network will then no longer need specific real handwriting samples to tune its model and introducing a new handwriting could be solved by a well-designed simulation. This approach is currently under study.

Acknowledgements. We warmly acknowledge the Auvergne-Rhône-Alpes Region for supporting the *Study* project on the *Research and Development Booster 2020–2023 Program* involving several teams of researchers and industrialists including the two French partner companies AMI and SuperExtraLab.

References

1. Augustin, E., Brodin, J., Carré, M., Geoffrois, E., Grosicki, E., Prêteux, F.: RIMES evaluation campaign for handwritten mail processing. In: Workshop on Frontiers in Handwriting Recognition, vol. 1, pp. 1–5 (2006)
2. Axler, G., Wolf, L.: Toward a dataset-agnostic word segmentation method. In: IEEE International Conference on Image Processing, 25th ICIP, pp. 2635–2639 (2018)

3. Bluche, T., Louradour, J., Messina, R.: Scan, attend and read: end-to-end handwritten paragraph recognition with MDLSTM attention. In: Proceedings of the International Conference on Document Analysis and Recognition, ICDAR, vol. 1, pp. 1050–1055 (2017). ISSN 15205363, https://doi.org/10.1109/ICDAR.2017.174
4. Chen, T.W., Chen, Y.L., Chien, S.Y.: Fast image segmentation based on K-means clustering with histograms in HSV color space. In: Proceedings of the 2008 IEEE 10th Workshop on Multimedia Signal Processing. MMSP, pp. 322–325 (2008). https://doi.org/10.1109/MMSP.2008.4665097
5. Chorowski, J., Bahdanau, D., Serdyuk, D., Cho, K., Bengio, Y.: Attention-based models for speech recognition. In: Advances in Neural Information Processing Systems, pp. 577–585 (2015). ISSN 10495258
6. Graves, A.: Generating sequences with recurrent neural networks, pp. 1–43 (2013)
7. Jia, Y.: Attention mechanism in machine translation. J. Phys. Conf. Seri. JOP **1314**(1) (2019). ISSN 17426596, https://doi.org/10.1088/1742-6596/1314/1/012186
8. Kang, L., Riba, P., Rusiñol, M., Fornés, A., Villegas, M.: Pay attention to what you read: Non-recurrent handwritten text-line recognition. arXiv (2020), ISSN 23318422
9. Marti, U.V., Bunke, H.: The IAM-database: an English sentence database for offline handwriting recognition. Int. J. Doc. Anal. Recogn. IJDAR **5**(1), 39–46 (2003). ISSN 14332833, https://doi.org/10.1007/s100320200071
10. Matcha, A.C.N.: Nanonets, how to easily do handwriting recognition using deep learning (2020). https://nanonets.com/blog/handwritten-character-recognition/
11. Pham, V., Bluche, T., Kermorvant, C., Louradour, J.: Dropout improves recurrent neural networks for handwriting recognition. In: Proceedings of International Conference on Frontiers in Handwriting Recognition, ICFHR, 2014-December, pp. 285–290 (2014). ISSN 21676453, https://doi.org/10.1109/ICFHR.2014.55
12. Wernersbach, J., Tracy, C.: East. Swimming Holes Texas, pp. 45–68 (2021). https://doi.org/10.7560/321522-007

Combining Image Processing Techniques, OCR, and OMR for the Digitization of Musical Books

Gonzalo Santamaría, César Domínguez(✉), Jónathan Heras, Eloy Mata, and Vico Pascual

Department of Mathematics and Computer Science, University of La Rioja, Logroño, Spain
{gonzalo.santamaria,cesar.dominguez,jonathan.heras,eloy.mata, vico.pascual}@unirioja.es

Abstract. Digitizing historical music books can be challenging since staves are usually mixed with typewritten text explaining some characteristics of them. In this work, we propose a new methodology to undertake such a digitization task. After scanning the pages of the book, the different blocks of text and staves can be detected and organized into music pieces using image processing techniques. Then, OCR and OMR methods can be applied to text and stave blocks, respectively, and the information conveniently stored using the *MusicXML* format. In addition, we explain how this methodology was successfully applied in the digitization of a book entitled "The Music in the Santo Domingo's Cathedral". In particular, we provide a new annotated database of musical symbols from the staves included in this book. This database was used to develop two new OMR deep learning models for the detection and classification of music scores. The detection model obtained a F1-score of 90% on symbol detection; and the classification model a note pitch accuracy of 98.4%. The method allows us to conduct text searches, obtain clean PDF files of music pieces, or reproduce the sound represented by the pieces. The database, models, and code of this project are available at https://github.com/joheras/MusicaCatedralStoDomingoIER.

Keywords: Image processing · OCR · OMR · Digitization · Music books

1 Introduction

Nowadays, there is an increasing need to digitize all kinds of printed documents [11]. A digital copy of a printed document enables electronic editing, quick and easy access, broad dissemination, in-document search, online reproduction, and use in different machine processes [32]. Among the types of books that are

This work was partially supported by Grant RTC-2017-6640-7; and by MCIN/AEI/10.13039/501100011033, under Grant PID2020-115225RB-I00.

© Springer Nature Switzerland AG 2022
S. Uchida et al. (Eds.): DAS 2022, LNCS 13237, pp. 553–567, 2022.
https://doi.org/10.1007/978-3-031-06555-2_37

worth digitizing are text books and music sheets. The digitization of the former type has a long tradition, and it is often successfully achieved by using techniques from a field called Optical Character Recognition (OCR) [11,32]. Several OCR tools, such as Tesseract OCR engine [33], have a good performance on scanned images of typewritten documents and are widely used.

The digitization of music sheets has also given rise to a new research field named Optical Music Recognition (OMR) [8]. This is an active research field that aims to develop techniques to teach computers to read music notation and store it in a digital structured format. Contrary to typewritten texts where the existing automatic systems are able to read and encode text with a high accuracy, there is not yet any software tool that is capable of interpreting complex music scores [8] and the field leaves space for improvement [30]. Although, OMR has previously been referred to as the OCR for music, OCR only deals with sequences of characters and words that are one-dimensional [30], whereas music makes use of a two-dimensional representation [2]. Music includes information for a more complex structure, with an ordered sequence of musical symbols together with their spatial relationships [8]. In this way, OMR methods must take into account not only the type of a symbol but also other aspects such as its specific vertical position on the staff, the variability in size of the symbols, or the rotations of some notes [2,8].

When the aim is to digitize a historical music book, wherein staves are mixed with typewritten text explaining some characteristics of them, methods from both fields, OMR and OCR, can be useful. The challenge is to integrate OCR and OMR techniques together with other computer vision methods in order to complete the digitizing process. The objective of this work is to develop a working methodology that can be applicable to the digitization of music books. The purpose is to generate a digital and structured version of a printed music book, in order to conduct text searches (for instance, to consult the works of different authors and/or different musical genres through a web page), to obtain a clean copy of the pieces, or to be able to reproduce the melody of the musical compositions. In this methodology, the combination and integration of OCR and OMR techniques with other methods coming from computer vision seems essential. Basically, computer vision methods can be used to distinguish between blocks of text and staves. Then, OCR and OMR techniques are used to obtain a structured representation of the blocks of text and staves, respectively. Finally, all the blocks can be properly recomposed and stored in a database for the different requested needs. Throughout this process, new OMR models could been produced to obtain structured representation of staves from historical music books. This methodology has been particularized for a book which published the musical archive of Santo Domingo's Cathedral (La Rioja, Spain) [21]. This is a book printed in 1988 wherein you can find the transcription of the historical music developed for this cathedral. It comprises different sacred musical compositions such as masses, carols, or psalms composed by different authors. This book did not have a digital version until now; therefore, its contents could only be accessed through one of the 500 printed copies. The final result can be found in the following link: https://domingo.unirioja.es/. More precisely, the contributions of this work are the following:

- We have proposed a new methodology for the digitization of music books which merge text with staves.
- We have produced a new annotated database of musical symbols coming from the historical music produced for the Cathedral of Santo Domingo.
- We have developed a detection and a classification deep learning OMR models for structured representation of historical music scores and conducted a thorough analysis of the results of the models.
- We have publicly released all the code, annotated images, and models developed in this work on the project website https://github.com/joheras/MusicaCatedralStoDomingoIER.

The rest of the paper is organized as follow. In the next section, we describe the book "The Music in the Santo Domingo's Cathedral" [21] which is used as use case in this work. Subsequently, in Sect. 3, we analyze different approaches to solve OCR and OMR task. The methodology and main features of our techniques are detailed in Sect. 4. A discussion of the obtained results is provided in Sect. 5. The paper ends with the conclusions.

2 The Music in the Santo Domingo's Cathedral Book

"The Music in the Santo Domingo's Cathedral" [21] is a book printed in 1988 that contains different religious musical works such as masses, carols, or psalms composed by different authors. This book did not have a digital version, so the only way to access its contents was to have access to one of the 500 existing copies. This book was the first volume within the project "Music in La Rioja" whose aim was to publish the musical archives from different religious sites of La Rioja (a region in the north of Spain). It is worth mentioning that the original music pieces were contained in historical manuscripts documents, and writing this type of books requires an arduous task of research, rewriting and organising the musical archives [21]. The music pieces are simple, including monophonic and homophonic music [7]. In addition, each music piece was enhanced with text comments. These comments include a brief explanation of each piece, the author, the type of piece, and so on. In Fig. 1, we show the cover and Page 144 of the book. This page includes the last staff of Piece 620, the complete Pieces 621, 622, and 623, and the beginning of Piece 624. Note that not all pieces start or end on the same page. The author of these four pieces was Manuel Pascual. We can also see, for example, that Piece 621 is a *Mass*. Finally, the book collected the restoration of ancient manuscript musical pieces, and it contains musical symbols that are out of use, such as C-clefs in first, or square notes.

3 Related Work

Optical Character Recognition (OCR) is defined as the automated conversion of images of typed, handwritten or printed text into machine-encoded text [11,32]. It has a long tradition as a research field and there is an extensive corpus of

JOSE LOPEZ-CALO

LA MUSICA

EN LA CATEDRAL DE

SANTO DOMINGO DE LA CALZADA

GOBIERNO DE LA RIOJA
Consejería de Educación, Cultura y Deportes

144

Placare, Christe, servulis

Manuel Pascual

621. *"Misa a solo y a 5":* S, SATB y acompañamiento al órgano. Sólo las particellas, manuscritas. 3/22

Kyrie

Et in terra pax

Patrem omnipotentem

Sanctus

Agnus Dei, qui tollis

622. *Beatus vir*. Salmo, a 5 v. (S, SATB) y acompañamiento. Sólo las particellas, manuscritas. 12/5

Beatus vir qui timet

623. *Laudate Dominum omnes gentes*. Salmo, a 5 v. (S, SATB), bajón y acompañamiento. El bajón es añadido y duplica al bajo del coro. Sólo las particellas, manuscritas. 10/24

Laudate Dominum

624. *Magnificat*, a 5 v. (S, SATB) y acompañamiento. Sólo las particellas, manuscritas. 1785. 20/15

Fig. 1. Book cover on the left and Page 144 of the book on the right.

work with many valuable tools [32]. OCR has been successfully applied to both typed and handwritten texts [23], in modern and historical documents [22], and in a wide variety of languages [17]. In addition, there exists different tools that correctly apply OCR to typewritten texts [32]. One of these tools is the Tesseract OCR engine [33], that performs a good work on scanned images of typewritten documents and it has been widely used.

Optical Music Recognition (OMR) is defined as a research field that seeks to develop techniques to teach computers to read music notations and to store them in a digital structured format [8]. This is an active research field, but, in contrast to OCR on typewritten texts, there is still no computer system that is capable to interpret complex music scores [8] and the field leaves space for improvement [30].

Music can be described as a structured assembly of notes defined by several parameters such as pith, duration, loudness, and timbre, which also have an onset or placement onto the axis of time preceded by a musical clef [8]. Notes and rests (periods without notes) can be grouped into graphical and logical hierarchies to form, for instance, beams. Although, OMR has previously been referred to as OCR for music, music scores carry information in a more complex structure, with an ordered sequence of musical symbols together with their spatial relationships [8]. In contrast, OCR deals with sequences of characters and words that are one-dimensional [30]. Therefore, OMR should take into consideration not only the type of symbol but also other aspects such as it specific vertical position on the staff, the variability in size of the symbols, the rotations of some notes and so on [2,8].

In a recent work, Calvo-Zaragoza et al. [8] detailed a taxonomy of OMR inputs, architectures, and outputs. This taxonomy could be useful to situate new OMR approaches. The inputs referred to the use of offline or online signals, typeset or handwritten symbols, the notation in which the music is represented (usually Common Western notation), the notational complexity of the music (i.e., monophonic, homophonic, polyphonic, or pianoform), the document quality (from perfect to degraded), and the image acquisition (from born-digital to scanned of varying quality). According to Calvo-Zaragoza et al. taxonomy, system architectures can be modular (with modules for pre-processing, object detection, notation assembly, and encoding) or end-to-end recognition system. Finally, outputs can range from a simple document metadata extraction, which seeks to answer simple questions such as whether or not a document contains music scores) to structured encoding, which aims to recognize the whole music score. The usual output formats are MusicXML [13] or MIDI [19]. Although both of them allow further editing by computers, the former is more focused on the encoding notation layout, and the latter on the replayability of the encoded notes [8,30].

Nowadays, the state-of-art approaches, both in OMR modular and end-to-end systems, are based on deep learning techniques. An example of modular system was developed in [12], which used autoencoders for the task of staff-line removal. An example of an end-to-end model was developed by Huang et al. [18] based on a deep convolutional neural network and feature fusion. The model was able to classify the music symbols, and the pitch and duration of the notes with a duration accuracy of 0.92 and a pitch accuracy of 0.96. Other model in this category was developed by Calvo-Zaragoza and Rizo [10]. This model combined a convolutional neural network and a recurrent neural network to apply OMR to monophonic scores. The model obtained a symbol error rate of 0.08. These methods usually work in specific scenarios and it is difficult to apply them to different contexts [10]. This is due to the internal diversity of the OMR field [8]. In the same vein, due to the lack of standards for the output of OMR results, their evaluation and comparison can be unfair [10]. The literature also includes some tools that attempt the structure encoding of musical scores; some commercial systems are, for instance, SmartScore [24] or PhotoScore [25], and there are also open-source solutions like Audiveris [3].

In spite of the existence of the aforementioned methods and tools, it worth mentioning the conclusion of the recent survey on OMR works by Shatri and Fazekas [30] that states that despite the introduction of deep learning the OMR field still has room for improvement. The most difficult obstacle that researches find when trying to apply deep learning algorithms is the lack of an appropriate and valid ground-truth data for the problem that they want to solve [10]—although there exists some publicly available datasets of annotated scores (see, for instance [9,14,31,35]). It also should be noted that models and tools that work well on certain types of scores, are not be able to conveniently process others [10]. That is the case, for instance, of Audiveris which did not properly work on some perfectly segmented monophonic staves under ideal conditions [10].

4 Methods

In this section, we propose a methodology, summarized in Fig. 2, for performing the task of digitizing a music book. After the digital scanning of each page of the book, staves, text blocks, and musical pieces (the latter are a combination of the previous two) are detected. Then, OCR or OMR techniques are used to extract the information from these blocks. Finally, the obtained information is properly recomposed and stored in a database. With this information it is possible to create, for instance, a web page to access the digital copy of the book through different types of searches. We also explain in this section how such a methodology has been particularized in the digitization of "The Music in the Santo Domingo's Cathedral" book. In particular, we detail the image processing techniques and deep learning models used to perform each task.

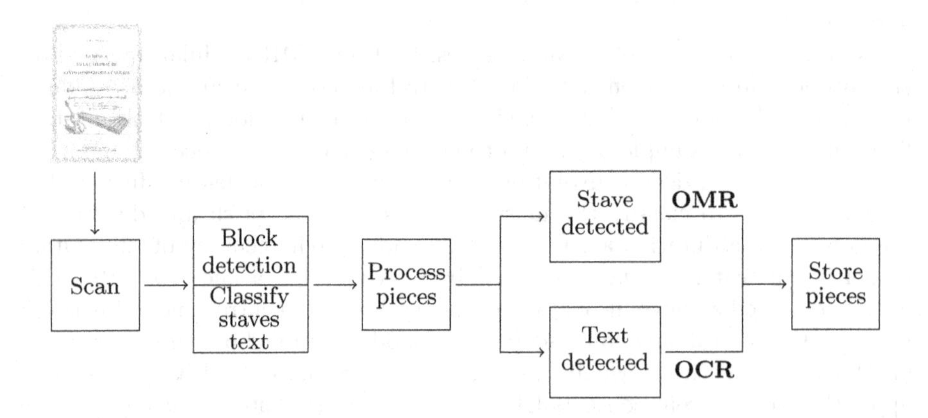

Fig. 2. Book digitization process.

If only a printed version of the book is available, the first step involves scanning each page of the book. The obtained images can be stored in a folder preserving the page order in the book; namely, the name of the file can be used for that purpose. In our case, we obtained a total of 377 images of size 1825 × 1200, one for each page of the book.

4.1 Block Detection

This step consists in detecting and classifying each block in each image. We have considered two types of blocks: text and staves. Subsequently, text and staves blocks are linked into musical pieces.

Image processing techniques can be used in this step. We have applied a thresholding process based on the Otsu method [37] combined with morphological operations and a contour detection process [29] provided by the *OpenCV* library [6]. Specifically, we have used a vertical dilation to locate the staves taking advantage of the staff lines and then selected the contours with greater area. Subsequently, to detect the text blocks, we first removed the staves, then dilated the image in all directions, and finally selected the contours with a minimum area. Figure 3 shows the two detection processes. Note that in the second row the staves have been removed to detect the text.

Fig. 3. Block detection with *OpenCV*. First row: staves; second row: text blocks.

Finally, the musical pieces were obtained by linking the segmented text blocks and stave blocks. To this end, we have noted that all the pieces begin with a number, the piece number, and end with the next piece or the beginning of a new section, which is marked by a new author or new musical genre. In order to detect the piece number, we have used the OCR provided by *PyTesseract* [28,33] and we have taken into account that these numbers follow an order (Piece 1, Piece 2, and so on). New sections are detected by taken into account that the text appearing in them is larger than usual. This method allowed us to detect musical pieces that begin on a page and ends on a different subsequent page. In Fig. 4, we show an example of piece detection.

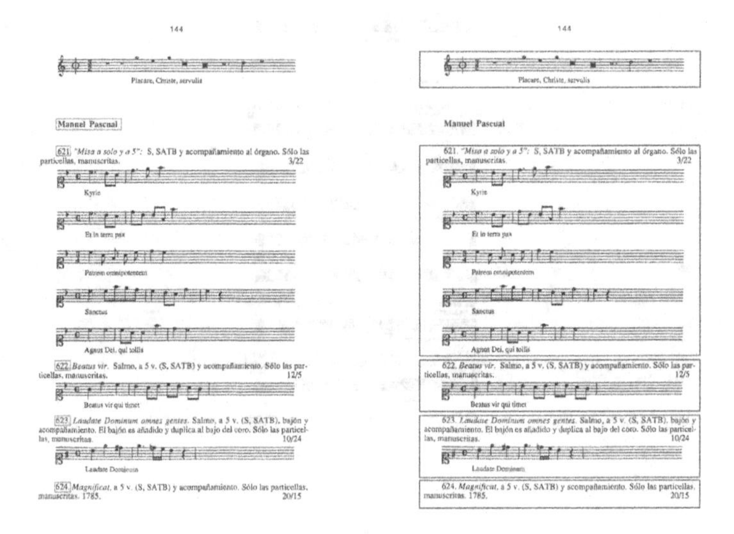

Fig. 4. Piece detection.

4.2 OCR

The text of a text block was easily extracted using *PyTesseract* library [28,33]. For each text block, the library returned a OCR of the text in a string. We have also used a second functionality of the library. This functionality returned all the text found in the image in the form of a dictionary whose keys are the words and values are the coordinates associated with them. This dictionary will be used to perform quick text searches within the book.

4.3 OMR

Contrary to OCR, in which *PyTesseract* performed an excellent work on the text blocks, we have developed our own OMR models for score recognition. In this subsection, we explain the steps we have followed to create such OMR models.

Specifically, we have built two models. The first was devoted to detect the musical symbols within the stave images, whereas the second model classified the note pitch of the notes previously detected by the first model. It is worth mentioning that the staves included in the book had out of use symbols which are not taken into consideration in other OMR models [1,18].

The first step to build our OMR models was the definition of two manually annotated datasets. The first model included the position of all the symbols included in the staves; and, the second, the pitch of the notes. For the construction of the first dataset, we manually annotated 300 of the 2308 staves in the book. To speed up the annotation process, we followed a semi-automatic procedure. This was an iterative process with the following steps: first, we manually annotated 20 staves which formed the set of annotated images; second, we trained a detection neural network model by using the set of annotated images, calculated the predictions of other set of 20 staves, and corrected the errors by hand; third, we added these staves to the set of annotated images and retrained the model with this new set; fourth, we repeated the previous two steps until we had 300 annotated staves. The second dataset was built in a similar way. In this process, we used *Faster R-CNN* [27] for the detection model, and *ResNet-18* [15] for the classification model because they were trained very quickly and produced reasonable results. A summary of both datasets can be found in Table 1. These datasets are freely available in the project *GitHub* repository.

Table 1. Datasets information.

Dataset	Training set	Test set	♯ classes
Symbol detection	225	75	34
Pitch classification	2015	504	17

These datasets have been used to build new models using a transfer learning approach [16]. In particular, the datasets have been split into 75% for training and 25% for testing. More concretely, for the detection task we have trained *Faster R-CNN* [27], *RetinaNet* [20], and *EfficientDet* [34] models from the *IceVision* library [36], and *YOLO v4* [5] model from the *Darknet* library [4]. The IceVision models have been trained using the environment provided by *Google Colaboratory*, and the YOLO model in an own server with four Nvidia RTX 2080 Ti GPUs. For the pitch note classification task, we have trained the *ResNet-18* [15] model from the *FastAI* library [16] using the Google Colaboratory environment.

We have compared the four detection models using the following metrics: *Precision*, *Recall*, *F1-score*, *mAP*, *mean Average Precision*, and *COCO-metric*. Table 2 shows the final results of the tested detection models, and we can see that the *YOLO v4* model obtained the best results and it was chosen for further use. In the case of the classification model, we used the *Accuracy* as metric; and the *ResNet-18* model obtained an accuracy of 98.4% on the classification of pitch notes – the training of this model took 56 s. The YOLO and ResNet models have been employed to process all the staves of the book and extract their information. These models together with the previously presented methods to process the text of the book have been employed to populate several databases.

Table 2. Results of the models for detecting musical symbols.

	Precision	Recall	F1-score	mAP	COCO	Training time
EfficientDet	0.28	0.14	0.19	13.86%	0.16	0.37 h
Faster R-CNN	0.86	0.77	0.82	**76.73%**	**0.59**	**0.21 h**
RetinaNet	0.73	0.15	0.25	14.73%	0.19	5.49 h
YOLO v4	**0.89**	**0.90**	**0.90**	68.25%	0.48	72 h

4.4 Data Storage

We have created different databases in order to offer a quick answer to the different uses of the obtained information from the music book. Obviously, other databases configurations can be easily added if other uses are required.

First, we have stored the scanned images of each page of the book in the *JPG* format. We have also stored each detected piece with the previously explained method, and concatenated them (if necessary) in this format. Second, we have stored the text of the book using a dictionary wherein keys are each word in the book and the values are the pages where those words appears. This dictionary has been stored in the *JSON* format [26] for later retrieval in order to perform quick searches. Third, we have also stored two dictionaries in *json* format in order to be able to consult the pieces by author or genre. In the first dictionary the keys were the authors and the associated values are the pieces by those authors. The second dictionary had a similar structure but indexed by musical genre.

Finally, all staves were automatically annotated with the best performing detection model and the classification model. Then, we integrated the predictions of our *ORM* models and the predictions of the OCR model using a *MusicXML* format, taking into account the vertical position of each block with respect to the image. We have stored all the book pieces using this *MusicXML* format. We have also made a conversion from *MusicXML* to *MIDI*, and from *MIDI* to *WAV* audio format, in order to be able to reproduce the sound represented by our pieces. From this *MusicXML* format we can also obtain a clean *PDF* file, thus eliminating the noise of the scanning of each piece.

5 Discussion

We have proposed a methodology for digitizing music books with text and staves on their pages. By applying a combination of computer vision methods to the scanned pages of the book, we detected and classified the different blocks that integrate every page. A summary of the whole process can be found in Fig. 5. Given a one-page image, we have defined a method that detects text blocks and staves separately, and then integrates them into pieces. Subsequently, OCR was applied to text blocks, and both detection and classification OMR models to the staves. In particular, a music symbol detection model based on *YOLO v4*, and a classification model of note pitch based on *ResNet-18*. Finally, we put everything together to generate *MusicXML* files with all the information. From this format, we can transform these files into sound or reconstruct the piece to a clean *PDF* file, thus eliminating the noise of the scanning.

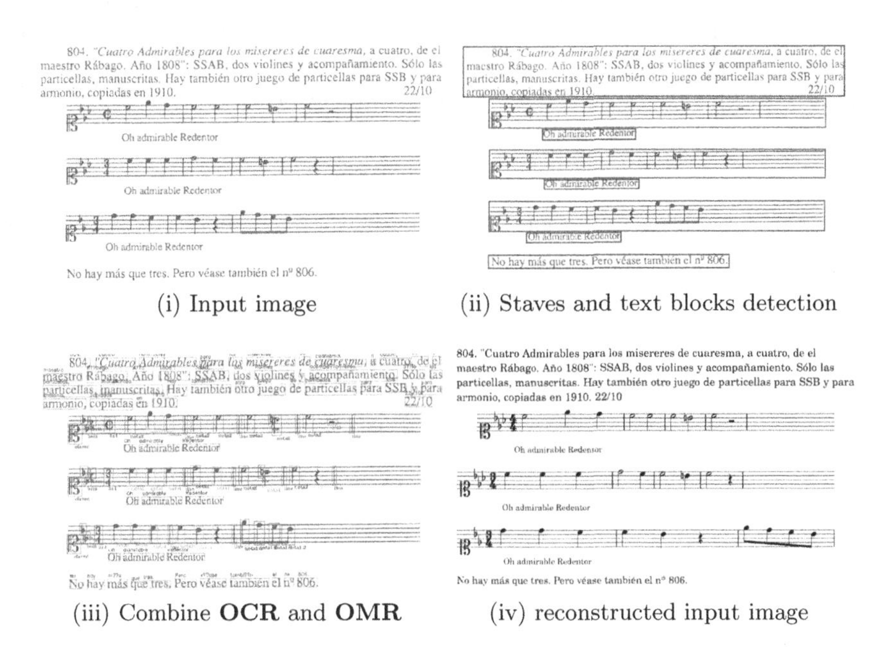

(i) Input image (ii) Staves and text blocks detection

(iii) Combine **OCR** and **OMR** (iv) reconstructed input image

Fig. 5. Score digitization process.

Whereas the application of OCR techniques was quite straightforward, the OMR of the staves required the development of new methods. Following the OMR taxonomy by Calvo-Zaragoza et al. [8], our OMR models involved as input offline, typeset, homophonic, using Common Western notation, scanned with good quality staves. The system architecture was an end-to-end recognition system which took the stave and detected, classified and assembled their components, and stored them using the MusicXML format that enabled them to be replayed and visualized in different formats.

Some state-of-art models for OMR were not suitable for our case. For instance, the model of Huang et al. [18] did not take into account some important symbols of the music book that are out of use, or the model by Calvo-Zaragoza and Rizo [10] was trained on perfectly segmented staves in ideal conditions at the document level. This situation was mentioned by other works that highlighted that there exist different models in the literature that were able to transform images from staves to music, but they usually were not precise enough when the context was different [10]. In a similar way, Audiveris [3], which was able to digitize both text and scores in MusicXML format, did not properly work in our case either. The results obtained by our models are comparable to the results obtained by other state-of-art deep learning based models defined in the literature. For instance Huang et al. model was able to classify music symbols as well as the pitch and duration of the notes with a duration accuracy of 92% and a pitch accuracy of 96% [18]; or Calvo-Zaragoza and Rizo model obtained a symbol error rate of 8% [10]. As previously mentioned, both models were not useful in our context. Figure 6 shows a comparison of the prediction of our model with other models available in the literature on an example extracted from the book.

It remains as future work to test our methodology and methods in other similar music books. It is reasonable to expect that our models will not work conveniently well in other settings and will have to be readjusted for those new scenarios.

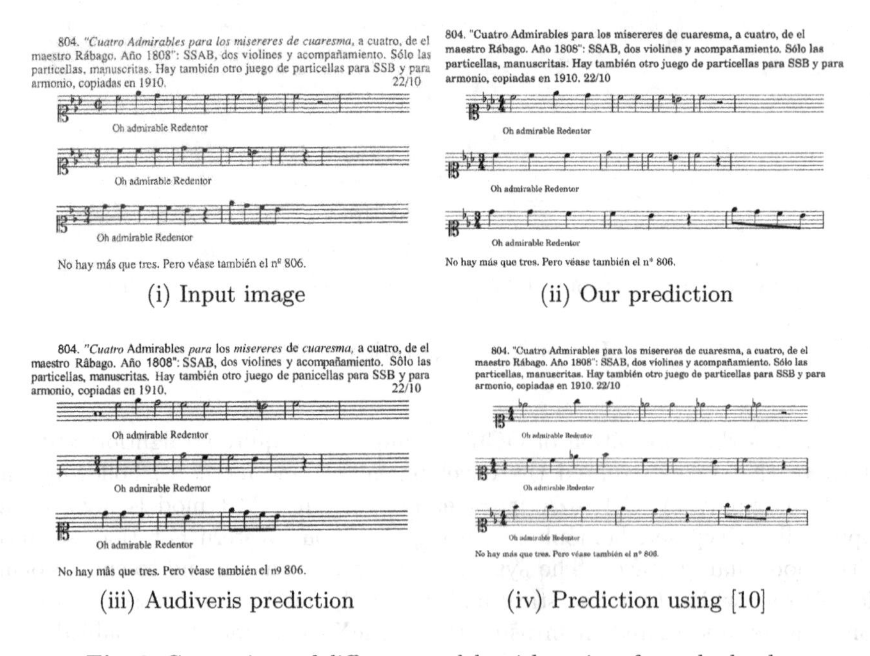

Fig. 6. Comparison of different models with a piece from the book.

6 Conclusions

In this work, we have proposed a new methodology for the digitization of music books. The methodology is designed to digitize book pages which contain both text and staves. After scanning each page of the book, the different blocks of text and staves are detected and organized into music pieces. OCR and OMR methods are applied to text and stave blocks and the information is conveniently stored. Finally, it is possible to perform text searches in the stored information, obtain clean PDF files of music pieces, or reproduce the sound represented by the pieces.

This methodology was applied to the digitization of the book The Music in the Cathedral of Santo Domingo. We provided a new annotated database of musical symbols from the historical music produced in this book. In addition, we have developed new OMR deep learning models for structured representation of historical music scores. These models obtained a F1-score of 90% on symbol detection, and a pitch notes accuracy of 98.4% on the scores of this book. Finally, we have publicly released all the code, annotated images, and models developed in this work on the project website.

References

1. Alfaro-Contreras, M., Calvo-Zaragoza, J., Iñesta, J.M.: Approaching end-to-end optical music recognition for homophonic scores. In: Morales, A., Fierrez, J., Sánchez, J.S., Ribeiro, B. (eds.) IbPRIA 2019. LNCS, vol. 11868, pp. 147–158. Springer, Cham (2019). https://doi.org/10.1007/978-3-030-31321-0_13
2. Alfaro-Contreras, M., Valero-Mas, J.J.: Exploiting the two-dimensional nature of agnostic music notation for neural optical music recognition. Appl. Sci. **11**(8), 3621 (2021)
3. Bitteur, H.: Audiveris (2004). https://github.com/audiveris
4. Bochkovskiy, A.: YOLO v4, v3 and v2 for Windows and Linux (2020). https://github.com/AlexeyAB/darknet
5. Bochkovskiy, A., Wang, C., Liao, H.M.: YOLO v4: optimal speed and accuracy of object detection (2020). https://arxiv.org/abs/2004.10934
6. Bradski, A.: Learning OpenCV, Computer Vision with OpenCV Library. O'Reilly Media, Sebastopol (2008)
7. Byrd, D., Simonsen, J.G.: Towards a standard testbed for optical music recognition: definitions, metrics, and page images. J. New Music Res. **44**(3), 169–195 (2015)
8. Calvo-Zaragoza, J., Hajič, J., Pacha, A.: Understanding optical music recognition. ACM Comput. Surv. **53**(4), 1–35 (2020). https://doi.org/10.1145/3397499
9. Calvo-Zaragoza, J., Rizo, D.: Camera-PrIMuS: neural end-to-end optical music recognition on realistic monophonic scores. In: Proceedings of the 19th ISMIR Conference, pp. 248–255 (2018)
10. Calvo-Zaragoza, J., Rizo, D.: End-to-end neural optical music recognition of monophonic scores. Appl. Sci. **8**(4) (2018). https://doi.org/10.3390/app8040606
11. Chandra, S., Sisodia, S., Gupta, P.: Optical character recognition-a review. Int. Res. J. Eng. Technol. **7**(04), 3037–3041 (2020)
12. Gallego, A.J., Calvo-Zaragoza, J.: Staff-line removal with selectional auto-encoders. Expert Syst. Appl. **89**, 138–148 (2017)

13. Good, M.: MusicXML: an internet-friendly format for sheet music. In: XML Conference and Expo, pp. 3–4 (2001). https://michaelgood.info/publications/music/musicxml-an-internet-friendly-format-for-sheet-music/
14. Hajic, J., Pecina, P.: In search of a dataset for handwritten optical music recognition: Introducing MUSCIMA++ (2017). http://arxiv.org/abs/1703.04824
15. He, K., Zhang, X., Ren, S., Sun, J.: Deep residual learning for image recognition (2015). https://arxiv.org/abs/1512.03385
16. Howard, J., Gugger, S.: FastAI: a layered API for deep learning. Information **11**(2), 108 (2020)
17. Huang, J., et al.: A multiplexed network for end-to-end, multilingual OCR. In: Proceedings of the IEEE/CVF Conference on Computer Vision and Pattern Recognition, pp. 4547–4557 (2021)
18. Huang, Z., Jia, X., Guo, Y.: State-of-the-art model for music object recognition with deep learning. Appl. Sci. **9**(13), 2645–2665 (2019). https://doi.org/10.3390/app9132645
19. Huber, D.M.: The MIDI Manual: A Practical Guide to MIDI within Modern Music Production. A Focal Press Book, Waltham (2020)
20. Lin, T.Y., Goyal, P., Girshick, R., He, K., Dollár, P.: Focal loss for dense object detection. In: Proceedings of the IEEE International Conference on Computer Vision, pp. 2980–2988 (2017)
21. López-Caro, J.: La Música en la Catedral de Santo Domingo de la Calzada. Vol. I: Catálogo del Archivo de Música (1988)
22. Lyu, L., Koutraki, M., Krickl, M., Fetahu, B.: Neural OCR post-hoc correction of historical corpora. Trans. Assoc. Comput. Linguist. **9**, 479–493 (2021)
23. Mursari, L.R., Wibowo, A.: The effectiveness of image preprocessing on digital handwritten scripts recognition with the implementation of OCR Tesseract. Comput. Eng. Appl. J. **10**(3), 177–186 (2021)
24. Musitek: SmartScore 64 (2021). https://www.musitek.com/
25. Neuratron: PhotoScore 2020 (2020). https://www.neuratron.com/photoscore.htm
26. Pezoa, F., Reutter, J.L., Suarez, F., Ugarte, M., Vrgoč, D.: Foundations of JSON schema. In: Proceedings of the 25th International Conference on World Wide Web, pp. 263–273 (2016)
27. Ren, S., He, K., Girshick, R.B., Sun, J.: Faster R-CNN: towards real-time object detection with region proposal networks (2015). http://arxiv.org/abs/1506.01497
28. Rosebrock, A., Thanki, A., Paul, S., Haase, J.: OCR with OpenCV, Tesseract and Python. PyImageSearch (2020)
29. Serra, J., Soille, P.: Mathematical Morphology and Its Applications to Image Processing. Springer Science & Business Media, Dordrecht (2012). https://doi.org/10.1007/978-94-011-1040-2
30. Shatri, E., Fazekas, G.: Optical music recognition: state of the art and major challenges (2020). https://arxiv.org/abs/2006.07885
31. Shatri, E., Fazekas, G.: DoReMi: first glance at a universal OMR dataset (2021). https://arxiv.org/abs/2107.07786
32. Singh, A., Bacchuwar, K., Bhasin, A.: A survey of OCR applications. Int. J. Mach. Learn. Comput. **2**(3), 314 (2012)
33. Smith, R.: An overview of the Tesseract OCR engine. In: Ninth International Conference on Document Analysis and Recognition, ICDAR 2007, vol. 2, pp. 629–633. IEEE (2007)
34. Tan, M., Pang, R., Le, Q.V.: EfficientDet: scalable and efficient object detection (2019). http://arxiv.org/abs/1911.09070

35. Tuggener, L., Satyawan, Y.P., Pacha, A., Schmidhuber, J., Stadelmann, T.: The DeepScoresV2 dataset and benchmark for music object detection. In: 2020 25th International Conference on Pattern Recognition (ICPR), pp. 9188–9195. IEEE (2021)
36. Vazquez, L.: IceVision: an agnostic object detection framework (2020). https://github.com/airctic/icevision
37. Yousefi, J.: Image binarization using Otsu thresholding algorithm (2015). https://doi.org/10.13140/RG.2.1.4758.9284

Evaluation of Named Entity Recognition in Handwritten Documents

David Villanova-Aparisi[1(✉)], Carlos-D. Martínez-Hinarejos[1], Verónica Romero[2], and Moisés Pastor-Gadea[1]

[1] PRHLT Research Center, Universitat Politècnica de València, Camí de Vera, s/n, València 46021, Spain
davilap@inf.upv.es

[2] Departament d'Informàtica, Universitat de València, València 46010, Spain

Abstract. Processing of handwritten documents is an important task for document analysis and cultural heritage. Among the most interesting processing tasks, Named Entity Recognition is one of the most important, since it allows to obtain relevant entities present in the handwritten documents. Many systems perform that in a decoupled way, automatically transcribing the document and, after that, applying classic Natural Language Processing techniques for Named Entity Recognition, but this approach is error-prone because of the noisy nature of the automatic transcription process. In this work we employ a single Deep Learning based model for coupled automatic transcription and Named Entity Recognition in historical handwritten texts. Such model leverages the generalization capabilities of recognition systems, combining Convolutional Recurrent Neural Networks (CRNN) and n-gram character models. The evaluation of that system is discussed and, as a consequence, two novel evaluation metrics, built upon the edit distance concept, are proposed. Additionally, we assess an alternative decoding process which includes syntactical constraints by exploring the n-best hypotheses.

Keywords: Named entity recognition · Historical documents · Evaluation metrics

1 Introduction

There is a multitude of reasons for which to work with Handwritten Text Recognition (HTR) [6]. Even with the appearance of new technologies, we still employ written text as a means to communicate and interact. Being able to obtain automatic transcriptions from such signal may be useful for a number of day-to-day tasks, and may be done via online [9] or offline approaches [12].

This work was supported by Grant RTI2018-095645-B-C22 funded by MCIN/AEI/10.13039/501100011033, by "ERDF A way of making Europe", by Grant ACIF/2021/436 funded by Generalitat Valenciana, by Generalitat Valenciana under the project GV/2021/072 and by Generalitat Valenciana under project DeepPattern (PROMETEO/2019/121).

© Springer Nature Switzerland AG 2022
S. Uchida et al. (Eds.): DAS 2022, LNCS 13237, pp. 568–582, 2022.
https://doi.org/10.1007/978-3-031-06555-2_38

Moreover, a large volume of historical archives are not digitized, which makes them vulnerable to degradation as time goes on. In order to preserve such information, an effort is being made to obtain digital versions of such documents. To expand on this issue, it must be mentioned that documents which have been digitized are mostly available as collections of scanned pages. Working with such format is a time consuming task, as the writing of many of the titles is obfuscated. Thus, many times the services of a paleographer are required to decipher the contents of the images, resulting in a significant monetary expense.

The main objective of historical HTR [14] is to build robust systems which are capable of obtaining accurate transcriptions from the provided scanned pages. Such task is an open area of study, as the present technology is capable of giving good results but still has to deal with the challenges associated to each corpus.

However, there are many collections of documents in which obtaining a perfect transcription is not necessarily the objective. In these cases, the goal is rather to perform some kind of information retrieval from the images, targeting specific fields within the records. A way to solve this problem is to rely on Named Entity Recognition (NER) [8], which is a process that allows to identify parts of text based on their semantic meaning, such as proper names or dates. The majority of NER technology employs Natural Language Processing (NLP) models. These models are usually trained from clean text, which makes them susceptible to input errors.

This work aims to not only obtain valid transcriptions, but also to perform NER in such process. The result of this combined approach is the transcription of the text along with the tags that help identify and categorize the Named Entities. The correct addition of such tags would allow further information extraction processes.

Our work approaches the task of historical HTR and NER over a multilingual corpus [3]. Our choice for such a corpus is motivated by the expected application of this kind of systems. In a real case, the provided solution should be able to deal with the writing style of different authors and different languages. The selected corpus can be seen as a valid sandbox on which to build such system.

The evaluation of the combined HTR and NER task is an open area of study. As we show later, classic metrics imported from HTR or NLP tasks fail to fully consider the nature of the task. One of the contributions of our work is the proposal of two novel metrics which have been designed with the purpose of correctly evaluating the combined task. We also improve the performance of the system with regards to such metrics by imposing constraints over the Named Entity syntax during the decoding process.

The rest of the paper is structured as follows. Section 2 reviews works that are related to this one. Section 3 overviews the employed architecture and describes the error correction strategies that have been considered. Section 4 discusses the current evaluation metrics being used and proposes two novel evaluation metrics. Section 5 describes the experimental methodology that has been followed and discusses the obtained results. Lastly, Sect. 6 concludes the paper remarking the key takeaways.

2 Related Work

The goal of HTR is to obtain the transcription from scanned documents while that of NER is to tag the Named Entities in digital text. When trying to perform HTR and NER over scanned documents, the first approach that may come to mind would be to first obtain the transcription and then to apply NER techniques over such output to obtain the tagged digital text. The alternative to this decoupled method is the development of a method that obtains the tagged output from the scanned documents in one step (coupled approximation).

Recent results indicate that the application of coupled models can improve the performance on such task [4] by avoiding error propagation, which arises as a consequence of the decoupled approach. These errors arise because the model which performs NER expects a clean input, similar to that with which it has been trained. However, the output of the HTR process is not exempt from mistakes, hindering the performance of the NER model. Such effect has also been seen in other tasks of similar complexity. In [5], both speech recognition and NER are performed via a coupled model, obtaining better results than those produced with a decoupled implementation.

This trend also seems to manifest in the results previously obtained in the corpus with which we are going to work. In [3], the "combined model" provides a stronger performance than that of the decoupled approach during the evaluation phase.

3 Framework

3.1 Characteristics of the Task

As we have already mentioned, our work approaches the task of HTR and NER over a multilingual corpus [3]. As such, we have a colored image in the input and the goal is to output the transcription of such image together with the corresponding tagging to identify the Named Entities present in the text. The number of different Named Entities varies according to the used corpus.

The chosen corpus employs parenthesized notation to tag the Named Entities, which imposes additional challenges. One of those challenges is the appearance of nested Named Entities. As an example of this event, we could have a birth date inside a proper noun. To correctly tag such kind of structures, the model must be able to incorporate some kind of syntactical knowledge in the decoding process. Another noteworthy challenge comes from Named Entities which span over several lines. In order to properly tag those Named Entities, the model should be able to have contextual information from lines prior to the one being transcribed. Since we work at line level, we will simplify the task by splitting such Named Entities upon line ending.

3.2 HTR and NER via a Coupled Model

Before introducing the chosen approach to deal with the problem, it may be useful to formalize its definition. Taking into account that the input to our system is an image corresponding to a text line, the HTR problem can be seen as the search for the most likely word sequence, $\hat{w} = \hat{w}_1\hat{w}_2...\hat{w}_t$, given the representation of the input line image, a feature vector sequence $x = x_1x_2...x_m$. This leads us to a search in the probability distribution $p(w \mid x)$:

$$\hat{w} = \underset{w}{\text{argmax}}\, p(w \mid x) \tag{1}$$

If we consider the sequences of NE tags $t = t_1t_2...t_t$ related with the word sequence w as a hidden variable in Eq. 1, we obtain the following equation:

$$\hat{w} = \underset{w}{\text{argmax}} \sum_{t} p(w, t \mid x) \tag{2}$$

If we follow the derivation presented in [3] and explicitly search for the most likely tagging sequence $\hat{t}$ during the decoding process, the obtained equation is:

$$(\hat{w}, \hat{t}) \approx \underset{w,t}{\text{argmax}}\, p(x \mid w, t) \cdot p(w, t) \tag{3}$$

If the information contained in t was added to the transcription w we would obtain the tagged transcription h, resulting in the following equation:

$$\hat{h} \approx \underset{h}{\text{argmax}}\, p(x \mid h) \cdot p(h) \tag{4}$$

Equation 4 resembles that of the original HTR problem, but taking into account that the hypothesis h to be generated is not only the transcription, but also the best tagging. We require, thus, a system that is capable of estimating both the optical probability, $p(x \mid h)$, and the syntactical probability, $p(h)$. The chosen architecture employs a Convolutional Recurrent Neural Network (CRNN) [17] to estimate the optical probability and a character n-gram to estimate the syntactical one. Both models are combined following the approach presented in [2].

3.3 Error Correction

Finding Syntactical Errors. Before introducing the error correction strategy with which we have experimented, it may be convenient to mention the type of errors that the system can produce. The system must be able to tolerate nested Named Entities as well as to avoid incurring in decoding mistakes.

However, these errors are not the only ones to account for. Via an analysis of the training and validation data, we observed that nested Named Entities follow specific structures. What we found was that two tags of the same type could not be nested one inside the other. Therefore, we decided to consider the production of these type of structures as a mistake.

The error correcting strategy that we have designed and implemented tries to discard sequences that are syntactically wrong at Named Entity level. As such, we will only deal with cases of incorrect closing of nested tags, tags that remain opened upon line ending, tags that were closed without a matching opening tag and sequences of nested Named Entities of the same type.

In order to find the errors that we have mentioned, we can perform a left-to-right analysis on the output. During such analysis, we employ an auxiliary stack that keeps track of the Named Entities that remain to be closed. Upon reading an opening or closing symbol, the auxiliary stack is consulted to determine if the tagging complies with the syntactical constraints and is updated accordingly. We also check if the stack is empty at the end of a line, in order to verify that all the Named Entities have been closed.

Exploration of the n-best Hypotheses. One possibility to increase the performance of the system is to improve the decoding process by restricting the output. The main idea here is to force the system to extract hypotheses which are syntactically sound at Named Entity level via an analysis of the output, discarding wrongful hypotheses and searching for correct ones.

The way in which such idea is implemented is simple. Instead of obtaining the best hypothesis during the decoding phase, we obtain a graph containing a sizeable number of hypotheses. This graph is called lattice and it can be directly obtained with the usage of Kaldi.[1] This graph can be pruned to obtain a smaller version containing only the n-best solutions, n being a parameter which we can adjust. From this reduced lattice we can directly obtain the sequence of the n-best hypotheses for an input line.

The decoding process for a specific line starts with an ordered exploration, from the 1-best hypothesis to the n-best. In such exploration, each hypothesis is evaluated by performing a left-to-right analysis with a stack that represents the state of the hypothesized Named Entities to detect syntactical errors. In case of finding an error, the current i-best hypothesis is disregarded and the exploration continues with the hypothesis $i + 1$. This search for the best valid hypothesis concludes when a hypothesis without syntactical errors is found.

It may also happen that, during the search, no valid hypothesis is found. This is an exception that will happen less often as the parameter n increases, but it should still be considered. In those cases, the applied policy will be to keep the 1-best hypothesis as the selected output. Other policies, such as selecting the hypothesis with the least amount of syntactical errors, may be considered. However, for consistency reasons, we are going to favor simplicity in the implementation.

The result of the customized decoding process will be the best valid hypothesis for each of the lines or, in case of not finding any correct output, the one with the most confidence. This error correction strategy includes the structural requirements during the decoding process.

[1] Kaldi's documentation is available at: http://kaldi-asr.org/doc/.

4 Evaluation Metrics

4.1 Character and Word Error Rates

The Character Error Rate (CER) and Word Error Rate (WER) are both based on the edit distance concept and serve the purpose of measuring the similarity between two text sequences. The usefulness of these metrics on the evaluation of the combined HTR and NER problem is debatable. It is clear that the edit distance can be used to compare two sequences of text. In our case, however, the focus does not rely on the quality of the transcription as a whole. Rather, the target is to identify the Named Entities correctly. The tags being employed are mere characters for the computation of the CER and the WER. As such, both metrics cannot evaluate the syntactical correctness of the output, thus limiting their usefulness on the evaluation of the proposed solution. In any case, we still use the aforementioned metrics to compare our work to that of [3].

4.2 Precision, Recall and F1-Score

As we have seen, we cannot fully evaluate the system in terms of CER and WER. The next step in the evaluation process is, thus, to consider a more specific metric that can take into account the syntactical correctness of the output. The approach that is followed starts with the extraction of the Named Entities from each of the hypothesized tagged transcriptions and the ground truth ones. Therefore, we take into account only the detected Named Entities (i.e., tags and contents) for each one of the lines. With this information we can compute the Precision, Recall and F1-Score for the whole corpus. Such metrics require the computation of the number of True Positives (TP), False Positives (FP) and False Negatives (FN).

To calculate the number of TP and FP, we can explore the hypothesized Named Entitities from each line and compare them to the reference data. If a hypothesized Named Entity is present on the ground truth file, then it is considered a TP. Otherwise, it is a FP. Another way to calculate the TP would be to inspect the Named Entities in the ground truth lines and to check if they are present in the corresponding hypothesis. A Named Entity which is present in a reference file and not in the output of the system is considered a FN.

These metrics evaluate the system regarding the detected Named Entities. However, their usage has some drawbacks worth mentioning. Firstly, order constraints are not being considered. If a Named Entity appears originally at the beginning of the sentence and, for some unknown reason, is hypothesized to appear at the end of the sentence, it would be considered as a TP.

Secondly, the number of appearances of a Named Entity cannot be easily taken into account with this computation. To give an example, consider that a proper name appears originally once in a line and that it is hypothesized that it appears a total of three times. In such a case, we may decide to account for three TP or just one, depending on the approach we follow to calculate them.

Lastly, there is also an issue regarding the strictness of the metric. A single character error in the contents of a Named Entity would lead to consider it as a FP. Even if it is theoretically true that it is a mistake and that such Named Entity was not detected correctly, we may still be able to make use of those mistakes in the expected final application of the system. By applying approximate search [1] in a document collection, we should be able to retrieve those documents with small transcription mistakes.

4.3 Entity CER and Entity WER

We have shown that the classic metrics are not exempt from drawbacks when it comes to the evaluation of the combined task. In order to face these drawbacks, a new evaluation metric is proposed. Such metric is based on the edit distance concept and it will evaluate the performance of the system via an analysis of the extracted Named Entities line by line, as it happened during the computation of both Precision and Recall.

More specifically, for each line of the testing corpus we are going to calculate the distance between the hypothesized Named Entities and the reference ones. This distance will be calculated similarly to how the CER and WER are calculated, except that in this case the minimal unit of information is a Named Entity, not a character nor a word. The edit operations that will be considered are the insertion, deletion and substitution of Named Entities. The associated cost to each of the operations for a pair of positions (i, j) can be expressed as:

$$\begin{aligned}
\mathrm{I}(i,j) &= 1 \\
\mathrm{D}(i,j) &= 1 \\
\mathrm{S}_{\mathrm{CER}}(i,j) &= \begin{cases} 2 & \text{if } E_i \neq E_j \\ 2 \cdot \mathrm{CER}(T_i, T_j) & \text{otherwise} \end{cases} \\
\mathrm{S}_{\mathrm{WER}}(i,j) &= \begin{cases} 2 & \text{if } E_i \neq E_j \\ 2 \cdot \mathrm{WER}(T_i, T_j) & \text{otherwise} \end{cases}
\end{aligned} \tag{5}$$

The cost of both the insertion, $\mathrm{I}(i, j)$, and deletion, $\mathrm{D}(i, j)$, is set to one. The cost of the substitution, $\mathrm{S}(i, j)$, depends on several factors. First of all, if the tagging of the Named Entities being compared do not match ($E_i \neq E_j$) the cost is 2, an absolute error of equal cost to an insertion and a deletion. Otherwise, the cost of the substitution depends on the similarity of the text contained within the Named Entities, T_i and T_j. This similarity can be measured as the CER or WER with saturated arithmetic, setting the upper bound to 1. In this way, the substitution cost is contained in the range $[0, 2]$.

If we consider a matrix of size $(|t| + 1) \times (|h| + 1)$, being $|t|$ the length of the reference Named Entities' list and $|h|$ the length of the hypothesized one, the Entity CER/WER (ECER/EWER) between two Named Entity lists is the cost of the best path that starts in position $i = j = 0$ and ends in $i = |t|, j = |h|$. The ECER is calculated according to the following recursive equation:

$$\mathrm{ECER}(t,h) = \mathrm{C}_{\mathrm{CER}}(|t|,|h|)$$
$$\mathrm{C}_{\mathrm{CER}}(i,j) = \begin{cases} i \,\mathrm{if}\, j = 0 \\ j \,\mathrm{if}\, i = 0 \\ \min \begin{cases} \mathrm{D}(i,j) + \mathrm{C}_{\mathrm{CER}}(i,j-1) \\ \mathrm{I}(i,j) + \mathrm{C}_{\mathrm{CER}}(i-1,j) \\ \mathrm{S}_{\mathrm{CER}}(i,j) + \mathrm{C}_{\mathrm{CER}}(i-1,j-1) \end{cases} \end{cases} \tag{6}$$

The EWER is calculated in a similar fashion, the difference being the consideration of the WER for the cost of the substitution. Figure 1 shows how the EWER could be calculated for an artificial sample.

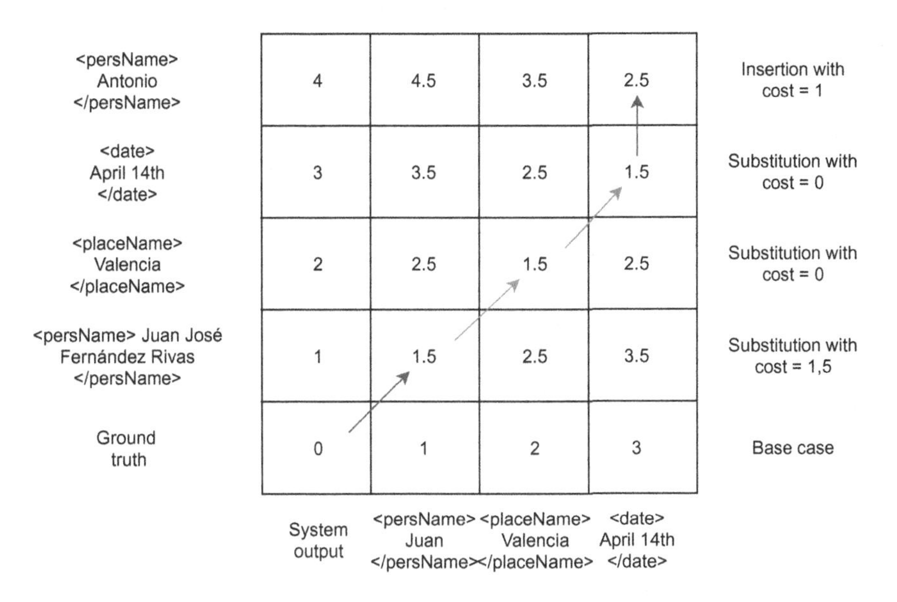

Fig. 1. An example on how the EWER is calculated. Note that the best sequence of edit operations is indicated by arrows.

To better understand how the metric is computed in the particular case shown in Fig. 1, we will analyze each of the edit operations in the optimal sequence of operations. The first operation that is considered is the substitution of the first detected Named Entity for the first Named Entity in the reference. As both of them have the same tags, the cost of the operation is twice the WER $(2 \cdot 0.75)$, which is calculated by comparing the transcriptions without any tagging. The two following operations in the optimal path are substitutions without cost, as both the tagging and the text of the Named Entities match. The final operation is to insert the Named Entity which was not detected by the system. This operation carries a cost of 1. As we can see, the best sequence of edit operations leads us to an EWER of 2.5.

In order to evaluate the performance of the model with several samples, we need to extend the definition of the metric. To tackle this issue, we perform a weighted average over each of the lines, l, of the corpus, C, by considering the associated hypothesis, h_l, and the reference, t_l:

$$\mathrm{ECER}(C) = \frac{\sum_{l \in C} \mathrm{ECER}(h_l, t_l)}{\sum_{l \in C} |h_l| + |t_l|} \tag{7}$$

$$\mathrm{EWER}(C) = \frac{\sum_{l \in C} \mathrm{EWER}(h_l, t_l)}{\sum_{l \in C} |h_l| + |t_l|} \tag{8}$$

We decided to employ the presented edit operation weights for two reasons. Firstly, we wanted to have, in the worst case, a substitution cost equal to both an insertion and a deletion so that the system did not prioritize any edit operation. We also set those weights in order to have the score fall within a certain range when evaluating the performance over a corpus. Given the operation costs and the way in which the metrics are computed, both the ECER and EWER are guaranteed to fall within the range [0, 1].

By definition, the proposed metrics deal with some of the drawbacks that Precision, Recall and F1-Score presented. Firstly, these metrics consider the order in which the Named Entities appear during the computation of the score for each hypothesized line. Secondly, the number of appearances is also taken into account during the computation of said score, as having more cases of a Named Entity in the output will result in a higher number of deletion operations. Lastly, the strictness of the metric can be adjusted by considering the CER or the WER for the cost of the substitutions.

5 Experimental Method

5.1 Dataset

Our experiments have been carried out on a corpus of handwritten medieval charters, which was presented and thoroughly described in [3]. This corpus consists of 499 letters written by different authors in three different languages: Latin, Czech and German. Even though the letters date from 1145 to 1491, their ink and paper are well preserved.

We followed an experimental scheme similar to that presented in [3]. Therefore, the available data is split into three parts: a training set containing 80% of the charters, a validation set with a 10% of the letters and a testing set with the remaining 10% of the samples. The details on how the data is partitioned are shown in Table 1.

Throughout the entirety of the corpus, we can only find three types of Named Entities: the name of a person, the name of a place and a date. As such, the model must be able to generate the corresponding opening and closing tags for each type of Named Entity. The number of different syntactical structures that can be generated, however, is potentially infinite due to the appearance of nested Named Entities.

Table 1. Experimental dataset split, equal to that proposed in [3].

	Number	Czech	German	Latin	All
Train	Pages	161	138	99	398
	Lines	2,905	2,556	1,585	7,046
	Tokens	52,708	60,427	28,815	141,950
	N. Entities	4,973	6,024	2,809	13,806
Validation	Pages	21	18	12	51
	Lines	300	252	150	702
	Tokens	5,997	5,841	2,467	14,305
	N. Entities	440	461	188	1,089
Test	Pages	20	17	13	50
	Lines	381	388	229	998
	Tokens	6,891	9,843	3,995	20,729
	N. Entities	467	744	295	1,506

During previous experimentation, the authors assumed two important simplifications over the original problem. First of all, Named Entities spanning over different lines (Continued Entities) were split so that their occurences are always contained within single lines. The other assumption regarded nested Named Entities. In this case, the authors decided to remove the nested entity. As an example, a date contained within a proper name would have its tag removed, so that there is no nesting.

Our work is based on a combined model that works at line level. Therefore, we are going to maintain the simplification assumed in the original work which consisted of splitting the Continued Entities as not doing so would lead to a noticeable performance decrease. However, we are going to forgo the other simplification and try to deal with nested Named Entities. Our reasoning for this choice is that the combined model should be capable of dealing with this kind of structural complexity without compromising the performance.

5.2 Implementation Details

The employed architecture is based on the combined approach that was used in [3]. Therefore, we scale the line images to a height of 64 pixels and apply contrast enhancement and noise removal as described in [16]. No additional preprocessing is applied to the input.

The optical model consists of a CRNN with four convolutional layers, where the n-th layer has $16n$ 3×3 filters, and a Bidirectional Long Short-Term Memory (BLSTM) unit of three layers of size 256 plus the final layer with a Softmax activation function. The values for the rest of the hyperparameters are the same that were used in [11]. Such model is implemented and trained with the PyLaia toolkit [7].

The employed language model is a character 8-gram with Kneser-Ney back-off smoothing. The estimation of its probabilities is done with the SRILM toolkit [15] by considering the tagged transcriptions in the training partition.

Both the CRNN and the character 8-gram are combined into a Stochastic Finite State Automata (SFSA) with the Kaldi toolkit [10]. Additional parameters such as the Optical Scale Factor (OSF) and Word Insertion Penalty (WIP) are, then, adjusted over the validation partition with the usage of the Simplex algorithm provided by the SciPy library.[2] Finally, the resulting SFSA is used to obtain the hypothesized tagged transcription for each line in the test partition. Our code is available at https://github.com/DVillanova/HOME-ECEREWER.

5.3 Obtained Results

Table 2 shows the best results obtained with each approach. Note that the first column corresponds to previous results, which were taken as a reference for the implementation of our model and evaluated with the proposed metrics. The second column reports the results obtained with our version of the combined approach. The third column shows the best results obtained with the application of the error correcting strategy.

Table 2. Reference results and evaluation scores of the best proposed models, with 95% confidence intervals.

Metric	Boroş, Emanuela et al. [3]	Combined model (nested NEs)	Combined model + 2500-best decoding
CER (%)	**8.00 ± 1.68**	9.23 ± 1.80	9.24 ± 1.80
WER (%)	**26.80 ± 2.75**	28.20 ± 2.79	28.14 ± 2.79
Precision (%)	**49.25 ± 3.10**	43.14 ± 3.07	40.05 ± 3.04
Recall (%)	37.08 ± 3.00	37.58 ± 3.00	**39.97 ± 3.04**
F1 (%)	**42.30 ± 3.07**	40.17 ± 3.04	40.01 ± 3.04
ECER (%)	34.48 ± 2.95	31.94 ± 2.89	**28.69 ± 2.81**
EWER (%)	52.79 ± 3.10	46.62 ± 3.10	**44.42 ± 3.08**

The results that we get with our implementation of the combined model show a drop in performance in almost every metric which was considered in the previous work, although it is not a statistically significant one. The consideration of nested named entities, however, improves the performance with respect to the proposed ECER and EWER metrics.

[2] The documentation for the employed Simplex implementation is available at: https://docs.scipy.org/doc/scipy/reference/optimize.linprog-simplex.html.

As we mentioned earlier, the exploration of the n-best hypotheses introduces a new parameter to adjust. In our experimentation, we have evaluated the effect that this parameter has by giving it different values and testing the performance of the resulting system. Such performance has been measured with each metric that has been discussed in Sect. 4, even though our focus is to optimize the proposed ECER and EWER metrics. During our experiments both the CER and WER remained unchanged independently of the value of n being considered. The evolution of the rest of the metrics with respect to the value of n is presented in Fig. 2.

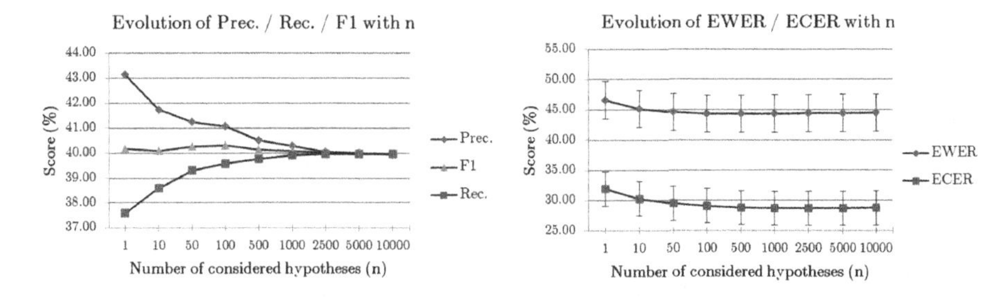

Fig. 2. Evolution of Precision, Recall and F1-score with different values of n. Evolution of EWER and ECER with different values of n considering a 95% confidence interval.

Our initial thought regarding this decoding strategy was that choosing less likely hypotheses would compromise the quality of the transcriptions in exchange for better ECER and EWER. The obtained results show that it is not the case because the differences between the hypotheses are minor, as we can see in Fig. 3. Moreover, the transcription errors that may happen due to the consideration of a less likely hypothesis end up being offset by the imposition of correct tagging.

<persName>Woita otewec Ugolt, Ruda<placeName>Pocimus</persName>. Si qua in futurum ecclesiastica sclarisue persona in factum irritum facere temptaverit, vinculo

<persName>Woita otewec Ugolt, Ruda Pocimus</persName>. Si qua in futurum ecclesiastica sclarisue persona in factum irritum facere temptaverit, vinculo

Fig. 3. Differences between the most likely hypothesis (top) and the 11-best hypothesis (bottom), which is syntactically correct, in one of the test lines.

It is worth mentioning that, in Fig. 2, we see how, with the increasing of n, the Precision worsens while the Recall keeps increasing, both in a logarithmic fashion. The reason behind this trade-off is that the increase in the number of considered hypotheses, n, results in a lower probability of obtaining a syntactically wrong output. By imposing correctness in the output, the number of Named Entities tagged by the system tends to rise as shown in Fig. 4. As a result, both the number of True Positives (TP) and False Positives (FP) increase, the second one at a faster rate than the first. As a consequence, Precision worsens and Recall improves. The F1 score, however, remains almost unaltered.

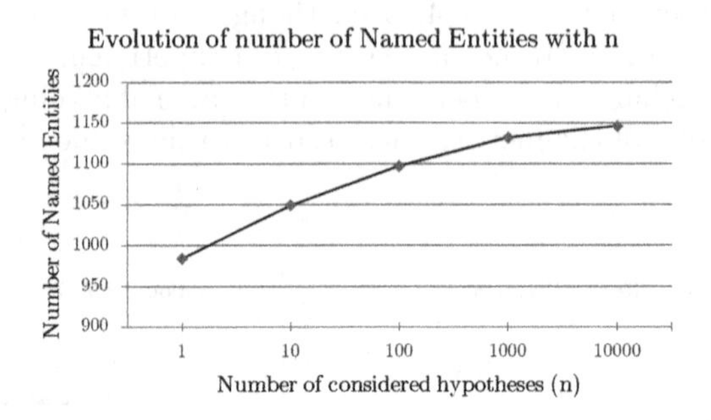

Fig. 4. Evolution of the number of Named Entities detected in the test set with the increase in the number of considered hypotheses.

This evolution in the number of tagged Named Entities affects the proposed ECER and EWER metrics. In Fig. 2 we can see the improvement of the performance with both metrics. It must be mentioned, however, that such improvement is not statistically significant considering a 95% confidence interval.

There are two reasons for this increase in performance in both metrics. Firstly, we have seen an increase in the number of TP. Perfectly recognized Named Entities are greatly rewarded in the proposed metric by being substitutions with no cost. Moreover, the increase that we saw in FP may not be harmful to the quality of the hypotheses. Some of the Named Entities considered as FP may be correctly tagged and have small character errors. Such cases qualify as beneficial for the performance of the system as long as the CER or WER of the transcription is lower than 0.5, depending on the metric being used.

The decrease in ECER and EWER is logarithmic with respect to the increase of n, as we can see in Fig. 2. Logarithmic increases in performance were also spotted in the evolution of Recall. To understand the reason behind this trend, we performed an analysis of the behavior of the system as the parameter n is increased.

The way in which such analysis was conducted is by recording, for each of the test lines, the index of the hypothesis which was selected as the output of the system. If the most likely hypothesis was the one being chosen because no valid hypothesis was found, the recorded index would be a 0, in contrast with the 1 that would be recorded if the 1-best hypothesis was valid. Having such data available, we calculated the frequency for each of the indexes and generated the histograms which can be seen in Fig. 5.

As we can see, by relying on the classical 1-best decoding, the system produces 248 syntactically incorrect outputs. With the usage of bigger values of n, the number of incorrect outputs decreases at a logarithmic rate. This trend

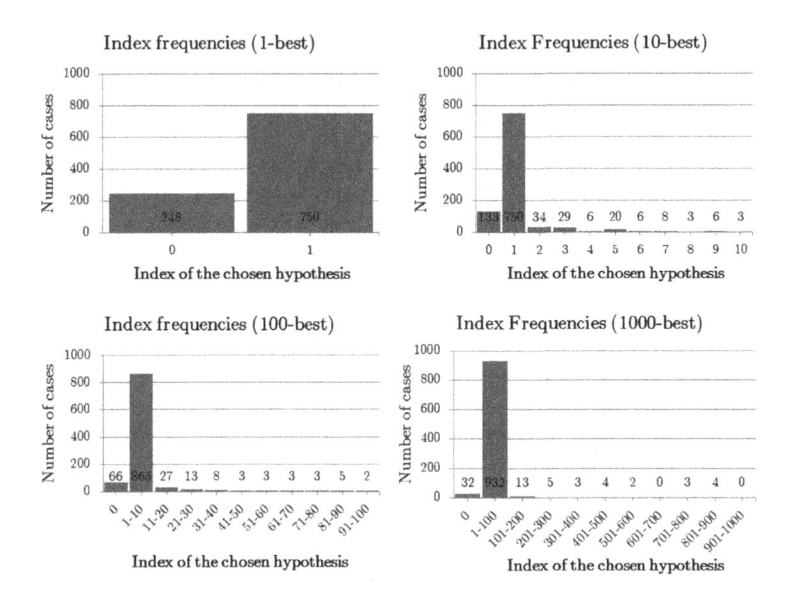

Fig. 5. Frequency histograms of the indexes of the hypotheses with n-best decoding.

justifies the logarithmic evolution that we have seen on the different evaluation metrics.

6 Conclusions

We have presented two novel metrics for the combined HTR and NER task which are based on the edit distance concept. Those metrics have been employed to evaluate a coupled model and the imposition of syntactical constraints during the decoding process. Even if the improvement brought by the n-best hypotheses exploration has not been statistically significant, we have managed to increase the number of syntactically correct outputs. Our model has worked over a simplified version of the corpus in which Named Entities did not span over several lines. A way to expand this work would be to forgo such simplification and to employ paragraph level decoding [13]. We could also opt to compare our approach to other previous works [4,5].

References

1. Andrés Moreno, J.: Approximate search for textual information in images of historical manuscripts using simple regular expression queries. Degree's Thesis, Universitat Politècnica de València (2020)
2. Bluche, T.: Deep Neural Networks for Large Vocabulary Handwritten Text Recognition. Ph.D. Thesis, Université Paris Sud-Paris XI (2015)

3. Boroş, E., et al.: A comparison of sequential and combined approaches for named entity recognition in a corpus of handwritten medieval charters. In: 2020 17th International Conference on Frontiers in Handwriting Recognition (ICFHR), pp. 79–84. IEEE (2020)
4. Carbonell, M., Villegas, M., Fornés, A., Lladós, J.: Joint recognition of handwritten text and named entities with a neural end-to-end model. In: 2018 13th IAPR International Workshop on Document Analysis Systems (DAS), pp. 399–404. IEEE (2018)
5. Ghannay, S., Caubrière, A., Estève, Y., Camelin, N., Simonnet, E., Laurent, A., Morin, E.: End-to-end named entity and semantic concept extraction from speech. In: 2018 IEEE Spoken Language Technology Workshop (SLT), pp. 692–699 (2018). https://doi.org/10.1109/SLT.2018.8639513
6. Kim, G., Govindaraju, V., Srihari, S.N.: An architecture for handwritten text recognition systems. Int. J. Doc. Anal. Recogn. **2**(1), 37–44 (1999). https://doi.org/10.1007/s100320050035
7. Mocholí Calvo, C.: Development and experimentation of a deep learning system for convolutional and recurrent neural networks. Degree's Thesis, Universitat Politècnica de València (2018)
8. Mohit, B.: Named entity recognition. In: Zitouni, I. (ed.) Natural Language Processing of Semitic Languages. TANLP, pp. 221–245. Springer, Heidelberg (2014). https://doi.org/10.1007/978-3-642-45358-8_7
9. Namboodiri, A.M., Jain, A.K.: Online handwritten script recognition. IEEE Trans. Pattern Anal. Mach. Intell. **26**(1), 124–130 (2004)
10. Povey, D., et al.: The Kaldi speech recognition toolkit. In: IEEE 2011 Workshop on Automatic Speech Recognition and Understanding. No. CFP11SRW-USB. IEEE Signal Processing Society (2011)
11. Puigcerver, J.: Are multidimensional recurrent layers really necessary for handwritten text recognition? In: 2017 14th IAPR International Conference on Document Analysis and Recognition (ICDAR), vol. 1, pp. 67–72. IEEE (2017)
12. Rajnoha, M., Burget, R., Dutta, M.K.: Offline handwritten text recognition using support vector machines. In: 2017 4th International Conference on Signal Processing and Integrated Networks (SPIN), pp. 132–136. IEEE (2017)
13. Rouhou, A.C., Dhiaf, M., Kessentini, Y., Salem, S.B.: Transformer-based approach for joint handwriting and named entity recognition in historical document. Pattern Recogn. Lett. (2021). https://doi.org/10.1016/j.patrec.2021.11.010
14. Sánchez, J.A., Bosch, V., Romero, V., Depuydt, K., De Does, J.: Handwritten text recognition for historical documents in the transcriptorium project. In: Proceedings of the First International Conference on Digital Access to Textual Cultural Heritage, pp. 111–117 (2014)
15. Stolcke, A.: SRILM-an extensible language modeling toolkit. In: Proceedings of 7th International Conference on Spoken Language Processing (ICSLP 2002), pp. 901–904 (2002)
16. Villegas, M., Romero, V., Sánchez, J.A.: On the modification of binarization algorithms to retain grayscale information for handwritten text recognition. In: Paredes, R., Cardoso, J.S., Pardo, X.M. (eds.) Pattern Recognition and Image Analysis, IbPRIA 2015. LNCS, vol. 9117, pp. 208–215. Springer, Cham (2015). https://doi.org/10.1007/978-3-319-19390-8_24
17. Xingjian, S., Chen, Z., Wang, H., Yeung, D.Y., Wong, W.K., Woo, W.C.: Convolutional LSTM network: a machine learning approach for precipitation nowcasting. In: Advances in Neural Information Processing Systems, pp. 802–810 (2015)

A Generic Image Retrieval Method for Date Estimation of Historical Document Collections

Adrià Molina(✉), Lluis Gomez, Oriol Ramos Terrades, and Josep Lladós

Computer Vision Center and Computer Science Department, Universitat Autònoma de Barcelona, Bellaterra/Cerdanyola, Catalunya, Spain
{amolina,lgomez,oriolrt,josep}@cvc.uab.cat

Abstract. Date estimation of historical document images is a challenging problem, with several contributions in the literature that lack of the ability to generalize from one dataset to others. This paper presents a robust date estimation system based in a retrieval approach that generalizes well in front of heterogeneous collections. We use a ranking loss function named smooth-nDCG to train a Convolutional Neural Network that learns an ordination of documents for each problem. One of the main usages of the presented approach is as a tool for historical contextual retrieval. It means that scholars could perform comparative analysis of historical images from big datasets in terms of the period where they were produced. We provide experimental evaluation on different types of documents from real datasets of manuscript and newspaper images.

Keywords: Date estimation · Document retrieval · Image retrieval · Ranking loss · Smooth-nDCG

1 Introduction

Universal access to historical archives and libraries has become a motivation to face innovative scientific and technological challenges. New services based in Document Analysis Systems are emerging for automatic indexing or extracting information from historical documents. In this context, the progress in Document Intelligence has brought the development of efficient methods for automatic tagging of such variety of documents. From a practical point of view, the association of metadata describing documents is extremely important in the early stages of archival processes, when archivists and librarians have to incorporate new collections (e.g. thousand new images of manuscripts, photographs, newspapers...). Metadata consists of annotated semantic tags that are used as keys in subsequent users' searches. One of the most relevant information terms, in particular in historical documents, is the time stamp. Automatic dating has been particularly addressed in historical photos [20,21]. Visual features are highly important in dating of historical photographs. Texture and color features are

© Springer Nature Switzerland AG 2022
S. Uchida et al. (Eds.): DAS 2022, LNCS 13237, pp. 583–597, 2022.
https://doi.org/10.1007/978-3-031-06555-2_39

good sources of information to accurately estimate when the image was taken, because photographic techniques have evolved. On another hand it is worth paying attention to the objects in the scene and their correspondence to some time periods (the clothes that people wear, their haircut styles, the overall environment, the tools and machinery, the natural landscape, etc.). But photographs are a particular type of historical assets, and the diversity is large. Date estimation is also interesting in mostly textual documents. A particular case is date estimation in manuscripts where writing style features and script recognition are crucial to classify a document within a time period or a geographic area (Hebrew, Medieval...) [6,7]. Hybrid methods combining multimodal features like appearance-based features and textual labels not only from the image itself but from its context when it appears in a document like a newspaper or a web page [19] have also demonstrated interesting performance. An interesting observation in the date is an ordinal feature, i.e. in some cases the date of a document can be better estimated by comparison to other ones. Ordinal classification techniques have been proposed so the task is to predict a ranked list of documents [19,20].

The existing date estimation methods have two main drawbacks. First, they are highly dependent of having a priori and contextual knowledge, and second they are conditioned by the data target. Technically, it means that a lot of training data, labeled by experts is required. On another hand, from a systemic perspective, i.e. date estimation as a service to scholars and archivists, there is a need of genericity and adaptability. In a practical situation, an archivist or a librarian receives diverse collections to classify with bunches of images. They require tools able to deal with heterogeneous data collections. Additionally, the response the archivist or social scientist is looking for, is not specific or affordable for any estimator. This is where retrieval plays a key role: Some professional may be looking for the date of an image, where an estimator makes its work properly; but other may be looking for cues, patterns or historical keywords (such as a word itself, a cool hat or a particular historical character) that belong to a certain period. Estimators fails at providing this service. Using ranking functions for this task, should pave the way to generate embedding spaces where communities are constituted by its contextual information such as the date instead of the visual and textual information.

For the discussed above reasons, in this paper we present a date estimation approach that follows a ranking learning paradigm in a retrieval scenario. We propose an incremental evolution of our previous work on date estimation for photographs collections [20]. In this work we proposed a model based in the Normalized Discounted Cumulative Gain (nDCG) ranking metric. In particular, the learning objective uses the *smooth-nDCG* ranking loss function. In the current work, we extend the functionality of the method, making it able to generalize to other types of documents. We will show its performance to estimate the date of two categories of documents: handwritten and historical newspapers joining the previously explored scanned photographs system in [20]. As additional feature of the proposed system, we integrate the human in the loop, so the user herself/himself if able to integrate her/his own feedback for fine tuning the model.

The overall contributions of our work can be summarized as follows:

- A flexible date estimation method based in a ranking loss function which is able to estimate the date of a document image within an ordering of the whole collection. The method is adaptable to different types of documents.
- The capability of generating embedding spaces according to contextual information such as dates instead of using textures, visual cues or textual ones.
- An application that is able to deal with document collections of different types. In particular we will show the performance on historical newspapers and manuscript documents.
- The inclusion of the user feedback in the process; making the application able to focus on those categories the user is currently interested on.
- The evaluation on different real datasets, some of them provided by the Catalan National Archives Department.

The rest of the paper is organized as follows: in Sect. 2 we review the state of the art in date estimation and we state the contribution of this paper with respect previous work. The data used for the experimental validation and the details of the peculiar loss function are exposed in Sects. 3, 4 respectively. In Sect. 5 the retrieval system is described. In Sect. 6 we present the resulting application architecture and the quantitative results in terms of regression. We conclude with Sect. 7 where we draw the conclusions.

2 Related Work

Before the widespread use of deep learning for document analysis tasks, automatic dating of historical manuscripts and printed documents was primarily based on hand-crafted features, carefully designed to capture certain characteristics of the handwriting style that are useful for identifying a particular historical period. Some examples of such hand-crafted features are Fraglet and Hinge features [3,13], Quill features [2], textural measures [8], histogram of strokes orientations [12], and polar stroke descriptors [10,14] among others.

In recent years, thanks to the latest developments in deep learning, there has been a greater tendency to use models in which features are learned directly from training data to perform various document analysis and recognition tasks [18]. In that regard, Li *et al.* [17] proposed a custom Convolutional Neural Network (CNN) consisting of two convolutional layers and two pooling layers for the task of publication date estimation for printed historical documents. They evaluated this model in two different tasks: date regression and century classification (with four classes that span for 100 years each).

Wahlberg *et al.* [29] used a more modern CNN based on the GoogleNet architecture [26] either for directly estimating the date of historical manuscripts or as a feature extractor in combination with other regression techniques. In particular, they explore the combination of CNN extracted features with a Gaussian

Processes Regressor and a Support Vector Regressor. In this case the model is pre-trained on the Imagenet dataset [5] and fine-tuned with a single output neuron for date regression. They extract 20 random crops of 256 × 256 from each manuscript, and at inference time they take the median of the 20 date estimates. A similar approach, using a pre-trained deep CNN model, was explored by Hamid *et al.* [7] and Studler *et al.* [24] but they evaluated several state of the art CNN architectures including VGG19 [23], GoogleNet [26], ResNet [9], Inceptionv3 [27], InceptionResnetv2 [25], and DenseNet [16]. Overall, their experiments demonstrated that ImageNet pre-training improves the performance of all the architectures, and that the patch-based ensemble (taking several crop patches of a manuscript) provides better results than using the whole document for both regression and classification tasks.

All the models mentioned above treat the problem of document dating either as a classification or a regression task. In this paper we propose to use different approach and recast the dating problem as a retrieval one. As shown in our previous work [20,22], it is possible to train a neural network to learn document representations in an embedding that preserves a desired metric as input; in this case, the proportion between different dates (years) is preserved in the output space. This new approach allows us to include in the results a retrieval document set ranked by date similarity, which, again, is contextual information rather than strictly visual; despite context is deducted by visual clues, we're not optimizing for looking certain clues, but a general embedding that better fits the representation of a certain period with respect other ones. This is specially desirable in the proposed application of a support tool for professional archivist and historical researchers (Sect. 6) since it doesn't just return a categorical result but a set of similar labeled data that may help to get the proper conclusions by comparing the query and output data.

3 Datasets

Several document dating databases exist for both historical manuscripts [1,4, 15] and printed documents [28]. In our experiments, we used two datasets: the Medieval Paleographic Scale (MPS) dataset and a custom historical newspaper dataset.

Manuscript Date Estimation

One of the problems we tested our method on, is the date estimation of Medieval Paleographic Scale manuscripts, MPS Dataset, presented by He Sheng *et al.* [15]. The dataset presents several manuscripts with Dutch and Flemish manuscripts from medieval mid period, years 1300 to 1550, divided in steps of 25 years. The dataset is not uniformly distributed with respect the years Fig. 1, which means that we lack of manuscripts from early and latest years.

As introduced by He *et al.*; we can observe a flagrant evolution in several characters in the dataset; leading to great results at character-level [11,12].

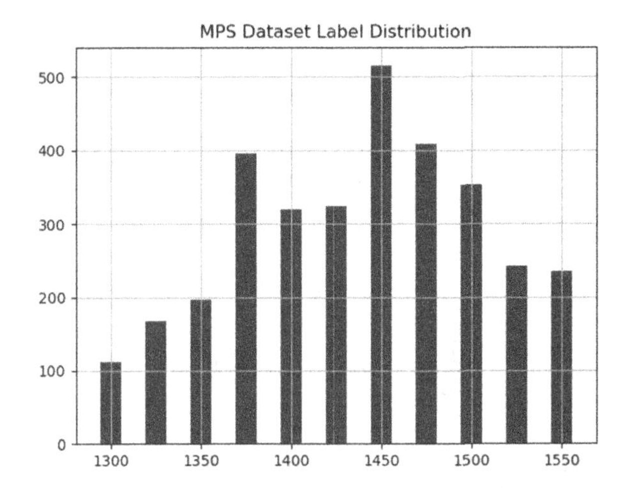

Fig. 1. MPS Dataset histogram of images per label.

Since the dataset provides directly the whole manuscript, we use this as input data 2 for a more general view.

Fig. 2. Sample of manuscripts extracted from the Medieval Paleographic Scale (MPS) dataset [15]. Images belong to years 1300, 1400 and 1500 respectively.

Newspaper Date Estimation

The second kind of document we have used to evaluate our method has been Newspaper date estimation Fig. 4. This data collection was gently yielded by the archivist from the Xarxa d'Arxius Comarcals[1]; belonging to the National Archives of Catalonia. This dataset consists of 10,001 newspaper pages from the year 1847 to 2021. Figure 3 shows some samples. In our experiments we gray-scaled all the images to prevent a color bias in the retrieval process.

Due historical and practical reasons the dataset is heavily unbalanced and sparsely distributed. Nevertheless almost every period contains some samples.

[1] XAC is a governmental archivist institution with url for further detail: https://xac.gencat.cat/en/inici/.

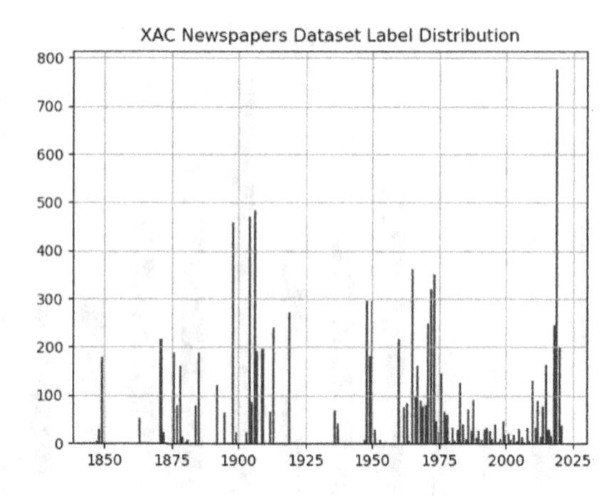

Fig. 3. XAC newspaper dataset: histogram of images per label. (Color figure online)

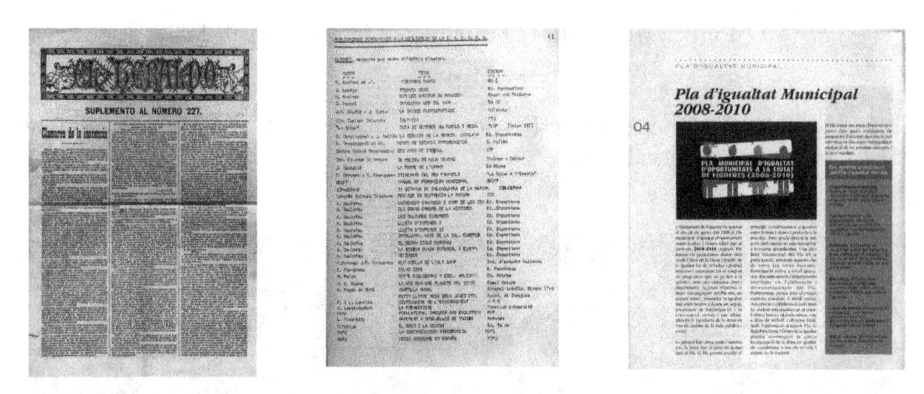
SUPLEMENTO AL NÚMERO 227.

Pla d'igualtat Municipal 2008-2010

04

Fig. 4. Example of different newspapers from the XAC-Newspapers dataset; years 1905 (left) 1982 (middle) and 2009 (right).

4 Learning Objectives

Following our previous work for word spotting [22] and scanned image retrieval [20], our document dating model uses the smooth-nDCG (Eq. 1) as its learning objective function. The smooth-nDCG is a differentiable information retrieval metric. Given a query q and the set of retrieved elements from the dataset Ω_q, the smooth-nDCG is defined as follows:.

$$\mathrm{DCG}_q \approx \sum_{i \in \Omega_q} \frac{r(i)}{\log_2\left(2 + \sum_{j \in \Omega_q, j \neq i} \mathcal{G}(D_{ij}; \tau)\right)} \tag{1}$$

where $r(i)$ is a graded function on the relevance of the i-th item in the returned ranked list, $\mathcal{G}(x;\tau) = 1/(1 + e^{\frac{-x}{\tau}})$ is the sigmoid function with temperature τ pointing out if element i is relevant with respect j as a mimic for binary step function, and $D_{ij} = s_i - s_j$, with s_i and s_j being respectively the cosine similarity between the i-th and j-th elements to the query.

This function measures the quality of retrieved content given a query but the retrieved content may be graded in a decreased scale. For example, in the case of document dating, it does not make sense to use a binary relevance function in which the documents of the same year as the query are considered relevant and the rest are considered not relevant. Instead, we would like to have a graded relevance function ($r(i)$) that ranks elements by their date-distance to the query (i.e. elements 1, 2 or 3 years away might be relevant in this order). This is reflected in Eq. 1 by the numerator $r(i)$ meaning the relevance of the document i with respect the query q. This can be considered the ground truth label for the problem to optimize. In Fig. 5 we illustrate graphically different relevance functions that prioritize retrieving images from closer dates (yellow and red indicates higher values for the relevance function $r(i)$ and darker colors indicates lower values).

Note that the relevance $r(i)$ given a query q can be represented by a relevance matrix where M_qi is the relevance of the document i with respect the query q. This can be saw as an attention matrix, what may be interesting to observe.

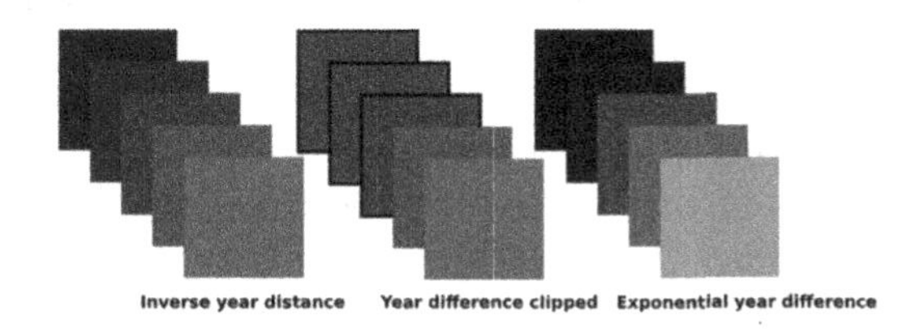

Fig. 5. Example of different relevance functions $r(i)$ that could be applied to a set of data. (Color figure online)

Since the document dates' annotations are $\mathbb{R}^1$ scalars (years), it makes sense to measure the distance between documents with their dates' absolute difference. In this paper we use various slightly different formulations of $r(i)$: γ-thresholded difference (Eq. 2), the logarithmic scaled difference (Eq. 3), or a combination of them. In order to keep the distances in a reasonable range; we used Eq. 2 for problems with a few categories such as the newspaper retrieval, where there's only a range of one century. When distances may return too big numbers, we scale it with Eq. 3 such as MPS dataset, which has a wider range $[min_{year}, max_{year}]$.

$$r(n; \gamma) = \max(0, \gamma - |y_q - y_n|) \tag{2}$$

$$r(n) = \log(1 + |y_q - y_n|) \tag{3}$$

5 Proposed Method

As mentioned above, to the best of our knowledge the existing methods for date estimation an retrieval can not be generalized to different types of document datasets. There's works that better meet this requirement such as He Sheng *et al.* [11,12] that requires a certain priory propositions; or Muller's work [21] that since it is based in categorical classification it can not predict if a manuscript classified of a certain period belongs to the early or late registers in it. Nevertheless it is notable that some methods such as Hamid's *et al.* [7] have shown an extraordinary performance in terms of strict date estimation without any a priory information. On the other hand, since this method relies on information retrieval; many applications and variations may emerge. This is the case of the application we are presenting.

Using the benefits of being able to effectively retrieve information from different layers of relevance, we can build a Human-in-the-loop system that uses user feedback to adapt the specificity of the use case Fig. 8. As it will be discussed in Sect. 6; the importance of the prediction error decays in favor of adaptability not only to different datasets but to focusing on different labels in the same one.

Given a labeled dataset, we train a Convolutional Neural Network (CNN) to learn a document projection to an embedding space where the distance between points is proportional to their distance in years, not specific visual features or content. This is made by maximizing the smooth-nDCG function; a normalization of Eq. 1. Algorithm 1 describes the training algorithm for the proposed model. Once the neural network is able to sort documents according to their actual dates; we can project a bunch of unlabeled ones expecting it to be properly clustered as a sorting Fig. 7.

With this method, if we have enough continuity and density of data so the "years" space is organized in cluster such as $Y = \{1700, ..., 1715\} \cup \{1730, ...\}$, the method can figure out that if the new bunch of data is placed between two clusters. Figure 6 illustrates this concept of inter-cluster embedding. We can use cosine distance to perform a weighted prediction for this new cluster. Note that given a new bunch of data many points could already belong to a certain cluster while many others could not, forming then new ones.

So, as it has been presented before, with smooth-nDCG date estimation any prior information is not required, and despite not having certain labels, we can find them out by properly exploring the cluster distribution with respect to the new data.

Algorithm 1. Training algorithm for the proposed model.

Input: Input data $\{\mathcal{X}, \mathcal{Y}\}$; CNN f; distance function D; max training iterations T
Output: Network parameters w

1: Initialize relevance matrix $R : |\mathcal{Y}| \times |\mathcal{Y}|$
2: **for** $y_q \leftarrow$ year in $\mathcal{Y}$ **do**
3:     **for** $y_n \leftarrow$ year in $\mathcal{Y}$ **do**
4:         Calculate relative relevance $R_{i,j} \leftarrow$ Eq. 2
5:     **end for**
6: **end for**
7: **repeat**
8:     Process images to output embedding $h \leftarrow f_w(\{x_i\}_{i=1}^{N_{batch}})$
9:     Get Distance matrix from embeddings, all vs all rankings $M \leftarrow D(h)$
10:     Get relevance from relevance matrix given a query q, $r(i) \leftarrow R_{i,q}$
11:     Using the relevance score, $\mathcal{L} \leftarrow$ Eq. 1
12:     $w \leftarrow w + \eta(\nabla_w \mathcal{L})$
13: **until** Max training iterations T

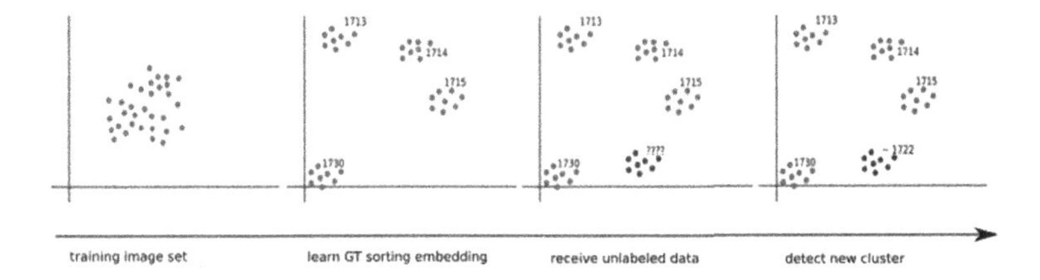

Fig. 6. Method pipeline for predicting years outside the training set. Since the clusters are sorted optimizing the ranking function; a new cluster found between two ones could be labeled as the mean between both clusters.

6 Application

6.1 Smooth-nDCG Human-in-the-Loop Architecture

As commented above we propose an application that is designed as a service to experts that deal with lot of unlabeled data that they have to sort, label, or extract some cues of information such as location or keywords. As illustrated in the application-level scheme in Fig. 9 our solution is generic and applicable to any document domain and dataset, the only module that needs to be changed is the relevance matrix of the data.

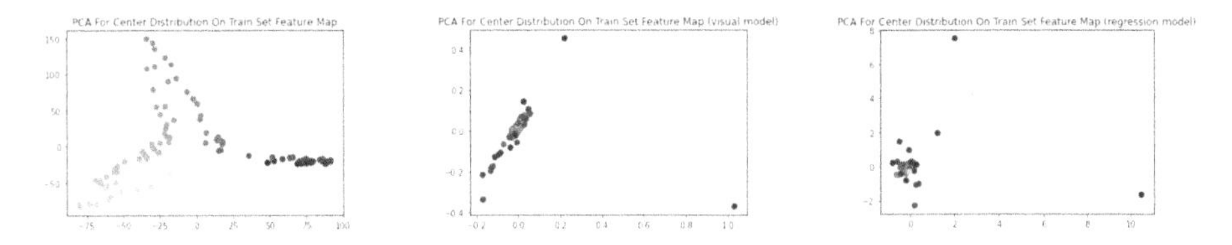

Fig. 7. PCA 2D Projection for Newspapers Dataset cluster centers on (from left to right): nDCG model, Imagenet weights, regression model from Table 2. Color gradient is proportional to label gradient (years). (Color figure online)

We appreciate that our main intention is being able to retrieve useful information for the user, this is where the retrieval function shines brighter than classification approaches. Once retrieved the information with which the user should be able to perform their task (such as comparative date estimation in this case) they can improve their own application instance by joining its newly labeled data to the set for the k-nn selection. As shown in Fig. 10 given a query we can suppose unlabeled, the top of retrieved documents should help the user to run a series of conclusions by comparison; since the entire response is labeled, they should not start from scratch, but from a good benchmark given by the k-nn approximation.

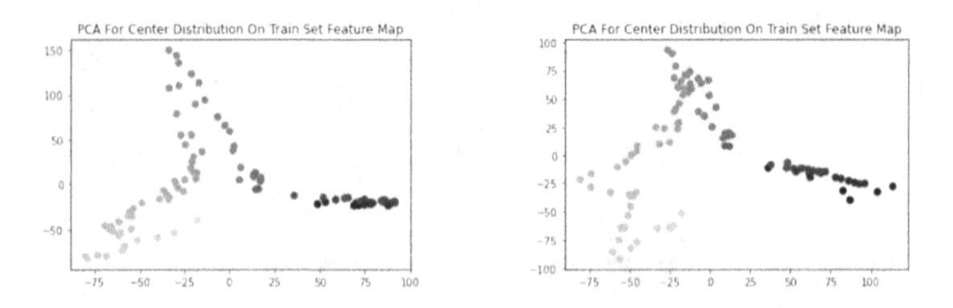

Fig. 8. Main 2D Embedding space (left) compared to an specialized in early years embedding space (right).

Additionally, as it was exposed in Algorithm 1, we are using a relevance matrix that tells how relevant is a category, or a certain data to another. In our application this property can be used to incorporate feedback from the user. For example, the user might not want to consider the distances in the early 20s period as relevant as the distances in the late 90s. If the user is not interested in studying data from further categories, this should allow the model to train again focusing on the categories the user is currently interested in. This Human-in-the-loop training allows the user to adapt the model to their specific needs in an intuitive manner, without any knowledge about deep learning or computer vision.

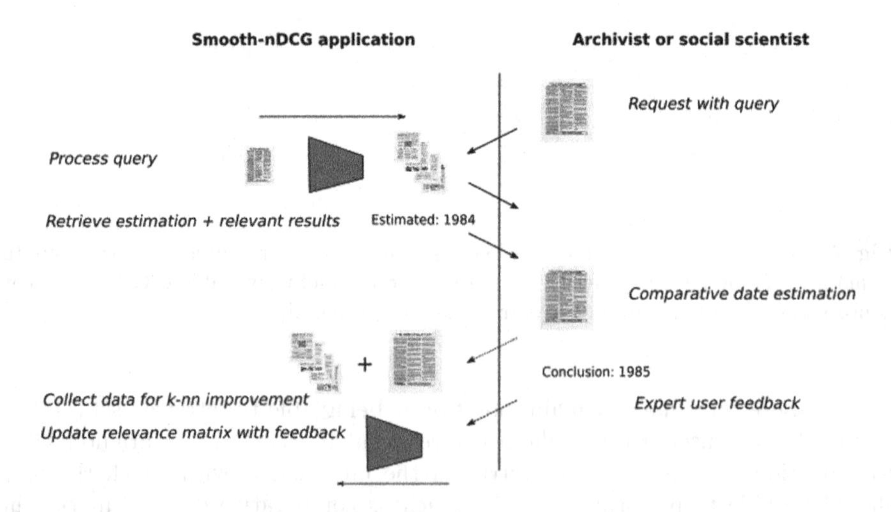

Fig. 9. Communication between user and a smooth-nDCG application. Observe that it's not an unilateral communication scheme.

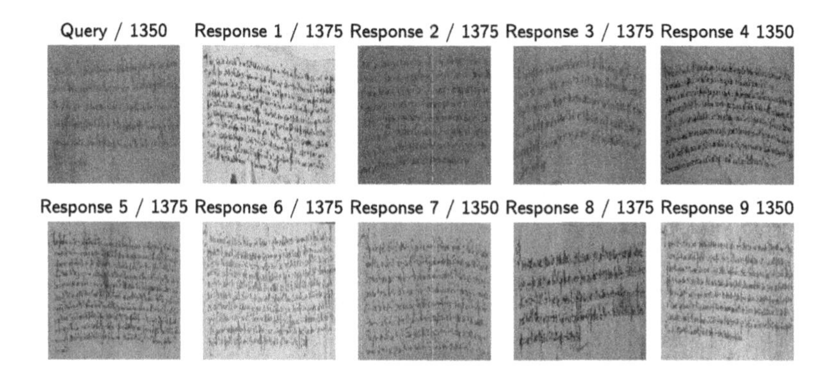

Fig. 10. Qualitative results for a query from the Medieval Paleographic Scale (MPS) test set [15].

6.2 Quantitaive Evaluation

We have trained and evaluated our document dating model in the two datasets described in Sect. 3. Tables 1 and 2 show the obtained results and a comparison with other existing methods and baselines in terms of Mean Average Error (MAE) and mean Average Precision (mAP). While in the MPS dataset (Table 1) our model performance is far from the state-of-the-art, we want to emphasise that our primarily goal in this paper is not to obtain a perfect MAE but into design a practical human-in-the-loop application. In that regard it is important to notice that no other model in Table 1 can directly provide a retrieval document set that allows the user to make more informed predictions and at the same time improve the model's performance in an easy way. The experiments ran on Tables 1 and 2 were performed with an Inception v3 pre-trained on ImageNet fine-tuning all the parameters. As it will be discussed on Sect. 7, further research is needed in terms of optimizing this CNNs. Using patches instead of the whole image as shown in [7] should make it perform better. Many clues in previous work [20] and Table 2 point that smooth-nDCG function should be able to perform with equivalent results to a regression model, with the added value of optimizing the ranking metrics.

We used as well an Inception v3 estimator for baseline in Table 2 with same fine-tuning criterion and a 1000×1 linear layer at the top of the network[2]. We added two frozen parameters: given the network $f : X \rightarrow \mathbb{R}^1$ and the frozen parameters $P = (w, b)$ the computed prediction h is the linear combination $h = (f(x), 1) \cdot P^T$. This parameters could be a learnable layer but for faster convergence we decided to let it frozen at $(10, 2000)$ as Muller *et al.* [21] froze the bias as the average of the dataset (1930) in the classifier. We used the whole dataset for training and testing the model.

---

[2] mAP approximated from training bacthes.

Table 1. Mean Absolute Error (MAE) comparison of our model with existing methods on the test set of the MPS dataset

Baseline	MAE	mAP
Fraglet and Hinge features [13]	35.4	-
Hinge features [3]	12.20	-
Quill features [2]	12.10	-
Polar Stroke Descriptor (PSD) [12]	20.90	-
PSD + temporal pattern codebook [10]	7.80	-
Textural features [8]	20.13	-
InceptionResnetv2 [7]	3.01	-
Smooth-nDCG Manuscript Retrieval (MPS)	23.8	0.43

Table 2. Results and benchmark for newspaper date estimation dataset

Baseline	MAE	mAP
Regression baseline (Inception v3)	3.5	0.24
Smooth-nDCG newspaper date estimation (Inception v3)	2.9	0.49

7 Conclusions

As exposed in Sects. 1 and 6; the retrieval task is a powerful tool for archivists and social scientists. Despite further research is needed in terms of this particular CNN optimization (as Hamid *et al.* [7] proved, Inception can outperform our current MAE error), the designed model provides lot of powerful interactions with the user. The most important idea of this work, is the fact commented in Sect. 1; the date estimation task is a very specific one as many others. This means that knowing the prediction of the year of a document may not be useful (which is indeed useful as cue for this specific task). Alternatively, given a query, retrieving a set of labeled images of the same context than the query, can be more useful in the study of a document collection.

In comparison to previous work reviewed in Sect. 2, the proposed method allows the user to evaluate topics under the dataset further than strict date estimation as it was designed to. In contrast to classic content based retrieval where visual features such as textures, patterns or keywords are used for indexing, our proposed system offers the functionality of indexing by historical context. Note that the context is something that emerges directly from the visual content, but by not biasing the model to look for one (or a combination of them) of the commented visual cues, the model gets an optimized representation of each year in the embedding space. This returns us to the previous statement, the fact that the model is learning a continuous efficient representation of each one of the periods (by using a graded retrieval function as loss) [20,22], the answers you can get by analyzing the retrieved content are wider than they seem.

The third key point of the application is the malleability and adaptability of the system to new requests and data. As shown in Sect. 5, the usage of a relevance matrix allows the users to provide feedback that improves the system just by tuning the values in the matrix to their convenience i.e. increasing the values of relevance for categories that fit their interest. In the case of date estimation, if there's an interest for a certain year, it's easy to just increase relevance near R_i. As illustrated in Fig. 9 the second part of this, actually Human-in-the-loop, architecture relies on the improvement of the k-NN estimator by feeding the database with recently labeled data.

In conclusion, this method should pave the way to may social sciences usages for big historical databases or institutions as national archives as mentioned in Sect. 1. Nevertheless, as it was demonstrated [7], the improvement of the hyper-parameters, usage of patches and, in general, better training is an immediate future work.

Acknowledgment. This work has been partially supported by the Spanish projects RTI2018-095645-B-C21, and FCT-19-15244, and the Catalan projects 2017-SGR-1783, the Culture Department of the Generalitat de Catalunya, and the CERCA Program/Generalitat de Catalunya.

References

1. Adam, K., Baig, A., Al-Maadeed, S., Bouridane, A., El-Menshawy, S.: Kertas: dataset for automatic dating of ancient Arabic manuscripts. Int. J. Doc. Anal. Recognit. (IJDAR) **21**(4), 283–290 (2018)
2. Brink, A.A., Smit, J., Bulacu, M., Schomaker, L.: Writer identification using directional ink-trace width measurements. Pattern Recogn. **45**(1), 162–171 (2012)
3. Bulacu, M., Schomaker, L.: Text-independent writer identification and verification using textural and allographic features. IEEE Trans. Pattern Anal. Mach. Intell. **29**(4), 701–717 (2007)
4. Cloppet, F., Eglin, V., Helias-Baron, M., Kieu, C., Vincent, N., Stutzmann, D.: ICDAR 2017 competition on the classification of medieval handwritings in latin script. In: 2017 14th IAPR International Conference on Document Analysis and Recognition (ICDAR), vol. 1, pp. 1371–1376. IEEE (2017)
5. Deng, J., Dong, W., Socher, R., Li, L.J., Li, K., Fei-Fei, L.: Imagenet: a large-scale hierarchical image database. In: 2009 IEEE Conference on Computer Vision and Pattern Recognition, pp. 248–255. IEEE (2009)
6. Dhali, M.A., Jansen, C.N., de Wit, J.W., Schomaker, L.: Feature-extraction methods for historical manuscript dating based on writing style development. Pattern Recogn. Lett. **131**, 413–420 (2020). https://doi.org/10.1016/j.patrec.2020.01.027
7. Hamid, A., Bibi, M., Moetesum, M., Siddiqi, I.: Deep learning based approach for historical manuscript dating. In: International Conference on Document Analysis and Recognition - ICDAR2019, pp. 967–972. IEEE (2019)
8. Hamid, A., Bibi, M., Siddiqi, I., Moetesum, M.: Historical manuscript dating using textural measures. In: 2018 International Conference on Frontiers of Information Technology (FIT), pp. 235–240. IEEE (2018)

9. He, K., Zhang, X., Ren, S., Sun, J.: Deep residual learning for image recognition. In: Proceedings of the IEEE Conference on Computer Vision and Pattern Recognition, pp. 770–778 (2016)
10. He, S., Samara, P., Burgers, J., Schomaker, L.: Historical manuscript dating based on temporal pattern codebook. Comput. Vis. Image Underst. **152**, 167–175 (2016)
11. He, S., Samara, P., Burgers, J., Schomaker, L.: Image-based historical manuscript dating using contour and stroke fragments. Pattern Recognit. **58**, 159–171 (2016)
12. He, S., Samara, P., Burgers, J., Schomaker, L.: A multiple-label guided clustering algorithm for historical document dating and localization. IEEE Trans. Image Process. **25**(11), 5252–5265 (2016). https://doi.org/10.1109/TIP.2016.2602078
13. He, S., Sammara, P., Burgers, J., Schomaker, L.: Towards style-based dating of historical documents. In: 2014 14th International Conference on Frontiers in Handwriting Recognition, pp. 265–270. IEEE (2014)
14. He, S., Schomaker, L.: A polar stroke descriptor for classification of historical documents. In: 2015 13th International Conference on Document Analysis and Recognition (ICDAR), pp. 6–10. IEEE (2015)
15. He, S., Schomaker, L., Samara, P., Burgers, J.: MPS Data set with images of medieval charters for handwriting-style based dating of manuscripts (2016)
16. Huang, G., Liu, Z., Van Der Maaten, L., Weinberger, K.Q.: Densely connected convolutional networks. In: Proceedings of the IEEE Conference on Computer Vision and Pattern Recognition, pp. 4700–4708 (2017)
17. Li, Y., Genzel, D., Fujii, Y., Popat, A.C.: Publication date estimation for printed historical documents using convolutional neural networks. In: Association for Computing Machinery, HIP 2015, New York, NY, USA, pp. 99–106 (2015). https://doi.org/10.1145/2809544.2809550
18. Lombardi, F., Marinai, S.: Deep learning for historical document analysis and recognition-a survey. J. Imaging **6**(10), 110 (2020)
19. Martin, P., Doucet, A., Jurie, F.: Dating color images with ordinal classification. In: Proceedings of International Conference on Multimedia Retrieval, pp. 447–450 (2014)
20. Molina, A., Riba, P., Gomez, L., Ramos-Terrades, O., Lladós, J.: Date estimation in the wild of scanned historical photos: an image retrieval approach. In: ICDAR (2021)
21. Müller, E., Springstein, M., Ewerth, R.: "When was this picture taken?"-image date estimation in the wild. In: Proceedings of the European Conference on Computer Vision, pp. 619–625 (2017)
22. Riba, P., Molina, A., Gomez, L., Ramos-Terrades, O., Lladós, J.: Learning to rank words: optimizing ranking metrics for word spotting. In: ICDAR (2021)
23. Simonyan, K., Zisserman, A.: Very deep convolutional networks for large-scale image recognition. arXiv preprint arXiv:1409.1556 (2014)
24. Studer, L., et al.: A comprehensive study of imagenet pre-training for historical document image analysis. In: 2019 International Conference on Document Analysis and Recognition (ICDAR), pp. 720–725. IEEE (2019)
25. Szegedy, C., Ioffe, S., Vanhoucke, V., Alemi, A.A.: Inception-v4, inception-resnet and the impact of residual connections on learning. In: Thirty-First AAAI Conference on Artificial Intelligence (2017)
26. Szegedy, C., et al.: Going deeper with convolutions. In: Proceedings of the IEEE Conference on Computer Vision and Pattern Recognition (2015)
27. Szegedy, C., Vanhoucke, V., Ioffe, S., Shlens, J., Wojna, Z.: Rethinking the inception architecture for computer vision. In: Proceedings of the IEEE Conference on Computer Vision and Pattern Recognition, pp. 2818–2826 (2016)

28. Vincent, L.: Google book search: document understanding on a massive scale. In: Ninth International Conference on Document Analysis and Recognition (ICDAR 2007), vol. 2, pp. 819–823. IEEE (2007)
29. Wahlberg, F., Wilkinson, T., Brun, A.: Historical manuscript production date estimation using deep convolutional neural networks. In: 2016 15th International Conference on Frontiers in Handwriting Recognition (ICFHR), pp. 205–210. IEEE (2016)

Combining Visual and Linguistic Models for a Robust Recipient Line Recognition in Historical Documents

Martin Mayr(✉), Alex Felker, Andreas Maier, and Vincent Christlein

Pattern Recognition Lab, Friedrich-Alexander-Universität Erlangen-Nürnberg, Erlangen, Germany
martin.mayr@fau.de

Abstract. Automatically extracting targeted information from historical documents is an important task in the field of document analysis and eases the work of historians when dealing with huge corpora. In this work, we investigate the idea of retrieving the recipient transcriptions from the Nuremberg letterbooks of the 15th century. This task can be solved with fundamentally different ways of approaching it. First, detecting recipient lines solely based on visual features and without any explicit linguistic feedback. Here, we use a vanilla U-Net and an attention-based U-Net as representatives. Second, linguistic feedback can be used to classify each line accordingly. This is done on the one hand with handwritten text recognition (HTR) for predicting the transcriptions and on top of it a light-wight natural language processing (NLP) model distinguishing whether the line is a recipient line or not. On the other hand, we adapt a named entity recognition transformer model. The system jointly performs the line transcription and the recipient line recognition. For improving the performance, we investigated all the possible combinations with the different methods. In most cases the combined output probabilities outperformed the single approaches. The best combination achieved on the hard test set an F1 score of 80% and recipient line recognition accuracy of about 96% while the best single approach only reached about 74% and 94%, respectively.

Keywords: Recipient recognition · Natural language processing · Semantic image segmentation · Handwritten text recognition

1 Introduction

In our modern society, almost all written texts are digitized and thus easy for machines to interpret and to further work with. By contrast, extracting information from older means of communication, such as letters, requires significantly more effort. In addition, the older the documents, the greater the problem of deterioration, such as water damage or bleaching. Due to high costs and lack of expert transcribers, there is a limited amount of training data available in

© Springer Nature Switzerland AG 2022
S. Uchida et al. (Eds.): DAS 2022, LNCS 13237, pp. 598–612, 2022.
https://doi.org/10.1007/978-3-031-06555-2_40

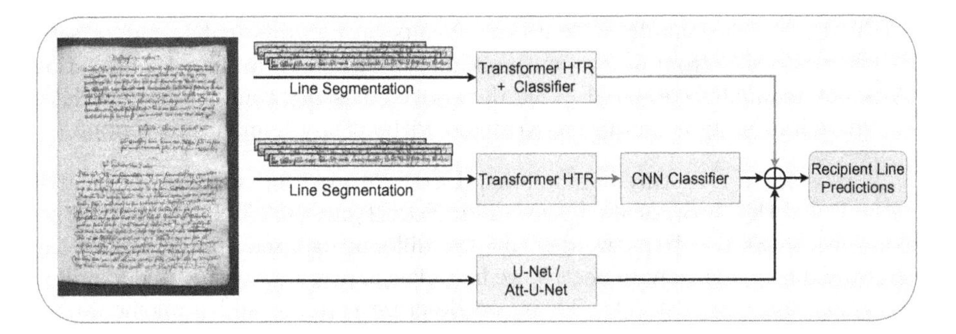

Fig. 1. Extraction of recipient information from Nuremberg Letterbooks with combining multiple different approaches. Methods analyzing the text content (Transformer HTR + Classifier and Transformer HTR + CNN Classifier) are reliant on a text line segmentation. U-Net and attention U-Net work on whole pages. The mean of all predictions is used as final output.

combination with the considerable variability in writing and document styles. In this work, we focus on the Nuremberg letterbooks of the 15th century, written by Nuremberg's small council, and investigate systems which automatically detect and transcribe the recipient lines. This task is important for local historians to better understand the role of Nuremberg during that period of time and also to the German Studies for evaluating Nuremberg's influence on the evolution of the written part of the New High German.

Simply detecting the headlines of the letters is not enough, because recipients are not only defined as the group of people which are directly addressed in the lines above the big text block, but as well as the people that received the letter's information, but were not directly addressed by the sender, named *similita*. The latter information can be found anywhere on the page and often looks like a normal text block.

For solving this task, we employ approaches from two completely different fields. On the one hand the problem can be solved with semantic image segmentation, which is performed with a vanilla [22] and an attention-based U-Net [20]. On the other hand extracting the recipients is done by using linguistic feedback. Here, one approach first transcribes all lines and then these text sequences are used as input for a CNN classification to filter out the recipient lines. Another option is to train a model for jointly recognizing and transcribing recipient lines. The latter one is adapted from a transformer-based named entity recognition model [23,28]. For achieving the best and most robust results we finally combine the predictions of all approaches, see Fig. 1.

In particular, we make the following contributions:

(1) We employ four different systems which use different modalities to detect direct addressees and similita on pages of the Nuremberg letterbooks.
(2) We show that the combination of multiple approaches lead to more robust and better results for this task.

(3) Finally, we thoroughly evaluate our results on two different test data sets, whereas one is from an unseen book and additionally its line segmentation is not manually corrected. This data set represents completely new data, where information should be extracted without any human intervention.

This work is structured as follows. First, we give an overview of related methods and the theoretical background in Sect. 2 and Sect. 3. The methodology is outlined in Sect. 4. Here, we describe the different utilized approaches for the recipient line recognition. In Sect. 5, we first give an overview of the used training, validation and two test sets. Then, we describe training and implementation details. Afterwards, we show the results for the different approaches and the performance gain when combining them. Finally, we discuss our results in Sect. 6 and conclude the paper with Sect. 7.

2 Related Work

The task of finding the important regions in documents can be solved either by relying on visual clues or by processing the text.

Semantic Segmentation. The most straight forward approach for detecting the recipients in a document is to exploit the exposed structure of most recipient lines, i. e., that we treat the recognition of recipient lines as a semantic segmentation. There are many possible approaches to achieve this, from the advent of fully-convolutional networks for semantic segmentation [15], refinement with CRF/MRF [5,14,31], dilated convolutions [30] or many more recent advancements with the use of transformers [32]. In our work we opt for a data-efficient and still very good performing model, the vanilla U-Net [22]. Further, to also use a more sophisticated model, we extend the U-Net architecture by integrating additive attention gates [20]. These approaches are solely based on visual features and do not incorporate any linguistic feedback. In principle, the U-Net architecture can be seen as an encoder-decoder model, where the encoder is a combination of multiple convolution-, max-pool- and batch-norm layers with ReLU activations. The main purpose of the encoder is to convert the input image into a more compact feature representation where no relevant information should be lost. The decoder is the exact counterpart. Instead of convolutions, transposed convolutions are used to recompose the compressed image. During this upsampling path, the spatial information, which is recreated, is partly imprecise. To address this problem, skip connections were introduced concatenating spatial information from the downsampling path [22]. The last layer of the model produces the final segmentation map with a 1×1 convolutional layer, where the feature maps match the number of classes of the task. Because of the fact that we are dealing with a pixel-wise binary segmentation, we need just a single feature map as output.

To further improve the information flow from every stage of the encoder, grid-based gating was proposed [20]. There, the purpose is to focus on local regions and scale the skip connections with attention coefficients [1]. It was empirically shown that by employing this mechanism, the network architecture can cancel out unnecessary activations, which provides a better generalisation power and more expressiveness compared to the vanilla U-Net. The attention gates get inputs x^l denoted as the data from the skip connection, and g denoted as the information from the previous block. Both vectors are then passed through a 1×1 convolution to bring them to the same number of channels, which are summed up afterwards. This result is then passed through a ReLU activation, another 1×1 convolution, and a sigmoid function. The outcome corresponds to a matrix with values between 0 and 1, where 1 marks the areas that are most important. This matrix is then resampled to the original size and multiplied by the skip connection input x^l to produce the final output of this block.

Named Entity Recognition. The most commonly used option for extracting specific information from documents is the use of named entity recognition (NER). These approaches can be mainly split into two groups. Group one first performs HTR and then uses this information as basis for NLP methods to obtain the desired text information [7,10,21]. Group two simultaneously performs the transcription of the content and the detection of the entities [3,4,26,27]. This has the advantage that the NLP system is not depending on almost error-free results from the HTR system.

In our case, we perform one approach out of each group. Out of the first group we utilize a sequence-to-sequence HTR system based on the transformer architecture [28] (described in Sect. 3) and combine it with a light-weight CNN [16] for detecting the recipient lines. This NLP model usually gets word embeddings as input, but due to having very versatile text and not having a lexicon, we decided to use the model on a character level. The architecture performs three 1D convolutions with different kernel sizes over the whole sequence (sentence+padding) to generate an n-gram-like feature extraction. The vectors generated in this way are then concatenated, passed through a dropout layer and through a fully connected layer with a single output neuron to obtain the logits for the classification. Out of the second group we extend the approach of Rouhou et al. [23] (described in more detail in Sect. 3) and integrate the recipient line recognition task into the end-of-sequence token of the HTR system.

3 Background: Handwritten Text Recognition with Transformers

Solely detecting recipient regions is only useful in combination with a good working transcription model. The HTR model is based on a transformer entity recognition system [23] that jointly performs HTR and NER. This is achieved by viewing the different entity tags as part of the alphabet and therefore also part of the output sequence to predict. For jointly performing HTR and recipient line recognition most of the sequence-to-sequence approaches [8,18] would have been appropriate, but the transformer architecture gives many hints on the decision

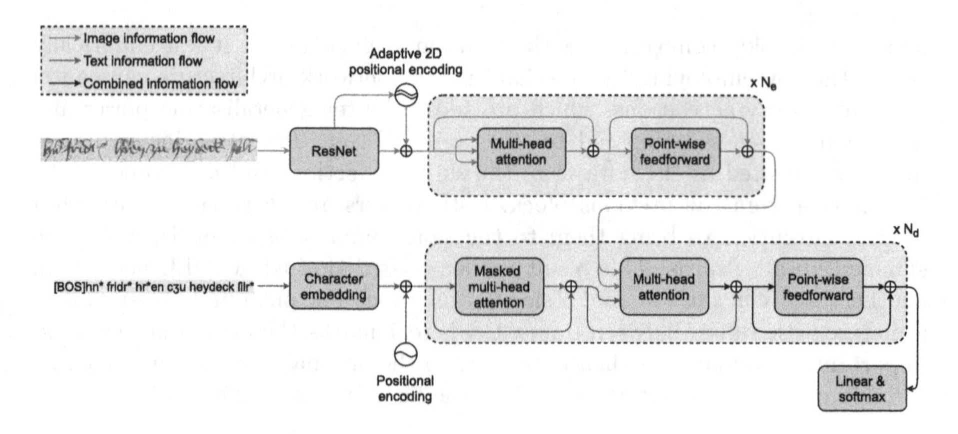

Fig. 2. Architecture of the transformer HTR model.

making and has a high modeling capabilities to simultaneously store all the necessary information for both tasks. The architecture is depicted in Fig. 2.

Visual Feature Extraction. First, the line image is fed into a pre-trained ResNet-50 [11] to better deal with the high computational costs of self-attention blocks. To further reduce the needed resources and to put more emphasis on the sequence part, we employ a smaller version with the ResNet-18 for extracting the features. The next step is to squeeze the 2-dimensional features maps into a sequence to make it accessible for the transformer encoder. Due to the position invariance of the self-attention layers, a positional encoding has to be added before reshaping the feature maps to a sequence. This can be done in different ways. One possible solution is to down-project the feature map in y-direction [12] and add sinusodial positional encoding to the sequence [28]. For being more robust to highly cursive writing, we use the adaptive 2D positional encoding [13]. A scaling parameter will be calculated based on the feature map and learned weights for each spatial dimension (α, β). These values are then used to add the positional encoding to the feature map F:

$$\tilde{F}_{h,w} = \{F_{h,w} + (\alpha(F)P_h^{\text{sinu}} + \beta(F)P_w^{\text{sinu}})\}, \tag{1}$$

where P is defined as $P_{p,2i}^{\text{sinu}} = \sin(p/10000^{2i/d})$ and $P_{p,2i+1}^{\text{sinu}} = \cos(p/10000^{2i/d})$. Here, p and i are the indices of the position and the hidden dimension and D is the size of the hidden dimension.

Refinement of Visual Features. The transformer encoder module is applied N_e times on the flattened output of $\tilde{F}$ to further condense the visual information. There, the multi-headed self-attention layer is used (for simplicity the splitting into multiple heads is not shown in the equation).

$$\hat{F} = \text{Softmax}\left(\frac{QK}{\sqrt{d}}\right)V, \tag{2}$$

where query $Q = \tilde{F}W_Q$, key $K = \tilde{F}W_K$ and value $V = \tilde{F}W_V$. W_Q, W_K and W_V are parameter matrices for the projections. A point-wise feed-forward layer attached to the attention mechanism further refines $\hat{F}$.

Fusion of Visual and Text Information. The next stage is to use all the already predicted text and the visual information from the encoder to output the next character. First, the already predicted characters will be embedded into a high-dimensional space. Again, to keep spatial information sinusodial positional encoding is added to the embeddings T. For training, this step can be parallelised with masking unseen parts of the sequence, which reduces training time. This is called teacher forcing. Note, during inference, there is of course no teacher forcing used and the model is predicting the outputs in an auto-regressive manner, i. e., in the first iteration, the model gets the start-of-sequence token and then, the best fit of every prediction will be added to the string until an end-of-sequence token appears. Afterwards, the transformer decoder will be applied N_d times. Here, multi-headed self-attention is applied on T to distill the context information. In the next step, the mutual-attention layer fuses visual information $\hat{F}$ with the textual information T by setting $Q = TW_Q$, $K = \hat{F}W_K$, and $V = \hat{F}W_V$ and apply the same formula as in Eq. (2). After running through the transformer decoder multiple times the output sequence is fed to a linear layer to match the alphabet size and then fed into a softmax layer to get the confidence scores of the next character.

4 Methodology

In this section, we outline the different approaches that we conducted to recognize and transcribe the recipient lines of the underlying historic corpus.

4.1 Semantic Segmentation of Recipient Lines

First, we are using only visual features and no linguistic feedback. For training the models, the shape of the image and the target mask have to be the same for every sample in a batch. Therefore, we scaled both to a fixed size. Due to the high class imbalance, a patch-based approach would not make sense and would lead to many blank patches. For deciding which lines to transcribe, the output of the model has to be scaled to the original size. Then, if the mean of all the pixels located in a specific line is over 0.5 the line is treated as a recipient line otherwise it is classified as non-recipient.

4.2 Handwritten Text Recognition and Recipient Line Classification with CNN

The first approach based on NLP consists of two steps: First, all the cropped line images are converted into text with an HTR system. Here, we use the model described in Sect. 3. For updating its weights, the training and validation splits are used. Second, the produced transcriptions are used to perform a classification if it is a recipient or a non-recipient line. Due to shape dependencies in the CNN classifier the produced outputs have to be padded to a fixed sequence length, i. e., that after one of the end-of-sequence tokens, no other characters should occur than the padding token. Afterwards, the end-of-sequence token is exchanged with a padding token to not bias the classifier.

For training the NLP classifier, we had to decide either to train it with line images and an HTR model in the beginning or with noisy teacher forcing. Initial experiments have shown that the latter training procedure is much more efficient and also works better. Even the use of larger batch sizes was no longer a problem. In the evaluation, this approach is denoted as CNN classifier.

4.3 Joint Recipient Line Recognition and Transcription

For this approach, we use the transformer HTR as the basis, but instead of predicting the default end-of-sequence token, the model has to predict either an end-of-sequence-recipient or an end-of-sequence-no-recipient token. In other words, the last token in a valid sequence is a classification token. When the model is not trained properly, there could be more than just one end-of-sequence token. In such cases, the first end-of-sequence token is taken for classification.

The training of the transformer models is in most cases done with teacher forcing. This is a problem when dealing with low amounts of data, like historical documents, because the model is very quickly over-fitting to the input sequences without using the visual features in the mutual attention layers. Therefore, we decided to perturb the input sequence of the decoder to make the model more robust to predicted errors and to force the model to rely on the encoder features.

In some cases during inference, the model does not output an end-of-sequence-recipient or an end-of-sequence-no-recipient token. There, the mean of the whole sequence is taken and fed into a softmax function. The end-token with the higher confidence score is taken. For the confidence of the classification, the score of the end-of-sequence-recipient token was divided by the sum of both end-of-sequence tokens.

The classification process of the joint recipient line recognition is further referred to as the HTR classifier.

4.4 Combination of Different Approaches

For fusing the outputs of M models, we are using the mean, which is similar to soft voting.

$$\overline{p} = \frac{1}{M} \sum_{m=1}^{M} \vec{p}_m, \tag{3}$$

where $\vec{p}_m$ are the output probabilities of a specific model m. Of course, there are more sophisticated procedures for fusing probabilities of multiple outputs, but for most of them an own classifier or re-ranking model has to be trained. Also, majority voting or hard voting is not considered, because it does not take into account the confidences of the predictions and the number of models is not high enough to get feasible results. There is also the possibility to perform weighted soft voting, but weighting coefficients have to be determined, which is very difficult for an imbalanced classification task. Thus, using the F1 score will lead to hit or miss for specific methods because of the high responsiveness of this metric.

5 Experimental Evaluation

This section shows the experimental evaluation. After describing the dataset, training parameters and metrics, the obtained results are presented.

5.1 Nuremberg Letterbooks

All the data is taken from the scans of the Nuremberg letterbooks from the 15th century. The line segmentation was done automatically in Transkribus[1] with CITLab advanced [19]. In total, there is annotated data for 3 books: Book 2, book 3 and book 4. Table 1 shows the number of pages for each book and each split. For books 3 and 4 non-appropriately detected lines were adjusted, for book 2 the line coordinates were not corrected. Transcriptions were manually annotated by experts for all three books. The recipient lines were extracted from the produced TEI[2] files, which are the basis for the hybrid edition of the letterbooks. Note, that book 2 was left out for training and validation to evaluate whether the trained models are also generalizing well on other data. We chose Book 2 for this task because the older the books are, the less legible they are. An example page from book 3 can be seen in Fig. 1, where the red boxes are indicating the recipient lines. Each page contains on average about 30 lines, including information such as the contained text and the recipient label. The alphabet used for the semantic feature extraction is constructed by all symbols that occur in the dataset with additional start-of-sequence, padding and end-of-sequence tokens.

5.2 Experiment Details

Both U-Nets are trained with pixel-wise binary cross-entropy loss. We weighted the loss function according to the class imbalance for more stable results. Other loss terms, like dice loss [25] or adversarial loss [9], were not showing better results in preliminary experiments. Early stopping monitoring the validation loss with a patience of 10 epochs is used. The model takes binarized pages as input, which are resized to a fixed width and height of 256 pixels. Sauvola binarization [24] with a window size of 55 is applied.

[1] https://transkribus.eu/lite/.
[2] https://tei-c.org/.

Table 1. Number of pages per book and in total for each used split. Every page consists of about 30 text lines.

	Number of pages			
	Book 2	Book 3	Book 4	Total
Training	–	375	201	576
Validation	–	53	29	82
Test 1	–	102	54	156
Test 2	48	–	–	48
Total	48	530	284	862

The recipient classifier uses noisy teacher forcing with a probability of 20% for each item of the input sequence to change to another random character. Early stopping is used with a patience of 15 epochs monitoring the validation loss. The batch size is set to 32 samples. The embedding employed a hidden dimension of 256. Filter sizes of the 1-D convolutional layers are 3, 4, and 5, each with 256 feature maps. Dropout is rather high with 0.5 because of fast over-fitting with this limited amount of training data.

For training the HTR, early stopping is applied with a patience of 20 monitoring the character error rate (CER) and not validation loss. Batch-size is 16 and learning rate is set to 0.0001. Probability for a random character with noisy teacher forcing is 20 %. The transformer encoder and decoder have a hidden dimension of 1024 with 4 heads and $N_e = N_d = 2$. Dropout is set to 0.1. The input image lines are scaled to a fixed height of 64 pixels and batch-wise zero padded on the right border. Binarization is applied by employing Sauvola thresholding [24]. For augmentation the same strategy is used as in [6,29], i. e., elastic distortion, random transform, dilation and erosion, contrast adjustment and random perspective.

For all models training is performed with the AdamW optimizer [17] and the model with the best monitored metric of early stopping is used for testing.

The PyTorch code of the used models is available at https://github.com/M4rt1nM4yr/recipient_line_detection_DAS22.

5.3 Metrics

Without a good HTR system all the recognized recipient lines would not be of any use. The metric for quantifying the performance is the character error rate (CER).

$$CER = \frac{S + D + I}{N}, \tag{4}$$

where S denotes the number of substitutions, D the number of deletions and I the number of insertions, which is basically the Levenshtein distance. This is divided by the number N of characters in the ground truth text.

The second metric captured is the word error rate (WER).

$$WER = \frac{S_w + D_w + I_w}{N_w}, \tag{5}$$

which is similar to the CER but on word level.

The accuracy is one of the main metrics to evaluate the recipient line recognition.

$$\text{Accuracy} = \frac{TP + TN}{TP + TN + FP + FN}, \tag{6}$$

where TP denotes the true positives, TN the true negatives, FP the false positives and FN the false negatives. But due to the strong class imbalance between recipient and non-recipient lines, we also report precision, recall and F1 score.

$$\text{Precision} = \frac{TP}{TP + FP}, \tag{7}$$

$$\text{Recall} = \frac{TP}{TP + FN}, \tag{8}$$

$$\text{F1} = 2 \cdot \frac{\text{Precision} \cdot \text{Recall}}{\text{Precision} + \text{Recall}}. \tag{9}$$

This metric reports the harmonic mean of precision and recall targeting the true recipient labels.

5.4 Results

The evaluation is split into two parts: (a) Checking the performance on a curated test data set, where the test set is taken from the same books as for the validation and train set. (b) Measuring the same models and model combinations on a more difficult test set, which is taken from an unseen and not curated book, i. e., the line segmentation is directly taken from the Transkribus tool and not manually corrected.

Test 1: Known and Curated Books, but Unseen Pages. Starting with the HTR model, it achieves a CER of 7.04% and a WER of 19.45% on test 1. This model outperforms our previously used model for this kind of data, where we used a Gated Convolutional Recurrent Neural Network [2] reaching on the same test data a CER of 7.36% and a WER of 22.17%, but being trained on over double the training data.

Table 2 shows the performances of all used methods and combinations of them. U-Net, CNN classifier and HTR classifier reach accuracies above 95.5% and F1 scores of above 81.5%. The HTR classifier has the best results with an accuracy of over 97% and F1 score of almost 88%. Only the attention U-Net under-performs compared to the other approaches. When combining the output probabilities of two methods the mix of both NLP-based methods performs best with about 98% and close to 91% accuracy and F1 score, respectively. Note, that

Table 2. Test 1 with the test data of books 3 and 4. Recipient recognition accuracy (Acc), precision (Pre), recall (Rec) and F1 score for the different single and combined approaches (in %). Δt gives the runtime in seconds per page. Best overall approach marked bold and best method of each section underlined. CNN classifier is denoted as CNN-Clf. and HTR classifier as HTR-Clf.

	U-Net	Att-U-Net	CNN-Clf.	HTR-Clf.	Acc↑	F1↑	Pre↑	Rec↑	Δt ↓
Single	✓				95.55	81.55	73.36	91.79	0.02
		✓			85.61	59.18	42.51	**97.34**	**0.01**
			✓		96.45	83.55	83.05	84.06	0.28
				✓	97.33	87.69	86.76	88.64	0.25
Double	✓	✓			89.34	65.84	50.13	95.89	0.03
	✓		✓		97.62	89.40	85.46	93.72	0.30
	✓			✓	97.67	89.46	86.82	92.27	0.27
		✓	✓		97.46	88.81	84.20	93.96	0.29
		✓		✓	97.64	89.43	86.13	93.00	0.26
			✓	✓	98.03	90.87	**90.43**	91.30	0.53
Triple	✓	✓	✓		97.20	88.13	80.85	96.85	0.31
	✓	✓		✓	97.26	88.33	81.17	96.86	0.28
	✓		✓	✓	98.11	91.32	89.93	92.75	0.55
		✓	✓	✓	98.19	91.71	90.00	93.48	0.54
All	✓	✓	✓	✓	**98.34**	**92.58**	89.06	96.38	0.56

the differences to the other doubles are very marginal, except for the combination of both vision approaches. For a fusion of three model outputs, the triple of both NLP models and the worst performing single method performs best with an F1 score of almost 92%. Finally, the best overall combination on test 1 is when using the predictions of all models. This leads to an accuracy of over 98% and F1 score of almost 93%.

Test 2 - Completely Unseen Books Without Manually Corrected Text Line Segmentation. On this less curated data the HTR model performs worse with a CER of 12.21% and a WER of 27.68%.

The recipient line recognition results of test 2 are depicted in Table 3. The order changed slightly and shows that U-Net is performing best on this unseen data with an F1 score of about 74%. The margin to the other approaches is bigger than in test 1. CNN classifier and attention U-Net are the best double combination with an accuracy of almost 96% and F1 score of close to 79%. The improvement from single to double approaches is higher than its improvement on test 1. Again, a combination of both visual models marks the worst double. But when incorporating the probabilities of the CNN classifier this triple achieves the best accuracy and F1 score in this category, almost 96% and 80%. As with the data of test 1, the best metrics are accomplished when combining all methods.

Table 3. Test 2 with the test data of book 2. Recipient recognition accuracy (Acc), precision (Pre), recall (Rec) and F1 score for the different single and combined approaches (in %). Δt gives the runtime in seconds per page. Best overall approach marked bold and best method of each section underlined. CNN classifier is denoted as CNN-Clf. and HTR classifier as HTR-Clf.

	U-Net	Att-U-Net	CNN-Clf.	HTR-Clf.	Acc↑	F1↑	Pre↑	Rec↑	Δt ↓
Single	✓				94.33	74.05	70.06	78.52	0.02
		✓			87.07	56.61	43.26	81.88	**0.01**
			✓		94.05	69.5	73.68	65.77	0.30
				✓	93.71	66.42	73.77	60.40	0.24
Double	✓	✓			91.01	66.32	54.01	**85.91**	0.03
	✓		✓		95.16	75.86	78.01	73.83	0.32
	✓			✓	94.47	71.22	76.74	66.44	0.26
		✓	✓		95.71	78.91	80.00	77.85	0.31
		✓		✓	94.40	71.17	75.76	67.11	0.25
			✓	✓	94.81	72.53	79.84	66.44	0.54
Triple	✓	✓	✓		95.85	**80.00**	79.47	80.54	0.33
	✓	✓		✓	94.95	75.09	76.39	73.83	0.27
	✓		✓	✓	95.71	77.70	83.72	72.48	0.56
		✓	✓	✓	95.57	76.81	83.46	71.14	0.55
All	✓	✓	✓	✓	**96.06**	**80.00**	**83.82**	76.51	0.57

But the best triple results are almost similar, except for the accuracy which is 0.21% better resulting in a detection rate of over 96%.

6 Discussion

The results reflect the hypothesis that fusing the output probabilities of multiple models is improving the performance and the robustness. This gets very apparent in this setting, because of the high class imbalance, where some single models tend to have either a high precision and low recall or the other way around. A bunch of outputs is regularizing this behaviour even though the approaches are not completely independent from each other, meaning they use the same data and the architecture of the different modalities differ only slightly. This gets more obvious when just looking at the mean of the top 3 approaches or combinations from each category of test 1 and test 2 (we used the Top 3 to regulate the effect of outliers, like attention U-Net in the single method category.) Due to the highly imbalanced data, we just check the F1 score and to better distinguish between them the hard test is chosen. Single U-Net, CNN classifier and HTR classifier have a mean score of about 70% which is almost 6% lower than the mean of the top 3 doubles (almost 76%). Adding one more system to the combinations leads to a top-3 F1 score of about 78%. The increase is not as big like as the one for the previous step, but still over 2%. Using all approaches is also just slightly better, but this can also be shown for test 1.

The last columns of Tables 2 and 3 show the runtime evaluated on an NVIDIA A40. The runtime of the approaches using the HTR system is higher than the runtime of both semantic segmentation models due to the auto-regressive nature of the HTR system. Note, this comparison only shows the time needed to get the specific information whether it is a recipient line, i. e., that for extracting the content of the target lines an HTR system is always necessary. In the setting of limited resources, a good solution would be to use one of the HTR based recognition models and combine it with both semantic segmentation models, because the extra runtime cost of the U-Net approaches is negligible and the performance is better and more robust.

7 Conclusion

We employed methods from NLP and from semantic image segmentation for detecting recipient lines in historical documents. The results of the individual models on their own lead to decent performance in terms of accuracy, but a closer look and measurement of the F1 score reveals the weaknesses of the trained models on the unbalanced data set. Therefore, we combined the outputs of different methods resulting in a highly improved performance and robustness regarding F1 score and accuracy. Investigating all possible permutations of the trained models, it seems that the more models are combined the better the performance regardless of the single performance. Using all of the methods at once showed great results on test data where the books are known and also on the pages of a completely unseen book. In the near future, the findings of this work can be used to detect the recipient lines of the newer books from the Nuremberg letterbooks.

References

1. Bahdanau, D., Cho, K., Bengio, Y.: Neural machine translation by jointly learning to align and translate (2016)
2. Bluche, T., Messina, R.: Gated convolutional recurrent neural networks for multilingual handwriting recognition. In: 2017 14th IAPR International Conference on Document Analysis and Recognition (ICDAR), vol. 01, pp. 646–651 (2017)
3. Carbonell, M., Fornés, A., Villegas, M., Lladós, J.: A neural model for text localization, transcription and named entity recognition in full pages. Pattern Recogn. Lett. **136**, 219–227 (2020)
4. Carbonell, M., Villegas, M., Fornés, A., Lladós, J.: Joint recognition of handwritten text and named entities with a neural end-to-end model. In: 2018 13th IAPR International Workshop on Document Analysis Systems (DAS), pp. 399–404 (2018)
5. Chen, L.C., Papandreou, G., Kokkinos, I., Murphy, K., Yuille, A.L.: Semantic image segmentation with deep convolutional nets and fully connected CRFs. In: International Conference on Learning Representations (2015)

6. Coquenet, D., Chatelain, C., Paquet, T.: SPAN: a simple predict & align network for handwritten paragraph recognition. In: Lladós, J., Lopresti, D., Uchida, S. (eds.) ICDAR 2021. LNCS, vol. 12823, pp. 70–84. Springer, Cham (2021). https://doi.org/10.1007/978-3-030-86334-0_5
7. Dinarelli, M., Rosset, S.: Tree-structured named entity recognition on OCR data: analysis, processing and results. In: Language Resources Evaluation Conference (LREC), Istanbul, Turkey (2012)
8. Doetsch, P., Zeyer, A., Ney, H.: Bidirectional decoder networks for attention-based end-to-end offline handwriting recognition. In: 2016 15th International Conference on Frontiers in Handwriting Recognition (ICFHR), pp. 361–366 (2016)
9. Gaál, G., Maga, B., Lukács, A.: Attention U-Net based adversarial architectures for chest X-ray lung segmentation (2020)
10. Hamdi, A., Jean-Caurant, A., Sidere, N., Coustaty, M., Doucet, A.: An analysis of the performance of named entity recognition over OCRed documents. In: 2019 ACM/IEEE Joint Conference on Digital Libraries (JCDL), pp. 333–334 (2019)
11. He, K., Zhang, X., Ren, S., Sun, J.: Deep residual learning for image recognition. In: Proceedings of the IEEE Conference on Computer Vision and Pattern Recognition (CVPR) (2016)
12. Kang, L., Riba, P., Rusiñol, M., Fornés, A., Villegas, M.: Pay attention to what you read: non-recurrent handwritten text-line recognition (2020)
13. Lee, J., Park, S., Baek, J., Oh, S.J., Kim, S., Lee, H.: On recognizing texts of arbitrary shapes with 2D self-attention. In: Proceedings of the IEEE/CVF Conference on Computer Vision and Pattern Recognition (CVPR) Workshops (2020)
14. Liu, Z., Li, X., Luo, P., Loy, C.C., Tang, X.: Semantic image segmentation via deep parsing network. In: Proceedings of the IEEE International Conference on Computer Vision (ICCV) (2015)
15. Long, J., Shelhamer, E., Darrell, T.: Fully convolutional networks for semantic segmentation. In: Proceedings of the IEEE Conference on Computer Vision and Pattern Recognition (CVPR) (2015)
16. Lopez, M.M., Kalita, J.: Deep learning applied to NLP (2017)
17. Loshchilov, I., Hutter, F.: Decoupled weight decay regularization. In: International Conference on Learning Representations (2019)
18. Michael, J., Labahn, R., Grüning, T., Zöllner, J.: Evaluating sequence-to-sequence models for handwritten text recognition. In: 2019 International Conference on Document Analysis and Recognition (ICDAR), pp. 1286–1293 (2019)
19. Muehlberger, G., et al.: Transforming scholarship in the archives through handwritten text recognition: Transkribus as a case study. J. Doc. **75**, 954–976 (2019)
20. Oktay, O., et al.: Attention u-net: learning where to look for the pancreas (2018)
21. Romero, V., Fornés, A., Granell, E., Vidal, E., Sánchez, J.A.: Information extraction in handwritten marriage licenses books. In: Proceedings of the 5th International Workshop on Historical Document Imaging and Processing, HIP 2019, pp. 66–71. Association for Computing Machinery, New York (2019)
22. Ronneberger, O., Fischer, P., Brox, T.: U-Net: convolutional networks for biomedical image segmentation. In: Navab, N., Hornegger, J., Wells, W.M., Frangi, A.F. (eds.) MICCAI 2015. LNCS, vol. 9351, pp. 234–241. Springer, Cham (2015). https://doi.org/10.1007/978-3-319-24574-4_28
23. Rouhou, A.C., Dhiaf, M., Kessentini, Y., Salem, S.B.: Transformer-based approach for joint handwriting and named entity recognition in historical document. Pattern Recogn. Lett. **155**, 128–134 (2022). ISSN 0167-8655
24. Sauvola, J., Pietikäinen, M.: Adaptive document image binarization. Pattern Recogn. **33**(2), 225–236 (2000)

25. Sudre, C.H., Li, W., Vercauteren, T., Ourselin, S., Jorge Cardoso, M.: Generalised dice overlap as a deep learning loss function for highly unbalanced segmentations. In: Cardoso, M.J., et al. (eds.) DLMIA/ML-CDS -2017. LNCS, vol. 10553, pp. 240–248. Springer, Cham (2017). https://doi.org/10.1007/978-3-319-67558-9_28
26. Toledo, J.I., Carbonell, M., Fornés, A., Lladós, J.: Information extraction from historical handwritten document images with a context-aware neural model. Pattern Recogn. **86**, 27–36 (2019)
27. Toledo, J.I., Sudholt, S., Fornés, A., Cucurull, J., Fink, G.A., Lladós, J.: Handwritten word image categorization with convolutional neural networks and spatial pyramid pooling. In: Robles-Kelly, A., Loog, M., Biggio, B., Escolano, F., Wilson, R. (eds.) S+SSPR 2016. LNCS, vol. 10029, pp. 543–552. Springer, Cham (2016). https://doi.org/10.1007/978-3-319-49055-7_48
28. Vaswani, A., et al.: Attention is all you need. In: Guyon, I., et a. (eds.) Advances in Neural Information Processing Systems, vol. 30. Curran Associates, Inc. (2017)
29. Yousef, M., Bishop, T.E.: Origaminet: weakly-supervised, segmentation-free, one-step, full page text recognition by learning to unfold. In: Proceedings of the IEEE/CVF Conference on Computer Vision and Pattern Recognition (CVPR) (2020)
30. Yu, F., Koltun, V.: Multi-scale context aggregation by dilated convolutions. In: International Conference on Learning Representations (2016)
31. Zheng, S., et al.: Conditional random fields as recurrent neural networks. In: Proceedings of the IEEE International Conference on Computer Vision (ICCV) (2015)
32. Zheng, S., et al.: Rethinking semantic segmentation from a sequence-to-sequence perspective with transformers. In: Proceedings of the IEEE/CVF Conference on Computer Vision and Pattern Recognition (CVPR), pp. 6881–6890 (2021)

Investigating the Effect of Using Synthetic and Semi-synthetic Images for Historical Document Font Classification

Konstantina Nikolaidou[1(✉)], Richa Upadhyay[1], Mathias Seuret[2], and Marcus Liwicki[1]

[1] Machine Learning EISLAB, Luleå University of Technology, Luleå, Sweden
{konstantina.nikolaidou,richa.upadhyay,marcus.liwicki}@ltu.se
[2] Pattern Recognition Lab, Friedrich-Alexander-Universität, Erlangen-Nürnberg, Erlangen, Germany
mathias.seuret@fau.de

Abstract. This paper studies the effect of using various data augmentation by synthetization approaches on historical image data, particularly for font classification. Historical document image datasets often lack the appropriate size to train and evaluate deep learning models, motivating data augmentation and synthetic document generation techniques for creating additional data. This work explores the effect of various semi-synthetic and synthetic historical document images, some of which appear as recent trends and others not published yet, on a font classification task. We use 10K patch samples as baseline dataset, derived from the dataset of Early Printed Books with Multiple Font Groups, and increase its size using DocCreator software and Generative Adversarial Networks (GAN). Furthermore, we fine-tune different pre-trained Convolutional Neural Network (CNN) classifiers as a baseline using the original dataset and then compare the performance with the additional semi-synthetic and synthetic images. We further evaluate the performance using additional real samples from the original dataset in the training process. DocCreator, and the additional real samples improve the performance giving the best results. Finally, for the best-performing architecture, we explore different sizes of training sets and examine how the gradual addition of data affects the performance.

Keywords: Historical document images · Synthetic image generation · Font classification · Convolutional neural networks · Generative Adversarial Networks · Document Image Analysis · Performance evaluation

1 Introduction

Deep learning is considered state-of-the-art in various fields. However, the performance highly depends on the amount and quality of training data. Obtaining

© Springer Nature Switzerland AG 2022
S. Uchida et al. (Eds.): DAS 2022, LNCS 13237, pp. 613–626, 2022.
https://doi.org/10.1007/978-3-031-06555-2_41

sufficient samples remains a challenge in many domains and tasks. More specifically, in Computer Vision, models not trained using adequate images suffer from generalizability issues and do not perform well on unseen data. To address these problems, research on image augmentation and generation is progressing rapidly.

In Historical Document Image Analysis, labeled images are scarce resources, making the application of Deep Learning methods challenging. Additional data with ground truth is most of the time hard and expensive to obtain. Moreover, expert knowledge from scholars in the humanities, such as paleographers or book historians, is often necessary. For this matter, research on image augmentation [24] and image generation [23] is progressing rapidly to address the lack of labeled training data.

In this work, we investigate the impact of semi-synthetic images generated by DocCreator and synthetic images created by GAN on the performance of different CNN architectures. The rest of the paper is organized as follows. In Sect. 2, we present the related work. Section 3 describes the dataset and methods used for data augmentation and image generation. In Sect. 4 we describe the experimental setup for the classification task and the architectures used, while in Sect. 5 we present our results and discuss future work. The code for data preprocessing and experiments, and the link to the generated data can be found on Github.[1]

2 Related Work

The advent of Deep Learning has set the field of document image analysis in demand for larger datasets, creating the need to synthesize these document images as obtaining new data is an expensive and time-consuming process. One way of achieving this is using software that helps to create document images. The DocEmul framework [2] combines data augmentation techniques and background extraction, along with Latex or XML structures to create synthetic documents. In [16], documents are generated by pasting characters, with some random variations, on a synthetic background. Several tools use defect models [1] and add degradations [20] to images in order to generate additional data. However, due to their complexity, semi-automatic behavior using such tools is tedious and requires human expertise making deep learning models widely employed for this purpose. DocCreator [10] is an open-source tool for synthetic and semi-synthetic document image generation that we aim to explore in this work. The work presented in [18] used several degradations of DocCreator as pseudo-ground truth for historical document image segmentation and examined whether they help improve neural network performance. We elaborate more on the software and its use in our work in Sect. 3.2.

In the past, Deep Neural Networks have been used not only for analysis and recognition of historical documents [14], but also for document synthesis. One such development is image-to-image translation using neural style transfer [5], which produces a synthetic output by mixing style from a source image and

[1] https://github.com/koninik/Font_Classification.

content from a target image. [19] employs this technique to generate synthetic historical documents. [15] applies both image style and writing style transfer to generate synthetic handwritten documents. Another method is image-to-image translation using Generative Adversarial Networks (GANs). GANs are generative models based on deep learning architectures [6]. These networks have the potential to generate synthetic instances of data, very similar to the real data. GANs use two simultaneously trained neural networks: the generator and the discriminator networks, which oppose each other, therefore the adversarial. The generator aims to fabricate new data samples that match the real data in the training set, while the discriminator learns to identify the fake data samples from the real ones. Each time the discriminator successfully rejects a generator-created fake image, this feedback helps the generator to increase the verisimilitude of its output. In order to create synthetic images of different modes or classes present in the training dataset, additional information of the class labels is necessary to direct the data generation. This category of GAN are the CGAN [17], where an additional feature vector related to the input labels guides the generator to generate class specific data. Pix2Pix GAN [9] transforms one image representation to another image representation, like converting a modern document image to a historical document image. The drawback of the above two algorithms is that they require paired images, so every source domain image must have a corresponding target image. The work in [19], handles the synthetic image generation similarly as an image-to-image translation problem where the goal is to transform modern documents generated from a CycleGAN [26] into historical documents without the need of paired images using Style Transfer as well. OpenGAN [4] uses features extracted from real samples to condition on the real samples and to generate synthetic images. As this system can generate images per sample, it can be directly compared with the per sample generation using DocCreator degradations. Hence, we use it to generate synthetic images in this work. More information can be found in Sect. 3.3.

3 Dataset and Image Generation

3.1 Dataset

We use the dataset of Early Printed Books with Multiple Font Groups [21]. The dataset consists of 24,866 images for training and 10,756 for testing and contains 12 font classes: Textura, Rotunda, Gotico-Antiqua, Bastarda, Schwabacher, Fraktur, Antiqua, Italic, Greek, Hebrew, "Other Font", and "Not a Font". Slightly more than 10% of the pages of this dataset have multiple labels (e.g., a German text printed in Fraktur, containing also a Latin sentence in Antiqua); for these pages, we consider only the main label of the page (Fraktur in the aforementioned example), as it is done in the paper presenting the dataset [21]. We only care about images containing text printed with a known font, and therefore we exclude the "Other Font" and "Not a Font" classes, thus concluding with 17,968 training images, 8,092 test images, and 10 classes. The dataset was further used in one of the tasks of the Historical Document Classification competition of the 16th International Conference of Document Analysis and Recognition [22].

The organizers introduced a new test set for the competition containing 5506 images (none with the Gotico-Antiqua font). In this work, we use the original test set presented in [21] as validation set and perform our test inference on the competition test set, meaning that we test and validate only on real samples and not on synthetic ones.

Pre-processing and Baseline: To create a baseline for our experiments, we applied several pre-processing steps to the training set in order to remove any unwanted noise and focus only in the center of the document which contains the text, as the goal of the task is the font classification, thus the text. First, we use binarization to detect and crop the out-of-page areas. Then, using Otsu binarization, we detect the text in the middle of each page and cut the image around this text. Finally, since the baselines presented in the dataset's original paper and competition use extracted patches of the images, we follow the same strategy and for every colored processed page image we create several non-overlapping patches of size 224 × 224. The patch extraction results in 206,402 training samples. The goal in this work is not to achieve the best performance, but to investigate how additional data derived from different sources can contribute to a small sized dataset performance; hence, we assume a small sized dataset of 10K samples which is quite common in Historical Document Image Analysis. To create the 10K sample baseline training set, we randomly select 1,000 patches per class. Figure 1 shows an example image (a) from the dataset with borders and background, then the cropped image (b) that comes up after the binarization steps that includes only text, and finally (c) 10 extracted patches from the processed image to use for our experiments. Both preprocessing and patch extraction code are available on the provided Github repository along with a link to the generated data we used.

3.2 Semi-synthetic Image Generation Using DocCreator

DocCreator is an open-source software that enables document image augmentation and generation, conditioning on the ground truth [10]. This tool offers both the option of synthetic image generation and data augmentation creating semi-synthetic images. For synthetic document generation, DocCreator offers several backgrounds and fonts, or gives the option to use particular backgrounds, create random text or use existing ones, and extract the desired font from a document image. Furthermore, one can generate semi-synthetic images applying degradation models provided by the software. By trying to generate fully synthetic images using DocCreator, we realized that the tool appeared to work better for Antiqua and Italic and, more specifically, English and French languages. This might happen because of the OCR system used in the software that performs better on Latin-based languages and characters. As we have mostly Gothic fonts in the used dataset, and two other fonts which do not contain Latin characters (Hebrew and Greek), we decide to proceed with the semi-synthetic image generation using the following degradations: characters degradation [11], frequent

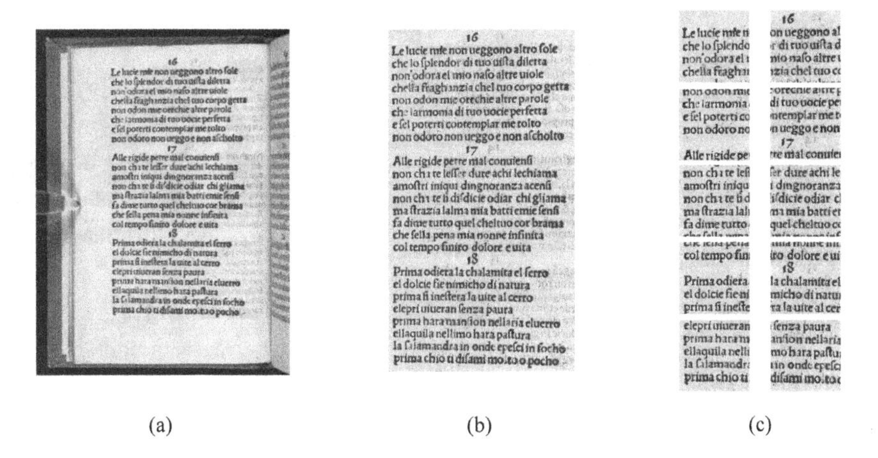

Fig. 1. Pre-processing and patch extraction results. (a) A document image from the training set, (b) the same image cropped around the text after the binarization pre-processing steps, and (c) 10 extracted patch samples from the processed image that we finally use.

phantom character[2], bleed-through, blur in specific zones, shadow binding, and finally, medium holes in the corners, border, and center of documents. We generate the DocCreator training set by creating samples per patch, thus increasing the training data by 60K samples. Figure 2, shows an original image and its corresponding generated semi-synthetic degradations.

3.3 Synthetic Image Generation Using Generative Adversarial Networks

In this work, for the synthetic image generation, we use a Conditional GAN named OpenGAN [4]. In OpenGAN, the Generator conditions on the real sample to generate the fake one. It uses a metric learning feature extractor that captures valuable semantic information and class-level information for every sample. The metric features extracted from the real images are fed to the generator and the discriminator through feature conditional normalization layers. As illustrated in Fig. 3, the generator gives as output a fake image, addressing two critical goals. The first is to mislead the discriminator in classifying the generated fake image as real, which is accomplished by the adversarial loss. The second is to obtain features $\overline{f}$ extracted from the fake image similar to the real image features f, which is acquired by the Mean Square Error (MSE) loss. As a result, the generator creates synthetic images that are semantically and visually similar to the real source image. OpenGAN enables the generation of diverse intra-class images. Therefore the generated samples can be appropriately used for data augmentation.

[2] Ink depot from the body of a type due to face erosion caused by frequent use of this type.

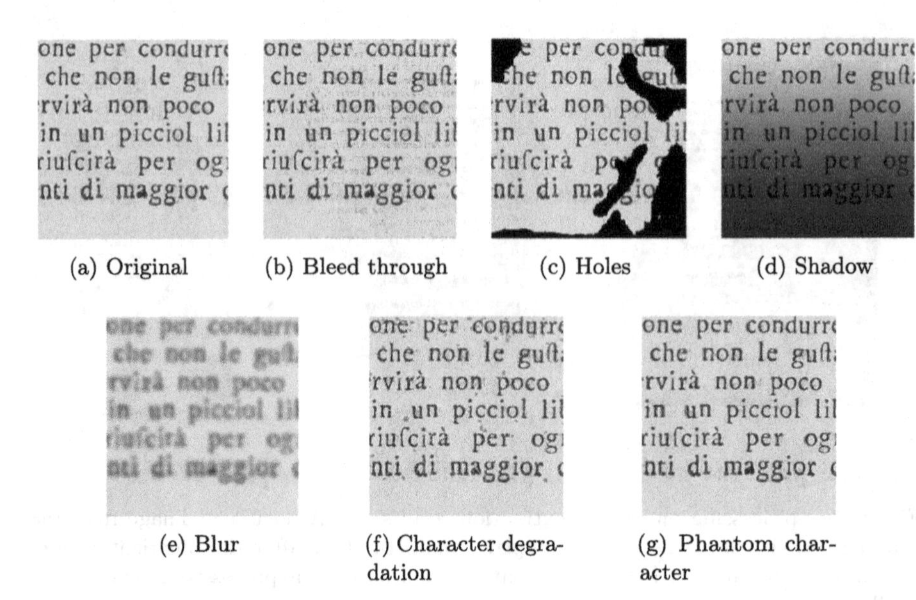

(a) Original (b) Bleed through (c) Holes (d) Shadow

(e) Blur (f) Character degradation (g) Phantom character

Fig. 2. The figure shows an patch image (a) from the original dataset and the generated images by DocCreator with the different degradations: (b) bleed-through, (c) holes around the borders and in the middle of the image, (d) shadow at the bottom of the image, (e) blur in the whole image, (f) character degradation, and (g) phantom character.

To generate the synthetic images from OpenGAN, we use the same set of hyperparameter values as the original paper [4] for feature extraction and GAN training. The images generated from Generative Adversarial Networks (GAN) are of lower resolution (256×256) than the original images, considering the computational capacity as well as time for training. Similar to DocCreator, we create 6 synthetic samples per real sample to end up with 60K synthetic samples. Figure 4 shows one example patch image generated using OpenGAN for each of the 10 font classes and their corresponding real sample they are conditioned on.

4 Experiments

To observe the effect of using synthetic images generated by DocCreator and GAN, twelve experiments were designed. These experiments were formed from four datasets used on three CNN pre-trained on ImageNet [3]. The datasets and the data distribution in non overlapping training, validation and test sets is described in Table 1. For evaluating the impact of the generated images with all CNN architectures, an equal number of synthetic images (from DocCreator and/or GAN) are added to the original dataset. The first dataset is the **Baseline** that contains 1,000 patches for each of the 10 classes derived and randomly selected from the original dataset. The **+DocCreator** dataset contains the **Baseline** samples along with the semi-synthetic samples generated

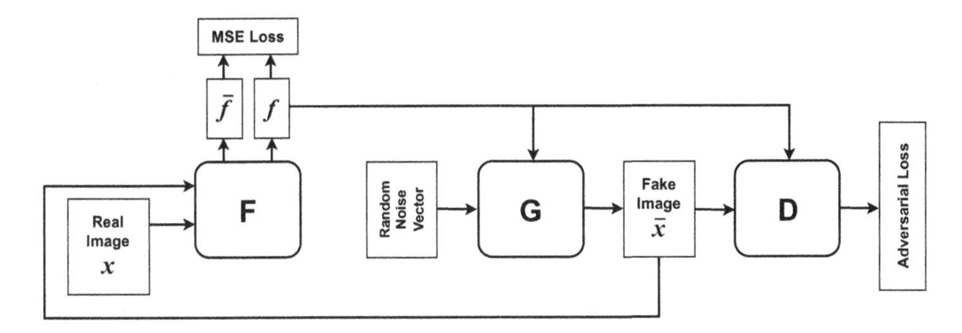

Fig. 3. Network Architecture of OpenGAN. This figure has three CNN architectures: a feature extractor F, a generator G, and a discriminator D. f and $\overline{f}$ are features derived from the metric feature extractor F from the real and the generated fake image, respectively.

by DocCreator. Similarly, in **+GAN** there are the **Baseline** patches with the synthetic GAN generated patches. Both synthetic and semi-synthetic datasets include same amount of samples per class. Finally, the **+Real** dataset is the balanced **Baseline** dataset combined with 60,000 randomly selected samples from the original dataset patches. It should be noted that the additional 60,000 real patches are not balanced across classes because of lack of training samples. Thus, this final new dataset is imbalanced. All experiments use a fixed validation and test set from real patch samples derived from the original dataset and only the training sets include synthetic images.

Table 1. Overview of the datasets used in the conducted experiments and their data distribution.

Dataset	N^o of patch samples			
	Total	Training	Validation	Test
Baseline	202,393	10,000	95,536	96,857
+DocCreator	262,393	70,000	95,536	96,857
+GAN	262,393	70,000	95,536	96,857
+Real	262,393	70,000	95,536	96,857

For the font type classification task, we evaluate 3 different CNN architectures: a 201-layer DenseNet [8], a residual network with 50 layers [7], and an EfficientNet-B0 [25]. All models are trained using Adam optimizer with initial learning rate of 0.001 and 0.00001 weight decay [12]. The learning rate is reduced by a factor of 0.1 in every epoch. The training images are downscaled to a size of 224×224 to match the input size of the networks. The training data is fed to the classifier in batches of 32 samples. The purpose of this work is to examine

(a) Textura† (b) Textura* (c) Rotunda† (d) Rotunda*

(e) Gotico Antiqua† (f) Gotico Antiqua* (g) Bastarda† (h) Bastarda*

(i) Schwabacher† (j) Schwabacher* (k) Fraktur† (l) Fraktur*

(m) Antiqua† (n) Antiqua* (o) Italic† (p) Italic*

(q) Greek† (r) Greek* (s) Hebrew† (t) Hebrew*

Fig. 4. Synthetic image samples of all classes generated by OpenGAN. Columns 1 and 3 show the original samples and columns 2 and 4 the synthetic samples conditioned on the original sample (on their left). For example, (g) depicts a baseline real sample of bastarda class (marked with $\dagger$) and (h) the generated synthetic sample (marked with $\star$) that is conditioned on (g).

the effect of the additional generated data and not the optimal performance of each model, thus we are not performing any hyper-parameter optimization. For every training process, we use early stopping when validation accuracy does not improve for 10 consecutive epochs. The training of the models in all cases using the early stopping is approximately 20 epochs. As for performance metrics, we compute the classification accuracy given by Eq. (1),

$$\text{Accuracy} = \frac{\text{Total number of correct predictions}}{\text{Total number of test instances}} \tag{1}$$

Since one of the datasets is imbalanced, accuracy alone may not be a correct representation of the model's performance. Therefore, we also use the multi-class F1 score given by Eq. (2), which is sensitive to the data distribution. It is the harmonic mean of precision and recall, and is suitable for imbalanced class distribution. Usually, F1 score is defined for binary classification. In a multi-class classification problem (such as this work), an F1 score per class is calculated using the one-vs-rest technique, and their average is considered as the multi-class F1 score. This is given by the following equation:

$$\text{F1 score} = \frac{1}{N} \times \sum_{i=1}^{N} \frac{\text{TP}_i}{\text{TP}_i + \frac{1}{2} \times (\text{FP}_i + \text{FN}_i)} \tag{2}$$

where N is the number of classes, and TP_i, FP_i, and FN_i are respectively the number of one-vs-rest true positives, false positives and false negatives for class i.

During inference, for every original document image we extract several patches and test on them. The final classification result is the most frequent value obtained from the patch classification of every image.

Finally, using the DocCreator generation method and the DenseNet architecture, that seem to achieve one of the highest performance according to the results, we investigate how the amount of generated training samples impact the classification result by gradually adding new training samples.

5 Results

Table 2 presents the resulting classification accuracy and F1 score of each trained model with the original data and the combination with the generated and real data. We further perform a significance test to conclude whether the additional data performance is statistically significant from the baseline. The statistics and p-values are presented in Table 3. The results suggest that the semi-synthetic images produces with DocCreator contribute to the improved performance. In the case of +GAN, there is no increase of the performance that is statistically significant in any model using the synthetic data. Thus, the low quality of the generated synthetic samples in terms of fine-grained details (words and characters that are not clear to the human eye) could not contribute in a better result. For the **+Real** dataset that contains additional real data to the baseline, we

Table 2. Classification performance on the test set for the different CNN architectures and the different data augmentation techniques. The resulted metrics are presented as **mean ± std %** over 5 runs for every model.

Dataset	ResNet50		DenseNet		EfficientNet	
	Accuracy	F1	Accuracy	F1	Accuracy	F1
Baseline	95.78 ± 0.55	95.46 ± 0.65	96.91 ± 0.41	96.78 ± 0.49	96.79 ± 0.10	96.70 ± 0.13
+DocCreator	**97.93 ± 0.21**	**97.97 ± 0.21**	**98.30 ± 0.69**	**98.53 ± 0.28**	**97.85 ± 0.24**	**97.88 ± 0.24**
+GAN	95.41 ± 0.33	95.04 ± 0.45	96.81 ± 0.43	96.73 ± 0.41	96.40 ± 0.42	96.33 ± 0.47
+Real	96.57 ± 0.23	96.48 ± 0.27	96.89 ± 0.54	96.85 ± 0.63	97.16 ± 0.12	97.14 ± 0.13

Table 3. Significance test among the Baseline and the three augmented training sets. The table presents the **statistic** and **p-value**. P-values larger than 0.05 suggest that the result of the augmented dataset is not significant.

Datasets	ResNet50		DenseNet		EfficientNet	
	Statistic	p-value	Statistic	p-value	Statistic	p-value
Baseline w/+DocCreator	−8.2183	0.00004	−9.2195	0.00002	−4.5174	0.00196
Baseline w/+GAN	1.3052	0.2281	0.4132	0.6902	1.994	0.0812
Baseline w/+Real	−2.9706	0.0179	−5.421	0.0006	−0.1099	0.9152

Table 4. Accuracy per class for the **+DocCreator** and **+Real** sets. The results are presented as mean ± std % over the 5 runs.

Class	ResNet50		DenseNet		EfficientNet	
	+DocCreator	+Real	+DocCreator	+Real	+DocCreator	+Real
Textura	95.81 ± 0.88	96.13 ± 0.88	96.77 ± 0	96.45 ± 1.35	96.45 ± 0.72	96.45 ± 0.72
Rotunda	99.75 ± 0	99.75 ± 0	99.75 ± 0	99.75 ± 0	99.75 ± 0	99.75 ± 0
Bastarda	75.74 ± 3.07	57.34 ± 3.06	83.94 ± 3.86	61.91 ± 6.49	74.15 ± 3.42	64.79 ± 2.33
Schwabacher	99.55 ± 0	99.55 ± 0	99.55 ± 0	99.55 ± 0	99.55 ± 0	99.55 ± 0
Fraktur	99.86 ± 0	99.86 ± 0	99.86 ± 0	99.86 ± 0	99.86 ± 0	99.86 ± 0
Antiqua	99.70 ± 0.11	99.75 ± 0	99.44 ± 0.11	99.75 ± 0	99.34 ± 0.53	99.75 ± 0
Italic	99.15 ± 0	99.15 ± 0	99.32 ± 0.16	99.15 ± 0	99.66 ± 0.13	99.26 ± 0.15
Greek	99.67 ± 0	98.54 ± 0.30	99.67 ± 0	98.93 ± 0.36	99.67 ± 0	99.07 ± 0.43
Hebrew	99.80 ± 0	99.80 ± 0	99.80 ± 0	99.80 ± 0	99.80 ± 0	99.80 ± 0

observe a performance increase in all network cases. However, the significance test results reveal that in the case of DenseNet the contribution is not significant, while it is in the case of ResNet50 and EfficientNet, though less significant compared to **+DocCreator** set.

While one would expect that additional real samples (**+Real**) would advance the performance more than semi-synthetic images (**+DocCreator**), the resulted metrics and significance test suggest that the **+DocCreator** dataset surpasses **+Real**. We present the accuracy per class for these two datasets in Table 4. The presented accuracy per class on the test set is the mean ± standard

Table 5. Class distribution of the random selection for baseline training set, the 60K additional images to create the **+Real** training set (2nd row), the validation, and the test set. The presented classes are the ones appearing in the test set. The validation and test set are fixed in all experiments.

Sets	N° of samples per class								
	Textura	Rotunda	Bastarda	Schwabacher	Fraktur	Antiqua	Italic	Greek	Hebrew
Baseline	1,000	1,000	1,000	1,000	1,000	1,000	1,000	1,000	1,000
+Real (training)	2,607	9,149	2,214	5,047	15,796	19,840	2,700	252	2,396
Validation	4,811	15,776	4,096	8,865	26,309	22,704	4,753	914	4,342
Test	1,984	17,156	6,812	7,378	9,531	12,666	9,388	9,836	22,106

deviation over the 5 runs that were presented in Table 2. The results for the classes **Rotunda**, **Schwabacher**, **Fraktur**, and **Hebrew** are identical for all models and both datasets showing no deviation over the different runs. For **Textura** and **Antiqua**, the results have slight differences, though very close, while for Greek all models perform the same using the **+DocCreator** dataset. The **Bastarda** class seems to be a special case, as the accuracy is much lower compared to the rest of the classes and reveals the difference in the performance for the two training datasets.

To analyze this observation more deeply, we show in Table 5 the class distribution of the additional real samples that we randomly selected to create the **+Real** training set. It is clear that the underrepresented classes, such as Textura, Bastarda, Italic, Greek, and Hebrew, are the ones with differences over the different training sets and runs. However, we would expect worse performance in classes such as Greek that have much fewer training samples than Bastarda. We can explain it with the inter-class variation: Bastarda is historically and visually closely related to the classes Fraktur, Schwabacher, and Rotunda; Greek does not share similarities with any of the other classes. We observe that the classifier recognizes a unique class with few samples better, than a class with more samples but high inter-class similarity.

We choose the best data generation method, **+DocCreator** and DenseNet, that gives the maximum performance to investigate further the impact of gradually adding data for this specific model. Figure 5 shows the accuracy and standard deviation on the test set for the different training set sizes over 3 runs. We can see that as we increase the number of training samples generated by DocCreator by 10K, we see a radical increase in the performance until size 40K. Above 40K samples, the performance does not seem to improve, showing several instabilities.

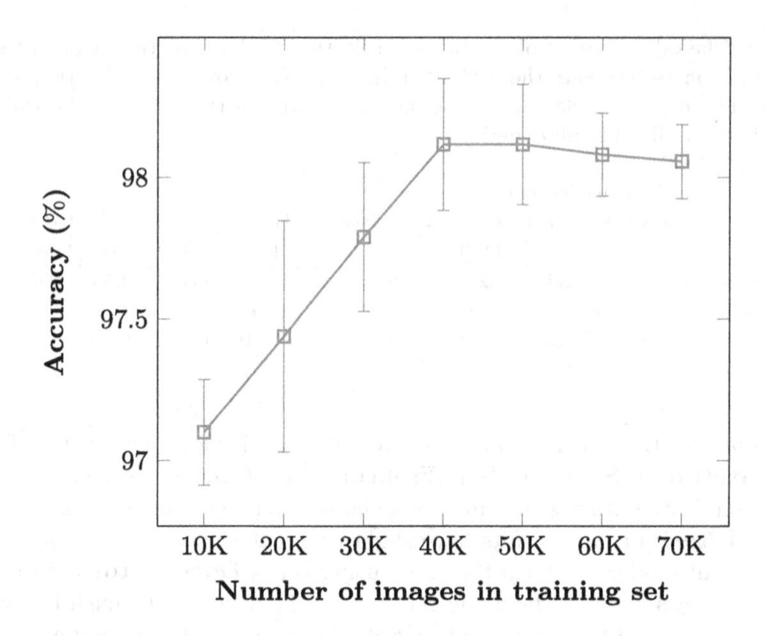

Fig. 5. Accuracy for different sizes of training set for DenseNet, starting with the 10K baseline samples, and with gradual addition of DocCreator samples.

6 Conclusion and Future Work

We examined the effect of semi-synthetic and synthetic images on a font classification task. We created images with various degradations using DocCreator and synthetic images using OpenGAN. We performed experiments on different pre-trained CNN, initially with the original dataset without any image augmentation, and then added the new generated and real sets separately. Between the two generation methods there is no difference in the classifier training in terms of time, however there is in the generation inference. On the one hand, DocCreator is slower at dealing with a large number of documents at once, and on the other hand, the GAN requires some training time initially. The demonstrated accuracy and F1 score indicate that the generated DocCreator and real samples lead to a nice and statistically significant improvement of the performance of the classifiers. In the case of GAN images, where the content is not understandable by humans, the performance does not increase. We experimented using the dataset generated by DocCreator and investigated the effect of different training set sizes. It would be beneficial to conduct the experiments using additional datasets and baselines that achieve worse performance with no data augmentation than in this case (~96%).

In the future, we plan to investigate synthetic image generation using advanced GAN architectures and improve the quality of generated images. Furthermore, other generative models such as Variational Autoencoders [13] can also create semi-synthetic or entirely synthetic images. Another possible direction for future work is to create fully synthetic images with DocCreator using existing text files from documents and Latin character fonts to investigate the performance further. Finally, a further investigation on the aspects of the data and recognition architecture is left for future work.

References

1. Baird, H.: Document image defect models and their uses. In: Proceedings of 2nd International Conference on Document Analysis and Recognition (ICDAR 1993), pp. 62–67 (1993). https://doi.org/10.1109/ICDAR.1993.395781
2. Capobianco, S., Marinai, S.: DocEmul: a toolkit to generate structured historical documents (2017)
3. Deng, J., Dong, W., Socher, R., Li, L.J., Li, K., Fei-Fei, L.: ImageNet: a large-scale hierarchical image database. In: 2009 IEEE Conference on Computer Vision and Pattern Recognition, pp. 248–255. IEEE (2009). https://doi.org/10.1109/CVPR.2009.5206848
4. Ditria, L., Meyer, B.J., Drummond, T.: OpenGAN: open set generative adversarial networks (2020)
5. Gatys, L.A., Ecker, A.S., Bethge, M.: Image style transfer using convolutional neural networks. In: Proceedings of the IEEE Conference on Computer Vision and Pattern Recognition (CVPR) (2016). https://doi.org/10.1109/CVPR.2016.265
6. Goodfellow, I.J., et al.: Generative adversarial networks (2014)
7. He, K., Zhang, X., Ren, S., Sun, J.: Deep residual learning for image recognition. In: 2016 IEEE Conference on Computer Vision and Pattern Recognition (CVPR), pp. 770–778 (2016). https://doi.org/10.1109/CVPR.2016.90
8. Huang, G., Liu, Z., Van Der Maaten, L., Weinberger, K.Q.: Densely connected convolutional networks. In: 2017 IEEE Conference on Computer Vision and Pattern Recognition (CVPR), pp. 2261–2269 (2017). https://doi.org/10.1109/CVPR.2017.243
9. Isola, P., Zhu, J.Y., Zhou, T., Efros, A.A.: Image-to-image translation with conditional adversarial networks (2018)
10. Journet, N., Visani, M., Mansencal, B., Van-Cuong, K., Billy, A.: DocCreator: a new software for creating synthetic ground-truthed document images. J. Imaging **3**(4), 62 (2017). https://doi.org/10.3390/jimaging3040062
11. Kieu, V., Visani, M., Journet, N., Domenger, J.P., Mullot, R.: A character degradation model for grayscale ancient document images. In: Proceedings of the 21st International Conference on Pattern Recognition (ICPR2012), pp. 685–688. IEEE (2012)
12. Kingma, D.P., Ba, J.: Adam: a method for stochastic optimization (2017)
13. Kingma, D.P., Welling, M.: Auto-encoding variational Bayes. arXiv preprint arXiv:1312.6114 (2013)
14. Lombardi, F., Marinai, S.: Deep learning for historical document analysis and recognition—A survey. J. Imaging **6**(10), 110 (2020). https://doi.org/10.3390/jimaging6100110
15. Mayr, M., Stumpf, M., Nicolaou, A., Seuret, M., Maier, A., Christlein, V.: Spatio-temporal handwriting imitation. In: Bartoli, A., Fusiello, A. (eds.) ECCV 2020. LNCS, vol. 12539, pp. 528–543. Springer, Cham (2020). https://doi.org/10.1007/978-3-030-68238-5_38

16. Mello, C.A.: Synthesis of images of historical documents for web visualization. In: Proceedings of the 10th International Multimedia Modelling Conference, pp. 220–226. IEEE (2004)
17. Mirza, M., Osindero, S.: Conditional generative adversarial nets (2014)
18. Pack, C., Liu, Y., Soh, L.K., Lorang, E.M.: Augmentation-based pseudo-groundtruth generation for deep learning in historical document segmentation for greater levels of archival description and access. ACM J. Comput. Cult. Herit. (2022)
19. Pondenkandath, V., Alberti, M., Diatta, M., Ingold, R., Liwicki, M.: Historical document synthesis with generative adversarial networks. In: 2019 International Conference on Document Analysis and Recognition Workshops (ICDARW), vol. 5, pp. 146–151 (2019). https://doi.org/10.1109/ICDARW.2019.40096
20. Seuret, M., Chen, K., Eichenbergery, N., Liwicki, M., Ingold, R.: Gradient-domain degradations for improving historical documents images layout analysis. In: 2015 13th International Conference on Document Analysis and Recognition (ICDAR), pp. 1006–1010 (2015). https://doi.org/10.1109/ICDAR.2015.7333913
21. Seuret, M., Limbach, S., Weichselbaumer, N., Maier, A., Christlein, V.: Dataset of pages from early printed books with multiple font groups. In: Proceedings of the 5th International Workshop on Historical Document Imaging and Processing, HIP 2019, pp. 1–6. Association for Computing Machinery, New York, NY, USA (2019). https://doi.org/10.1145/3352631.3352640
22. Seuret, M.: ICDAR 2021 competition on historical document classification. In: Lladós, J., Lopresti, D., Uchida, S. (eds.) ICDAR 2021. LNCS, vol. 12824, pp. 618–634. Springer, Cham (2021). https://doi.org/10.1007/978-3-030-86337-1_41
23. Shamsolmoali, P., et al.: Image synthesis with adversarial networks: a comprehensive survey and case studies. Inf. Fusion **72**, 126–146 (2020)
24. Shorten, C., Khoshgoftaar, T.M.: A survey on image data augmentation for deep learning. J. Big Data **6**(1), 1–48 (2019)
25. Tan, M., Le, Q.V.: EfficientNet: rethinking model scaling for convolutional neural networks (2020)
26. Zhu, J.Y., Park, T., Isola, P., Efros, A.A.: Unpaired image-to-image translation using cycle-consistent adversarial networks. In: 2017 IEEE International Conference on Computer Vision (ICCV), pp. 2242–2251 (2017). https://doi.org/10.1109/ICCV.2017.244

Poster Session 2

3D Modelling Approach for Ancient Floor Plans' Quick Browsing

Wassim Swaileh[1,2](✉), Michel Jordan[2], and Dimitris Kotzinos[2]

[1] University of Rennes, INSA of Rennes, IRISA, CNRS, 35000 Rennes, France
wassim.swaileh@irisa.fr

[2] ETIS, UMR 8051, CY Cergy Paris Université, ENSEA, CNRS, 95000 Cergy, France
{michel.jordan,dimitrios.kotzinos}@cyu.fr

Abstract. Although 2D architectural floor plans are a commonly used way to express the design of a building, 3D models provide precious insight into modern building usability and safety. In addition, for a historical monument like the Palace of Versailles, the 3D models of its ancient floor plans help us to reconstruct the evolution of its buildings over the years. Such old floor plans are hand made and thus present some problems in automatic creation of a 3D model due to the drawing style variability. In this paper, we introduce a fully automatic and fast method to compute 3D building models from a set of architectural floor plans of the Palace of Versailles dated to the 17th and 18th century. First, we detect and localise walls in an input floor plan image using a statistical image segmentation model based on the U-net convolutional neural network architecture and a binary wall mask image is obtained. Secondly, using the generated wall mask image, the 3D model is built upon the linear edge segments representing the detected wall sides in the mask image. In order to cope with the lack of accurate ground truth information for the 3D models of ancient floor plans, we use a dedicated semi-automatic software to build a set of reference 3D models that describe plans' wall projections from three sides of view. We evaluate the performance of our approach on an input floor plan image by measuring the overlapping between the 3D reference model and our 3D model. Our fast and fully automatic approach performs efficiently and produces quite accurate 3D models with 84.2% of IoU score in average. Furthermore, its performance surpasses the performance of the state of the art approach in the wall detection task.

Keywords: 3D modelling · Convolutional neural network · Ancient architectural archives

1 Introduction

Since the ancient architectural plans of a historical monument, such as the Palace of Versailles, reflect the evolution of the monument's buildings over the years,

First and second authors contributed equally to this research.

© Springer Nature Switzerland AG 2022
S. Uchida et al. (Eds.): DAS 2022, LNCS 13237, pp. 629–643, 2022.
https://doi.org/10.1007/978-3-031-06555-2_42

the conversion of these 2D plans into 3D models automatically, allows the visualisation of certain disappeared or unrealised places of the monument. Such old plans present several difficulties and challenges regarding drawing styles and paper format variabilities, in addition to foreground and background variable colour contrast. Due to these variabilities, building a 3D model from such floor plan images is not trivial. Actually, even creating the 3D model manually is not trivial either, and requires skill and time [23] In the literature, most of the related works detect walls in modern architectural drawings for building representative 3D models. Such drawings respect a set of standard protocols and symbols. However, architectural drawings in ancient floor plans show sometimes shapeless walls of variable drawing style [20]. The navigation through a virtual 3D environment becomes a common practice in order to access the past of the historical monuments and is nowadays commonly used in various domains and applications such as cultural heritage, archaeology, virtual tourism, etc. (see for instance [6,10]). Virtual reality and augmented reality, combined with semantic analysis, allow professionals, as well as the general audience, to visualise any kind of data including cultural heritage and historic data [13].

In this article, we present a 3D modelling approach that provides access to historical architectural archives by building 3D models in a fast and fully automated way. The paper is organised as follows: related works are presented in the following section. The proposed approach is then introduced in Sect. 3. Finally, we present the evaluation protocol and system performance evaluation results in Sect. 4 before introducing the conclusion and future works.

2 Related Work

In the literature we observe that the 3D modelling process of scanned floor plans consists of two main steps: the wall detection process and the 3D model generation process. Indeed, while the wall detection is not the main task but can be a sub-task, there are some works which are only focused on wall detection and we discuss those works in the following section.

2.1 Wall Detection

Traditionally, the wall detection task consists of several low-level image processing techniques including floor plan image noise reduction, graphics/text splitting, and drawing vectorisation. For detecting walls in ancient floor plan images, [21] used mathematical morphology operators in order to detect the main walls' drawings in a given input floor plan image, then built the corresponding 3D model. The proposed method requires human intervention to tune a threshold parameter in order to accurately detect thick walls and thin ones. In the same context, [20] introduced a U-Net convolutional neural network model for wall segmentation in ancient floor plan images. The model is trained to produce main wall mask images from grayscale floor plan images with no need for human intervention. We are using this work in our fully automatic 3D modelling processing chain.

In modern architectural drawings, many works are reported on wall detection and/or room segmentation. [15] used Hough transform to detect walls and doors. Subsequently wall polygons are created by the Hough lines and are partitioned iteratively into rooms, assuming convex room shapes. [1] first segment wall footprints in high-resolution images according to their line thickness. Then, they used geometrical reasoning to find room segments. SURF descriptors are used to detect doors. Convolutional neural networks were recently adopted to achieve the wall detection task. [4] used a simple neural network for denoising the plan image then used a fully convolutional neural network (FCN) for detecting walls. This approach is one of the state-of-the-art methods and performs well on multi-color contemporary floor plan images. [14] used a combination of deep neural networks and integer programming, where they first identify junction points in a given floor plan image and then join the junctions to locate walls in the floor plan. The method can only handle walls aligned with the two major axes of the floor plan image. Therefore, it can recognise layouts with only rectangular rooms and walls of uniform thickness, this is a critical limitation in the context of historical layouts where the shapes of the rooms are not always rectangular and often round shaped rooms are used. [22] also trained an FCN to label pixels in a floor plan image with multiple classes. The classified pixels formed a graphic model and were used to recover houses of similar structures.

As stated earlier all these works use different methods ranging from low level image processing to heuristics and furthermore to deep learning. However, the success of these methods in modern architectural documents cannot be replicated in the case of ancient architectural documents.

2.2 3D Modelling

Nowadays, 3D representations are a useful tool for architects because they provide an intuitive view on the architects' work and help them to present their projects to their clients. Thus, the research about 3D model computation from architectural drawings is an active research field since more than 20 years [23]. Contemporary architectural floor plans are composed of geometric shapes (straight lines, curves) representing the building structure (external and inner walls), while various symbols represent openings, stairs, heating and furniture elements, *etc.*. A lot of textual annotations provide information about the building's use, its rooms, dimensions, *etc.*. This standard drawings give a strong basis for a lot of research works that succeeded in the floor plan analysis and automatic 3D modelling tasks. In [5], authors from the LORIA lab present a system for the analysis of architectural drawings, with the aim of reconstructing in 3D the represented buildings. Authors follow a set of graphics recognition steps including image processing and feature extraction. This system demonstrates some robustness though it requires moderate human assistance. Siu-Hang Or *et al.* [18] developed a system to solve a slightly simplified problem that considers only walls, doors, and windows. The system distinguishes walls as inner structures from building outlines and uses it to match neighbouring floors; during vectorisation, the system extracts outlines of black pixels in the raster image and matches them with walls of various shapes. An improved example-driven symbol recognition algorithm is proposed for CAD engineering drawings in [11]. Firstly, in order to represent the structure of symbols, Guo *et al.* involve the text entity as one of the basic elements and redefine the relation representation mechanism. Then, the structure graph and a constrained tree is established for the target symbol, using the knowledge acquisition algorithm. In the recognition process, the nodes with the same type as the key features are located first in the drawing. The authors of [12] propose a database for structural floor plan analysis and the associated ground truth tool. The ground truth tool especially focuses on wall segmentation using the Hough line transform combined with an alignment heuristic fitting, and room detection.

Beside the 3D model computation, another field of interest concerning floor plan analysis is the need to refer to reference drawings when designing new buildings and to find solutions for similar architectural problems. In [2], authors present a system for semantic search from an architect's sketch: the semantic information of reference floor plans is extracted thanks to a "divide and conquer" strategy that separates text from graphics, and then graphics between thick lines (mainly walls) and thin lines (symbols). A further structural analysis allows us to perform the room detection and labelling. The DANIEL architecture presented in [19] is a deep learning model that learns the representative floor plan features using a convolutional neural network (CNN); the existing datasets provide the examples for the learning step. The authors of this paper proposed the ROBIN dataset that they used for evaluation.

The historical floor plans such as the ones we are working with have some drawing and digitizing specific features that do not allow their processing by one of the system described in previous section: the symbols for openings, doors, stairs, furniture, *etc.*, are different depending of the epochs and architects, and the digitization retains marks of the document state (for instance creases, since these documents have been kept in folders during long periods) or stamps that are on the drawing's reverse side. For these reasons, some of us developed in a previous project a system which copes with these particularities and help to build an accurate 3D model from the floor plans, thanks to a limited number of user interactions [21]. This system demonstrates a good ability to perform this task, but is not appropriate when processing a lot of drawings.

In the field of cultural heritage and architecture, some projects deal with text and image extraction from archived documents [17], other projects intend to build 3D models from various data sources such as photographs, laser scanning or point clouds [9]. The European Time Machine project intends to build 3D models of historical European cities across the centuries and analyses for this purpose various types of data including texts and graphics such as cadastral data; this allows to build city 3D models but not to enter the building interiors. In fact, it seems that the architectural drawing collection describing the palace of Versailles between 17th and 18th centuries is exceptional and lead to particular problems and particular algorithmic solutions.

3 Proposed Approach

Our approach relies on three main steps: the floor plan digitization step, the wall mask generation step and the 3D modelling step. There is no need for human intervention during the whole processing chain and the result is obtained in a few seconds. Such fast and automatic 3D model construction system allow fast access to disappeared or non-realised architectural projects that interest specialists as well as the general public (Sects. 3.1, 3.2 and 3.3). Figure 1 illustrates our processing pipeline from the floor plan image as input to the resulting 3D model.

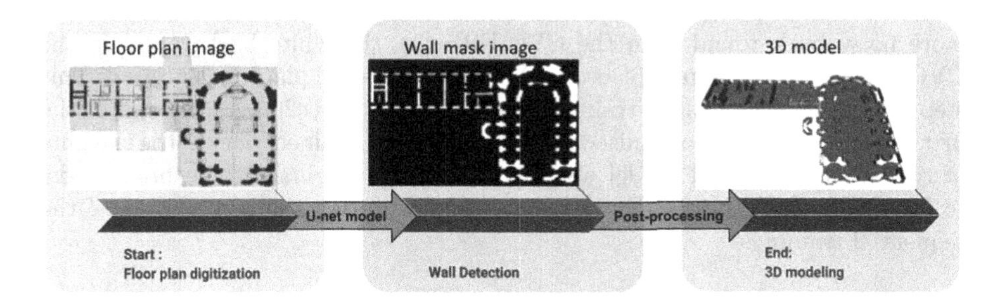

Fig. 1. Summary scheme of the processing pipeline

3.1 Floor Plan Digitization

The ancient floor plans of the Palace of Versailles have various sizes and drawing styles. This is due to the lack of standards for the drawing styles and paper formats at that age, and to the long period (about 120 years) covered by these plans. The images of these floor plans, including those of Versaille-FP dataset, were captured in a very high resolution. Our approach receives a multi-colour floor plan image as input. Then, it converts it into grayscale in order to reduce the colour noise and increase the intensity between the foreground and background colours at the input image. After that, it scales the grayscale image down to images of 512 $\times$ 512 pixels. To maintain the aspect ratio of the original input image, we square padded the input image to its largest dimension before applying the scaling process. Once the scaled image is obtained, we apply the wall segmentation model based on the U-net convolutional neural network to obtain the wall mask image. The wall mask image is a binary image where the foreground colour represents the wall regions detected by the segmentation model in the scaled input image.

3.2 Wall Mask Generation

At the second step of our proposed approach we used a U-Net architecture of convolutional neural network (CNN) to generate the wall mask image from the input image. We used the same CNN model architecture introduced by [20] and illustrated in Fig. 2. We choose this architecture due to its precious proprieties of high speed and performance with small amount of data, according to [8]. The CNN model consists of two symmetric paths connected at a bottle neck layer. The first path namely called contracted path extracts contextual and semantic information from the input image. The second path (expanding path) achieves the accurate localization, by copying, cropping and concatenating information that are observed by the contracting path toward the expanding path. The model is trained in two steps of sequential learning where data augmentation is applied. The first step consists in learning an initial model on the modern CVC floor plan dataset samples [12]. As the Palace of Versailles dataset samples have a more noisy background than the CVC-FP ones, the white background in the CVC original samples are replaced by Versailles coloured plans background. This step provides a good initial training for the final model [20]. At the end of the first learning step, an optimised initial model is obtained. Next, the weights of the optimised initial model are used as initial weights for the final model to be trained on Versailles-FP dataset[1] samples during the second step of the sequential training.

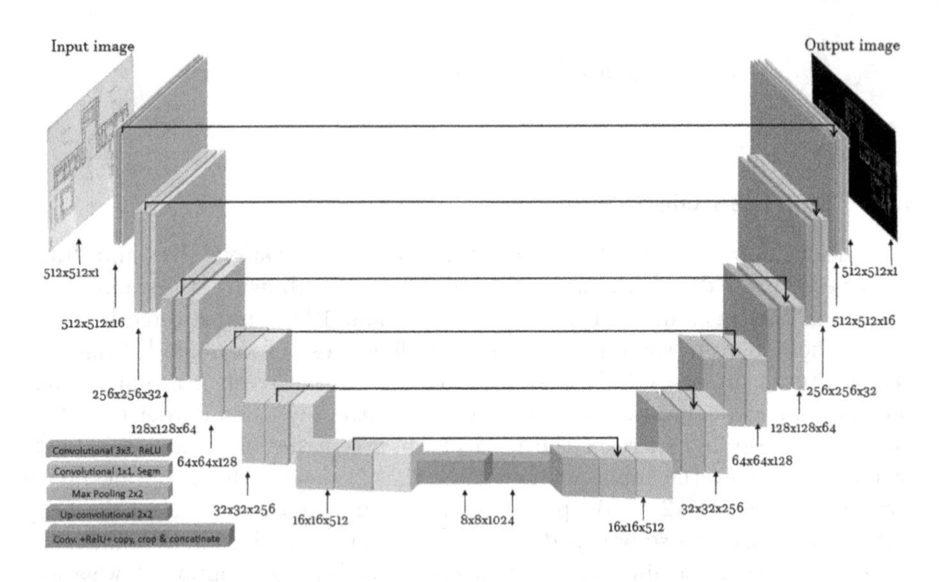

Fig. 2. U-net architecture for detecting walls in ancient floor plans

The loss function L_D of Dice [16] is used to train the model. The Dice Coefficient is an overlapping measure between two samples. It ranges from 0 to 1 where a Dice coefficient of 1 denotes perfect and complete overlap. The Dice loss is simply equals to 1− *Dice coefficient*. The L_D loss value is calculated as in Eq. 1.

[1] https://www.etis-lab.fr/versailles-fp/.

$$L_D = 1 - \frac{2 \times |A \cap B| + 1}{|A| + |B| + 1} \tag{1}$$

where A is the neural network prediction about the segmentation and B is the ground truth value. We compared the performance of the U-net model using different loss functions including the Dice loss with which we obtained the best results. Similarly, we selected the Rmsprop optimisation algorithm for learning our model. We used a learning rate that start with $10e^{-4}$ with batch size of 25 and we selected the best model during the learning process. We applied learning rate decay whenever validation accuracy does not improve. Although our model training is guided by the validation datasets, learning rate adjustment based on validation accuracy still works well according to our experience. It is worthy to mention that our model differs from the state art model [20] by the models' hyper-parameters fine tuning strategies, specially with regards to the batch size and the learning rate decay. The model training is achieved in 4 days using 2 x GPU Nvidia GTX 1080 11Go. We used early stopping with patience of fifty epochs.

To produce wall mask image for an input floor plan image, we applied the final CNN model on the scaled grayscale image at the input. The generated wall mask image is then feed to the next process of the pipeline in order to produce the corresponding 3D model. In the next section, we describe in detail the 3D model construction process and its related sub-processes.

3.3 3D Model Generation

We develop in this section an algorithm for a fast computation of 3D models from a wall mask image such as the one provided by the deep learning architecture in Sect. 3.2. This algorithm includes the following steps:

- pruning of the residual small structures in the wall mask image;
- computation of edges;
- polygonalization of the edges;
- computation of the 3D model based on the wall polygons.

These steps consist of an arrangement of well-known image processing algorithms, whose parameters are fixed.

Small Structures Pruning. Though the wall image computation is an accurate and carefully designed process, it may persist some pixels either isolated or in thin structures. The first step of our algorithm is a morphological opening with a rectangular 3×3 structuring element. We heuristically fixed the morphological opening number of iterations according to the wall image size. This operation allows not only to discard noisy pixels in the wall image, but also to remove thin structures such as dividing walls (*cf.* Fig. 3).

Fig. 3. Discarding thin structures: original wall image (left), after processing (center), difference image (right).

Edge Detection and Polygonalization. The center image of Fig. 3 represents the basis for the 3D model extrusion. The walls in the 3D model have inner and outer sides, since in future steps they could be covered with more or less realistic textures. We compute wall contours by means of the Canny edge detector [3] with σ parameter 0.1, and then perform edge polygonalization using a probabilistic Hough line detector [7] (*cf.* Fig. 4).

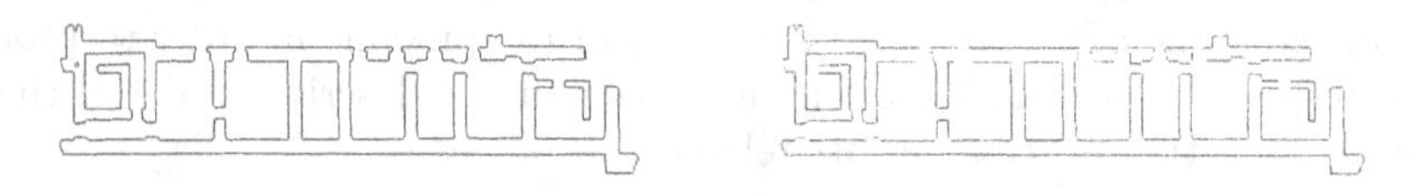

Fig. 4. Edge (left) and straight line (right) images computed from center image in Fig. 3.

3D Model Building. The 3D model is computed as an extrusion upon the straight lines of Fig. 4, at an arbitrary height that we fixed as 0.15 of $\max(H, W)$ where H and W are the initial image dimensions, in order to get a clearly visible 3D model. This height will be easily modified in further processes, depending on the final objective of the 3D modelling. Each line resulting from the polygonalization step gives an edge and two vertices for the 3D model, two other vertices are positioned vertically upon the first two, and these four vertices finally shape a vertical face (*cf.* Fig. 5).

4 Results and Evaluation

In this section, we present the experimental framework we followed to evaluate the performance of our fast and automatic 3D modelling approach. First, we introduce the ancient floor plan dataset that we used. Then, we present the evaluation protocol and results for a set of 15 3D models obtained by our approach and evaluated against reference 3D models that were produced using the VERSPERA semi-automatic 3D modelling tool [21].

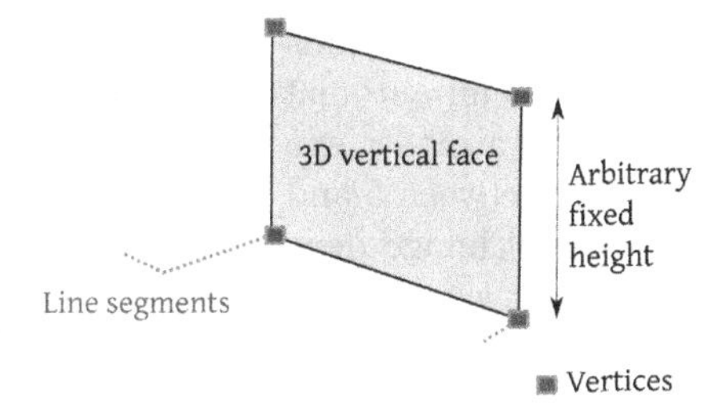

Fig. 5. A 3D face is computed upon a polygonal line.

4.1 Dataset

In this study, we used the ancient floor plan *"Versailles-FP"* dataset which is based on the french VERSPERA research project[2]. The dataset consists in 500 annotated images collected from scanned floor plans that belongs to the Palace of Versailles constructions dated of 17th and 18th century. The VERSPERA research project was started in 2013 with the aim of digitizing a large amount of graphical documents related to the construction of the Palace of Versailles during the 17th and 18th centuries. There is a large corpus of floor plans including elevations and sketches present in the collection of French National Archives. The total number of documents in this archive is around 6,500 among which about 1,500 are floor plans. An ambitious project to digitize this varied corpus started in 2014, extraordinary technical capability is needed to achieve this task due to fragile and varied nature of the paper documents (for example some document can be as big as $3\,\text{m} \times 4\,\text{m}$). The digitized plans of the Palace of Versailles consist of graphics that illustrate the building architecture such as walls, stairs, halls, king rooms, royal apartments, *etc.* in addition to texts and decorations. Since the palace of Versailles digitized floor plans cover 120 years (1670–1790) of architectural design, different drawing styles clearly appear in the corpus.

4.2 Evaluation Protocol

Since most of the used ancient floor plans do not refer to real buildings of the Palace of Versailles of today, a straightforward numerical evaluation of the produced 3D models is not applicable. In fact, there is no absolute 3D ground truth for these data: 3D measurements in real rooms is not ever possible, since these floor plans represent parts of the monument that have been rebuilt or destroyed; even in the case of the current state of the monument is the same as the plan drawings, we can not infer that a floor plan of the 18th century is a perfectly scaled representation of the monument. In fact, the only absolute reference would be an entirely manual processing of the floor plan images by computer graphics professionals, which is a time consuming and expensive task. We tried thus to compare some of these 3D models with the ones generated

[2] https://verspera.hypotheses.org/.

from the same floor plan images by means of the VERSPERA software [21]. This software is an interactive semi-automatic application that we developed in a previous research project; for a rather experienced user, building a 3D model with this application requires between 5 and 10 min. The VERSPERA software involves the following steps: *(i)* image denoising and preprossessing, including downsizing to a roughly similar size for all plans; *(ii)* image binarization; *(iii)* wall detection through mathematical morphology tools and manual thresholding; *(iv)* wall footprint extrusion for 3D modelling. The VERSPERA software allows some other processings such as staircase detection and wall decoration, that we do not consider in this study. In the following sections, we compare our method and the VERSPERA software method by means of two criteria: computation time and 3D model accuracy.

4.3 Results

Wall Detection Evaluation. To evaluate the performance of the wall detection model we distinguish between two training schemes; 1) training from scratch where the model weights are initialised arbitrary for training the model on the Versailles-FP dataset samples directly, 2) sequential training in two phases; I) training from scratch the model on the CVC modern floor plan dataset, and II) continue the training of the model on the Versailles-FP training set. For the wall detection task, we used the same 5 fold cross validation protocol introduced in the Versailles-FP dataset paper [20]. We provide in Fig. 6 a set of good and bad wall detection images (images with black background). Bad results (lower row of the figure) occur when sketch scale differs too much from the learning set (lower left example) or sketch technique is different (pencil—lower right example—against ink in the learning set).

We applied our software to floor plans of other monuments, freely available in the Gallica database from the French national Library (Fig. 7); results appear to be not so good for these low resolution images, it will be interesting to add some of these examples in a transfer learning set to add more versatility to our models.

Figure 8 presents a comparison of the wall detection task evaluations using the accuracy, IoU and Dice coefficient scores for both cases of "learning from scratch" (blue lines, with best Dice score of 93%), and the "sequential learning" (red lines with best Dice score of 94%). From this two learning tasks, we observe that the sequential learning of the pre-trained model is better than learning from scratch. In addition, our model outperforms the state of the art model. The slit performance enhancement is obtained thanks to the applied training parameter settings that are different of the state of the art ones. We think that the pre-trained model is learning better when using fine tuned learning rate with bigger batch size, recalling that the state of the art batch size is 16 since we used batch size of 25.

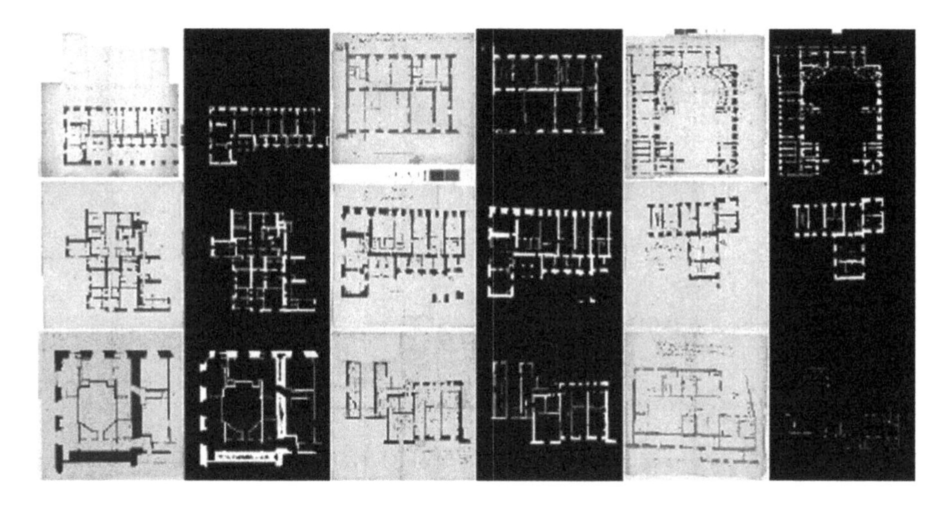

Fig. 6. Illustration of wall detection good and bad results.

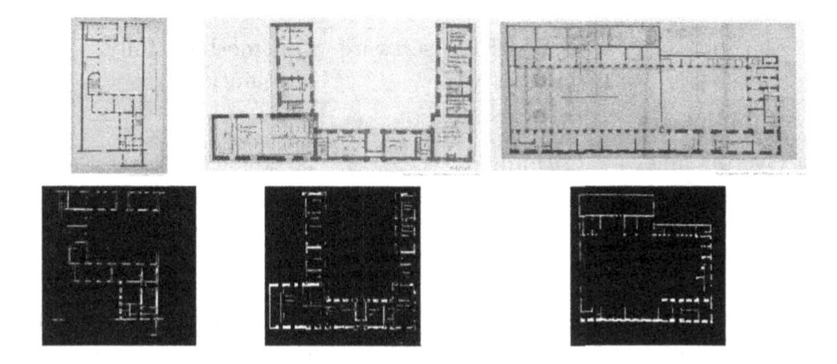

Fig. 7. Some examples of wall mask computation for other monuments.

Computation Time. Figure 9 presents some 3D model examples of floor plans we computed using the approach described in Sect. 3.3. Computing 3D models for the 500 wall mask images of the Versailles-FP dataset took around 20 min on a laptop with a *Intel Xeon CPU E3-1505M* processor, which gives an average of 2.4 s for each image. In comparison with the computation time of the reference 3D models constructed using the interactive VERSPERA software, we observe that our approach is 25 time faster than the interactive one. We give on Table 1 a comparison of computation times for both methods.

3D Model Accuracy. We compare first our models by a visual examination and comparison of the resulting 3D models for both automatic and interactive methods (see Fig. 10). Then, in order to get a more accurate and quantitative evaluation, we overlap the polygonalized wall mask image resulting of our approach with the vertical ground projection of the 3D model resulting from the VERSPERA software (Fig. 10, right column); in fact, these two images are

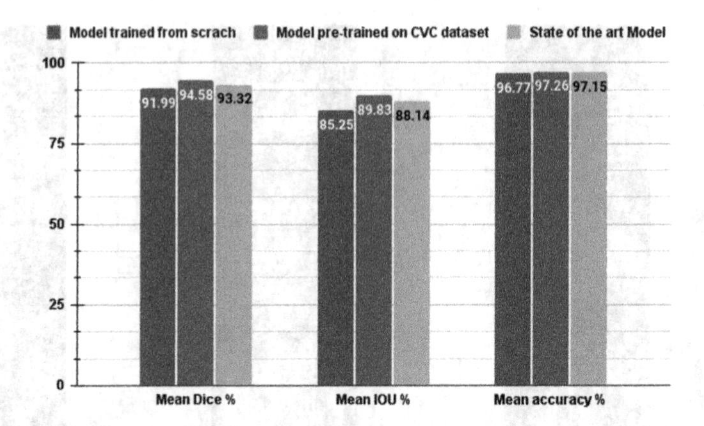

Fig. 8. Wall detection evaluation using pre-trained/trained from scratch model and comparison with the state-of-the-art. (Color figure online)

Table 1. Comparison of computation times.

Method	VERSPERA software (interactive)	Our method (fully automatic)
Nb of images	15	500
Overall time	120 min	20 min
Time *per* image	8 min	2.4 s

the basis of the 3D model extrusion, and comparing them gives a rather good approximation of fidelity between both 3D models since, as we discussed, a real reference 3D model is missing when dealing with the ancient floor plans. The right column images show small differences (non-white pixels) between the two masks that do not lead to drastic 3D model differences. For a numeric evaluation, we computed the min, max and average values and standard deviation of the IoU scores for the overlapping masks on a set of 15 models; Table 2 sums up these values and shows a 84.2% average value, which is a rather good value depending on the final objective of the 3D modelling.

Table 2. IoU scores computed on 15 models.

Min	Max	Average	Std dev.
0.780	0.906	0.842	0.041

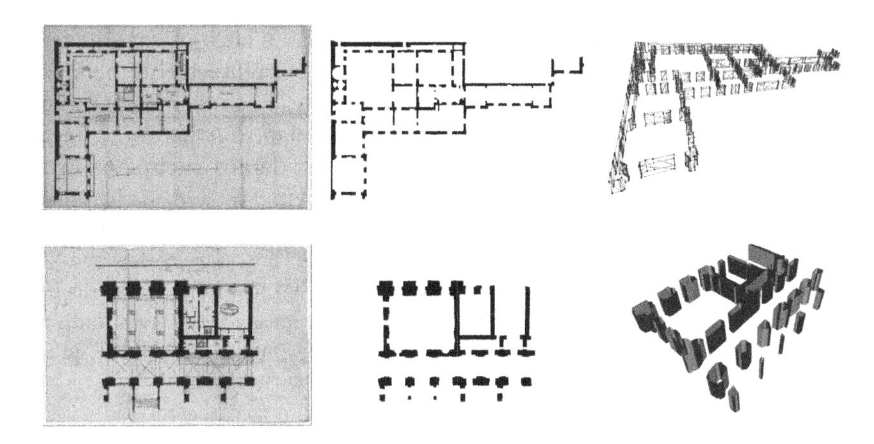

Fig. 9. Some 3D models (wireframe or solid screen capture on right) computed from the floor plan (left) and wall mask images (middle).

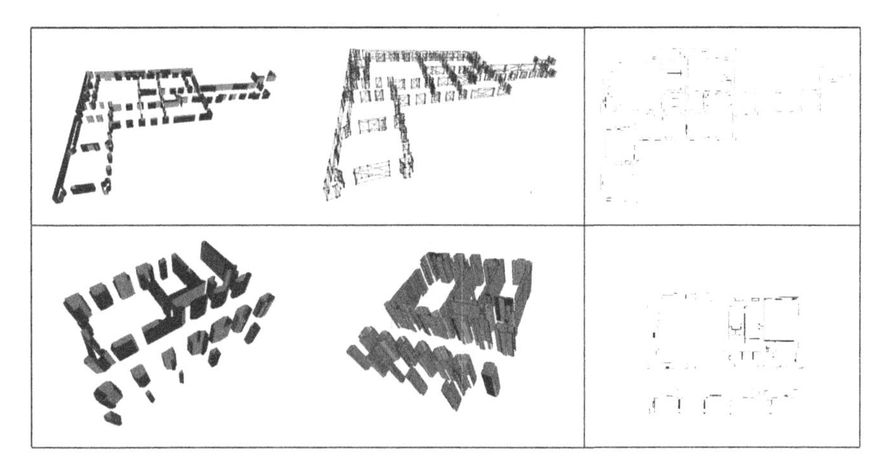

Fig. 10. Comparison of 3D models: VERSPERA semi-manual software (left), our automatic and fast approach (center), and wall mask difference images (right).

5 Perspectives and Conclusion

We presented in this paper a fast 3D model computation approach for historical floor plan quick browsing. The proposed approach uses the U-net architecture of the convolutional neural networks for detecting the wall regions in an input floor plan image, resulting in a binary wall mask image. The resulting image is filtered in a way that only the main walls remain, corresponding to the main structures of the monument. Finally, a 3D model is computed by extruding 3D faces upon linear segments corresponding to the wall edges. We applied this method to build 3D models from the 500 floor plan images of the Versailles-FP dataset (floor plans representing the palace of Versailles between 17th and 18th

centuries), and were able to build 3D models very fast. This tool gives historians and archivists a mean to have a fast and quick but complete 3D perspective at the floor plans, and thus to get an easy perception and understanding of the volumes of the monument represented on the plan. Further work includes the integration in 3D models of other architectural details such as staircases, by means of a dedicated machine learning approach. We will also apply our algorithms to other historical datasets of architectural plans, if available.

Acknowledgements. We thank our colleagues of the VERSPERA research project in the Research Center of Château de Versailles, French national Archives and French national Library, and the Fondation des sciences du patrimoine which supports VERSPERA.

References

1. Ahmed, S., Liwicki, M., Weber, M., Dengel, A.: Automatic room detection and room labeling from architectural floor plans. In: 2012 10th IAPR DAS, pp. 339–343 (2012)
2. Ahmed, S., Weber, M., Liwicki, M., Langenhan, C., Dengel, A., Petzold, F.: Automatic analysis and sketch-based retrieval of architectural floor plans. Pattern Recognit. Lett. **35**, 91–100 (2014). Frontiers in Handwriting Processing
3. Canny, J.: A computational approach to edge detection. IEEE Trans. Pattern Anal. Mach. Intell. **PAMI–8**(6), 679–698 (1986)
4. Dodge, S., Xu, J., Stenger, B.: Parsing floor plan images. In: MVA, May 2017
5. Dosch, P., Tombre, K., Ah-Soon, C., Masini, G.: A complete system for the analysis of architectural drawings. Int. J. Doc. Anal. Recogn. **3**, 102–116 (2000)
6. Fiorucci, M., Khoroshiltseva, M., Pontil, M., Traviglia, A., Del Bue, A., James, S.: Machine learning for cultural heritage: a survey. Pattern Recogn. Lett. **133**, 102–108 (2020)
7. Galambos, C., Kittler, J., Matas, J.: Progressive probabilistic hough transform for line detection. In: Proceedings of 1999 IEEE Computer Society Conference on Computer Vision and Pattern Recognition (Cat. No PR00149), vol. 1, pp. 554–560 (1999)
8. Gao, H., Yuan, H., Wang, Z., Ji, S.: Pixel deconvolutional networks. arXiv preprint arXiv:1705.06820 (2017)
9. Grilli, E., Özdemir, E., Remondino, F.: Application of machine and deep learning strategies for the classification of heritage point clouds. Int. Arch. Photogr. Remote Sens. Spat. Inf. Sci. **XLII-4/W18**, 447–454 (2019)
10. Grilli, E., Menna, F., Remondino, F.: A review of point clouds segmentation and classification algorithms. Int. Arch. Photogr. Remote Sens. Spat. Inf. Sci. **42**, 339 (2017)
11. Guo, T., Zhang, H., Wen, Y.: An improved example-driven symbol recognition approach in engineering drawings. Comput. Graph. **36**(7), 835–845 (2012). Augmented Reality Computer Graphics in China
12. de las Heras, L.-P., Terrades, O.R., Robles, S., Sánchez, G.: CVC-FP and SGT: a new database for structural floor plan analysis and its groundtruthing tool. Int. J. Doc. Anal. Recognit. (IJDAR) **18**(1), 15–30 (2015)

13. Lampropoulos, G., Keramopoulos, E., Diamantaras, K.: Enhancing the functionality of augmented reality using deep learning, semantic web and knowledge graphs: a review. Vis. Inform. **4**(1), 32–42 (2020)
14. Liu, C., Wu, J., Kohli, P., Furukawa, Y.: Raster-to-vector: revisiting floorplan transformation. In: Proceedings of the IEEE ICCV, pp. 2195–2203 (2017)
15. Macé, S., Locteau, H., Valveny, E., Tabbone, S.: A system to detect rooms in architectural floor plan images. In: Proceedings of the 9th IAPR DAS, pp. 167–174 (2010)
16. Milletari, F., Navab, N., Ahmadi, S., V-net: fully convolutional neural networks for volumetric medical image segmentation. In: Proceedings of the 2016 Fourth International Conference on 3D Vision (3DV), pp. 565–571 (2016)
17. Oliveira, S.A., Seguin, B., Kaplan, F.: dhSegment: a generic deep-learning approach for document segmentation. CoRR abs/1804.10371 (2018)
18. Or, S., Wong, K., Yu, Y., Chang, M.: Highly automatic approach to architectural floorplan image understanding & model generation. In: Vision, Modeling & Visualization 2005, pp. 25–32. Erlangen, Germany (2005)
19. Sharma, D., Gupta, N., Chattopadhyay, C., Mehta, S.: DANIEL: a deep architecture for automatic analysis and retrieval of building floor plans. In: 2017 14th IAPR International Conference on Document Analysis and Recognition (ICDAR), vol. 01, pp. 420–425 (2017)
20. Swaileh, W., Kotzinos, D., Ghosh, S., Jordan, M., Vu, N.-S., Qian, Y.: Versailles-FP dataset: wall detection in ancient floor plans. In: Lladós, J., Lopresti, D., Uchida, S. (eds.) ICDAR 2021. LNCS, vol. 12821, pp. 34–49. Springer, Cham (2021). https://doi.org/10.1007/978-3-030-86549-8_3
21. Tabia, H., Riedinger, C., Jordan, M.: Automatic reconstruction of heritage monuments from old architecture documents. J. Electron. Imaging **26**(1), 011006 (2016)
22. Yang, S.T., Wang, F.E., Peng, C.H., Wonka, P., Sun, M., Chu, H.K.: DuLa-net: a dual-projection network for estimating room layouts from a single RGB panorama. In: Proceedings of the IEEE/CVF Conference on CVPR, pp. 3363–3372 (2019)
23. Yin, X., Wonka, P., Razdan, A.: Generating 3D building models from architectural drawings: a survey. IEEE Comput. Graph. Appl. **29**(1), 20–30 (2008)

A Comparative Study of Information Extraction Strategies Using an Attention-Based Neural Network

Solène Tarride[1,2(✉)], Aurélie Lemaitre[1], Bertrand Coüasnon[1], and Sophie Tardivel[2]

[1] Univ. Rennes, IRISA, CNRS, Rennes, France
solene.tarride@irisa.fr
[2] Doptim, Cesson-Sévigné, France

Abstract. This article focuses on information extraction in historical handwritten marriage records. Traditional approaches rely on a *sequential* pipeline of two consecutive tasks: handwriting recognition is applied before named entity recognition. More recently, *joint* approaches that handle both tasks at the same time have been investigated, yielding state-of-the-art results. However, as these approaches have been used in different experimental conditions, they have not been fairly compared yet. In this work, we conduct a comparative study of sequential and joint approaches based on the same attention-based architecture, in order to quantify the gain that can be attributed to the joint learning strategy. We also investigate three new joint learning configurations based on multi-task or multi-scale learning. Our study shows that relying on a joint learning strategy can lead to an 8% increase of the complete recognition score. We also highlight the interest of multi-task learning and demonstrate the benefit of attention-based networks for information extraction. Our work achieves state-of-the-art performance in the ICDAR 2017 Information Extraction competition on the Esposalles database at line-level, without any language modelling or post-processing.

Keywords: Document image analysis · Historical documents · Information extraction · Handwriting recognition · Named entity recognition

1 Introduction

In recent years, many European libraries, museums and archives have undertaken to digitize their collections of historical documents [20], as a way to ensure the preservation of our cultural heritage. Indeed, digital documents do not need to be physically handled, thus reducing the potential damage to fragile collections, and can be quickly and easily accessed from any location. However, searching for a specific document among millions of digitized entries remains a challenge. As a result, there is a need for data indexation, which would allow users to browse

© Springer Nature Switzerland AG 2022
S. Uchida et al. (Eds.): DAS 2022, LNCS 13237, pp. 644–658, 2022.
https://doi.org/10.1007/978-3-031-06555-2_43

a collection and retrieve specific documents using keyword queries. A practical application is the search of specific information contained in population records. These documents hold a strong value for genealogists around the world, as they provide precious information about our ancestors. The ability to search for a specific name in a collection of records would substantially ease genealogical research. Additionally, a macro-analysis of such documents could provide an interesting perspective for historians and demographers.

This raises the question of how to extract information from large collections of historical records. Several platforms have been proposed for collaborative manual annotation, such as Transcribathon[1] and CrowdHeritage[2]. Yet, even with collaborative tools, manual extraction of information from large collections still requires a lot of effort. Recently, the advances in computer vision have opened the way for automatic document understanding using computational models. These systems can be trained to recognize a sequence of handwritten characters, and to assign a semantic label to each predicted word. However, these approaches are not entirely reliable when applied to challenging documents, such as historical documents. Indeed, historical records can have various layouts and handwriting styles, and often feature paper and ink degradations, abbreviations, inter-line annotations, or crossed-out words.

The interest for the task of information extraction in historical documents has been reinforced by the 2017 Information Extraction in Historical Handwritten Records (IEHHR) competition [12], based on Esposalles database [24]. The database is a collection of historical handwritten marriage records from the Archives of the Cathedral of Barcelona from the 17th century. The aim of the competition is to extract relevant information about the wife and husband and their parents, such as their name, occupation and place of origin, as illustrated in Fig. 1.

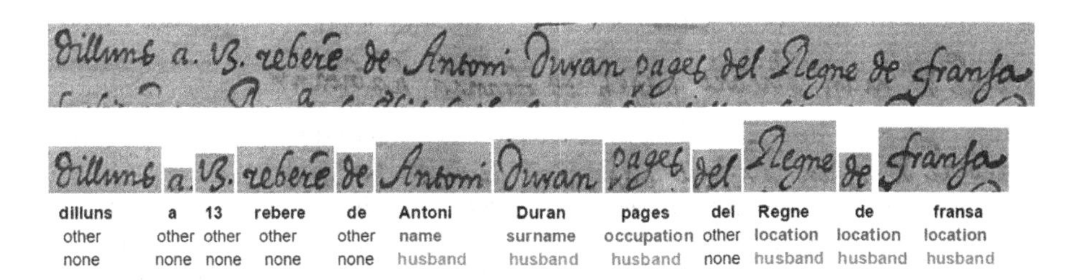

Fig. 1. Example from the database proposed for the IEHHR competition [12]. Each image is associated with a transcription, and two semantic labels are associated with each word: *category* and *person*.

Most of the approaches submitted to this competition are based on a sequential approach, in which the handwriting recognition task is performed upstream of the named entity recognition task [12]. However, several researchers have highlighted the possibility of combining these two tasks, by training a model to output

[1] https://www.transcribathon.com/en/.
[2] https://crowdheritage.eu/en.

characters and contextual tags [7,25]. These joint approaches yield competitive results, however it is impossible to assert whether the improvement comes from the architecture used by the authors or from the joint learning strategy.

In this work, we conduct a comparative study of these two training strategies using the same attention-based model, in order to quantify the improvement linked to the joint learning strategy. We also introduce three additional joint learning configurations based on multi-task and multi-scale learning.

This article is organized as follows. Section 2 introduces the works related to automatic information extraction and handwriting recognition in historical documents. Section 3 introduces the attention-based sequence-to-sequence architecture proposed in this work. Section 4 presents our comparative study and introduces three joint learning strategies based on multi-task and multi-scale learning. The results of our experiments are presented and discussed in Sect. 5. Finally, we summarize this work and propose future directions.

2 Related Works

Automatic document understanding is a challenging research area that includes document layout analysis, handwriting recognition, and information extraction. As we work on pre-segmented line images, we focus this study on handwritten text recognition (HTR) and information extraction (IE).

2.1 Handwriting Recognition

State-of-the art HTR models are currently based on deep neural networks. One of the most popular architectures is composed of a convolutional neural network (CNN) to extract features from the image, and a recurrent neural network (RNN) to capture sequential information. This architecture is often referred to as CRNN-CTC, as it is trained using the Connectionist Temporal Classification (CTC) loss function [13]. This architecture has become very popular over the recent years [5,14,23]. More recently, attention-based sequence-to-sequence (seq2seq) networks have been investigated for HTR. Contrary to the CRNN-CTC architecture, attention-based models learn to align image pixels with the target sequence. As a result, the network learns to focus on a small relevant part of the feature vector to predict each token [19,22]. Another strength of this architecture is that the recurrent decoder learns an implicit language model at character-level. The seq2seq architecture has also demonstrated its ability to handle multiple tasks at once [16], which is interesting in the context of information extraction. Finally, the Transformer architecture is also gaining a lot of attention from the HTR community [15], although this architecture requires many training images.

2.2 Information Extraction

The task of information extraction (IE) consists in extracting semantic information from documents. For structured documents, such as tables, forms or invoice, semantic information can be derived from the word localization. In this case, it is common to build end-to-end models that localize each word, transcribe them, and derive context from localization features [21,30]. But for semi-structured documents, such as marriage records, context can only be derived from phrasing. For example, a surname generally comes after a name. In this scenario, the first step is to localize textual zones, such as paragraphs, text-lines or words. Then, handwriting recognition and named entity recognition (NER) is performed.

Three main strategies have been considered so far. The traditional strategy is a *sequential approach* where handwriting recognition is performed before named entity recognition [12]. Another *sequential approach* consists in classifying each word into semantic categories, then applying handwriting recognition techniques [27]. The last strategy is a *joint approach*, where both tasks are tackled at the same time, using an end-to-end model [6,7].

Transcription Before Word Semantic Classification. The most common approach relies on a HTR model to predict a transcription, and on natural language processing (NLP) techniques to classify each word into named entities using textual features. The drawback of this sequential approach is that there is no contextual information during the transcription stage. Rather, the context is used in the post-processing stage, using category-based language modelling. Several methods have been proposed in the ICDAR2017 competition on Information Extraction [12]. The Hitsz-ICRC team developed an approach at word-level. First, a bi-gram based CNN is trained on word images to recognize characters, without any language model. Then, words are classified using the CRF sequence tagging method. The CITlab ARGUS team developed an approach at line-level. Images are passed into a CRNN-CTC architecture for handwritten text recognition, then regular expressions are used to decode and classify each word.

Word Semantic Classification Before Transcription. This approach consists in labelling each word before predicting the transcription. The interest is that knowing the semantic category beforehand helps the system to predict the right transcription. It is particularly suited when dealing with word image, as each image can be easily classified before being transcribed. Toledo et al. [27] extract semantic context from word images, and obtain the transcription of each word using semantic context of precedent words. The authors observe that their system benefits from knowing the semantic category. For example, if it makes sense to read a male name, the word "John" will be more likely than the word "born", even if the handwritten word looks more like "born". The main drawback of this approach is that it requires word bounding box annotation.

Joint Transcription and Word Semantic Classification. This approach consists in producing a transcription and a semantic category at the same time. A method has been proposed in [6], where an end-to-end model is used to localize word bounding boxes and jointly classifying and transcribing them. A major drawback is that it requires word-level segmentation for training. Another approach based on line-level images [7] relies on a CRNN-CTC network to predict a transcription with semantic categories, using tags located before important words. If this approach is very promising, we believe that it could benefit from using an attention-based network, which would allow the network to learn the appropriate features to predict the tags. More recently, this joint approach has been successively used with a Transformer network at line-level and record-level [25]. We believe that this combined approach is very promising, and benefits from an attention-based network to contextual features that are relevant to predict the semantic tags.

2.3 Our Statement

In this work, we address different aspects of information extraction at line-level.

First, we want to assess the interest of the joint approach for information extraction. Our intuition is that combining handwriting recognition and named entity recognition should be helpful, as both tasks are related and could benefit from shared contextual features. Indeed, semantic context can be derived from the transcription, but knowing the semantic context beforehand should also make the transcription more reliable. Recently, competitive results have been achieved using this joint approach [7,25], which indicates its relevance for information extraction. However, it is challenging to assert if the improvement comes from the neural network architecture or from the combination of HTR and NER. To validate this intuition, we compare sequential and joint approaches using the same neural network. Our study aims to measure the improvement that can be attributed to the learning strategy.

Secondly, we present a simple seq2seq architecture with attention, and demonstrate the interest of attention-based models for the task of information extraction. Indeed, the attention mechanism learns to focus on relevant zones of the image at each step, which should facilitate the recognition of named entities based on visual features. Moreover, this architecture can learn an implicit language model, which is convenient for semi-structured documents. To this end, we compare its performance with other neural network architectures, mainly CRNN-CTC [7] and Transformer [25].

Finally, we investigate three additional joint learning strategies based on our seq2seq architecture using multi-task and multi-scale learning. Indeed, we believe that these strategies can be helpful for joint handwriting recognition and named entity recognition, as they have been proved efficient for neural translation and image captioning [16].

Our contributions are summarized as follows:

- We show the benefit of joint strategies for information extraction and handwriting recognition.
- We highlight the interest of attention-based models for joint handwriting and named entity recognition.
- We propose three additional joint strategies for HTR and NER using multi-task and multi-scale learning strategies.

- We obtain state-of-the-art results on the IEHHR competition benchmark at line-level [12], without any post-processing.

3 The Attention-Based Seq2seq Architecture

The seq2seq architecture is an encoder-decoder network which was initially proposed for automatic translation [26]. It recently gained popularity for speech recognition, image captioning and neural translation [3,8,29]. The architecture has since been adapted for HTR [19,22], as illustrated in Fig. 2. In this work, we use an architecture adapted from [19].

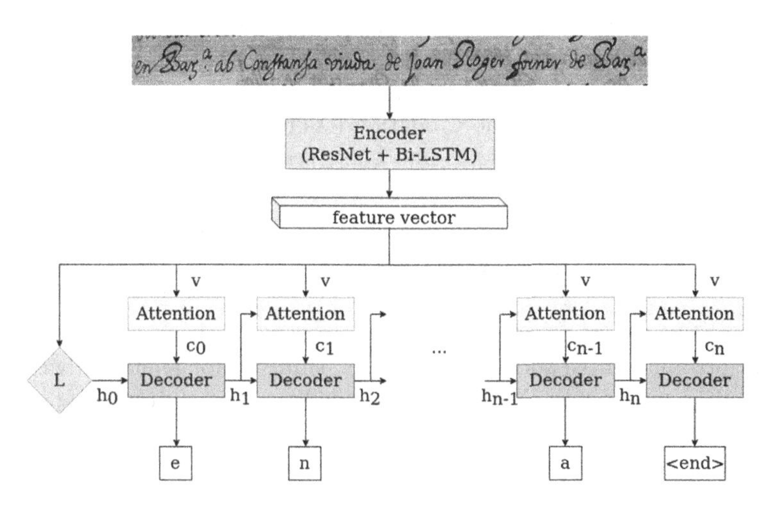

Fig. 2. Seq2seq architecture used in this work, where v is the feature vector, L is a linear function, (c_i) and (h_i) are the context vector and the decoder's hidden state at time step i, and n is the number of decoding steps.

Input. Line images are pre-processed and augmented before entering the encoder. First, they are resized, padded and normalized. Then, random augmentations are applied on the fly, including DPI adjusting, random perspective, random transform, elastic distortion, dilation, erosion, sign flipping, brightness and contrast adjustment.

Encoder. The role of the encoder is to extract features from the image. Our encoder is a convolutional recurrent neural network (CRNN) based on a ResNet-101 backbone, which is pre-trained on ImageNet [10]. Only the last two convolutional blocks are trained to reduce the number of training parameters. We add to this architecture a bi-dimensional Long Short Time Memory (BLSTM) network to capture the sequential information. The final feature vector is then fed to the attention mechanism and the decoder.

Attention. The attention mechanism has been introduced by Bahdanau et al. [3] as a way to focus on parts of the feature vector, without any manual segmentation. For every step, the attention mechanism allows the network to rely on contextual features that are useful for the task at hand. Several attention mechanisms have been proposed over the years [3,8,17]. In this work, we use Chorowski attention (hybrid) [8], as it yields better results for HTR [19].

Decoder. The role of the decoder is to generate a textual sequence from the attention-aware feature vector. The decoder is a simple LSTM cell trained with an embedding layer. For each time step, the output is passed through a softmax layer to obtain probabilities for each character of the alphabet.

Training Settings. The seq2seq is trained using the hybrid loss proposed in [19], and using the Adam optimization algorithm. We use teacher forcing in the decoder during training. Early stopping is used to stop the training if no improvement is observed on the validation loss for 20 epochs. During inference, we use the best validation weights and beam search decoding with a beam size of 5.

4 Strategies for Information Extraction

In this section, we present the two main approaches compared in this work. We also introduce three additional learning strategies for joint handwriting and named entity recognition. No language modelling or post-processing is used in this work.

4.1 Comparing the *Sequential* and *Joint* Approaches

We present the two main strategies for information extraction: the *sequential approach* and the *joint approach*. These two approaches are illustrated in Fig. 3 and will be compared using the same seq2seq architecture in Sect. 5.

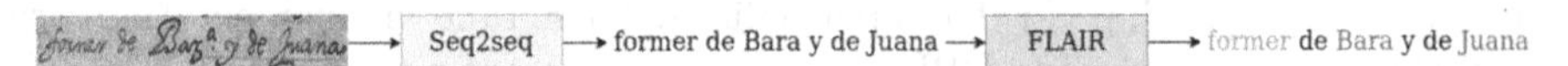

(a) *Sequential approach*: the seq2seq network is used to predict characters and FLAIR is used to classify predicted words into named entities.

Seq2seq → <occupation_WF>former de <location_WF>Bara y de <name_WM>Juana

(b) *Joint approach*: the seq2seq network is used to predict characters and tags

Fig. 3. Illustration of the two main strategies compared in this work. Legend: wife's father occupation, wife's father location, wife's mother name. Figure best viewed in color. (Color figure online)

Sequential Approach. This strategy is the most popular approach for information extraction [12]. Handwriting recognition and named entity recognition are addressed as separate subsequent tasks, as illustrated in Fig. 3a. Once handwritten text recognition is performed, each predicted word is classified into multiple semantic categories, based on textual features. Our implementation of this approach relies on our seq2seq network for handwriting recognition, and the FLAIR system [1] for named entity recognition, trained using catalan word embedding and FLAIR embedding.

Joint Approach. This approach, illustrated in Fig. 3b, was initially proposed in [7] using a CRNN-CTC neural network. In this strategy, the network is trained to predict tags as well as characters, with tags being located before each word of interest. In this work, we propose to reproduce this experiment using our seq2seq architecture trained to predict characters and tags. We believe that using an attention-based architecture is appropriate, as the network learns to focus on relevant parts of the feature vector to predict the tags. The tags are designed to encode the semantic category and the person relative to each word. We use the combined tags proposed in [7], e.g. <wife_name> is used to represent the wife's name.

4.2 Exploring Additional *Joint* Learning Configurations

Finally, we propose and evaluate three original *joint* configurations using multi-task and multi-scale training. Our intuition is that learning multiple tasks, or from multiple scales, could help the encoder to extract richer contextual features.

Joint Multi-task Strategy Without Tags. In this first learning configuration, three decoders are connected to the same encoder, as illustrated in Fig. 4a. Each decoder is trained for a specific task:

- the *htr* decoder predicts the sequence of characters
- the *category* decoder predicts the sequence of categories (6 classes: name, surname, location, occupation, state, other)
- the *person* decoder predicts the sequence of persons (8 classes: wife, father, wife's father, wife's mother, husband's father, husband's mother, other person, none)

Each decoder relies on a specific attention mechanism that focuses on relevant features for each subtask. The network is trained using a single loss function that is computed as the mean of the three individual loss functions. During inference, the sequences are merged to assign a category and a person to each predicted word.

However, this approach has two possible issues. The first is a potential convergence issue, as the *htr* decoder typically learns slower than the other decoders. The second issue comes from the alignment of the predicted sequences: an error at the beginning of a sequence can potentially offset the entire prediction.

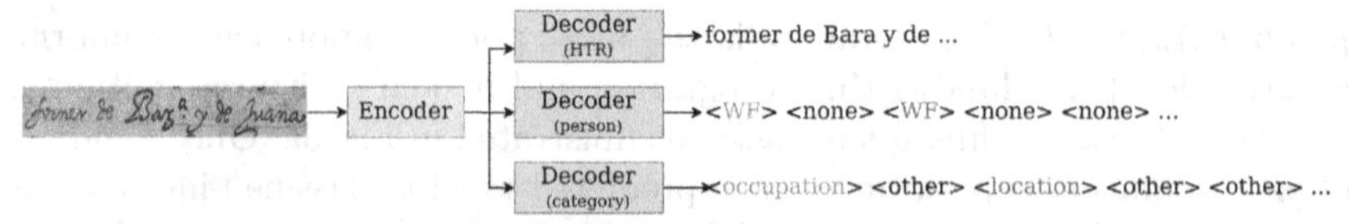

(a) ***Joint multi-task without tags*: A single encoder is connected to three decoders. Each decoder is specialised for the prediction of characters, categories and persons.**

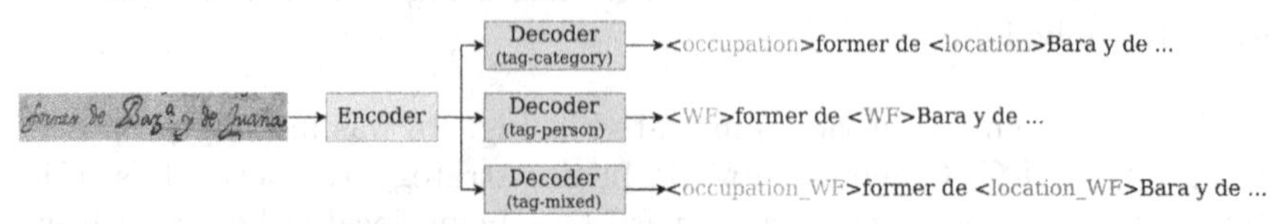

(b) ***Joint multi-task with tags*: A single encoder is connected to three decoders. Each decoder is specialised for the prediction of category-based tags, person-based tags, and mixed tags.**

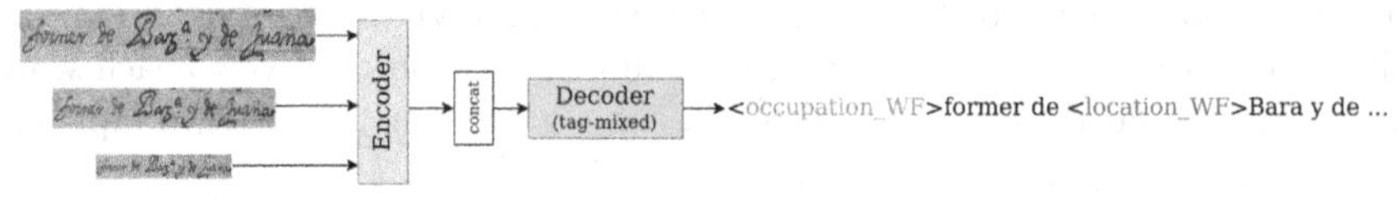

(c) ***Joint multi-scale strategy*: the image is fed to the encoder at multiple scales.**

Fig. 4. Illustration of three *joint* strategies proposed in this work. Legend: WF for wife's father

Joint Multi-task Strategy with Tags. This second learning configuration is designed to overcome the issues of the last strategy. It is illustrated in 4b. Three decoders are connected to the same encoder, and each decoder is trained using a specific semantic encoding:

- *tag-category* is a single tag for the category, e.g. <name>
- *tag-person* is a single tag for the person, e.g. <wife>
- *tag-mixed* is a combined tag for both category and person, e.g. <wife_name>

Each decoder relies on a specific attention mechanism that focuses on relevant features for each subtask. The network is trained using a single loss function that is computed as the mean of the three individual loss functions. During inference, only the *tag-mixed* branch is evaluated to ensure a fair comparison with the single-task *joint* strategy. In this scenario, the subtasks tackled by each decoder are balanced, which ensure proper convergence.

Joint Multi-scale Strategy with Tags. The multi-scale learning configuration is based on the *joint* strategy and is illustrated in Fig. 4c. In this scenario, the image is passed through the encoder at different scales to get contextual information at different levels. The three feature vectors are concatenated and passed through the attention network and the decoder.

5 Experiments

In this section, we evaluate our contributions for information extraction. First, we present results achieved by our attention-based network for handwriting recognition, showing that it is competitive. Then, we conduct a comparative study of *sequential* and *joint* approaches for information extraction based on our attention-based network, and evaluate three additional joint learning configurations. Finally, we compare this work with other participants in the IEHHR competition.

5.1 Handwriting Recognition Using Seq2seq

First, we demonstrate the interest of our seq2seq architecture for HTR, as it is used for the *sequential strategy.* The evaluation is carried out on two public databases at line-level: IAM [18] (modern, English) and RIMES [2] (modern, French). We use the Character Error Rate (CER) as the evaluation metric. We compare our architecture with other state-of-the-art methods, without any post-processing or language model, in Table 1. Our architecture yields state-of-the-art results on the IAM database. When comparing our seq2seq (encoder-decoder) with the corresponding CRNN-CTC (encoder-CTC) architecture, we observe that the seq2seq architecture is more efficient. Indeed, when using the attention-based model, the CER drops from 6.1% to 5.2% on the IAM database, and from 5.2% to 4.4% on the RIMES database. This is likely due to attention mechanism and the implicit language model learned by the decoder. Consequently, we use the seq2seq architecture in the following experiments.

Table 1. Benchmark comparison of methods for handwriting recognition at line-level on the test set and without language modeling or post-processing.

System	Method	IAM	RIMES
CRNN-CTC	Wigington et al. [28]	6.4	**2.1**
CRNN-CTC	Puigcerver [23]	5.8	2.3
CRNN-CTC	Dutta et al. [11]	5.2	5.1
CRNN-CTC	*Ours*	6.1	5.2
Seq2seq	Poulos et al. [22]	16.6	12.1
Seq2seq	Chowdhury et al. [9]	8.1	3.5
Seq2seq	Bluche [4]	7.9	2.9
Seq2seq	Michael et al. [19]	**5.2**	–
Seq2seq	*Ours*	**5.2**	4.4
Transformers	Kang et al. [15]	7.6	–

5.2 Information Extraction Using Seq2seq

We analyse the four approaches described in Sect. 4 for information extraction on the Esposalles database. Then, we compare our best approaches with state-of-the-art methods.

Evaluation Protocol. Two tasks can be evaluated using the Esposalles database [24]. The first task is full text recognition, which is evaluated using the Character Error Rate (CER) and the Word Error Rate (WER). The second task is information extraction, which can be assessed using the metrics introduced in the ICDAR 2017 competition on Information Extraction in Historical Handwritten Records (IEHHR) [12]. The objective of information extraction is to go beyond full text recognition, by assigning two semantic labels to each word. For the basic track, the aim is to recognize the category among *name*, *surname*, *occupation*, *location*, *civil state* and *other*. The score associated to the basic task is $100 - CER$ if the category is correctly identified, 0 otherwise. For the complete track, the role of the person must also be identified among: *wife*, *husband*, *wife's father*, *wife's mother*, *husband's father*, *husband's mother*, *other person*, *none.* The score associated to the complete task is $100 - CER$ if both labels are correctly identified, 0 otherwise. It should be noted that only relevant words (e.g. not classified as *other*/*none*) are taken into account in this evaluation.

Comparing the Sequential and Joint Approaches. The results presented in Table 2 compare the two main strategies presented in this work. The results highlight the interest of the *joint* strategy for information extraction task, as it yields a 3% increase on the basic score and an 8% increase on the complete score of the IEHHR competition.

We observe that the *sequential* approach tends to propagate and amplify transcription errors. Indeed, we have observed that words with errors are more likely to be misclassified: the complete score drops from 98.5% when FLAIR is applied to ground truth words to 86.7% when applied to predicted words. Another weakness of the *sequential* approach can be observed on the complete score. The score drops by 5% when the evaluation of persons is considered, suggesting that this task is particularly problematic for the FLAIR model at line level.

Overall, the strength of the *joint* approach comes from the implicit knowledge regarding the context of each word during the prediction. We observe that this knowledge helps the handwriting recognition task, as the CER drops from 2.82% when predicting only characters to 1.81% when predicting characters and tags.

Evaluation of Additional Joint Learning Configurations. We observe that the *multi-task without tags* approach obtains a high error rate, which confirms our intuition on the asymmetric convergence of the three branches: the two *ner* decoders converge much quicker than the HTR decoder. This causes either an overfitting of the two *ner* decoders or an underfitting of the *htr* decoder. Moreover, difficulties related to the alignment of the three predicted sequences may explain the low scores on the IEHHR competition. An interesting contribution of our study is the

Table 2. Comparison of various learning strategies for information extraction on the Esposalles database at line-level. The first table presents results for handwriting recognition and information extraction using scores from the IEHHR competition. The second table details these scores for each semantic category.

	CER	WER	Basic score	Complete score
Sequential	2.82	8.33	91.2	86.7
Joint	1.81	6.10	94.7	94.0
Joint multi-task without tags	7.75	17.38	61.8	48.1
Joint multi-task with tags	**1.74**	**5.38**	**95.2**	**94.4**
Joint multi-scale with tags	5.61	15.13	83.0	80.3

	Name	Surname	Location	Occupation	State
Sequential	94.6	85.9	91.7	92.3	90.4
Joint	96.1	91.0	93.7	**95.3**	**97.8**
Joint multi-task without tags	63.2	41.0	48.2	69.8	86.7
Joint multi-task with tags	**97.0**	**92.6**	**94.5**	**95.3**	96.7
Joint multi-scale with tags	87.4	61.1	84.2	87.5	96.9

joint multi-task with tags strategy. This learning strategy yields better results than the *joint* strategy, even though we only evaluate the *tag-mixed* branch. This highlights that the network benefits from learning different semantic representations at once. This result also confirms the observation of Luong et al. [16] who observed that multi-task learning improves the performance of seq-to-seq models for neural translation. However, we must relativize the small improvement of this strategy with its high computational cost. Finally, the *joint multi-scale with tags* strategy does not meet our expectations. Our intuition was that extracting features at different scales would help to get more contextual information. However, we observe that this makes it more difficult for the attention network to select relevant features. As a result, the final performance is quite poor.

Benchmarking Information Extraction (IEHHR). We now compare our work with other participants in the IEHHR competition in Table 3. Our *joint multi-task with tags* strategy achieves state-of-the-art results at line-level, without any post-processing and language model. This result shows the interest of multi-task learning for information extraction, despite a high computational cost.

Another interesting point comes from the comparison of our *joint* strategy with the articles that rely on the same strategy using a different architecture, mainly CRNN-CTC [7] and Transformer [25]. When compared to the CRNN-CTC [7], our seq2seq model is able to boost the complete score from 89.40% to 94.0% using the same methodology. When compared to the Transformer [25],

Table 3. Benchmark comparison of the IEHHR competition on the Esposalles database at line level.

Method	Model	Strategy	Basic score	Complete score
Baseline HMM [12]	HMM+LM	Sequential	80.2	63.1
CITlab-ARGUS-1 [12]	CRNN-CTC	Sequential	89.5	89.2
CITlab-ARGUS-2 [12]	CRNN-CTC	Sequential	91.9	91.6
CITlab-ARGUS-3 [12]	CRNN-CTC	Sequential	91.6	91.2
CVC (tags) [7]	CRNN-CTC	Joint	90.6	89.4
InstaDeep (tags) [25]	Transformer	Joint	**95.2**	93.3
Joint (ours)	Seq2seq	Joint	94.7	94.0
Joint multi-task with tags (ours)	Seq2seq	Joint	**95.2**	**94.4**

our seq2seq model obtain competitive results, with a lower basic score but a higher complete score. This observation highlights the interest of attention-based architectures, such as seq2seq or Transformer, for information extraction.

Another valuable observation comes from the analysis of the attention maps. We observe that the attention is very narrow and well focused on the corresponding pixels when predicting characters. However, when the network predicts semantic tags, the attention spreads over the previous words, showing that it learns to attend over relevant visual features to predict semantic categories.

It should be noted that a work submitted on the IEHHR competition website[3] outperforms our approaches, although the methodology has not been published yet. Moreover, the joint approach based on a Transformer model [25] achieves state-of-the-art results at record-level, as the network benefits from a larger bi-dimensional context.

6 Conclusion

This study compares joint and sequential approaches for information extraction. Our results demonstrate the interest of using *joint* approaches, as we show that training a network for handwriting recognition and named entity recognition increases performance on both tasks. Compared to the traditional *sequential* approach, our *joint* strategy yields to an 8% increase in complete recognition score and a significant decrease of the Character Error Rate (from 2.82% to 1.81%). In addition, this work highlights the interest of seq2seq architectures. Indeed, we obtain a substantial performance increase when the *joint* strategy is applied using a seq2seq network, as compared to the CRNN-CTC approach [7]. This is because seq2seq networks rely on an attention mechanism to extract relevant visual features, as well as a recurrent decoder to learn the implicit language model. Its performance is comparable to the Transformer proposed in [25], although our approach yield a better complete score. Finally, we explore different joint learning configurations and observe that multi-task learning from

[3] https://rrc.cvc.uab.es/?ch=10.

multiple semantic encodings helps the network to extract relevant features for each task. Indeed, our *joint multi-task with tags* approach yields a complete score of 94.4% on the IEHHR competition at line-level. As a consequence, we believe that multi-task seq2seq architectures should be investigated in more depth. We obtain state-of-the-art results on the IEEHR competition [12], without any post-processing or external language modeling. We believe that future work should also focus on information extraction at paragraph-level to take advantage of the recurrent phrasing at record-level. Recent work shows that relying on incremental learning strategies could ease information extraction at record-level [25].

Acknowledgements. Solène Tarride is partly funded by the CIFRE ANRT grant No. 2018/0896.

References

1. Akbik, A., Bergmann, T., Blythe, D., Rasul, K., Schweter, S., Vollgraf, R.: FLAIR: an easy-to-use framework for state-of-the-art NLP. In: NAACL Annual Conference of the North American Chapter of the Association for Computational Linguistics (Demonstrations), pp. 54–59 (2019)
2. Augustin, E., Brodin, J.M., Carré, M., Geoffrois, E., Grosicki, E., Prêteux, F.: RIMES evaluation campaign for handwritten mail processing. In: Proceedings of the Workshop on Frontiers in Handwriting Recognition (2006)
3. Bahdanau, D., Cho, K., Bengio, Y.: Neural machine translation by jointly learning to align and translate (2016)
4. Bluche, T.: Joint line segmentation and transcription for end-to-end handwritten paragraph recognition (2016)
5. Bluche, T., Messina, R.: Gated convolutional recurrent neural networks for multilingual handwriting recognition. In: 2017 14th IAPR International Conference on Document Analysis and Recognition (ICDAR), vol. 01, pp. 646–651 (2017)
6. Carbonell, M., Fornés, A., Villegas, M., Lladós, J.: A neural model for text localization, transcription and named entity recognition in full pages. Pattern Recogn. Lett. **136**, 219–227 (2020)
7. Carbonell, M., Villegas, M., Fornés, A., Lladós, J.: Joint recognition of handwritten text and named entities with a neural end-to-end model. CoRR abs/1803.06252 (2018)
8. Chorowski, J., Bahdanau, D., Serdyuk, D., Cho, K., Bengio, Y.: Attention-based models for speech recognition. In: Proceedings of the 28th International Conference on Neural Information Processing Systems, NIPS 2015, vol. 1, p. 577–585. MIT Press, Cambridge (2015)
9. Chowdhury, A., Vig, L.: An efficient end-to-end neural model for handwritten text recognition. CoRR abs/1807.07965 (2018)
10. Deng, J., Dong, W., Socher, R., Li, L.J., Li, K., Fei-Fei, L.: ImageNet: a large-scale hierarchical image database. In: 2009 IEEE Conference on Computer Vision and Pattern Recognition, pp. 248–255. IEEE (2009)
11. Dutta, K., Krishnan, P., Mathew, M., Jawahar, C.: Improving CNN-RNN hybrid networks for handwriting recognition. In: 16th International Conference on Frontiers in Handwriting Recognition, pp. 80–85 (2018)

12. Fornés, A., et al.: ICDAR2017 competition on information extraction in historical handwritten records. In: 2017 14th IAPR International Conference on Document Analysis and Recognition (ICDAR), vol. 01, pp. 1389–1394 (2017)
13. Graves, A., Fernández, S., Gomez, F., Schmidhuber, J.: Connectionist temporal classification: labelling unsegmented sequence data with recurrent neural networks. In: Proceedings of the 23rd International Conference on Machine Learning, ICML 2006, pp. 369–376. Association for Computing Machinery, USA (2006)
14. Graves, A., Liwicki, M., Fernández, S., Bertolami, R., Bunke, H., Schmidhuber, J.: A novel connectionist system for unconstrained handwriting recognition. IEEE Trans. Pattern Anal. Mach. Intell. **31**(5), 855–868 (2009)
15. Kang, L., Riba, P., Rusiñol, M., Fornés, A., Villegas, M.: Pay attention to what you read: non-recurrent handwritten text-line recognition. CoRR abs/2005.13044 (2020)
16. Luong, M.T., Le, Q., Sutskever, I., Vinyals, O., Kaiser, L.: Multi-task sequence to sequence learning. In: Proceedings of ICLR, San Juan, Puerto Rico, November 2015
17. Luong, M.T., Pham, H., Manning, C.D.: Effective approaches to attention-based neural machine translation (2015)
18. Marti, U.V., Bunke, H.: The IAM-database: an English sentence database for offline handwriting recognition. Int. J. Doc. Anal. Recogn. **5**, 39–46 (2002)
19. Michael, J., Labahn, R., Grüning, T., Zöllner, J.: Evaluating sequence to sequence models for handwritten text recognition. CoRR abs/1903.07377 (2019)
20. Nauta, G.J., Heuveland, W.V.D., Teunisse, S.: Europeana DSI 2-access to digital resources of European heritage - report on ENUMERATE core survey 4 (2017)
21. Palm, R.B., Laws, F., Winther, O.: Attend, copy, parse end-to-end information extraction from documents. In: 2019 International Conference on Document Analysis and Recognition (ICDAR), pp. 329–336 (2019)
22. Poulos, J., Valle, R.: Character-based handwritten text transcription with attention networks. CoRR abs/1712.04046 (2017)
23. Puigcerver, J.: Are multidimensional recurrent layers really necessary for handwritten text recognition? In: 14th IAPR International Conference on Document Analysis and Recognition (ICDAR), vol. 01, pp. 67–72 (2017)
24. Romero, V., et al.: The ESPOSALLES database: an ancient marriage license corpus for off-line handwriting recognition. Pattern Recogn. **46**, 1658–1669 (2013)
25. Rouhou, A.C., Dhiaf, M., Kessentini, Y., Salem, S.B.: Transformer-based approach for joint handwriting and named entity recognition in historical document. Pattern Recogn. Lett. **155**, 128–134 (2021)
26. Sutskever, I., Vinyals, O., Le, Q.V.: Sequence to sequence learning with neural networks. In: Proceedings of NIPS, Montreal, CA (2014)
27. Toledo, J.I., Carbonell, M., Fornés, A., Lladós, J.: Information extraction from historical handwritten document images with a context-aware neural model. Pattern Recogn. **86**, 27–36 (2019)
28. Wigington, C., Tensmeyer, C., Davis, B.L., Barrett, W., Price, B.L., Cohen, S.: Start, follow, read: end-to-end full-page handwriting recognition. In: ECCV (2018)
29. Xu, K., et al.: Show, attend and tell: neural image caption generation with visual attention. CoRR abs/1502.03044 (2015)
30. Yu, W., Lu, N., Qi, X., Gong, P., Xiao, R.: PICK: processing key information extraction from documents using improved graph learning-convolutional networks. In: 2020 25th ICPR, pp. 4363–4370 (2021)

QAlayout: Question Answering Layout Based on Multimodal Attention for Visual Question Answering on Corporate Document

Ibrahim Souleiman Mahamoud[1,2](✉), Mickaël Coustaty[1], Aurélie Joseph[2], Vincent Poulain d'Andecy[2], and Jean-Marc Ogier[1]

[1] La Rochelle Université, L3i Avenue Michel Crépeau, 17042 La Rochelle, France
{ibrahim.souleiman_mahamoud,mickael.coustaty,jean-marc.ogier}@univ-lr.fr
[2] Yooz, 1 Rue Fleming, 17000 La Rochelle, France
{aurelie.joseph,vincent.poulaindandecy}@getyooz.com

Abstract. The extraction of information from corporate documents is increasing in the research field both for its economic aspect and a scientific challenge. To extract this information the use of textual and visual content becomes unavoidable to understand the inherent information of the image. The information to be extracted is most often fixed beforehand (i.e. classification of words by date, total amount, etc.). The information to be extracted is evolving, so we would not like to be restricted to predefine word classes. We would like to question a document such as "which is the address of invoicing?" as we can have several addresses in an invoice. We formulate our request as a question and our model will try to answer. Our model got the result 77.65% on the Docvqa dataset while drastically reducing the number of model parameters to allow us to use it in an industrial context and we use an attention model using several modalities that help us in the interpertation of the results obtained. Our other contribution in this paper is a new dataset for Visual Question answering on corporate document of invoices from RVL-CDIP [8]. The public data on corporate documents are less present in the state-of-the-art, this contribution allow us to test our models to the invoice data with the VQA methods.

Keywords: Visual question answering · Multimodality · Attention mechanism

1 Introduction

Imagine a near future where you will not process any document but your digital clone will do it for you. It will summarize and/or extract the relevant and useful information for you. For some, this future seems close while for others it is a simple illusion. Some companies have already started this transition such as the

© Springer Nature Switzerland AG 2022
S. Uchida et al. (Eds.): DAS 2022, LNCS 13237, pp. 659–673, 2022.
https://doi.org/10.1007/978-3-031-06555-2_44

New Zealand company UneeQ [16], which created an avatar of Albert Einstein that you can interact with on various topics.

If we go back to the present, companies try to provide easy-to-use solutions to extract information from the corporate documents. These corporate documents are varied in both content and form (i.e. invoices, order form, resume, pay slip etc.). Thus the customer processing are evolving everyday, and if they today only focus on the extraction of information from invoice, tomorrow they will ask for more and more intelligent process of documents (i.e. for instance the automatic processing of collaborators resume, or the automatic linking of file folders). We would therefore like to have a method to extract information with few data and able to adapt to another type of document while taking into account the changes in layout and content to be extracted.

To answer these problems, various state-of-the-art papers have tried to provide a solution. Some few-shot learning [19] methods were trained only a corpus composed of a hundred of documents. These methods have all shown their weakness when used with a large dataset. Some other methods, based on deep-learning techniques such as Lambert [6] or LayoutLM [22], have appeared recently. Their performances were impressive but with the drawback that these methods require a lot of data to converge. A very large dataset of annotated documents is not always available and in order to deal with this limitation, some recent works proposed to use incremental methods [18] which can evolve over time.

We propose in this paper a method to process different types of documents and can also extract the different information by the customer for a question answering model. The usefulness of visual question answering on corporate documents unlike predefined extraction is to allow to extract more general information that can be adapted to a new corpus of data.

In general when we read a paper whether it is scientific or other, some questions come to mind. For example for our paper, if you try to understand our contribution you quickly read the abstract or propose method to have an answer. This natural ability to focus on a part of the information from questions is innate to us we look at what is essential to solve a problem. It is by being inspired by this that the mechanism of attention has seen the day several papers such as [17] are sold the merit of mechanism of attention. The attention mechanism is used in several situations, some to optimize the performance of their model, others to be able to interpret the model and thus brings some element of answer to the behavior of this black box. The visual question answering is known to be very popular in the community because many problems can be solved with it, from simple questions on images of natural scenes [7] to medical assistance [11] to help specialties to better understand some images thanks to the strength of accumulating a large amount of information to synthesize and then to keep it in its memory of artificial neural networks.

The visual questions answering approaches on corporate documents are minimal and the data also concerns this subject. In this article we will describe our method based on the Qanet architecture [23]. The key motivation behind the design of our model is as follows: The convolution layer helps to capture the local structure of the image and text. The co-attention layer allows to have a global relationship between the inputs in order to define the positive or negative impact of one in relation to the other. We have evaluated our model on the Squad [14] and Docvqa [13] datasets. The choice of these two datasets is that one is only textual, it will allow us to test and optimize the textual part of our model while the Docvqa is a corpus of documents of different type, it will allow us to test if our model will manage to use well this multimodal corpus.

The contributions of this paper are summarized as follows:

- We have made available a corpus of Visual question answering data VQA-CD for corporate documents, this corpus is to our knowledge the first in the state-of-the-art. We have annotated 3 thousand questions from ~693 documents.
- We propose a multi-modal co-attention model for visual question answering. This template will use the visual and textual content features of this document and the layout features for each word. This model learns the best way to use cross-modality to predict the answer of a question. We use this self-attention to focus our network on common features from the input. This will allow us to exploit the context and query correlation at the initial stage.

2 Related Work

The approaches proposed for VQA are generally distinguished in three categories. Some use a single modality like Qanet [23]. Although these methods show good performances by using some transformer-based model [17], they remain less effective when the multimodal understanding is necessary.

The second category of methods proposes to rely on multimodal architectures in order to be able to deal with the visual and the textual content at the same time [20]. These methods have then be designed to include a second modality in the proposed architecture. Even if only these models require better performances, this can only be done by using large dataset and they require to re-train the model each time new kind of input is used. This relies on the fact that the document layout may vary in a significant way and this is not taken into account. For example, if you train a model on a dataset where the addresses are generally located at the top of documents, this one will have difficulties to find them elsewhere (addresses could be located at the bottom, or in the middle of new documents and dataset). Such kind of models then need to be re-trained on new dataset which should come with its annotations.

To overcome this problem, some recent models like LayoutLM [22] have been introduced to be able to add a third modality representing the layout information of documents. The most recent and popular model from this third category includes the famous LayoutLM [22], LamBERT [6] and ViBERTgrid [10]. LAMBERT [6] proposed a model based on the Transformer encoder architecture RoBERTa [12]. The main contribution of this paper was to propose a general-purpose language model that views text not simply as a sequence of words, but as a collection of tokens on a two-dimensional page by applying relative attention bias. In their industrial context the use of text, bounding box and therefore not image allows to eliminate an important performance factor in industrial systems. They have conducted several evaluations on several public datasets The Kleister NDA and Kleister Charity, SROIE and CORD. A deep experimental study allows comparing it to other state-of-the-art methods such as their baseline Roberta, LayoutLM [22], and LayoutLMv2 [21]. Although lambert is a good method, it does not take into account the whole image and the correlation that can exist between the layout information and the textual content. ViBERTgrid [10] proposed a new multi-modal network by combining the best of BERTgrid and CNN to generate a more powerful grid-based document representation. It simultaneously encode the textual, layout and visual information of a document in a 2D feature map.

In order to go one step further, and to reduce the gap between users' need and the documents, we propose to integrate a question answering approach in this kind of architecture. To the best of our knowledge, and as discussed before, different 2D representation of documents have been proposed in the litterature but none of them integrate a question answering in the process and no works have proposed to include an attention mechanism mixing the question and the 2D representation.

3 Problem Definition

The main objective of our proposal is to provide solutions for the automatic comprehension of documents that requires both a visual analysis and a semantic understanding of their content. The visual analysis remains essential to capture some contextual information. In parallel, the document layout is necessary to extract the correct word in the document (like when a human distinguish two similar tokens based on their visual context). The other hand, automatic comprehension means being able to provide an answer to a question about the content of the document. Starting from a visual context with D documents, we can define the document image $I = i_1, i_2, .., i_d$, its associated bounding boxes (one for each word) $B = b_1, b_2, .., b_m$ and its textual information based on the extracted text (using an OCR) T $= t_1, t_2, .., t_m$ with m being the number of words from the document. We also propose to define the query sentence (with k words) Q $= q_1, q_2, ..; q_k$ which then correspond to the request that a user could submit to the system (what he/she is looking for). The questions are related to the corporate document field (i.e. What is the total amount of this invoice? What is this type of document?) and the answer varies according to the question they can be categorical, numerical or textual.

$$I \in R^d, B \in R^{d \times m}, T \in R^{d \times m}\, and\, Q \in R^{d \times k} \quad (1)$$

The questions refer to the document's content. The answers are therefore information extracted from the text present in the document. The objective of our proposed prediction task is then to predict the present beginning and ending word of the answer present in the text. The general assumption is that an unknown function correlates the samples of the questions with the prediction of the beginning word $s1$ and the end word $s2$, i.e. $(s1, s2) = f(Q)$. The goal of the learning process is to provide an approximation of this unknown function. To better approximate this function f, we need to know the correlation between the question and the answer from the visual and textual features and from the similarity between the questions.

4 Proposed Approach

In order to set up our proposed model, we got inspired by state-of-the-art models such as Qanet [23] and LayoutLM [22] to propose an architecture based on textual and visual attention models. As we tend to address industrial tasks, where the time needed to process each document must remain low, we only use convolutional and self-attention mechanisms, discarding recurrent neural network (known as slower architectures as they can not process the input tokens in a parallel way).

Another limitation from the most recent and relevant state-of-the-art models relies on their large number of parameters (i.e. 300M parameters for LayoutLMV2). Setting up a model mixing visual content, the semantic, the layout and the question would then require more than 600 million data to converge. We would like to remember that in industrial context, no large annotated dataset could be available, and they would then require a huge effort of annotation to link question and content. We then propose to develop a model able to learn with few data (e.g. hundreds or thousands of annotated documents) while maintaining the execution time as low as possible (generally, companies may not devote more than a second to each document).

Several studies have been conducted to know if we could establish a correspondance between the number of training data and the number of model parameters. The paper [24] conducted several tests on hypothesis related to the generalization capacity of neural networks. The authors demonstrated that "Theoretically, a simple two-layer neural network with $2n + d$ parameters is capable of perfectly fitting any dataset of n samples of dimension d". We can then observe that state-of-the-art approaches have much more than $2n + d$ parameters and one of our hypothesis is to propose a model with much less parameters (less than 10 Millions). In practice, finding the optimal number of parameters is not an easy matter. It depends on the diversity and number of data available in the training set and on the selected architecture. Models with many parameters are likely to overfit while they have the advantage of being able to model much more complicated knowledge (like some latent relationship between the data). Methods with a few numbers of parameters (i.e. less than 20M) are not much studied in the state-of-the-art especially applied on the corporate document. Our second contribution is then to propose a model having a maximum number of 8 millions parameters and which having almost similar results to the state-of-the-art method thanks to our proposed attention mechanisms.

In a first step, we will present the global architecture. Then we will describe the used encoder architecture, and finally we will discuss the proposed co-attention.

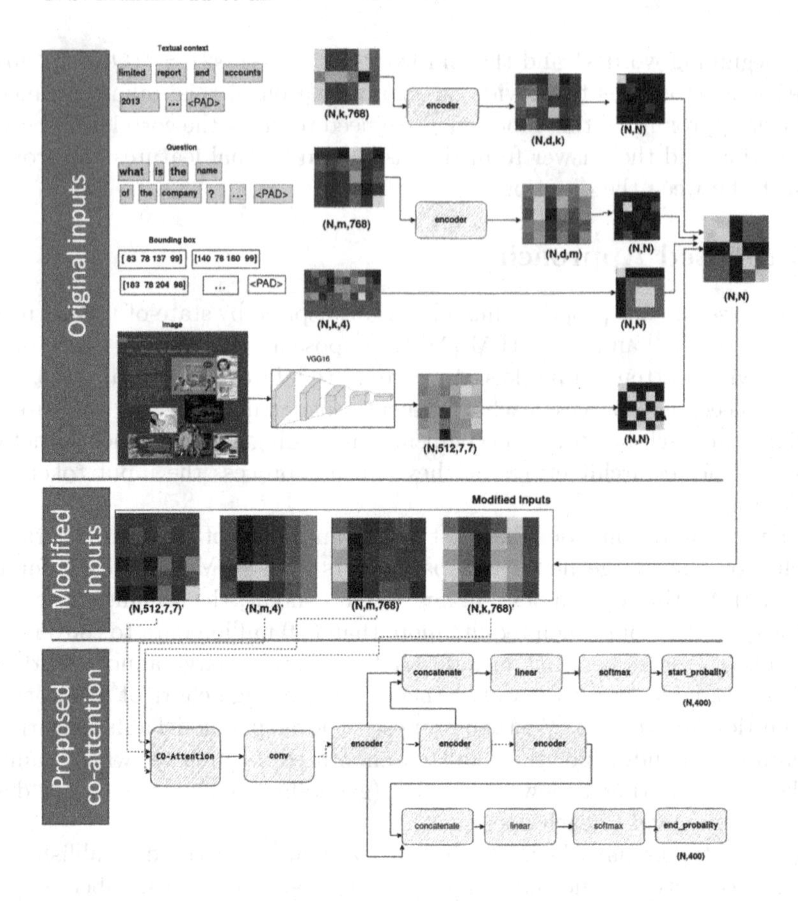

Fig. 1. Description of the QAlayout model. The inputs are the textual context extracted from the image, the visual context is whole image. Also the layout information (i.e. bounding box of each word) and finally the question. We will try to predict two probality the beginning and end of the answer from the word present in the textual content. We have a self-attention and co-attention mechanism to learn multimodal comprehension

4.1 Global Description

The global architecture, presented in Fig. 1, describes the inputs, the intermediate layers and the prediction layer. This last step predicts the probality of having the beginning and the end of the answer. The model has different input features (denoted T, Q and B and I in the Eq. 2).

The features of each sample are decomposed as follows: $T \in R^{D\times m\times d_1}$ for the textual context, $B \in R^{D\times m\times d_2}$ is the dimension of bounding box for each word of the textual content, Q is the question input with dimension $R^{D\times k\times d_1}$ and I is the image of the document with the dimension $R^{D\times d_3}$. The dimensions d_1, d_2 and d_3 are the respective feature dimensions for the textual content, the bounding boxes and the visual parts. D is the document size of the dataset (i.e. $1 \leq i \leq D$).

The probality of the beginning and the end of the answers corresponding the label as $y_i \in R^{2\times m}$. The beginning and the end correspond to the first and last words that the answer will contain, which is a selection from the words present in the document. The symbole m corresponds the total number of word in the textual context.

The textual content of a document is different, we can have a document with ten words while others exceed 500 words. So that our entries are of the same dimension we keep the prepocessing method described in Qanet to keep one size for all entries. For the semantic embedding of each document, we chose to keep the 400 first words. If the number of word is lower than this, we add some padding tags. We justify this choice of 400 words as we observed that the average numbers of words in corporate documents is around 300. This parameters could be obviously modified to other context, but we use this value even for larger datasets like DocVQA. In a similar way, we chose to limite the question size to the 50 first words. Once the text has been extracted, we use the classical BERT embedding features [5], where each word is represented by 768 (d_1) values. We then obtain a total size of (400, 768) for the textual context, and a total size of (50, 768) for questions.

For the bounding boxes, we used the outlines of the word from the original image and we normalized them to the width and height of the image. We then obtained a two dimensions vector d_2 composed of their cartesian coordinates $(x_{max}, y_{max}, x_{min}, y_{min})$.

The last part of the documents input relies on the visual content. To this end, we resized document images to a dimension vector $d_3 = (224, 224)$. We then use a VGG16 convolutional network to embedded the visual content to obtain a (512, 7, 7) feature dimension vector. This vector is then provided as an input to our co-attention model.

$$\begin{aligned} I &= [i_1, i_2, ..., i_n] \in \mathbf{R}^{D\times d_3} \\ T &= [[t_1, t_2, ..., t_m], ..] \in \mathbf{R}^{D\times m\times d_1} \\ B &= [[b_1, b_2, ..., b_m], ..] \in \mathbf{R}^{D\times m\times d_2} \\ Q &= [[q_1, q_2, ..., q_k], ..] \in \mathbf{R}^{D\times k\times d_1} \end{aligned} \tag{2}$$

As described in the Fig. 1, it's following end-to-end operations:

- In a first step, the input described in the equation goes through models such as Bert and Vgg16 to extract the relevant embedding for our model.
- Then all the input text tensors are passed to the embedding encoding layer which is a single encoding block with 4 conv layers. 8 attention heads are used in the auto-attention module which is the same for all encoding blocks of the model.
- Finally we use self-attention to transform our input into new input, allowing us to exploit the correlation between question and context (e.g. textual context, bounding box and image) at the initial stage.

The self-attention mechanism used in this paper is inspired by [3]. They propose a self-attention mechanism on the input image to consider the inherent correlation (attention) between the input features themselves, and then use a graph neural network for the classification task.

Self-attention allows us to transform T, B, Q, I into T', B', Q', I' which will be the inputs to our co-attention model. To do that, we follow several steps, the first one consists in computing the correlation matrices between the sample and the label.

$$\begin{aligned} C^t &= softmax(TT^T) \\ C^i &= softmax(II^T) \\ C^b &= softmax(BB^T) \\ C^q &= softmax(QQ^T) \end{aligned} \tag{3}$$

Here softmax(·) denotes a softmax operator. The inputs of this function have all the same dimension where BB^T, QQ^T, II^T *and* $TT^T \in R^{N \times N}$

The self-attention module exploits C^t, C^i, C^b, C^q (see Eq. 3).

$$C^m = fusion([C^t, C^i, C^b, C^q]) \in R^{N \times N} \tag{4}$$

where $[C^t, C^i, C^b, C^q]$ denotes the attention map concatenation. This fusion function $fusion$ is equivalent e.g. $C^m = w_1 C^t + w_2 C^i + w_3 C^b + w_4 C^q$, where the weighted parameters w_1, w_2, w_3 and w_4 are learned adaptively and N is batch size.

$$\begin{aligned} T' &= TC^m \\ I' &= IC^m \\ B' &= BC^m \\ Q' &= QC^m \end{aligned} \tag{5}$$

These modified inputs as indicated in the Eq. 5 will be reused in the co-attention part of the model. In the following sections we will detail the encoder part and the co-attention part.

4.2 Encoder

As presented in Fig. 2, the encoder is composed of several convolution layers and a attention layer. Unlike the traditional convolution layer, we use depthwise separable convolution [4]. In this paper [4] depthwise describes that this convolution layer is memory efficient and has better generalization. That help us because this model will be used in a production system and it should be light and fast. For this work, we chose to set up the kernel size to 7, the number of filters to d = 128 and the number of convolutional layers within a block to 3. The output of this convolutionnal model is then transfered to an attention layer. We adopt the multi-headed attention mechanism defined in [17] which, for each position in the input, called a query, computes a weighted sum of all positions, or keys, in the input based on the similarity between the query and the key, as measured by the system. The scalar product.

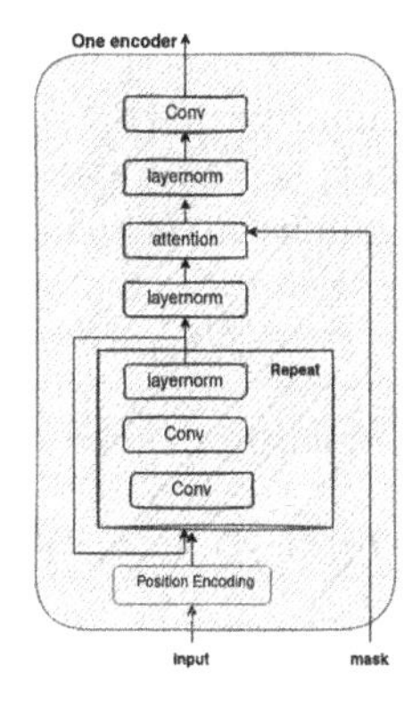

Fig. 2. One encoder

As one can see in the Fig. 2, the model involves residual connections, layer normalizations and dropouts too. For each input x and a given operation f, the f is defined as $f = (layernorm(x)) + x$ [1] of identity at the input and output of each block repeat. This block is repeated 7 times.

4.3 Co-attention

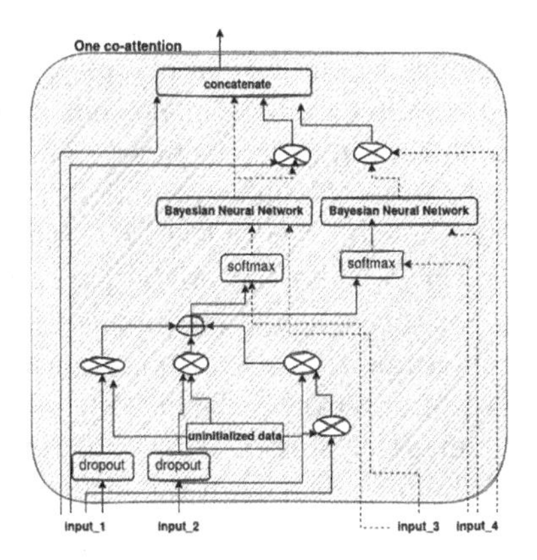

Fig. 3. Description of the one co-attention

The proposed "Co-attention" step proposed in our work is inspired by the attention flow layer from the BIDAF architecture [15] (see Fig. 3. It calculates attention in two directions. Context-query attention tells us what query words are the most relevant to each context word like descripted in Eq. 6 and 7. The input_1 (i_1) represents the textual context, the bounding box of each word is input_3(i_3) and input_2 (i_2) represents the question finally input_4 (i_4)the whole box and encoded context and query respectively. Given that the context length is j and query length is m, a similarity matrix is calculated first. The similarity matrix captures the similarity between each pair of context and query words. It is denoted by S and is a n-by-m matrix. The similarity matrix is calculated as,

$$S = f(i_1, i_2) \tag{6}$$

where f is a trilinear similarity function defined as,

$$f(i_1, i_2, i_3, i_4) = W_0[i_1; i_2; i_1 \cdot i_2, i_3 \cdot i_4] \tag{7}$$

5 Experiments

5.1 Dataset

Proposed Dataset. VQA-CD is a new public dataset containing 3000 questions extracted ~693 documents from RVL-CDIP [9]. To the best of our knowledge, no public visual question answering dataset exists for corporate documents. We then decided to annotate and to share our work with the scientific community. This dataset is based on the public RVL-CDIP [9] dataset. This document extracted from RVL-CDIP [9] invoice class does not only contain invoices but also other types of document such as purchase order.

The documents found in VQA-CD dataset contains some documents are well structured while others contain some semi-structured or raw text. This heterogenity of content mimics the industrial context.

When we annotated, we were careful to make sure that each question could be repeated on other documents in order to have a balanced corpus. This balanced corpus is necessary because the inherent understanding of language forces the model to focus on the question rather than the image. For example, if the question is "How much is the ..." and the prediction of the model is a number even if this number does not correspond to the expected one, the model assigns a high probability to it, the same for other type of question ("who is ...", "is it ...?", etc.) If for example the question is "What is the total amount" and this question is found in two images I and I/. If the answer related to two images differs then the model will be forced to learn the visual feature to distinguish it.

In order to ensure a kind of compatibility with the DocVQA dataset, we use the same organization of the dataset. We separated our questions for each document randomly. We then divided our 3000 questions into train, test and validation. We took 50% for train and 25% for validation and test.

SQuAD Dataset. Stanford Question [14] Answering Dataset is a reading comprehension dataset consisting of questions posed by crowdworkers on a set of Wikipedia articles. The answer to each question is a segment of text.

This dataset consists of 100,000 questions. The corpus content is only text and no layout or image information is available.

DocVQA Dataset. is to our knowledge the most complete dataset in both content and number of samples. The dataset consists of 50,000 questions defined on 12,000 document images. In the Fig. 4 you will notice the most recurrent words in the questions and also in the answers. So we can see our questions are often linked to a date or an amount inside in the document.

Fig. 4. Word clouds of words in answers (left) and questions (right).

Table 1. This table contains the results of the proposed QAlayout model and the results of the state-of-the-art method (LayoutLm, Bert).

Modality	Data		SQUAD	DOCVQA
Method		Param	F1-SCORE	ANLS
Text only	*Bert*	~110M	74.430%	45.57%
	QAlayout(Only_Text)	~1M	82.19%	48.63%
Text + Layout + Image	*LayoutLMv2*	~426M	–	86.72
	LayoutLM	~160M	–	68.93
	QAlayout(All_inputs)	~8M	–	77.65%

Implementation Details. The training phase of our model was run with the ADAM optimizer with $\beta 1 = 0.8, \beta 2 = 0.999$ and a batch size of 64, an initial learning rate of 10−3 scaled from 0.1 every 3 epochs without improvement in validation loss and an early stopping after 5 epochs without improvement.

5.2 Performance Evaluation

To evaluate the performance of our models we will use three metrics (ANLS, F1-SCORE, EM). Average normalized Levenshtein (anls) measure the distance between two string (q_k, p_k) see the Eq. 8. Where Q is the total number of questions. Each question can have 3 answers and k words, q_k the ground truth answers and p_k is the prediction of model. $NL(q_k, p_k)$ is the Normalized Levenshtein distance between ground truth and the prediction. Then a threshold $\tau = 0.5$ to filter NL values larger than τ by returning a score of 0 (Table 2).

$$ANLS = \max_{1..3} s(q_k, p_k)$$
$$s(q_k, p_k) = \begin{cases} (1 - NL(q_k, p_k)) & \textbf{if} \quad NL(q_k, p_k) < \tau \\ 0 & \textbf{if} \quad NL(q_k, p_k) \geq \tau \end{cases} \tag{8}$$

Table 2. The result for the different categories of question-answer in the Dovqa

Method	Figure/Diagram	Form	Table/List	Layout	Free_text
Bert	22.33%	52.59%	26.33%	51.13%	77.75%
QAlayout(Only_Text)	18.53%	56.12%	35.85%	53.42%	74.75%
QAlayout(All_inputs)	39.20%	73.21%	86.21%	63.20%	71.96%
Method	Image/Photo	Handwritten	Yes/No	Others	
Bert	48.59%	35.65%	3.45%	5.778%	
QAlayout(Only_Text)	30.10%	37.26%	17.24%	34.50%	
QAlayout(All_inputs)	44.66%	62.82%	59.07%	67.49%	

$$precision = \frac{1 * same\_word}{tail(p_k)} \quad recall = \frac{1 * same\_word}{tail(q_k)} \quad f1 = \frac{2 * precision * recall}{(precision + recall)} \tag{9}$$

The metric F1-SCORE is descripted in 9. Where same_word count the number of similar words between GT and the prediction.

In the Table 1, we have the results on the corpus squad containing only text and the corpus Docvqa containing image and text.

First we tried to compare the performance of the text-only part using the context text and the question only (QAlayout(Only_Text)) with Bert [5].

When we compare our QAlayout(Only_Text) to the state-of-the-art Bert [5] model on the squad corpus we get better results as described in the section. These good performances also add up to a much faster training time. The different attention mechanisms that we have detailed in Sect. 4.1 have largely contributed to these results. QAlayout(Only_Text) despite its good performance has limitations. These limitations in a text-only corpus may be due to not understanding the syntactic structure. For example if the question is *"What is the name of the Bungie Inc. founder who is also a university graduate?"* and the context contains the following words: *"In the arts and entertainment, minimalist composer Philip Glass, dancer, choreographer and leader in the field of dance anthropology Katherine Dunham, Bungie founder and developer of the Halo video game series Alex Seropian, ..."* our model predict "Katherine Dunham" while the correct prediction is "Alex Seropian". Although Katherine is close to the word founder this does not grant her the status of founder. QAlayout(Only_Text) also has other limitations related to the question related to the document layout. For example in the result table, we notice that the QAlayout(Only_Text) is good

Table 3. The results VQA-CD dataset with different metrics

Method	ANLS	F1-SCORE	Exact
QAlayout(Only_Text)	36.29%	29.16%	25.58%
QAlayout(All_inputs)	42.54 %	35.92%	33.01%

when the question is related to free_Text and has trouble with the question related to Figure or Form.

The QAlayout method using the image and the characteristic bounding box in addition to the QAlayout(Only_Text) is that allowed to provide answers to the question related to visual structure of the text. The performance has greatly improved on the question related to figure/diagram or form.

In the Table 3 we also get the performance of our model on the VQA-CD dataset.

In our VQA for corporate documents, we have either ~80% of the answers with one word and ~17% of contain only two words. Unlike to the other VQA task where the answer size is longer. These one-word answers will certainly result in a kiss on the F1-SCORE. You will notice that the multimodal model is better than the QAlayout(Only_Text) model 5% in all scores. For the VQA-CD corpus we have managed to compute several scores because we have the GT.

The metrics ANLS, F1-SCORE and EM have a value of 100% each if the correspondance between the prediction and GT are totally similar. Their difference comes if the prediction is different from the GT in this case for the em score will have 0% (i.e. either we have all or nothing for this score). For example if the model predicts 1200 and the GT is 100 for the metric ANLS we would have **1 - transformation cost** so NL is $1 - 0.25 = 0.75$ and as the threshold is equal to 0.5 (i.e. the same as the paper [2]) this result it will be taken into account in the final results. In a document this two numbers (1200, 100) can be two different amounts or just a case where the ocr can't extract the character 2 in the image. This score ANLS remains an approximate value that can help in some cases to limit the impact of ocr errors. Finally for the metric F1-SCORE as it is based on the token, we have calculated two tokens one at the word level and another at the character level (i.e. in Table 3 it is the token at the word level). The F1-SCORE that we obtained at character level is ~60%. It remains clearly higher than the token words. Nevertheless this F1-SCORE metric is not adapted in our task VQA for corporate document because either we take a token at word level and as the majority of our tokens are one word we end up with a very low score or we take at character level and therefore the ANLS score would be more accurate.

Although our model has a good performance, in most cases it confuses the GT amount with another amount or extracts only a part of this answer. The other errors are often OCR errors as the images VQA-CD are old with low-resolution

Also the limitations of QAlayout are also numerous. Sometimes, these limitations are due to a bad understanding of the visual part (i.e. difficulty to correlate the elements inside the image). Also the multimodality help us on some cases its performance is not yet the desired one.

6 Conclusion and Future Work

Visual question answering is a task that requires a good understanding of both visual and textual information by correlating this information with the question.

We propose a fast and accurate end-to-end method QAlayout that uses visual information the whole image document or layout as well as textual information. Our QAlayout method uses an attention mechanism to take into account the inherent correlation between the question and its visual or textual context at the input of the model with self-attention or after with co-attention. Compared to some state-of-the-art models we have much better results while having less parameters (8M). The limit number of parameters corresponds to the expectation of the industrial context which requires a fast training and prediction time while keeping a reasonable performance. We also contributed to annotate a new dataset VQA-CD containing 3000 questions on corporate documents. Despite the good performance of our model, some limitations exist and we will try to provide a solution. For example, we will build a graph system to establish links between words in the textual content or between areas in the visual content.

Acknowledgment. This research has been funded by the LabCom IDEAS under the grand number ANR-18-LCV3-0008, by the French ANRT agency (CIFRE program) and by the YOOZ company.

References

1. Ba, J.L., Kiros, J.R., Hinton, G.E.: Layer normalization (2016)
2. Biten, A.F., et al.: ICDAR 2019 competition on scene text visual question answering (2019)
3. Cheng, H., Zhou, J.T., Tay, W.P., Wen, B.: Attentive graph neural networks for few-shot learning (2020)
4. Chollet, F.: Xception: deep learning with depthwise separable convolutions (2017)
5. Devlin, J., Chang, M.W., Lee, K., Toutanova, K.: BERT: pre-training of deep bidirectional transformers for language understanding (2019)
6. Garncarek, L., et al.: LAMBERT: layout-aware language modeling for information extraction. In: Lladós, J., Lopresti, D., Uchida, S. (eds.) ICDAR 2021. LNCS, vol. 12821, pp. 532–547. Springer, Cham (2021). https://doi.org/10.1007/978-3-030-86549-8_34
7. Goyal, Y., Khot, T., Summers-Stay, D., Batra, D., Parikh, D.: Making the V in VQA matter: elevating the role of image understanding in Visual Question Answering. In: Conference on Computer Vision and Pattern Recognition (CVPR) (2017)
8. Harley, A.W., Ufkes, A., Derpanis, K.G.: Evaluation of deep convolutional nets for document image classification and retrieval. In: International Conference on Document Analysis and Recognition (ICDAR) (2015)
9. Harley, A.W., Ufkes, A., Derpanis, K.G.: Evaluation of deep convolutional nets for document image classification and retrieval. In: 2015 13th International Conference on Document Analysis and Recognition (ICDAR), pp. 991–995. IEEE (2015)
10. Lin, W., Gao, Q., Sun, L., Zhong, Z., Hu, K., Ren, Q., Huo, Q.: VIBERTgrid: a jointly trained multi-modal 2D document representation for key information extraction from documents (2021)
11. Lin, Z., et al.: Medical visual question answering: a survey (2021)
12. Liu, Y., et al.: RoBERTa: a robustly optimized BERT pretraining approach (2019)
13. Mathew, M., Karatzas, D., Jawahar, C.V.: DocVQA: a dataset for VQA on document images (2021)

14. Rajpurkar, P., Jia, R., Liang, P.: Know what you don't know: unanswerable questions for squad (2018)
15. Seo, M., Kembhavi, A., Farhadi, A., Hajishirzi, H.: Bidirectional attention flow for machine comprehension (2018)
16. TOMSETT, D.: https://digitalhumans.com
17. Vaswani, A., et al.: Attention is all you need (2017)
18. Wang, W., Zhang, J., Li, Q., Hwang, M.Y., Zong, C., Li, Z.: Incremental learning from scratch for task-oriented dialogue systems (2019)
19. Wang, Y., Yao, Q., Kwok, J., Ni, L.M.: Generalizing from a few examples: a survey on few-shot learning (2020)
20. Wu, J., Lu, J., Sabharwal, A., Mottaghi, R.: Multi-modal answer validation for knowledge-based VQA (2021)
21. Xu, Y., et al.: LayoutLMv2: multi-modal pre-training for visually-rich document understanding (2022)
22. Xu, Y., Li, M., Cui, L., Huang, S., Wei, F., Zhou, M.: LayoutLM: pre-training of text and layout for document image understanding. In: Proceedings of the 26th ACM SIGKDD International Conference on Knowledge Discovery and Data Mining, July 2020. https://doi.org/10.1145/3394486.3403172, http://dx.doi.org/10.1145/3394486.3403172
23. Yu, A.W., et al.: QANet: combining local convolution with global self-attention for reading comprehension (2018)
24. Zhang, C., Bengio, S., Hardt, M., Recht, B., Vinyals, O.: Understanding deep learning requires rethinking generalization (2017)

Is Multitask Learning Always Better?

Alexander Mattick(✉), Martin Mayr, Andreas Maier, and Vincent Christlein

Pattern Recognition Lab, Friedrich-Alexander University Erlangen-Nürnberg, Nuremberg, Germany
alex.mattick@fau.de

Abstract. Multitask learning has been a common technique for improving representations learned by artificial neural networks for decades. However, the actual effects and trade-offs are not much explored, especially in the context of document analysis. We demonstrate a simple and realistic scenario on real-world datasets that produces noticeably inferior results in a multitask learning setting than in a single-task setting. We hypothesize that slight data-manifold and task semantic shifts are sufficient to lead to adversarial competition of tasks inside networks and demonstrate this experimentally in two different multitask learning formulations.

Keywords: Multitask learning · Document analysis · Deep learning · Document classification

1 Introduction

Multitask learning is a technique that has been employed for decades to improve the results learned by artificial neural networks [3]. Empirical observations have shown that training a singular model on multiple tasks has a subadditive effect on all tasks. The reason for this could be interpreted as a form of "significance hacking" of random correlations inside the network. This increases the likelihood of a certain sub-network being good at least at one of the tasks.

If we assume that the tasks are at least in some form related we can generally observe a boost in performance for all tasks from the improvement of just one of them. This is presumably caused by a single task bootstrapping of off their own success and shaping increasingly larger shared subnetworks that then give the other tasks an easier time [3]. Since this effect is not one-sided, one can obtain improvements on all tasks simultaneously.

We divide the space of multitask learning into two types of network architectures: Multihead-based architectures and fullbody-based architectures. The former is considered more generally, in particular by dynamic reweighting algorithms like [22] that attempt to automatically find optimal weights between tasks. Mathematically, these architectures can be seen as two-stage algorithms that first transform the features into a latent space, where they branch off into

© Springer Nature Switzerland AG 2022
S. Uchida et al. (Eds.): DAS 2022, LNCS 13237, pp. 674–687, 2022.
https://doi.org/10.1007/978-3-031-06555-2_45

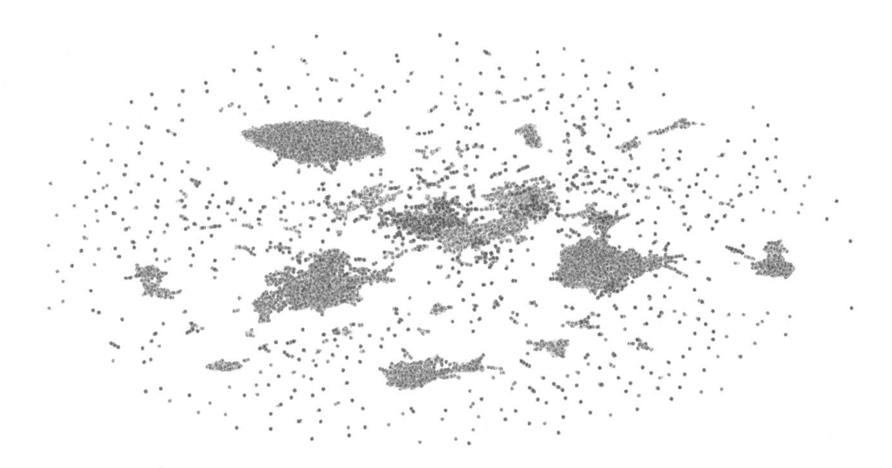

Fig. 1. UMAP projection of the 2048-D latent space. One can clearly see the distinct clustering of font (green), the mixing of the common parts of the date (orange) and script (red) datasets in the center, and the small clusterings of location (blue). Most notably: we do not observe a uniform mixing of the different tasks. The model learns distinct submodels rather than a joint model. (Color figure online)

multiple per-task networks. These algorithms are commonly trained end-to-end to achieve higher performance on multiple tasks that form the body network. Let $(x, y_0, \ldots, y_n) \in \mathcal{D}$ be the dataset with multiple labels,

$$l = \text{body}(x), \tag{1}$$

$$k_i = \text{head}_i(l). \tag{2}$$

The loss is now represented by a weighted sum of the individual task losses. It is very common to either hand-engineer these weights or simply weight all tasks equally, although there are other options [14,22] (at the cost of some overhead):

$$\mathcal{L}(\vec{k}, \vec{y}) = \alpha_0 \mathcal{L}_0(k_0, y_0) + \alpha_1 \mathcal{L}_1(k_1, y_1) + \cdots + \alpha_n \mathcal{L}_n(k_n, y_n). \tag{3}$$

The second type of architecture is more recent and the one preferred in autoregressive natural language modeling (e.g., [2,28]). This framework calls for a more general dynamic adaptation to each task by querying a jointly learned model for the information required. These task queries can be statically learned index vectors or dynamically generated via natural language queries, such as in natural-language question-answering problems. This query model is often realized via the use of Transformer networks [27] since Transformers are more generally applicable due to the dynamic generation of fast weights (see [20,21]). Another example of this (and the one more closely considered later) is the Perceiver [12] which embeds the inputs into a more compact latent space. The multitask effect is here obtained via querying the latent space using task-embeddings and cross-attention. Mathmatically, this can be framed as

$$k_i = \text{body}(x, t_i), \tag{4}$$

where t_i is a task-specific embedding that queries the proper prediction from the joint body directly. In practice, we often also need to have at least a small projection head to adapt to the different amount of classes in each task, but this should be considered less as actual task-specific processing and more as dimensional adaptation. In fact, we found negligible differences between using multiple layers for the output heads and using only a single bias-free linear projection to adjust the dimension.

Even though all of these models have been used with great success on multitask modeling problems, we show a practical scenario in which multitask learning is actively harmful to the final model's performance.

We hypothesize that multitask learning is actively harmful to the models performance in settings where the features are drawn from even only slightly different data-manifolds, such as multiple datasets each with a single label rather than the commonly assumed "one-dataset, multiple labels" setting. This style of multitask learning is often referred to as "heterogeneous" multitask learning, in contrast to "homogeneous" learning where a single dataset with multiple labels is present [29]. Additionally, there exists the equivalent notion of "joint learning" [30] that views heterogeneous multitask learning as simultaneously learning multiple weight-sharing networks jointly, rather than the multitask learning viewpoint of training a single feature extractor as a prior to smaller classifiers. A color-coded UMAP projection of the latent space of the different datasets can be seen in Fig. 1. It shows the behavior of the different datasets in the latent space.

Specifically, we make the following contributions:

(1) We compare the performance of single-against multitask learning in the setting of historical document analysis.
(2) We show that multitask learning can have adversarial results when trained on even subtly shifted data-manifolds.
(3) We further show that this may be mitigated to some degree by semantically closer tasks.

We proceed as follows: First, we introduce the datasets in Sect. 3 and the models in Sect. 4. Then, we present our experiments proving that multiple tasks compete (Sect. 5). Finally, we will discuss our findings with practical considerations when to use multitask learning (Sect. 6).

2 Related Work

There have been lots of works discussing multitask-learning in different contexts. The first mention of multitask learning was in [3], which explained the efficacy of multitask learning through the lens of statistical amplification and inductive biases. Related to this is the statistical phenomenon of Stein's paradox, which shows that there exists a combined estimator of three or more parameters that exceeds the performance of any three independent estimators on average [25]. [18] is a meta-review of evolutionary algorithms and Bayesian optimization and finds a novel combination of both. "A Survey on Multitask Learning" [29] discusses five different categories of multitask learning and benchmarks them head-to-head. Multitask learning with attention [16] builds a global "feature pool" from which different attention-heads draw information. The used Perceiver [13] and PerceiverIO [12] models can be seen as an extreme case of multitask learning, where one tries to marry independent domains into a common general perception model. An adjacent field of study is automatic task-weighting. Sener *et al.* [22] use multi-objective optimization theory to find pareto-optima between tasks in the network while Kendal *et al.* [14] automatically weight different tasks based on uncertainty estimates for each task. Random loss weighting [15] tries to use multiple tasks to escape local optima by randomly permuting task weightings. Another piece of related work is, of course, the "ICDAR 2021 Competition on Historical Document Classification" [24] associated with the dataset used and the methods included within. Similar to us, most approaches used a ResNet style architecture. However, most systems train on the individual tasks rather than multitask training a single network.

3 Datasets

We will use the group of datasets introduced for the "ICDAR 2021 Historical Document Classification Competition" [24] consisting of five datasets across four tasks.[1] Task 1a covers font classification across the "font group" dataset [23]. It contains images in different resolutions and from ten different classes. Similar to the final evaluation of the challenge, we do not consider the alternative classes for multi-font pages, as well as filtering out pages with "other-font" or "not a font". Task 1b covers scripts across the ICFHR16 [4] and ICDAR17 [5] datasets featuring twelve different classes and different document resolutions. Task 2 involves date classification using the ICDAR21-date [24] and ICDAR17 [5] datasets. Last but not least, task 3 covers is about location classification using the ICDAR21-location dataset [24] with 13 different classes as well as different resolutions for each image.

Figure 2 exemplifies the vast differences in datasets. It is essential to perform some preprocessing to make the models trainable and at least partially align the datasets. Specifically, we first scale the images to maintain roughly the same font size and take eight centered random 224×224 crops. We do this by first binarizing the image using Otsu's algorithm [17] and then performing an "erosion" mathematical operation with a small, quadratic structuring element and use this to mask out noise from the original image (mask 1 in Fig. 3). We then crop empty space around the documents introduced during scanning and binarize again using Otsu [17]. On this binarized image, we then perform an "opening" mathematical morphological operation with a long $(48, 1)$ rectangular structuring element. This is used to produce continuous bars from each text-line (mask 2 in Fig. 3). We then scan over the image vertically and save the median

[1] We consider the tasks 1a and 1b as independent, as they feature different datasets and labels.

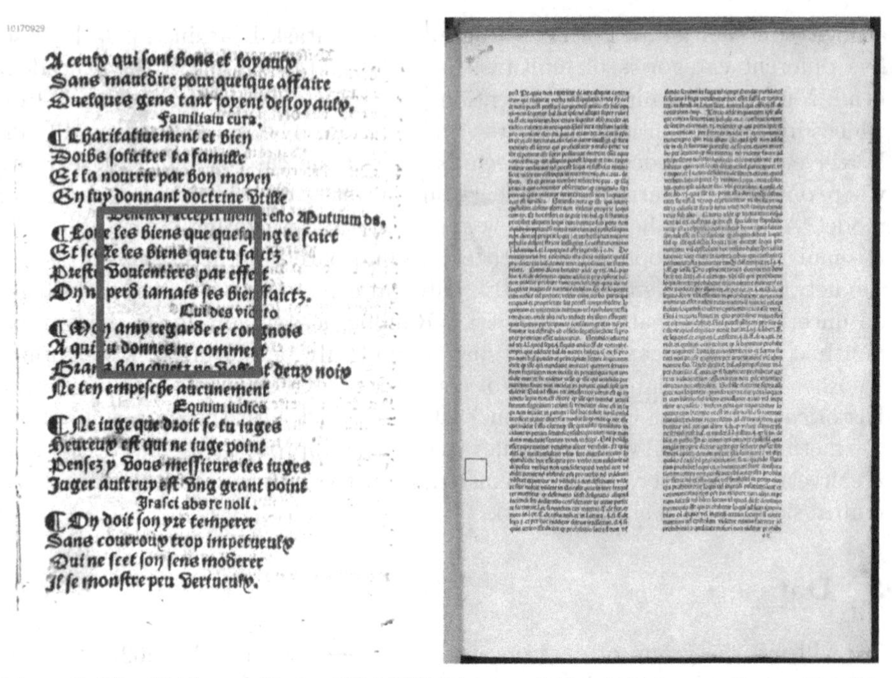

(a) small 607 × 874 image, filesize 227.8 KiB (b) large 4500 × 6624 image, filesize 36 MiB

Fig. 2. Two images with 224 × 224 pixel crops (red), the stroke width of the drawn windows is 5px in both images. Right image: the bottom crop on the border shows an example of an empty crop detected by our empty-patch detector. (Color figure online)

amount of continuous black pixels. The font height is then determined by computing the median across all scanlines (line height prediction in Fig. 3). In a last step, we scale the images to have a constant font height across all samples. This is necessary to achieve convergence and handle the potentially large size discrepancies between images. The scaling allows us to maintain approximately the same amount of content in each patch, even if the amount of text varies greatly across images even within the same dataset. The crops are done by picking a random translation across the middle of the image to decrease the likelihood of the crop not selecting any text. We further also ensure the measurement of the total amount of variation inside a patch by computing the sum of all edges and discarding patches below a minimal threshold to remove empty patches, cf. Fig. 2.

Aside from image normalization, we also normalize the different labeling formats and, most notably, convert the "dating" task into a classification problem. The ICDAR17 [5] dataset already features date classes that describe from-to date ranges. To homogenize the different datasets, we also convert the ICDAR21-date dataset [24] into date classes. This allows us to treat every task as a classification problem, which eases the comparability between the tasks as it allows us

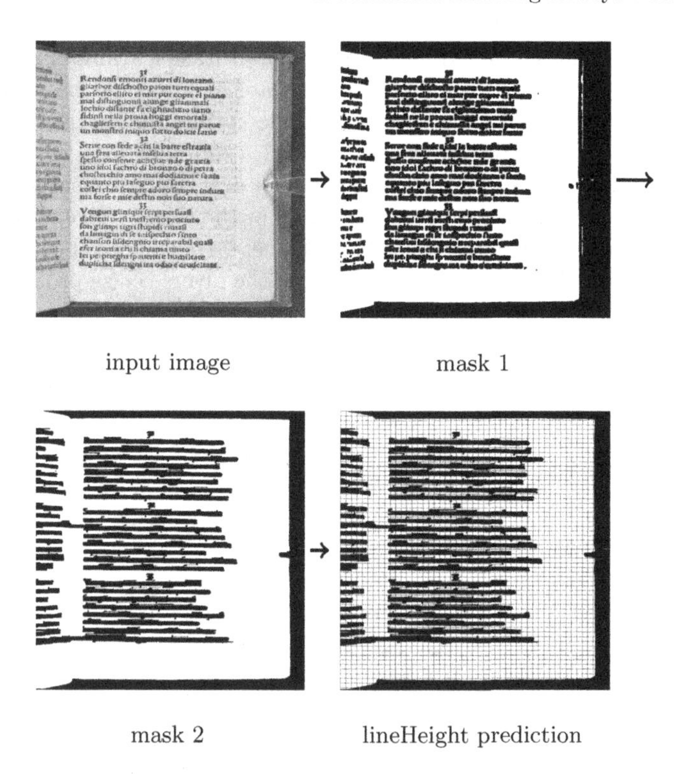

Fig. 3. Samples after each algorithm stage. Note that the scanlines on the left are only spaced according to the font size, but are not aligned to the bottom of each line. Also note that in the final output, the pincer on the right has barely any influence on the normalization process.

to use the same class re-weighting techniques used in all other tasks. To include the temporal structure inside the date classification problem, we decided to add label-smoothing in the form of a truncated normal-distribution around the date, i.e.:

$$\text{Smooth}_{c,r}(x) \sim \begin{cases} 0 & x < c - r \\ \frac{1}{\sigma\sqrt{2\pi}} \exp(-\frac{1}{2}(\frac{x - \text{class}}{\sigma})^2) & x \in (c - r, c + r) \\ 0 & x > c + r \end{cases}, \qquad (5)$$

where c denotes the class with range r. The goal of this is to recover some of the hierarchical nature of dates inside the categorical space: A date-error of 25 years is not as bad as one of 500 years and our label-smoothing helps to reflect that. Further, while we re-use the ICDAR17 [5] classes, we also decided to add additional ones below the year 1000, as ICDAR17 considers all of those years as the same class. We made sure that those classes have approximately the same number of samples and have approximately the same time range as the ICDAR17 ones. While the choice of modeling the dates as classes may not be optimal, we ultimately only want to reason about the relative performance advantages/disadvantages introduced through multitask learning. Instead of converting ICDAR21-date [24] to categorical variables, one could have also chosen to convert the ICDAR17 [5] dataset to continuous time-ranges. However, due to the fact that continuous variables would need a different loss function with a different numerical range, we decided to model everything as categorical as to not accidentally over- or under-emphasize any of the tasks. Since there is a high degree of intra- and inter-dataset class imbalance, we normalize the classes within each dataset by weighting the loss according to the inverse class-probabilities and normalize the datasets by oversampling the smaller ones.

4 Methodology

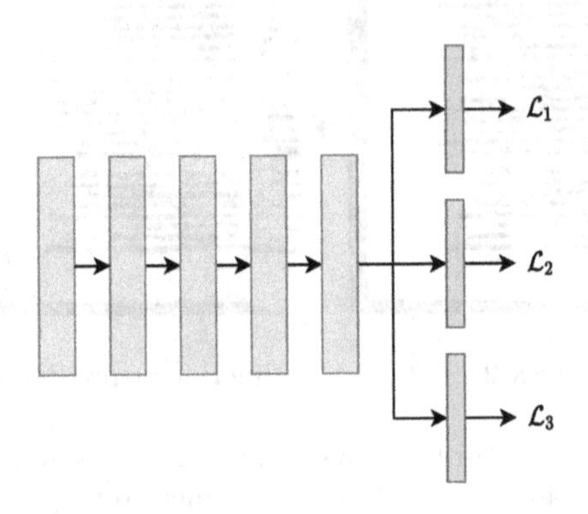

Fig. 4. Illustration of the Mulitheaded approach. The blue backbone is often a pre-trained off-the shelf network like a ResNet-50 [10] while the heads are considerably smaller, often just a single linear layer. (Color figure online)

We study two different architectures. Our main object of study is the ResNet [10] as it is a common high performance image recognition technique. Specifically, we use a common pretrained ResNet-50 with four independent heads. The ResNet model family consists of multiple stages, all of which are comprised of ResNet-Blocks. At the beginning of each stage the input is downsampled to halve the resolution while the number of channels is doubled. After the initial downsampling, the stage maintains the same resolution and number of channels to facilitate skip-connections inside each ResNet-Block. The ResNet-Block is characterised by two stacked convolutional layers with batch normalization [11] that are additionally connected with skip-connections that allow for unobstructed gradient flow in deep networks. We use a ResNet-50 that returns a 2048-D representation of the input. Independent heads are used to classify the different tasks. All heads consist of a single linear layer, as we did not notice any beneficial effect when using multiple layers on top of the ResNet-50 body. We use a learning rate of 10^{-4}, which is a standard learning rate for training ResNets. We also trialed several other learning rates, all of which lead to consistently worse results.

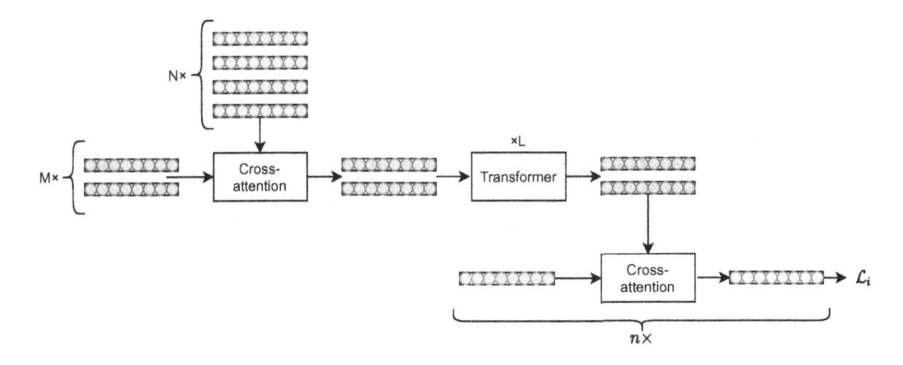

Fig. 5. The Perceiver as an example of a fullbody architecture. Inference here is performed by querying the latent state for each task. The perceiver inputs a N-dimensional input (blue) that is downprojected into a smaller M-dimensional space via cross-attention to a set of M learned vectors (red). Afterwards L standard transformer layers are added [27]. At the end of the model, each of the N tasks are queried from the latent space (purple) via a task-embedding vector (green) (Color figure online)

Additionally we use color jitter, grayscaleing, horizontal mirroring, random noise from a $\mathcal{N}(0, 0.1)$ distribution and random-cropping for data augmentation.

To show that our findings are architecture independent, we also train a PerceiverIO [12] with a single common body. The PerceiverIO [12] and Perceiver [13] (used interchangeably in this paper) models have an entirely different learning mechanism than the ResNet, opting for self-attention rather than convolutions to achieve globally, rather than locally, connected transformations of the data. The PerceiverIO learning can be roughly split into 3 stages. First, a potentially very long sequence of input tokens N is projected into the more compact space M via "cross-attention" [13]. This can be seen as a soft-assignment of the N inputs to M cluster centers by projecting into a coordinate system relative to the M statically learned centers. From this point on, one is in a sufficiently compact space to perform the usually memory intensive $M \times M$ self-attention operation [27] in the form of stacked transformer layers (in our case 6). This constitutes stage two and is the main workhorse of the Perceiver architecture. The major difference between Perceiver and PerceiverIO becomes visible in the last stage. While Perceiver simply accumulates the M output vectors into one mean/sum vector used for further processing, PerceiverIO applies another cross-attention operation to retrieve targeted information from the M latent vectors, cf. Fig. 5. This has the advantage that we are not bound to a static number of N heads and that the number of tokens retrieved can change for each task (especially interesting for e.g., joint classification and semantic segmentation training). We chose to use a PerceiverIO architecture as it encapsulates the "querying a joint model" paradigm as described in Fig. 1. As input, we split the 224×224 pixel images into 64 smaller 28×28 micro-patches which we run through the initial stages of a ResNet-18 to get a 128-D representation for the 64 token. This gives us the embeddings used for the cross-attention input (Fig. 5). We use the same learning rate of 10^{-4} as in the ResNet models, as it proved to be effective during a preliminary hyperparameter search. We also use the same augmentations as were used for training the ResNets, as additional augmentation did not prove to be beneficial to model performance.

We decided to test the Perceiver rather than a transformer based model like ViT [6] because of the significantly higher sample-efficiency present in the Perceiver when compared to other pure-attention models which tend to be trained on terabytes of data (i.e., imagenet-21k [19] or JFT-300M [26]).

5 Evaluation

5.1 ResNet

Table 1. Comparison of the peak accuracies of single vs. multitask learning using the ResNet architecture. Note that all tasks refer to individual optima rather than joint optima.

Model	Location	Date	Font	Script
Location only	**0.77**	–	–	–
Date only	–	**0.49**	–	–
Font only	–	–	**0.98**	–
Script only	–	–	–	**0.90**
All tasks	0.66	0.32	0.97	0.88

At first, we compare the joint training on all four tasks with the training for each individual task. We train the ResNet model featuring all four datasets with a batchsize of 16 per dataset, resulting in an effective batchsize of $4 \cdot 16 = 64$. For the runs with a single task, we fix the batchsize to the effective batchsize of 64 and keep all other hyperparameters equal. All models were trained with an early stopping threshold of 16 epochs to ensure convergence and a reduce-on-plateau learning rate scheduler with patience of 5 epochs.

The individual tasks outperform the joint training in all cases. In fact, the results of Table 1 may give a too optimistic impression: We chose to pick the individual optima for the multitask trained model, rather than the joint peak of all tasks simultaneously. While technically possible, this is rarely done in practice. By reporting the individual optima, we give an upper bound on the task accuracies that may in practice be weighted based on problem specific requirements. When deployed classically, i.e., with the jointly optimal model, one can expect even worse performance from the multitask model.

Naturally, one might ask the question whether there is a single dataset that poisons the multitask model or whether the mere existence of multiple data streams already results in degraded performance. The columns of Table 2 show that we can observe noticeable performance degradation in both "location" and

Table 2. Comparing the peak accuracies of dual vs. multitask learning using the ResNet architecture. Note that like before, we always consider individual rather than joint optima.

Model	Location	Date	Font	Script
Date and location	**0.67**	0.25	–	–
Script and font	–	–	**0.98**	0.89
Script and location	**0.67**	–	–	0.87
Script and date	–	**0.32**	–	**0.91**
Font and date	–	0.24	0.97	–
Font and location	**0.67**	–	0.97	–
All tasks	0.66	**0.32**	0.97	0.88

"date" tasks. This makes sense since these are the datasets that most significantly break the assumption of "one-dataset, multiple labels" present in most analyses of multitask learning. The "font" task retains most of its performance, which can be attributed to the sheer size of the font-groups dataset in contrast to the others. Noteworthy, the combination that retains 100% of the single task performance is the one that combines the most similar tasks: font and script classification. Finally, we can see the expected multitask learning performance increases in the "script" column. However, this does align with our hypothesis as the winning combination "script and date" is precisely the one combination that does partially follow the "one-dataset, multiple labels" system since script and date share the common dataset ICDAR17, cf. Sect. 3. We further note that the very same model also performs best in the "date" task, though with some performance degradation when compared with the single task setting. This can be explained by the addition of the ICDAR21 dataset into the "date" task, which leads to some domain shift. This domain shift does not take place in the "script" task since the ICHFR16 and ICDAR17 datasets are very similar and, in fact, share parts of the training set.

5.2 Perceiver

So far, we only studied the more common single-body multi-heads approach, but not the full-body adaptation setting outlined in Sect. 4. To cover this case and demonstrate that our results extend to vastly different methods of learning (i.e., CNNs vs. self-attention), we train another batch of models on the PerceiverIO [12] architecture. From the results in Table 3, we can draw a very similar conclusion than for the ResNet model. The differences are not as pronounced as in the single-body multi-heads setting, though this probably has more to do with the decreased performance of the Perceiver when compared to the ResNet. The dual-task models also perform similarly, see Table 4. We still have a general decrease of performance across all task combinations aside from "font and date" which manages a surprising 1% improvement above the "font"-only setting

Table 3. Comparing the peak accuracies of single vs. multitask learning using the Perceiver architecture. Note that all tasks refer to individual optima rather than joint optima.

Model	Location	Date	Font	Script
Location only	**0.67**	–	–	–
Date only	–	**0.46**	–	–
Font only	–	–	**0.96**	–
Script only	–	–	–	**0.82**
All tasks	0.64	0.23	0.95	0.78

Table 4. Comparing the peak accuracies of dual vs. multitask learning using the Perceiver architecture. Note that all tasks refer to individual optima rather than joint optima.

Model	Location	Date	Font	Script
Date and location	0.60	0.19	–	–
Script and font	–	–	0.96	**0.80**
Script and location	**0.65**	–	–	0.78
Script and date	–	**0.24**	–	**0.80**
Font and date	–	0.17	**0.97**	–
Font and location	0.58	–	0.95	–
All tasks	0.64	0.23	0.95	0.78

though at the cost of catastrophic performance in the "date" task. Once again, the only combination that was able to unilaterally improve over the multitask model was "date" and "script", which share part of the dataset, though they still perform significantly worse than the single-task models.

This result is interesting for an additional reason: The original motivation of PerceiverIO [12] was incorporating multi-modal information sources into a single model. The Perceiver architecture was thought to be particularly useful for this as the "querying the latenspace" style of training forces the model to align all information sources in its latent space. The results in Table 3 show that this is not generally the case, even with information that appears to fall within the scope of "historical document analysis". This may also be connected with the small amounts of data used to train when compared to the massive datasets usually used for attention-based models, such as JFT-300M [26] or imagenet-21k [19]. However, in this setting one would still expect the joint dataset to outperform the individual ones on size alone, even with imperfect task alignment (consider that JFT-300M also only features noisy labels). Therefore, we conclude that the tasks themselves are adversarial to each other.

6 Discussion

Though our experiments showed quite convincingly that different data manifolds are detrimental to this model's performance, this does not mean that multitask learning is a bad idea in general. However, we argue, it makes sense to consider the domains of reasoning necessary to accomplish a certain task and relate this to the available datasets. For example, we saw little degradation of performance when joining the semantically similar tasks of font and script classification (see Tables 2 and 4) in comparison to joining semantically unrelated tasks, like font and date. This makes sense when viewed through the lens of multitask learning as an inductive bias, as semantically unrelated tasks may induce conflicting biases, while related ones narrow the search space of possible parameter weights.

On the other hand, we also found that semantically unrelated tasks, like script and date classification can be combined beneficially, if they share the same dataset (see script in Table 2). This can be explained through the lens of statistical significance: odds are that if a certain randomly initialized feature detector works for task A, it is at least not detrimental to task B if both come from the same distribution. In cases where a feature is not as evident in task B as it is in task A, we get an effective boost for task B as it gets the detector for the hard to discover feature for free. This however only works if the data manifolds are overlapping, as statistical features drawn from one data distribution may appear differently in a different distribution.

In an ideal world, one would have both a common dataset and related tasks, but this is often not a realistic assumption. However, one may be able to align the data-distributions by preprocessing the dataset, such as giving the model normalized line-segments (like the PERO team did in [24]) for script and font classification. Specific to note here is the "CLUZH" team who found that combining script and fontgroup training could be beneficial for the script task, which is one of the cases where we found the least amount of performance degradation, see Table 2. We hypothesize that aligning the datasets for semantically related tasks through normalization would give boosts in both tasks as we could benefit from both the semantic and statistical effects of multitask learning. Such normalizations though are, by definition, task-dependent as they aim to remove statistical confounders to make the relevant signals of the task apparent through the dataset-sampling induced noise. This effectively amounts to proving the causal independence of the dataset from the confounders (for a more explicit treatment of confounders and causality, see e.g., [9]).

7 Conclusions

We demonstrate a realistic scenario in which multitask learning is detrimental to the overall performance of each task. We suggest that the likely culprit is the semantic and statistical shift between the datasets by comparing the results of training on a single task vs. multiple tasks simultaneously for two fundamentally different architectures. Our analysis demonstrates that multitask learning on one dataset with multiple labels produces the best results, closely followed by semantically related tasks with independent datasets. Due to the validity of this observation in both ResNets and Perceiver architectures, we argue that the blame cannot be placed on the chosen model, but rather is an intrinsic attribute of the datasets. We conclude that one cannot, in general, expect increased performance when combining multiple tasks derived from different datasets, even if they all share a common domain.

A promising direction for future work could be incorporating methods that reduce the domain shift of neural network at training time, such as gradient reversal networks [8] or other techniques from domain adaptation [7]. This may enable a more sound usage of multiple datasets with related tasks, as they reduce the chance of segmenting the latent space by dataset source (cf. Fig. 1). Another avenue for improvement could be the extension of automatic task-weighting algorithms, like [22], which currently often rely on having a common dataset with multiple labels, rather than having completely independent task-dataset pairs. From the point of view of specific domains (such as document analysis) one could also seek to find equivariances or invariances to prevent certain directions of embedding misalignment [1]. The inclusion of such constraints on the model could allow the use of more diverse data sources by removing certain confounders from the model's search space.

References

1. Bronstein, M.M., Bruna, J., Cohen, T., Velivckovi'c, P.: Geometric deep learning: grids, groups, graphs, geodesics, and gauges. ArXiv abs/2104.13478 (2021)
2. Brown, T.B., et al.: Language models are few-shot learners. ArXiv abs/2005.14165 (2020)
3. Caruana, R.: Multitask learning. Mach. Learn. **28**(1), 41–75 (1997)
4. Cloppet, F., Eglin, V., Stutzmann, D., Vincent, N., et al.: ICFHR 2016 competition on the classification of medieval handwritings in latin script. In: 2016 15th International Conference on Frontiers in Handwriting Recognition (ICFHR), pp. 590–595. IEEE (2016)
5. Cloppet, F., Eglin, V., Helias-Baron, M., Kieu, C., Vincent, N., Stutzmann, D.: ICDAR 2017 competition on the classification of medieval handwritings in latin script. In: 2017 14th IAPR International Conference on Document Analysis and Recognition (ICDAR), vol. 01, pp. 1371–1376 (2017)
6. Dosovitskiy, A., et al.: An image is worth 16x16 words: transformers for image recognition at scale. ArXiv abs/2010.11929 (2021)
7. Farahani, A., Voghoei, S., Rasheed, K.M., Arabnia, H.R.: A brief review of domain adaptation. ArXiv abs/2010.03978 (2021)
8. Ganin, Y., et al.: Domain-adversarial training of neural networks. J. Mach. Learn. Res. (2016)
9. Greenland, S., Pearl, J., Robins, J.M.: Confounding and collapsibility in causal inference. Stat. Sci. **14**(1), 29–46 (1999)
10. He, K., Zhang, X., Ren, S., Sun, J.: Deep residual learning for image recognition. In: 2016 IEEE Conference on Computer Vision and Pattern Recognition (CVPR), pp. 770–778 (2016)

11. Ioffe, S., Szegedy, C.: Batch normalization: accelerating deep network training by reducing internal covariate shift. ArXiv abs/1502.03167 (2015)
12. Jaegle, A., et al.: Perceiver IO: a general architecture for structured inputs & outputs. ArXiv abs/2107.14795 (2021)
13. Jaegle, A., Gimeno, F., Brock, A., Zisserman, A., Vinyals, O., Carreira, J.: Perceiver: general perception with iterative attention. In: ICML (2021)
14. Kendall, A., Gal, Y., Cipolla, R.: Multi-task learning using uncertainty to weigh losses for scene geometry and semantics. In: 2018 IEEE/CVF Conference on Computer Vision and Pattern Recognition, pp. 7482–7491 (2018)
15. Lin, B., Ye, F., Zhang, Y.: A closer look at loss weighting in multi-task learning. ArXiv abs/2111.10603 (2021)
16. Liu, S., Johns, E., Davison, A.J.: End-to-end multi-task learning with attention. In: 2019 IEEE/CVF Conference on Computer Vision and Pattern Recognition (CVPR), pp. 1871–1880 (2019)
17. Otsu, N.: A threshold selection method from gray-level histograms. IEEE Trans. Syst. Man Cybern. **9**(1), 62–66 (1979)
18. Ponti, A.: Multi-task learning on networks. ArXiv abs/2112.04891 (2021)
19. Russakovsky, O., et al.: ImageNet large scale visual recognition challenge. Int. J. Comput. Vis. **115**(3), 211–252 (2015). https://doi.org/10.1007/s11263-015-0816-y
20. Schlag, I., Irie, K., Schmidhuber, J.: Linear transformers are secretly fast weight programmers. In: ICML (2021)
21. Schmidhuber, J.: Learning to control fast-weight memories: an alternative to dynamic recurrent networks. Neural Comput. **4**(1), 131–139 (1992)
22. Sener, O., Koltun, V.: Multi-task learning as multi-objective optimization. In: NeurIPS (2018)
23. Seuret, M., Limbach, S., Weichselbaumer, N., Maier, A., Christlein, V.: Dataset of pages from early printed books with multiple font groups, August 2019
24. Seuret, M., et al.: ICDAR 2021 competition on historical document classification. In: Lladós, J., Lopresti, D., Uchida, S. (eds.) ICDAR 2021. LNCS, vol. 12824, pp. 618–634. Springer, Cham (2021). https://doi.org/10.1007/978-3-030-86337-1_41
25. Stein, C.: Inadmissibility of the usual estimator for the mean of a multivariate normal distribution. In: Proceedings of the Third Berkeley Symposium on Mathematical Statistics and Probability, vol. 1, pp. 197–206 (1956)
26. Sun, C., Shrivastava, A., Singh, S., Gupta, A.: Revisiting unreasonable effectiveness of data in deep learning era. In: Proceedings of the IEEE International Conference on Computer Vision (ICCV), October 2017
27. Vaswani, A., et al.: Attention is all you need. ArXiv abs/1706.03762 (2017)
28. Yang, Z., Dai, Z., Yang, Y., Carbonell, J.G., Salakhutdinov, R., Le, Q.V.: XLNet: generalized autoregressive pretraining for language understanding. In: NeurIPS (2019)
29. Zhang, Y., Yang, Q.: A survey on multi-task learning. ArXiv abs/1707.08114 (2017)
30. Zhang, Z., Huang, X., Huang, Q., Zhang, X., Li, Y.: Joint learning of neural networks via iterative reweighted least squares. ArXiv abs/1905.06526 (2019)

SciBERTSUM: Extractive Summarization for Scientific Documents

Athar Sefid[1(✉)] and C. Lee Giles[1,2]

[1] Computer Science and Engineering, Pennsylvania State University, University Park, PA 16802, USA
atharsefid@gmail.com, clg20@psu.edu

[2] Information Sciences and Technology Department, Pennsylvania State University, University Park, PA 16802, USA

Abstract. The summarization literature focuses on the summarization of news articles. The news articles in the CNN-DailyMail are relatively short documents with about 30 sentences per document on average. We introduce SciBERTSUM, our summarization framework designed for the summarization of long documents like scientific papers with more than 500 sentences. SciBERTSUM extends BERTSUM to long documents by 1) adding a section embedding layer to include section information in the sentence vector and 2) applying a sparse attention mechanism where each sentences will attend locally to nearby sentences and only a small number of sentences attend globally to all other sentences. We used slides generated by the authors of scientific papers as reference summaries since they contain the technical details from the paper. The results show the superiority of our model in terms of ROUGE scores. (The code is available at https://github.com/atharsefid/SciBERTSUM).

1 Introduction

Automatic summarization frameworks condense an input document into shorter text consisting of the main points in that document. Neural networks have achieved state of the art results for both paradigms of abstractive summarization [4, 19] and extractive summarization [15, 18]. While extractive models are factually more consistent with the content in the input document, abstractive models can be novel and less redundant. Most of the existing methods are used on news datasets [8, 16] where the input document is relatively short and normally less than 30 sentences long. Summarization of long documents such as scientific papers is different from a short article summarization since it requires more memory and computational power to encode the full document and model the relationship between the sentences.

Natural language processing applications have completely been revolutionized with the advent of pre-trained models. Pre-trained language models are easy to incorporate and don't require much-labeled data to deal with, which makes it appropriate for many problems such as prediction, transfer learning, and feature extraction. Bidirectional Encoder Representations from Transformers (BERT) [6] have combined both word and sentence representations into a single very large Transformer [23]. This has

© Springer Nature Switzerland AG 2022
S. Uchida et al. (Eds.): DAS 2022, LNCS 13237, pp. 688–701, 2022.
https://doi.org/10.1007/978-3-031-06555-2_46

shown superior results on many NLP tasks such as question answering and text generation. BERT was trained on large amounts of data with the objective of predicting the masked tokens and the next sentence and it can be fine-tuned for various task-specific objectives [14].

Language models such as BERT [6] or SciBERT [1] have improved many language based tasks, especially with SciBERT for science related documents. The impact of BERT on extractive summarization was due to BERTSUM. BERTSUM extended BERT from a two-sentence language model to one that covers all sentences in a document. BERTSUM model with a full attention layer can capture the document-level features. However, full attention is not efficient for the summarization of long documents such as scientific papers which have more than 500 sentences. Here we propose an extractive transformer based summarizer for longer documents such as scientific articles with multiple sections.

The contributions of our model are:

- Design a section embedding layer to the embedding module of BERTSUM where all tokens in the same section are embedded with the same embedding token. This is crucial for the embedding of long documents with multiple sections in a hierarchical structure.
- Employ a sparse inter-sentence attentional model with local and global attention schemes where each sentence will attend locally to nearby sentences and some random sentences attend globally to all other sentences in the document.
- Devise summarization modules for scientific articles using the presentation slides as the ground-truth summaries. The slides contain the technical details from the paper and usually follow the structure of the papers.

2 Related Work

2.1 Summarization

We believe summarizing scientific articles is more challenging than summarizing generic text since such articles have a hierarchical structure [9]. They contain technical terms and formulas [24], and much valuable content can be embedded in figures, tables, and algorithms [3].

Scientific article summarization has been less investigated compared to news articles summarization [5, 19]. This seems to be mainly due to the lack of training data for full scientific articles. Types of reference summaries for scientific articles are:

- Abstract: Most of the traditional summarization methods use the abstract as the reference summary of the paper. However, abstracts are extremely compressed versions of a papers and usually do not have enough space to include all of the contributions [7].
- Citation-based: These types of summaries integrate the authors' highlights in the abstract of the paper with citation context of the citing papers which in some ways reflects the impact in the paper of the research community [24].

- Speaker Transcript: Many conference proceedings/workshops require the authors to verbally present their work. TalkSum [10] uses the transcript of these presentations as a summary of the scientific article. However, the transcripts in the TalkSum data set are often noisy and can not be readily used as reference summaries.

Presentation slides for a paper are a different class of summaries that intend to cover in some way the important content of the entire paper, sometimes section by section. They contain the main highlights and also valuable images/tables. They are not as noisy as speaker transcripts and are becoming more available as more conferences are providing slides that go with their papers. We used the PS5k dataset [20,21] to build our summarizer.

2.2 Transformer Based Summarization

Pre-trained language models such as BART [11] produce state-of-the-art results on the summarization tasks. However, they are often used on short news articles such as XSum [17] or CNN-DailyMail [16] datasets. These models are not designed for scientific articles and their space/computational complexity grows quadratically with the size of the input.

HIBERT [25] is an extractive summarizer that learns context aware sentence representations using multiple layers of transformers. Here, 15% of the full sentences are masked (replaced with a single [mask] token) with the goal to predict the sentence embedding of the masked sentences. BERTSUM [12] is another BERT style extractive summarizer that extends BERT to multiple sentences by expanding the positional embedding and using interval segmentation embeddings to distinguish multiple sentences within a document. Sotudeh et al. [22] added section information to the objective function of BERTSUM so it could optimize both the sentence prediction and section prediction tasks in a multi-task setting. However, most of these transformer-based extractive summarizers do not scale for long documents with thousands of tokens nor can they be applied to many full scientific documents.

3 Method - SciBERTSUM

Most of the previous language models such as BERT are employed as encoders for short pieces of text usually covering at most 2 sentences. The summarization task besides other NLP tasks (e.g. predicting entailment relationship, question answering) requires a wide coverage of the full document containing multiple sections and many sentences. We propose a document encoder based on BERT. Our encoder model will help build sentence representations by stacking multiple transformer layers on top of sentence vectors to model the inter sentence relations in the full document.

Our SciBERTSUM model is an extension of BERTSUM and can generate sentence embeddings for all sentences in a full document with multiple sections. Our model applies a linear sparse attention mechanism between sentences to represent inter sentence relations and it outperforms BERTSUM on our dataset.

4 Language Model Architecture

To explain the architecture of our language model, we first explain how we generate the sentence embeddings by adding section information to sentences and then we explain how our sparse attention mechanism helps us process the full document efficiently.

4.1 Embedding Layer

The embedding layer of BERT [6] applies the byte-pair encoding to tokenize the text. It adds [CLS] tokens to the beginning of the sequences and [SEP] tokens to separate the first and second sentences in the sequence. The embedding representation of the [CLS] token is the representation of the full sequence and is used for sentence classification tasks.

BERT combines 1) Semantic embedding (the meaning of the token) 2) Positional embedding (the position of the token in the sequence), 3) Segmentation embedding (the embedding layer to distinguish the first and the second sentence in the sequence) to form the embedding of a token in a sequence.

BERTSUM [12] extends BERT to multiple sentences by adding [CLS] tokens to the beginning of all sentences. It changes the segmentation embedding to distinguish odd and even sentences. The embedding model is depicted in Fig. 1. The green boxes are the segmentation embeddings. Light greens are the embedding for odd sentences and dark green boxes are for even sentences. BERTSUM extended the positional embedding of BERT beyond 512 tokens to cover all tokens of the input document.

The sentence embeddings are the embedding of the [CLS] tokens which are the combination of semantic, segment, and position embeddings. The Positional encoding is the sinusoid embedding from Vaswani et al. [23].

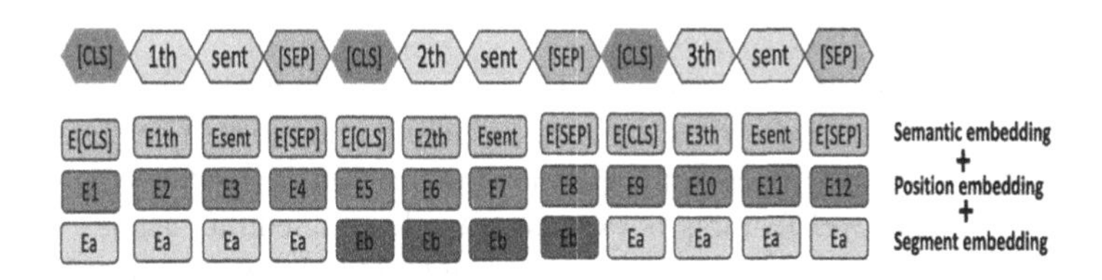

Fig. 1. BERTSUM architecture covering multiple sentence. Each sentence has a [CLS] token at the beginning.

Long documents especially long scholarly articles contain multiple sections. The section of the document is important in the selection of salient sentences. For instance, the sentences in the '*acknowledgment*' section are less important compared to other sections like '*abstract*' or '*results*'. We enhance BERTSUM by adding section embedding as shown in Fig. 2. The sentence embeddings (E_{sents}) are the combination of the section, semantic, position, and segmentation embeddings. The section embeddings are the blue boxes in Fig. 2. All of the tokens of the sentences in the first section are embedded by dark blue and the tokens of sentences in the second section are embedded in light blue. Each section has the same segmentation embedding as in BERTSUM.

$$E_{sents} = Semantic + Position + Segment + Section \tag{1}$$

To overcome the memory issue where we cannot load the full document with the full position embedding in the memory. We get the sentence vectors section by section. We can load the maximum 3072 tokens to the memory based on experiments on an Nvidia GPU with 11,019 MiB memory capacity.

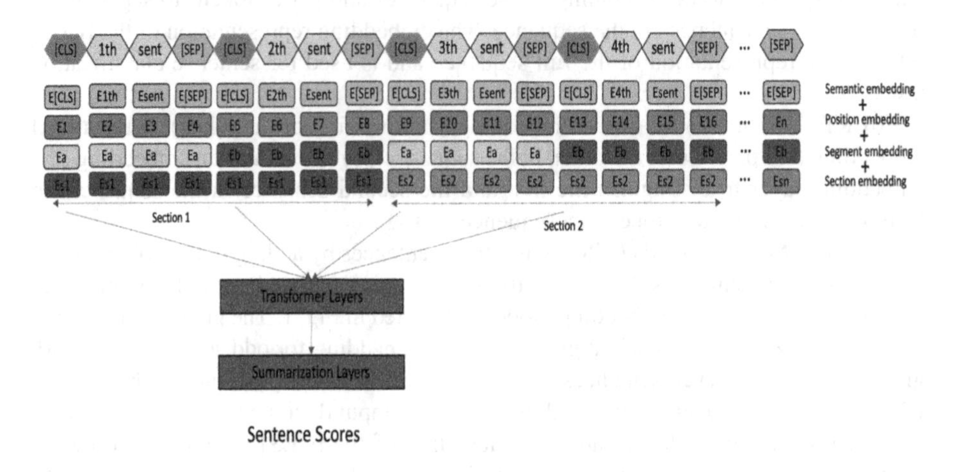

Fig. 2. SciBERTSUM architecture

4.2 Attention Mechanism

BERTSUM applies multiple layers of full dot-product attention. However, applying full attention on the sentence vectors is expensive where the number of sentences in documents is large. Scientific documents can have more than 500 sentences. Therefore, extracting document-level features is expensive for long scientific articles with a full attention layer. Therefore, we introduce a lightweight attention mechanism inspired by LongFormer [2]. The LongFormer language model applies sparse attention between tokens to learn the embedding of the masked tokens. We apply their attention mechanism at the **sentence level** where each sentence will fully attend locally to the nearby sentences and some sentences will attend globally to all sentences in the document.

This attention mechanism will help model select salient sentences locally from the window and at some random and selected positions sentences will attend to all other sentences to identify the salient sentences that are globally important regardless of the section they belong to. The attention window in Fig. 3 is 2 which means each sentence will attend to 2 sentences before and after it and in Fig. 4 sentences 2 and 7 (marked with *) are attending to all other sentences.

Applying window-based local attention requires a few preprocessing steps. We list the main steps in the following section.

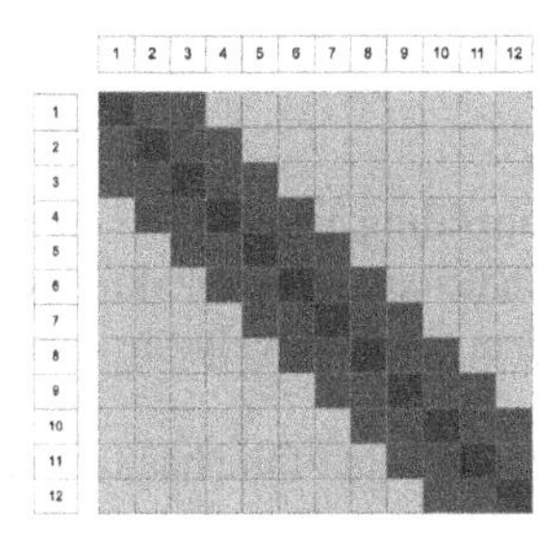

Fig. 3. Local attention

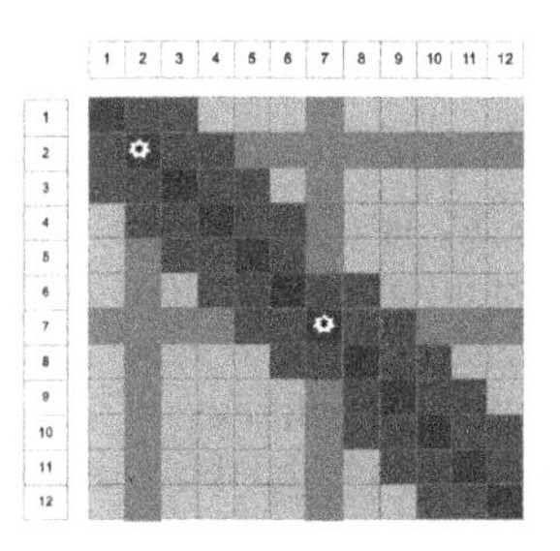

Fig. 4. Local and global attention

Building the Attention Matrix. Since we are processing multiple documents in batch mode and each document has a different length, we fix the number of sentences and make the length of the documents to be a multiple of the attention window. Therefore the following steps are required to process the document:

1. Padding to document size: the document size is fixed to 500 sentences for scientific documents in our corpus.
2. Padding to attention window: the length of the document must be a multiple window size to be able to apply the sliding window attention mechanism.
3. Building attention matrix: the attention matrix has a value of 0 for the padded sentences, a value of 1 for the local attention, and 2 specifies the combination of both local and global attention. Figure 5 shows an attention matrix for a batch of size 3. This batch contains 3 documents with 6, 2, and 6 sentences respectively. For example, the first document attends locally at positions [1, 2, 3, 4, 5, 6] and attends globally at position 4.

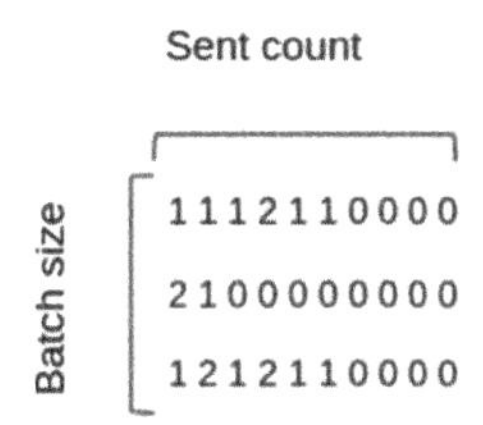

Fig. 5. Attention mask matrix

Calculating Attention Value. Here we list the steps for calculating the local attention. The global attention follows the same approach by adjusting the sentences vectors.

1. Three linear layers are applied on the sentence vectors to generate the query, key and value vectors

$$E_q = W_q * E_s + b_q \tag{2}$$

$$E_k = W_k * E_s + b_k \quad (3)$$

$$E_v = W_v * E_s + b_v \quad (4)$$

where E_s is the sentence embedding from the embedding layer that embeds the section information. Here b is the bias term and W matrices are learned in the training phase in order to generate E_q, E_k, and E_v which are respectively the query, key, and value embeddings.

2. For the second step the query is normalized by the square root of the head dimension

$$E_q = E_q / \sqrt{heads}. \quad (5)$$

3. The attention scores are calculated by a sliding query and key matrix multiplication on all chunks of the attention-window size

$$S_{attn}[i] = E_q[i] * E_v[i] \quad (6)$$

where $E_q[i]$ and $E_v[i]$ are the query and value embeddings of window i, and $s_{attn}[i]$ is the attention score for that window.

4. The values of the attention scores at the padding positions are set to 0 to ignore values at these locations so that

$$S_{attn}[padIndex] = 0. \quad (7)$$

5. $Softmax$ is applied to attention scores to generate the attention probabilities

$$P_{attn} = Softmax(S_{attn}). \quad (8)$$

6. Finally, the attention probabilities are multiplied by the value vectors chunk by chunk in a sliding window

$$out[i] = P_{attn}[i] * E_v[i]. \quad (9)$$

4.3 Transformer Layer

Our sparse attention mechanism is applied in each layer of transformation [23] and the inputs to the first layer of transformer are the sentence embeddings that include the section information

$$\widetilde{h}^l = h^{l-1} + normalize(SparseAttention(h^{l-1})) \quad (10)$$

$$h^l = PositionwiseFeedForward(\widetilde{h}^l) \quad (11)$$

where $h^0 = E_{sents}$. We apply a sparse attention mechanism here instead of full attention.

5 Sentence Extractor

To generate the final sentence score, we combined the sentence embedding from the language module with a list of features necessary for the score prediction. Section 5.1 elaborates on the list of features applied to generate the sentence scores. These features depend on the document embedding calculated as in Sect. 5.2.

5.1 Sentence Features

The features used to predict the final scores are:

1. Length: number of characters in the sentence i

$$E_{length} = ReLU(Linear(Embedding(length[i]))). \tag{12}$$

2. Position: position of the sentence (i) in the document

$$E_{position} = ReLU(Linear(Embedding(i))). \tag{13}$$

3. Section: section of the sentence i in the document

$$E_{section} = ReLU(Linear(Embedding(section[i]))). \tag{14}$$

   Each of the embedding layers is a simple lookup table that stores embeddings of a fixed dictionary and size. The size of the Embedding layers is d. The linear layer applies a linear transformation to the input data x ($y = Wx^T + b$).
4. Correlations: the sentence correlations embed the correlation between sentences. The correlation embeddings help the model to identify sentences with a high degree of correlation to other sentences and then exclude them.

$$Correlation = tanh(E_{sents} \times W_c \times E_{sents}^T), \tag{15}$$

$$E_{correlation} = ReLU(Linear(Correlation \times E_{sents})) \tag{16}$$

   where $W_c \in R^{d \times d}$ is the learned correlation matrix and $Correlation \in R^{n \times n}$ and $E_{Correlation} \in R^{n \times d}$.
5. Saliency: the saliency embedding will embed the importance of sentence vectors with respect to the document embedding. The saliency weight matrix $W_s \in R^{d \times d}$ is learned in the training phase.

$$Saliency = tanh(E_{sents} \times W_s \times E_D^T), \tag{17}$$

$$E_{Saliency} = ReLU(Linear(Saliency * E_{sents})) \tag{18}$$

   where $W_s \in R^{d \times d}$ is the learned saliency matrix and $Saliency \in R^{n \times 1}$. The document embedding ($E_D =$) is the weighted average of the sentence embeddings as explained in Sect. 5.2

5.2 Document Embedding

The document encoder is a simply the weighted average of the sentence vectors:

$$Weight = Softmax(E_{sents} \times W_{sents}) \tag{19}$$

where $E_{sents} \in R^{n \times d}$ and $W_{sents} \in R^{d \times 1}$. The wights are initialized randomly and will be learned during the training process. Therefore, $Weight \in R^{n \times 1}$ are the weights of the sentences.

Therefore, the embedding of document D is:

$$E_D = \frac{1}{n} \sum_{i=1}^{n} Weight[i] * E_{sents[i]} \tag{20}$$

where the terms are defined above.

5.3 Score Predictor

The score prediction module concatenates all of the features and feeds them to a linear layer to generate the final scores. Our cross-entropy loss evaluates the difference between the prediction and the ground truth scores. We also evaluated the loss factored by the rewards to see if the model makes better predictions using reinforcement learning (in Sect. 6)

$$\begin{aligned} p(y_i) = Linear(&E_{sent} + E_{length} + E_{position} \\ &+ E_{section} + E_{correlation} + E_{saliency}) \end{aligned} \tag{21}$$

where $p(y_i)$ is the probability of adding sentence i to the summary and the linear layer format is $Wx^T + b$ and $x \in R^{1 \times d}$ and $w \in R^{1 \times d}$.

6 Reinforcement Learning

Ground truth summaries are abstractive summaries that cover the important content from the input documents. Extracive summarization frameworks need conversion of abstractive summaries to extractive 0/1 labels and they maximize the likelihood of the 0/1 ground truth labels for sentences. The Objective function is to minimize this negative log likelihood

$$loss = -\sum_{i=1}^{n} log[p(y_i)]. \tag{22}$$

The objective in Eq. 22 maximizes the correct prediction of 0/1 labels where $p(y_i)$ is the probability of label y_i for sentence i. However, the evaluation of summaries are based on the similarity of selected sentences to the abstractive summaries evaluated by ROUGE scores. Therefore, in the training phase we are minimizing the cross entropy loss while in the test phase we evaluate the ROUGE scores [18].

To mitigate the discrepancy between the train and test objectives, Narayan et al. [18] suggest using ROUGE scores in the a reinforced setting to factor the pure cross-entropy loss

$$\nabla loss \simeq -r(y) \sum_{i=1}^{n} \nabla log[p(y_i)] \tag{23}$$

where $r(y)$ is the average of ROUGE-F1 and ROUGE-F2 scores.

Since there are multiple collection of sentences or candidate summaries that all could have reasonably high ROUGE scores, they suggest training the model with a selection of good candidate summaries. Therefore, if a candidate summary (made by extractive labels) have high overlap with the abstractive summary, the model wants to predict those labels since the $loss$ will be higher.

7 Experimental Results

7.1 Hardware

Three NVIDIA GPUs (GeForce RTX 2080 Ti) with 11019 MiB on memory were used. The batch size is set to 1 because of the size of the input document. Since we could not have a large batch size, we accumulate the gradients for 10 steps and then update the parameters. Learning rate is one of the most important hyper-parameters to tune. We used the NOAM scheduler to adjust the learning rate while training and apply gradient-clipping to prevent exploding gradients.

7.2 Experiments

Table 1 shows different values for the size of the local attention window and the ratio of the sentences that will attend globally to all other sentences in the document. The results show that increasing the window size for local attention and global attention ratio will improve the ROUGE recall scores. We can set the size of global and local attentions based on the hardware available at hand. Our model converges faster with a larger attention window and more global attentions.

The effect of reinforcement learning on our model is shown in Table 2. The reinforcement learning does not improve the results on our dataset mainly due to reducing the bias toward the position and length of the sentences.

Our results outperform many of the tested extractive and abstractive models as seen in Table 3. The sentence scores of BERTSUM in Table 3 are generated chunk by chunk since this model is not designed for extractive summarization of long documents. The BART and T5 summaries are generated section by section since they were developed for short sequences and have problems with long documents.

Table 4 shows the effect of trigram blocking in our dataset. If we block adding a sentence if it has a shared trigram, the results are not improved (first row). We also tried allowing some shared trigram in the sentence. For example, the third row in Table 4 only blocks sentences if there are more than 5 shared trigrams with the current summary. We see that the results get worse with trigram blocking. It shows that the scores predicted by the model are good enough to understand if adding a sentence with shared tokens can improve the ROUGE scores.

Table 1. Tuning the local attention window size and ratio of global attentions. The evaluation is based on the ROUGE recall scores. The summary limit is 20% of the size of the input document

Local window size	Global ratio (%)	ROUGE-1	ROUGE-2	ROUGE-L
6	–	56.854	19.692	41.210
10	–	58.854	20.392	41.810
20	–	59.06	20.77	42.00
30	–	58.989	20.664	42.031
40	–	58.97	20.44	41.91
50	–	59.408	21.099	42.232
40	20	59.47	21.11	42.34
40	40	59.72	21.45	42.77
50	20	**59.829**	21.479	42.973
50	40	59.714	**21.498**	**43.057**

Table 2. Applying the reinforcement learning suggested in [18] does not improve ROUGE scores.

Local window size	Global ratio (%)	Reinforced	ROUGE-1	ROUGE-2	ROUGE-L
20	–	No	59.06	20.77	42.00
20	–	Yes	55.27	16.40	38.54
30	–	No	58.989	20.664	42.031
30	–	Yes	55.38	16.57	38.72

Table 3. Comparison with baselines based on ROUGE recall.

MODEL	ROUGE-1	ROUGE-2	ROUGE-L
Lead20%	37.68	6.62	15.90
TextRank [13]	38.87	9.28	19.75
SummaRuNNer [15]	45.04	11.67	23.03
BART (section-based)	46.34	11.14	29.85
T5 (section-based)	44.72	10.23	29.63
BERTSUM	52.34	15.06	36.87
SciBERTSUM	**59.714**	**21.498**	**43.057**

Table 4. ROUGE results for tri-gram blocking.

Local window size	Global ratio (%)	Tri-gram count	ROUGE-1	ROUGE-2	ROUGE-L
40	40	Tri-grams > 0	51.27	14.57	33.34
40	40	Tri-grams > 3	57.28	19.10	39.49
40	40	Tri-grams > 5	58.49	20.18	40.90
40	40	No-blocking	59.72	21.45	42.77

8 Conclusions and Future Work

We created an extractive summarization framework, SciBERTSUM, based on BERT-SUM for long documents with multiple sections (e.g. scientific papers). We generate sentence vectors based on their sections. The section information is important for the summarization task since sentences in the abstract or method sections are more important compared to the acknowledgement parts. To build a computationally efficient model that scales linearly with the number of sentences in the document, we employed the sparse attention mechanism of LongFormer [2] to embed the inter sentence relations. All sentences attend to a limited number of sentences before and after the current sentence and only a small number of random sentences attend globally to all other sentences. Our model is computationally efficient and improves the ROUGE scores on the dataset of paper-slide pairs.

Future work could be applying our model on existing summarization datasets and other long scholarly documents. It would also be interesting to see whether the SciBERT language model, which is pre-trained on scientific text, will give improved performance.

Acknowledgement. Partial support from the National Science Foundation is gratefully acknowledged.

References

1. Beltagy, I., Lo, K., Cohan, A.: SciBERT: a pretrained language model for scientific text. In: Proceedings of the 2019 Conference on Empirical Methods in Natural Language Processing and the 9th International Joint Conference on Natural Language Processing (EMNLP-IJCNLP), pp. 3615–3620. Association for Computational Linguistics, Hong Kong, November 2019. https://doi.org/10.18653/v1/D19-1371, https://aclanthology.org/D19-1371
2. Beltagy, I., Peters, M.E., Cohan, A.: Longformer: the long-document transformer. arXiv preprint arXiv:2004.05150 (2020)
3. Bhatia, S., Mitra, P.: Summarizing figures, tables, and algorithms in scientific publications to augment search results. ACM Trans. Inf. Syst. **30**(1), 1–24 (2012). https://doi.org/10.1145/2094072.2094075
4. Celikyilmaz, A., Bosselut, A., He, X., Choi, Y.: Deep communicating agents for abstractive summarization. In: Proceedings of the 2018 Conference of the North American Chapter of the Association for Computational Linguistics: Human Language Technologies, Volume 1 (Long Papers), pp. 1662–1675. Association for Computational Linguistics, New Orleans, June 2018. https://doi.org/10.18653/v1/N18-1150, https://aclanthology.org/N18-1150
5. Cheng, J., Lapata, M.: Neural summarization by extracting sentences and words. In: Proceedings of the 54th Annual Meeting of the Association for Computational Linguistics (Volume 1: Long Papers), pp. 484–494. Association for Computational Linguistics, Berlin, August 2016. https://doi.org/10.18653/v1/P16-1046, https://www.aclweb.org/anthology/P16-1046
6. Devlin, J., Chang, M.W., Lee, K., Toutanova, K.: BERT: pre-training of deep bidirectional transformers for language understanding. In: Proceedings of the 2019 Conference of the North American Chapter of the Association for Computational Linguistics: Human Language Technologies, Volume 1 (Long and Short Papers), pp. 4171–4186. Association for Computational Linguistics, Minneapolis, June 2019. https://doi.org/10.18653/v1/N19-1423, https://www.aclweb.org/anthology/N19-1423

7. Elkiss, A., Shen, S., Fader, A., Erkan, G., States, D., Radev, D.: Blind men and elephants: what do citation summaries tell us about a research article? J. Am. Soc. Inform. Sci. Technol. **59**(1), 51–62 (2008)
8. Grusky, M., Naaman, M., Artzi, Y.: Newsroom: a dataset of 1.3 million summaries with diverse extractive strategies. In: Proceedings of the 2018 Conference of the North American Chapter of the Association for Computational Linguistics: Human Language Technologies, Volume 1 (Long Papers), pp. 708–719. Association for Computational Linguistics, New Orleans, June 2018. https://doi.org/10.18653/v1/N18-1065, https://aclanthology.org/N18-1065
9. Ibrahim Altmami, N., El Bachir Menai, M.: Automatic summarization of scientific articles: a survey. J. King Saud Univ. - Comput. Inf. Sci. (2020)
10. Lev, G., Shmueli-Scheuer, M., Herzig, J., Jerbi, A., Konopnicki, D.: TalkSumm: a dataset and scalable annotation method for scientific paper summarization based on conference talks. arXiv preprint arXiv:1906.01351 (2019)
11. Lewis, M., et al.: BART: denoising sequence-to-sequence pre-training for natural language generation, translation, and comprehension. In: Proceedings of the 58th Annual Meeting of the Association for Computational Linguistics, pp. 7871–7880. Association for Computational Linguistics, July 2020. https://doi.org/10.18653/v1/2020.acl-main.703, https://www.aclweb.org/anthology/2020.acl-main.703
12. Liu, Y., Lapata, M.: Text summarization with pretrained encoders. In: Proceedings of the 2019 Conference on Empirical Methods in Natural Language Processing and the 9th International Joint Conference on Natural Language Processing (EMNLP-IJCNLP), pp. 3730–3740. Association for Computational Linguistics, Hong Kong, November 2019. https://doi.org/10.18653/v1/D19-1387, https://www.aclweb.org/anthology/D19-1387
13. Mihalcea, R., Tarau, P.: TextRank: bringing order into text. In: Proceedings of the 2004 Conference on Empirical Methods in Natural Language Processing, pp. 404–411 (2004)
14. Mosbach, M., Andriushchenko, M., Klakow, D.: On the stability of fine-tuning BERT: misconceptions, explanations, and strong baselines. arXiv preprint arXiv:2006.04884 (2020)
15. Nallapati, R., Zhai, F., Zhou, B.: SummaRuNNer: a recurrent neural network based sequence model for extractive summarization of documents. In: Thirty-First AAAI Conference on Artificial Intelligence (2017)
16. Nallapati, R., Zhou, B., dos Santos, C., Gulçehre, Ç., Xiang, B.: Abstractive text summarization using sequence-to-sequence RNNs and beyond. In: Proceedings of The 20th SIGNLL Conference on Computational Natural Language Learning, pp. 280–290. Association for Computational Linguistics, Berlin, August 2016. https://doi.org/10.18653/v1/K16-1028, https://www.aclweb.org/anthology/K16-1028
17. Narayan, S., Cohen, S.B., Lapata, M.: Don't give me the details, just the summary! Topic-aware convolutional neural networks for extreme summarization. In: Proceedings of the 2018 Conference on Empirical Methods in Natural Language Processing. Brussels, Belgium (2018)
18. Narayan, S., Cohen, S.B., Lapata, M.: Ranking sentences for extractive summarization with reinforcement learning. arXiv preprint arXiv:1802.08636 (2018)
19. See, A., Liu, P.J., Manning, C.D.: Get to the point: summarization with pointer-generator networks. In: Proceedings of the 55th Annual Meeting of the Association for Computational Linguistics (Volume 1: Long Papers), pp. 1073–1083. Association for Computational Linguistics, Vancouver, July 2017. https://doi.org/10.18653/v1/P17-1099, https://www.aclweb.org/anthology/P17-1099
20. Sefid, A., Mitra, P., Giles, L.: SlideGen: an abstractive section-based slide generator for scholarly documents. In: Proceedings of the 21st ACM Symposium on Document Engineering, pp. 1–4 (2021)

21. Sefid, A., Mitra, P., Wu, J., Giles, C.L.: Extractive research slide generation using windowed labeling ranking. In: Proceedings of the Second Workshop on Scholarly Document Processing, pp. 91–96. Association for Computational Linguistics, June 2021. https://doi.org/10.18653/v1/2021.sdp-1.11, https://aclanthology.org/2021.sdp-1.11
22. Sotudeh Gharebagh, S., Cohan, A., Goharian, N.: GUIR @ LongSumm 2020: learning to generate long summaries from scientific documents. In: Proceedings of the First Workshop on Scholarly Document Processing, pp. 356–361. Association for Computational Linguistics, November 2020. https://doi.org/10.18653/v1/2020.sdp-1.41, https://www.aclweb.org/anthology/2020.sdp-1.41
23. Vaswani, A., et al.: Attention is all you need. In: Guyon, I., et al. (eds.) Advances in Neural Information Processing Systems, vol. 30, pp. 5998–6008. Curran Associates, Inc. (2017). http://papers.nips.cc/paper/7181-attention-is-all-you-need.pdf
24. Yasunaga, M., et al.: ScisummNet: a large annotated corpus and content-impact models for scientific paper summarization with citation networks. In: Proceedings of the AAAI Conference on Artificial Intelligence, vol. 33, pp. 7386–7393 (2019)
25. Zhang, X., Wei, F., Zhou, M.: HIBERT: document level pre-training of hierarchical bidirectional transformers for document summarization. arXiv preprint arXiv:1905.06566 (2019)

Using Multi-level Segmentation Features for Document Image Classification

Panagiotis Kaddas[1,2](✉) and Basilis Gatos[1]

[1] Computational Intelligence Laboratory, Institute of Informatics and Telecommunications, National Center for Scientific Research Demokritos, 153 10 Agia Paraskevi, Athens, Greece
{pkaddas,bgat}@iit.demokritos.gr

[2] Department of Informatics and Telecommunications, University of Athens, 157 84 Athens, Greece

Abstract. Document Image classification is a crucial step in the processing pipeline for many purposes (e.g. indexing, OCR, keyword spotting) and is being applied at early stages. At this point, textual information about the document (OCR) is usually not available and additional features are required in order to achieve higher recognition accuracy. On the other hand, one may have reliable segmentation information (e.g. text block, paragraph, line, word, symbol segmentation results), extracted also at pre-processing stages. In this paper, visual features are fused with segmentation analysis results in a novel integrated workflow and end-to-end training can be easily applied. Significant improvements on popular datasets (Tobacco-3482 and RVL-CDIP) are presented, when compared to state-of-the-art methodologies which consider visual features.

Keywords: Document image classification · Document image segmentation · Convolutional Neural Network · Deep Learning

1 Introduction

Digitization of Documents has already become a necessity in order to assist daily tasks and transactions. Moreover and throughout the globe, historical documents can be accessed on-line and information can be exploited by the community for any purpose. To this end, several methodologies are being applied on scanned document images in order to convert them to their digital twins. This leads to the application of image processing techniques such as Optical Character Recognition (OCR), automatic indexing and keyword-spotting for searching documents in huge databases.

Unfortunately, most of the techniques that have been mentioned do not always apply successfully due to the vast diversity of document types. For example, a historical handwritten document must be processed using a completely

© Springer Nature Switzerland AG 2022
S. Uchida et al. (Eds.): DAS 2022, LNCS 13237, pp. 702–712, 2022.
https://doi.org/10.1007/978-3-031-06555-2_47

different workflow when compared to processing a scanned invoice. So, it is clear that a prior processing step must be applied in order to: a) classify document images to their corresponding type (class) and b) select a suitable system for processing based on document class. Motivated by this observation, this paper addresses the problem of document classification, based on the visual features of document images.

Classic approaches [1–3] towards the document classification problem focus on extracting image features for defining structural similarity of documents. During the past ten years, the rise of Deep Learning has been proven suitable to match the need of classifying huge document image databases [4] with high intra-class and low inter-class variability by only using the document image as input and leveraging the advantages of transferring knowledge from similar domains [5,6]. Recent works [7–9] have shown that textual information can be combined with image features in order to improve classification accuracy.

The proposed work focuses on the Document Image Classification problem by combining a Convolutional Neural Network (CNN) architecture with multi-level information provided by image segmentation techniques (text block, paragraph, line, word, symbol segmentation results). Textual information is not considered in the proposed work, under the assumption that document image classification usually takes place in pre-processing stages where textual information (OCR) is not available.

The contributions of this paper are as follows: a) A novel integrated architecture is described and end-to-end training can be applied by only using a document image and one or more image masks that correspond to the segmentation levels that are mentioned above. b) An experimental study is being conducted in order to determine which segmentation levels should be used and decide whether multi-level segmentation features contribute to the task at hand. c) We present competitive results when compared to the state-of-the-art techniques, evaluated on commonly-used datasets (Tobacco-3482 [3], RVL-CDIP [4]). d) An additional proof-of-concept is presented for a new private dataset from The Library of the Piraeus Bank Group Cultural Foundation (PIOP)[1] used in the *CULDILE*[2] project.

The rest of the paper is organized as follows. Section 2 presents related works, Sect. 3 introduces the proposed architecture, Sect. 4 demonstrates experimental results and Sect. 5 presents the conclusion of this work.

2 Related Work

Several methodologies that leverage the advantages of Deep Neural Networks over document images have been proposed over the last decade for the document image classification problem. In [10], a basic CNN architecture is proposed in order to learn features from raw image pixels instead of relying on hand-crafted

[1] https://www.piop.gr/en/vivliothiki.aspx.
[2] http://culdile.bookscanner.gr.

features. In [4], it was shown that a CNN can extract robust features from different parts of the image (header, footer left and right body) and by training such different networks and using them in an ensemble scheme, significant classification or retrieval accuracy improvements are achieved. Moreover, reduction of feature space by using Principal Component Analysis (PCA) is important, while the performance is not affected significantly.

Further experiments for document image classification where presented in [5,11,12], where many neural network architectures where compared (e.g. *AlexNet* [13], *VGG-16* [14], *ResNet50* [15], *GoogleNet* [16]) under different scenarios (transfer learning [5], data augmentation [17]). The advantages of learning using some kind of spatial information from parts of the document (holistic, header, footer, left and right body) and parallel VGG-16 based systems were presented in [6]. Furthermore, inter-domain and intra-domain transfer learning schemes are described and finally they present a comparison of possible meta-classifier techniques on the stacked output of the parallel sub-systems.

In addition to the methodologies described above which try to learn and make use of visual features of document images, there are recent techniques [7–9] that fuse visual and textual information in parallel systems. In [7], two classifiers are trained in parallel. The first is a classic visual-based CNN and the second takes as input text embeddings, extracted using open-source Tesseract OCR Engine[3] and FastText[4]. Similar approach is also presented in [8], where text embeddings proposed in [18] are considered as a textual feature selection scheme. Fernando et al. [9] proposed the use of *EfficientNet* models [19] as a lighter alternative to classic CNN architectures for the visual feature extraction and combined it with the well-known *BERT* model [20] as a textual transformer. Finally, Xu et al. [21] proposed the LayoutLM model, where layout and image embeddings extracted from *Faster R-CNN* [22] are integrated into the original BERT architecture and work together for feature extraction.

3 Proposed Method

As mentioned in Sect. 1, this work does not focus on using textual information, under the assumption that document image classification usually takes place in pre-processing stages where textual information (OCR) is not available or reliable. Instead of this, we try to embed segmentation features of multiple levels (e.g. text block, paragraph, line, word, symbol segmentation results), which are usually extracted during pre-processing stages (Fig. 1).

3.1 Integrated CNN Architecture

The proposed system relies on two kinds of input: At first, a classic CNN flow is considered, using *ResNet50* as a backbone architecture. We chose *ResNet50*,

[3] https://github.com/tesseract-ocr/tesseract.
[4] https://github.com/facebookresearch/fastText.

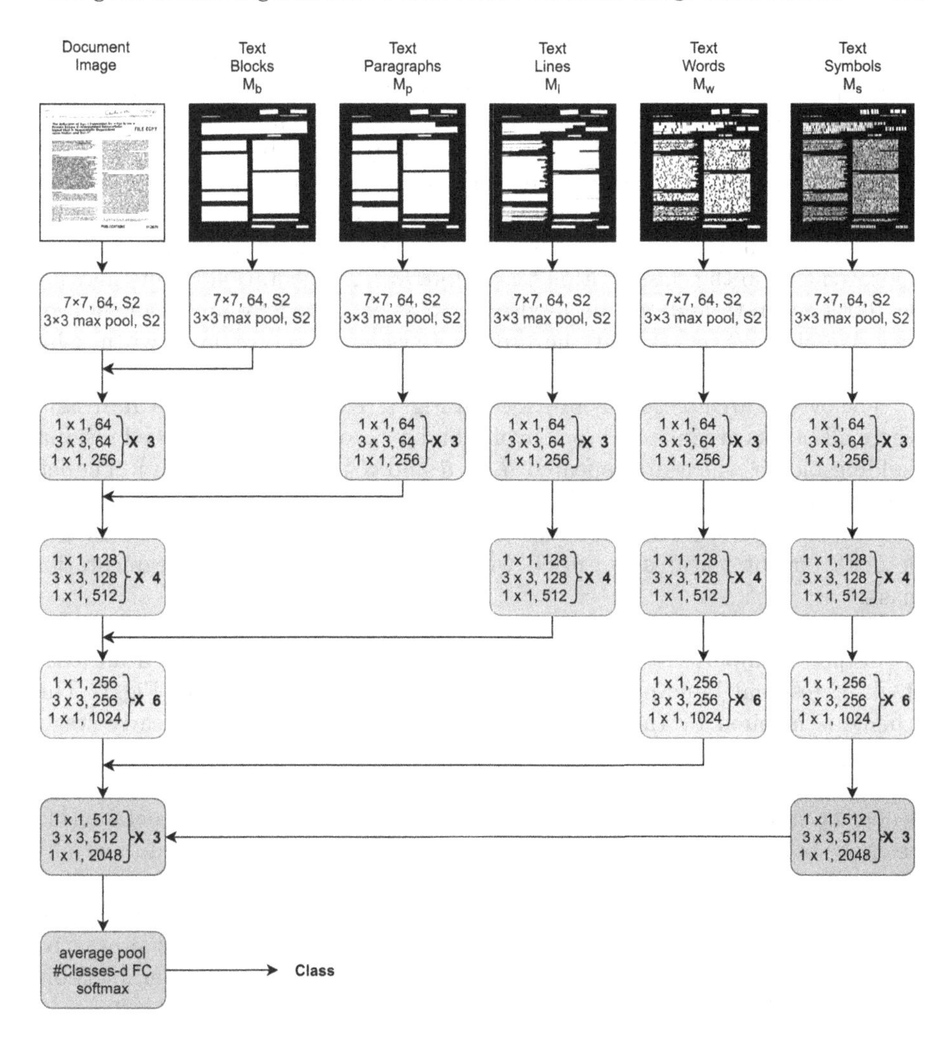

Fig. 1. Overview of the proposed network. Document image and segmentation masks are forwarded in parallel network streams using *ResNet50* as backbone network. Each segmentation stream is "deeper" than the previous one in order to be able to learn higher level of information. Each output stream is added to the corresponding layer of the backbone (left branch) and finally a Fully Connected layer yields class probabilities. (Color figure online)

over other backbones (e.g. *VGG-16*, *GoogleNet*) for reasons such as: advantages of residual connections [15], simplicity in architecture, number of weight parameters. We do not conduct experiments using other CNN backbones and this is out of the scope of this paper. In addition, our main goal is to demonstrate accuracy improvements of our proposed system over similar state-of-the-art techniques mainly based on *ResNet50*.

The second kind of inputs to our system are document image binary masks that represent segmentation information at various text levels, namely M_b : *block*, M_p : *paragraph*, M_l : *line*, M_w : *word*, M_s : *symbol*. A pixel of each mask has an "on" value if it is contained in a detected polygon of this level. So, we consider binary masks that correspond to the detected polygons *(x,y coordinates)* of multiple segmentation levels as inputs. As described in our experiments (Sect. 4), not all segmentation masks are required in order to achieve accuracy improvements in the classification task. Moreover, segmentation results may not be 100% accurate and can be used exactly as extracted by any segmentation tool. We claim (Sect. 4) that the proposed system is robust even when using noisy segmentation results.

The overall architecture is provided in Fig. 1. The left branch is a *ResNet50*, where all blocks are included (a convolution block followed be 4 stacked residual building blocks which are repeated *3*, *4*, *6* and *3* times respectively). The left branch is considered as the backbone of the proposed system. When forwarding to the next type of a residual block (illustrated with different colors in the left branch), output features are fused with those extracted from a segmentation mask.

The other five branches take as input the segmentation masks. Each branch is "deeper" compared to the previous one and is forwarded through an extra stack of residual building blocks, following the scheme shown in Fig. 1. This scheme is inspired by the fact that convolution layers applied at early stages learn abstract layouts and shapes (such as spatial position of text blocks in our case, which are considered the higher level of information), in contrast to deeper layers which can handle more complex visual elements and details (like positions of text lines, words and even symbols, the lowest level of information). So, we handle higher levels of information with less layers and we increase the depth of a branch considering the level of details in the input segmentation mask. The proposed architecture was also verified after trying many alternative schemes (e.g. forward high level segmentation masks through "deeper" network branches) which yielded less accurate results.

3.2 Implementation Details

In general, our TensorFlow[5] implementation of the proposed model follows [15]. We use "bottleneck" residual blocks for convolutions, followed by Batch Normalization (*BN*) and *ReLU* activation layers. The main differences from [15] are that, during training, we use input images of size 256×256 as long as cyclic learning rate [23] with Stochastic Gradient Descent (*SGD*), with values ranging in $[0.0001, 0.1]$. Before the final Fully Connected (*FC*) layer, we apply *Dropout* with skipping ratio of 0.5. Finally, our inputs are augmented using random cropping (80% of the original size at most) and mirroring over $y - axis$. Weight initialization comes from ImageNet weights [24].

[5] https://www.tensorflow.org.

Training can be easily applied over the integrated network. If information is not available at a certain segmentation level (e.g. paragraphs), the corresponding branch can be discarded. We used an NVIDIA GTX 1080 Ti 11 GB GPU with batch size 16 at most cases.

For segmenting the documents, the Google Cloud Vision API[6] is considered.

4 Experiments

4.1 Datasets

For our experiments, we use three datasets. The smallest one is Tobacco-3482 [3] which consists of ten document classes. As there is no official split in train-val-test subsets, we average over five random splits, following the same logic as in other state-of-the-art methods [4]. We use this dataset in order to evaluate the performance of our proposed network and to investigate the contribution of segmentation information for the document classification task.

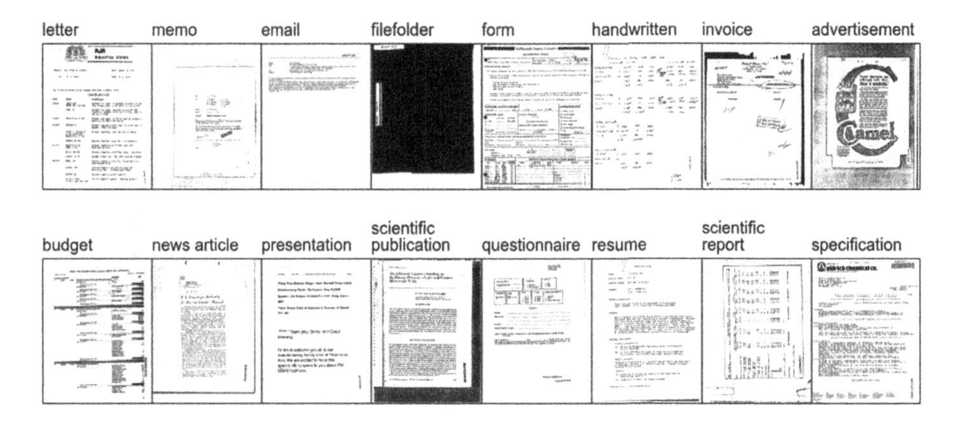

Fig. 2. Example of the RVL-CDIP dataset, consisting of 16 document classes.

Secondly, we use the large-scale Ryerson Vision Lab Complex Document Information Processing (RVL-CDIP) dataset [4]. It consists of 320, 000 training images and a validation and test dataset with 40, 000 images each. This dataset has 16 document classes (see Fig. 2) and is considered the most challenging dataset for Document Image Classification.

Finally, we introduce a new dataset, obtained from The Library of the Piraeus Bank Group Cultural Foundation (PIOP)[7] and used in the *CULDILE*[8] project. We selected pages belonging to four classes (Mail, Contracts, Financial,

[6] https://cloud.google.com/vision.

[7] https://www.piop.gr/en/vivliothiki.aspx.

[8] http://culdile.bookscanner.gr.

Architecture Plans) and each class has about 9,000 images. We split the dataset randomly (60% train, 20% validation and 20% test). Exemplar images of this dataset are given in Fig. 3.

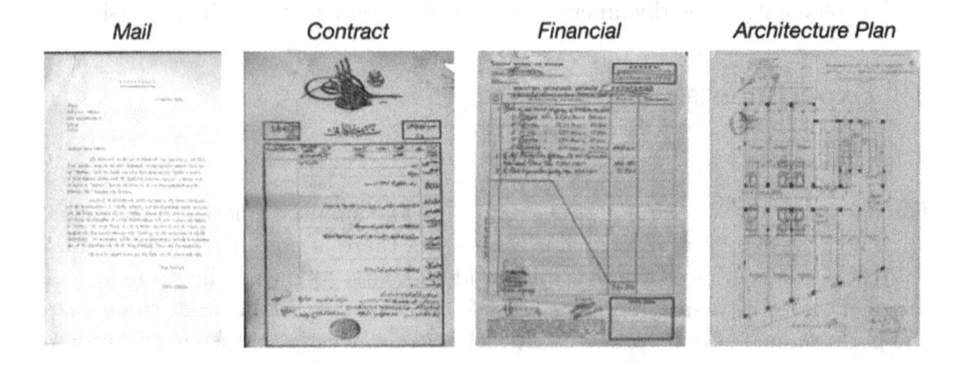

Fig. 3. Exemplar images of the PIOP dataset, consisting of 4 document classes.

4.2 Experimental Results

As a first experiment, we investigate the contribution of using multi-level segmentation features to the document classification problem. For this reason, we use the Tobacco-3482 and train using different schemes. At scenario $Baseline_A$, we train a single $ResNet50$. At $Baseline_B$, we train several classic $ResNet50$ models using only segmentation information (and not the original document image), by stacking masks in a single image of depth n, where n is the number of the stacked masks. We do this for every possible combination over the segmentation masks. At $Baseline_C$, we just concatenate the output probabilities of already trained models of $Baseline_A$ and $Baseline_B$ for every combination of the latter. Finally, at $Baseline_D$, we train $Baseline_A$ and $Baseline_B$ models (all combinations again) in a parallel scheme, where we concatenate the convolution outputs for both models and we use an FC layer of 1024 neurons before the output FC layer. This can be considered as a simple ensemble scheme.

The investigation mentioned above help us to decide which levels of segmentation to keep in the proposed architecture. Table 1 demonstrates the best results for each baseline scenario. The best combination uses the initial document image, line, word and symbol segmentation masks. For completeness, we summarize the results of previous works for the Tobacco-3482 dataset that use visual features (we do not include methods that use textual features). As shown in Table 1, our proposed method outperforms (80.64%) all ResNet50-based models ($Baseline_A$ and [5]), as long as other architectures that use AlexNet, VGG-16 or GoogleNet as backbones [4,5]. We note that we do not compare with methods that use weight initialization from models trained on the much larger RVL-CDIP dataset.

Table 1. Accuracy of combinations over multi-level segmentation masks for document image classification using F-Measure for Tobacco-3482 (%).

Method	Image	Block	Line	Word	Symbol	Accuracy (%)
$Baseline_A$	✓					68.78
$Baseline_B$			✓	✓	✓	75.40
$Baseline_C$	✓		✓	✓	✓	79.86
$Baseline_D$	✓	✓	✓	✓		78.63
Proposed method	✓		✓	✓	✓	**80.64**
Harley et al. - Ensemble of regions [4]	✓					79.90
Afzal et al. - VGG-16 [11]	✓					77.60
Afzal et al. - ResNet50 [5]	✓					67.93
Audebert et al. - MobileNetV2 [7]	✓					**84.50**
Fernando et al. - EfficientNet [9]	✓					**85.99**

Furthermore, from all baseline experiments that were conducted, it was clear that segmentation information contributes significantly to classification tasks (almost an additional 11% in accuracy for $Baseline_C$). In fact, even when using only segmentation masks instead of the original document image, accuracy increases remarkably ($Baseline_B$). We found out that Line and Word and Symbol segmentation masks play the most important role in most combinations and yielded better results (not included in Table 1, for convenience), in contrast to Paragraph and Block masks that are of less importance. Finally, our proposed method does not outperform methods that depend on more recent backbones (MobileNetV2 [7] and EfficientNet [9]). We believe that applying our proposed scheme in a future work, using such backbones, will improve accuracy even more, when compared to [7] and [9].

Our second experiment concerns the evaluation of our proposed architecture over the RVL-CDIP and PIOP datasets. Again, we do not consider textual methods in our comparison (Table 2). As in our first experiment, we use the best input combination (document image, line, word and symbol segmentation masks) as the proposed system. We observe that, concerning the RVL-CDIP dataset, the proposed method outperforms all other techniques no matter the backbone architecture that is used (92.95%). Finally, the PIOP dataset is another proof that our proposed scheme can improve accuracy results when applied on a ResNet50 backbone.

Table 2. Accuracy of combinations over multi-level segmentation masks for document image classification using F-Measure for RVL-CDIP and PIOP datasets (%).

Method	Accuracy on RVL-CDIP (%)	Accuracy on PIOP (%)
Harley et al. - Ensemble of regions [4]	89.80	–
Csurka et al. - GoogleNet [12]	90.70	–
Afzal et al. - ResNet50 [5]	90.40	–
Afzal et al. - VGG-16 [5]	90.97	–
Das et al. - Ensemble of VGG-16 models [6]	92.21	–
Fernando et al. - EfficientNet [9]	92.31	–
$Baseline_A$	90.55	84.28
Proposed method	**92.95**	**86.31**

5 Conclusion

In this paper we proposed a novel integrated architecture in which document images and multi-level segmentation features can be fused for document classification. We showed in our experiments that segmentation is considered to be useful for improving image-based classification methods, even when we use noisy multi-level masks. This also introduces accuracy improvements for the RVL-CDIP dataset, as presented in our experiments. Moreover, we conducted an investigation on Tobacco-3482 in order to define which segmentation levels can yield better results and found that more detailed segmentation (at line, word and symbol) is of greater importance rather than segmentation of higher levels (blocks, paragraphs).

Acknowledgements. This research has been co-financed by the European Union and Greek national funds through the Operational Program Competitiveness, Entrepreneurship and Innovation, under the RESEARCH-CREATE-INNOVATE call (project code: T1EDK-03785 and acronym: CULDILE) as well as by the program of Industrial Scholarships of Stavros Niarchos Foundation[9].

References

1. Shin, C.K., Doermann, D.S.: Document image retrieval based on layout structural similarity. In: Proceedings of the 2006 International Conference on Image Processing, Computer Vision, & Pattern Recognition (IPCV), Las Vegas, Nevada, USA, pp. 606–612 (2006)

[9] https://www.snf.org/en/.

2. Chen, S., He, Y., Sun, J., Naoi, S.: Structured document classification by matching local salient features. In: 21st International Conference on Pattern Recognition (ICPR), Tsukuba Science City, Japan, pp. 1558–1561 (2012)
3. Kumar, J., Ye, P., Doermann, D.: Structural similarity for document image classification and retrieval. Pattern Recogn. Lett. **43**, 119–126 (2016)
4. Harley, A.W., Ufkes, A., Derpanis, K.G.: Evaluation of deep convolutional nets for document image classification and retrieval. In: 13th International Conference on Document Analysis and Recognition (ICDAR), Nancy, France, pp. 991–995 (2015)
5. Afzal, M.Z., Kölsch, A., Liwicki, S.A.M.: Cutting the error by half: investigation of very deep CNN and advanced training strategies for document image classification. In: 14th International Conference on Document Analysis and Recognition (ICDAR), Kyoto, Japan, pp. 883–890 (2017)
6. Das, A., Roy, S., Bhattacharya, U., Parui, S.K.: Document image classification with intra-domain transfer learning and stacked generalization of deep convolutional neural networks. In: 24th International Conference on Pattern Recognition (ICPR), Beijing, China, pp. 3180–3185 (2018)
7. Audebert, N., Herold, C., Slimani, K., Vidal, C.: Multimodal deep networks for text and image-based document classification. In: Cellier, P., Driessens, K. (eds.) ECML PKDD 2019. CCIS, vol. 1167, pp. 427–443. Springer, Cham (2020). https://doi.org/10.1007/978-3-030-43823-4_35
8. Asim, M.N., Khan, M.U.G., Malik, M.I., Razzaque, K., Dengel, A., Ahmed, S.: Two stream deep network for document image classification. In: 15th International Conference on Document Analysis and Recognition (ICDAR), Sydney, Australia, pp. 1410–1416 (2019)
9. Ferrando, J., et al.: Improving accuracy and speeding up document image classification through parallel systems. In: Krzhizhanovskaya, V.V., et al. (eds.) ICCS 2020. LNCS, vol. 12138, pp. 387–400. Springer, Cham (2020). https://doi.org/10.1007/978-3-030-50417-5_29
10. Kang, L., Kumar, J., Ye, P., Li, Y., Doermann, D.: Convolutional neural networks for document image classification. In: 22th International Conference on Pattern Recognition (ICPR), Stockholm, Sweden, pp. 3168–3172 (2014)
11. Afzal, M.Z., et al.: DeepDocClassifier: document classification with deep convolutional neural network. In: 13th International Conference on Document Analysis and Recognition (ICDAR), Nancy, France, pp. 1273–1278 (2015)
12. Csurka, G., Larlus, D., Gordo, A., Almazan, J.: What is the right way to represent document images?. arXiv preprint arXiv:1603.01076 (2016)
13. Krizhevsky, A., Sutskever, I., Hinton, G.E.: ImageNet classification with deep convolutional neural networks. In: 26th Conference on Neural Information Processing Systems (NIPS), Harrah's Lake Tahoe, USA, pp. 1097–1105 (2012)
14. Simonyan, K., Zisserman, A: Very deep convolutional networks for large-scale image recognition. arXiv preprint arXiv:1409.1556 (2014)
15. He, K., Zhang, X., Ren, S., Sun, J.: Deep residual learning for image recognition. In: Proceedings of the IEEE Conference on Computer Vision and Pattern Recognition (CVPR), Las Vegas, Nevada, USA, pp. 770–778 (2016)
16. Szegedy, C., et al.: Going deeper with convolutions. In: Proceedings of the IEEE Conference on Computer Vision and Pattern Recognition (CVPR), Boston, USA, pp. 1–9 (2015)
17. Tensmeyer, C., Martinez, T.: Analysis of convolutional neural networks for document image classification. arXiv preprint arXiv:1708.03273 (2017)

18. Noce, L., Gallo, I., Zamberletti, A., Calefati A.: Embedded textual content for document image classification with convolutional neural networks. In: Proceedings of the ACM Symposium on Document Engineering (DocEng), Vienna, Austria, pp. 165–173 (2016)
19. Tan, M., Le, Q.: EfficientNet: rethinking model scaling for convolutional neural networks. In: Proceedings of the 36th International Conference on Machine Learning (PMLR), Long Beach, California, pp. 6105–6114 (2019)
20. Devlin, J., Chang, M.W., Lee, K., Toutanova, K.: BERT: pre-training of deep bidirectional transformers for language understanding. In: Proceedings of the 2019 Conference of the North American Chapter of the Association for Computational Linguistic (NAACL), Mineapolis, Minesota, USA, pp. 4171–4186 (2019)
21. Xu, Y., Li, M., Cui, L., Huang, S., Wei, F., Zhou, M.: LayoutLM: pre-training of text and layout for document image understanding. In: Proceedings of the 26th ACM SIGKDD International Conference on Knowledge Discovery (SIGKDD), pp. 1192–1200 (2020)
22. Ren, S., He, K., Girshick, R., Sun, J.: Faster R-CNN: towards real-time object detection with region proposal networks. IEEE Trans. Pattern Anal. Mach. Intell. **39**(6), 1137–1149 (2017)
23. Smith, L.N.: Cyclical learning rates for training neural networks. In: IEEE Winter Conference on Applications of Computer Vision (WACV), Santa Rosa, California, USA, pp. 464–472 (2017)
24. Deng, J., Dong, W., Socher, R., Li, L.J., Li K., Fei-Fei, L.: ImageNet: a large-scale hierarchical image database. In: IEEE Computer Vision and Pattern Recognition (CVPR), Miami, Florida, USA, pp. 248–255 (2009)

Eye Got It: A System for Automatic Calculation of the Eye-Voice Span

Mohamed El Baha[1], Olivier Augereau[3(✉)], Sofiya Kobylyanskaya[2], Ioana Vasilescu[2], and Laurence Devillers[2]

[1] IMT Atlantique, Brest, France
mohamed.el-baha@imt-atlantique.net
[2] Université Paris-Saclay, CNRS LISN, Paris, France
sofiya.kobylyanskaya@limsi.fr
[3] Lab-STICC CNRS UMR 6285, ENIB, Brest, France
augereau@enib.fr

Abstract. Over the past decade, eye movement has been widely looked into for describing and analyzing several cognitive processes and especially for human-document interaction, such as estimating reading ability and document understanding. Most of the existing applications have been done for silent reading but we propose to explore reading aloud interaction through a powerful measurement named the "eye-voice span" which measures the distance between the eyes and the voice. In this paper we present an open-source platform named "Eye got it" and the underlying algorithms that can be used for processing eye-tracking and voice data in order to compute automatically the eye-voice span.

Keywords: Eye tracking · Eye-voice span · Human-document interaction · Voice analysis

1 Introduction

Studying the way people are interacting with documents can give insightful information about the documents and the readers. For example, it can be used for detecting if a document is hard to understand (for all readers) or if a reader is struggling to read the document and needs help. In the same way it can be used to detect if a document is interesting [3] or has some emotional content [12] or if a reader is interested or feels something while reading the document.

Eye movement research related to reading starts to attract attention thanks to the development of more affordable eye-trackers, which are able to obtain accurate measurements. Eye-tracking is the process of measuring either the point of gaze (the position where we look) or the movement of the eyes. The tracking is done by an eye tracker that measures the position and movement of the eyes. A great deal of research studies has been done with eye trackers. The goal is

This work is supported by ANR (ref ANR-20-IADJ-0007), DFG and JST.

© Springer Nature Switzerland AG 2022
S. Uchida et al. (Eds.): DAS 2022, LNCS 13237, pp. 713–725, 2022.
https://doi.org/10.1007/978-3-031-06555-2_48

generally to estimate the gaze position when doing an activity, such as driving a car, shopping, or for marketing and industrial applications, etc. In our case, we are considering education applications and more specifically we focus on L2 (second language) learning paradigms. We aim thus to provide valuable feedback about L2 English learning by adults, such as English for non-native speakers. It has been shown that the pattern of the eye movement can be used to predict the level of English of a reader [2] and even predict the reader's TOEIC score after reading some texts [1].

Most of this research has been done for silent reading but some researchers also looked into reading aloud. By simultaneously recording the eye movements and the voice while reading aloud, we can see how these two measurements are related to each other and how they vary with time which can provide more information about the reader and help, for example, to detect dyslexia [7]. According to Laubrock and Kliegl, when we read a text aloud, our eyes are generally looking ahead from our voice [10]. This distance is what we call the eye-voice span (EVS). The EVS has been a center of various research since 1920 where Buswell found patterns that describe eye movement during oral reading similar to silent reading such as forward and backward saccades, fixations and word skipping [5]. According to several studies, the EVS can be an accurate indicator of the reader's reading skills and text understanding. According to Buswell, a skilled reader will tend to maintain a significant average span between the eye and the voice, while a novice reader will tend to keep the eye and voice very close together, in many cases not moving the eye from a word until the voice has pronounced it [6]. Silva et al. [17] demonstrate in their experience that the word familiarity and the word length have a strong effect on the eye-voice span. Laubrock and Kiegl showed that the EVS is constantly regulated [10] thought the reading time according to cognitive, oculomotor, and articulatory demands and that the EVS can be used to predict regression, fixations, and saccades which are related to the reading skill.

The EVS is commonly measured either with a time reference such as millisecond (it is then called the temporal EVS), or with a spatial reference such as the number of letters or words (in which case it will be called the spatial EVS). In his study, Buswell found out an average spatial distance of 15 letters for college students [5] reading a text in their native language. In more recent study, Inhoff et al. [9] reported a average temporal EVS of 500ms for standard reader and, De Luca et al. showed that dyslexic readers have an average spatial EVS of 8.4 letters whereas standard readers have an average spatial EVS of 13.8 letters [7].

The present work is part of a large-scale research project aiming at building an experimental paradigm dedicated to the learning of a second language using multimodal information such as the eye movement, the voice, and facial expressions. The part about facial expression will not be described in this paper. For this purpose and to help researchers to collect data, we built a system named "Eye Got it". It is open source and available on GitHub[1]. Our system integrates several algorithms for eye movement and voice processing and analysis, and com-

[1] https://github.com/oaugereau/Eye_Got_It.

putes the EVS automatically. In this paper, we will present the architecture of this system, the underlying algorithms and describe preliminary data in terms of voice/eye movements recordings.

In the following section, we will start by defining the "Eye Got it" system and present the process of computing the EVS step by step. In the next section we will present an experiment that we set up in order to test and validate the system. Finally, we will conclude by explaining the limitations of the program and presenting possible improvements.

2 Eye Got It

We built an easy-to-use system called "Eye got it" that will allow researchers to record the voice and eye movement of people reading aloud texts in a second language with different levels. Our platform is used for displaying a text, recording eye tracking and voice data and displaying multiple choice questions to assess the user's understanding.

After recording the data, the same system processes eye tracking and audio features and calculates the EVS automatically. The EVS will then be integrated into the analysis permitting to evaluate the reader's level of understanding and language skills. The overview of the system is described in Fig. 1.

2.1 Eye Tracking

Our system is compatible with stationary eye trackers such as Tobii Pro Nano[2]. These eye trackers are fixed with mounting plates on a computer screen and give us as a raw output the coordinates (in pixels) on the screen of the eye gaze through time. The first step consists in obtaining the words' positions of the screen and in processing the raw eye-tracking data by computing the fixations and saccades from the raw eye gaze data. When we are reading, the eye movement is not a smooth movement but a sequence of fixations (when our eyes maintain a position on a single location) and saccades which are quick movements of the eyes between fixations.

For detecting fixations and saccades, we implemented two existing algorithms into our system: a) the Buscher et al. algorithm [4], and b) the Nystrom and Holmqvist algorithm [13]. Buscher et al. algorithm detects fixations when neighboring gazes are closed to each other (based on the two-dimensional location on the screen) whereas Nystrom and Holmqvist algorithm detect saccades based on the angular rotation speed of the eyes. The output of the fixation-saccade processing is a list of fixations. Each fixation has a starting time, a duration and a center. Between a pair of fixations are the saccades. The result of the output of such an algorithm computed by our system is displayed in the Fig. 2.

It should be noted that the quality of the eye-tracking recordings depends on several factors such as the movements of the recorded person and especially

[2] https://www.tobiipro.com/product-listing/nano/.

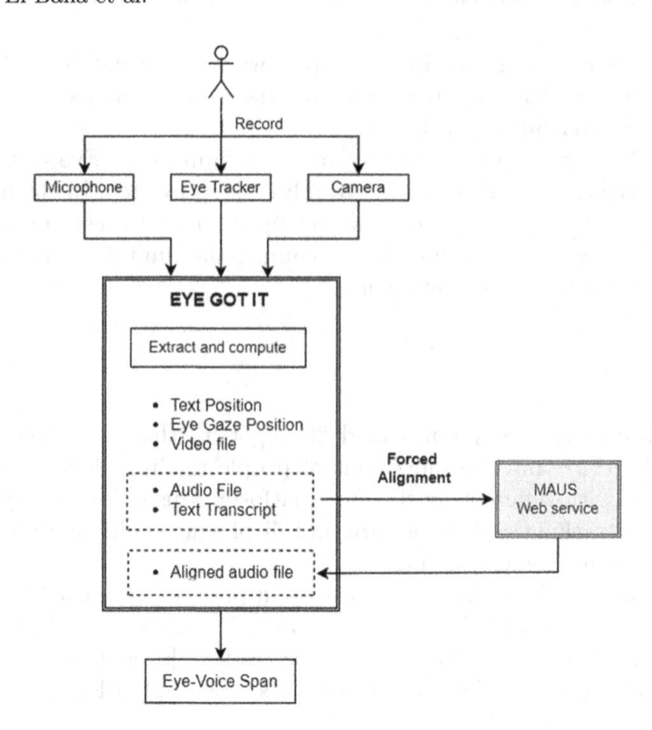

Fig. 1. The computing process of the EVS thought "Eye Got it" system. The system records the reader's voice with a microphone, the eye movement with an eye tracker and the facial expression with a camera. The position of the text and words on the screen is known by the system.

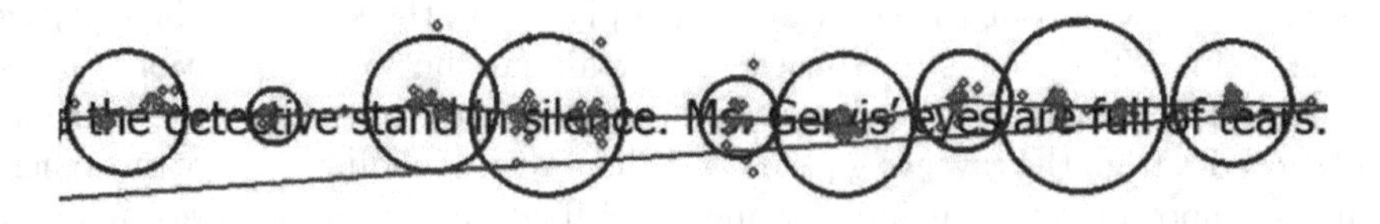

Fig. 2. Result of the Buscher et al. [4] fixation-saccade algorithm. The red dots are the eye gazes, i.e. the raw output of the eye tracker. The blue circles correspond to the fixations: the diameter of the circles is proportional to the duration of the fixation. The purple lines are the saccades. Long backward saccades are observed when the eyes of the reader jump from the end of one line to the next one. (Color figure online)

the head movements, the brightness of the room, the presence of other infrared lights than the eye trackers, the use of glasses or contact lenses, etc. For these reasons, it is important to frequently control the calibration of the eye-tracker and proceed to re-calibration if necessary. Even after careful calibration, it can happen to obtain an eye gaze recording that is inaccurate (such as the ones in Fig. 3). If such inaccurate recordings are used to compute the EVS, the results will not be correct.

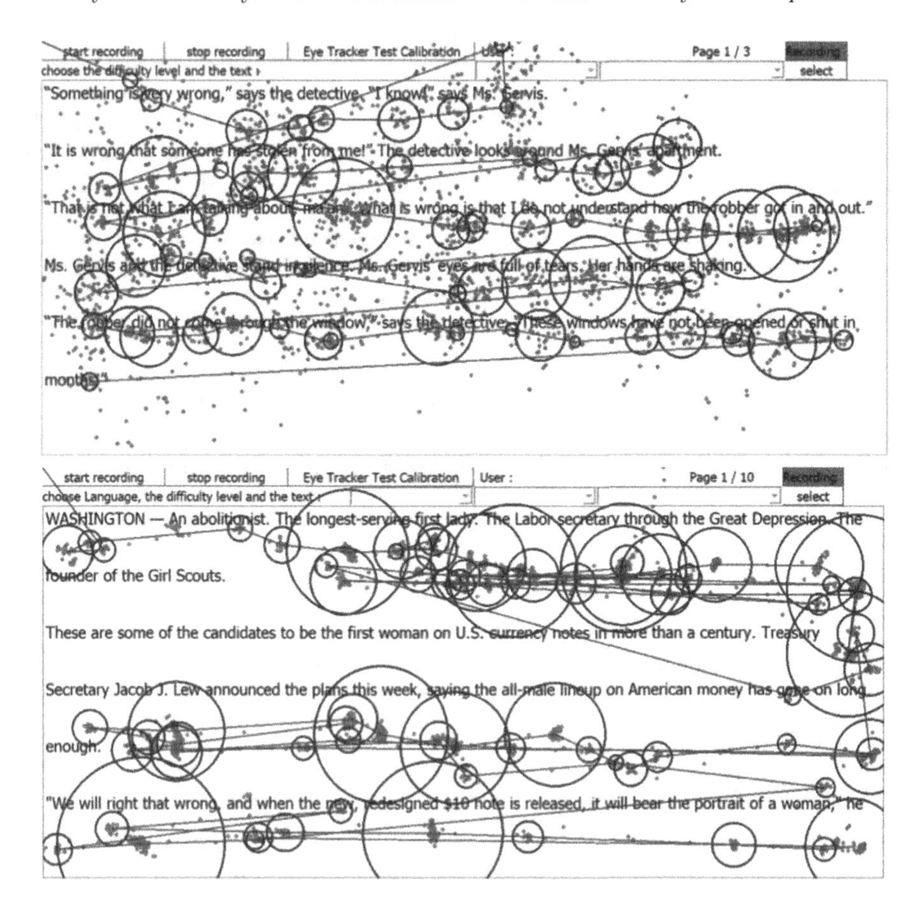

Fig. 3. Eye tracking recordings with low quality. On the top we can see that eye gazes are far from each other and this phenomenon is not possible (since the eyes cannot have such a movement), so we know that it is a problem from the eye tracker. On the bottom we can see another problem: the fixations seem natural this time, but they are not aligned with the text. This might be due to a calibration problem.

In order to avoid low quality recordings/processing errors, we set up three metrics to estimate the quality of a recording based on known eye tracking patterns from the literature:

- the percentage of fixations whose center is inside the bounding box of a word. When reading a text, the eyes are moving from word to word. Some fixations can be outside of the bounding box of a word, especially at the beginning and end of the recording or when the reader's eyes jump from one line to another but most of the time they should be aligned with the text;
- the percentage of words with at least one fixation. We know that there is not necessarily a fixation per word (for skilled readers, the eyes typically move about seven to nine letter spaces with each saccade [14]) but still, it is

not possible to read a text without a significant percentage of words with a fixation;
- the percentage of eye gazes used to process a fixation. Except during the saccades, most of the eye gazes are near each other and will form the fixations. A high number of isolated eye gazes should not be obtained.

A threshold was experimentally defined for each metric. If the value of each metric is higher than the threshold, then the recording is validated for processing the EVS. The three thresholds can be found in Table 1.

Table 1. We introduce in our system three metrics to control the quality of the eye tracking recordings. The thresholds were selected experimentally and are dependent on the quality of the eye-tracker and on its calibration. If the values of the three metrics are higher than the thresholds then the recording is considered to be processed by our system.

Metrics	Thresholds
% Fixations in words	80%
% Words with at least a fixation	60%
% Eye gazes in a fixation	70%

2.2 Speech Processing

Speech is used by many researchers as complementary information to eye-tracking features in different fields such as linguistics, psychology and cognitive sciences. For example, eye fixations combined with the analysis of the word stress used by the speaker can give salient information about word recognition strategy implemented by the speaker during a listening task [15].

The analysis of the correlation of the latency (the pause before starting to pronounce a word and the start of the fixation on the word that is to be read) with the eye gaze in object naming task allows to get information about linguistic planning: the fixation on a word is lasting until the phonological form of the word is recognized. This explains why the fixations on the longer words are longer than on shorter words [8]. In the object naming task, the analysis of eye fixations combined with speech helped to understand that two successive words can be processed in parallel by the speakers under the assumption that both words are known and easy to name by the speakers [8].

In order to compute the EVS, we need to know which word has been pronounced at what time. To do this we align the voice signal, recorded with a microphone and the read text. The Munich AUtomatic Segmentation system (MAUS) computes the phonetic labeling and segmentation of a speech signal based on a given phonological pronunciation, i.e. a forced alignment between a voice and a text [16]. MAUS is based on Hidden Markov Model and supports 21 languages. It provides an accurate annotation and will help us to save an important amount of time that could be spent on manual annotation. However, the presence of disfluencies (such as pauses and hesitations) especially in L2 production influenced by L1 specificity can worsen the performance of automatic speech recognition systems [18]. That is why manual annotation can be needed to adjust the frontiers of the words and phonemes detected automatically.

In spontaneous interactive speech, disfluencies reflect various speaker's intentions such as maintain their speaking turn, end the interaction, mention a new piece of information [19]. We suppose that the alignment of disfluencies with eye gaze during the reading aloud experiment can provide aid in formulating hypotheses about cognitive processing of the text such as an attempt to pronounce correctly a word or a combination of letters, to get the sense of the sentence by returning the eye to the beginning of the sentence or of the text, to understand the user's interface.

2.3 EVS Computation

To compute the EVS, we will combine the two results that we have obtained on a single axis of time, namely the audio file aligned with the script and the position of the eyes, which will allow us to calculate this distance between the spoken word, and the gazed word. To get the EVS, a user will need to perform the four following steps with "Eye got it": (1) recording the eye movements and the voice while reading aloud, (2) processing the fixations and saccades, (3) aligning the voice with the text from the other and (4) computing the distance between the eye gaze and the pronounced word. We detail these three steps as follows.

1. The user starts a recording session and reads a text aloud displayed on the screen via the "Eye Got it" interface. During this time a microphone and an eye tracker start recording. Optionally a camera can be used for recording the facial movement but this will not be detailed further in this paper. The program then organizes all the collected data in the following format: the coordinates of each word and their corresponding bounding box ("Text position"), the voice recording ("Audio file"), the text file read during the session ("Text input") and the coordinates of the gaze indexed by time during the reading session ("Gaze position").
2. The fixations and the saccades are computed. If the center of a fixation is inside the bounding box of a word, the fixation is associated with that word. This way, we can deduce the specific moment when a word is actually being looked at by the user. The quality of the recording is computed with three metrics, if the quality is not high enough then the other steps will not be processed.
3. The voice is aligned with the text based on the MAUS web service. The recorded audio file and the text are given as an input and MAUS returns a file which contains audio snippets with words from the text read by the user.
4. Using the aligned audio file with the text to read, the system checks if a word has been looked at when the user pronounced the word. If this is the case, it

will compute the distance between the word pronounced and the word looked at. This distance can be represented as a time duration and the number of words between the word pronounced and the word looked at.

3 Experiment

In this section we present a preliminary experiment that we set up for testing and validating the different parts of the system and the user experience of "Eye Got it".

3.1 Participants

Five French native participants volunteered to participate in the experiment (average age 25 years, one woman and four men). All had normal or corrected-to-normal vision and had English as a second language (L2 speakers). The participants were asked to read three texts aloud while their eye movements and voice were being recorded. They all signed a consent form and were free to stop the experiment at any time.

3.2 Apparatus and Material

The texts are displayed on a 22-in. computer screen with a resolution of 1280 × 960 pixels. The voice was recorded using an AKG Perception Wireless 45 Sports Set Band-A 500–865 MHz microphone, connected to the computer via a USB Jack cable. Video recording is done via the computer's internal web camera, but this data has not been used. The eye-tracking recording is done using a professional eye tracker, a Tobii Pro Nano with a sampling rate 60 Hz, fixed underneath the screen. All the devices are controlled by the Eye-Got-It software on a standard PC. The texts displayed are from the corpus "English For Everyone"[3]. Three texts of different levels are displayed (one text per level): beginner (170 words), intermediate (260 words), and advanced (237 words). All texts are spread over three pages with a font size of 25 pts.

3.3 Procedure

Firstly, we calibrated the eye tracker and tested if all devices were correctly functioning. The calibration is done with Tobii "Eye Tracker Manager" and is checked after each recording via the recording interface. Secondly, each participant creates a session by entering their information (name, first name, sex, date of birth, and language level). After that, the participants are asked to read aloud the text that is displayed on the screen. They will go through the three texts of different levels (beginner, intermediate, and advanced) that are displayed in a random order. Thirdly, after the recording session, all data collected during the

[3] https://englishforeveryone.org/.

session is saved and organized automatically by "Eye Got It". Before processing the EVS, the quality of the eye tracking recordings is estimated with the three proposed thresholds. If the quality seems to be not high enough, we advise to not compute the EVS. We observe that we sometimes found low quality recordings (as previously shown in Fig. 3), despite meticulous calibration and care during the recording. Finally, the system processes and plots the EVS for each recording of each participant.

3.4 Results

The format of the output is a graph such as the one displayed on Fig. 4, where the y-axis is the EVS defined by the number of words and in the x-axis is the total duration of the lecture (in seconds). In this example, the maximum value of the EVS is seven words, which might correspond to the easier part of the text (for this reader). When the EVS reaches zero, it means that the reader is pronouncing the same word that he or she is looking at. The variation of the EVS is correlated to the cognitive process of each word and thus the understanding of that word. The average of the EVS will reflect the language skill of the reader. In general, the preliminary results seem to be consistent with the tendencies described in [4,9,17], but further recordings will be needed to confirm this outcome.

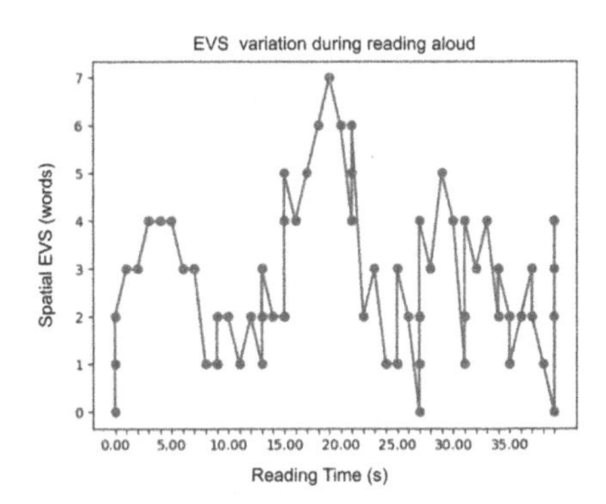

Fig. 4. An example of an automatically generated EVS graph of a participant while reading a text aloud. The variation of the EVS reflects the reader's reading skill.

4 Discussion

In this section, we comment on the preliminary results obtained during the first experiment setup conducted among 5 participants and analyze the two aspects of the system: from the eye tracking and speech modality processing issues. The described results permit us to verify the correct functioning of the "Eye Got It" platform and will serve us as a baseline for the system's quality improvement as well for further data collection. We also present the limitations of the systems and their possible remediation.

4.1 Eye Tracking

A pillar factor of an accurate EVS computation is eye movement recordings, and as explained in the sections above, eye-tracking is very sensitive and depends on many parameters. Nevertheless, we can obtain decent results by following good practices during the recording, such as controlling the brightness of the room, using multiple calibration points and asking the participants to restrain their body movement while recording. However, this will not guarantee the quality of the recorded eye gaze, and we still might end up with some recordings that are not usable. In this case, it is important to have a way to automatically estimate the quality of the recordings. Our proposition with three thresholds must be refined in the future. In the cases where all the fixations are shifted vertically, some algorithms have been proposed by some research to automatically correct the recordings [11,20]. For example, Lima Sanches et al. proposed a vertical error correction based on Dynamic Time Warping and match the text lines and the fixations lines [11].

4.2 Audio

From the audio perspective, the precision of the calculation of EVS depends on the quality of the audio provided to the system and on the reading strategies of the participants. We identified two main difficulties that need to be taken into account when calculating EVS.

The first one will happen if an audio file which does not correspond to the text to read is used in MAUS. As MAUS is based on a forced alignment, it cannot detect this kind of error and will still give us an output. But the output of MAUS will lead to a totally incorrect EVS. The Fig. 5 illustrates two following situations: one where the EVS is calculated with an audio file corresponding to the text input and one where the audio does not correspond to the text input. As we can see, the EVS where a different text was read than the expected one, the number of points is much lower which could indicate that there is a problem. Another parameter that could also influence MAUS alignment is the accent of the reader. In our experiment, the participants were French native speakers but their English pronunciation was correct enough to be aligned with MAUS. But if a reader has a very strong accent, the alignment might fail (in the same way as if a different audio file was used) and will also lead to an incorrect computation of the EVS.

The second possible difficulty is the use of disfluencies (such as pauses, hesitations, stuttering, repetitions, etc.) by the speakers while reading aloud. As we discussed earlier, the disfluencies are crucial elements in verbal communication, so we cannot ignore them while analyzing the experiment, but they also can lead to errors in speech-text alignment and thus in computing the EVS. For instance, the Fig. 6 shows the difference between two EVS computed with and without stuttering when the same text is read by the same speaker.

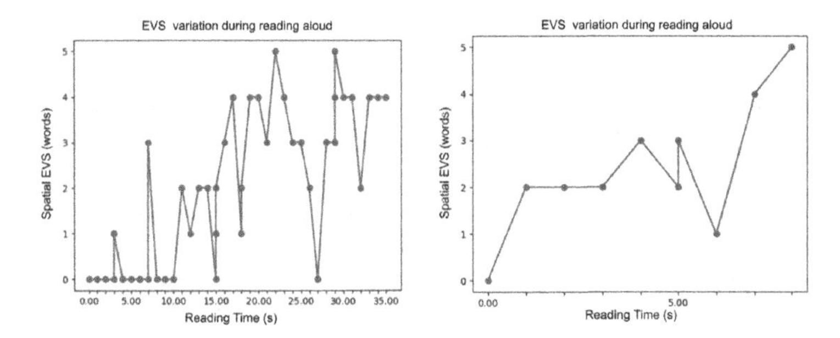

Fig. 5. Two EVS graphs: on the left the audio input was corresponding to the text input; on the right the audio file was not corresponding to the text input (i.e. the reader was not reading the displayed text). We can see that the discrepancy between the text input and the audio will be reflected by fewer points on the EVS graph.

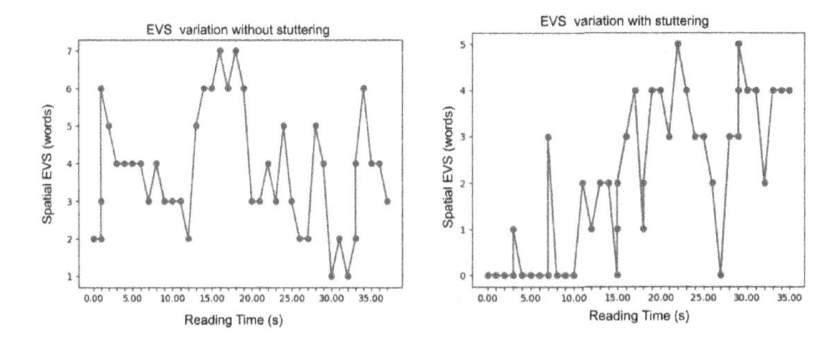

Fig. 6. The EVS graph of a reader reading the text with few disfluencies (left), and the EVS graph of same text but with stuttering. The stuttering tends to generate more zero values.

As we can observe, the EVS often reaches zero when the speaker is stuttering, which is not the case without it. One possible solution to this problem is to add disfluencies to the text to input and to correct MAUS output manually before the EVS calculation.

5 Conclusion

Current approaches to measure the EVS when reading aloud are not fully automatic and require manual intervention in several parts of the process. We propose in this paper "Eye Got It", an automatic system for computing the eye-voice span, ready-to-use, free, and easy to manipulate. The accuracy of the results expected from this system is highly dependent on the quality of the input data. In particular, the quality of the recordings, whether it is the eye tracking recording or the audio recording. Thus, the proposed system finally provides an aggregation of several tools and methods to automate the computation of several features that helps to describe and analyze the skill of a second language reader.

We also presented a preliminary experiment conducted on five participants to test and to validate that the system is working correctly.

Many points of improvement were raised, notably, the need for a metric to evaluate the accuracy of the audio alignment, since MAUS processes an alignment even if the audio and text input are not corresponding. Furthermore, MAUS is designed for native speakers, so readers with a strong accent cannot use the system or they might obtain an EVS that does not reflect their reading behavior. A possible improvement to solve this problem is to integrate an alignment model for non-native readers. We are able to classify native and non-native readers, via a convolutional neural network trained on spectrograms, the next step would be to train an alternative version of MAUS for the mother language of the reader in order to take into account their accent.

Acknowledgment. We would like to thanks the students from ENIB who participated to the development of the "Eye got it" system: Axel NOUGIER, Victor MENARD, Marine LE GALL, Yohan MAUPAS, Nicolas MENUT, Maelie MIGNON, Alexandre TROFIMOV and Asma NAIFAR.

References

1. Augereau, O., Fujiyoshi, H., Kise, K.: Towards an automated estimation of English skill via TOEIC score based on reading analysis. In: 2016 23rd International Conference on Pattern Recognition (ICPR), pp. 1285–1290. IEEE (2016)
2. Augereau, O., Kunze, K., Fujiyoshi, H., Kise, K.: Estimation of English skill with a mobile eye tracker. In: Proceedings of the 2016 ACM International Joint Conference on Pervasive and Ubiquitous Computing: Adjunct, pp. 1777–1781 (2016)
3. Buscher, G., Dengel, A., Biedert, R., Elst, L.V.: Attentive documents: eye tracking as implicit feedback for information retrieval and beyond. ACM Trans. Interact. Intell. Syst. (TiiS) **1**(2), 1–30 (2012)
4. Buscher, G., Dengel, A., van Elst, L.: Eye movements as implicit relevance feedback. In: CHI 2008 Extended Abstracts on Human Factors in Computing Systems, pp. 2991–2996 (2008)
5. Buswell, G.T.: An experimental study of the eye-voice span in reading, No. 17. University of Chicago (1920)
6. Buswell, G.T.: The relationship between eye-perception and voice-response in reading. J. Educ. Psychol. **12**(4), 217 (1921)
7. De Luca, M., Pontillo, M., Primativo, S., Spinelli, D., Zoccolotti, P.: The eye-voice lead during oral reading in developmental dyslexia. Front. Hum. Neurosci. **7**, 696 (2013)
8. Huettig, F., Rommers, J., Meyer, A.S.: Using the visual world paradigm to study language processing: a review and critical evaluation. Acta Physiol. (Oxf) **137**(2), 151–171 (2011)
9. Inhoff, A.W., Solomon, M., Radach, R., Seymour, B.A.: Temporal dynamics of the eye-voice span and eye movement control during oral reading. J. Cogn. Psychol. **23**(5), 543–558 (2011)
10. Laubrock, J., Kliegl, R.: The eye-voice span during reading aloud. Front. Psychol. **6**, 1432 (2015)

11. Lima Sanches, C., Augereau, O., Kise, K.: Vertical error correction of eye trackers in nonrestrictive reading condition. IPSJ Trans. Comput. Vis. Appl. **8**(1), 1–7 (2016)
12. Matsubara, M., Augereau, O., Sanches, C.L., Kise, K.: Emotional arousal estimation while reading comics based on physiological signal analysis. In: Proceedings of the 1st International Workshop on coMics ANalysis, Processing and Understanding, pp. 1–4 (2016)
13. Nyström, M., Holmqvist, K.: An adaptive algorithm for fixation, saccade, and glissade detection in eyetracking data. Behav. Res. Methods **42**(1), 188–204 (2010)
14. Rayner, K., Chace, K.H., Slattery, T.J., Ashby, J.: Eye movements as reflections of comprehension processes in reading. Sci. Stud. Read. **10**(3), 241–255 (2006)
15. Reinisch, E., Jesse, A., McQueen, J.M.: Early use of phonetic information in spoken word recognition: lexical stress drives eye movements immediately. Q. J. Exp. Psychol. **63**(4), 772–783 (2010)
16. Schiel, F.: A statistical model for predicting pronunciation. In: ICPhS (2015)
17. Silva, S., Reis, A., Casaca, L., Petersson, K.M., Faísca, L.: When the eyes no longer lead: familiarity and length effects on eye-voice span. Front. Psychol. **7**, 1720 (2016)
18. Tomokiyo, L.M.: Linguistic properties of non-native speech. In: Proceedings of the 2000 IEEE International Conference on Acoustics, Speech, and Signal Processing, (Cat. No. 00CH37100), vol. 3, pp. 1335–1338. IEEE (2000)
19. Vasilescu, I., Rosset, S., Adda-Decker, M.: On the role of discourse markers in interactive spoken question answering systems. In: LREC (2010)
20. Yamaya, A., Topić, G., Martínez-Gómez, P., Aizawa, A.: Dynamic-programming–based method for fixation-to-word mapping. In: Neves-Silva, R., Jain, L.C., Howlett, R.J. (eds.) Intelligent Decision Technologies. SIST, vol. 39, pp. 649–659. Springer, Cham (2015). https://doi.org/10.1007/978-3-319-19857-6_55

Text Detection and Post-OCR Correction in Engineering Documents

Mathieu Francois[1,2](✉), Véronique Eglin[1], and Maxime Biou[2]

[1] Univ Lyon, INSA Lyon, CNRS, UCBL, LIRIS, UMR5205, 69621 Villeurbanne, France
veronique.eglin@insa-lyon.fr
[2] Orinox, Vaulx-en-Velin, France
{francois.mathieu,biou.maxime}@orinox.com

Abstract. As the amount of born-analog engineering documents is still very large, the information they contain can not be processed by a machine or any automatic process. To overcome this, a whole process of digital transformation must be implemented on this type of documents. In this paper, we propose to detect and recognize all textual entities present on this type of documents. They can be part of technical details about a technical diagrams, bill of material or functional descriptions, or simple tags written in a standardized format. These texts are present in the document in an unstructured way, so that they can be located anywhere on the plan. They can also be of any size and orientation. We propose here a study allowing the text detection and recognition with or without associated semantics (symbolic annotations and dictionary words). A solution coupling a text detector based on a deep learning architecture, an open-source OCR for string recognition and an OCR post-correction process based on text clustering is proposed as a first step in the digital transformation process of industrial plans and P&ID schemes. The results applied to a database of 30 images of industrial maps and plans from different industries (oil, gas, water...) are very promising and close to 84% of correct detection and 82% of correct tags (and lexicon-free words) recognition after post-correction.

Keywords: Deep learning text detection · Post-OCR correction on tags and lexicon-free words · Affinity clustering

1 Introduction

Today, a large number of engineering documents are stored either in paper format (as printed version of a native digitized document or as unique representation medium), in weakly structured digital representation (pdf-like documents), or in their native digitized format (as images). The digital transformation of these contents is a challenging research topic in the industrial world. Companies involved in such digital transformation hope an access to numerical resources in

© Springer Nature Switzerland AG 2022
S. Uchida et al. (Eds.): DAS 2022, LNCS 13237, pp. 726–740, 2022.
https://doi.org/10.1007/978-3-031-06555-2_49

an automatic, indexed and enriched way (with for instance new semantic relations between digitized content, information retrieval in indexed and annotated datasets, smart edition of P&ID schemas and maps). The engineering documents or other technical documentation of industrial installations are composed of a large number of elements of different nature, see Fig. 1. One can find for example texts with or without semantics, graphical symbols and plots... These documents can be very dense in information, these can even overlap. For a man with engineering knowledge, it is quite obvious to distinguish between these information but what about a machine?

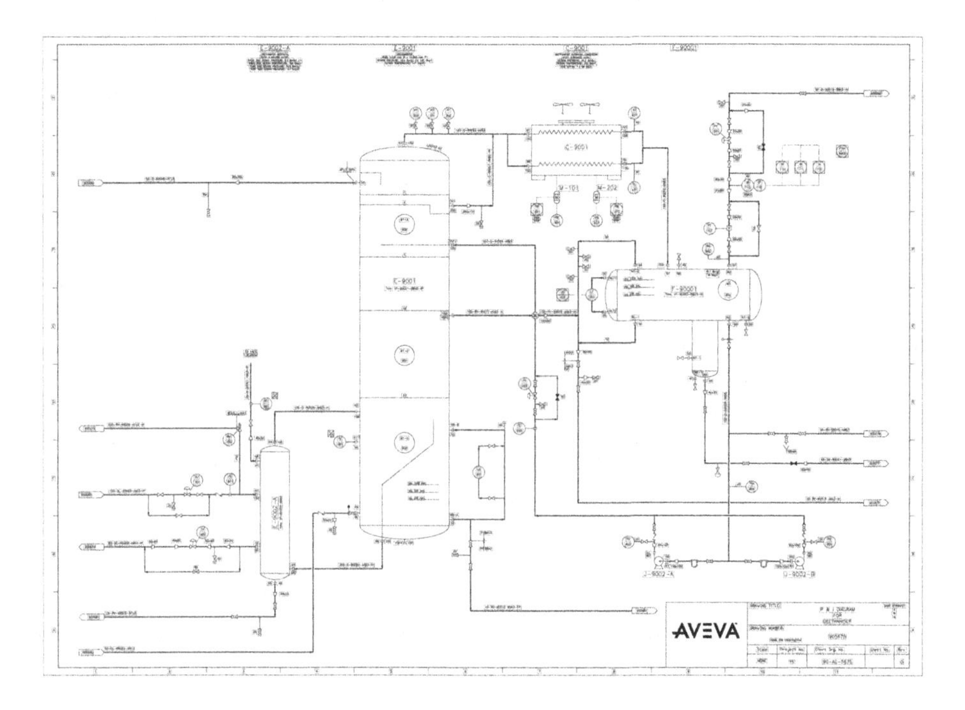

Fig. 1. Example of engineering document with multi-oriented texts and symbols.

To extract information from such documents, detection and recognition of textual entities is firstly required. However, most of those documents are not qualified for any classic OCR process that would produce frustrating full of errors transcriptions. First of all, we must be aware that these documents are often very large (A1-A0 format, which represents the equivalent of several hundred Million pixels to process) and the quality of the scans can lead to serious problems for the detection (blurring effects, noise distribution...). Moreover, in the field of industrial map and plan design, standards of representation produce unstructured documents: this means that the texts can be represented in any orientation and can have different sizes and colors on the same map. Thus, in this study where we are interested in the digital transformation of industrial engineering maps, we will try to answer two main fundamental questions. First, we will question FCN-type techniques and their ability to produce a suitable detection of all types of texts present on the maps. Those can be multi-oriented, of large sizes, having or not linguistic semantics and some structural regularities (font, size, and structure for annotations and nomenclature texts, also called *tags*). Then, we will question the accuracy of a transcription produced by OCR, knowing that the texts have neither semantics nor link with any lexicon or even less any language model.

In this paper, we address these two challenges with an end-to-end pipeline that performs text recognition on very large scale industrial P&ID drawings. This pipeline consists of two main steps: a text detection according to a model adapted from the Efficient and Accurate Scene Text Detector (EAST [2]), technique which allows an efficient localization of text zones in natural scenes; followed by an OCR step (by Tesseract [9]) and an OCR post-correction based on a sequences clustering *by affinity propagation* [13].

The paper is organized as follows: in Sect. 2, we present related works in connection with unconstrained text detection techniques focusing printed text and recent innovations for post-OCR correction, especially on those based on isolated token features and sequence clustering. In Sect. 3, we present our contribution to an end-to-end scenario for text detection and recognition through the deep adapted extractor EAST and a post-OCR correction for lexicon-free words and tags using the affinity clustering. In Sect. 4, we report then the experimental results of the end-to-end pipeline on a real industrial annotated dataset composed of thirty large engineering P&ID maps. It shows the first results of text detection and recognition followed by the original proposition of tags post-correction. The results are quite promising with F1-score for the detection near 84% which corresponds to a coverage of 90% of lexicon-free words (tags) and 88% of the nomenclature texts and other textual descriptions. The post-OCR correction with sequences clustering shows no less than 82% of correct (entire) tag recognition, i.e. an improvement of more than 7% compared to the rough OCR results.

2 Related Work

The domain of semi-structured document and image map analysis also dealing with engineering documents is very challenging. As developed by Moreno-García et al. in [7], this domains must face almost three core problems: text extraction, irregular strings or words clustering, text recognition and semantic interpretation. The core problems concerning the identification of printed textual information in engineering documents must face three research domains: the text detection in unconstrained environments, the text recognition and the post-OCR correction. As we made the choice to use open-source OCR recognition engine for the text transcription, we only relate here text detection and post-OCR correction approaches.

2.1 Text Detection in Unconstrained Documents

Text detection is the first step to process when automatically reading text in documents. Since existing methods often fail when applied to unconstrained document images, they still exist an active research in that challenging domain. Recently, Jamieson et al. [1] carried out a complete study on text detection and recognition for engineering documents exploiting the potential of deep learning solutions. For that, they used the EAST detector fully described in [2] for the text areas localization. EAST text detector is identified today as one of the more efficient fully connected network (FCN) for scene text detection. Its FCN-like architecture can be divided into a feature extraction part, a feature fusion part (to improve text detection) and an output layer, see Fig. 2. The text predictions are finally sent to the non-maximal suppression layer to obtain the final results.

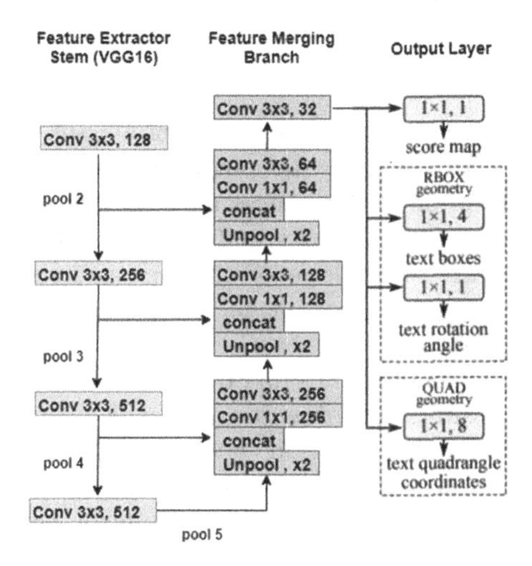

Fig. 2. The EAST architecture, [2].

DUET as *Detector Utilizing Enhancement for Text* is a neural network model for text detection of scanned documents developed by Jung et al., [10]. Authors proposed a method for text detection reducing noise of input images and enhancing text region appearance as well as its detection. They introduce an enrichment of the training data with automatically labeled synthetic images allowing to train the model with few data.

Baek et al. recently proposed in [11] a method named CRAFT (Character Region Awardness for Text Detection) that detect each character and the affinity between them. Two scores are predicted (Region score and Affinity score), the first one allows to localize the character in the image and the second one allows to gather the characters belonging to the same word. This detector is basically used in natural scenes contexts.

One difficulty with most of these systems is the complexity of training step which very often requires specific adjustments (pre-training, transfer learning, introduction of synthetic data...). To that extend, Yoshihashi et al. [12] proposed a method to simplify the text detection system. To achieve it, they use only simple convolutions and some post-processing. For the text areas detection, they proposed an adaptation of CRAFT and for the characters grouping, only simple image processing techniques. Anyway, models developed for scene text detection or text detection in the Wild are not of the same complexity as if they are employed to detect text in more homogeneous environments such as maps. For these last more simpler cases, the main challenge is to overcome difficulties of multi-oriented text, superposition of text with graphics, and in some extend geometrical distortions of schemas. So, given the intrinsic qualities of the EAST detector for multi-oriented texts, we will show how we have adapted them to face the specific challenges of P&ID documents.

2.2 Post-OCR Correction

Today, Post-OCR correction has become an essential issue to enhance the results of text recognition on documents. Although OCR engines have continuously improved their transcription performances and can efficiently work on most of modern printed texts, they still lack specific training data when dealing with technical documents with non-generic specificities of their texts. An OCR can make errors for many reasons (poor image, unknown font...). Applying a Post-OCR correction to remedy this has become very common, either by integrating some lexicon or language model for literary or semantic texts, either by proposing to analyse brute OCR results according to the frequency of transcribed terms. In our industrial context, the transcription concerns isolated *tags*: identifiers of the technical components. The tags do not have any linguistic semantic but a contextual significance due to their closeness to symbolic objects (graphic objects of the plan). Consequently, without any specific information on the technical documentation of the P&ID document, it is not possible to make any fair Post-OCR correction of mis-transcribed sequences. In the context of post-OCR correction on isolated words, there is a very large diversity of techniques that rely on the characteristics of isolated tokens. They can be classified into several groups of methods that can rely on merging of OCR results, on lexicon-based approaches, on error models, on specialized language models, and other models mainly focusing on sequence clustering, [6]. Here, we will not detail the different affordable techniques as they are not suitable for our post-correction problem, but we only focus here on those based on isolated token features and sequence clustering.

In [8], Das et al. proposed a solution for Post-OCR correction on a large number of documents using clustering. They compared several clustering algorithms such as k-means or MST clustering. They exploit the collection of documents as a whole, with its inner structure leading to detect repetition of words. Such words efficiently grouped together and corrected lead to significant reduction of post-OCR correction cost. An other clustering based approach has been proposed by Hakala et al. in [5] who train character models using OpenNMT translation system to correct OCR errors, [4]. For that, they create training data based on text duplication and search repeated texts to cluster. For each cluster of more than 20 sequences, they group similar words based on Levenshtein distance. An alignment with the most common character for each position is then realized and a replacement occurrence is done when necessary. In the same vein, the research of common motifs between strings through clustering is an interesting way to correct sequences without any associated lexicon. Our Post-OCR proposition is part of this context (Sect. 4.2.2).

3 Our Approach for Lexicon-Free Text Recognition

3.1 EAST-Based Text Detection

To train the detection system, the dataset is a batch of engineering documents that the Orinox company owns. These files are currently confidential and cannot be shared. These documents come from several fields of activity and do not have the same form. They include P&IDs, sales documents, isometrics and others, providing a great diversity in our dataset.

Before using EAST detector on our large sized document images, data must be prepared. First, images are binarized by simple Otsu thresholding to enhance light color texts. Then, oriented versions (at 90° rotation) are produced to improve the detection of vertical texts. For each couple of images from these two batches, 16 zoomed sub-images are created to improve the detection of the smallest text areas. The procedure also miss text areas if they are located at the place where the image is cut (tolerance of 1% error).

For our purpose, we use the FCN model EAST [2] with pre-trained Resnet V1. A fine-tuning step is also realized with 330 fully annotated engineering documents from our dataset. The part where the predictions are sent to the non-maximal suppression are also removed. We did not use *Non Maximum Suppression* to filter the predictions because this method reacted badly with long strings (see Fig. 3). Indeed, it cut off the beginning and end of each long string and several words were not detected.

To perform realistic detections, we identified all interlocking rectangles that had similar $y0$ and $y1$ coordinates with a margin of error of 5 pixels. We then joined these rectangles by taking the minimum and maximum of all these coordinates. Then, we cleaned up the rectangles to make them as close to reality as possible. So, all the rectangles being only composed of white pixel column at the beginning and at the end of the text area are readjusted. Then, the rectangles representing several very close text areas are split, this happens especially in the tables. For this, when we detect a column composed only of black pixels, we split the text area in two parts.

By strictly applying this procedure, the system also detects potential false positives. To overcome this effect, we separate each text area character by character and check that there is no line only composed of white pixels. If this is the case, we check that this character is not an "i, j, é ...". If it is not one of these characters we delete the text zone. Finally, we delete the rectangles that

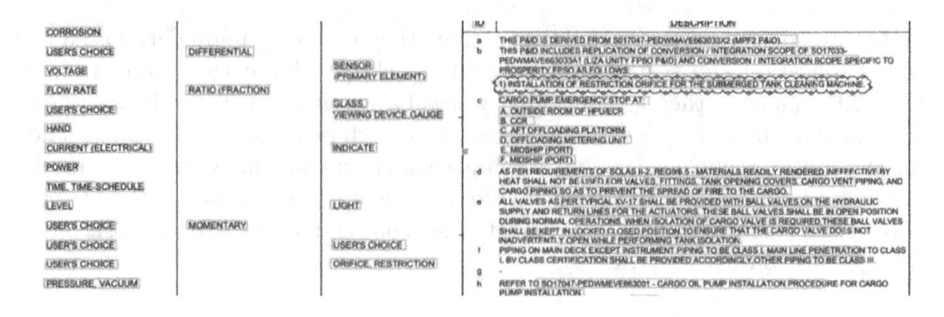

Fig. 3. Example of errors made by EAST on long strings.

are already completely or almost completely included in a larger text area. After these steps, we have a prediction of the text zones present on the documents and this is what will be given as input to the OCR for recognition part.

3.2 Open-Source Engine for Text Recognition

For the recognition stage, we used the open-source Tesseract OCR based on a LSTM neural network architecture [3,9]. We also configured Tesseract to assume that each image given as input corresponds to a single line of text. Each previously detected image area serves as input to the OCR. From brute-force OCR, 74% of the input text was correctly transcribed and some recurring errors has been identified. In Table 1, we find the details of recognition based on the criteria: strings written in horizontal or vertical and strings representing tags or other.

Table 1. Result of the brute-force Tesseract OCR for text written vertically or horizontally, text representing tags or others and total.

	Recognised	Total	Percentage
Horizontal	9343	12298	76.0%
Vertical	632	1183	53.4%
Tag	923	1224	75.4%
Other	9052	12257	73.9%
Total	9975	13481	74.0%

These results highlight several weaknesses: first, during the detection of the vertical zones, we didn't know if the text had been detected at 90° or −90°. For each area of vertical text, we proposed to the OCR the two images and kept the prediction with the highest percentage of confidence. In some cases, the prediction of the image text upside down has a higher confidence index than the one upright. So it can wrong as we can see in the Fig. 4. Another recurrent problem concerns situations where drawings are to close to the text areas. The detection system may have included a small part of these drawings in the text area. Therefore, during the recognition phase, we can see characters added at the beginning of strings, see Fig. 4. From the results detailed in Table 1, we understand that the texts written in vertical have less accurate results. A part concerning the post-OCR correction will help to solve this kind of transcription errors and others (confusion between a digit and a letter, omission of a character or a portion of a string).

Fig. 4. Examples of errors made by the recognition system.

3.3 Post-OCR Correction of Tags and Lexicon-Free Worlds

Tags are the identifiers of technical elements present on documents. They are not linked to any lexicon nor language model. They are mostly composed of an uppercase letter, a number and a separator character (example: AA-1234-BB), and because of these properties, they are frequently the target of transcription errors. This part aims to improve the predictions initially proposed by the OCR for this kind of sequences.

We propose here a solution based on sequence clustering to improve the tags and lexicon-free words transcription. Following the predictions of OCR, we grouped the textual entities into several clusters using *Affinity propagation* approach as described in [13] and [14].

3.3.1 Affinity Propagation for Sequence Clustering

Affinity propagation can be used in different contexts such as human face recognition or text analysis. Compared to other clustering approaches (i.e. K-Means) requiring a selection of the number of clusters and the choice of the initial set of points, *Affinity Propagation* takes, as input, measures of similarity between pairs of data points, and simultaneously considers all data points as potential exemplars (future centroïds of the clusters). In our context, *Affinity propagation* takes as input a set of pairwise similarities between data points and defines clusters by maximizing the score of similarity between data points. A matrix representing the Levenshtein distances between each tag (the exemplars) is given as input to the clustering and will serve as similarity measure between pairs of points. Formally, let x_i to x_n be a set of data points (the text sequences), with no assumptions made about their internal structure. Let $\mathcal{L}$ be the *Levenshtein* function that quantifies the dissimilarity between any two points, such that $\mathcal{L}$(i, j) $\geq$ $\mathcal{L}$(i, k) if and only if x_i is more similar to x_k than to x_j.

The diagonal of $\mathcal{L}$ (i.e. $\mathcal{L}$(i, i)) represents the instance preference, meaning how likely a particular instance is to become an exemplar.

Each x_i will communicate with the other elements to let them know if they are similar or not. These will respond by specifying the availability of the two elements to associate. Each time affinity coefficients are updated and these steps are repeated until there is no more change. The two steps are represented by two matrices. The algorithm proceeds by alternating between two message-passing steps, which update two matrices:

1. The *Responsibility* matrix $\mathcal{R}$ has values $\mathcal{R}$(i, k) that quantify how well-suited x_k is to serve as the exemplar for x_i, relative to other candidate exemplars for x_i.
2. The *Availability* matrix $\mathcal{A}$ contains values $\mathcal{A}$(i, k) that represent how "appropriate" it would be for x_i to pick x_k as its exemplar, taking into account other points' preference for x_k.

This method works then in an iterative way, updating in turn the values of the responsibility and availability matrices. One of the advantages of affinity propagation that is particularly interesting in our case is that it is not necessary to give the number of clusters as input and the algorithm is able to identify outliers instead of exemplars, also useful for our purposes (even not fully investigated here) (Fig. 5).

3.3.2 Clusters Exploration

When the predictions are divided into several clusters, we define whether the cluster data is a tag group or a word group and we only keep clusters containing tags. First, we checked whether the vertically oriented texts had been detected in the right direction. To do this, we retrieved the prediction that had not been selected by the OCR and compared it with the data from the different tag clusters using the Levenshtein distance. If the distance is very close with a cluster and the tag selected by the OCR is quite isolated, then we update the new tag and integrate it into the corresponding cluster.

Then, we studied, cluster by cluster, the structure of the tags. For 'C' corresponding to a number and 'M' for an uppercase letter, we have a large majority in one form (CC-MMM-CC) and that some tags have a very close structure with a difference (CM-MMM-CC). We replay the tag in Tesseract by recovering character by character all the assumptions it makes, if the OCR predicts a character allowing the tag to have the same structure as the majority group then we replace this character.

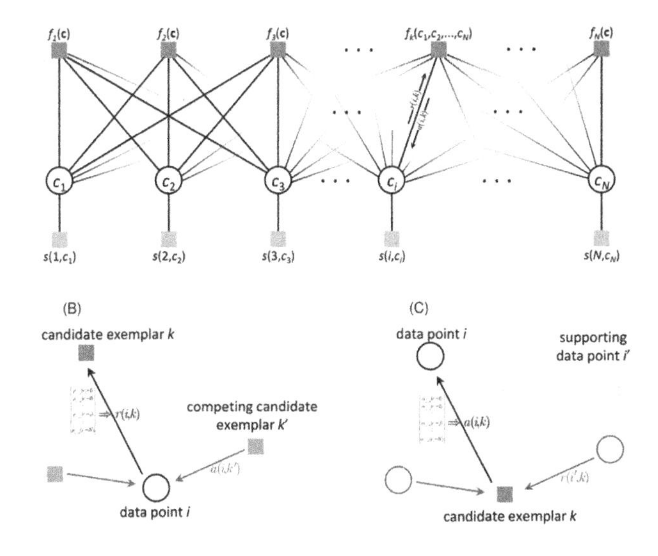

Fig. 5. Affinity propagation structure [13].

4 Experimentation and Discussions

4.1 Dataset

Datasets from engineering documents are mostly confidential and open-source datasets of P&ID documents do not exist at efficient scale. We therefore chose to create a fully annotated dataset to qualify our results and compare them with related works only.

Orinox. Company has access to several hundred thousand of documents, but as we wanted to automate the annotation process, we have privileged pdf files with selectable text for 330 engineering documents in various forms (technical document, drawings, component legend...). There is an average of 440 text boxes per document. 40 are tags and the other 400 are either words, technical details (measurement, size...) or isolated characters. We used a dedicated library (*PDFBox*) to extract the coordinates of each character of the pdf files. Then, we developed a small tool allowing the merging of characters sequences so as to create a single rectangle per text zone. This program also converts the input data into an image file and transforms the coordinates to match. This annotation returns a csv file containing the coordinates of the texts which will act as ground-truth (Fig. 6).

For the datasets preparation, we separated it in two parts: 300 files for the training and 30 for the evaluation, which makes a ratio of 91% for the training dataset and 9% for the evaluation.

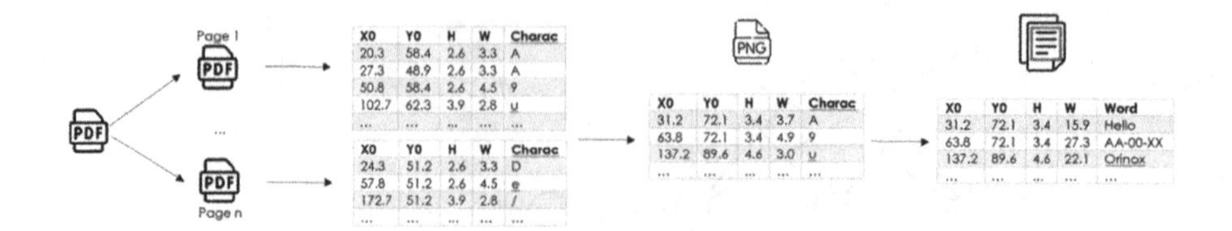

Fig. 6. Diagram summarizing the method used for data annotation.

4.2 Results

4.2.1 Text Detection

In this section, we present the results of our detection system based on the EAST. We considered that a text area is well detected if all its coordinates correspond exactly to the ground-truth coordinates with a margin of error of 10 pixels. For the evaluation, we used the classical precision, recall and F1-score metrics:

$$Precision = TP/(TP + FP) \tag{1}$$

$$Recall = TP/(TP + FN) \tag{2}$$

$$F1Score = 2 * (Precison * Recall)/(Precison + Recall) \tag{3}$$

To determine which confidence threshold of the text field predictions are the most suitable for our problem, we tested several values as we can see in the Fig. 7. These values represent the limit where the detection system can validate or not the propositions of text zones thanks to the confidence index. If, for example, the threshold is set to 0.10 then all bounding boxes predicted with a confidence of more than 10% will be selected.

Logically, we notice that the lower the threshold, the closer the recall is to 1 because it will have selected most of the true text areas. However, it will also have selected many more false positives and the precision will be low. On the contrary, the higher the confidence threshold, the more true text areas can not be selected by the detection system and the recall becomes lower. But the false positives will also decrease in a consequent way and so the precision will be closer to 1.

We also notice that the values of the F1 score curve is at its maximum when the confidence threshold is equal to 0.10 and 0.15. We select the threshold value equal to 0.10 because it seemed more beneficial offering a better recall rather than a good precision. This choice is also motivated by future works, based on co-occurrences of symbols and text, allowing to remove some false textual positives representing symbols.

On the basis on these general configuration components, we compared the original text detection method of EAST and our adaptation. As a reminder, we removed the post-processing part of the prediction made by EAST (NMS) and replaced it by our own method (Table 2).

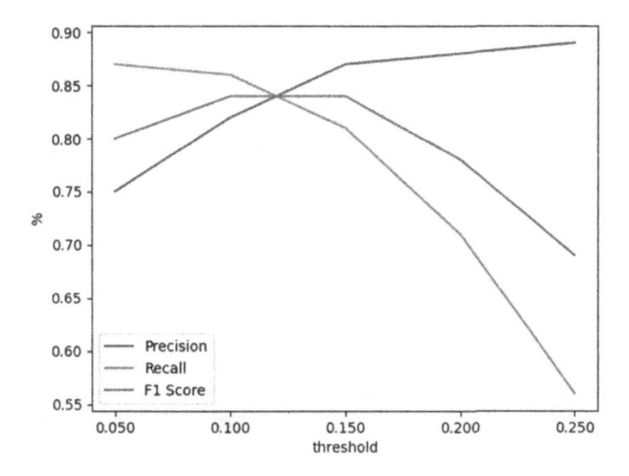

Fig. 7. Curve showing the values of recall, precision and F1-score of our detection system according to the confidence threshold.

Table 2. Comparison of the detection results of the original version of EAST [2] and our approach.

	Precision	Recall	F1 Score
EAST (original)	0.818	0.619	0.705
Our Approach	0.824	0.863	0.843

We can see that 82% of the detected texts are actually text. The majority of the errors are valve symbols or other symbols detected as text by the system.

We also have 86% of the text present in the document correctly detected. The 14% missing are mostly small text areas in symbols that are not or incorrectly detected by the system and several text areas grouped together into one (Fig. 8).

Whether it is for false positives representing symbols or small areas of undetected text, future work will improve the detection results of text areas. Indeed, the graphic symbols containing text inside are very often simple shapes (circle, hexagon, square...). We will then be able to detect these shapes and then recover the undetected text (this part has not been presented here).

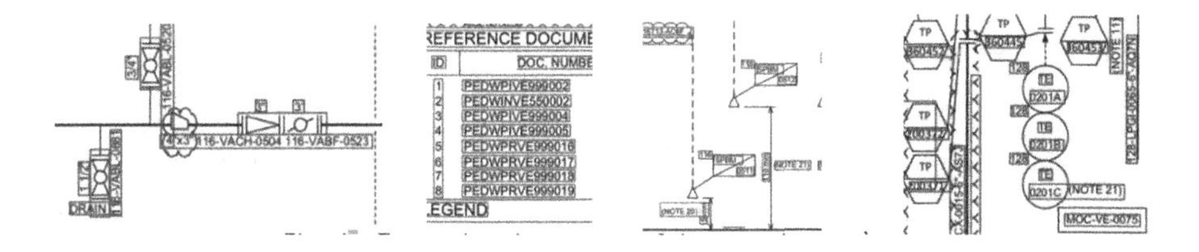

Fig. 8. Examples of errors made by the detection system.

4.2.2 Post-OCR Correction

As announced in the objectives of the paper, an OCR post-correction step is essential to compensate for the lack of knowledge and lexicon. The algorithm realized for this phase has been summarized below. For each map, the number of clusters generated by *Affinity propagation* is not fixed. Within our dataset with 30 documents, the clusters vary between 15 and 50 groups.

After defining which cluster is a tag group, we found a total of 1224 tags in the 20 documents making up the evaluation dataset. This represents about 40 tags per document. The predictions made by the OCR resulted in a 75% (923/1224) rate of well recognized tags. The affinity propagation-based method resulted in a gain of 7% and after the Post-OCR correction, the tag recognition results amount to 82% (1009/1224). This method allowed us to correctly correct 86 tags.

Algorithm 1. Post-OCR Correction for tags

```
Require: prediction
  clusters_all ← AffinityPropagation(prediction)
  cluster ← clusters_all[tag]
  if cluster[tag].orientation() == rotate then
    if distance(cluster[tag], cluster) < distance(cluster[tag_reverse], cluster) then
      cluster[tag] ← cluster[tag_reverse]
    end if
  end if
  if struture(cluster[tag]) ≠ structure(cluster) then
    diff ← difference(struture(cluster[tag]),structure(cluster))
    predic_multi ← ocr_multi(diff)
    if struture(difference) == structure(predic_multi) then
      cluster[tag][diff] ← predic_multi
    end if
  end if
```

The main errors remaining after the Post-OCR correction on the tags are errors with the same structure ('7' instead of '1' or 'D' instead of 'O'). The incorrect tags are not detected and therefore not updated. Also, the predicted tags having errors on the separator character are also badly detected by the current method.

Following the clustering part, we performed some processing based on the correction of the vertically written tags and on tags structure. By analyzing in more detail the content of the tags of each cluster, we plan to perform text and image correlations. Indeed, if some tags have differences while all the other data of the cluster are identical on this characteristic, then text & image association could remove decision ambiguity and guide characters corrections with well-qualified hypotheses.

The links between tags and symbols are also a very interesting track. We will be able to predict the similarities of the tags according to the component it represents. And if some tags representing the same symbol do not have the same similarities, then we can also propose corrections.

5 Conclusion

In this paper, we intended to demonstrate whether FCN-type technologies are suitable for text detection of multi-oriented texts of large size variability, and intended to improve an OCR-based transcription for tags and lexicon-free words in P&ID document.

For text detection, we proposed an approach based on the EAST architecture by introducing our technique for selecting the best bounding box. Concerning the Post-OCR correction, we based it on the *Affinity propagation clustering* algorithm to sort the tags having potentially the same structure. The results of these two parts are promising and proved that these methods can be used for engineering documents.

As future investigations, we plan to focus on graphic symbols present in this kind of document and to study relationships between small text areas in the neighborhood of graphic symbols so as to improve even more the post-OCR correction.

References

1. Jamieson, L., Moreno-Garcia, C.F., Elyan, E.: Deep learning for text detection and recognition in complex engineering diagrams. In: 2020 International Joint Conference on Neural Networks (IJCNN), pp. 1–7 (2020)
2. Zhou, X., et al.: EAST: an efficient and accurate scene text detector. In: 2017 IEEE Conference on Computer Vision and Pattern Recognition (CVPR), pp. 2642–2651 (2017)
3. Hochreiter, S., Schmidhuber, J.: Long short-term memory. Neural Comput. **9**(8), 1735–1780 (1997)
4. Klein, G., Kim, Y., Deng, Y., Senellart, J., Rush, A.: OpenNMT: open-source toolkit for neural machine translation. In: Proceedings of the Association for Computational Linguistics on System Demonstrations, ACL 2017, pp. 67–72 (2017)
5. Hakala, K., Vesanto, A., Miekka, N., Salakoski, T., Ginter, F.: Leveraging text repetitions and denoising autoencoders in OCR post-correction. CoRR abs/1906.10907 (2019)
6. Huynh, V.-N., Hamdi, A., Doucet, A.: When to use OCR post-correction for named entity recognition? In: Ishita, E., Pang, N.L.S., Zhou, L. (eds.) ICADL 2020. LNCS, vol. 12504, pp. 33–42. Springer, Cham (2020). https://doi.org/10.1007/978-3-030-64452-9_3
7. Moreno-García, C.F., Elyan, E., Jayne, C.: New trends on digitisation of complex engineering drawings. Neural Comput. Appl. **31**(6), 1695–1712 (2018). https://doi.org/10.1007/s00521-018-3583-1

8. Das, D., Philip, J., Mathew, M., Jawahar, C.V.: A cost efficient approach to correct ocr errors in large document collections. In: 2019 International Conference on Document Analysis and Recognition (ICDAR), pp. 655–662 (2019)
9. Smith, R.: An overview of the Tesseract OCR engine. In: Ninth International Conference on Document Analysis and Recognition (ICDAR 2007), pp. 629–633 (2007)
10. Jung, E.-S., Son, H., Oh, K., Yun, Y., Kwon, S., Kim, M.S.: DUET: detection utilizing enhancement for text in scanned or captured documents. In: 2020 25th International Conference on Pattern Recognition (ICPR), pp. 5466–5473 (2021)
11. Baek, Y., Lee, B., Han, D., Yun, S., Lee, H.: Character region awareness for text detection. In: 2019 IEEE/CVF Conference on Computer Vision and Pattern Recognition (CVPR), pp. 9357–9366 (2019)
12. Yoshihashi, R., Tanaka, T., Doi, K., Fujino, T., Yamashita, N.: Context-Free TextSpotter for real-time and mobile end-to-end text detection and recognition. In: Lladós, J., Lopresti, D., Uchida, S. (eds.) ICDAR 2021. LNCS, vol. 12822, pp. 240–257. Springer, Cham (2021). https://doi.org/10.1007/978-3-030-86331-9_16
13. Dueck, D.: Affinity propagation: clustering data by passing messages. Ph.D. dissertation. Citeseer (2009)
14. Frey, B.J., Dueck, D.: Clustering by passing messages between data points. Science **315**(5814), 972–6 (2007)
15. Refianti, R., Mutiara, A.B., Syamsudduha, A.A.: Performance evaluation of affinity propagation approaches on data clustering. Int. J. Adv. Comput. Sci. Appl. (IJACSA) **7**(3), 420–429 (2016)
16. Volk, M., Furrer, L., Sennrich, R.: Strategies for reducing and correcting OCR errors. In: Sporleder, C., van den Bosch, A., Zervanou, K. (eds.) Language Technology for Cultural Heritage. TANLP, pp. 3–22. Springer, Heidelberg (2011). https://doi.org/10.1007/978-3-642-20227-8_1
17. Mittendorf, E., Schäuble, P.: Information retrieval can cope with many errors. Inf. Retrieval **3**, 189–216 (2000)
18. Drobac, S., Lindén, K.: Optical character recognition with neural networks and post-correction with finite state methods. Int. J. Doc. Anal. Recogn. (IJDAR) **23**(4), 279–295 (2020)
19. Nguyen, T., Jatowt, A., Nguyen, N., Coustaty, M., Doucet, A.: Neural machine translation with BERT for Post-OCR error detection and correction. In: Proceedings of the ACM/IEEE Joint Conference on Digital Libraries in 2020, pp. 333–336. Association for Computing Machinery (2020)
20. Bazzo, G.T., Lorentz, G.A., Suarez Vargas, D., Moreira, V.P.: Assessing the impact of OCR errors in information retrieval. In: Jose, J.M., et al. (eds.) ECIR 2020. LNCS, vol. 12036, pp. 102–109. Springer, Cham (2020). https://doi.org/10.1007/978-3-030-45442-5_13

TraffSign: Multilingual Traffic Signboard Text Detection and Recognition for Urdu and English

Muhammad Atif Butt[1(✉)], Adnan Ul-Hasan[2], and Faisal Shafait[1,2]

[1] School of Electrical Engineering and Computer Science (SEECS), National University of Science and Technology (NUST), Islamabad, Pakistan
matifbutt@outlook.com

[2] Deep Learning Laboratory, National Center of Artificial Intelligence (NCAI), Islamabad, Pakistan
{adnan.ulhassan,faisal.shafait}@seecs.edu.pk

Abstract. Scene-text detection and recognition methods have demonstrated remarkable performance on standard benchmark datasets. These methods can be utilized in human-driven/self-driving cars to perform navigation assistance through traffic signboard text detection and recognition. Existing datasets include scripts of numerous languages like English, Chinese, French, Arabic, German, etc. However, traffic navigation signboards in Pakistan and many states of India are written in Urdu along with the English translation to guide human drivers. To this end, we present Deep Learning Laboratory's Traffic Signboards Dataset (DLL-TraffSiD) to develop multi-lingual text detection and recognition methods for traffic signboards. In addition, we present a pipeline for multi-lingual text detection and recognition for an outdoor road environment. The results show that our presented system signified better applicability in text-detection and text recognition, and achieved 89% and 92.18% accuracy on the proposed dataset (The proposed dataset along with implementation is available at https://github.com/aatiibutt/TraffSign/).

Keywords: Multi-lingual (Urdu and English) Traffic Signboards Data-set · Multi-lingual text detection · Multi-lingual text recognition

1 Introduction

Traffic signboards are considered as the main source of guidance for human drivers to navigate on the roads. Various types of traffic signboards are placed along highways and roads displaying warnings, speed limits, and directions (written mostly in native languages) to assist human drivers [7]. With the advancement of deep learning, scene-text recognition has drawn a significant attention of the computer vision community, which is evident by large-scale datasets along with scene-text detection and recognition methods [18]. Recently, RoadText-1K [23], COCO-Text

© Springer Nature Switzerland AG 2022
S. Uchida et al. (Eds.): DAS 2022, LNCS 13237, pp. 741–755, 2022.
https://doi.org/10.1007/978-3-031-06555-2_50

[29], and Total-Text [13] datasets consisting of English, Arabic, German, French, Italian, Japanese, and Korean language scripts have been proposed to perform text detection and recognition tasks. However, it is observed from the literature that no significant contribution is made in developing scene text detection and recognition dataset consisting of traffic signboards in Urdu language.

There has been some progress in past couple of years to collect traffic signboard data in Urdu language., Chandio et al. [8] and Arafat and Iqbal [3] have contributed Urdu text detection and recognition datasets; however, one [3] is based on synthetic data where Urdu words are typed on scenic background images whose performance can be influenced on Urdu-text unseen outdoor natural text images. The second dataset [8] consists of text written on random shop boards and wall-choking in non-symmetric font styles and variable text sizes. On the contrary, Urdu-text written on traffic signboards is purely written in sharp-cursive Nastaleeq and Naskh scripts in symmetrical font size and appropriate alignment. In addition, traffic signboards placed on the roads in Pakistan and many states of India (including west Bengal, Uttar Pradesh, Bihar, Jharkhand, etc.) are in Urdu and English languages [3]. These limitations are an indicative of a need for a multi-lingual scene text detection and recognition dataset of traffic signboards along with multi-lingual scene text detection and recognition methods.

To address the above-discussed shortcomings, we have made the following contributions in this research work.

- A multi-lingual (English and Urdu) scene-text detection and recognition dataset - DLL-TraffSiD - has been proposed. This dataset consists of 2,600 text-enriched traffic signboard images such as directional boards, distance boards, instructions, and warning boards, speed limit boards, and location boards with 9,051 bounding boxes annotations containing 13,481 words.
- We have also proposed a scene-text detection and recognition pipeline to perform multi-lingual traffic signboard text detection and recognition.

The remainder of this paper is comprised of the following sections. Section 2 summarizes the related work for scene text detection and recognition and benchmark scene text datasets. Section 3 describes the different steps of developing a DLL-TraffSiD dataset. Section 4 explains the employed methodology for TraffSign and Sect. 5 delineates the experiments and results. Section 6 concludes the paper with some future directions.

2 Related Work

Scene-text understanding methods generally consist of two step processes: (i) text detection, and (ii) text recognition, which are briefly discussed in the following subsections.

2.1 Text Detection

In the early era of ML, researchers introduced handcrafted features-based scene text detection methods. For instance, Hossain et al. [15] proposed Maximally

Stable Extremal Regions (MSER) based road sign text detection from scenic images. Similarly, Basavaraju et al. [5] presented a Laplacian-component analysis fusion-based approach for multilingual text detection from image/video data. However, text recognition under different luminous conditions is still a challenging task. In this regard, Tian et al. [28] presented a Co-occurrence Histogram (Co-HOG) of oriented gradient and convolutional Co-HOG feature descriptor-based scene text detection. Co-HOG is employed to encode context-aware spatial information in neighbor pixels to detect text regions from an input image. However, these methods require a sequence of steps of manual feature extraction along with the optimization in the classifiers that make these methods structurally complex and inefficient.

With the advancement in computer vision, end-to-end representation learning based text detection methods have been proposed. In this regard, Panhwar et al. [21] proposed an Artificial Neural Network (ANN) based text detection method to detect text regions in signboards. Liao et al. [17] presented a Differentiable Binarization (DB) based segmentation network for text localization in scene images. In another research work, Seha et al. [24] introduced an end-to-end Stroke Width Transform (SWT) and Maximally Stable Extremal Regions (MSER) based text detection method in an outdoor environment. However, these methods only focused on unilingual text detection that can be employed in a very limited scale application. To address these issues, Chandio and Pickering [9] proposed a multilingual text-detection method by combining the convolutional layers and the VGG network to perform text detection on Urdu and Arabic text scripts.

2.2 Text Recognition

Convolutional neural networks have demonstrated better applicability in scene-text recognition in various scripts. Bains et al. [4] proposed a text recognition method to recognize Gurumukhi text regions in signboard images. Aberdam et al. [1] presented a Sequence-to-sequence based Contrast Learning method (SeqCLR) for text recognition using the attention-driven technique. However, one main limitation of attention-driven text recognition methods is that the performance can be easily influenced through minor attention drift. To cater with this issue, Cheng et al. [12] proposed a ResNet-based Focusing Attention Network (FAN) to perform scene text recognition tasks in an outdoor environment. Similarly, Lu et al. [19] proposed a multi aspect transformer-based network to perform scene text recognition.

Though deep neural networks can learn robust representations of image artifacts and text style changes, they still run into challenges while coping with scene texts having a pattern and curvature distortions. To cater with this limitation, Shi et al. [25] proposed an end-to-end image-based sequence recognition method for scene text recognition. Chen [11] proposed transformation, attention-driven, and rectification-based networks to improve the text recognition performance in natural scene images. In another research work, Arafat and Iqbal [3] have proposed customized Faster-RCNN based scene text detection, recognition, and ori-

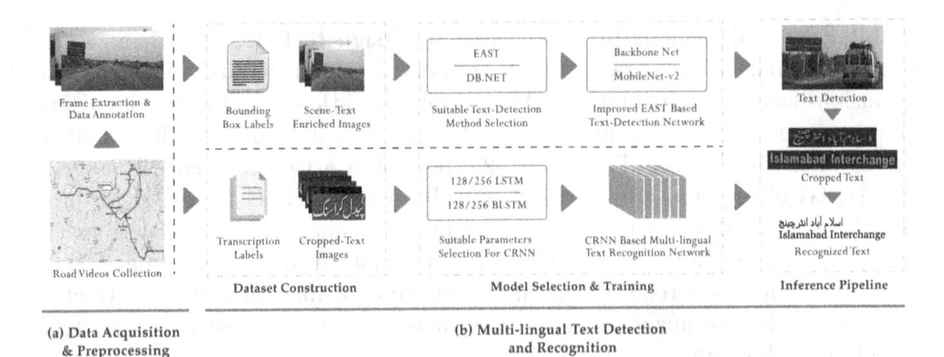

Fig. 1. The Method: Firstly, 900-KM road video sequences are captured, and text-enriched signboard frames are extracted. Secondly, bounding box and transcription annotations are generated to form datasets. In the next step, scene text detection and recognition methods are fine-tuned on proposed datasets. Based on the analysis, best-performing methods are further improved to achieve maximum performance.

entation prediction of Urdu ligatures in outdoor images. The authors employed customized regression residual neural networks for the orientation prediction of text ligatures. Busta et al. [6] proposed a fully connected network-based end-to-end method for multi-language text detection and recognition.

2.3 Standard Benchmark Datasets for Text Detection and Recognition

In recent years, numerous datasets are presented to pave the way towards the generalization of text detection and recognition methods. For instance, Karatzas et al. [16] presented ICDAR 2015 benchmark for incidental scene text detection comprising of $1,600$ images. However, the images are captured using google glasses without considering the importance of image quality. Therefore, the models trained over such data get exposed to unseen data vulnerabilities which influence the performance of models in real-world applications. Sun et al. [27] presented Chinese Street View Text (C-SVT) consisting of two chunks; one is completely labeled including bounding boxes of words and characters, while the other consists of annotations of dominant text instances only. Yuan et al. [30] presented Chinese Text in the Wild (CTW) dataset comprising of $32,285$ images along with transcription-based annotations. Chandio et al. [10] developed Urdu characters-based dataset containing outdoor scene text images. Gupta et al. [14] presented a synthetic-Synth90K dataset which contains 9 million synthetic images constructed via synthetic text generation engine. Ali et al. [2] presented an Urdu characters-based dataset for the cursive Urdu text recognition in natural outdoor scene images. Shi et al. [26] have presented an ICDAR2017 dataset for text localization and recognition tasks. Nayef et al. [20] introduced ICDAR2019-MLT datasets for multi-lingual scene text detection in an outdoor environment.

Fig. 2. Sample dataset images: Demonstrating the diversity of signboards such as, directional boards, distance boards, location boards, and warning boards.

3 Dataset Preparation

We develop a large-scale text-detection and recognition dataset—DLL-TraffSiD, to overcome the shortcomings in existing text detection and recognition methods, as discussed in Sect. 1. It is worth mentioning that collecting large-scale data through driving videos postures numerous challenges including capturing device configuration, data filtration, and suitable frame selection, ground truth generation alongside the development of multi-lingual text-detection and recognition methods. To accomplish these gigantic tasks, we carefully designed and followed the pipeline, as shown in Fig. 1.

3.1 Data Acquisition and Pre-processing

Driving Platform Setup. A high resolution cameras—GoPro Hero 8, mounted over the dashboard of a standard vehicle to capture the front field of view, is used for data acquisition. The installed camera is configured to 4K resolution with a 16:9 super-wide aspect ratio to capture the ultimate width of the road from the driver's perspective. It is important to mention that the main reason for considering monocular vision over stereo vision is because traffic signboards are placed straight oriented towards the driving road direction [22]. Moreover, the installed camera is stabilized by using a mounting device to avoid vibration effects of the vehicle at varying acceleration and deceleration speed patterns.

Video Sequence Collection. Keeping the limitations of existing datasets in view, we collected driving video sequences of 928 KM covering general traffic roads, highways, and motorways of Khyber-Pakhtunkhwa (KP), Punjab, and Azad Jammu and Kashmir (AJK), Pakistan. We primarily focused on diverse direction boards, warning and distance boards, navigation boards placed alongside the roads, as shown in Fig. 2.

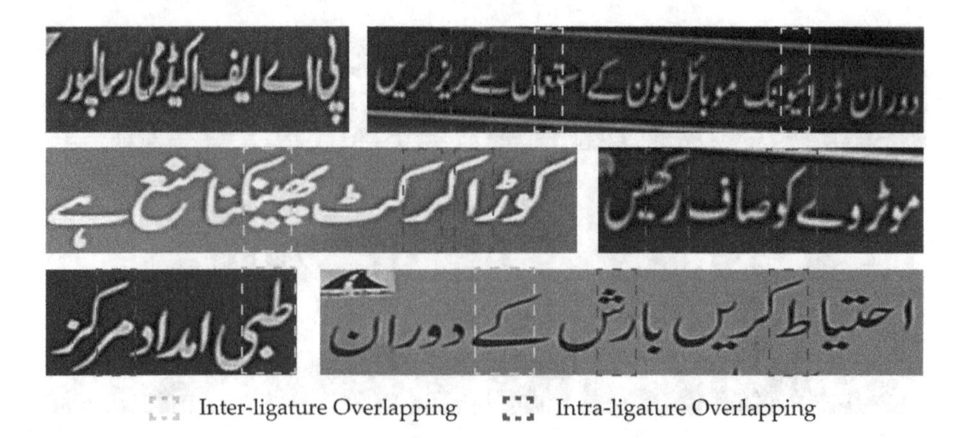

Fig. 3. Dataset-Complexities: Sample images demonstrating inter-ligature and intra-ligature overlapping in Urdu-text instances.

3.2 Multi-lingual Text Detection and Recognition

Dataset Construction. After completing the data collection, the most important task is to align the video sequences to extract frames with signboards and to exclude non-relevant frames. Firstly, we used a frame extraction tool to extract the frames at 30 fps from the video sequences. In the next step, the frames with signboards were manually selected to get a representative subset of signboards considering the diverse text font size, complex scenic backgrounds, writing styles, aspect ratios, and context-sensitivity. Consequently, a subset of 2,600 images is formed as shown in Fig. 2.

Dataset Complexities: Unlike English script, Urdu is a bidirectional cursive script, which makes text detection and recognition tasks more challenging in scene images. In addition, Urdu text is written from right to left in the Nastaleeq script, which is considerably different than the Naksh script that is primarily used to write Arabic. Therefore, detection and recognition of such text become more challenging due to non-uniform inter-ligature overlaps between the letters of the same or two ligatures, as shown in Fig. 3. Moreover, Urdu scripts also include joining and non-joining letters. The joining letters change their shape while merging into the sequence of alphabets within a word, as shown in Fig. 4.

Text Detection Dataset: The proposed dataset contains 2,600 images of traffic signboards. The initial step in annotating text data is to detect the text regions in each frame. In this regard, all the text regions are manually annotated with an enclosed bounding box and four-cornered polygons. The reason for annotating the text regions with polygons is to annotate the tilted text regions accurately. In addition, the legible text regions i.e., single word and illegible text regions i.e., sequences are annotated with one bounding box. Resultantly, 12,179 bounding boxes are generated to perform multi-lingual text detection as shown in Table 1.

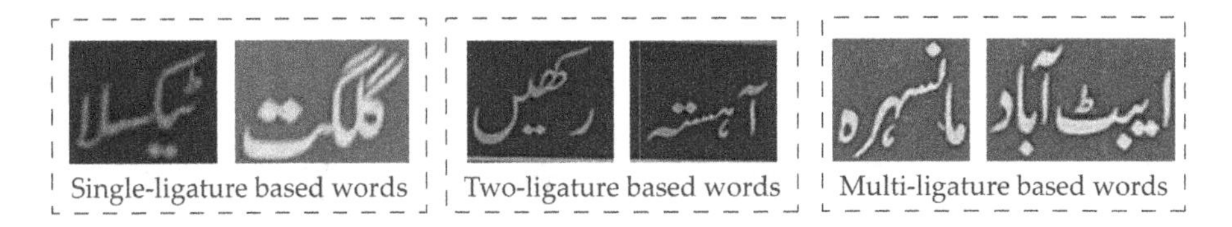

Fig. 4. Types of ligatures in Urdu words: Urdu-text instances with single-ligature, two-ligature, and multi-ligature words.

Text Recognition Dataset. We constructed a text recognition dataset containing 9,120 images along with the corresponding multi-lingual transcriptions. It is worth mentioning that the text regions have been cropped from the original images and the corresponding annotations have been generated in UTF-8 encoded text files. All the words/sequences of words are manually annotated and verified for text recognition purposes. In addition to annotations files, two lexicon text files consisting of 50,000 and 80,000 common Urdu and English words are created to address the contextual error while word recognition.

4 The Methodology

This section describes three sub-parts of the TraffSign system. As described in the pipeline, our proposed system includes (i) a multi-lingual text detection network and (ii) a multi-lingual text recognition network. Each of these architectures is briefly elaborated in the below subsections.

4.1 Multi-lingual Text Detection Architecture

The size of word regions varies in the dataset images; therefore, it is important to extract the features from the late stage of the feature extraction block. To choose suitable network for multi-lingual text detection, we fine-tuned existing text detectors, namely, Efficient and Accurate Scene Text (EAST) [31] and Differentiable Binarization (DB) [17] networks.

Table 1. Statistics of the bounding boxes, words, and characters for each script in the DLL-TraffSiD text detection and recognition dataset

Script type	No. of bounding boxes	No. of words	No. of characters
English	3,269	5,828	11,754
Urdu	5,782	7,653	38,521

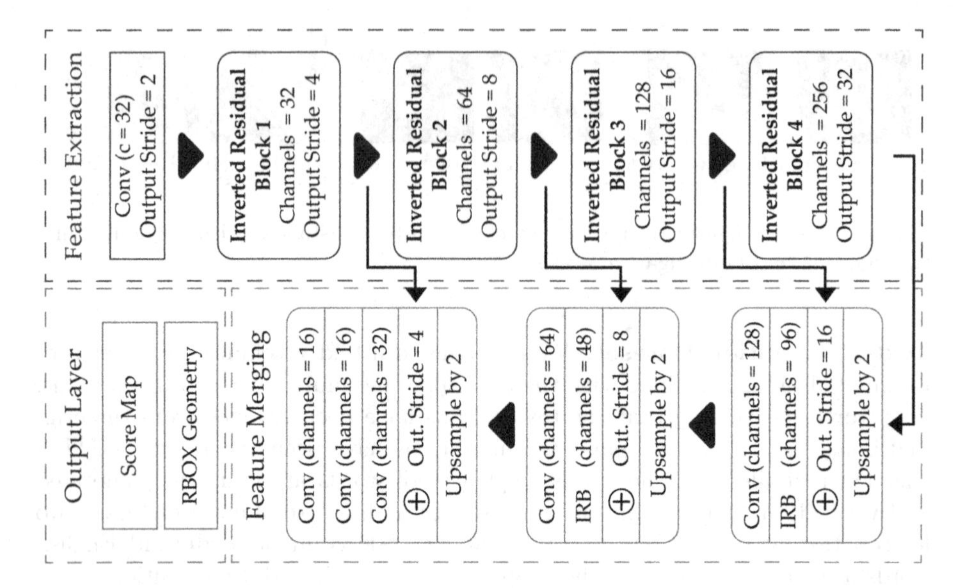

Fig. 5. Improved EAST Architecture with MobileNet-v2 as Backbone for Multi-lingual Text Detection. MobileNet-v2, consisting of four residual blocks is used to perform feature extraction. Each residual block is concatenated with EAST network in feature merging block. Lastly, output layer is used to generate bounding box and score map.

Based on the better performance, EAST architecture has been further improved as shown in Fig. 5. The presented architecture consists of (i) feature extraction, (ii) feature merging, and (iii) output blocks. In the existing EAST based text detection implementations, PA-net, VGG-16, and ResNet-50 architectures [18,31] have been used as feature extraction networks. However, the EAST detector with these feature extraction networks has not performed well in terms of accuracy and precision on our proposed dataset.

To improve upon this shortcoming, we present an improved MobileNet-v2 feature extraction network comprising of four inverted residual blocks including convolution, batch normalization, and pooling layers. Initially, convolution is performed on an input image and the resultant feature map is passed towards the inverted residual blocks. The inverted residual blocks process the feature maps in a feed-forward fashion while also passing the feature merging blocks to concatenate multi-layered information.

In the next step, feature merging block comprising of three blocks connected in a bottom-up fashion. The output feature map of the feature extraction network is up-sampled by a factor of 2 and a series of convolutions are performed with various output numbers of channels over the concatenated feature map. Consequently, the final feature map is passed to the output block to produce the predictions. The output block contains 1×1 convolution projecting 32 channels to generate the score map along with the geometric output containing coordinate-based information of the corresponding text regions.

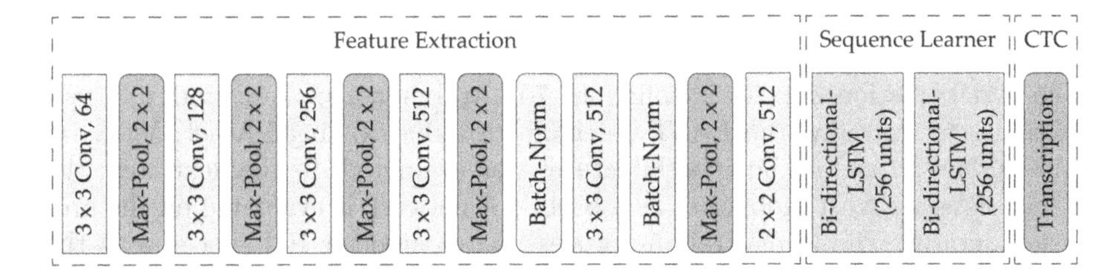

Fig. 6. Bi-Directional LSTM based CRNN Architecture For Multi-lingual Text Recognition. Feature Extraction block contains custom convolution layers followed by max-pooling and activation layers. BLSTMs with 256 units, followed by CTC layer are employed to perform sequence learning and transcription generation.

4.2 Text Recognition Architecture

Multi-lingual text recognition in scenic image data is a challenging task due to variable font size, style, and orientation. To this end, firstly, a base convolutional recurrent neural network (CRNN) has been selected to perform multi-lingual text recognition using both unidirectional and bi-directional Long Short-term Memory (LSTM) with 128 and 256 units, respectively. CRNN is a combination of two categorial neural networks i.e., convolutional and recurrent neural networks. CNN layers in CRNN perform feature extraction from the visual input, whereas RNN layers perform sequence classification. Based on the performance, we further optimized bi-directional LSTM (256-units) based CRNN by inserting batch-normalization followed by max-pooling layer prior to base convolution layer in feature extraction block to achieve maximum performance. As depicted in Fig. 6, a feature extraction block consisting of six convolution layers followed by max-pooling and batch normalization layers is used to extract useful information from the text regions. In the next step, two LSTMs with unidirectional and bi-directional units are used in the architecture to evaluate the text recognition performance on our proposed dataset. Based on better performance, two bi-directional LSTMs with 256 hidden units are used to perform sequence classification. Lastly, Connectionist Temporal Classification (CTC) layer is used to generate final transcription along with the associated confidence score of prediction.

5 Experiments and Results

The experiments of proposed scene text detection and recognition methods are performed on the high-performance machine equipped with the core i9 - 9900k CPU with 32GB RAM, and RTX 2080TI, 11GB DDR5 GPU, having a 64bit windows 10 operating system. Pytorch 1.8.0 library is utilized for data preprocessing, training, and evaluation of the proposed methods. The implementation guide along with training and evaluation codes are available at https://github.com/aatiibutt/TraffSign/.

5.1 Evaluation of Multi-lingual Text Detection Methods

The text detection dataset is split into 70% training, 20% test, and 10% validation sets. Initially, the EAST detector with the existing PA-net, VGG-16, and ResNet-50 as backbone architectures was evaluated. In the next step, the EAST detector with MobileNet as a backbone network is evaluated on the proposed dataset. The whole training process is executed for 100 epochs with the validation patience of 4. In addition, the Stochastic Gradient Descent (SGD) optimizer is set to 0.9, while the piece-wise learning rate and batch size are configured at 0.001 and 32, respectively. Moreover, the validation is performed after each epoch to monitor the learning progress. Performance has been in terms of precision, recall, and F-score of the evaluated networks.

Table 2. Comparison of text detection networks on proposed multilingual dataset demonstrating performance metrics. DB-Net and EAST with four backbone networks, PA-Net, VGG-16, ResNet-50, and improved MobileNet based architecture are evaluated after training with variable learning rates including piece-wise, 0.0001, and 0.001.

Model	Backbone	Learning rate	Precision	Recall	F-score
DB Net	FCN	Piece-wise	0.68	0.72	0.69
		0.0001	0.63	0.69	0.65
		0.001	0.61	0.66	0.62
EAST	PA-net	Piece-wise	0.51	0.39	0.44
		0.0001	0.54	0.42	0.46
		0.001	0.52	0.40	0.43
	VGG-16	Piece-wise	0.43	0.27	0.33
		0.0001	0.41	0.25	0.31
		0.001	0.39	0.24	0.29
	ResNet-50	Piece-wise	0.67	0.71	0.68
		0.0001	0.70	0.74	0.72
		0.001	0.64	0.68	0.69
	Mobile-Net	Piece-wise	**0.86**	**0.93**	**0.89**
		0.0001	0.82	0.89	0.85
		0.001	0.77	0.84	0.86

Discussion: It can be seen from Table 2 that DB-Net achieved 68% accuracy with piece-wise learning rate. Whereas, the EAST detector with PA-net which is base architecture, achieved 51% precision. However, it is worth noting that the same EAST detector with VGG-16 backbone architecture performed worst. Similarly, a base EAST detector with ResNet-50 architecture improved detection performance in terms of precision, recall, and F-score. While on the other side, our proposed improved EAST architecture with MobileNet outperformed the above-mentioned base architectures in terms of precision and recall i.e., 86% and

Fig. 7. Qualitative examples of proposed text detection network. The precise bounding boxes over Urdu and English text regions show that proposed network achieved significant performance in spotting multi-lingual text in test data.

93%, respectively. It is also important to mention that the learning rate played a crucial role in maximizing the performance of the networks. For instance, both architectures, i.e., DB and EAST performed well with piece-wise learning rates, while our presented MobileNet based EAST architecture achieved maximum precision. Some of the sample predictions of our presented network are depicted in Fig. 7.

5.2 Evaluation of Multi-lingual Text Recognition Methods

The proposed text recognition dataset is split into 70% training, 20% test, and 10% validation sets. CRNN-based architectures with LSTM and BLSTM layers comprising of variable hidden units have been employed while transfer learning is used to fine-tune these baseline models on our proposed dataset. The performance of the presented networks on our proposed dataset is measured in terms of Word Recognition Rate (WRR), as statistically elaborated in Table 3. It is worth mentioning that WRR is calculated by mapping the predicted characters over ground truth transcriptions and dividing the sum of correctly predicted characters with the sum of ground truth.

Discussion: It can be seen from Table 3 that baseline CRNN based text recognition methods achieved 71.22%, and 68.10%, WRR on English and Urdu scripts respectively. Moreover, baseline CRNN with BLSTM consisting of 256 hidden units achieved 78.34% and 76.41% WRR on English and Urdu scripts respectively. The proposed CRNN with BLSTM with 256 hidden units achieved 92.18%

Table 3. Performance of proposed CRNN Architectures on proposed multi-lingual text recognition dataset.

Architecture	Sequence learner	Language	Hidden units	WRR (%)
Baseline CRNN	LSTM	English	128	71.22
		Urdu		68.10
	BLSTM	English		74.87
		Urdu		72.29
	LSTM	English	256	75.41
		Urdu		72.66
	BLSTM	English		78.34
		Urdu		76.41
Improved CRNN	LSTM	English	128	82.64
		Urdu		79.28
	BLSTM	English		86.71
		Urdu		85.56
	LSTM	English	256	89.07
		Urdu		86.43
	BLSTM	English		92.18
		Urdu		90.85

and 90.85% WRR on English and Urdu respectively. Some of the sample predictions are shown in Table 4 to provide more insight into the influential factors. Among these, similar alphabets or special characters influenced the performance of proposed methods. For instance, referencing *output 79* in Table 4, our model is influenced by the English special character "**backward slash: /**", and predicted as Urdu alphabet " ا ". Moreover, in output 77 in Table 4, the proposed model predicts English special character **"exclamation mark: !"** as Urdu alphabet " ا " due to similarity. Another observation is that the proposed method demonstrated better applicability in recognizing text scripts of both languages; however, its performance is influenced by the presence of inter-ligature and intra-ligature overlapping. For example, in *output 21* in Table 4, our model ignored the Urdu alphabet " ت " in the word " رفتار " due to distant epigrams, written on next alphabet. Similarly, in *output 19*, proposed model missed " ن " in " پھینکا " and misrecognized " ع " in " منع " because of inter-ligature overlapping in multi-ligature based words. Whereas, it performed well with single ligature and two ligature-based words, as shown in *output 58, 39, 82.*

5.3 Evaluation of Proposed Text-Detection and Recognition Models as an End-to-End Pipeline

To evaluate the proposed methods as an end-to-end method, we combined both, the text detection and recognition models in a pipeline to evaluate in terms

Table 4. Qualitative examples of proposed CRNN architecture, evaluating on test set of proposed text recognition dataset.

Input	Recognized Text	WRR
	Out [74]: پیدل روڈ کراسنک والے	97.21%
	Out [21]: رفار آہستہ رکھیں	98.60%
	Out [19]: کوڑاکرکٹ پھیکنا مغ ہے	88.57%
	Out [79]: پشاور ا اسلام آباد	98.12%
	Out [77]: خبردار ا	98.54%
	Out [33]: Hakla to DI Khan	100%
	Out [27]: PAF Academy Risalpur	100%

of inference time, taken in detecting and recognizing the text in test images. Overall, the proposed pipeline achieved 97.3% precision in text detection, 97.9% word recognition rate, with an average inference time of 0.17 s. Some of the sample predictions along with the recognized text and inference time are shown in Table 5. It can be observed from the table that the text detection network has demonstrated significant performance in spotting English and Urdu text regions. Precision in text detection is one of the most influential factors in the performance of text recognition networks. Subsequently, the text-recognition network also recognized the multi-lingual text with a minimal error rate with less inference time. However, the performance of text recognition is influenced to

Table 5. Performance of proposed text-detection and recognition models as an end-to-end pipeline (Matrices – Text-Detection: Precision)

Input Image	Text-Detection	Recognized Text	WRR (%)	Inf. Time (s)
	98.3%	اسلام آباد Islamabad Chakri چکری Lahore لاہور	100%	0.11 + 0.08
	98.7%	راولپندی Rawalpindi Peshawar پشاور Lahore لاور اسلام آباد Islamabad	98.83%	0.09 + 0.06

some extent while recognizing Urdu alphabets with epigrams, written in overlapping and tilt orientation. These issues can be addressed by introducing text orientation-awareness in text recognition models.

6 Conclusions

In this paper, we present a large-scale multi-lingual dataset—DLL-TraffSiD for text detection and text recognition for Urdu/English traffic signboards. We also proposed a high-performance pipeline for multi-lingual text detection and text recognition tasks in scene images. The results show that the proposed text detection and recognition method outperformed the existing base architectures on our proposed dataset. As a future direction, we are aiming to extend this research work to text orientation assessment and context development.

References

1. Aberdam, A., et al.: Sequence-to-sequence contrastive learning for text recognition. In: CVPR, pp. 15302–15312 (2021)
2. Ali, A., Pickering, M., Shafi, K.: Urdu natural scene character recognition using convolutional neural networks. In: IEEE 2nd International Workshop on Arabic and Derived Script Analysis and Recognition, pp. 29–34. IEEE (2018)
3. Arafat, S.Y., Iqbal, M.J.: Urdu-text detection and recognition in natural scene images using deep learning. IEEE Access **8**, 96787–96803 (2020)
4. Bains, J.K., Singh, S., Sharma, A.: Dynamic features based stroke recognition system for signboard images of Gurmukhi text. Multimed. Tools Appl. **80**(1), 665–689 (2021)
5. Basavaraju, H.T., Manjunath Aradhya, V.N., Guru, D.S.: A novel arbitrary-oriented multilingual text detection in images/video. In: Satapathy, S.C., Tavares, J.M.R.S., Bhateja, V., Mohanty, J.R. (eds.) Information and Decision Sciences. AISC, vol. 701, pp. 519–529. Springer, Singapore (2018). https://doi.org/10.1007/978-981-10-7563-6_54
6. Bušta, M., Patel, Y., Matas, J.: E2E-MLT - an unconstrained end-to-end method for multi-language scene text. In: Carneiro, G., You, S. (eds.) ACCV 2018. LNCS, vol. 11367, pp. 127–143. Springer, Cham (2019). https://doi.org/10.1007/978-3-030-21074-8_11
7. Butt, M.A., Riaz, F.: CARL-D: a vision benchmark suite and large scale dataset for vehicle detection and scene segmentation. Signal Process.: Image Commun. **104**, 116667 (2022)
8. Chandio, A.A., Asikuzzaman, M., Pickering, M., Leghari, M.: Cursive-text: a comprehensive dataset for end-to-end Urdu text recognition in natural scene images. Data Brief **31**, 105749 (2020)
9. Chandio, A.A., Pickering, M.: Convolutional feature fusion for multi-language text detection in natural scene images. In: 2019 2nd International Conference on Computing, Mathematics and Engineering Technologies, pp. 1–6. IEEE (2019)
10. Chandio, A.A., Pickering, M., Shafi, K.: Character classification and recognition for Urdu texts in natural scene images. In: 2018 International Conference on Computing, Mathematics and Engineering Technologies, pp. 1–6. IEEE (2018)

11. Chen, X., Jin, L., Zhu, Y., Luo, C., Wang, T.: Text recognition in the wild: a survey. ACM Comput. Surv. (CSUR) **54**(2), 1–35 (2021)
12. Cheng, Z., Bai, F., Xu, Y., Zheng, G., Pu, S., Zhou, S.: Focusing attention: towards accurate text recognition in natural images. In: ICCV, pp. 5076–5084 (2017)
13. Ch'ng, C.K., Chan, C.S.: Total-text: a comprehensive dataset for scene text detection and recognition. In: 14th IAPR ICDAR, vol. 1, pp. 935–942. IEEE (2017)
14. Gupta, A., Vedaldi, A., Zisserman, A.: Synthetic data for text localisation in natural images. In: CVPR, pp. 2315–2324 (2016)
15. Hossain, M.S., Alwan, A.F., Pervin, M.: Road sign text detection using contrast intensify maximally stable extremal regions. In: 2018 IEEE Symposium on Computer Applications & Industrial Electronics, pp. 321–325. IEEE (2018)
16. Karatzas, D., et al.: ICDAR 2015 competition on robust reading. In: 13th ICDAR, pp. 1156–1160. IEEE (2015)
17. Liao, M., Wan, Z., Yao, C., Chen, K., Bai, X.: Real-time scene text detection with differentiable binarization. In: AAAI, vol. 34, pp. 11474–11481 (2020)
18. Long, S., He, X., Yao, C.: Scene text detection and recognition: the deep learning era. Int. J. Comput. Vision **129**(1), 161–184 (2021)
19. Lu, N., et al.: MASTER: multi-aspect non-local network for scene text recognition. Pattern Recogn. **117**, 107980 (2021)
20. Nayef, N., et al.: ICDAR 2019 robust reading challenge on multi-lingual scene text detection and recognition—RRC-MLT-2019. In: 2019 ICDAR, pp. 1582–1587. IEEE (2019)
21. Panhwar, M.A., Memon, K.A., Abro, A., Zhongliang, D., Khuhro, S.A., Memon, S.: Signboard detection and text recognition using artificial neural networks. In: 2019 IEEE 9th International Conference on Electronics Information and Emergency Communication, pp. 16–19. IEEE (2019)
22. Rasib, M., Butt, M.A., Riaz, F., Sulaiman, A., Akram, M.: Pixel level segmentation based drivable road region detection and steering angle estimation method for autonomous driving on unstructured roads. IEEE Access **9**, 167855–167867 (2021)
23. Reddy, S., Mathew, M., Gomez, L., Rusinol, M., Karatzas, D., Jawahar, C.: RoadText-1K: text detection & recognition dataset for driving videos. In: 2020 ICRA, pp. 11074–11080. IEEE (2020)
24. Saha, S., et al.: Multi-lingual scene text detection and language identification. Pattern Recogn. Lett. **138**, 16–22 (2020)
25. Shi, B., Bai, X., Yao, C.: An end-to-end trainable neural network for image-based sequence recognition and its application to scene text recognition. IEEE Trans. Pattern Anal. Mach. Intell. **39**(11), 2298–2304 (2016)
26. Shi, B., et al.: ICDAR 2017 competition on reading Chinese text in the wild (RCTW-17). In: 14th IAPR ICDAR, vol. 1, pp. 1429–1434. IEEE (2017)
27. Sun, Y., Liu, J., Liu, W., Han, J., Ding, E., Liu, J.: Chinese street view text: large-scale Chinese text reading with partially supervised learning. In: ICCV, pp. 9086–9095 (2019)
28. Tian, S., et al.: Multilingual scene character recognition with co-occurrence of histogram of oriented gradients. Pattern Recogn. **51**, 125–134 (2016)
29. Veit, A., Matera, T., Neumann, L., Matas, J., Belongie, S.: COCO-text: dataset and benchmark for text detection and recognition in natural images. arXiv (2016)
30. Yuan, T.L., Zhu, Z., Xu, K., Li, C.J., Mu, T.J., Hu, S.M.: A large Chinese text dataset in the wild. J. CS Technol. **34**(3), 509–521 (2019)
31. Zhou, X., et al.: EAST: an efficient and accurate scene text detector. In: CVPR, pp. 5551–5560 (2017)

Read While You Drive - Multilingual Text Tracking on the Road

Sergi Garcia-Bordils[1,3], George Tom[2], Sangeeth Reddy[2], Minesh Mathew[2], Marçal Rusiñol[3], C. V. Jawahar[2], and Dimosthenis Karatzas[1](✉)

[1] Computer Vision Center (CVC), UAB, Barcelona, Spain
{sergi.garcia,dimos}@cvc.uab.cat
[2] Center for Visual Information Technology (CVIT), IIIT Hyderabad, Hyderabad, India
{george.tom,sangeeth.battu,minesh.mathew,jawahar}@research.iiit.ac.in
[3] AllRead Machine Learning Technologies, Barcelona, Spain

Abstract. Visual data obtained during driving scenarios usually contain large amounts of text that conveys semantic information necessary to analyse the urban environment and is integral to the traffic control plan. Yet, research on autonomous driving or driver assistance systems typically ignores this information. To advance research in this direction, we present RoadText-3K, a large driving video dataset with fully annotated text. RoadText-3K is three times bigger than its predecessor and contains data from varied geographical locations, unconstrained driving conditions and multiple languages and scripts. We offer a comprehensive analysis of tracking by detection and detection by tracking methods exploring the limits of state-of-the-art text detection. Finally, we propose a new end-to-end trainable tracking model that yields state-of-the-art results on this challenging dataset. Our experiments demonstrate the complexity and variability of RoadText-3K and establish a new, realistic benchmark for scene text tracking in the wild.

Keywords: Scene text · Tracking · Multilingual · Driving videos

1 Introduction

There is text in about 50% of the images in large-scale datasets such as MS Common Objects in Context [32], and the percentage goes up sharply in urban environments. Specific activities, such as making a purchase, using public transportation or finding a place in the city, are highly dependent on understanding textual information in the wild, and driving is a prime example.

Text on traffic signs is an integral part of the traffic control plan as it provides the driver with information on the upcoming situation. Nevertheless, textual information is currently not exploited by Advanced Driver Assistance Systems (ADAS) or autonomous driving systems. Automatic road text understanding could allow introducing new driving instructions in the route, updating maps automatically, and identifying target locations in the street. At the same time,

© Springer Nature Switzerland AG 2022
S. Uchida et al. (Eds.): DAS 2022, LNCS 13237, pp. 756–770, 2022.
https://doi.org/10.1007/978-3-031-06555-2_51

Fig. 1. Sample frames from the new RoadText-3K dataset taken from different locations containing multilingual text. The top-left frame was captured in the US (English text), the top-right frame in Spain (Spanish/Catalan) while bottom row frames are taken from India videos (Telugu and Hindi). Transcriptions are shown only for some of the bounding boxes to avoid clutter.

much text on the road is a distraction for the driver. For example, 71% of Americans consciously look at billboard messages while driving [34]. As a matter of fact, drivers who detected more traffic signs also detected more advertisements [31], as text naturally attracts bottom-up human attention [6] (Fig. 1).

The lifetime of text objects while driving is quite short. At normal city driving speeds (30 km/h), a road text instance enters (becomes readable) and exits the scene within 3–5 s. Thus a reading system for driving is required to detect, track and recognise text early on, at the initial instances of its occurrence, while the text is typically far from the vehicle. This requires a fast model, tolerant to occlusions, which can deal with tiny text instances, typically affected by motion blur and important perspective distortions, especially in the case of roadside text. While object tracking is a well explored area of research, there have only been a few attempts at extending these ideas to text tracking. Our dataset contains high-resolution videos where many of the text instances have a small size, suffer significant perspective changes during the sequence and present visual artifacts such as blurred or out of focus text. This makes the extension of object tracking to road-text tracking non-trivial.

In this work, we introduce a significant quantitative and qualitative extension to the RoadText-1k dataset [26] and a comprehensive study of baseline tracking methods before introducing a new tracking model that yields better performance at high speeds compared to the baselines.

Specifically our contributions are the following:

- We extend the existing RoadText-1K dataset by adding 2000 more videos. The extended RoadText-3K dataset contains videos captured from three countries

that contain text instances in three scripts—Latin (English/Spanish/Catalan), Telugu and Devanagari.
- We provide a detailed study of various tracking methods and compare their performance on RoadText-3K. We build multiple trackers based on state-of-the-art scene text detectors and highlight the key aspects that influence tracking in each of the approaches.
- We propose a new tracking approach using a CenterNet [37] based text detector. Our approach outperforms other trackers in terms of MOTA and MOTP metrics while maintaining real-time speed.

2 Related Work

2.1 Datasets for Text Spotting in Videos

Existing datasets for spotting text in videos are ICDAR Text in Videos dataset [15], YouTube Video Text (YVT) [24], RoadText-1K [26], Large-scale Video Text dataset (LSVTD) [7] and Bilingual Open World Video Text (BOVText) [35].

ICDAR Text in Videos dataset was introduced as part of 2013–2015 Robust Reading Challenge in ICDAR. It contains 51 videos (28k frames) of varying lengths, captured in different scenarios such as highways, shopping in a supermarket or walking inside buildings. The videos are captured using a handheld device or a head-mounted camera. YVT contains 30 videos (13k frames) in total, sourced from Youtube. The videos contain scene text and born digital overlay text such as captions, titles or logos. RoadText-1K [26] has 1,000 driving videos (300k frames) with annotations for text detection, recognition and tracking. The videos contain only text in English since all the videos are captured from the United States. LSVTD [7] has 100 videos (65k frames) captured in 13 indoor and 8 outdoor scenarios. BoVText is a recently introduced dataset with 2021 videos (1,750k frames). The dataset contains both born digital overlay text and scene text instances. The videos are harvested from video-sharing platforms, and consequently, there are videos captured from different parts of the world.

The newly introduced RoadText-3K contains driving or road videos and it is an extension of the existing RoadText-1K. 2000 new videos from two different geographical locations are added to the existing RoadText-1K to make the new RoadText-3K dataset. Among the existing datasets, BoVText, a work that is concurrent to ours is the only dataset that has more number of videos and text instances in it compared to RoadText-3K. Compared to BOVText, which has text instances in Chinese and English, RoadText-3k has annotated text instances in Latin (English, Spanish and Catalan), Telugu and Devanagari (Hindi).

2.2 Text Detection

Text detection approaches can be classified into two types—regression-based and segmentation-based. TextBoxes++ [18] and CTPN [30] are examples for regression-based methods. TextBoxes++ generates proposals using a quadrilateral representation of the bounding boxes. CTPN uses vertical anchors of fixed width to predict the location of text. The model combines the output of a Convolutional Neural Network (CNN) with a Recurrent Neural Network (RNN) to build more meaningful text proposals. Segmentation based methods include EAST [38], which generates dense pixel-level proposals. It employs non-maximal suppression to filter the proposals. FOTS [20] generates pixel-level predictions, outputting a confidence score, the distance, and the rotation of the bounding box the pixel belongs to. Models such as CRAFT [2] focus on detecting curved text. CRAFT uses a more unconventional bottom-up approach and learns to output individual character predictions and their affinity (whether they belong to the same word or not). Since most datasets do not include character-level annotations, CRAFT uses a weakly-supervised method to generate ground truth.

2.3 Text Tracking

Text tracking methods cover both main families of tracking approaches—Tracking by Detection (TbD) and Detection by Tracking (DbT). In TbD, text instances in each individual frame are detected. Then, subsequent detections that correspond to the same text instance are linked to form a track [7,25,33]. For example, [7,25] use spatio-temporal redundancy between frames to track text instances across frames.

In the case of DbT, text detection is performed in an initial frame and these detections are then propagated to the subsequent frames using a propagation algorithm. Detection is then repeated at set intervals to update the trackers. In [10] text regions are extracted using a Maximally Stable External Regions (MSER)-based detector [9] every 5 frames, and are then propagated for the next 5 frames using MSER propagation. Snooper-track [22] uses a similar strategy, where text is tracked using a particle filter system. In [28] a combination of TbD and DbT is used (spatio-temporal learning and template matching) to improve text tracking. The authors use the Hungarian algorithm to do the final association of the detections. In [29] TbD and DbT are explored for the problem of tracking and recognizing embedded captions in online videos.

In addition to the above two categories of approaches, there have been some end-to-end approaches that do detection and tracking simultaneously. In [36] a Convolutional Long Short Term Memory (ConvLSTM) is used in the detection branch to capture spatial structure information and motion memory. Yu et al. [36] proposed an end-to-end tracking model where a two branch network is used to detect and track text instances simultaneously.

2.4 Multiple Object Tracking Metrics

Evaluation of multiple object tracking has evolved extensively over the past years [3,21]. In this work, we evaluate the tracking of text instances using four standard metrics—MOTA, MOTP, IDF1 and IDs. Multiple Object Tracking Accuracy(MOTA) takes into account false positives, false negatives and ID switches at track level. MOTP (Multiple Object Tracking Precision) measures the similarity between the true positive detections and their corresponding ground truth objects (in this case, the similarity is measured in terms of the Intersection over Union). IDF1 is similar to the F-score metric used in binary classification, it reports the harmonic mean of identification precision and recall. ID switches (IDs) indicates the number of re-identifications of a tracked object.

3 RoadText-3K Dataset

We introduce the RoadText-3K dataset, an extension of the existing RoadText-1K [26] dataset. We extend the former by adding new 2000 videos captured from two different geographical locations and containing text in 6 languages, including English. RoadText-1K has videos captured from the United States. In the new 2000 videos, 1000 are captured from Spain, and the remaining 1000 are from India. The dataset can be downloaded from https://datasets.cvc.uab.es/roadtext3k/.

Table 1. RoadText-3K is an extension of RoadText-1K. The new dataset has 2000 more videos that are captured in locations in two continents making it ideal for detection and tracking of multilingual text on roads.

Dataset	RoadText-1K [26]	RoadText-3k
Source	Car-mounted	Car-mounted
Videos	1,000	3,000 (2,000 new videos)
Length (seconds)	10	10
Resolution	1280×720	1280×720
Annotated frames	300,000	927,974
Text Instances	1,280,613	4,039,250
Tracks	28,280	88,427
Unique words	8,263	22,115
Location	US	US, Europe and India
Scripts	Latin	Latin, Telugu and Devanagari

Similar to RoadText-1K, the new videos are annotated with bounding boxes of text tokens in each frame, transcriptions for the text tokens and track information. Videos from each of the three locations are split in 50:20:30 ratio to train, validation and test splits respectively.

3.1 Videos

Videos are collected from various geographical locations to accommodate the diversity of scripts scenes and geographies. Out of the 2,000 new videos, 1,000 videos are collected from India and the other 1,000 are collected from Europe. Videos are captured with a camera mounted on a vehicle.

3.2 Annotations

We follow the same annotation approach as RoadText-1K [26]. Annotations include text bounding box, text transcription and track id. Text bounding boxes are added at line level as in RoadText-1K [26]. The videos from Europe have text in Spanish, Catalan and English. Videos captured in India include text in English, Telugu and Hindi. Few text instances do not belong to any of these languages and are labelled as "Others". Transcriptions are not provided for text instances in this category.

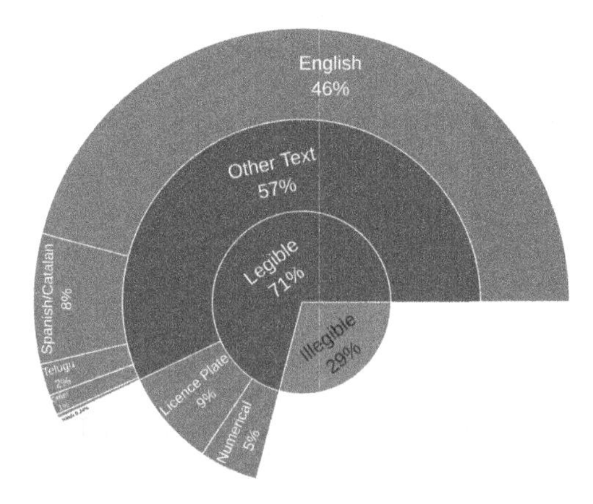

Fig. 2. Distribution of text instances in Roadtext-3K based on legibility, type and language.

3.3 Analysis

Basic statistics of the dataset in comparison with Roadtext-1K are shown in Table 1. The distribution of text instances based on their legibility, text type and language is shown in Fig. 2. Around 29% of the text instances are illegible. This is expected since texts in driving videos are subject to various artifacts including low resolution, motion blur, glare and perspective distortions. Indian roads and highways are dominated by English text, leading to a low percentage of Indian language text instances. The distribution of track lengths in the dataset is shown in Fig. 3. The lifetime of text instances in the driving videos is generally short, with most tracks having a duration of <1 s (<30 fps). Figure 4 shows a word cloud of the most common text tokens in the dataset. It can be seen that tokens like "P" and "30" are among these, suggesting that there are many text tokens from traffic/road boards.

4 Methodology

We first evaluate state-of-the-art scene text detection models on individual frame of the videos in RoadText-3K to identify efficient detectors for our tracking

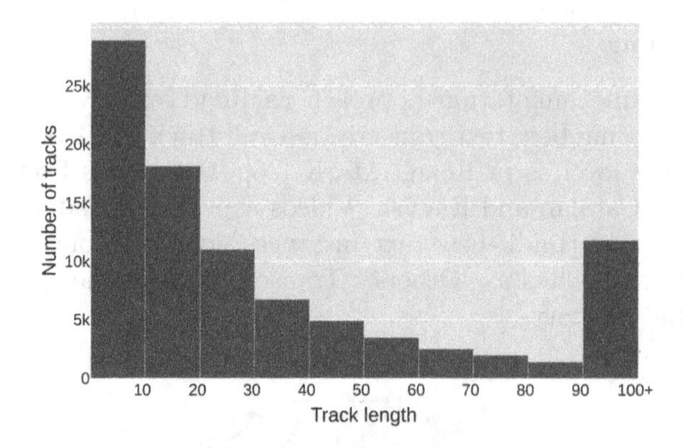

Fig. 3. Distribution of track lengths in RoadText-3K.

methods. For text tracking, multiple baseline trackers are built using these text detectors in both TbD and DbT paradigms. Finally, we propose a new TbD approach that uses CenterNet for detections.

4.1 Text Detection

To test the performance of modern text detectors on RoadText-3K, we have evaluated CTPN [30], EAST [38], FOTS [20] and CRAFT [2] on the test split of our dataset. Since consecutive frames contain similar text instances, we evaluate these detectors on every 10th frame in a video. All four detection models were originally trained to detect single words. Since text instances in our dataset are annotated at line level, we have fine-tuned these models on the train split of the RoadText-3K. For CTPN, EAST and FOTS, we have used implementations available online, which include training and evaluation code[1]. The authors of CRAFT provide an implementation of their model but do not include the training code, so we have used their provided pre-trained model on SynthText [11], ICDAR 2013 [16] and ICDAR 2015 [15] datasets. Since we have many small text instances, each frame is resized to an input size of 1280×720.

4.2 Text Tracking

Both TbD and DbT trackers are built using the above mentioned text detection models. In addition, we propose a new TbD model that uses CenterNet for temporally aware text detection. We quantitatively evaluate our methods using the

[1] For CTPN, EAST and FOTS we have used unofficial implementations of the original methods, for CRAFT we have used the author's released implementation:

- CTPN: https://github.com/eragonruan/text-detection-ctpn
- EAST: https://github.com/argman/EAST
- FOTS: https://github.com/jiangxiluning/FOTS.PyTorch
- CRAFT: https://github.com/clovaai/CRAFT-pytorch.

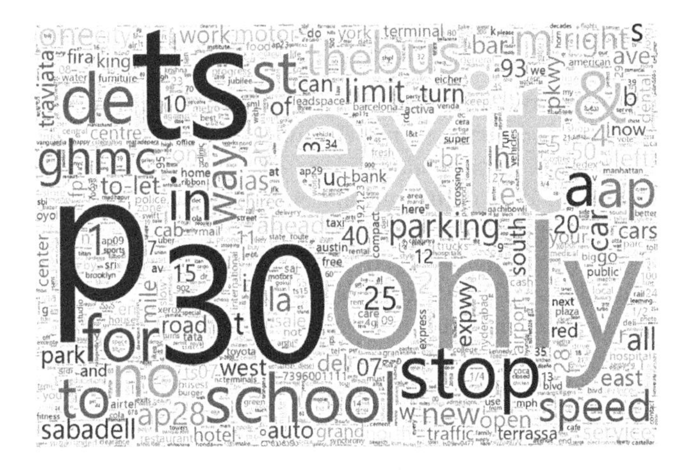

Fig. 4. WordCloud of the text tokens in the dataset. The most common text comes from road signs (for example, "P", "stop", "30" and "exit") and the common prefixes on Indian licence plates (for example "ts" and "ap").

previously introduced MOT metrics[2], which have also been used in the ICDAR 2013 [16] and ICDAR 2015 [15] challenges.

Tracking by Detection (TbD). In the TbD paradigm, trackers seek to associate detections between frames using temporal and visual information. These methods perform text detection on every frame, which can make the text tracking system slower. Since detections are made independently on each frame, flickering, detection merging might occur, while the final tracking result is affected by any detection failures. In order to evaluate TbD on our dataset, we use text detection from the scene text detectors discussed in the previous section and use SORT [4] to associate the instances across frames. SORT uses Kalman filters to predict the position of the objects (text instances in our case) in consecutive frames. Using the IoU as the distance, it solves the matching problem between predicted positions and frame detections using the Hungarian algorithm [17].

Detection by Tracking (DbT). In DbT every single object is explicitly tracked by a different instance of the tracker. This technique can employ visual and temporal information to find the location of the object in consecutive frames, while keeping the inference times down. One of the major drawbacks is that, in the case of multiple object tracking, we need to initialize a new instance of the tracker for every new object. We have opted for a setup similar to the ones proposed in [10,22]. Every 5 frames, we perform text detection using one of the text detection models that we discuss in Sect. 4.1. For every new text instance found, we launch a tracker that will follow the text instance for the next 5 consecutive frames. In the final, frame we compare the location of the tracked objects

[2] We used the implementation given in https://github.com/cheind/py-motmetrics.

with the detections of the text detector on the current frame. When the IoU between a tracked instance and one of the detections surpasses 0.5, we consider this a match, and the tracker continues following the text instance for the next 5 frames. When no tracked instances match any of the detections, we launch a new tracker. Finally, if no detection matches a tracked instance, we keep tracking it for 5 more frames, but if no detections match it 5 frames later the tracker stops. This avoids relying too much on the detections of the detector, since motion blur or temporal occlusion can introduce detection failures. This also introduces the risk of increasing the number of false positives, since a text instance that leaves the scene or gets permanently occluded may still get tracked for a few frames.

FOTS obtained the highest F-score and recall from all the detection methods, as well as a relative high inference speed. For this reason, we have used its detections to start and stop the trackers. To perform the tracking of the text instances we use CSR-DCF [1], KCF [13] and MedianFlow [14]. These trackers use traditional approaches to object tracking such as correlation filters and kernels.

4.3 CenterNet-Based Detection and Tracking

Our dataset features high resolution images with many small text instances. While downsampling frames allows faster inference, the detection of small instances requires working at a higher resolution. Nevertheless, the complexity of many modern text detectors results in slow inference speeds on high-resolution images. We have tried to simplify the approach towards scene text detection and tracking on RoadText-3k, and we have tailored a framework that focuses on real-time inference on high-resolution inputs. Our approach is based on CenterNet [37], and we use it for both text detection and tracking. We made the code and the weights publicly available at https://github.com/Sergigb/roadtext3k-baselines.

CenterNet is an object detection model that represents objects as a single point at their bounding box center, and then regresses the width and height at the center of the location. The centers of the objects are represented as a Gaussian kernel on a heatmap and focal loss [19] is used to learn this representation. To achieve better performance, we adopt ResNet-18 [12] as a backbone to our networks, which offers a good balance between performance and real-time inference. Inspired by YOLOv4 [5], we replace ReLU with the Mish activation function [23]. We evaluated our approach on single frame object detection in the same fashion as the previous object detectors.

For our tracking model, we have tried to leverage temporal information into our CenterNet-based text detector. We add temporal awareness to the model by adding one convolutional GRU cell before upsampling the feature map. This cell which is similar to the convolutional LSTM model presented in [27], but uses the GRU cell [8] layout. In the decoder of the network we apply transpose convolutions to upscale the latent feature map. Similar to the other TbD approaches we discuss in Sect. 4.2, we use SORT to perform the object association between the frames. Figure 5 shows an overview of the architecture.

5 Results

Results of the detection and tracking experiments are presented in this section. All experiments results are reported on the test split of the Roadtext-3K dataset.

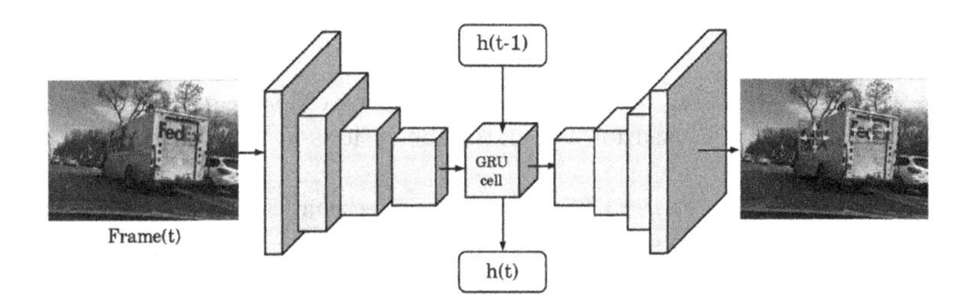

Fig. 5. Our method extends the CenterNet architecture with a convolutional GRU cell at the output of a ResNet-18 to aggregate spatial and temporal information. $h(t)$ represents the hidden state of the GRU cell in the frame t.

5.1 Frame Level Text Detection

We consider a detection to be a true positive if it overlaps with any ground truth instance with an IoU of over 0.5. Table 2 shows results of the frame level text detection. FOTS obtains the best F-score and the highest recall. Our CenterNet-based method gets competitive performance while being the fastest method we have tested, obtaining an inference speed of 44 FPS. Despite not being fine-tuned on our dataset, CRAFT obtains the highest precision but has the lowest FPS.

Table 2. Results of frame level text detection. FOTS has the highest F-score and recall, while our method obtains the fastest inference speed.

Detector	Precision (%)	Recall (%)	F-score (%)	FPS
CTPN	34.62	32.74	33.65	13
EAST	32.14	29.51	31.27	17
FOTS	42.77	**50.74**	**46.41**	19
CRAFT	**54.3**	37.21	44.15	5
CenterNet	50.3	39.1	43.9	**44**

5.2 Tracking

TbD: Results of the TbD on the test split of RoadText-3K are shown in Table 3. Our CenterNet+GRU model and CRAFT obtain similar scores. Both methods have high precision and similar recall, and both models produce a low number of ID switches. Using GRU with CenterNet reduces the number of ID switches (column IDs) by a large margin and improves the MOTA score. We hypothesize that the GRU cell helps reducing the flickering and improves the consistency of the tracking. Speed is reported based on the combined time taken to run the detector and associate detections using SORT. Our CenterNet-based approaches have the highest inference speeds, reaching 40 FPS when we do not use the GRU cell and 31 when we use it.

Table 3. Results of Tracking by Detection (TbD) using different detectors. In all cases the SORT algorithm [4] is used for associating detections in two consecutive frames.

Detector	MOTA (%)	IDF1 (%)	Recall (%)	Precision (%)	IDs	MOTP (%)	FPS
CTPN	24.93	51.60	52.00	66.80	15524	65.11	11
EAST	25.20	51.33	49.20	68.00	13178	69.47	12
FOTS	28.47	**56.57**	**59.73**	66.33	12883	71.82	14
CRAFT	35.40	54.77	46.20	**81.73**	**6328**	70.59	5
CenterNet	33.80	54.80	53.00	74.50	15032	72.44	**40**
CenterNet+GRU	**36.00**	54.80	47.60	81.30	8896	**72.74**	31

DbT: Table 4 shows results of the three different DbT methods we evaluated. It can be seen that precision is much lower for all the three approaches compared to TbD results shown in Table 3. Lower precision is partly due to a higher number of false positives. One possible reason for this is the fact that even text instances that have disappeared are tracked until the next set of detections are available. However recall scores for DbT approaches are comparable to that of TbD. For example CSR-DCF has a recall of 54.10%, second only to FOTS in the TbD setup. DbT is usually a good choice if inference speed is a constraint. Note nevertheless that the proposed CenterNet-based TbD approach still yields competitive speeds (35 FPS).

Table 4. Results of detection by tracking (DbT) approach using various trackers. FOTS [20] is used for detections in all cases.

Tracker	MOTA (%)	IDF1 (%)	Recall (%)	Precision (%)	IDs	MOTP (%)	FPS
CSR-DCF	**−33.70**	**35.30**	**54.10**	**38.70**	**16136**	71.66	25
KCF	−51.80	31.10	49.60	33.40	19534	71.8	76
MEDIANFLOW	−53.00	28.10	48.40	32.90	22614	**71.81**	**83**

5.3 Qualitative Analysis

Visually inspecting the results of the different methods gives us a hint of how they behave under different conditions. For example, the two approaches we tested behave very differently in cases of temporal occlusions. We can see one of such cases in Fig. 6, where the vehicle's windscreen wiper temporally occludes the text being tracked. Since the detector does not have any sort of temporal awareness, in TbD the occluded text fails to be localized. In this case, SORT managed to recover from the occlusion and correctly reassigned the IDs in the following frames. Our model displays a similar behaviour to the TbD approach, but SORT recovers again from the occlusions. In DbT, the CSR-DCF algorithm manages to keep tracking the text instances even under partial occlusion. This robustness to temporal occlusions minimizes the probabilities of ID switches. However, the two text instances in the upper left side of the board ("SOUTH" and "678") appear to have slightly shifted bounding boxes after the occlusion. Since we check the detections every 5 frames to start and stop trackers, a shifted bounding box can result in ID switches if it does not match any detection. This could be attributed to the fact that CSR-DCF uses the last known visual appearance of the object to find it in the next frame. Even though TbD generally offers more precise and reliable detection overall, distant or blurry text can introduce flickering. As seen in Fig. 7a, the two lower text instances disappear for a few frames and then reappear, increasing the chances of ID switches. In DbT (Fig. 7b), the proposals keep being reliably tracked. Our model displays a more conservative behaviour, one of the smaller instances is not tracked but the others are consistently tracked.

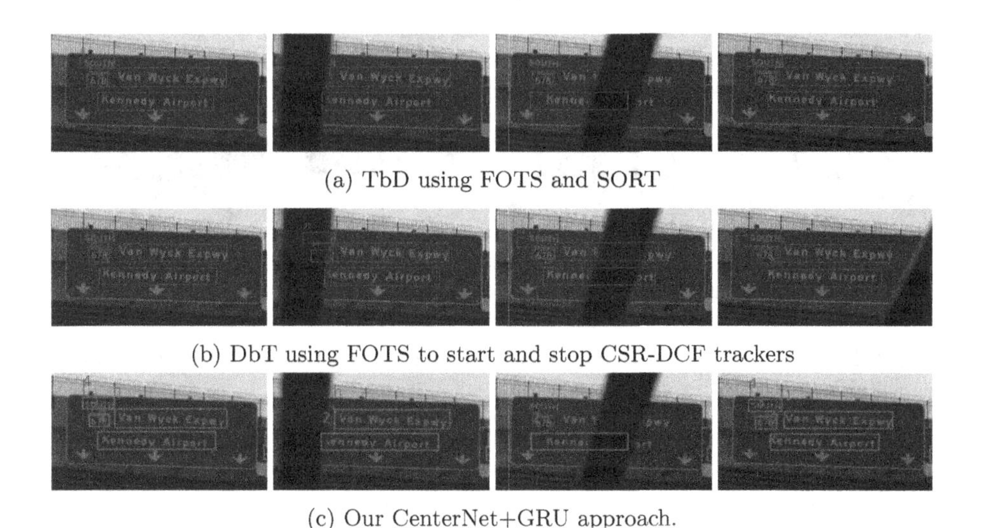

(a) TbD using FOTS and SORT

(b) DbT using FOTS to start and stop CSR-DCF trackers

(c) Our CenterNet+GRU approach.

Fig. 6. Results of TbD, DbT, and our CenterNet-based model in case of a temporal occlusion. Numbers shown alongside the boxes are track numbers.

Judging by the quantitative results, the TbD approach seems to obtain better overall tracking results (best MOTA, lowest ID switches, etc.), but qualitative results suggest that DbT can be advantageous in cases with occlusions or detector failures. The biggest drawback of DbT is the increase in false positives, and this is primarily due to the inability to immediately stop tracking when a text instance leaves the scene. The lower recall in DbT (5.64% between FOTS + CSR-DCF and FOTS + SORT) can be partially explained by the fact that we check for new detections every 5 frames, which can delay starting a new tracker and increase the amount of false negatives.

(a) TbD using FOTS and SORT

(b) DbT using FOTS to start and stop CSR-DCF trackers.

(c) Our CenterNet+GRU method.

Fig. 7. Performance of the various models under a case of detection flickering. Numbers shown along with the boxes are track numbers.

6 Conclusions

We have introduced RoadText-3K, an extension to the existing RoadText-1K dataset with an additional 2000 driving videos captured in different geographical locations and containing text in different scripts and languages. We evaluated several state-of-the-art detectors in this dataset and employed them to construct tracking by detection and detection by tracking methods. Results demonstrate that driving videos are especially challenging. Finally, we have presented a new simple and efficient approach for tracking by detection which incorporates temporal information in the detection branch. Our method yields competitive tracking results while obtaining real-time inference speeds.

Acknowledgements. This work has been supported by the Pla de Doctorats Industrials de la Secretaria d'Universitats i Recerca del Departament d'Empresa i Coneixement de la Generalitat de Catalunya; Grant PDC2021-121512-I00 funded by MCIN /AEI/ 10.13039/501100011033 and the European Union NextGenerationEU/PRTR; Project PID2020-116298GB-I00 funded by MCIN/ AEI /10.13039/501100011033; Grant PLEC2021-007850 funded by MCIN/AEI/10.13039/501100011033 and the European Union NextGenerationEU/PRTR; Spanish Project NEOTEC SNEO-20211172 from CDTI and CREATEC-CV IMCBTA/2020/46 from IVACE and IHub-Data at IIIT-Hyderabad.

References

1. Lukežič, A., Vojíř, T., Čehovin, L., Matas, J., Kristan, M.: Discriminative correlation filter tracker with channel and spatial reliability. IJCV **126**, 671–688 (2018)
2. Baek, Y., Lee, B., Han, D., Yun, S., Lee, H.: Character region awareness for text detection. In: CVPR (2019)
3. Bernardin, K., Stiefelhagen, R.: Evaluating multiple object tracking performance: the CLEAR MOT metrics. EURASIP J. Image Video Process. **2008**, 1–10 (2008)
4. Bewley, A., Ge, Z., Ott, L., Ramos, F., Upcroft, B.: Simple online and realtime tracking. In: ICIP (2016)
5. Bochkovskiy, A., Wang, C.-Y., Liao, H.-Y.M.: YOLOv4: optimal speed and accuracy of object detection. arXiv preprint arXiv:2004.10934 (2020)
6. Cerf, M., Frady, E.P., Koch, C.: Faces and text attract gaze independent of the task: experimental data and computer model. J. Vision **9**, 10 (2009)
7. Cheng, Z., et al.: FREE: a fast and robust end-to-end video text spotter. IEEE Trans. Image Process. **30**, 822–837 (2020)
8. Cho, K., et al.: Learning phrase representations using RNN encoder-decoder for statistical machine translation. arXiv preprint arXiv:1406.1078 (2014)
9. Donoser, M., Bischof, H.: Efficient maximally stable extremal region (MSER) tracking. In: CVPR (2006)
10. Gomez, L., Karatzas, D.: MSER-based real-time text detection and tracking. In: ICPR (2014)
11. Gupta, A., Vedaldi, A., Zisserman, A.: Synthetic data for text localisation in natural images. In: CVPR (2016)
12. He, K., Zhang, X., Ren, S., Sun, J.: Deep residual learning for image recognition. In: CVPR (2016)
13. Henriques, J.F., Caseiro, R., Martins, P., Batista, J.: Exploiting the circulant structure of tracking-by-detection with kernels. In: Fitzgibbon, A., Lazebnik, S., Perona, P., Sato, Y., Schmid, C. (eds.) ECCV 2012. LNCS, vol. 7575, pp. 702–715. Springer, Heidelberg (2012). https://doi.org/10.1007/978-3-642-33765-9_50
14. Kalal, Z., Mikolajczyk, K., Matas, J.,: Forward-backward error: automatic detection of tracking failures. In: ICPR (2010)
15. Karatzas, D., et al.: ICDAR 2015 competition on robust reading. In: ICDAR (2015)
16. Karatzas, D., et al.: ICDAR 2013 robust reading competition. In: ICDAR (2013)
17. Kuhn, H.W.: The Hungarian method for the assignment problem. Naval Res. Logistics Q. **2**, 83–97 (1955)
18. Liao, M., Shi, B., Bai, X.: TextBoxes++: a single-shot oriented scene text detector. TIP **27**, 3676–3690 (2018)
19. Lin, T.-Y., Goyal, P., Girshick, R., He, K., Dollár, P.: Focal loss for dense object detection. In: ICCV (2017)

20. Liu, X., Liang, D., Yan, S., Chen, D., Qiao, Y., Yan, J.: FOTS: fast oriented text spotting with a unified network. In: CVPR (2018)
21. Milan, A., Leal-Taixé, L., Reid, I., Roth, S., Schindler, K.: MOT16: a benchmark for multi-object tracking. arXiv preprint arXiv:1603.00831 (2016)
22. Minetto, R., Thome, N., Cord, M., Leite, N.J., Stolfi, J.: SnooperTrack: text detection and tracking for outdoor videos. In: ICIP (2011)
23. Misra, D.: Mish: a self regularized non-monotonic neural activation function. arXiv preprint arXiv:1908.08681 (2019)
24. Nguyen, P.X., Wang, K., Belongie, S.: Video text detection and recognition: dataset and benchmark. In: WACV (2014)
25. Petter, M., Fragoso, V., Turk, M., Baur, C.: Automatic text detection for mobile augmented reality translation. In: ICCV Workshops (2011)
26. Reddy, S., Mathew, M., Gomez, L., Rusinol, M., Karatzas, D., Jawahar, C.V.: RoadText-1K: text detection & recognition dataset for driving videos. In: ICRA (2020)
27. Shi, X., Chen, Z., Wang, H., Yeung, D.-Y., Wong, W.K., Woo, W.: Convolutional LSTM network: a machine learning approach for precipitation nowcasting. In: NeurIPS (2015)
28. Tian, S., Pei, W.-Y., Zuo, Z.-Y., Yin, X.-C.: Scene text detection in video by learning locally and globally. In: IJCAI (2016)
29. Tian, S., Yin, X.-C., Ya, S., Hao, H.-W.: A unified framework for tracking based text detection and recognition from web videos. IEEE Trans. Pattern Anal. Mach. Intell. **40**(3), 542–554 (2017)
30. Tian, Z., Huang, W., He, T., He, P., Qiao, Yu.: Detecting text in natural image with connectionist text proposal network. In: Leibe, B., Matas, J., Sebe, N., Welling, M. (eds.) ECCV 2016. LNCS, vol. 9912, pp. 56–72. Springer, Cham (2016). https://doi.org/10.1007/978-3-319-46484-8_4
31. Topolšek, D., Areh, I., Cvahte, T.: Examination of driver detection of roadside traffic signs and advertisements using eye tracking. Transp. Res. Part F: Traffic Psychol. Behav. **43**, 212–224 (2016)
32. Veit, A., Matera, T., Neumann, L., Matas, J., Belongie, S.: COCO-text: dataset and benchmark for text detection and recognition in natural images. arXiv preprint arXiv:1601.07140 (2016)
33. Wang, X., et al.: End-to-end scene text recognition in videos based on multi frame tracking. In: ICDAR (2017)
34. Williams, D.: The Arbitron National In-Car Study. Arbitron Inc., Columbia (2009)
35. Wu, W., et al.: A bilingual, OpenWorld video text dataset and end-to-end video text spotter with transformer. In: NeurIPS 2021 Track on Datasets and Benchmarks (2021)
36. Yu, H., Huang, Y., Pi, L., Zhang, C., Li, X., Wang, L.: End-to-end video text detection with online tracking. PR **113**, 107791 (2021)
37. Zhou, X., Wang, D., Krähenbühl, P.: Objects as points. arXiv preprint arXiv:1904.07850 (2019)
38. Zhou, X., et al.: EAST: an efficient and accurate scene text detector. In: CVPR (2017)

A Fair Evaluation of Various Deep Learning-Based Document Image Binarization Approaches

Richin Sukesh, Mathias Seuret(✉), Anguelos Nicolaou, Martin Mayr, and Vincent Christlein

Friedrich-Alexander-Universität Erlangen-Nürnberg, Erlangen, Germany
{richin.sukesh,mathias.seuret}@fau.de

Abstract. Binarization of document images is an important preprocessing step in the field of document analysis. Traditional image binarization techniques usually rely on histograms or local statistics to identify a valid threshold to differentiate between different aspects of the image. Deep learning techniques are able to generate binarized versions of the images by learning context-dependent features that are less error-prone to degradation typically occurring in document images. In recent years, many deep learning-based methods have been developed for document binarization. But which one to choose? There have been no studies that compare these methods rigorously. Therefore, this work focuses on the evaluation of different deep learning-based methods under the same evaluation protocol. We evaluate them on different Document Image Binarization Contest (DIBCO) datasets and obtain very heterogeneous results. We show that the DE-GAN model was able to perform better compared to other models when evaluated on the DIBCO2013 dataset while DP-LinkNet performed best on the DIBCO2017 dataset. The 2-StageGAN performed best on the DIBCO2018 dataset while SauvolaNet outperformed the others on the DIBCO2019 challenge. Finally, we make the code, all models and evaluation publicly available (https://github.com/RichSu95/Document_Binarization_Collection) to ensure reproducibility and simplify future binarization evaluations.

Keywords: Binarization · Deep learning · Evaluation

1 Introduction

Image binarization is a process that converts a color or grayscale image into an image whose pixels can have only two different values, usually black and white. In the domain of document image analysis, binarization typically consists in separating the text (foreground) from its support (background), e. g., the paper. While it became less popular for text recognition, it remains an important preprocessing step in many other tasks, such as writer identification [4,5], word

© Springer Nature Switzerland AG 2022
S. Uchida et al. (Eds.): DAS 2022, LNCS 13237, pp. 771–785, 2022.
https://doi.org/10.1007/978-3-031-06555-2_52

spotting or optical character recognition (OCR) [10]. In traditional global binarization, the grayscale intensity frequency histogram of an image is analyzed and an appropriate threshold is set, e. g., Otsu's thresholding [17]. Alternatively, binarization is applied locally using statistics such as mean and standard deviation like the popular Sauvola method [27]. However, these methods have problems with ink bleed-through artifacts and other artifacts such as stains, blurring, faint characters and noise [15]. An error that may be generated through incorrect binarization may propagate forward and lead to performance reduction in subsequent tasks. Document binarization also acts as a means to filter out these undesirable features. A thorough overview of binarization techniques, datasets, and metrics is given in a survey by Tensmeyer and Martinez [31].

In recent years, rather than relying on traditional image binarization techniques, many studies have been conducted that employ deep learning models to binarize document images. The advent of deep learning has brought a multitude of changes to the domain of computer vision and image processing. Convolutional neural networks (CNNs) identify features automatically by learning from training data. The image features are discovered at multiple layers and are learned gradually from lower-levels to higher-levels. This multi-layered architecture performs a series of convolutions on the input image. A training process is implemented to adjust the parameters of the network to achieve the desired output.

In the past decade, there have been immense progress in the field of binarization of contemporary and historical documents using deep learning techniques. Although many approaches using deep learning for document binarization have been put forward, it is difficult to identify which among these models perform best when compared to one another. The root cause of this problem is the fact that most of these models have never been trained and tested on a common dataset using the same evaluation protocol. This paper aims to resolve this disparity by training and testing some well-known binarization models [2,8,10,13,28,29,32] on common datasets from the well-known Document Image Binarization Contests (DIBCO) [6,16,18–25]. While we evaluated the results of the models using four metrics, we omitted investigations on the relationship between result quality and processing time as Lins *et al.*. did [11]. Our evaluations draw a very heterogeneous picture. All four evaluation datasets have a different winner. Overall, DE-GAN ranks best across the four chosen DIBCO test datasets while metric-wise, the 2-Stage GAN outperforms the other models.

The following Sect. 2 of the paper provides a brief overview on the network architectures and methodologies used in the different binarization models that would be compared against one another. Section 3 gives a detailed description on the various datasets, validation metrics and on how all the models were trained. Section 4 shows the results of evaluating all models on the various test datasets and provides a brief discussion on the outcome of the experiments.

2 Overview of Evaluated Binarization Methods

2.1 Document Enhancement Generative Adversarial Network

The work presented by Souibgui *et al.* [28] models the document binarization problem as an image-to-image translation task. The Document Enhancement Generative Adversarial Network (DE-GAN) model basically consists of a generator and a discriminator. The generator follows a U-Net architecture [26] and its objective is to generate a clean image given the original degraded image. The goal of the discriminator is then to determine if the image shown is a fake image generated by the generator or the original binarized ground truth. An adversarial loss function is employed for training the model [28]:

$$\begin{aligned} L_{GAN}(\phi_G, \phi_D) = E_{I^W, I^{GT}} log[D_{\phi D}(I^W, I^{GT})] \\ + E_{I^W, I^{GT}} log[1 - D_{\phi D}(I^W, G_{\phi G}(I^W))], \end{aligned} \tag{1}$$

where $G_{\phi G}$ and $D_{\phi D}$ are the generator and discriminator functions respectively, I^W is the degraded image and I^{GT} is the ground truth. After a few epochs, the network is able to generate images similar to the ground truth. To maintain a good text quality and to improve training speed an additional log loss function is added. The objective is that the text output from the generator is identical to the ground truth text [28]:

$$L_{\log}(\phi G) = E_{I^{GT}, I^W}[-(I^{GT} \log(G_{\phi G}(I^W)) + ((1 - I^{GT}) \log(1 - G_{\phi G}(I^W)))]. \tag{2}$$

The overall loss of the network is denoted as [28]:

$$L_{\text{net}}(\phi G, \phi D) = \min_{\phi G} \max_{\phi D} L_{\text{GAN}}(\phi G, \phi D) + \lambda L_{\log}(\phi G), \tag{3}$$

where L_{GAN} is the adversarial loss function used to train the cGAN and λ is a hyper-parameter that is set to 500 for document binarization. The generator follows an encoder-decoder structure. The encoder performs down-sampling of the given input up to a certain layer and the decoder then up-samples the encoder output. The discriminator used is a simple Fully-Connected Network (FCN) with 6 convolutional layers. To train the DE-GAN model, overlapped patches of size 256×256 pixel are obtained from the degraded images and fed as input to the generator.

2.2 SauvolaNet

Inspired by the traditional Sauvola thresholding algorithm [27], the work by Li *et al.* [10] presents a deep learning approach to learn the Sauvola parameters, called the "SauvolaNet". The network aims to making the model computationally efficient. The model also comprises of an attention mechanism that aims to estimate the required Sauvola window sizes for each pixel location. One main drawback of the traditional Sauvola thresholding approach is that the algorithm achieves its highest performance only when the right hyperparameters are manually tuned for each input image (window size, estimated level of document degradation and dynamic range of input image intensity). SauvolaNet uses three modules, the Multi-Window Sauvola (MWS), Pixelwise Window Attention (PWA), and Adaptive Sauvola Threshold (AST) to learn an auxiliary threshold estimation function.

The MWS module takes an image as input and uses the Sauvola algorithm to estimate the local thresholds for different window sizes. The PSA module also takes the same image as input to estimate the window sizes for each pixel location. The AST module then predicts the final threshold for each pixel location by fusing the thresholds of different windows from the MWS and weights from the PWA modules. The SauvolaNet function is modelled as [10]:

$$T = g_{SauvolaNet}(D), \tag{4}$$

where, T is the output, $g_{SauvolaNet}$ is the auxiliary threshold estimation function and D is the input image. The PWA uses instance normalization instead of batch normalization in order to avoid overfitting when training with a small dataset. When training the SauvolaNet, the input image D is normalized to values in the range (0,1) and a modified hinge loss was developed [10]:

$$\text{loss}[i,j] = \max(1 - \alpha \cdot (D[i,j] - T[i,j]) \cdot B[i,j], 0), \tag{5}$$

where B is the binarization ground truth with values -1 for foreground and $+1$ for the background. i and j are indices that specify the location of a pixel. α is a parameter to control the margin of the decision boundary and only the pixels close to the decision boundary are used in gradient-backpropagation.

2.3 Two-Stage GAN

The work presented by Suh *et al.* proposes a two-stage color document image binarization deep learning architecture using generative adversarial neural networks (GANs) [29]. The GAN architecture generally consists of two networks, i.e., the generator and the discriminator. For this model, the EfficientNet [30] was used as the generator on account of its efficiency in the domain of image classification. In the case of the discriminator, the discriminator network from the PatchGAN [9] was implemented.

The first part of the network consists of four color independent generators that are trained with the red, green, blue, and gray channels in order to generate an enhanced image by removing background color information. The resulting channel images and corresponding ground truths first concatenated and then fed to the discriminator network. The binarization in the first stage is performed using local predictions in small patches. In order to cater to regions with larger backgrounds, the second stage of the network performs global binarization with the resized original input image and local binarization using the first stage output. Except for the input image channels, the structure for the generators in the second stage is identical to that of the first stage. During training, the images are divided into patches of 256 × 256 pixels resolution without scaling. When training GANs in general, it is common to observe an instability in loss function convergence [29]. To solve this issue, the Wasserstein GAN with penalty was used which implements the Wasserstein K-distance as the loss function. Further, instead of the typically used L1 loss, pixel-wise binary cross-entropy is defined as the additional loss term for the generator update.

2.4 Robin U-Net Model

The implementation by Mikhail Masyagin [13] presents the Robust Documentation Binarization (ROBIN) tool. ROBIN makes use of a simple U-Net model [26] to perform document binarization. The U-Net model was originally developed for the purpose of semantic segmentation of medical images. The U-Net architecture can be described as an encoder-decoder network. The input image is first fed into the encoder network, where multiple convolution blocks are applied followed by a maxpool downsampling layer. The idea here is to encode the input image into feature representations at multiple levels. The output from the encoder is then sent to the decoder where the activation map undergoes upsampling or deconvolution. Skip connections are also introduced between the encoder and decoder structure such that the deep and shallow features can be combined.

When training the model, the input images are split into patches of 128 × 128 px resolution. The learning rate was set to 0.0001 with the Adam optimizer. The training is trained using the dice coefficient loss and run for 250 epochs with an early stopping criteria.

2.5 DP-LinkNet

The DP-LinkNet is a segmentation model introduced by Xiong *et al.*. It makes use of the D-LinkNet [33] and LinkNet [3] models with a pre-trained encoder as the backbone.

The model consists of: 1) an encoder, 2) a hybrid dilated convolution module, 3) a spatial pyramid pooling (SPP) module, and 4) a decoder [32]. Firstly, the input image is fed to the encoder where the text stroke features are extracted. The series of convolutions and down-sampling occurring at the encoder causes a reduction in the resolution of the obtained feature map. To counter this effect, dilated convolutions are introduced into the model. Dilated convolutions help in exponentially increasing the size of the receptive field without affecting the spatial resolution. An issue that still persists here is the fact that the dilated convolution module may still find it difficult to identify objects of different sizes with a fixed-sized field-of-view. To counter this effect, the spatial pyramid pooling is employed. This helps to present the input feature maps at different scales. Lastly, the decoder performs transposed convolution. Skip connections between the decoder and encoder structure are present to combine the shallow-level and high-level features, helping to compensate any loss encountered by convolution and pooling operations. When training the model, the binary cross entropy and dice coefficient losses are used. The input images were split into patches of size

128×128 px. The adam optimizer was set with an initial learning rate of 2×10^{-4}. The model was trained for 500 epochs with an early stopping criteria to avoid overfitting.

2.6 Selectional Auto-Encoder

The work presented by Calvo-Zaragoza *et al.* [2] uses an auto-encoder network topology to perform an image-to-image processing task. Such a task results in higher computational efficiency since all pixels in the input image are processed at the same time. Generally, an auto-encoder network is trained to learn the identity function. However, in the selectional auto-encoder (SAE), the network is trained to learn a selectional map over a $w \times h$ image, preserving the input shape. The activation of each pixel depends on whether the pixel belongs to the foreground or the background. When training the SAE, the images along with their corresponding ground-truth (binarized image) are fed as input to the network. Auto-encoders are feed-forward networks and generally consist of two sections, i.e., the encoder and decoder. The encoder learns to extract the latent representation given an input image, downsampling the image until an intermediate representation is achieved. The output from the encoder is then upsampled and reconstructed to the original input image dimensions by the hidden layers of the decoder. The last layer consists of a set of neurons and a sigmoid activation layer which then gives an output prediction between the range of 0 and 1.

Since the binarized output image should consist of pixel values being 0 or 1 and not in between, a thresholding process is implemented to decide whether the certain pixel belongs to the background or foreground. The encoder and decoder both consisted of 5 layers each and the sampling operators were fixed at 2×2. Network weights were initialized using Xavier initialization [7]. Optimization is handled with stochastic gradient descent and a mini-batch size of 10. The initial learning rate is set to 0.001 and the network is trained for 200 epochs with an early stopping criteria kept in place.

2.7 DeepOtsu

The work presented by He *et al.* [8] proposes an iterative deep learning approach to obtain binarized images called the DeepOtsu model. However, unlike the aforementioned methods in this section, the deep learning network in this case aims to remove artifacts and generate a non-degraded version of the input image. The degraded input image $\mathbf{x}$ is modeled as:

$$\mathbf{x} = \mathbf{x}_{\mathrm{u}} + \mathbf{e}, \tag{6}$$

where $\mathbf{x}_{\mathrm{u}}$ is the latent uniform image and e is the degradation. The aim of the deep learning network is to ultimately obtain $\mathbf{x}_{\mathrm{u}}$.

The network was trained with images split into patches of size 256×256. The patches are first fed to the CNN model and the obtained output is then compared to the ground truth, which in this case should be representative of the uniform, clean version of the input image. To obtain this ground truth, the degraded input image is compared to the already available binarized images from the dataset. Then, the ground truth image is computed as the average pixel value with the same label within the image patch. Once the non-degraded, uniform version of the input image is obtained, the binarized version of the image can be easily obtained using Otsu thresholding [17]. The basic U-Net model [26] is used for learning the degradation. The down-sampling path of the network consisted of 5 convolutional layers with a 3×3 kernel size, followed by a leaky-ReLU activation [12] and 2×2 max pooling. The batch size was set to 8 and the learning rate set to 10^{-4}.

3 Materials and Methods

3.1 Datasets

All models mentioned in the previous section are trained and tested on document images from the DIBCO dataset. To keep the comparison between the models fair and precise, the training set and validation set remain the same for all models. The training set consists of the DIBCO2009, DIBCO2010, DIBCO2011, DIBCO2012, DIBCO2014, and DIBCO2016 datasets. The models are evaluated on DIBCO2013, DIBCO2017, DIBCO2018, and DIBCO2019 datasets. The four test sets were chosen based on the unique properties present in the three sets. DIBCO2013 consists of both handwritten and printed documents. The images from DIBCO2017 had more textual content in them. The DIBCO2018 dataset consisted of images of textual content present towards the borders or corners of the papers and higher intensity of bleed-through artifacts. The DIBCO2019 dataset had large variations in the types of images. Note that we used only track A since track B, containing text content on papyri, are not present in any training data which lead to rather poor learning-based results. Evaluations based on these four datasets give an idea of how well the models are able to generalize on different types of unseen images. Figure 1 shows some samples of images that belong to the DIBCO datasets used for validating the models.

3.2 Metrics

Our evaluation of the various models is based on the standard evaluation metrics used in the DIBCO challenges: (1) F-measure (FM), (2) pseudo F-measure (pFM), (3) peak signal to noise ratio (PSNR), and (4) distance reciprocal distortion (DRD). The FM and pFM reach their best value at 1 and worst at 0 (Eqs. (7) and (8)). PSNR describes how close the binarized and ground truth images are (Eq. (9)). The higher the PSNR, the better is the binarized result.

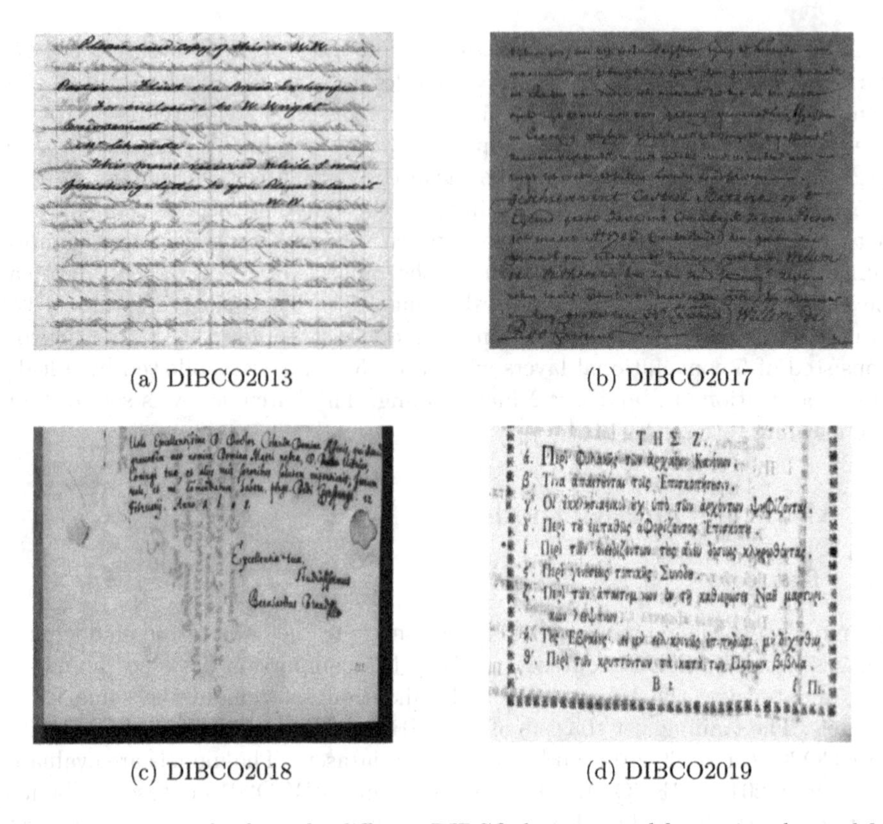

(a) DIBCO2013 (b) DIBCO2017

(c) DIBCO2018 (d) DIBCO2019

Fig. 1. Image examples from the different DIBCO datasets used for testing the models.

The DRD is based on the reciprocal of distance, matching well to subjective evaluation by human visual perception (Eq. (10)).

$$\mathrm{FM} = \frac{2 \times \mathrm{Recall} \times \mathrm{Precision}}{\mathrm{Recall} + \mathrm{Precision}}, \tag{7}$$

where, Recall $= \frac{TP}{TP+FN}$ and Precision $= \frac{TP}{TP+FP}$. TP, FP and FN denote true positive, false positive and false negative values respectively.

$$\mathrm{pFM} = \frac{2 \times \mathrm{pRecall} \times \mathrm{pPrecision}}{\mathrm{pRecall} + \mathrm{pPrecision}}, \tag{8}$$

where, pRecall and pPrecision, respectively the pseudo-recall and the pseudo-precision, are metrics weighted based on the distance to the contours of the foreground in the ground truth. For the pseudo-recall, pixels around strokes have weights starting from 1, and reaching 0 at a distance corresponding to the stroke's width, and pixels inside of the strokes have a weight of 1. For the pseudo-precision, pixels outside strokes but not further than the stroke's thickness have a weight of 1, and inside the stroke their weight increase toward the center, where they reach a value of 2.

$$\text{PSNR} = \log_{10}\left(\frac{C^2}{\text{MSE}}\right), \tag{9}$$

where, $\text{MSE} = \frac{\sum_{x=1}^{m}\sum_{y=1}^{n}(L(x,y)-L'(x,y))^2}{mn}$. The terms m and n denote the dimensions of the image. C denotes the difference present between the text and background.

$$\text{DRD} = \frac{\sum_k \text{DRD}_k}{\text{NUBN}}, \tag{10}$$

where DRD_k is the distortion of the kth flipped pixel and NUBN is the number of non-uniform 8×8 blocks in the ground truth image.

3.3 Training

All models are trained on the DIBCO datasets as mentioned in the previous sections. Based on the configuration of the models, the degraded images along with the accompanying ground truths are first split into patches of size 256×256 pixel or 128×128 pixel resolutions. The patches are further augmented by random horizontal flipping, vertical flipping and rotations. The number of epochs for training each model is set based on the recommendation of the authors for each model, along with an early stopping criteria to monitor any possibility of overfitting the models. If the validation loss of the model does not show significant changes for 15 consecutive epochs, the training would stop and the model would be saved. Certain pre-processing and post-processing operations on the images exclusive to specific models have also been implemented. Such an example is the application of Otsu's thresholding on the output of the DeepOtsu method. The hyper-parameters for the models are optimized using the python library "optuna" [1].

4 Evaluation

The results of testing each model on the different test DIBCO datasets are as shown in the following tables. Table 1a shows the results of validating the models on the DIBCO2013 dataset. The DIBCO2013 dataset contains images that have a good representation of the training data, without any major artifacts or degradation present. All methods display comparable performance with the DE-GAN performing best. For reference, we also show the DIBCO winners of the respective challenge. Note that the participants of 2017 and later potentially used more data for training.

Table 1b shows the results of validating the models on the DIBCO2017 dataset. Here, the performance of the models start to fluctuate more when compared to Table 1a. This might be due to the fact that the DIBCO2017 dataset contains more images that have more densely packed textual content. The DP-LinkNet model outperforms the other models in terms of PSNR, FM and DRD whereas the DE-GAN model has a higher performance in terms of pFM. However, it can be observed that the DRD value for DE-GAN is quite high, indicating

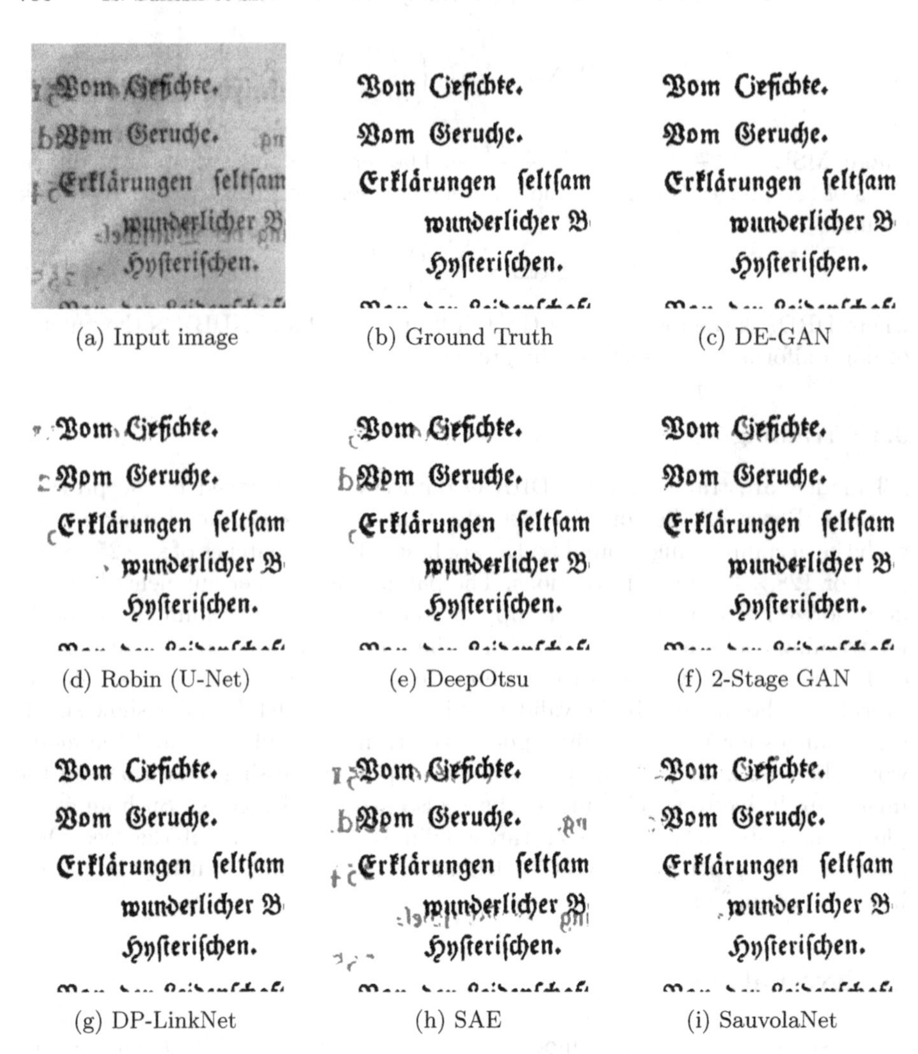

(a) Input image (b) Ground Truth (c) DE-GAN

(d) Robin (U-Net) (e) DeepOtsu (f) 2-Stage GAN

(g) DP-LinkNet (h) SAE (i) SauvolaNet

Fig. 2. Illustration of some results for an image from DIBCO-2017. Pixels in cyan are false positives. The few pixels in orange are false negatives. Pixels in white or black match the ground truth. (Color figure online)

that the resulting binarized images have higher rate of distortions. This may be attributed to the training process of the DE-GAN model, which may have introduced distortions to the generated images. Qualitative results for a randomly chosen sample from DIBCO2017 can be seen in Fig. 2.

The results for the DIBCO2018 dataset is shown in Table 1c. The winner is clearly the 2-Stage GAN approach, outperforming all other methods in each metric. For the pFM and the DRD metrics, the DE-GAN ranks second. Interestingly, the DP-LinkNet struggles with black page borders, see Fig. 3b. While

Table 1. Results of different image binarization methods on the (a) DIBCO2013, (b) DIBCO2017, (c) DIBCO2018, and (d) DIBCO2019 datasets. Note that the winners of the respective DIBCO2017, DIBCO2018 and DIBCO2019 challenge had more data available.

Model	PSNR↑	FM↑	pFM↑	DRD↓
DE-GAN	**24.08**	**97.68**	**98.09**	**1.11**
Robin (U-Net)	22.81	95.07	95.82	1.99
DeepOtsu	21.19	93.46	95.99	2.25
2-Stage GAN	22.60	95.75	96.40	1.46
DP-LinkNet	23.63	96.49	97.24	1.10
SAE	20.88	93.35	94.44	3.17
SauvolaNet	23.41	96.31	97.53	1.28
Winner [21,24]	20.68	92.12	94.19	3.10

(a) DIBCO2013

Model	PSNR↑	FM↑	pFM↑	DRD↓
DE-GAN	18.31	96.23	**98.10**	3.22
Robin (U-Net)	19.99	92.05	94.06	2.23
DeepOtsu	18.02	89.01	91.84	3.50
2-Stage GAN	20.89	95.56	96.54	1.33
DP-LinkNet	**22.84**	**97.92**	97.94	**0.77**
SAE	16.73	87.59	90.41	5.60
SauvolaNet	19.40	93.33	96.26	2.20
Winner [21,24]	18.28	91.04	92.86	3.40

(b) DIBCO2017

Model	PSNR↑	FM↑	pFM↑	DRD↓
DE-GAN	15.98	76.21	83.29	8.01
Robin (U-Net)	15.78	78.80	81.11	12.20
DeepOtsu	12.72	66.60	68.83	42.52
2-Stage GAN	**19.93**	**92.40**	**94.90**	**2.67**
DP-LinkNet	15.73	78.56	80.70	13.72
SAE	14.48	73.45	76.33	15.45
SauvolaNet	16.03	77.94	81.92	10.41
Winner [22,25]	19.11	88.34	90.24	4.92

(c) DIBCO2018

Model	PSNR↑	FM↑	pFM↑	DRD↓
DE-GAN	15.12	70.86	70.69	6.23
Robin (U-Net)	14.39	65.55	65.34	7.36
DeepOtsu	14.82	70.81	70.91	7.59
2-Stage GAN	12.87	65.09	65.72	12.71
DP-LinkNet	14.20	61.84	61.55	7.58
SAE	12.50	62.17	61.90	13.43
SauvolaNet	**15.83**	**72.04**	**71.59**	**5.55**
Winner [22,25]	14.48	72.88	72.15	16.24

(d) DIBCO2019

it wins for the 2017 dataset that does not have borders, it performed poorly on images that have borders that are present in the DIBCO2018 dataset, cf. Fig. 1c.

While SauvolaNet ranks behind these two methods in the DIBCO2018 challenge, it outperforms both methods on the DIBCO2019 dataset, see Table 1d. The 2-Stage GAN, which performs very well for the 2013 to 2018 datasets had some difficulties to deal with the squared paper (check paper, quadrille paper) of the 2019 dataset, which can be observed in Fig. 3d. When we average all metrics for all different evaluated datasets, see Table 2a, the 2-Stage GAN seems to be on average the most suitable binarization method appearing to be consistent in terms of performance. Interestingly, computing the average rank over all metrics, i.e., the average over all 16 ranks for each method, it falls behind DE-GAN and SauvolaNet, cf. Table 2b.

We also evaluated the runtime, reported as throughput, i.e., images per second in the last column of Table 2a. The best throughput has the Robin binarization method. Note, however that we evaluated the methods on a small-sized

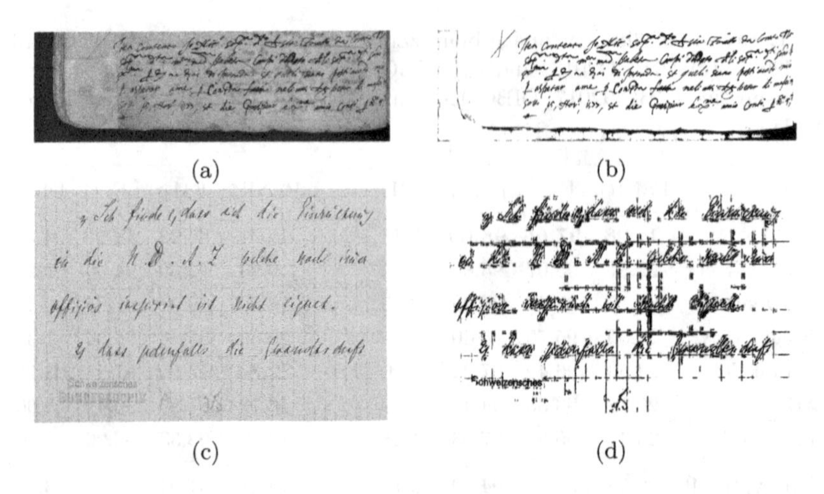

Fig. 3. Qualitative examples of failure modes: (b) shows that DP-LinkNet binarizes the large black borders present in images of DIBCO2018 to white; (d) shows that the 2-Stage GAN struggles with the squared paper given in images of DIBCO2019, and additionally produces halo-artifacts.

Table 2. Average over (a) all metrics and (b) all ranks. Runtimes evaluated using an NVIDIA RTX 2060 GPU (12 GB RAM). Note that DeepOtsu and 2-Stage GAN were limited by the available memory.

Model	PSNR↑	FM↑	pFM↑	DRD↓	img/sec↑
DE-GAN	18.37	85.25	87.54	4.64	0.67
Robin (U-Net)	18.24	82.87	84.08	5.95	1.99
DeepOtsu	16.69	79.97	81.89	13.96	0.01
2-Stage GAN	19.07	**87.20**	**88.39**	**4.54**	0.01
DP-LinkNet	**19.10**	83.70	84.36	5.79	0.49
SAE	16.15	79.14	80.77	9.41	0.68
SauvolaNet	18.67	84.91	86.83	4.86	0.37

(a) Average metrics

Avg. rank↓
2.44
4.19
5.50
3.25
3.38
6.63
2.63

(b) Average ranks

GPU (NVIDIA RTX 2060) with 12 GB GPU-RAM. Unfortunately, this affected the throughput of DeepOtsu and 2-Stage GAN because multiple images of the DIBCO 2013 dataset contain very large images, e. g., image sizes of 4161×1049.

5 Conclusion

In this paper, we thoroughly evaluated seven deep learning-based methods in a fair evaluation where we fixed the data and augmentation used. We evaluated the methods using all ten available DIBCO datasets. Therefore, we used six datasets for training and the remaining four datasets for testing. Our evaluations show that the results are very diverse on the four different tested datasets and no clear winner could be established. Overall, the DE-GAN approach achieved the best rank averaged over all four different datasets followed by SauvolaNet. When we compare the metrics individually, then the 2-Stage GAN approach performed best followed by the DE-GAN. In the very different DIBCO2019 dataset, however, the SauvolaNet outperformed these methods.

For future work, we would like to evaluate the methods also with a different protocol. In particular, we would like to simulate the DIBCO scenario of each year's challenge to be comparable with the single DIBCO papers, i. e., training with the datasets 2015–2016, then evaluating with 2017, adding 2017 to the training set, re-train and evaluate on 2018, and so on. The use of additional augmentation techniques as well as additional training datasets is also worth investigating and might have huge impact on the overall performance of the binarization methods. Furthermore, pixel-based evaluation is not optimal [31]. While the pFM metric incorporates the distance to the script contour, it might be worth investigating indirect measures, such as OCR/HTR accuracy or purely skeleton-based metrics [14]. From a practical point of view, the inference time is also worth investigating. This has been mainly studied in the competitions on time-quality document image binarization.

References

1. Akiba, T., Sano, S., Yanase, T., Ohta, T., Koyama, M.: Optuna: a next-generation hyperparameter optimization framework. In: 25th ACM SIGKDD International Conference on Knowledge Discovery & Data Mining, KDD 2019, pp. 2623–2631. Association for Computing Machinery, New York, NY, USA (2019)
2. Calvo-Zaragoza, J., Gallego, A.J.: A selectional auto-encoder approach for document image binarization. Pattern Recogn. **86**, 37–47 (2019)
3. Chaurasia, A., Culurciello, E.: LinkNet: exploiting encoder representations for efficient semantic segmentation. In: 2017 IEEE Visual Communications and Image Processing (VCIP), pp. 1–4 (2017)
4. Christlein, V., Bernecker, D., Hönig, F., Maier, A., Angelopoulou, E.: Writer identification using GMM supervectors and exemplar-SVMs. Pattern Recogn. **63**, 258–267 (2017)
5. Christlein, V., Gropp, M., Fiel, S., Maier, A.: Unsupervised feature learning for writer identification and writer retrieval. In: 2017 14th IAPR International Conference on Document Analysis and Recognition (ICDAR), vol. 01, pp. 991–997 (2017)
6. Gatos, B., Ntirogiannis, K., Pratikakis, I.: ICDAR 2009 document image binarization contest (DIBCO 2009). In: 2009 10th International Conference on Document Analysis and Recognition, pp. 1375–1382 (2009)
7. Glorot, X., Bengio, Y.: Understanding the difficulty of training deep feedforward neural networks. In: Teh, Y.W., Titterington, M. (eds.) Proceedings of the Thirteenth International Conference on Artificial Intelligence and Statistics. Proceedings of Machine Learning Research, 13–15 May 2010, vol. 9, pp. 249–256. PMLR, Chia Laguna Resort, Sardinia, Italy (2010)
8. He, S., Schomaker, L.: DeepOtsu: document enhancement and binarization using iterative deep learning. Pattern Recogn. **91**, 379–390 (2019)

9. Isola, P., Zhu, J.Y., Zhou, T., Efros, A.A.: Image-to-image translation with conditional adversarial networks. In: 2017 IEEE Conference on Computer Vision and Pattern Recognition (CVPR), pp. 5967–5976 (2017)
10. Li, D., Wu, Y., Zhou, Y.: SauvolaNet: learning adaptive Sauvola network for degraded document binarization. In: Lladós, J., Lopresti, D., Uchida, S. (eds.) ICDAR 2021. LNCS, vol. 12824, pp. 538–553. Springer, Cham (2021). https://doi.org/10.1007/978-3-030-86337-1_36
11. Lins, R.D., Bernardino, R.B., Smith, E.B., Kavallieratou, E.: ICDAR 2021 competition on time-quality document image binarization. In: Lladós, J., Lopresti, D., Uchida, S. (eds.) ICDAR 2021. LNCS, vol. 12824, pp. 708–722. Springer, Cham (2021). https://doi.org/10.1007/978-3-030-86337-1_47
12. Maas, A.L.: Rectifier nonlinearities improve neural network acoustic models (2013)
13. Masyagin, M.: Robust document image binarization. https://github.com/masyagin1998/robin. Accessed 1 Apr 2022
14. Monteiro Silva, A.C., Hirata, N.S.T., Jiang, X.: Skeletal similarity based structural performance evaluation for document binarization. In: 2020 17th International Conference on Frontiers in Handwriting Recognition (ICFHR), pp. 37–42 (2020)
15. Mustafa, W.A., Kader, M.M.M.A.: Binarization of document images: a comprehensive review. J. Phys.: Conf. Ser. **1019**, 012023 (2018)
16. Ntirogiannis, K., Gatos, B., Pratikakis, I.: ICFHR 2014 competition on handwritten document image binarization (H-DIBCO 2014). In: 2014 14th International Conference on Frontiers in Handwriting Recognition, pp. 809–813 (2014)
17. Otsu, N.: A threshold selection method from gray-level histograms. IEEE Trans. Syst. Man Cybern. **9**(1), 62–66 (1979)
18. Pratikakis, I., Gatos, B., Ntirogiannis, K.: H-DIBCO 2010 - handwritten document image binarization competition. In: 2010 12th International Conference on Frontiers in Handwriting Recognition, pp. 727–732 (2010)
19. Pratikakis, I., Gatos, B., Ntirogiannis, K.: ICDAR 2011 document image binarization contest (DIBCO 2011). In: 2011 International Conference on Document Analysis and Recognition, pp. 1506–1510 (2011)
20. Pratikakis, I., Gatos, B., Ntirogiannis, K.: ICFHR 2012 competition on handwritten document image binarization (H-DIBCO 2012). In: 2012 International Conference on Frontiers in Handwriting Recognition, pp. 817–822 (2012)
21. Pratikakis, I., Gatos, B., Ntirogiannis, K.: ICDAR 2013 document image binarization contest (DIBCO 2013). In: 2013 12th International Conference on Document Analysis and Recognition, pp. 1471–1476 (2013)
22. Pratikakis, I., Zagori, K., Kaddas, P., Gatos, B.: ICFHR 2018 competition on handwritten document image binarization (H-DIBCO 2018). In: 2018 16th International Conference on Frontiers in Handwriting Recognition (ICFHR), pp. 489–493 (2018)
23. Pratikakis, I., Zagoris, K., Barlas, G., Gatos, B.: ICFHR 2016 handwritten document image binarization contest (H-DIBCO 2016). In: 2016 15th International Conference on Frontiers in Handwriting Recognition (ICFHR), pp. 619–623 (2016)
24. Pratikakis, I., Zagoris, K., Barlas, G., Gatos, B.: ICDAR 2017 competition on document image binarization (DIBCO 2017). In: 2017 14th IAPR International Conference on Document Analysis and Recognition (ICDAR), vol. 01, pp. 1395–1403 (2017)
25. Pratikakis, I., Zagoris, K., Karagiannis, X., Tsochatzidis, L., Mondal, T., Marthot-Santaniello, I.: ICDAR 2019 competition on document image binarization (DIBCO 2019). In: 2019 International Conference on Document Analysis and Recognition (ICDAR), pp. 1547–1556 (2019)

26. Ronneberger, O., Fischer, P., Brox, T.: U-Net: convolutional networks for biomedical image segmentation. In: Navab, N., Hornegger, J., Wells, W.M., Frangi, A.F. (eds.) MICCAI 2015. LNCS, vol. 9351, pp. 234–241. Springer, Cham (2015). https://doi.org/10.1007/978-3-319-24574-4_28
27. Sauvola, J., Pietikäinen, M.: Adaptive document image binarization. Pattern Recogn. **33**(2), 225–236 (2000)
28. Souibgui, M.A., Kessentini, Y.: DE-GAN: a conditional generative adversarial network for document enhancement. IEEE Trans. Pattern Anal. Mach. Intell. **44**(3), 1180–1191 (2022)
29. Suh, S., Kim, J., Lukowicz, P., Lee, Y.O.: Two-stage generative adversarial networks for document image binarization with color noise and background removal. CoRR abs/2010.10103 (2020). https://arxiv.org/abs/2010.10103
30. Tan, M., Le, Q.: EfficientNet: rethinking model scaling for convolutional neural networks. In: Chaudhuri, K., Salakhutdinov, R. (eds.) Proceedings of the 36th International Conference on Machine Learning. Proceedings of Machine Learning Research, 09–15 June 2019, vol. 97, pp. 6105–6114. PMLR (2019)
31. Tensmeyer, C., Martinez, T.: Historical document image binarization: a review. SN Comput. Sci. **1**(3), 1–26 (2020). https://doi.org/10.1007/s42979-020-00176-1
32. Xiong, W., Jia, X., Yang, D., Ai, M., et al.: DP-LinkNet: a convolutional network for historical document image binarization. KSII Trans. Internet Inf. Syst. **15**(5), 1778–1797 (2021)
33. Zhou, L., Zhang, C., Wu, M.: D-LinkNet: LinkNet with pretrained encoder and dilated convolution for high resolution satellite imagery road extraction. In: 2018 IEEE/CVF Conference on Computer Vision and Pattern Recognition Workshops (CVPRW), pp. 192–1924 (2018)

Correction to: How Confident Was Your Reviewer? Estimating Reviewer Confidence from Peer Review Texts

Prabhat Kumar Bharti, Tirthankar Ghosal, Mayank Agrawal, and Asif Ekbal

Correction to: Chapter "How Confident Was Your Reviewer? Estimating Reviewer Confidence from Peer Review Texts" in: S. Uchida et al. (Eds.): *Document Analysis Systems*, LNCS 13237, https://doi.org/10.1007/978-3-031-06555-2_9

In the version of this paper that was originally published one crucial acknowledgement was missing. This has now been corrected.

The updated version of this chapter can be found at
https://doi.org/10.1007/978-3-031-06555-2_9

© Springer Nature Switzerland AG 2022
S. Uchida et al. (Eds.): DAS 2022, LNCS 13237, p. C1, 2022.
https://doi.org/10.1007/978-3-031-06555-2_53

Author Index